T0188828

Lecture Notes in Computer Science 1099

Edited by G. Goos, J. Hartmanis and J. van Leeuwen

Advisory Board: W. Brauer   D. Gries   J. Stoer

Springer
Berlin
Heidelberg
New York
Barcelona
Budapest
Hong Kong
London
Milan
Paris
Santa Clara
Singapore
Tokyo

F. Meyer auf der Heide   B. Monien   (Eds.)

Automata, Languages and Programming

23rd International Colloquium, ICALP '96
Paderborn, Germany, July 8-12, 1996
Proceedings

 Springer

Series Editors

Gerhard Goos, Karlsruhe University, Germany

Juris Hartmanis, Cornell University, NY, USA

Jan van Leeuwen, Utrecht University, The Netherlands

Volume Editors

Friedhelm Meyer auf der Heide
Burkhard Monien
Heinz Nixdorf Institut und FB Mathematik-Informatik
Universität-GH Paderborn
D-33095 Paderborn, Germany

Cataloging-in-Publication data applied for

Die Deutsche Bibliothek - CIP-Einheitsaufnahme

Automata, languages and programming : 23rd international
colloquium ; proceedings / ICALP 96, Paderborn, Germany,
July 8 - 12, 1996. F. Meyer auf der Heide ; B. Monien (ed.). -
Berlin ; Heidelberg ; New York ; Barcelona ; Budapest ; Hong
Kong ; London ; Milan ; Paris ; Santa Clara ; Singapore ;
Tokyo : Springer, 1996
  (Lecture notes in computer science ; Vol. 1099)
  ISBN 3-540-61440-0
NE: Meyer auf der Heide, Friedhelm [Hrsg.]; ICALP <23, 1996,
  Paderborn>; GT

CR Subject Classification (1991): F, E, G.2, I.2

ISSN 0302-9743
ISBN 3-540-61440-0 Springer-Verlag Berlin Heidelberg New York

This work is subject to copyright. All rights are reserved, whether the whole or part of the material is
concerned, specifically the rights of translation, reprinting, re-use of illustrations, recitation, broadcasting,
reproduction on microfilms or in any other way, and storage in data banks. Duplication of this publication
or parts thereof is permitted only under the provisions of the German Copyright Law of September 9, 1965,
in its current version, and permission for use must always be obtained from Springer-Verlag. Violations are
liable for prosecution under the German Copyright Law.

© Springer-Verlag Berlin Heidelberg 1996
Printed in Germany

Typesetting: Camera-ready by author
SPIN 10513283   06/3142 – 5 4 3 2 1 0   Printed on acid-free paper

Foreword

The International Colloquium on Automata, Languages, and Programming (ICALP) is the annual conference series of the European Association for Theoretical Computer Science (EATCS). It is intended to cover all important areas of theoretical computer science, such as: computability, parallel and distributed computation, automata, formal languages, term rewriting, analysis of algorithms, computational geometry, computational complexity, symbolic and algebraic computation, cryptography, data types and data structures, theory of data bases and knowledge bases, semantics of programming languages, program specification, transformation and verification, foundations of logic programming, theory of logical design and layout, theory of concurrency, theory of robotics.

ICALP '96 was held at Paderborn University, Germany, from July 8th to July 12th, 1996.

Previous colloquia were held in Szeged (1995), Jerusalem (1994), Lund (1993), Wien (1992), Madrid (1991), Warwick (1990), Stresa (1989), Tampere (1988), Karlsruhe (1987), Rennes (1986), Nafplion (1985), Antwerpen (1984), Barcelona (1983), Århus (1982), Haifa (1981), Amsterdam (1980), Graz (1979), Udine (1978), Turku (1977), Edinburgh (1976), Saarbrücken (1974), and Paris (1972).

ICALP '97 will be held in Bologna, Italy, from July 7th to July 11th, 1997.

The Program Committee selected 52 papers from a total of 172 submissions. Authors of submitted papers came from 38 countries on 5 continents. Each submitted paper was sent to at least four Program Committee members, who were often assisted by their referees. The Program Committee meeting took place in Paderborn on February 2nd and 3rd, 1996. This volume contains the 52 papers selected at the meeting plus four invited papers.

We would like to thank all the Program Committee members and the referees who assisted them in their work.

The members of the Program Committee and the Organizing Committee, as well as further members of the theory groups and the "Paderborn Center for Parallel Computing" (PC^2) in Paderborn deserve our gratitude for their contributions throughout the preparations.

We gratefully acknowledge support from the Deutsche Forschungsgemeinschaft, Stiftung Westfalen, Paderborn University, PESAG Paderborn, Volksbank Paderborn e. G., Sparkasse Paderborn, Weidmüller Interface GmbH, Miele & Cie. GmbH & Co., and the City of Paderborn.

April 1996   Friedhelm Meyer auf der Heide and Burkhard Monien
                                                  Paderborn University

Invited Lecturers

Harald Ganzinger, Saarbrücken
Véronique Bruyère, Mons, and Michel Latteux, Lille
Michael Rabin, Jerusalem and Harvard
Abhiram Ranade, Berkeley and Bombay
Alexander Razborov, Moscow

Program Committee

J. Balcazar, Barcelona
J. Berstel, Paris •
R. Freivalds, Riga •
Z. Galil, New York
J. Karhumäki, Turku •
R. Kemp, Frankfurt •
W. Maass, Graz
A. Marchetti-Spaccamela, Roma •
J. Matoušek, Praha •
F. Meyer auf der Heide, Paderborn (Co-Chairman) •
B. Monien, Paderborn (Co-Chairman) •
I. Munro, Waterloo
L. Pacholski, Wrocław
A. Pnueli, Rehovot •
A. Rosenberg, Amherst
D. Sannella, Edinburgh •
S. Skyum, Århus •
P. Spirakis, Patras •
P. Vitányi, Amsterdam •

Members marked with • participated in the PC meeting.

Organizing Committee

Bernard Bauer
Birgit Farr
Erich Köster
Friedhelm Meyer auf der Heide
Burkhard Monien
Christian Scheideler
Willy-B. Strothmann
Walter Unger
Rolf Wanka

List of Referees

P. Alimonti
T. Altenkirch
F. d'Amore
S. Anderson
P. Arenas-Sánchez
L. Arge
A. Arnold
J.-M. Autebert
J. Bang Jensen
G. Bhat
D. A. Mix Barrington
F. Barthélémy
M.-P. Béal
D. Beauquier
P. Berenbrink
S. Bezrukov
S. L. Bloom
L. Boasson
H. Bodlaender
D. J. B. Bosscher
C. Bouras
J. Bradfield
D. Breslauer
C. Brown
H. Buhrman
D. Caucal
O. Čepek
J.-M. Champarnaud
W. Charatonik
R. Cleaveland
B. Courcelle
R. Cramer
P. Crescenzi
M. Crochemore
K. Culik II
P.-L. Curien
A. Czumaj
I. Damgaard
O. Danvy
M. Dauchet
Z. Dayar
T. Decker
J. Diaz
A. Dicky

R. Dicosmo
V. Diekert
R. Diekmann
W. Dittrich
M. Dietzfelbinger
D. Dubashi
P. van Emde Boas
P. Fischer
M. Flammini
D. Fotakis
P. G. Franciosa
G. Frandsen
D. Frigioni
J. Gabarró
P. Gastin
J. von zur Gathen
R. Gavaldà
R. Giaccio
E. Goubault
S. Grigorieff
J. Grundy
I. Guessarian
D. Guijarro
T. Harju
M. Hermo
J. G. Henriksen
T. Hildebrandt
J. Hillston
Y. Hirshfeld
J.-H. Hoepman
M. Hofmann
J. Honkala
J. Hromkovič
M. Hühne
H. Hüttel
T. Husfeldt
Z. Huzar
K. Indermark
B. Jenner
M. Jerrum
M. Jourdan
E. Jurvanen
T. Jurdziński
S. Kahrs

T. Karvi
P. Kilpelainen
N. Klarlund
R. Klasing
M. Klazar
H. C. M. Kleijn
L. Knudsen
T. Knuutila
A. Kościelski
J. Kratochvíl
M. Krause
M. Kutyłowski
K. G. Larsen
M. Latteux
U. Lechner
H. Lefmann
D. Le Metayer
S. Leonardi
A. Lepistö
J. Levy
M. Li
I. Litovsky
K. Loryś
A. Lozano
R. Lyngsø
J. van Maanen
C. Marche
J. Marcinkowski
U. Martin
A. Mateescu
J. Mazoyer
M. Młotkowski
E. Moggi
F. Moller
G. Morrill
M. Morvan
P. D. Mosses
F. Mráz
M. Nebel
F. Nielson
H. R. Nielson
M. Nielsen
R. Nieuwenhuis
N. Nishimura

D. Niwiński
P. Ochsenschläger
F. Orejas
J. van Oosten
F. Otto
S. E. Panitz
G. Pantziou
F. Parisi-Presicce
M. Paterson
G. Paun
W. Penczek
M. Penttonen
G. Persiano
A. Petit
J. F. Peyre
M. Piotrów
J. Pokorný
J. Power
M. Protasi
C. Queinnec
A. Rauzy
D. Reed
A. Renvall
B. Ritchie
M. Roettger
A. Rubio
J. Sakarovitch
A. Salomaa

K. Salomaa
A. Sandholm
C. Scheideler
O. Schmidt
M. Schmidt-Schauß
G. Schnitger
J. Schulze
M. Schwartzbach
R. Schwarz
P. Séébold
H. Seidl
E. Sekerinski
K. Sere
D. Sieling
R. Silvestri
H. U. Simon
E. Speckenmeyer
L. Staiger
P. Steckler
M. Steinby
P. Stevens
C. Stirling
C. N. Storm Pedersen
W.-B. Strothmann
K. Sunesen
A. Szałas
B. Tampakas
A. Tarlecki

B. Terhal
P. Thanisch
L. Thimonnier
W. Thomas
J. Tiuryn
C. Tofts
J. Torán
A. Törn
L. Torenvliet
V. Triantafillou
T. Truderung
D. N. Turner
J. Underwood
W. Unger
P. Urzyczyn
F. Vaandrager
U. Vaccaro
R. Wanka
I. Walukiewicz
A. Weber
I. Wegener
P. Weil
R. Werchner
G. Winskel
T. Wierzbicki
J. von Wright
M. Zawadowski

Table of Contents

Session 14: Parallel Algorithms

Session 15: Distributed Systems

Session 16: Algorithms II

Saturation-Based Theorem Proving
(Abstract)

Harald Ganzinger

Max-Planck-Institut für Informatik
Im Stadtwald, D-66123 Saarbrücken, Germany
hg@mpi-sb.mpg.de

Saturation is one of the major techniques for automated theorem proving in first-order logic. Saturation means to compute the closure of a given set of formulas under a given set of inference rules. One of the most prominent examples of theorem proving by saturation is the resolution method. Resolution theorem proving essentially means to compute the closure of a given set of clauses under the resolution inference rule. Resolution is refutationally complete in the sense that if a set of clauses is closed under resolution then it is inconsistent if and only if it contains the empty clause. Knuth/Bendix completion for presentations of equational theories is another major example of saturation. Here the given set of equations is closed under certain paramodulation rules. Knuth/Bendix completion is a refutationally complete method for solving word problems. The main drawback of saturation-based theorem proving is a lack of goal orientation. In extreme cases one enumerates all consequences of the theory until, by mere luck, the theorem one wants to prove is found.

Restricting forward computation as much as possible is the main motivation behind recent refinements of saturation-based methods. Two concepts are helpful in this regard: orderings and selection. Orderings restrict inferences to instances in which the main atoms and/or terms of the inference are maximal in the respective premises. Conceptually, orderings represent the progress that one makes when computing a new inference. Satisfying the ordering restrictions, the conclusion of an inference is smaller than its premises, with the formula to be proved (usually the empty clause) being minimal in the ordering. For a satisfiable set of formulas, an ordering identifies a unique minimal model, and ordered inferences approximate truth with respect to this model.

Selection exploits the asymmetry in computing with negative and positive literals of clauses. While the choice between positive literals in a clause leads to (ordering-guided) search, the negative literals represent unification problems which can be solved in a don't-care nondeterministic fashion, hence can be "selected" at random. It has been shown that respective refinements of resolution and paramodulation remain refutationally complete.

Ordering restrictions and selection constrain inferences locally in that they depend only on the premises of the inference. Constraint inheritance propagates this information to all descendants. Nevertheless they cannot avoid that, when formulas have more than one proof, many of these proofs are enumerated simultaneously, although finding one proof would be sufficient. Simplification techniques are a means to reduce this form of global redundancy. We have been able to show that practically all simlification techniques that are used in practice can be justified by a simple abstract notion of redundancy that is largely independent of the particular inference system and mainly depends on the ordering that one employs for constraining the inference system. Essentially, formulas are redundant in a particular set of formulas if they follow from smaller formulas in that set. Saturation up to redundancy, which means closure of a set of formulas under non-redundant inferences, remains refutationally complete.

There has always been strong practical evidence of the fact that simplification and redundancy elimination form an essential ingredient of any theorem proving method. With the abstract notion of saturation up to redundancy we now have means to also theoretically analyse the effects of simplification. Requiring saturation up to redundancy is strictly weaker than requiring the closure under a given inference system. Saturation up to redundancy, if sufficiently powerful techniques for redundancy detection are employed, terminates in many non-trivial examples of first-order theories, while the closure under the respective inference system is usually infinite. This has many applications. Saturation can be used as a decision procedure for many decidable fragments of first-order logic, including the monadic class with equality and Maslov's class K. Being able to effectively saturate a theory provides a strong normal form for proofs of entailment with respect to that theory. Such proofs are linear in that inferences in which not at least one of the premises descends from the goal to be proved need not be considered: in a saturated theory they are redundant. This obvious fact is, among others, the basis for refined saturation calculi for first-order logic over certain algebraic theories such as abelian groups or rings which do posses saturated presentations in the form of canonical rewrite systems.

Saturation of a theory yields an improved presentation of that theory and hence can be viewed as an optimizing compiler. A well-known result is that Knuth/Bendix completion, if it terminates, provides us with a decision procedure for the word problem. More theoretical results that indicate why saturated presentations are better behaved computationally have becoming available more recently. Saturation under ordered basic paramodulation, if it terminates, proves that the respective theory has a finitary and decidable unification problem. Moreover, saturation up to redundancy by ordered resolution can be used as a method for automated complexity analysis. If saturation terminates, one may derive a strong relationship between the complexity of certain entailment problems for that theory and the cardinality of downward closed sets of ground terms in the given term ordering.

References

1. L. Bachmair and H. Ganzinger. Rewrite-based equational theorem proving with selection and simplification. *Journal of Logic and Computation*, 4(3):217–247, 1994.

2. L. Bachmair and H. Ganzinger. Rewrite techniques for transitive relations. In *Proc. 9th IEEE Symposium on Logic in Computer Science*, pages 384–393. IEEE Computer Society Press, 1994.
3. L. Bachmair, H. Ganzinger, Chr. Lynch, and W. Snyder. Basic paramodulation. *Information and Computation*, 121(2):172–192, 1995.
4. L. Bachmair, H. Ganzinger, and U. Waldmann. Superposition with simplification as a decision procedure for the monadic class with equality. In G. Gottlob, A. Leitsch, and D. Mundici, editors, *Proc. of Third Kurt Gödel Colloquium, KGC'93*, volume 713 of *Lecture Notes in Computer Science*, pages 83–96, Berlin, 1993. Springer-Verlag.
5. D. Basin and H. Ganzinger. Automated complexity analysis based on ordered resolution. Research Report MPI-I-95-2-006, Max-Planck-Institut für Informatik, Saarbrücken, 1995. Extended abstract to appear in Proc. LICS'96.
6. R. Nieuwenhuis. Basic paramodulation and decidable theories. In *Proc. 11th IEEE Symposium on Logic in Computer Science*. IEEE Computer Society Press, 1994. To appear.

Bandwidth Efficient Parallel Computation

Abhiram Ranade

Computer Science Division, University of California, Berkeley
and
Department of Computer Science and Engineering, I.I.T. Bombay

Abstract. We believe that for the next few years, the most pressing research question in parallel computation will concern communication bandwidth: Can we design fast algorithms for parallel computers that only support low bandwidth communication (such as most existing parallel computers). An alternative formulation of the question is, can we design parallel algorithms that have communication locality? While good locality preserving techniques are known for application problems with regular, predictable dataflow, few theoretical results have been developed for irregular problems e.g. problems involving sparse graphs, or problems that adapt to data distribution dynamically. And yet, since most existing parallel computers only offer low communication bandwidth, it is necessary to either develop techniques to live with low bandwidth, or provide arguments in favor of building parallel computers with high bandwidth communication systems.

This paper provides a rough sketch of a research plan for rigorously answering some of these questions. First, we propose a formal definition of what it means to exploit locality, e.g. to be able to decide whether it is possible to exploit locality for a given problem, and if so, to what extent a given implementation is successful in it. Using our formal notion of locality, we describe some preliminary work regarding the development of strategies to exploit locality. Finally, our formal definition opens up the possibility of formally proving that a given problem does not have locality (though it may have parallelism), i.e. it is impossible to design fast algorithms for the problem without having high communication bandwidth. We give examples of some such problems.

1 Introduction

"Is there enough parallelism in my application? Can I get some speedup at all, if not linear? Do there exist application problems that are inherently sequential?" These are some of the questions that have driven the research in parallel computation over the last 15-20 years. Today, intuitively satisfactory answers are known for all of these questions at least in theory. Indeed, most theoreticians as well as practitioners will grant that most applications have the potential of beeing speeded up using parallel computers. The questions driving the theoretical research for the next decade will not center around "is there enough parallelism?", but will concern techniques necessary to efficiently exploit available parallelism. The purpose of this paper is to speculate on what the fundamental obstacles are in efficiently exploiting available parallelism– and prioritize the main theoretical research questions.

A number of factors make it difficult to efficiently utilize available parallelism. Most existing parallel computers have high communication latency (10s or 1000s of cycles) as compared to the cost of executing a local computation (1 cycle). Most existing computers also have low bandwidth, e.g. on the Connection Machine CM-5, each node can send out data at the rate of around 2 Mwords per second, while each node can perform upto 128 million floating point operations per second. Another problem is issue overhead, the time that a processor must spend in issueing (packetizing, generating headers, managing buffers, operating system overhead) a message. This again is in the range from 10s to 1000s of cycles for most machines. Finally, most existing machines are asynchronous. This complicates communication protocols and related issues such as buffer management, as well as programming in general.

Nearly every difficult issue in exploiting parallelism can be traced back to one of these four: latency, bandwidth, issue overhead and asynchrony. This can be seen by considering a model in which these factors have been abstracted away: the PRAM model. Programming on the PRAM is easy; issues such as processor coordination and load balancing can be dealt with gracefully. In fact it is reasonable to consider the PRAM to be a formal vehicle for answering the question "Is there enough parallelism?", since all other issues except that of identifying parallelism are abstracted away on the PRAM. And in this sense, the question has been satisfactorily answered.

The question of efficiently exploiting available parallelism can thus be naturally formulated as "Given that a problem has a fast PRAM algorithm, what does it take to design a fast algorithm for a real machine?", where a real machine is one afflicted by the four factors mentioned above[10].

1.1 Asynchrony, Latency and Overhead

The problems of asynchrony and latency can be overcome if we have *parallel slackness*[37], i.e. if the problem size is substantially larger than the number of processors in the machine. This assumption is expected to hold in practice. While future generations of parallel computers will have a large number of processors, each processor will also have a large local memory, and such machines are expected to be put to work on very large problems[10]. As we will see, this simplifies the problem of exploiting available parallelism.

Consider asynchrony first. If we have a n processor standard (synchronous) PRAM algorithm, it can be simulated to run with nearly the same efficiency on a p processor asynchronous PRAM if n is reasonably larger than p[14]. Since most parallel algorithms on PRAMs typically use nearly as many processors as the problem size, it is reasonable to think of n also as the problem size. In other words, if the problem size is very large as compared to the number of processors, then we can effectively ignore the problem of asynchrony. Parallel slackness can also overcome long latency using techniques such as multithreading[7], or *bulk synchrony*[37]. In several cases, fairly simple algorithmic transformations can drastically reduce the parallel slackness required[14].

The question of issue overhead has received substantial attention. For message passing machines, techniques such as *active messages*[38] substantially reduce issue overhead. For machines with shared address spaces and coherent caches, the overhead is already quite small.

1.2 Bandwidth limitation

Parallel slackness can also be used to alleviate bandwidth limitation. This is well known for several applications (e.g. dense matrix multiplication) in which the dataflow is highly structured. While in the past parallel computers have focussed largely on applications with regular dataflow, they are increasingly called upon to work on problems in which the dataflow is irregular, and even dynamically depends upon the data.

Unlike the case with latency and asynchrony, parallel slackness cannot be directly used to overcome bandwidth limitation, in any general sense. We could try a similar strategy and attempt to design a general efficiency preserving simulation of PRAMs on bandwidth limited models, but this is not possible. For example, consider any general simulation of a single step of the PRAM with n processors on a p processor bandwidth limited machine such as a mesh of processors. It is easily proved that such a simulation must take time at least $\Omega((n/p)\sqrt{p})$. This is in contrast to results regarding asynchrony[14], where simulation time was $O(n/p)$ for reasonably large n.

1.3 The Challenge

The next generation of parallel processors will inevitably be called upon to solve very large unstructured problems. For such endeavours bandwidth limitation will pose a severe challenge.

There are really two options available to overcome this challenge: (1) build machines with high bandwidth communication systems and use general PRAM simulation ideas. This approach has the advantage of very low programming overhead. This approach is taken in the machines such as the TERA[5] and the Saarbrucken PRAM[1]. The obvious drawback is the high cost of the extra hardware. (2) Devise algorithms that are efficient even on parallel computers with low bandwidth communication. In this case we cannot use general PRAM simulations, but must devise techniques for exploiting locality. In other words, we need techniques for distributing data among the memories of a parallel computer such that most of the time each processor only needs data from its local memory; data generated by other processors is required fairly infrequently, so that low bandwidth networks suffice to perform the required data movement without much loss of efficiency.

The second approach is especially relevant for the "Networks of Workstations" proposal for designing parallel processors. Networks of workstations of today are very severely bandwidth limited. With advances in processor design being faster than network design, the networks of tomorrow will very likely be even more bandwidth limited for the purpose of parallel computing. Thus how to exploit locality will be crucially important.

1.4 Overview

In this paper we explore the second approach outlined above. We begin by developing a framework for understanding bandwidth limitation (Section 2). Previous work is considered in Section 2.3. We then consider a number of applications in the context of our framework and either point out locality exploitation strategies, or give proofs that locality cannot be exploited, or point out open problems. We comment on the general methodology in section 3. We consider applications such as convex hulls in two dimensions (section 4), sparse matrix multiplication (section 5), relational joins (section 6), connected component labelling (section 7), combinatorial search (section 8), parallel data structures (section 9), and the n-body problem (section 10). Section 11 concludes.

2 Framework

We begin by formalizing the notion of "bandwidth limited". Consider a parallel computer model M having p processors. Let $B_p(M)$ denote the bisection bandwidth of M, i.e. the minimum number of words that can be exchanged in unit time between any set of $p/2$ processors and the remaining $p/2$ processors of M. We shall say that M is a bandwidth limited model (BLM) if $B_p(M) = o(p)$, i.e. if the bisection bandwidth is strictly smaller than the number of processors. We will assume the wordsize to be $O(\log p)$ bits. Throughout this paper we will assume that that the problem size n is polynomial in p, so that the wordsize will also be $O(\log n)$ bits.

Clearly, most standard (distributed memory) network models[21][1] such as Butterflies, constant dimensional meshes, shuffle exchange networks, fat-trees are BLMs. It may be noted that standard hypercubes are not BLMs, since their bisection bandwidth is proportional to p. Our definition also applies to models such as logP[10]. The bisection bandwidth of the logP model is p/g, where g is the *gap* parameter of the model. Realistic machine families have $g = \omega(1)$ and hence these would then become BLMs. We note that PRAMs have bisection bandwidth of p, and thus PRAMs are not BLMs as per the definition. This is as it should be.

Most existing parallel computers can be considered to be BLMs, since typically, their bisection bandwidth is substantially smaller than the number of processors. Some machines currently under construction do not fit this definition, e.g. the Saarbrucken PRAM[1] and the TERA[5] that support a bisection bandwidth as large as the number of processors. For these machines, exploiting locality is not important; however such machines are few.

---

[1] In a standard network model, processors may execute one instruction every time step, and exchange one word of data with neighbors in the network in every time step.

2.1 Exploiting Locality

Let $T_p(M)$ denote the best time for solving a problem on a model M of parallel computation having p processors. Let $S_p(M_1/M_2)$ denote the time to simulate a single step of model M_1 on model M_2, both having p processors. Then we will say that the program P that achieves the time $T_p(M)$ exploits locality iff

$$T_p(M) = o(T_p(PRAM)S_p(PRAM/M))$$

We will say that P fully exploits locality iff

$$T_p(M) = O(T_p(PRAM))$$

In other words, in order to claim that you exploit locality, you need to do something better than the simulation of the PRAM algorithm. In the rest of the paper we will drop the subscript denoting the number of processors, and it will be implicitly assumed to be p in all cases.

As an example, consider the problem of computing Fast Fourier transforms. A p processor PRAM can compute the FFT on p points in $O(\log p)$ steps, and so can a p processor Butterfly[21]. The Butterfly implementation thus exploits locality fully. On a p processor mesh, however, the best known time is $O(\sqrt{p})$, which is worse than the PRAM time, but better than simulating the PRAM on the mesh, which would have taken time $O(\sqrt{p}\log p)$. The mesh algorithm thus exploits locality, but not fully. More complex examples will follow shortly.

Our definition of exploiting locality is quite robust. This is because there are *bandwidth efficient simulations* between most known models (e.g. Butterflies, meshes, logP) i.e. $S(M_1/M_2) = O(1 + \frac{B(M_1)}{B(M_2)})$. For example, it is possible to simulate PRAMS[2] on Butterfly networks with $S(PRAM/Butterfly) = O(B(PRAM)/B(Butterfly)) = O(\log p)$, or simulate a 3 dimensional mesh on a 2 dimensional mesh with $S(3D/2D) = O(B(3D)/B(2D)) = O(p^{1/6})$. We will assume that such simulations are available.

Theorem 1. *If locality can be fully exploited for one BLM M_1 then it can be exploited for another BLM M_2, provided $B(M_1) > B(M_2)$, and fully exploited if $B(M_2) > B(M_1)$.*

Proof: First consider the case $B(M_1) > B(M_2)$. Then we have $S(M_1/M_2) = B(M_1)/B(M_2)$. Further, since M_1 is a BLM, we know that $B(M_1) = o(p)$.

Let P be the program that fully exploits locality on M_1, taking time $T(M_1) = O(T(PRAM))$. When we simulate P on M_2, the time taken is $T(M_1)S(M_1/M_2) = O(T(PRAM))\frac{B(M_1)}{B(M_2)} = o(T(PRAM)\frac{p}{B(M_2)}) = o(T(PRAM)S(PRAM/M_2))$. Thus locality is exploited on M_2 as well.

If $B(M_1) < B(M_2)$, then $S(M_1/M_2) = O(1)$. Thus the time taken on M_2 is $S(M_1/M_2)T(M_1) = O(T(PRAM))$, i.e. locality is fully exploited. □

---

[2] We implicitly allow randomized simulations and algorithms in this paper.

Theorem 2. *If locality can be exploited for one BLM M_1 then it can be exploited for another BLM M_2, provided $B(M_1) > B(M_2)$.*

Proof: Omitted. □

As an example of theorem 1 note that a p processor Butterfly can be simulated in $O(\sqrt{p}/\log p)$ time in a bandwidth efficient manner on a p processor 2 dimensional mesh. As per theorem 1, simulating the Butterfly FFT algorithm on the mesh takes time $O(\sqrt{p})$. This is precisely the time for the standard algorithm, and exploits locality as mentioned earlier.

2.2 Lack of locality

We next consider the question of lower bounds, i.e. how to prove that locality cannot be exploited. The main idea behind lower bounds is the communication complexity of the problem being solved, which places a lower limit on the amount of communication that is inherently necessary.

Gross Locality Simply stated, the gross locality of a problem is the ratio: parallel work $W = pT$ required to solve the problem using the best PRAM algorithm, divided by the two processor communication complexity of the problem. The ratio of the amounts of computation to communication performed by a program is commonly used to analyze its quality, especially in scientific computing. The intuition behind our measure is similar; the difference being that we are seeking to characterize the intrinsic locality of problems rather than of particular algorithms.

Our notion of communication complexity[24, 36] is standard, except we require input and output to be *balanced*, as described below. Our model consists of two processors connected together by a communication link. We assume that the processors have unlimited computing power and that local computation at each processor is free. However we are charged for each word ($O(\log n)$ bits, for problem size n) that is communicated along the link. The inputs and outputs to the problem being solved must be specified in a where and when oblivious manner (i.e. where and when each input bit is read and output generated is specified in advance, independent of the value of the bits). In addition, we require that the inputs and the outputs be balanced, i.e. each processor inputs half the bits and outputs half the bits. Which bits to input/output at each processor is at the discretion of the algorithm designer, but this must be specified in advance. The number of bits transmitted across the link will depend upon the algorithm used, as well as the problem instance. The communication complexity of a problem is the minimum over all possible algorithms of the maximum (over problem instances) number of words transmitted by the algorithm to solve the problem. The notion of randomized communication complexity has also been studied[36, 24]; in this case the processors are only required to produce the correct answer with probability strictly larger than 1/2.

Let $C(n)$ denote the randomized communication complexity for the given problem, with problem size n. Let $W(n)$ denote the work required to solve the problem on a PRAM. Then the gross locality of the problem is $L(n) = W(n)/C(n)$.

Theorem 3. *Consider a problem with gross locality $L(n)$. Then for any model M,*

$$T(M) = \Omega(T(PRAM)(1 + \frac{p}{L(n)B(M)}))$$

If we have a bandwidth efficient simulation of PRAMs on M, then

$$T(M) = \Omega(T(PRAM)(1 + \frac{S(PRAM/M)}{L(n)}))$$

Proof: The communication time on M must be at least

$$\frac{C(n)}{B(M)} = T(PRAM)\frac{C(n)}{pT(PRAM)}\frac{p}{B(M)} = T(PRAM)(\frac{p}{L(n)B(M)})$$

Adding the computation time which must at least be $T(PRAM)$ gives the first result. The second follows since $B(M) = \theta(p/S(PRAM/M))$. □

Corollary 4. *Suppose the gross locality $L(n)$ of a problem P is $O(1)$, and suppose that a bandwidth efficient PRAM simulation is available for a given network. Then the simulation of the fastest PRAM algorithm on the network is the fastest algorithm for P on the network.*

Proof: Since $L(n) = O(1)$, the time on M is at least $\Omega(T(PRAM)S(PRAM/M))$. But this is achieved by PRAM simulation. □

Informally this corollary characterizes problems without locality. In particular, if a problem has at most constant gross locality, then the best way of implementing it on most networks is just to use the brute force simulation of PRAMs on that network. We will see several examples of this in the following sections.

2.3 Related Work

Several researchers have noted the need to model bandwidth limitation and have proposed a number of different models, the most discussed among these are perhaps logP[10] and BSP[37]. The work on most of these models has dealt only with exploiting locality in very regular and structured computations. Also, little has been said in these papers regarding lower bounds: the idea that a problem might be inherently nonlocal has not been formally explored.

The work that comes closest to ours is that of Aggarwal, Chandra, and Snir[2]. They present techniques for exploiting locality on their *LPRAM* model, which is bandwidth limited in the sense we defined above. They also present techniques for lower bounding the amount of communication for solving a problem on the LPRAM. Finally, they describe relationships between LPRAMs and meshes and LPRAMs and butterfly networks, and thus implicitly argue that exploiting locality on LPRAMs is equivalent to exploiting locality on meshes.

Our formulation is more general, and more explicit in defining locality as a (robust) property of problems. Second, the work in [2] deals only with computations expressible as (possibly irregular) directed acyclic graphs. In fact, they mention graph algorithms and data structure manipulation as future research directions–we have analyzed these in the present paper.

Also noteworthy is the work by Leiserson and Maggs[22]. They consider pointer based algorithms. Given any pointer based data structure stored in the memories of different processor in a parallel computer, they (in spirit) define its communication load as the time required to chase all pointers exactly once in parallel. They then develop algorithms for several problems whose communication time is within polylog factors of the communication load of the input pointer structure. This approach is attractive in that it attempts to get the best performance possible for each input instance, unlike our model which only considers the performance for the worst input instance. On the other hand, our model is more general and works with problems that cannot be expressed naturally in terms of pointers (e.g. convex hulls).

3 General Strategies

A natural strategy for designing algorithms for BLMs is as follows. We start with the best PRAM algorithm for the problem and augment it with a *customized data layout*. The data layout specifies how the PRAM data structures are distributed in the local memories of the processors in a BLM. The processors execute the same program as in the PRAM model. The hope is that with a good layout processors find the data that they need in their local memories most of the time; and only infrequently would they need to fetch data from other processors. This will lead to high efficiency.[3]

This approach is supported by programming languages like CMFortran and others which provide language constructs for defining data layout. This approach has been very successfully used for applications with regular data flow, such as several applications in Scientific Computing. An important idea is that of *blocking*– wherebye related data elements are grouped together and are stored on a single processor. The idea of *blocking* has been successfully used for several numerical computations with regular dataflow such as banded matrix multiplication, LU and other decompositions, and transforms such as FFT[20]. Our main focus in this paper is on applications with irregular dataflow. Our goal is to design strategies akin to blocking for such applications.

We should point out that not every (efficient) PRAM algorithm can be augmented with a customized layout to produce an efficient algorithm for BLMs. As we will see in the section on convex hulls (section 4), the standard PRAM algorithm will require provably large data movement for any layout whatsoever.

---

[3] Notice that this strategy is similar to PRAM simulation. The crucial difference is that in PRAM simulation, the data structures of the PRAM program are distributed in an oblivious manner, whereas this strategy chooses a layout designed to minimize data movement (based on an analysis of the data access pattern of the program).

In this case, we need to look for a different algorithm, possibly avoiding algorithmic primitives that require high data movement. Section 4 will show how this can be done for convex hull computation.

Of course, when the application problem has $O(1)$ locality, i.e. is inherently nonlocal, it is futile to look for good layouts or good PRAM algorithms. The algorithm designer also needs to keep this possibility in mind; while looking for good layouts and appropriate PRAM algorithms it is useful to also simultaneously attempt to estimate the locality of the application problem. As we will see in the following sections, not all problems will have good layouts, for some it will indeed be possible to prove lack of locality.

A note on communication complexity: While evaluating communication complexity, we have avoided being too formal. In particular, our proofs are not described at the level of crossing sequences and fooling sets[36]. It should be clear however, that these formalisms could be incorporated, but we have avoided them for brevity. Second our estimates are also applicable for randomized protocols using ideas like "one-way information"[36].

4 Convex Hulls

Computation of convex hulls is a basic problem in Computational Geometry. Several parallel algorithms are known for the two dimensional case (e.g. [6]). We show that locality can be fully exploited for this problem. Besides serving as an illustration of locality exploitation techniques, we also improve the result in [11].

The simplest approach to designing bandwidth efficient convex hull algorithms is to provide a customized data layout for the data structures of a suitable PRAM algorithm, e.g. [6]. Unfortunately, this does not work, since these algorithms repeatedly use a step that is equivalent to vector compaction. Vector compaction, as is easily seen, is inherently nonlocal.

Our bandwidth efficient algorithm is similar in structure to the PRAM algorithms, but it avoids the compaction step. Our algorithm runs on a p processor Butterfly network and computes the convex hull of n points in time $O(n \log n/p)$, for $n > p \log^3 p$). Our algorithm gives linear speedup, as do the PRAM algorithms, and thus fully exploits locality.

We describe an algorithm to compute the upperhull (i.e. the hull obtained by including a point at $y = -\infty$). The lower hull can be computed likewise, and will together yield the convex hull. The algorithm has three parts. In the first, the points are input by the processors and sorted by the x-coordinate. At the end of this step, the points are stored in an array $Q[1..n]$, with each processor i holding elements $Q[i(n/p) + 1..(i+1)n/p]$. To describe the remaining two parts, it is best to consider the p processors to be at the leaves of a complete binary tree of processors with p leaves. This is only for ease of description; it will be obvious that the p processor Butterfly can also accomplish the same result in the same time to within constant factors. The rest of the algorithm is as follows:

1. Each processor computes upperhulls of points it holds locally. Let j denote the number of points on the upperhull computed by processor i. Processor

i stores its upperhull in $Q[i(n/p) + 1..i(n/p) + j]$. Processor i also initializes 3 variables $First[i] = 1$, $Last[i] = j$, and $Precede[i] = 0$. First and last are used to point to the subvector which contain points in the upperhull. $Precede[i]$ is of use during subsequent execution, when a single upperhull might be stored across several leaf processors. In this case, $Precede[i]$ gives the number of points in the hull preceding those stored in leaf i.

2. Let the levels of the tree be numbered 0 through $\log p$ from the leaves to the root. This step has $\log p$ phases. Phase i is executed independently in subtrees of level i nodes, which we will call *masters*. In phase i each master computes the joint upperhull of the upperhulls stored in its left and right subtrees. This has two parts:

    (a) *Determine which points survive to the joint hull.* This is done by the divide-and-conquer procedure of [26]. The procedure is executed entirely by each master; when the master needs data regarding the points in the two hulls it broadcasts requests of the form "Examine point i in the right (or left) hull". Each leaf processor determines whether it has the requested point, and if so sends it to the master. Notice that the information in First, Last and Precede is sufficient for this purpose. Each communication takes $2i < 2\log p$ steps for upward and downward traversal in the tree. The procedure of [26] requires $O(\log(n/p))$ such queries, and does constant work per query, for a total time of $O(\log p \log(n/p))$ for the entire step.

    (b) *Update data structures for subsequent mergeing steps.* The joint hull consists of some left prefix of the left hull concatenated with some right prefix of the right hull. Thus, we need to update First, Last and Precede in each leaf in the subtree accordingly. It is easily seen that this can be done using a constant number of parallel prefix operations in time $O(i)$, parallelly in each subtree.

Step 2 takes time $O(\log^2 p \log(n/p))$, and step 1 takes $O(n/p \log(n/p))$, so that the total time is $O((n \log n)/p)$ for $n = \Omega(p \log^3 p)$.

Theorem 5. *A p processor Butterfly network of processors can compute the convex hull of n points in time $O((n \log n)/p)$ for $n = \Omega(p \log^3 p)$.*

Proof: Notice that the tree algorithm described above can be executed in any $O(p)$ leaf tree of height $O(\log p)$ in time $O((n \log n)/p)$ for $n = \Omega(p \log^3 p)$. A p processor Butterfly contains such a tree and will also execute it in the same time. The sorting step can be finished in time $O(n \log n/p)$ using an algorithm such as Flashsort[33]. □

5   Sparse Matrix-vector Multiplication

We show that general sparse matrix multiplication is inherently non-local. We make this claim in a very strong sense: we will show that there exist sparse matrices for which matrix vector multiplication requires large data movement even if the sparsity pattern is known beforehand. We will also considered sparse matrix vector multiplications that arise in Finite element computations. As we will see, these matrices possess some structure which can be exploited to produce locality. Dense matrix operations, as might be expected, possess high locality, as may be seen from well-known algorithms[20].

5.1 General Sparse Matrices

The sparse matrices we consider are derived from expander graphs.

Consider a bounded degree bipartite graph $G = (X, Y, E)$, with vertex sets $X = [X_1, \ldots, X_n]$ and $Y = [Y_1, \ldots, Y_n]$. The edges E are selected so that G has an *expansion* property. In particular, we require that for any $X' \subset X$ such that $|X'| \leq n/2$, X' has a large neighborhood, i.e. $|\{y | x \in X', (x, y) \in E\}| \geq 3 |X'| /2$. It is known that such graphs exist[4], even with the degree of vertices bounded by an appropriate constant d.

Say a matrix A is G-sparse if $A[i, j]$ is nonzero iff $(X_i, Y_j) \in E$. Consider the problem of multiplying a G-sparse matrix A by a vector x, for any fixed G having the expansion property as described above. The input to the problem consists of $m + n$ numbers, first n of which are the elements $x_1, x_2, \ldots, x_n$ of x. The remaining m specify the nonzero entries in A. Clearly $m = O(n)$ since G has bounded degree. The output consists of n numbers $y_1, \ldots, y_n$ which are the elements of the vector $y = Ax$.

Theorem 6. *The problem of multiplying a G-sparse matrix by a vector has locality $O(1)$.*

Proof: We show that the problem has communication complexity of $C(n) = \Omega(n)$ words. Since the work $W(n) = O(m) = O(n)$, the result will then follow.

Any two processor implementation of the problem induces an embedding of G onto the processors as follows: embed vertex X_i onto the processor that reads x_i, and vertex Y_j onto the processor that generates y_j. Let X' denote the vertices of X placed on the first processor, and Y' the vertices of Y placed on the second processor. By the balance requirement, we know that $|X'| = |Y'| = n/2$. We also know that X' is connected to at least $3n/4$ vertices in Y by the expansion property. Thus X' is connected to at least $n/4$ vertices in Y'. Because the degree of G is a constant, it follows that there exists a matching consisting of $\Omega(n)$ edges such that each edge has an endpoint on both processor. Thus by setting A appropriately, we can force $\Omega(n)$ values to flow between the processors. $\square$

5.2 FEM matrices

The Finite Element Method (FEM) is commonly used to solve partial differential equations such as those arising from structural mechanics. An important step in the FEM is the solution of a sparse linear system $Ax = y$, for which iterative methods are often used. This requires the multiplication of the (sparse) matrix A by a vector x. The sparsity pattern of A is derived from a graph G almost exactly as described in the previous section. A minor difference is that these graphs are not bipartite; $A[i,j]$ is nonzero iff (i,j) is an edge in G whose vertices are labelled 1 through n. Finite element graphs do possess bounded degree, but do not have expansion, instead they have small *separators*[25].

Definition 7. A subset of vertices C of a graph G with n vertices is an $f(n)$-separator that δ-splits if $|C| \leq f(n)$ and the vertices of $G - C$ can be partitioned into two sets A and B such that there are no edges from A to B, and $|A|, |B| \leq \delta n$, where $0 < \delta < 1$ and f is a function.

Miller et al[25] show that graphs arising from 3 dimensional finite element problems have $O(n^{2/3})$ separators that $1/2$ split. Further, this property applies to the graphs obtained after the separation and so on. From this it follows that an n vertex finite element graph can be embedded onto p processors such that each processor receives $O(n/p)$ vertices, and further only $O((n/p)^{2/3})$ edges leave each processor.

We consider the p processors to be connected in a Butterfly network, and show that the Butterfly network will perform the matrix vector multiplication in time $O(n/p)$. The algorithm is defined by the embedding mentioned above. Each processor that holds vertex i is responsible for reading in x_i, as well as computing y_i. It also holds the values of nonzero $A[j,i]$ for all j. Each processor computes $A[j,i] * x_i$ and sends it to the processor responsible for computing y_j. Each processor responsible for computing y_i receives all the necessary terms from the other processors, and adds them up. Since G has bounded degree, each processors does $O(n/p)$ operations, and sends out $O((n/p)^{2/3})$ values. The communication time is $O((n/p)^{2/3} \log p) = O(n/p)$ for $\log p = O((n/p)^{1/3})$. Thus the total time is $O(n/p)$ whenever $n = \Omega(p \log^3 p)$. But the PRAM time cannot be better than this, since this is linear speedup. Thus locality is fully exploited.

6 Relational Join

For distributed databases, an intensely studied problem is the *fragmentation* problem, viz. how to distribute the tuples in each relation among the available processors. This is analogous to the data layout problem facing us. While much research in databases focusses on how to reduce disk I/O operations (or how to encourage parallel I/O), minimizing data movement between processors is also important. This is our central question: Given a database and a set of possible join operations that can potentially be performed on its relations, is their a fragmentation strategy that guarantees small data movement for all the joins?

The answer will obviously depend upon the database and the joins allowed; we only show here that a universally good fragmentation strategy does not exist, even for a database consisting of a small number of relations and allowed joins. In particular, we will show that there exists a database with $r = O(1)$ relations and $O(n)$ triples in each relation such that given any fragmentation strategy some join operation will require $\Omega(n)$ data to be moved. Since any join operation will require $W(n) = O(n)$, this shows that the join is inherently nonlocal.

Our database is derived from a d-regular graph G having n vertices, where d is a suitable constant. We require that the graph not have an $(n/10)$ separator that $1/2$ splits. It is easily proved that such graphs exist using the probabilistic method [4]. We assign each edge of the graph a color in the range 1 through d such that edges incident at any vertex have distinct colors. Further we assign a weight to each edge also. We now define the database: each relation corresponds to edges of a single color. In particular relation i consists of tuples (u, v, w) where (u, v) is an edge in G of color i, and weight w.

Consider any particular scheme for storing the database on two processors. We will assume that each tuple is stored integrally on one processor. Call a vertex *live* if it appears in tuples stored on both processors.

Lemma 8. *At least $n/10$ vertices are live.*

Proof: Removal of the live vertices splits the graph into two subgraphs, one stored on each of the two processor. How large can these subgraphs be? All the tuples involving a non-live vertex must be stored on a single processor. Since the graph is d-regular, and because we know that the tuples must be stored in a balanced manner among the two processors, it is clear that each processor holds all the tuples for at most $n/2$ non-live vertices. Thus the live vertices comprise a separator that $1/2$ splits the graph. But we know that the graph does not have an $n/10$ separator that $1/2$ splits. Thus there must be at least $n/10$ live vertices. □

For each live vertex v, let p_v be a tuple containing v stored on the first processor, and let q_v be a tuple containing v stored on the second processor.

Clearly, there exist relations R_1 and R_2 and a set L of live vertices such that $|L| > n/(10r^2)$ and for every $v \in L$ we have (i) $p_v \in R_1$ and (ii) $q_v \in R_2$.

Now consider a join of relations R_1 and R_2. It is clear that $\Omega(n)$ words must be moved no matter where the result of the join is stored.

7 Connected Component Labelling

We consider 3 variations of the basic problem. The first is the general sparse version, in which the number of edges is proportional to number of vertices. The second is the dense version. In the third version the graph is constrained to be a subgraph of a grid.

7.1 General sparse graphs

JaJa[18] showed that the communication complexity of determining whether or not an n vertex graph is connected is $\Omega(n \log n)$ bits, or $\Omega(n)$ words. This is true even for sparse graphs, i.e. those having $m = O(n)$ edges. The best deterministic algorithm for sparse graphs is due to Cole and Vishkin[9] and requires time $T = O(\log n)$ and uses $n\alpha(m, n)/\log n$ processors, where $\alpha(m, n)$ is the inverse Ackerman function. Thus $W(n) = n\alpha(m, n)$. The problem thus has locality $\alpha(m, n)$.

By theorem 3, even with full exploitation of locality, the time on any model M would have to be at least $\Omega(TS(PRAM/M)/\alpha(m,n))$. This is hardly better than $O(TS(PRAM/M))$, the time acheived using PRAM simulation.

We conjecture that the randomized communication complexity of connectivity is also $O(n)$ words, in which case the locality of the sparse graph connectivity problem is $O(1)$ using the efficient (randomized) parallel algorithm by Gazit[13], i.e. the problem is inherently nonlocal. In this case it is clear that a direct network implementation will be no more efficient than a straightforward simulation of the PRAM algorithm.

7.2 Dense Graphs

In this case the number of edges is $\theta(n^2)$, with n the number of vertices. Driscoll et al[12] show how $n/\log n$ processors connected in a tree may be used to compute a minimum spanning forest in time $O(n\log n)$. This algorithm can be easily adapted to find connected components without changing the time and processor bounds. Thus we get linear speedup, which is the best possible even on a PRAM, giving full exploitation of locality.

We do not know if the number of processors can be increased while maintaining the linear speedup that the above algorithm provides. This is an interesting open question.

7.3 Grid graphs

The so called region-labelling problem[3] is as follows. Given an $n \times n$ array of pixels, assign a distinct label to contiguous pixels of the same color. This is really a connected component labelling problem on subgraphs of two dimensional grids. Similar problems also arise in the simulation of Ising models[35].

These problems also possess locality. We sketch how to solve a region labelling problem on $n \times n$ pixels using a $p = n^2/\log^2 n$ processor butterlfy network. We assign regions of size $\log n \times \log n$ to each processor. Each processor finds connected components locally in time $O(\log^2 n)$, generating at most $4\log n$ components which might be connected to pixels outside its region. At this stage we have the task of mergeing $q = 4\log n(n^2/\log^2 n) = O(n^2/\log n)$ components generated by all processors together. This is done using Gazit's algorithm[13] simulated on the Butterfly. We know that a $q/\log n$ processor PRAM would find the components in time $O(\log n)$. This algorithm when simulated on $n^2/\log^2 n = O(q/\log n)$ processor Butterfly would finish in time $O(\log^2 n)$. Thus the over all time is $O(\log^2 n)$. Thus we have linear speedup, and we have fully exploited locality.

8 Combinatorial Search

Backtrack search and branch-and-bound are commonly used strategies for solving combinatorial search/optimization problems. Parallel algorithms have been developed for both these strategies, and some work has been done on exploiting locality.

A combinatorial search problem may be thought of as the problem of exploring an implicitly specified tree. The input consists of the root of the tree, and a procedure, which when applied to any tree node generates its children if any. Each tree leaf is associated with a cost c, and the goal is to find the least cost leaf. This models iterative refinement strategies: the root corresponds to the least refined solution (with several problem constraints unsatisfied), and leaves to most refined solution (all constraints satisfied). Optionally, internal nodes of the tree might also have an associated cost with the constraint that the cost function is monotone increasing, i.e. $c(\text{parent}(v)) < c(v)$, for all nodes v except the root. In this case the cost function can be used to prune the search space, resulting in branch-and-bound methods. If a monotonic cost function is not available, then the entire tree must be generated and the least cost leaf reported (backtrack search).

Karp and Zhang[19, 30] give an optimal PRAM algorithm for performing branch-and-bound searches. Their algorithm uses the following simple strategy. Each processor maintains a local priority queue to store tree nodes in increasing order of the cost. At each step, each processor picks the least cost node from its queue, expands it, and sends the resulting children to randomly chosen processors in the PRAM. They show that their algorithm is competitive with the sequential Best-first heuristic: if the height of the tree explored by the sequential heuristic is h and the number of nodes explored is n, then their parallel algorithm essentially accomplishes the same and finishes in time $O(n/p+h)$. For backtrack search they propose several strategies to finish in the optimal time $O(n/p + h)$ where n is the number of nodes in the tree and h the height.

Locality preserving algorithms are known for both problems. Ranade[32] shows how the time of $O(n/p + h)$ can be achieved on a Butterfly network of processors using the following strategy: whenever a new child is generated, it is sent to the queue of a randomly chosen *nearby* processor. Zhang has recently shown[40] how to exploit locality using the "idle initiates" heuristic of [19]. With this, it is possible to acheive time $O(n/p + h)$ for backtrack search on several other networks besides the Butterfly.

Jain[17] has recently proposed a strategy for exploiting locality in branch-and-bound computation. Jain uses a simple extension of Karp and Zhang's idea: instead of sending every node to a randomly chosen processor, send only a small fraction of the nodes, retain others in the local priority queue. Jain shows that the time of $O(n/p + h)$ can be achieved on a Butterfly network of p processors, but p cannot be larger than $2^{O(\sqrt{\log n})} = o(n^\epsilon)$ for any constant $\epsilon > 0$. Thus full exploitation of locality is achieved, though fairly small parallelism is used.

9 Data Structure Maintenance

We survey locality exploitation strategies in designing parallel priority queues and parallel analogues of 2-3 trees.

Pinotti and Pucci[28] define a parallel priority queue data structure. This supports 2 main operations: insert and deletemin. The insert operation allows p items to be simultaneously inserted; the deletemin retrieves the p items having the smallest key values. On a p processor PRAM they show that both operations can be executed in time $O(\log p)$. Ranade et al[29] show how parallel priority queue operations can be executed with partial exploitation of locality on d dimensional processor arrays and Butterfly networks.

Paul, Vishkin and Wagener[27] show how to implement 2-3 tree like data structures on a PRAM. They provide an insert operation which inserts p elements. They also provide a Find(X) operation where X is a vector of p elements. The result of the operation is a vector Y of p elements. Each X[i] is a key; if an item with the same key exists in the datastructure, then Y[i]=X[i]. If such an item does not exist, then Y[i] is required to be the smallest element in the datastructure larger than X[i]. Both operations can be implemented in time $O(\log p)$ using a p processor PRAM. It is also possible to delete elements rather than just find them, in the same amount of time. Ranade[31] presents a data structure using which 2 dimensional processor arrays can support the same operations in time $O(\sqrt{p})$. This is partial exploitation of locality. A data structure is also presented using which a $p/\log p$ processor Butterfly can execute p find operations in time $O(\log^2 p)$. This corresponds to full exploitation of locality, though insert operations are not supported, and the number of processors is reduced somewhat.

10 N-body problem

The n-body problem is as follows: given the positions and magnitudes of n masses compute the gravitational force exerted on each by all of the others. This problem is central in many areas such as astrophysics, electrostatics and fluid dynamics.

A direct method that evaluates every pairwise interaction separately and then adds up the contributions is easily seen to take $O(n^2)$ time; this is prohibitively large when n is large, as is often necessary. The Fast Multipole Method of Greengard and Rokhlin[8, 16, 15] completes these computations in time $O(n)$. We sketch an approach for exploiting locality in parallel implementations of the fast multipole method.

The principle data structure in the fast multipole method is an oct-tree (quad-tree in the two dimensional version). The oct-tree holds the particles/charges whose interaction is being computed. The structure of the oct-tree is adaptive; it is deeper where particles are distributed densely. The algorithm has three phases. The first phase is akin to a data reduction; data flows from leaves towards the root, with $O(1)$ words flowing on each edge. In the second phase nodes exchange $O(1)$ words of information with their siblings. The third phase is the reverse of the first, with data flowing from the root towards leaves.

Exploiting locality in these computations requires a good embedding of the octtree (augmented with sibling connections) among the processors in the parallel machine. In particular, it is desirable that each processor in the machine receives the same number of tree nodes, and has only a small number of oct-tree edges (or sibling edges) leaving to nodes on other processors. Existing implementations use data layout heuristics such as the *costzones* heuristic[34], or *orthogonal recursive bisection*[23, 39]. These can be shown to work well with uniform distribution of particles, but no rigorous analysis of these is known for the case of non-uniform distribution which are common in practice. There is some experimental evidence that the heuristics may work well for the (non-uniform) particle distributions that researchers have experimented with, however, distributions that will lead to bad communication performance are known.

It is possible to show that n node octtrees have $O(n^{2/3})$ separators. From this it follows, using the argument in section 5.2 that we can partition the nodes of an octtree among p processors such that each processor only receives $O(n/p)$ nodes, and number of edges leaving each processor is $O((n/p)^{2/3})$. Using this, it is possible to show that a p processor Butterfly network can execute phase 2 in time $O(n/p)$ whenever $n = \Omega(p \log^3 p)$. Phases 1 and 3 would require time at most $O(n/p + h \log p)$, so that linear speedup would be obtained for $h \log p = O(n/p)$, which is likely to be true. Thus we would also have full exploitation of locality.

The above solution is theoretically unsatisfactory in that it does not take into account the overhead of computing separators. In practice this may be fine because the n-body method is itself known to be extremely compute intensive as well as communication intensive (especially phase 2), thus the time spent in computing separators may well be negligible. But this needs more analysis. Second, it would be desirable to eliminate the requirement that $h \log p = O(n/p)$.

11 Concluding Remarks

One goal of this paper was to present a framework for designing bandwidth efficient algorithms. We have presented such a framework and illustrated its use for several application problems; in the process presenting several new results as well as interpreting old results in the context of designing bandwidth efficient algorithms. Our larger goal really is to highlight this as an important research area. We believe that the entire range of PRAM algorithms needs to be reexamined in this context and bandwidth efficient algorithms designed for all problems for which we have good PRAM algorithms (or failing which, lack of locality established).

A discouraging implication of what we have presented is that to understand locality it is necessary to examine each application problem anew. Essentially, it is necessary to understand in great detail the data dependencies implied in the problem, understand their graphical structure and then devise good data layouts. Thus, unlike problems like latency and asynchrony, there is no easy universal solution (such as parallel slackness) that can be used. The only universal solution is to build high bandwidth networks, as is being done by groups like the Saarbrucken PRAM project and the TERA project. This solution is not entirely satisfactory either. It is possible that a few computers based on high bandwidth networks will be built and be commercially viable and successful. But there will continue to exist low bandwidth networks (such as those targeted by the "Networks of Workstations" initiatives) which will demand bandwidth efficient algorithms.

Acknowledgements

I am grateful to Wolfgang Paul for suggesting to me to examine locality issues in sparse matrix multiplication.

References

1. F. Abolhassan, R. Drefenstedt, J. Keller, W. Paul, and D. Scheerer. On the Physical Design of PRAMS. *The Computer Journal*, 36(8), 1993.
2. Alok Aggarwal, Ashok Chandra, and Marc Snir. Communication Complexity of PRAMS. *Theoretical Computer Science*, pages 3–28, March 1990.
3. A. Agrawal, L. Nekludova, and W. Lim. A parallel $o(\log n)$ algorithm for finding connected components in planar images. In *Proceedings of the International Conference on Parallel Processing*, pages 783–786, June 1987.
4. N. Alon, J. Spencer, and P. Erdos. *The Probabilistic Method*. 1992.
5. Robert Alverson, David Callahan, Daniel Cummings, et al. The TERA Computer System. In *Proceedings of Supercomputing 90*, pages 1–6, 1990.
6. M. Atallah and M. Goodrich. Efficient parallel solutions to some geometric problems. *Journal of Parallel and Distributed Computing*, 3:492–507, 1986.
7. Robert Boothe and Abhiram Ranade. Improved multithreading techniques for hiding communication latency in multiprocessors. In *Proceedings of the Nineteenth Annual Symposium on Computer Architecture*, pages 214–223, May 1992.
8. J. Carrier, L. Greengard, and V. Rokhlin. A fast adaptie multipole algorithm for particle simulations. *SIAM Journal of Scientific and Statistical Computing*, July 1988.
9. R. Cole and U. Vishkin. Approximate and exact parallel scheduling with application to list, tree and graph problems. In *Proceedings of the IEEE Annual Symposium on The Foundations of Computer Science*, pages 478–491, 1986.
10. D. Culler, R. Karp, D. Patterson, A. Sahay, K. Schauser, E. Santos, R. Subramonian, and T. Eicken. LogP: Towards a realistic model of Parallel Computation. In *Principles and Practice of Parallel Programming*, 1992.
11. F. Dehne, A. Fabri, and A. Rau-Chaplin. Scalable Parallel Geometric Algorithms for Coarse Grained Multicomputers. In *Proceedings of the ninth annual symposium on Computational Geometry*, pages 298–307, May 1993.
12. James R. Driscoll, Harold N. Gabow, Ruth Shrairman, and Robert E. Tarjan. Relaxed heaps: An alternative to fibonacci heaps with applications to parallel computation. *Communications of the ACM*, 31(11):1343–1354, November 1988.
13. H. Gazit. An optimal randomized parallel algorithm for finding connected components in a graph. In *Proceedings of the IEEE Annual Symposium on The Foundations of Computer Science*, pages 492–501, 1986.
14. Phillip B. Gibbons. *The Asynchronous PRAM: A Semi-Synchronous Model for Shared Memory MIMD Machines*. PhD thesis, University of California, Berkeley, 1989.

15. Leslie Greengard and Vladimir Rokhlin. A fast algorithm for particle simulation. *Journal of Computational Physics*, (73):325–348, 1987.

16. Leslie F. Greengard. *The rapid evaluation of potential fields in particle systems.* MIT Press, 1988.

17. Sanjay Jain. Branch and Bound on the Network Model. In *Proceedings of the Symposium on Foundations of Software Technology and Theoretical Computer Science*, 1995. To appear.

18. Joseph Ja'Ja'. The VLSI Complexity of Selected Graph Problems. *Journal of the ACM*, 31:377–391, April 1984.

19. Richard Karp and Yanjun Zhang. Randomized parallel algorithms for backtrack search and branch-and-bound computation. *Journal of the ACM*, 40(3):765–789, 1993.

20. V. Kumar, A. Grama, A. Gupta, and G. Karypis. *Introduction to Parallel Computing.* The Benjamin/Cummings Publishing Company, 1994.

21. F. T. Leighton. *Introduction to parallel algorithms and architectures.* Morgan-Kaufman, 1991.

22. C. E. Leiserson and B. M. Maggs. Communication-efficient parallel graph algorithms for distributed random-access machines. *Algorithmica*, 3:53–77, 1988.

23. P. Liu and S. Bhatt. Experiences with Parallel N-body Simulation. In *Proceedings of the ACM Symposium on Parallel Algorithms and Architectures*, 1994.

24. Laszlo Lovasz. Communication complexity: A survey. In *Paths, Flows and VLSI Layout.* Springer–Verlag, 1989.

25. G. Miller and W. Thurston. Separators in Two and Three Dimensions. In *Proceedings of the ACM Annual Symposium on Theory of Computing*, pages 300–309, 1990.

26. M. H. Overmars and J. van Leeuwen. Maintenance of configurations in the plane. *Journal of Computer and System Sciences*, 23:166–204, 1981.

27. W. Paul, U. Vishkin, and H. Wagener. Parallel dictionaries on 2-3 trees. In J. Diaz, editor, *ICALP 83: Automata Languages and Programming.* Volume 154 of *Lecture Notes in Computer Science*, pages 597–609. Springer Verlag, New York, NY, 1983.

28. Maria Pinotti and Geppino Pucci. Parallel priority queues. Technical Report TR-91-016, International Computer Science Institute, March 1991.

29. A. Ranade, S. Cheng, E. Deprit, J. Jones, and S. Shih. Parallelism and Locality in Priority Queues. In *IEEE Symposium on Parallel and Distributed Processing*, 1994.

30. Abhiram G. Ranade. A Simpler Analysis of the Karp-Zhang Parallel Branch-and-Bound Method. Technical Report UCB/CSD 90/586, University of California, August 1990.

31. Abhiram G. Ranade. Maintaining dynamic ordered sets on processor networks. In *Proceedings of the ACM Symposium on Parallel Algorithms and Architectures*, pages 127–137, June-July 1992.

32. Abhiram G. Ranade. Optimal speedup for backtrack search on a butterfly network. *Mathematical Systems Theory*, (27):85–101, 1994. Also in the Proceedings of the Third ACM Sympsium on Parallel Architectures and Algorithms, July 1991, pages 40–48.

33. John Reif and Leslie Valiant. A logarithmic time sort for linear size networks. *Journal of the ACM*, 34(1):60–76, January 1987.

34. J. Singh, C. Holt, T. Totsuka, A. Gupta, and J. Hennessy. Load Balancing and Data Locality in Hierarchical N-body Methods. Technical Report CSL-TR-92-505, Stanford University, 1992.
35. R. H. Swendsen and J-S. Wang. Nonuniversal critical dynamics in monte carlo simulations. *Physical Review Letters*, 58:86–88, 1987.
36. J. D. Ullman. *Computational aspects of VLSI*. Computer Science Press, 1984.
37. L. G. Valiant. A Bridging Model for Parallel Computation. *Communications of the ACM*, 33(8):103–111, August 1990.
38. T. von Eicken, D. Culler, S. Goldstein, and K. Schauser. Active Messages: a Mechanism for Integrated Communication and Computation. In *Proceedings of the International Symposium on Computer Architecture*, 1992.
39. M. Warren and J. Salmon. Astrophysical N-body Simulations Using Heirarchical Tree Data Structures. In *Supercomputing*, 1992.
40. Yanjun Zhang. Personal communication.

Variable-Length Maximal Codes*

Véronique Bruyère[1] and Michel Latteux[2]

[1] University of Mons-Hainaut, 15 Avenue Maistriau,
B-7000 Mons, Belgium.
vero@sun1.umh.ac.be
[2] University of Lille, L.I.F.L., Bâtiment M3,
F-59655 Villeneuve d'Ascq, France.
latteux@lifl.fr

Abstract. In this survey, we present some of the main open problems
in the theory of variable-length codes, together with the major advancements recently realized to solve them.

1 Introduction

The beginnings of both coding and information theory go back to the pioneer
works of C.E. Shannon in 1948. Coding theory has later developed into two independent directions. One is the design of constant-length codes for the problems
of error detection and correction. This theory is a nice application of commutative algebra [34, 35, 51]. The other direction, initiated by M.-P. Schützenberger
in 1955 [45], has lead to the theory of variable-length codes. This theory is now
clearly a part of theoretical computer science, strongly connected to formal languages, combinatorics on words, automata theory and the theory of semigroups
[2, 4, 27, 32, 43, 49].

Maximal codes play a central role in the theory of variable-length codes.
They reflect extremal properties of codes when they are considered as communication tools. In some sense, maximal codes make use of the whole capacity of the
transmission channel. Despite a lot of research activities and important progress,
some open problem on maximal codes are still open today, whose cornerstone
is the so-called *commutative equivalence conjecture*. This conjecture belongs to
the list of six old open problems about regular languages presented by J. Brzozowski at the International Symposium on Formal Language Theory, Santa
Barbara, California, 1979 [11]. It states the optimality of prefix codes in a way
that reinforces the Kraft-McMillan inequality well known in information theory.

In this paper, we want to survey some of the major developments on regular
codes for the last ten years, made around the commutative equivalence conjecture
and the related open problems. We particularly emphasize on the variety of the
approaches. We sketch the proofs when they are simple; we provide examples in
case of more complex arguments.

---

* This work was partially supported by a cooperation project CGRI-FNRS-CNRS.

The paper is organized as follows. In Sect. 2, we recall the basic notions on codes and the way to characterize them by simple properties of finite automata. For this section, our reference book is mainly [2]. In Sect. 3, we enunciate the commutative equivalence conjecture and its partial results. It has stimulated two other open problems, the problem of completion of finite codes discussed in Sect. 4 and the factorization conjecture studied in Sect. 5. It appears that partial answers to both questions display a strong relation with factorizations of cyclic groups. In Sect. 6, we explain what are the consequences on codes of some properties and open problems on factorizations of cyclic groups. The last section deals with the problem of embedding a code into a maximal one sharing common properties, for which recent important advancements have occurred.

2 Codes

2.1 Definitions

Assume that we are given a *source alphabet B* and a *channel alphabet A*, and we want to code messages v over B into coded messages w over A. The letters of B are put in correspondence with the words of a *code X* over A. Any letter of the source message v is then replaced by the corresponding word of X. We require that the decoding of the resulting coded message $w \in X^*$ does not lead to ambiguities, i.e., any product of words of X is factorized in a unique fashion.

More formally, $X \subseteq A^*$ is a *code* if for any $x_1, \ldots, x_n, y_1, \ldots, y_m \in X$, $n, m \geq 1$,

$$x_1 \cdots x_n = y_1 \cdots y_m \quad \Rightarrow \quad n = m \text{ and } x_i = y_i, \forall i .$$

Example 1. The Morse code X_M is a code over the alphabet $\{., \wedge, -\}$ used to code the letters $a, b, \ldots, z$ (see Fig. 1). The blank symbol $\wedge$ only appears at the end of words of X_M, ensuring that X_M is a code. Another example is the ASCII code X_A over the binary alphabet. Whereas the words of X_M have variable length, all the words of X_A have constant length 8.

It is reasonable to require for a code X that the decoding of a coded message can begin without waiting for the whole reception of the message. Such codes are said to have *bounded decoding delay*, that is, there exists $d \geq 0$ such that for any $x, x' \in X$, $y' \in X^*$ and y prefix of a word of X^* (see Fig. 2),

$$xy \text{ prefix of } x'y' \text{ and } |y| \geq d \quad \Rightarrow \quad x = x' . \tag{1}$$

The least integer d satisfying (1) is called the *delay* of X.

Codes with decoding delay 0 are the most interesting ones. They are also called *prefix* codes, or *instantaneous* codes. In this case, no word of the code is prefix of another one. A prefix code can then be represented as a tree with arity $|A|$,– the size of the alphabet. The leaves correspond to the words of the code (see Fig. 1). This shows that prefix codes are easy to construct.

a	$. - \wedge$	j	$. - - - \wedge$	s	$... \wedge$
b	$- ... \wedge$	k	$-. - \wedge$	t	$- \wedge$
c	$-. - . \wedge$	l	$. - .. \wedge$	u	$.. - \wedge$
d	$-.. \wedge$	m	$- - \wedge$	v	$... - \wedge$
e	$. \wedge$	n	$-. \wedge$	w	$. - - \wedge$
f	$.. - . \wedge$	o	$- - - \wedge$	x	$-.. - \wedge$
g	$- - . \wedge$	p	$. - - . \wedge$	y	$-. - - \wedge$
h	$.... \wedge$	q	$- - . - \wedge$	z	$- - .. \wedge$
i	$.. \wedge$	r	$. - . \wedge$		

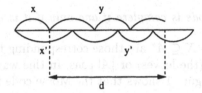

Fig. 1. Morse code.

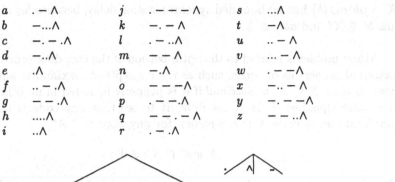

Fig. 2. Code with bounded decoding delay d

Example 2. The Morse code and the ASCII code are prefix codes. The code $X = \{a, aab\}$ is not prefix, it has a decoding delay 2. The code $X = \{a, ab, bb\}$ has no bounded decoding delay, because the coded message ab^n can be decoded only after its complete reception.

Notice that the latter example is a prefix code if one looks at its words from right to left. Such codes are called *suffix*, and a code which is both prefix and suffix is a *bifix* code.

Another interesting class is the one of codes with bounded synchronization delay. A code $X \subseteq A^*$ has a *synchronization delay* s if

$$uxyv \in X^*, \quad \text{with} \quad x, y \in X^s, \ u, v \in A^* \quad \Rightarrow \quad ux, yv \in X^* .$$

In practice, the presence inside a coded message $uxyv$ of the synchronization pair xy forces the decoding to pass between x and y. In this way, an error occurring before x does not affect the decoding of yv.

Example 3. The Morse code X_M has synchronization delay 1, since the end of each word of X_M is "marked" by the blank symbol $\wedge$ (see Fig. 1). The bifix code

$X = ab^*c \cup \{b\}$ has no bounded synchronization delay, because for all $s \in \mathbb{N}$, one has $b^s \in X^s$ and $ab^sc \in X$.

Many problems related to the optimization of the encoding process deal with extremal properties of codes, such as the concepts of maximality and completeness. A code $X \subseteq A^*$ is *maximal* if it is properly included in no other code over the same alphabet A. It is not difficult to see that any code is included in a maximal one. A code X is *complete* if for any word $w \in A^*$,

$$A^* w A^* \cap X^* \neq \emptyset .$$

Roughly, any word of A^* occurs as part of a coded message. In this sense, a complete code makes use of the whole capacity of the transmission channel.

A fundamental result of M.-P. Schützenberger [45] states the equivalence of these two notions for finite codes and more generally for the class of regular codes. We recall that a set of words is *regular* or *rational* if it is recognizable by a finite automaton [22, 26, 38].

Theorem 1. *A regular code is complete if and only if it is maximal.*

Complete prefix codes $X \subseteq A^*$ are those corresponding to trees of which all nodes have either 0 sons (the leaves) or $|A|$ sons. In this way, any word $w \in A^*$ verifies $wA^* \cap X^* \neq \emptyset$. Figure 1 shows that the Morse code is not complete.

2.2 Automata

In this paper, we focus on maximal codes which are finite or regular. The automata that we consider are strongly connected directed graphs $\mathcal{A} = (S, 1, T)$ with a set S of vertices or *states*, a particular state $1 \in S$ and a set T of edges or *transitions* labeled by an alphabet A. Notation $(r, a, s) \in T$ means that a labels an *edge* going from state r to state s. It is extended to paths: $(r, w, s) \in T^*$ means that w is the label of a *path* from r to s. The set $X_{\mathcal{A}} \subseteq A^*$ of *first returns* is defined as the set of labels of all paths going from state 1 to state 1, without passing through 1 (see Fig. 3). If the automaton $\mathcal{A}$ is finite, this set $X_{\mathcal{A}}$ is regular.

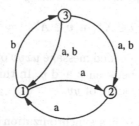

Fig. 3. Automaton $\mathcal{A}$ and set $X_{\mathcal{A}} = \{aa, ba, baa, bba, bb\}$.

Code properties are easily checked on automata. We recall that an automaton is *deterministic* if

$$(r, a, s_1), (r, a, s_2) \in T \quad \Rightarrow \quad s_1 = s_2 \ .$$

More generally, an automaton is *unambiguous* if for all $r, s \in S$ and $w \in A^*$, there is at most one path from r to s labeled by w (see Fig. 3).

Proposition 2. *A regular set X is a code if and only if $X = X_A$ for some unambiguous finite automaton A. Moreover X is prefix if and only if A can be chosen deterministic.*

Completeness of a code $X = X_A$ is translated by the following property of the automaton A: any word $w \in A^*$ labels some path of A. In the case of a complete prefix code, A is deterministic and complete, i.e.,

$$\forall r \in S, \forall a \in A, \exists! s \in S, \quad (r, a, s) \in T.$$

Finally, a regular code $X = X_A$ is finite if and only if all cycles of A go through the particular state 1.

2.3 Composition of Codes

We now introduce the operation of composition of codes for constructing more complicated codes from simpler. Beginning with prefix or suffix codes, this operation can lead to codes neither prefix nor suffix.

Let $Z \subseteq A^*$ and $Y \subseteq B^*$ be two codes together with a bijection $\varphi : B \to Z$. Then the set $X = \varphi(Y)$ obtained by coding the letters of Y by the corresponding words of Z is again a code over the alphabet A. We say that X is obtained by *composition* of Y and Z and we denote $X = Y \circ Z$.

Example 4. Let $Z = \{a, ba, bb\}$ over $A = \{a, b\}$ and $Y = \{\alpha\alpha, \beta, \beta\alpha, \gamma\alpha, \gamma\}$ over $B = \{\alpha, \beta, \gamma\}$, then $X = Y \circ Z = \{aa, ba, baa, bba, bb\}$ with φ a bijection such that $\varphi(\alpha) = a$, $\varphi(\beta) = ba$ and $\varphi(\gamma) = bb$.

Notice that in Example 4, Z is prefix, Y is suffix, but X is neither prefix nor suffix. However, if the codes used in the composition are all prefix (resp. suffix), then the resulting code is still prefix (resp. suffix). This is still true for complete codes: the composition of complete codes always gives a complete code (see Example 4).

A code which is constructed by a finite number of compositions of prefix and/or suffix codes only, is called a *prefix-suffix composed code*. Example 4 is such a code. Until the seventies, people thought that the composition of codes was a powerful tool, in the sense that any finite maximal code was prefix-suffix composed. In 1974, Y. Césari [13] found the first counterexample with a total number of 73 words over a 4-letter alphabet. J.-M. Boë [5] later proposed smaller counterexamples of size 11, such as the following code

$$X = \{b, aba, ab^2, a^4, ba^2b, aba^3, ab^2a^2, ba^3b, ba^2ba^2, ba^3ba^2, ba^6\}$$

which is not prefix, not suffix and not decomposable into simpler codes.

These results show that finite maximal codes are complex objects. In the next sections, we will enunciate some of the major conjectures on codes. These conjectures would all be true if every finite maximal code was prefix-suffix composed.

3 Optimality of Prefix Codes

In information theory, Shannon's first theorem relates the average length of a code X to the entropy of the alphabet B encoded by X [25]. More precisely, assume that X is a code over the channel alphabet A, and let $p : B \to [0, 1]$ be a probability distribution of the source alphabet B. The *entropy* of B is then defined as

$$H(B) = - \sum_{b \in B} p(b) \log_{|A|} p(b) \ ,$$

and the *average length* of the code X is equal to

$$L(X) = \sum_{b \in B} p(b) |x_b|$$

such that x_b is the word of X encoding the letter b. One has the following inequality

$$H(B) \le L(X)$$

which can be proved thanks to the Kraft-McMillan inequality:

Theorem 3. *Let $X \subseteq A^*$ be a code with length-distribution $(f_n)_{n \in \mathbb{N}}$, i.e., f_n is the number of words in X with length n. Then*

$$\sum_{n \in \mathbb{N}} \frac{f_n}{|A|^n} \le 1 \ . \tag{2}$$

Moreover, there exists a prefix code $Y \subseteq A^*$ with the same length-distribution.

The first part of Theorem 3 shows that if a set has too many too short words, then it cannot be a code. For regular codes, one can prove that equality with 1 in (2) holds exactly when the code is maximal. The second part of the theorem shows the optimality of prefix codes. This remarkable property is used by the well-known Huffman algorithm which constructs a finite maximal prefix code X with minimal average length $L(X)$.

One central conjecture in the theory of codes states the optimality of prefix codes in a stronger way. Let us give an example.

Example 4 (continued). Consider the code $X = \{aa, ba, baa, bba, bb\}$ with length-distribution $f_2 = 3, f_3 = 2$. By Theorem 3, there exists a prefix code with the same length-distribution, namely $Y = \{aa, ab, baa, bab, bb\}$. This set Y satisfies a stronger property: it has been obtained by a rearrangement of the letters inside each word of X.

We say that two codes X and Y are *commutatively equivalent* if there exists a bijection between them such that a word and its image only differ by the ordering of their letters.

Conjecture 1. Let $X \subseteq A^*$ be a finite maximal code. Then X is commutatively equivalent to a finite maximal prefix code $Y \subseteq A^*$.

Clearly every suffix code is commutatively equivalent to a prefix code. More generally, any prefix-suffix composed code is commutatively equivalent to a prefix code (just replace each suffix code used in the composition by a commutatively equivalent prefix code,– see Sect. 2.3). See Example 4 above.

Conjecture 1 can be translated into a divisibility property of polynomials. Given a set $X \subseteq A^*$, we denote by $\underline{X}$ (resp. $\underline{\underline{X}}$) its characteristic series in non commutative (resp. commutative) variables $a \in A$. For example, if $X = \{ab, ba\}$, then $\underline{X} = ab + ba$ and $\underline{\underline{X}} = 2ab$. Given a finite maximal prefix code $Y \subseteq A^*$, we can write

$$\underline{\underline{X}} - 1 = \underline{\underline{P}}(\underline{\underline{A}} - 1)$$

where P is the set of proper prefixes of words of Y. Hence, if a code X is commutatively equivalent to Y, then $\underline{\underline{X}} = \underline{\underline{Y}}$ and $\underline{\underline{X}} - 1 = \underline{\underline{P}}(\underline{\underline{A}} - 1)$. More generally, one can prove the following proposition.

Proposition 4. A finite maximal code $X \subseteq A^*$ is commutatively equivalent to a prefix code if and only if the quotient $T = (\underline{\underline{X}} - 1)/(\underline{\underline{A}} - 1)$ is a polynomial with coefficients in $\mathbf{N}$.

Example 4 (continued). For $X = \{aa, ba, baa, bba, bb\}$, T is the polynomial $1 + a + b + ab$.

A first step in this direction is proved in [46]: if $X \subseteq A^*$ is a finite maximal code, then the quotient T is a polynomial with integer coefficients; moreover if T is irreducible, then X is prefix or suffix.

Two interesting results shed different light on Conjecture 1. The first one states that any code with bounded synchronization delay (see Sect. 2.1) is commutatively equivalent to a prefix code [39]. The second one is due to P. Shor [48]:

Theorem 5. The code

$$X_S = b\{1, a, a^7, a^{13}, a^{14}\} \cup \{a^3, a^8\}b\{1, a^2, a^4, a^6\} \cup a^{11}b\{1, a, a^2\} \tag{3}$$

is not commutatively equivalent to a prefix code.

Of course, code (3) is not maximal. It is an open problem whether it can be embedded into a finite maximal code (see Sect. 4). If yes, Conjecture 1 would be false.

Up to now, Conjecture 1 is widely open. It has stimulated other open problems that we discuss in the next two sections: the problem of completion of finite codes and the factorization conjecture.

4 Finite Codes

4.1 A Decision Problem for Finite Codes

Another main open problem in the theory of variable-length codes deals with the completion of finite codes (see Sect. 7 for the general problem of completion). A code X is said to have a *finite completion* if it can be embedded into a finite complete code.

Problem 2. Is it decidable whether a finite code has a finite completion?

As mentioned in Sect. 3, if it exists, such a decision procedure could be applied to Shor's code X_S (see (3)). In the case of a positive answer, Conjecture 1 would be false. There exists another code for which we do not known whether it has a finite completion. It is the 4-word code [41]

$$X_4 = \{a^5b, a^2b, ba, b\} \ . \tag{4}$$

Some classes of finite codes always have a finite completion. For instance, a finite prefix (resp. suffix) code X can be embedded into a finite complete prefix (resp. suffix) code. Just add to the tree associated with X the lacking leaves (see Sect. 2.1). More generally, any finite prefix-suffix composed code has a finite completion: any prefix or suffix code used in the composition is replaced by a complete prefix or suffix code (see Sect. 2.3).

Example 5. The code $X = \{aa, ba\} = \{\alpha\alpha, \beta\} \circ \{a, ba\}$ has the finite completion $\{\alpha\alpha, \beta, \beta\alpha, \gamma\alpha, \gamma\} \circ \{a, ba, bb\}$.

However it is known [33, 40] that some finite codes have *no* finite completion, such as the small code

$$\{a^5, a^2b, ba, b\} \ . \tag{5}$$

Notice that this code and code X_4 above only differ by one word.

Of course, code (5) is not prefix-suffix composed, otherwise it would have a finite completion. However each code with 1 or 2 words has a finite completion because it is always prefix-suffix composed [41]. To fill the gap between 2-word codes and the 4-word code (5), the authors of [41] have conjectured that any code with 3 elements was prefix-suffix composed. D. Derencourt [15] has recently found a family of counterexamples whose smallest one is the not prefix-suffix composed code

$$\{b, bab, abab^2a\} \ .$$

However, the following problem remains open, since all these counterexamples have a finite completion [16].

Problem 3. Is there a 3-word code which has no finite completion?

4.2 Codes with no Finite Completion

There exists a simple method to construct finite codes with no finite completion, which is related to factorizations of cyclic groups.

Let $\mathbb{Z}_n = \{0, 1, \ldots, n-1\}$ be the cyclic group of order $n \geq 1$. We say [23] that (R, S), with R, S two finite *subsets* of N, is a *factorization* of $\mathbb{Z}_n$ if and only if each element of $\mathbb{Z}_n$ can be expressed uniquely as the sum, modulo n, of an element of R and an element of S:

$$\mathbb{Z}_n = R \oplus S \bmod n \ .$$

More generally [31], given $m \geq 1$, a pair (R, S) of finite *multisets* of N is called a *m-factorization* of $\mathbb{Z}_n$ if

$$\underbrace{\mathbb{Z}_n \cup \ldots \cup \mathbb{Z}_n}_{m \text{ times}} = R \oplus S \bmod n \ .$$

Example 6. If n is a prime number, then $\mathbb{Z}_n$ only admits the trivial factorization (R, S) (up to a translation) with $R = \{0\}$ and $R = \mathbb{Z}_n$ (since $|R| \cdot |S| = n$). The cyclic group $\mathbb{Z}_6$ admits several factorizations, such as $(\{0, 1\}, \{0, 2, 4\})$ or $(\{0, 5\}, \{0, 2, 4\})$. It also admits the 2-factorization $(\{0, 0, 1, 5\}, \{0, 2, 4\})$ since

$$\{0, 0, 1, 5\} \oplus \{0, 2, 4\} = \{0, 0, 1, 1, 2, 2, 3, 3, 4, 4, 5, 5\} \bmod 6 \ .$$

The next proposition shows how m-factorizations of cyclic groups appear in the framework of finite complete codes. It uses the property that in a finite complete code X over A, X contains a unique power of each letter $a \in A$.

Proposition 6. Let X be a finite complete code over the binary alphabet $\{a, b\}$. Let $n, m \geq 1$ such that $a^n, b^m \in X$. Define the multisets

$$R_X = \{p \in \mathbb{N} \mid a^p b^i \in X, \text{ with } i \geq 1\} \ , \tag{6}$$
$$S_X = \{q \in \mathbb{N} \mid b^i a^q \in X, \text{ with } i \geq 1\} \ .$$

Then (R_X, S_X) is a m-factorization of $\mathbb{Z}_n$.

Example 7. For $X = \{a^6, aba^4, ba^4, aba^2, ba^2, a^5b, aba^3b, ba^3b, abab, bab, b\}$, we obtain the factorization $(\{0, 5\}, \{0, 2, 4\})$ of $\mathbb{Z}_6$. For the code $\{aa, ba, baa, bba, bb\}$ of Example 4, one gets the 2-factorization $(\{0\}, \{0, 1, 1, 2\})$ of $\mathbb{Z}_2$.

Proof. We just sketch the proof of Proposition 6 when $m = 1$ [41]. The general case is solved in [31]. The sum $R \oplus S$ is unambiguous, otherwise using the hypothesis that $b \in X$, one can find a word which ambiguously decomposes over the code X. To show that $\mathbb{Z}_n$ is contained in $R \oplus S \bmod n$, we use the completeness of the code: any word $b^{i'} a^m b^{j'}$ is factor of some word of X^*. If i', j', m are chosen large enough, as the code is finite, we obtain a decomposition $b^i a^q (a^n)^t a^p b^j$ with $m = p + q \bmod n$ and $i \leq i', j \leq j'$ (see Fig. 4). $\qquad\square$

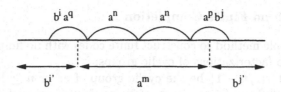

Fig. 4. Decomposition over X.

We are now ready to describe an infinite family of finite codes which cannot be embedded into a finite complete code [40]. The smallest example is the code $\{a^5, a^2b, ba, b\}$ given above (see (5)).

Theorem 7. Let $X = \{a^n\} \cup a^{R_X}b \cup ba^{S_X}$ be a finite code, where sets R_X, S_X both contain 0 (that is $b \in X$). If $|R_X| \geq 2$, $|S_X| \geq 2$ and if n is a prime number, then X has no finite completion.

Proof. Assume that X can be embedded into a finite complete code Y. By Proposition 6, the pair (R_X, S_X) is included into the factorization (R_Y, S_Y) of the cyclic group $\mathbb{Z}_n$. But n is prime, and neither R_Y nor S_Y is equal to $\{0\}$. $\square$

Codes of Theorem 7 are all subsets of $a^* \cup a^*b \cup ba^*$ containing both a^n and b. A collection of results show that inside the class of codes $X \subseteq a^* \cup a^*b \cup ba^*$ such that $a^n, b \in X$, there exist codes with no finite completion if and only if $n \neq 1, 2, 3, 4, 6$ [18, 19, 30, 40, 41]. Very recently, N.H. Lam [31] has used m-factorizations to get examples of codes with no finite completion, containing b^m instead of b. Among these examples, one finds the code $\{a^8, b^4, ba^8, b^4a^3, ba^3, a^8b, ab^4, ab, a^2b^4, a^2b\}$.

Methods to construct codes with no finite completion all use properties of m-factorizations of cyclic groups. We do not know other methods. Concerning Problem 3, we think that m-factorizations are not helpful. Indeed, a pair (R_X, S_X) associated with a code X gives information only if $|R_X| \geq 2$ and $|S_X| \geq 2$. These lower bounds are not achieved by 3-word codes.

4.3 Codes with Unknown Finite Completion

We now come back to finite codes of Sect. 4.1, namely Shor's code X_S and code $X_4 = \{a^5b, a^2b, ba, b\}$ (see (3) (4)) .

The latter code is very close to the code $\{a^5, a^2b, ba, b\}$ having no finite completion (see (5)). The only difference is about word a^5b replacing a^5, in a way that hypotheses of Theorem 7 are no longer satisfied. In other words, Theorem 7 is helpful only if one knows the order n of the cyclic group, i.e., the exponent n such that a^n belongs to the code.

People think (or hope) that codes X_S, X_4 have no finite completion. In this spirit, the next conjecture is formulated in [41] in relation with results of Sect. 4.2.

Conjecture 4. Let $p, q \geq 2$ such that $\gcd(p, q) = 1$. Then the pair (R, S) with $R = \{0, p, q\}$ and $S = \{0, 1\}$ is not embeddable into a factorization of a cyclic group.

Example 8. The pair (R_X, S_X) (see (6)) associated with Shor's code X_S is equal to $(\{0, 3, 8, 11\}, \{0, 1, 7, 13, 14\})$. If Conjecture 4 is true, then X_S cannot be embedded into a finite complete code Y. Otherwise $\{0, 3, 8\} \subseteq R_Y$, $\{0, 1\} \subseteq S_Y$ where (R_Y, S_Y) is the factorization associated with Y by Proposition 6. A similar argument holds for the pair $(\{0, 2, 5\}, \{0, 1\})$ of the code X_4.

5 Factorization Conjecture

5.1 Factorizing Codes

A finite code $X \subseteq A^*$ is called *factorizing* if there exist two finite sets $P, Q \subseteq A^*$ such that

$$\underline{A}^* = \underline{Q} \underline{X}^* \underline{P} , \tag{7}$$

i.e., any word over A uniquely factorizes as qxp, with $q \in Q$, $x \in X^*$ and $p \in P$. Equivalently, X is factorizing if and only if

$$\underline{X} - 1 = \underline{P}(\underline{A} - 1)\underline{Q} . \tag{8}$$

Notice that (7) implies that the code X is complete. Definition (8) shows that if X is factorizing, then X is commutatively equivalent to a prefix code (see Proposition 4). So Conjecture 1 is a particular case of the following conjecture.

Conjecture 5. Any finite maximal code is factorizing.

Again, finite maximal codes which are prefix-suffix composed, are factorizing. Indeed, any finite maximal prefix code $X \subseteq A^*$ is factorizing, since $\underline{X} - 1 = \underline{P}(\underline{A} - 1)$ with P the set of proper prefixes of words of X. Symmetrically for suffix codes, we get $\underline{X} - 1 = (\underline{A} - 1)\underline{Q}$. Finally, a composition of factorizing codes is still a factorizing code.

Example 4 (continued). Let $X = Y \circ Z$ such that $X = \{aa, ba, baa, bba, bb\}$, $Y = \{\alpha\alpha, \beta, \beta\alpha, \gamma\alpha, \gamma\}$ and $Z = \{a, ba, bb\}$. Then $\underline{Y} - 1 = (\alpha + \beta + \gamma - 1)(1 + \alpha)$, $\underline{Z} - 1 = (1 + b)(a + b - 1)$ and $\underline{X} - 1 = (1 + b)(a + b - 1)(1 + a)$.

Definition (7) is "dual" to the notion of complete code, since a code X is complete if any word over A is factor of a word of X^* (see Sect. 2.1). The next results make this relationship more precise [4, p. 136][42].

Theorem 8. Let $X \subseteq A^*$ be a regular maximal code. Then, there exist an integer $d \geq 1$ and $d + 1$ words $y, z_1, \ldots, z_d \in A^*$ such that

$$\forall w \in A^*, \exists! i \in \{1, \ldots, d\}, \, ywz_i \in X^* .$$

The integer d in Theorem 8 is unique; it is the *degree* of the code X. Codes with degree 1 are called *synchronous*. See [12] for interesting properties on unambiguous finite automata $\mathcal{A}$ where code $X_\mathcal{A}$ is synchronous.

Corollary 9. *Let $X \subseteq A^*$ be a finite maximal code. Then, there exist finite subsets P, Q, R of A^* such that*

$$\underline{A^*} = \underline{QX^*P} + \underline{R} \ .$$

Proof. Apply Theorem 8 to each word w of A^*. Either the decomposition over X of the word ywz_i "cuts" w, or w is strictly inside a word $x \in X$ of the decomposition. All words w verifying the second case are put in a set R. This set is finite because X is finite. In the first case, w is cut and factorized as

$$w = qxp$$

with q (resp. p) a proper suffix (resp. prefix) of some word of X, and x a word of X^* (see Fig. 5). All such possible words q define set Q, similarly with words p and set P. Finally, any $w \in A^* \setminus R$ uniquely factorizes as qxp with $q \in Q$, $x \in X^*$ and $p \in P$ because of the unicity of word z_i in Theorem 8. □

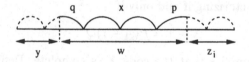

Fig. 5. Decomposition of ywz_i over X.

So, one has to erase set R in Corollary 9 to prove Conjecture 5. In [7], another interesting property is proved for finite maximal codes with degree 1: there exist finite sets P, Q, R such that

$$\underline{A^*} = \underline{QX^*P} - \underline{R} \ .$$

Example 9. Code $X = \{aa, ba, baa, bba, bb\}$ of Example 4 is synchronous, because

$$bwaa \in X^* \tag{9}$$

for all $w \in A^*$. A simple method to verify (9) is to make deterministic the automaton $\mathcal{A}$ of Fig. 3 by beginning with state 1. The new automaton is depicted on Fig. 6. We see that feeding b at state 1 leads to the strongly connected component $\{(3), (1, 2)\}$, and that inside it all paths labeled by aa lead to $(1, 2)$ which contains the particular state 1. Hence, for any word w, there is a path $(1, bwaa, 1)$ in $\mathcal{A}$, which means that $bwaa \in X^*$.

Now, as in the proof of Corollary 9, we get $\underline{A^*} = \underline{QX^*P} + \underline{R}$ with $Q = \{a, aa, b, ba\}$, $P = \{1, b\}$ and $R = \{1\}$.

Notice that we have also $bbwaa \in X^*$ for all $w \in A^*$. This choice is better because R becomes empty and thus $\underline{A^*} = \{1, a\}\underline{X^*}\{1, b\}$.

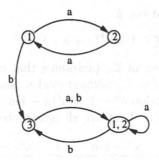

Fig. 6. Determinization of $\mathcal{A}$.

5.2 Reutenauer's Theorem

Up to now, the best result approaching Conjecture 5 is due to C. Reutenauer [4, 42]. It uses polynomials of $\mathbb{Z}[A]$ with coefficients in $\mathbb{Z}$ and non commutative variables in A.

Theorem 10. *Let $X \subseteq A^*$ be a finite maximal code. Then there exist polynomials $P, Q \in \mathbb{Z}[A]$ such that*

$$\underline{X} - 1 = P(\underline{A} - 1)Q . \tag{10}$$

More precisely, if d is the degree of X, there exist polynomials $P, Q, R \in \mathbb{Z}[A]$ such that

$$\underline{X} - 1 = P\left(d(\underline{A} - 1) + (\underline{A} - 1)R(\underline{A} - 1)\right)Q .$$

Therefore, to get Conjecture 5, one has to prove that polynomials P and Q can be chosen with coefficients $0, 1$ in (10). This idea has been developed for codes over $A = \{a, b\}$ having a bounded number of b inside their words [17, 20, 40]:

Theorem 11. *Let $X \subseteq \{a, b\}^*$ be a finite maximal code with at most 3 occurrences of letter b inside its elements. Then, any polynomials P, Q as in (10) have necessarily coefficients $0, 1$. In particular, X is a factorizing code.*

Proof. We only give the proof for codes with at most one b in each word. The general case is rather elaborated. By (10), $\underline{X} - 1 = PaQ + PbQ - PQ$. By hypothesis, P and Q are polynomials in $\mathbb{Z}[a]$ instead of $\mathbb{Z}[a, b]$. Hence PbQ is the only term dealing with words of X containing letter b, and P, Q only have coefficients $0, 1$. □

Theorem 11 cannot be extended. For example, the maximal code $\{b^6\}$ verifies (10) with $P = 1 - b + b^2$, $Q = 1 + 2b + 2b^2 + b^3$ [20]. However, it is factorizing since other polynomials exist which have coefficients $0, 1$ only: $P = 1$ and $Q = 1 + b + b^2 + b^3 + b^4 + b^5$.

As observed in Sect. 4.2, finite maximal codes and factorizations of cyclic groups cannot be separated. Factorizations of cyclic groups also appear in the context of codes X described in Theorem 11:

- If words in X have at most one b:

$$\underline{X} - 1 = a^I(a + b - 1)a^J \tag{11}$$

where (I, J) is a factorization of $\mathbb{Z}_n$ (assuming that $a^n \in X$). This factorization is particular since $I \oplus J = \mathbb{Z}_n$ *without* mod n (see Sect. 4.2). Indeed, (11) restricted to a gives the relation $a^n - 1 = a^I(a - 1)a^J$. Such factorizations are called *Krasner's factorizations*. They are all constructed as follows. Given a decomposition

$$\frac{a^n - 1}{a - 1} = \frac{a^n - 1}{a^{n_1} - 1}\frac{a^{n_1} - 1}{a^{n_2} - 1} \cdots \frac{a^{n_k} - 1}{a - 1} \ ,$$

separate factors $(a^{n_l} - 1)/(a^{n_{l-1}} - 1)$ into two groups, in a way to get $\frac{a^n - 1}{a - 1} = a^I a^J$ [29].

- If words in X have at most two b:

$$\underline{X} - 1 = a^I(a + b - 1)(a^J + \sum_{j \in J} a^{M_j} ba^j)$$

where (I, J) is a Krasner's factorization of $\mathbb{Z}_n$ and for all $j \in J$, (I, J_j) is a factorization of $\mathbb{Z}_n$ such that

$$a^{J_j} = a^J(1 + a^{M_j}(a - 1)) \ . \tag{12}$$

We will come back on these factorizations in Sect. 6.

5.3 Separating Boxes

Another attempt in solving Conjecture 5 uses combinatorics on words instead of polynomials. The starting point is Theorem 8 and its Corollary 9. In [6], the concept of *separating* synchronous code X is introduced, i.e., there exist $y, z \in A^*$ such that any $w \in A^*$ admits a decomposition

$$w = w'w'' \text{ such that } yw', w''z \in X^* \ . \tag{13}$$

Looking at the proof of Corollary 9, one sees that set R is empty. It follows that any finite maximal code which is separating synchronous is factorizing (see end of Example 9). In [6], J.-M. Boë asked whether all synchronous codes were separating. Y. Césari [2, p. 419] gave a negative answer (see Fig. 7). This remarkable finite maximal code X over $A = \{a, b\}$ is a synchronous code such that $X^* = Y^* \cap Z^*$ where $Y = (A^2 \setminus \{b^2\}) \cup b^2 A$ is a maximal prefix code and $Z = A^2 a \cup \{b\}$ is a maximal suffix code (see [9, 28] for codes constructed in this way). Nevertheless, Césari's code is factorizing because

$$\underline{X} - 1 = (1 + \underline{A} + b^2)(\underline{A} - 1)(1 + ab + \underline{A}a + a\underline{A}^2 a) \ . \tag{14}$$

Recently, L. Zhang [52] has proposed the notions of box and separating box for codes. A code $X \subseteq A^*$ has a *box* (U, V), with U, V two finite subsets of A^*, if for any $w \in A^*$, there exist a unique $u \in U$ and a unique $v \in V$ such that

$$uwv \in X^* \ .$$

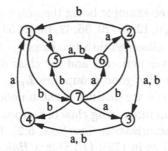

Fig. 7. A synchronous code which is not separating.

This box is *separating*, if

$$uwv \in X^* \text{ with } w = w'w'' \text{ such that } uw', w''v \in X^* .$$

By Theorem 8, any regular maximal code has a box $(\{y\}, \{z_1 \ldots z_d\})$. Separating synchronous codes all have a separating box $(\{y\}, \{z\})$ (see (13)). As for separating synchronous codes, one can easily verify that finite maximal codes having a separating box are factorizing.

An interesting fact is that Césari's code has a separating box $(\{b^3\}, \{1, a^5\})$ leading to (14). Such a box can be read in the automaton of Fig. 7, in the manner briefly outlined in Example 9. Another separating box is $(\{1, a^4, a^4b\}, \{1, aba\})$. There is also a box which is not separating like $(\{a, a^2, a^3\}, \{1, a^5\})$.

Hence, an open question is the following.

Problem 6. Has any finite maximal code a box which is separating?

A first step in this direction appears in [53] for finite maximal code $X \subseteq \{a, b\}^*$ such that $b^n \in X$ with $n = 4$ or n a prime number: if the number of b's inside words of X is bounded by n, then X is factorizing (see also Theorem 11). The proof is simple and a separating box is constructed thanks to factorizations of cyclic groups.

6 Factorization of Cyclic Groups

In Sect. 4 and 5, several results have pointed out connections between finite maximal codes and factorizations of cyclic groups. The general structure of factorizations of cyclic groups is unknown, despite a lot of efforts. This open problem has consequences on the conjectures mentioned before for codes. We briefly explain them (for more details and further references, see the survey [18]).

G. Hajós was the first to introduce the notion of factorization of a cyclic group, by solving a conjecture proposed by H. Minkowski [23] (see Sect. 4.2). He has conjectured that any factorization (R, S) of a cyclic group $\mathbb{Z}_n$ was *periodic*, that is, $R + m = R \bmod n$ for some $m \in \mathbb{Z}_n \setminus \{0\}$. The conjecture was false, the smallest counter-example being the non periodic factorization [14] $(\{0, 8, 16, 18, 26, 34\}, \{0, 1, 5, 6, 12, 25, 29, 36, 42, 48, 49, 53\})$ of $\mathbb{Z}_{72}$.

Cyclic groups have been classified into two classes,– the class of *good groups* for which *all* factorizations are periodic, and the class of *bad groups*. As a consequence of several papers, cyclic good groups are $\mathbb{Z}_{p^n q}$, $\mathbb{Z}_{p^2 q^2}$, $\mathbb{Z}_{p^2 qr}$, $\mathbb{Z}_{pqrs}$ and their subgroups, where p, q, r, s are distinct prime numbers [23].

G. Hajós [24] has found an interesting class of factorizations, whose construction is based on Krasner's factorizations (see Sect. 5.2). In [18], it is proved that they are exactly constructed as in (12): (R, S) is a *Hajós' factorization* of $\mathbb{Z}_n$ if and only if there exist a Krasner's factorization (I, J) of $\mathbb{Z}_n$ and two finite sets $M, N \subset \mathbb{N}$ such that

$$a^R = a^I(1 + a^M(a - 1)) \quad a^S = a^J(1 + a^N(a - 1)) \ .$$

It is unknown whether other factorizations than Hajós' ones appear in the structure of factorizing codes (see (11) (12)).

Hajós' factorizations is a subclass of the periodic factorizations, but it contains all the factorizations of the good groups. They have the following nice property [18].

Proposition 12. *If (R, S) is a Hajós' factorization of $\mathbb{Z}_n$, with R, S both containing 0, then there exists an integer $k \geq 2$, $k|n$, such that $R = 0 \bmod k$ or $S = 0 \bmod k$.*

It is an open question whether this property holds for every factorization [44]. A positive answer to this question would solve Conjecture 4.

Proposition 12 implies some consequences on Shor's code X_S and code $X_4 = \{a^5 b, a^2 b, ba, b\}$ (see (3) (4)), whose finite completion is an open problem (see Sect. 4.3). Sets $R = \{0, p, k\}$ and $S = \{0, 1\}$ of Conjecture 4 cannot be embedded into a Hajós' factorization. Hence, codes X_S and X_4 are contained in no finite maximal code such that $a^n \in X$, with $\mathbb{Z}_n$ a good group. In [18], it is also proved that X_S and X_4 are contained in no finite maximal factorizing code.

As $\mathbb{Z}_{72}$ is the smallest bad group, if X_S or X_4 have a finite completion, one must add to them the word a^n with n at least equal to 72. This lower bound can be increased thanks to the following result [50].

Proposition 13. *Let $m, n \in \mathbb{N}$ be such that $\gcd(m, n) = 1$. If (R, S) is a factorization of $\mathbb{Z}_n$, then (R, mS) is also a factorization of $\mathbb{Z}_n$.*

Assume that the pair $(\{0, 2, 5\}, \{0, 1\})$ associated with code X_4 is contained in a factorization (R, S) of $\mathbb{Z}_n$. If $\gcd(2, n) = 1$, then $(R, 2S)$ is still a factorization, which is impossible since $0 + 2 = 2 + 0$ in $R + 2S$. Thus $2|n$. In the same way, one proves that 3, 5 and then 30 divides n. For the pair $(\{0, 3, 8, 11\}, \{0, 1, 7, 13, 14\})$ associated with Shor's code, similar computations lead to $330|n$.

7 Completion of Codes

7.1 Embedding into a Complete Code

In Sect. 4, the problem of embedding a finite code into a complete one has been investigated. This question is part of a more general program.

Problem 7. Given a code $X \subseteq A^*$ with Property $\mathcal{P}$, is there a procedure to embed X into a complete code $Y \subseteq A^*$ sharing the same property?

The answer is simple for codes whose Property $\mathcal{P}$ is to be finite prefix, or regular prefix. In both cases, the answer is "yes". The finite case has already been discussed in Sect. 4.1: a complete finite prefix code Y is obtained by adding the lacking leaves to the tree associated with X. The regular case is solved similarly. Given a deterministic finite automaton $\mathcal{A} = (S, 1, T)$ associated with X (see Sect. 2.2), for any state s and letter a such that no transition (s, a, r) exists, add a new transition $(s, a, 1)$. The automaton remains deterministic, is now complete, and the related code Y is a regular complete prefix code containing X.

Embedding procedures are similarly defined for codes which are finite suffix, regular suffix, finite prefix-suffix composed, regular prefix-suffix composed.

When Property $\mathcal{P}$ is being a finite code, we recall that some codes has no finite completion (see Sect. 4.2). The smallest known example is the 4-word code $\{a^5, a^2b, ba, b\}$. However, for the class of regular codes, any regular code can be embedded into a complete one, and the embedding procedure is simple [21].

Theorem 14. Any regular code can be effectively included in a regular code which is complete.

Recently, the statement of this theorem has been reinforced: any regular code X can be realized as $X^* = Y^* \cap Z^*$ with Y, Z two regular maximal codes [9].

Problem 7 has made important progress for two families of codes, the bifix codes and the codes with bounded decoding delay. However, this problem is open for codes with bounded synchronization delay (see Sect. 2.1).

Problem 8. Is any regular code with bounded synchronization delay included in a maximal one?

7.2 Codes with Bounded Decoding Delay

Codes with bounded decoding delay naturally generalize prefix codes which have delay 0 (see Sect. 2.1). A central result due to M.-P. Schützenberger [47] states that a finite complete code has either delay 0 or an infinite delay.

Theorem 15. Any finite complete code with bounded decoding delay is prefix.

In other words, there is no hope to embed a finite code with bounded decoding delay into a complete one, except if the code is prefix. However, if the code is regular instead of finite, such a procedure exists which keeps the same decoding delay [10]. The procedure of Theorem 14 is not suitable because the constructed complete code Y has always an infinite delay.

Theorem 16. *Any regular code with decoding delay d can be effectively included in a regular complete code with the same decoding delay d.*

Original proofs of both theorems use combinatorics on words; they are long and complex. Automata lead to simpler proofs, we want to sketch them.

In the same way deterministic automata correspond to prefix codes (Proposition 2), automata with bounded delay exactly describe codes with bounded decoding delay. Let $d \geq 0$, an automaton has a *bounded delay d* if [1, 8]

$$\begin{array}{c} (r, a, s_1) \in T, (s_1, w, t_1) \in T^* \\ (r, a, s_2) \in T, (s_2, w, t_2) \in T^* \end{array} \text{ with } |w| = d \quad \Rightarrow \quad s_1 = s_2 \ . \qquad (15)$$

When $d = 0$, this is the definition of a deterministic automaton. The next proposition is proved in [8] for finite codes only.

Proposition 17. *A regular code X has a bounded decoding delay d if and only if $X = X_A$ for some finite automaton A with bounded delay d.*

Proof. If A has bounded delay d, then it is not difficult to see that the code X_A has bounded decoding delay d. For the converse, we have to construct a particular automaton $A = (S, 1, T)$ called *literal*. It is a "nearly deterministic" automaton in the sense that

$$(r, a, s_1), (r, a, s_2) \in T, \ s_1 \neq s_2 \quad \Rightarrow \quad s_1 = 1 \ \text{or} \ s_2 = 1 \ .$$

Such an automaton always exists, with $S = \{w^{-1}X \mid w \in A^*\} \setminus \{\emptyset, \{1\}\}$ and state 1 equal to X. Transition (r, a, s) is in T if $a^{-1}r = s$ or if $a \in r$ and $s = 1$. Notice that this automaton is trim. Now, assume that X has decoding delay d. With the notations of (15), either s_1 or s_2 equals 1 as the automaton is literal. Suppose $s_1 = 1$. As A is trim, there exists some paths $(1, u, r)$ and $(t_2, v, 1)$. Reading the two paths of Fig. 8, we get (1) with $x = ua$ showing that $s_2 = 1 = s_1$. $\square$

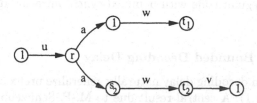

Fig. 8. Two paths with $|w| = d$.

Now, one can prove [8] that given an automaton A with delay d and a code $X = X_A$ with bounded decoding delay d, X is complete if and only if A is d-complete. An automaton A is *d-complete* if

$$(r, w, s) \in T^* \text{ with } |w| = d \quad \Rightarrow \quad \forall a \in A, \exists t \in S, \ (r, wa, t) \in T^* \ .$$

When $d = 0$, a d-complete automaton with delay d is exactly a complete deterministic automaton.

A simple way to prove Theorem 15 is proposed in [8]. Let $\mathcal{A}$ be a finite d-complete automaton with delay d associated with the finite complete code X with bounded decoding delay d. As X is finite, one proves that all pairs of states s_1, s_2 such that $(r, a, s_1), (r, a, s_2) \in T$ can be merged without modifying X. The resulting automaton is deterministic showing that X is prefix.

In [1], it is shown how to make d-complete an automaton with delay d by adding it a *finite* number of new states and new transitions. This procedure leads to a simple proof of Theorem 16. It works as follows.

Example 10. Let $X = \{a, aab\}$ be a code with decoding delay 2. A literal automaton $\mathcal{A}$ for X is depicted on Fig. 9 (see the proof of Proposition 17). In a first step, add a "tree" of height $d = 2$ whose root is state 1 (see Fig. 10). In a second step, if $(r, w, s) \in T^*$ but $(r, wa, t) \notin T^*$ with $|w| = 2$, $a \in A$ (showing that $\mathcal{A}$ is not 2-complete), add a transition (r, b, q_u) such that $wa = bu$. This procedure keeps the delay 2 of the automaton, and makes it 2-complete. By Proposition 17, this new automaton leads to a regular code Y with delay 2 which is complete and contains X.

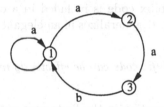

Fig. 9. A literal automaton for the code $\{a, aab\}$.

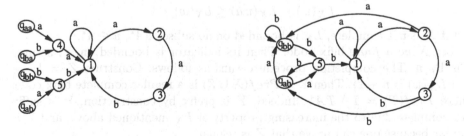

Fig. 10. The two steps for completing $\{a, aab\}$.

In [1], one also finds a nice procedure which embeds a finite automaton with bounded bi-delay (d, d') into a (d, d')-complete one. By bi-delay (d, d'), we mean a (left) delay d as before and a right delay d'. See also [36] for the completion of deterministic local automata.

7.3 Bifix Codes

Bifix codes are particular prefix codes, since they are also suffix. Though the embedding of a prefix code into a complete one is easy, this is more complicated for bifix codes due to the two-sided condition defining them. For example, contrarily to prefix codes, there exist finite bifix codes not embeddable into a finite complete one.

Proposition 18. *The bifix code* $X = \{a^n, b^m\}$, *with* $n \neq m$, *is included in no finite complete bifix code.*

Proof. Assume that $n < m$ and that X is included in a finite complete bifix code Y. Thus, as a complete prefix code, Y satisfies (see Sect. 2.1)

$$\forall w \in A^*, \ wA^* \cap Y^* \neq \emptyset \ .$$

Hence for all i, $1 \leq i < m$, there exists $k_i \geq 1$ such that $b^i a^{k_i} \in Y$ (remember that Y is finite). The k_i's are pairwise distinct since Y is suffix. But $a^n \in Y$ with $n < m$, and a^n is a proper suffix of some $b^i a^{k_i}$. □

However, every regular bifix code is included in a complete one. The proof has been found recently [54]. It generalizes in an elegant way the proof proposed in [37] for finite bifix codes.

Theorem 19. *Any regular bifix code can be effectively included in a regular complete bifix code.*

Proof. We outline the construction in the finite case only. Given a bifix code X, its *indicator* L_X is defined on A^* by

$$L_X(w) = card\{(u, x, v) \mid w = uxv, u \in A^* \setminus A^* X, x \in X^*, v \in A^* \setminus X A^*\} \ .$$

The indicator is increasing

$$L_X(w) \leq L_X(wa) \leq L_X(w) + 1$$

and if X is not complete, L_X is bounded on no subset uA^*, $u \in A^*$.

Let X be a *finite* bifix code. Then its indicator is bounded on X; assume it is by n. The completion procedure runs as follows. Construct $Z = \{w \in A^* \mid L_X(w) = n + 1\}$. Then $Y = Pref(X \cup Z)$ is a regular complete bifix code, where $Pref(T) = T \setminus TA^+$. Indeed, Y is prefix by construction, Y is suffix and complete due to the increasing property of L_X mentioned above, and Y is regular because one can prove that Z is regular.

In the case X is a *regular* bifix code, L_X is no longer bounded on X and the previous method fails. However, it can be applied "locally" to get a regular bifix code Y which is not complete but strictly contains X. The construction is thus repeated on Y and after a finite number of steps, one gets a regular bifix code containing X, which is now complete. □

Example 11. The bifix code $X = \{a, b^2\}$ cannot be embedded into a finite complete one by Proposition 18. Its completion into a regular bifix code is depicted on Fig. 11 with $n = 2$ (value of $L_X(w)$ is indicated inside the node corresponding to w). This leads to the complete code $\{a\} \cup ba^*b$.

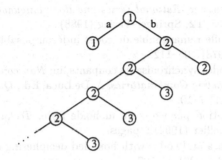

Fig. 11. Completion of the bifix code $\{a, b^2\}$.

Notice that the completion of automata with bounded bi-delay mentioned at the end of Sect. 7.2 does not lead to another proof of Theorem 19. Indeed, for some regular bifix codes X, there exists no finite automaton $\mathcal{A}$ with bi-delay $(0, 0)$ such that $X = X_{\mathcal{A}}$. It is then natural to try to generalize Theorem 19 to the class of regular codes with bounded decoding bi-delay.

Problem 9. Is any regular code with decoding bi-delay (d, d') included in a regular complete code with the same decoding bi-delay?

Problem 2 is solved for bifix codes, since it is decidable whether a finite bifix code has a finite completion into a bifix code [3]. However the decision procedure does not give a real characterization of the subsets of finite complete bifix codes.

8 Conclusion

In this survey, we have focused on finite and regular maximal codes. They play a central role in the theory of variable-length codes. We have emphasized the richness and the variety of tools introduced to study the difficult conjectures on maximal codes. The important developments obtained during the last ten years seem promising to solve the still open problems.

Acknowledgments

We thank J. Berstel, D. Perrin and A. Restivo for stimulating discussions, and we are grateful to C. De Felice for numerous comments on a preliminary version of this paper. We also wish to thank H. Jürgensen for providing helpful references.

References

1. J. Ashley, B. Marcus, D. Perrin, S. Tuncel, Surjective extensions of sliding block codes, *SIAM J. Discrete Math.* **6** (1993) 582–611.
2. J. Berstel, D. Perrin, *Theory of Codes*, Academic Press (1985).
3. J. Berstel, D. Perrin, Trends in the theory of codes, *Bull. EATCS* **29** (1986) 84–95.
4. J. Berstel, C. Reutenauer, *Rational series and their languages*, EATCS Monogr. Theoret. Comput. Sci. **12**, Springer-Verlag (1988).
5. J.-M. Boë, Une famille remarquable de codes indécomposables, *Lecture Notes in Comput. Sci.* **62** (1978) 105–112.
6. J.-M. Boë, Sur les codes synchronisants coupants, in: *Non-commutative Structures in Algebra and Geometric Combinatorics*, A. De Luca, Ed., *Quaderni de la Ricerca Scientifica* **109** (1981) 7–10.
7. J.-M. Boë, Factorisation par excès du monoïde libre, *Technical Report* **94-005**, University of Montpellier (1994) 6 pages.
8. V. Bruyère, Automata and codes with bounded deciphering delay, *Lecture Notes in Comput. Sci.* **583** (1992) 99–107.
9. V. Bruyère, D. Derencourt, M. Latteux, The meet operation in the lattice of codes *submitted* (1996) 16 pages.
10. V. Bruyère, L. Wang, L. Zhang, On completion of codes with finite deciphering delay, *European J. Combin.* **11** (1990) 513–521.
11. J. Brzozowski, Open problems about regular languages, in: *Formal Language Theory: Perspectives and Open Problems*, R.V. Book, Ed., Academic Press (1980) 23–45.
12. A. Carpi, On synchronizing unambiguous automata, *Theoret. Comput. Sci.* **60** (1988) 285–296.
13. Y. Césari, Sur l'application du théorème de Suschkevitch à l'étude des codes rationnels complets, in: *Automata, Languages and Programming, Lecture Notes in Comput. Sci.* (1974) 342–350.
14. N.G. de Bruijn, On the factorization of cyclic groups, *Indag. Math.* **15** (1953) 258–264.
15. D. Derencourt, A three-word code which is not prefix-suffix composed, to appear in *Theoret. Comput. Sci* (1996) 15 pages.
16. D. Derencourt, personal communication (1996).
17. C. De Felice, A partial result about the factorization conjecture for finite variable-length codes, *Discrete Math.* **122** (1993) 137–152.
18. C. De Felice, An application of Hajós factorizations to variable-length codes, to appear in *Theoret. Comput. Sci.* (1996) 31 pages.
19. C. De Felice, A. Restivo, Some results on finite maximal codes, *RAIRO Inform. Théor. Appl.* **19** (1985) 383–403.
20. C. De Felice, C. Reutenauer, Solution partielle de la conjecture de factorisation des codes *C. R. Acad. Sci. Paris* **302** (1986) 169–170.
21. A. Ehrenfeucht, G. Rozenberg, Each regular code is included in a regular maximal code, *RAIRO Inform. Théor. Appl.* **20** (1985) 89–96.
22. S. Eilenberg, *Automata, Languages and Machines*, Vol. A, Academic Press (1974).
23. L. Fuchs, *Abelian Groups*, Pergamon Press (1960).
24. G. Hajós, Sur la factorisation des groupes abéliens, *Casopis Pest. Mat. Fys.* **74** (1950) 157–162.
25. R.W. Hamming, *Coding and Information Theory*, Prentice-Hall (1986).

26. J. Hopcroft, J. Ullman, *Introduction to Automata Theory, Languages and Computation*, Addison-Wesley (1979).

27. H. Jürgensen, S. Konstantinidis, Codes, in: *Handbook of Formal Languages*, G. Rozenberg, A. Salomaa, Eds, Springer-Verlag, to appear (1996).

28. R. König, Lectures on codes, *Technical Report* **93-3**, University of Erlangen (1993) 66 pages.

29. M. Krasner, B. Ranulac, Sur une propriété des polynômes de la division du cercle, *C. R. Acad. Sci. Paris* **240** (1937) 397–399.

30. N.H. Lam, On codes having no finite completion, *Lecture Notes in Comput. Sci.* **775** (1994) 691–698.

31. N.H. Lam, A property of finite maximal codes, *preprint* (1996) 8 pages.

32. D. Lind, B. Marcus, *Symbolic Dynamics and Coding*, Cambridge University Press (1996).

33. A.A. Markov, An example of an independent system of words which cannot be included in a finite complete system (in Russian), *Mat. Zametki* **1** (1967) 87–90.

34. R.J. McEliece, *The Theory of Information and Coding*, Enc. of Math. **3**, Addison-Wesley (1977).

35. F.J. McWilliams, N.J.A. Sloane, *The Theory of Error-Correcting Codes*, North-Holland (1977).

36. R. Montalbano, Local automata and completion, *Lecture Notes in Comput. Sci.* **665** (1993) 333–342.

37. D. Perrin, Completing biprefix codes, *Theoret. Comput. Sci.* **28** (1984) 329–336.

38. D. Perrin, Finite automata, in: *Handbook of Theoretical Computer Science*, vol. B, J. Van Leeuwen, Ed., Elsevier (1990) 2–57.

39. D. Perrin, M.-P. Schützenberger, Un problème élémentaire de la théorie de l'information, *Théorie de l'Information, Colloques Internat. CNRS* **276**, Cachan (1977) 249–260.

40. A. Restivo, On codes having no finite completions, *Discrete Math.* **17** (1977) 309–316.

41. A. Restivo, S. Salemi, T. Sportelli, Completing codes, *RAIRO Inform. Théor. Appl.* **23** (1989) 135–147.

42. C. Reutenauer, Non commutative factorization of variable length codes, *J. Pure Appl. Algebra* **36** (1985) 157–186.

43. A. Salomaa, *Jewels of Formal Language Theory*, Washington, D.C.: Computer Science Press (1981).

44. A.D. Sands, On a conjecture og G. Hajós, *Glasgow Math. J.* **15** (1974) 88–89.

45. M.-P. Schützenberger, Une théorie algébrique du codage, *Séminaire Dubreil-Pisot* 1955-56, exposé no 15 (1955).

46. M.-P. Schützenberger, Sur certains sous-monoïdes libres, *Bull. Soc. Math. France* **93** (1965) 209–223.

47. M.-P. Schützenberger, On a question concerning certain free submonoids, *J. Combin. Theory* **1** (1966) 437–442.

48. P. Shor, A counterexample to the triangle conjecture, *J. Combin. Theor. Ser. A* **38** (1983) 110–112.

49. H.J. Shyr, *Free Monoids and Languages*, Hon Min Book Company, Taichung, second ed. (1991).

50. S. Szabo, personal communication to A. Restivo (1992).

51. J.H. Van Lindt, *Introduction to Coding Theory*, Graduate Texts in Math. **86**, Springer-Verlag (1982).

52. L. Zhang, Every finite maximal code is a factorizing code, *manuscript* (1993) 19 pages.
53. L. Zhang, C.K. Gu, Two classes of factorizing codes – (p, p) codes and $(4, 4)$ codes, In: *Words, languages and combinatorics II*, M. Ito, H. Jürgensen, Eds., World Scientific (1994) 477–483.
54. L. Zhang, Z. Shen, Completion of recognizable bifix codes, *Theoret. Comput. Sci.* **145** (1995) 345–355.

Lower Bounds for Propositional Proofs and Independence Results in Bounded Arithmetic

Alexander A. Razborov*

Steklov Mathematical Institute
Vavilova 42, 117966, GSP-1, Moscow, RUSSIA

Abstract. We begin with a highly informal discussion of the role played by Bounded Arithmetic and propositional proof systems in the reasoning about the world of feasible computations. Then we survey some known lower bounds on the complexity of proofs in various propositional proof systems, paying special attention to recent attempts on reducing such bounds to some purely complexity results or assumptions. As one of the main motivations for this research we discuss provability of extremely important propositional formulae that express hardness of explicit Boolean functions with respect to various non-uniform computational models.

1. Propositional proofs as feasible proofs of plain statements

Interesting and viable logical theories do not appear as result of sheer speculation. Conversely, they attempt to summarize and capture a certain amount of reasoning of a certain style about a certain class of objects that had existed in the math community before the mathematical logics entered the stage. For example, the set theory ZF was developed to distill those partial cases of the comprehension scheme (contradictory in full generality) that are really used in the common day work of a "practical mathematician". The aim of Peano Arithmetic PA was to capture that part of reasoning about integers which involves only finite objects, or at least can be in principle reduced to this form like some deep results using trigonometrical sums.

For many computer scientists, however, the world is usually even more restricted. Namely if we perform certain computation with an input binary string a, then all objects viewed during this computation not only are finite, but usually are required to have the bit length bounded a priori in terms of the length of a by some simple function like a polynomial or a quasipolynomial. Respectively, as a logical basis for the reasoning about this bounded world we would like to have a formal theory capable exactly of formalizing "common" arguments that may involve only those finite objects whose length is bounded in the length of a, and that, moreover, can be efficiently computed from this parameter. The exact

* Supported by the grant # 96-01-01222 of the Russian Foundation for Fundamental Research

meaning of efficient computability is determined by the class of computations we are trying to emulate in the context of Proof Theory.

A large variety of such theories appeared in the literature, most of them under the generic name *Bounded Arithmetic*. Naturally, the theories corresponding to (supposedly) different complexity classes are (also supposedly) different. One important empiric observation is that, vice versa, theories differently defined to capture the same class of computations tend to be isomorphic. For example, theories corresponding to the most basic world of time-bounded computations were independently defined in at least three different forms: Cook's equational theory PV [13], first order theories S_k^1 and second order theories V_k^1 [10]. It turned out, however, that PV is in some sense equivalent to S_2^1 [10, Chapter 6], and that S_k^1 is simply isomorphic to (an inessential modification of) V_{k-1}^1 [41, 29]. This robustness additionally suggests that we are on the right track in our quest for the theories of "feasible reasoning about feasible objects".

It is a common place in the Proof Theory that when we are interested in the provability of some formulae, the first question to ask is what is their logical complexity. This parameter typically measures the number of alternations of connectives or quantifiers and usually dictates which kind of techniques we should look for in our proof-theoretical studies. In the case of the theories of Bounded Arithmetic this helps us to draw a rather clear and natural distinction between what we are and what we are not going to do in this paper. Namely, there exists a rich and powerful *witnessing technique* for studying the provability of formulae that contain non-trivial bounded quantifiers, non-triviality meaning that they quantify over the full domain of objects whose length is comparable to the length of initial parameters. This technique is *not* considered in this paper, and the interested reader is referred e.g. to the monographs [10, 17, 22].

The only formulae we are dealing with in this paper are Σ_0^b-formulae. These are essentially the formulae in which all quantifiers are *sharply bounded* i.e., quantify over some domain whose *size* is comparable to the length of initial parameters.

The bad news about these formulae is that the previously known witnessing technique can not distinguish between their truth and provability and hence can be hardly used for studying them proof-theoretically. The good news is that the provability of Σ_0^b-formulae in theories of Bounded Arithmetic is closely related to the existence of short propositional proofs for certain propositional tautologies associated with the formula.

We do not give here exact definitions or details, they can be found e.g. in [22]. The best way for a complexity-oriented reader to imagine this correspondence is to invoke the familiar analogy with uniform vs. non-uniform computational models. A Σ_0^b-formula corresponds to a language, provability in Bounded Arithmetic is analogous to computability within specified amount of computational resources in some uniform model, and the length of propositional proofs for associated tautologies corresponds to the size of circuits computing restrictions of our language onto words of prescribed length.

The main purpose of the introductory part above was to convey to nonspecialists at least some feeling about the theory of feasible provability, and about the role played in this area by propositional proofs. We are not going to return to this topic (with a few minor exceptions). In comparison with first or second order formal theories, the model of propositional proof systems is much simpler to formulate and much cleaner combinatorially. Hence, similarly to Boolean circuits in the context of Computational Complexity, propositional proof systems provide a convenient and elegant framework for contemplating over lower bounds arguments, and explaining to a sufficiently broad audience the main ideas of what has been done. The latter task is exactly what we will try to do in the rest of the paper. Once again, the reader should bear in mind that our original motivation is to study the provability of plain ($= \Sigma_0^b$) statements by the amount of reasoning allowed in the world of feasibly computed objects ($=$ in certain fragments of Bounded Arithmetic), and that this is basically equivalent to the study of complexity of propositional proofs.

2. Some concrete propositional proof systems

The most general definition of a propositional proof system was given in [14]:

Definition 1. Let C be a certain class of propositional formulae in variables $p_1, \ldots, p_n, \ldots$, and TAUT_C be the set of all tautologies from the class C (if C is not specified, we assume by default that it consists of all bounded-fanin formulae in the standard language $\{0, 1, \neg, \lor, \land\}$ and abbreviate TAUT_C to TAUT). A *propositional proof system* (p.p.s.) for the class C is a poly-time computable function P from $\{0, 1\}^*$ *onto* TAUT_C. For a tautology $\phi \in \text{TAUT}_C$, any string w such that $P(w) = \phi$ is called a *P-proof of* ϕ.

Let $s_P(\phi)$ be the minimal possible length $|w|$ of a P-proof w of ϕ. A p.p.s. P is called *optimal* if $s_P(\psi)$ is bounded from above by a polynomial in $|\phi|$, uniformly for $\phi \in \text{TAUT}_C$. The following easy result, also from [14], indicates that in its full generality this definition is just a reformulation of the standard characterization of $co - \textbf{NP}$ and has only little to do with the "real" proof theory:

Theorem 2. $\textbf{NP} = co - \textbf{NP}$ *if and only if there exists an optimal propositional proof system.*

In this paper, however, we will consider only natural proof systems, and are primarily interested in proving their non-optimality. This is equivalent to obtaining lower bounds on $s_P(\phi_n)$ for *some* sequence of tautologies $\{\phi_n\}$. As we will see many times throughout this paper, there is a striking analogy between this task (and approaches to it) and the similar task of obtaining lower bounds on the complexity of Boolean functions. Here is one place where this analogy fails: for propositional proofs there does not seem to exist any analogue of the statement that almost all Boolean functions are hard, and the question whether the sequence ϕ_n should consist of "explicit" or arbitrary tautologies does not seem to be too relevant.

The following definition allows us to compare different p.p.s. by their strength:

Definition 3. Let P_0, P_1 be propositional proof systems for some classes C_0, C_1 of formulae respectively, and assume that $C_0 \subseteq C_1$. We say that P_1 *polynomially simulates* P_0 if there is a poly-time computable function $f : \{0, 1\}^* \longrightarrow \{0, 1\}^*$ such that $P_0(w) = P_1(f(w))$ for all $w \in \{0, 1\}^*$. Two proof systems for the same class of formulae are *polynomially equivalent* if they polynomially simulate each other.

Notice that if P_0 is polynomially simulated by P_1 then it has a *polynomial speed-up* over P_1, that is $s_{P_1}(\phi) \leq s_{P_0}(\phi)^{O(1)}$, $\phi \in \text{TAUT}_{C_0}$. Vice versa, in all known cases when one natural p.p.s. has a polynomial speed-up over another p.p.s., this is due to the existence of some (also natural) polynomial simulation.

As the most basic example of a p.p.s. take the ordinary Hilbert-style propositional calculus from your favourite textbook in mathematical logics. It is based upon some finite number of axiom schemes and inference rules like modus ponens and can be easily converted into a p.p.s. F in the sense of Definition 1 as follows. Let $F(w)$ be the final tautology in w if w is some (encoding of) legal inference in our calculus. If w is a meaningless word, we let $F(w) \rightleftharpoons 1$. This p.p.s. F is called *Frege proof system*.

At the first sight, this definition is somewhat ambiguous as there are several variants of the propositional calculus (almost as many as textbooks). However, the following theorem [35] shows that the differences between them are actually inessential:

Theorem 4. *If the language of some Frege system F_1 contains the language of another Frege system F_0, then F_1 polynomially simulates F_0. In particular, every two Frege systems in the same language are equivalent.*

Coming back for a moment to the discussion from the previous section, Frege proofs correspond to (non-uniform) class $\mathbf{NC}^1$ in the sense that they allow plain reasoning about $\mathbf{NC}^1$-predicates (= predicates expressible by small propositional formulae). All p.p.s. considered in this paper follow the same pattern for different complexity classes, and always allow straight-line proofs (that is, every deduced formula can be used more than once in forthcoming inferences). Missing details of definitions as well as more information about these p.p.s. and their modifications can be found e.g. in [22, 44].

Resolution. This proof system was introduced in [8] and further developed in [16, 38]. Let C be the class of all disjunctive normal forms. A *resolution proof* of a tautology $\phi = (K_1 \vee \ldots \vee K_m)$ from TAUT_C is actually a resolution refutation of the set $\{D_1, \ldots, D_m\}$, where D_i is the *clause* (elementary disjunction) that is the negation of the elementary conjunction K_i. In turn, a *resolution refutation* of a set of clauses is an inference of the empty clause 0 from this set in the calculus with the only *resolution rule*

$$\frac{D \vee p \qquad E \vee (\neg p)}{D \vee E}.$$

We denote this proof system by R.

Bounded depth Frege systems. Let $d \geq 2$ be a fixed constant. F_d is the p.p.s. defined similarly to the ordinary Frege system F with the exception that it operates with (unbounded fan-in) formulae of depth at most d over the standard basis $\{0, 1, \neg, \vee, \wedge\}$. Axioms and inference rules are modified in a natural way so that they take care of unbounded fan-in.

We will also consider extensions of F_d with axiom schemes $Count_m$, where m is some fixed integer, expressing that no universe of cardinality not divisible by m can be partitioned into m-sets. An even stronger system is obtained if we append to our language the connective MOD_m of counting modulo m, along with natural axioms and inference rules expressing its basic properties. This system will be denoted by $F_d(MOD_m)$.

Cutting planes ([15]). These generalize resolutions and are defined similarly with the exception that this time K_i (and hence D_i) have the form of arbitrary inequalities of the form $a_1 p_1 + \cdots + a_n p_n \geq T$, where the coefficients $a_1, \ldots, a_n, T$ are integer numbers, and p_i are treated as $(0, 1)$-valued variables. This system has a number of obvious axioms and inference rules, and just one less straightforward *division rule*

$$\frac{a_1 c p_1 + \cdots + a_n c p_n \geq T}{a_1 p_1 + \cdots + a_n p_n \geq \lceil T/c \rceil},$$

where $a_1, \ldots, a_n, c, T$ are integers and $c > 0$ (its non-triviality stems from rounding up rather than down). By analogy with threshold circuits (see e.g. [30]), cutting planes come in two different versions. Namely, we can write the coefficients $a_1, \ldots, a_n, T$ in unary (the case of *small weights*) or in binary (*arbitrary weights*). The corresponding p.p.s. will be denoted by CP^*, CP respectively.

Extended Frege system, denoted EF [13]. We enlarge the ordinary Frege system by allowing it to introduce *extension axioms* of the form $(p \equiv A)$, where A is a formula, and p is a new *extension atom* that did not occur previously in the proof. This is an extremely powerful and probably the most important propositional proof system. The reason is that it corresponds to the complexity class $\mathbf{P}/poly$: roughly speaking, what extension atoms and axioms do, they allow us to evaluate arbitrary Boolean circuits.

There are many polynomial simulations between natural propositional proof systems, their numerous modifications etc. For p.p.s. introduced above there are no real surprises on the map of known simulations between them: it completely mimics the map of known containments between underlying complexity classes, see Figure 1. [44, Figure 1] contains more remarkable landmarks in this neighborhood.

3. Algebraic and combinatorial machinery borrowed from Boolean complexity

When superpolynomial lower bounds are proved for some class of Boolean circuits, the machinery developed for that purpose usually conveys much more

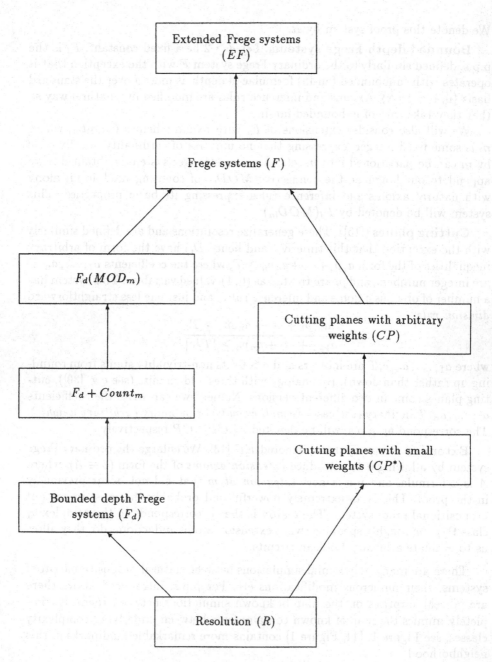

Fig. 1. Map of known simulations

information. In many cases this lead us to a rather clear understanding of what, how and why can or cannot be computed by small circuits from this class. Thus, it is natural to try to apply this or similar machinery to propositional proofs that operate with tautologies expressible by circuits already understood in this way.

In some cases this analogous approach to proving lower bounds on the complexity of propositional proofs turned out to be fruitful. Exponential lower bounds were consequently established for resolutions [42, 18, 43, 12], bounded depth Frege systems [1, 7, 21, 25, 26, 2, 6, 36] and their extensions $F_d + Count_m$ [3, 5, 37, 11].

We do not present here these important results in more details as this is already quite successfully done e.g. in [22, 44]. We only remark that our experience gathered in this field strictly suggests that the problem of proving lower bounds on the complexity of propositional proofs is usually much harder than the similar problem for the underlying circuit class, and there are no obvious generalizations to the case of p.p.s.

Lower bounds for resolutions are treated as deep and ingenious results, whereas lower bounds for DNF is something completely trivial. Even with Håstad Switching Lemma in hand [20] it took a lot of work to adopt it for constant-depth propositional proofs, and even more work to prove lower bounds for the extension $F_d + Count_m$. Moreover, looking closely at Figure 1, we might expect lower bounds for cutting planes and the system $F_d(MOD_p)$, p a prime, as such bounds are known for the corresponding circuit models.

Lower bounds for cutting planes will be presented in the next section, but they are based on ideas totally different from the machinery used in the theory of threshold circuits. No "combinatorial" proof of lower bounds for cutting planes is currently known. The situation with the system $F_d(MOD_p)$ is even worse: the question of its optimality is still open. In the recent paper [11] this question was reduced to lower bounds on the degree of coefficients in a certain "extended" version of one partial case of Hilbert's Nullstellensatz. This reduction emulates in the context of propositional proofs the main lemma from [28, 40] on approximating $ACC^0[p]$-circuits by low degree polynomials. The remaining step, however (proving lower bounds on the degrees for this algebraic problem) is still elusive.

We conclude this section with mentioning another recent paper [19] that provides one of the rare examples of influence in another direction. Namely, in that paper combinatorial ideas previously used for resolutions in [18] were successfully employed in Boolean complexity to provide a new elegant method of obtaining exponential lower bounds on the monotone circuits size of explicit Boolean functions.

4. Interpolation theorems and disjoint NP-pairs

Given our inability to prove superpolynomial lower bounds on the size of Boolean formulae or circuits computing explicit Boolean functions, and observations made in the previous section, we can hardly expect a "combinatorial" proof of non-optimality for strong p.p.s. like F or EF. In this section we present another approach that is more akin to the witnessing technique mentioned in Section 1 and looks more promising.

The general idea of this approach is not to try to analyze directly potential propositional proofs but rather to extract from them some unlikely algorithmic consequences in the real world. This, if successful, would give us *conditional* statements about non-optimality of p.p.s. modulo complexity assumptions saying that the consequences extracted from the proofs actually do not take place. Roughly speaking, if lower bounds for propositional proofs must necessarily be at least as hard as complexity lower bounds (and the latter are currently inaccessible), let us at least show that they are *just* as hard!

One successful scheme fulfilling this idea for substantially weaker p.p.s. appeared, implicitly and independently, in [32, 9]. More explicit and slightly different treatments of this scheme were later given in [23, 33], and in our presentation we follow an intermediate course between them.

Let $U, V \subseteq \{0,1\}^*$ be two disjoint **NP**-sets, and let us arbitrarily fix their **NP**-representations

$$\left. \begin{array}{l} x \in U \equiv \exists y \in \{0,1\}^{p(n)} A_n(x,y), \\ x \in V \equiv \exists z \in \{0,1\}^{p(n)} B_n(x,z), \end{array} \right\} \tag{1}$$

$n = |x|$, where $A_n(\mathbf{p}, \mathbf{q})$, $B_n(\mathbf{p}, \mathbf{r})$ are propositional formulae of polynomial in n size. Then $U \cap V = \emptyset$ implies that for every n, $\neg A_n(\mathbf{p}, \mathbf{q}) \vee \neg B_n(\mathbf{p}, \mathbf{r})$ is a tautology and hence provable in the propositional calculus. This tautology can be re-written in the form $B_n(\mathbf{p}, \mathbf{q}) \supset \neg A_n(\mathbf{p}, \mathbf{r})$, and Craig's Interpolation Lemma (which, although, amounts to something completely trivial in the propositional case) guarantees us the existence of a formula $C_n(\mathbf{p})$ called an *interpolant* that depends on **p**-variables only and such that both $B_n(\mathbf{p}, \mathbf{q}) \supset C_n(\mathbf{p})$ and $C_n(\mathbf{p}) \supset \neg A_n(\mathbf{p}, \mathbf{r})$ are provable. Semantically this means that the set $L \rightleftharpoons \{x \mid C_n(x_1, \ldots, x_n), n = |x|\}$ has the property $U \cap L = \emptyset$, $V \subseteq L$, i.e., it *separates* U and V.

Suppose now we additionally know that the tautology $\neg A_n(\mathbf{p}, \mathbf{q}) \vee \neg B_n(\mathbf{p}, \mathbf{r})$ has a *short* proof in some p.p.s. P (informally speaking, this means that P can easily prove the disjointness of U and V). Then we might expect that in this case something could be additionally said about the *complexity* of the interpolant C_n, i.e., we would have some constructive form of the interpolation theorem. And then we could argue another way around: if U and V can *not* be separated by any set L of the complexity prescribed by the interpolation theorem, then the corresponding tautologies $\neg A_n(\mathbf{p}, \mathbf{q}) \vee \neg B_n(\mathbf{p}, \mathbf{r})$ do not have short P-proofs, and, thus, P is not optimal.

Lower bounds based upon this idea independently appeared in [32] (for the system R in the uniform framework), and in [9] (for CP^*). [27] established the interpolation theorem for cutting planes with arbitrary coefficients, and this is the strongest system for which it is currently known:

Theorem 5. *There exists a polynomial time algorithm which does the following. Given a CP-refutation of some set of clauses*

$$\{A_i(\mathbf{p}, \mathbf{q}) \mid i \in I\} \cup \{B_j(\mathbf{p}, \mathbf{r}) \mid j \in J\},$$

where all variables occurring in A_i, B_j are explicitly displayed, the algorithm produces a Boolean circuit $C(\mathbf{p})$ that computes an interpolant for $A(\mathbf{p}, \mathbf{q}) \rightleftharpoons \bigwedge_{i \in I} A_i(\mathbf{p}, \mathbf{q})$ and $B(\mathbf{p}, \mathbf{r}) \rightleftharpoons \bigwedge_{j \in J} B_j(\mathbf{p}, \mathbf{r})$ in the above sense.

In particular, if the size of the original CP-refutation is polynomial in the number n of variables in $\mathbf{p}$, the same is true for the size of $C(\mathbf{p})$. Therefore, if there exist two disjoint **NP**-sets U and V that can not be separated by any set in **P**/*poly*, then the cutting planes system (and any forthcoming p.p.s. for which we might manage to prove a similar interpolation theorem) is not optimal. The sequence of hard tautologies would have the form $\neg A_n(\mathbf{p}, \mathbf{q}) \vee \neg B_n(\mathbf{p}, \mathbf{r})$, where A_n, B_n are arbitrary CNF of polynomial size representing U, V in the sense of (1).

The existence of such pairs (U, V) follows from either **NP**$\cap$co$-$**NP** $\not\subseteq$ **P**/*poly* (for trivial reasons) or **UP** $\not\subseteq$ **P**/*poly* (see [39, Theorem 9]). In the next section we will discuss some concrete pairs that are likely to fulfill this assumption.

But the monotone analogue of this assumption *is* known, although the proof is rather hard [4]. Quite remarkably, a *monotone* version of Theorem 5 allows us to obtain *unconditional* lower bounds for cutting planes based upon this knowledge.

To be more specific, assume that the set U is an ideal w.r.t. the natural partial ordering on $\{0, 1\}^*$ given by $x \leq y \rightleftharpoons (|x| = |y| \ \& \ \forall i \leq |x|(x_i \leq y_i))$, and V is a filter w.r.t. the same ordering. Assume also that the representation (1) witnesses these monotonicity properties in the sense that $A_n(\mathbf{p}, \mathbf{q})$ contains $\mathbf{p}$-variables only negatively, and $B_n(\mathbf{p}, \mathbf{r})$ contains them only positively. Then one might hope that the short circuit whose existence is guaranteed by Theorem 5 can be additionally guaranteed to be monotone.

This is indeed so for the restricted version CP^* of cutting planes [9] which immediately implied in that paper exponential lower bounds on $s_{CP^*}(\phi_n)$ for some tautologies ϕ_n. More specifically, let propositional variables p_{ij} ($1 \leq i < j \leq n$) encode in a natural way an undirected graph on n vertices, and let $Colour_n(\mathbf{p}, \mathbf{q})$ be a poly-sized CNF that contains $\mathbf{p}$ only negatively and expresses that $\mathbf{q}$ is a proper $(m-1)$-colouring of this graph, $m \rightleftharpoons \lfloor \frac{1}{8}(n/\log n)^{2/3} \rfloor$. Similarly, let $Clique_n(\mathbf{p}, \mathbf{r})$ contain $\mathbf{p}$ only positively and say that $\mathbf{r}$ is an m-clique in the graph represented by $\mathbf{p}$. It is the main result of [4] that every monotone circuit separating the sets $\{x \mid |x| = n \ \& \ \exists y \ Colour_n(x, y)\}$ and $\{x \mid |x| = n \ \& \ \exists z \ Clique_n(x, z)\}$ must have size $2^{\Omega((n/\log n)^{1/3})}$. Hence:

Theorem 6. *Every CP^*-proof of the tautology $\neg Clique_n(\mathbf{p}, \mathbf{q}) \vee \neg Colour_n(\mathbf{p}, \mathbf{r})$ must have size $2^{\Omega((n/\log n)^{1/3})}$.*

The situation with cutting planes proofs having arbitrary coefficients is slightly more complicated. For this case it is not known whether there always exists an interpolant $C(\mathbf{p})$ computable by ordinary monotone circuits of poly-nomial size. However, Pudlák [27] introduced some natural generalization of Boolean monotone circuits which he called *monotone circuits over reals*, and showed that a poly-sized interpolant for CP-proofs can be constructed in this broader class. The second main result of [27] (which is also of independent interest in "pure" Complexity Theory) is a generalization of the Alon-Boppana bound to the case of monotone circuits over reals. Altogether this implies the extension of Theorem 6 to cutting planes with arbitrary coefficients. No "direct" combinatorial proof of non-optimality for CP, or even for CP^* is currently known.

5. Tautologies expressing hardness of Boolean functions

So far we were interested in general methods for showing lower bounds on the complexity of propositional proofs paying only little attention to the look of tautologies for which these bounds are attained. On the contrary, in this section we study the proof complexity for quite concrete and extremely important propositional formulae expressing major results and central open problems in Computational Complexity.

Fix some parameters n and $t = t(n) \leq 2^n$. Let $\mathbf{p} \rightleftharpoons (p_x \mid x \in \{0,1\}^n)$ be a vector of propositional variables of length 2^n encoding the truth-table of a Boolean function in n variables, and $Circuit_{t,n}(\mathbf{p}, \mathbf{q})$ be a CNF of size $2^{O(n)}$ expressing that $\mathbf{q}$ encodes a Boolean circuit of size at most t that computes this function. As we require the size to be polynomial only in 2^n, not in n, constructing such a CNF presents no difficulties. For example, $\mathbf{q}$ can simply encode all instructions of the circuit as well as truth-tables of all intermediate results. Like in Computational Complexity, exact details of this encoding are unimportant for our purposes.

Now, $\mathbf{NP} \overset{?}{\subseteq} \mathbf{P}/poly$ is equivalent to the question if $\neg Circuit_{t(n),n}(s_n, \mathbf{q})$ are tautologically true for some function $t(n)$ not bounded by any polynomial, and an arbitrary $\mathbf{NP}$-complete function $\{s_n\}$ (for definiteness, SATISFIABILITY). Moreover, what Boolean complexity has been basically doing over last 30 years was trying to prove, with the help of the algebraic and combinatorial methods mentioned in Section 3, propositional formulae of the form $\neg Circuit^*_{t(n),n}(f_n, \mathbf{q})$ for explicit functions f_n, integer-valued $t(n)$ that are as large as possible, and $Circuit^*_{t,n}$ corresponding to some restricted class of Boolean circuits.

One more face of the already strong connections existing between propositional proofs and Boolean circuits was observed in [31], where I proposed the thesis that *in all known cases of success achieved in proving propositional formulae of the form $\neg Circuit^*_{t(n),n}(f_n, \mathbf{q})$, this proof is actually an EF-proof[2] of size $2^{O(n)}$.* This immediately gives raise to the following natural ques-

---

[2] Strictly speaking, [31] deals with some second order theories $V_1^0(\delta)$ of Bounded Arithmetic, but all considerations from there translate to propositional proof systems as sketched at the end of Section 1. In fact, only a few proofs in Boolean complexity use the full strength of the Extended Frege system: many major results already have short F-proofs, and some of them even F_d-proofs for a fixed reasonable d.

tion: since we are currently unable to prove things like $\neg Circuit_{n^2,n}(s_n, \mathbf{q})$ or $\neg Formula_{n^{10},n}(s_n, \mathbf{q})$, then perhaps we should think in another direction and try to show that these supposed tautologies significantly differ from their already proven restricted versions and do not possess short propositional proofs at all. And the first natural thing to wonder about this question: is there any hope to *prove* something intelligent about it provided we *believe* that in reality things like $\mathbf{NP} \not\subseteq \mathbf{P}/poly$ take place but are extremely hard to prove?

The hope to show with a direct combinatorial or algebraic argument that $\mathbf{NP} \not\subseteq \mathbf{P}/poly$ does not have short EF-proofs or F-proofs seems to be even more vain than the hope to prove non-optimality of these proof systems using arbitrary tautologies. Indeed, it finally may happen (although it seems more and more unlikely – see [34] for one possible explanation) that we significantly underestimate the potential of this machinery, and it will eventually prove $\mathbf{NP} \not\subseteq \mathbf{P}/poly$ and then non-optimality of EF. But even in that case, given the main thesis from [31], the first step in this program would probably give a *short EF-proof* of $\mathbf{NP} \not\subseteq \mathbf{P}/poly$, and with all our nice machinery in hands we could not establish strong lower bounds on $s_{EF}(\neg Circuit_{t,n}(s_n, \mathbf{q}))$ simply for the reason these bounds are not true!

On the other hand, it seems that "witnessing" arguments from Section 4, if successful at all, tend to be perfectly applicable to supposed tautologies like $\neg Circuit_{t,n}(s_n, \mathbf{q})$ since the latter are of "universal" nature and have a lot of useful structure hidden inside. Let us show, for example, how to associate to these tautologies some natural pair of $\mathbf{NP}$-sets that presumably can not be separated by a set in $\mathbf{P}/poly$.

Fix any sufficiently constructive super-polynomially growing function $t(n)$, and let the $\mathbf{NP}$-set $SIMPLE_t$ consist of all truth-tables f_n of Boolean functions in n variables that are computable by a circuit of size at most $t(n)$. Then

$$f_n \in SIMPLE_t \equiv \exists y \in \{0,1\}^{p(n)} Circuit_{t(n),n}(f_n, y)$$

for some polynomial $p(n)$. Let $s = \{s_n \mid n \in \omega\}$ be a sufficiently constructive sequence of Boolean functions, and consider also the shifted version $SIMPLE_t^{\oplus s}$ given by

$$f_n \in SIMPLE_t^{\oplus s} \equiv \exists z \in \{0,1\}^{p(n)} Circuit_{t(n),n}(f_n \oplus s_n, z).$$

The following is an easy consequence of the main result from [34] observed in [32]:

Theorem 7. *If there exists a pseudorandom number generator $G_k : \{0,1\}^k \longrightarrow \{0,1\}^{2k}$ in $\mathbf{P}/poly$ that is secure against attack by $2^{k^{\epsilon}}$-sized circuits for some fixed $\epsilon > 0$, then $SIMPLE_t$ and $SIMPLE_t^{\oplus s}$ can not be separated by any set computable by circuits of quasipolynomial size.*

In combination with Theorem 5, this implies lower bounds $2^{n^{\omega(1)}}$ on the length of any CP-proof of the formula

$$\neg Circuit_{t(n),n}(\mathbf{p},\mathbf{q}) \vee \neg Circuit_{t(n),n}(\mathbf{p}^{\oplus s_n},\mathbf{r}),$$

where $\mathbf{p}^{\oplus s_n}$ has the obvious meaning[3], modulo the same cryptographic assumption as in Theorem 7. This formula, however, has an easy propositional proof from an instance of $\neg Circuit_{4t(n)+3,n}(s_n,\mathbf{q})$ obtained by a formalization of the obvious construction that takes XOR of one $t(n)$-sized circuit $\mathbf{q}$ computing some function $\mathbf{p}$ with another such circuit computing $\mathbf{p}^{\oplus s_n}$ giving a $(4t(n)+3)$-sized circuit that computes s_n. Putting things together, and observing that our argument does not actually use the constructiveness of $t(n), s_n$, we get:

Theorem 8. *If there exists a pseudorandom number generator $G_k : \{0,1\}^k \longrightarrow \{0,1\}^{2k}$ in $\mathbf{P}/poly$ that is secure against attack by 2^{k^ϵ}-sized circuits for some fixed $\epsilon > 0$, then for every super-polynomially growing function $t(n)$ and every sequence of Boolean functions $\{s_n \mid n \in \omega\}$, the formulae $\neg Circuit_{t(n),n}(s_n,\mathbf{q})$ do not possess CP-proofs of size $2^{n^{O(1)}}$.*

This conditional lower bound is true for any sequence $\{s_n\}$. As observed by Avi Wigderson (in the context of Natural Proofs – see the journal version of [34]), for some specific s_n we can obtain *unconditional* results of this sort. Suppose for example that B_n is a hard bit of the discrete logarithm problem. Then for sufficiently large $t(n)$ (certainly, $t(n) = 2^{n^\epsilon}$ would do for any fixed $\epsilon > 0$) the formulae $\neg Circuit_{t(n),n}(B_n,\mathbf{q})$ do not possess quasipolynomial size CP-proofs, *without any unproven complexity assumptions*. The reason, roughly speaking, is that if the discrete logarithm function is hard, we apply Theorem 8 with G_k based on this function, and if it is easy then $\neg Circuit_{t(n),n}(B_n,\mathbf{q})$ are not tautologies and do not have any CP-proof at all. Of course, this argument is highly non-constructive, essentially uses the law of the excluded middle and does *not* imply non-optimality of CP since we do not know for sure whether $\neg Circuit_{t(n),n}(B_n,\mathbf{q})$ are really tautologies.

6. Some directions for future research

As we observed in Section 3, there still remains one natural p.p.s., namely $F_d(MOD_p)$, p a prime, for which the traditional algebraic and combinatorial machinery has very good chances to succeed in proving non-optimality. That would be very nice to obtain such a result as this would practically equalize the achievements of this machinery in Computational Complexity and in Proof Theory.

One bad news about the interpolation theorem is that its extension to the system EF turned out to be highly unlikely. Namely, [24] gave an example of

---

[3] Notice that if the function s_n is actually easy, then this formula has no CP-proof at all for the trivial reason it is not a tautology.

two disjoint **NP**-sets such that the fact of their disjointness *has* poly-size *EF*-proofs but which can not be separated by a set in **P**/*poly* if the cryptosystem RSA is secure. Hence, the analogue of Theorem 5 for *EF* could be used for a successful cryptoattack to break RSA. Despite this discouraging fact, the general approach of extracting some unlikely algorithmic consequences from potential propositional proofs still looks (in my opinion) promising.

For example, suppose we manage to show that every poly-size *EF*-proof (or *F*-proof) of any tautology of the form $\phi(\mathbf{q}) \vee \psi(\mathbf{r})$ implies the *existence* of a poly-size *EF*-proof for either $\phi(\mathbf{q})$ or $\psi(\mathbf{r})$ (but in general we can neither construct this proof efficiently nor even indicate for which of the two parts it exists). Notice that this is true if *exactly* one of the components $\phi(\mathbf{q})$, $\psi(\mathbf{r})$ is a tautology, simply by substituting into the original proof an arbitrary falsifying assignment for another component, and that at the moment this assumption does not seem to contradict any piece of our complexity intuition. Then the optimality of *EF* would imply, similarly to Section 4, that every two disjoint **NP**-sets can be separated by a disjoint pair of $co - \mathbf{NP}$ sets. This would have almost as striking consequences for the tautologies $\neg Circuit_{t,n}(f_n, \mathbf{q})$ discussed in Section 5 as the "strong" version of the interpolation theorem for *EF*.

References

1. M. Ajtai. The complexity of the pigeonhole principle. In *Proceedings of the 29th IEEE Symposium on Foundations of Computer Science*, pages 346–355, 1988.
2. M. Ajtai. Parity and the pigeonhole principle. In S. R. Buss and P. J. Scott, editors, *Feasible Mathematics*, pages 1–24. Birkhauser, 1990.
3. M. Ajtai. The independence of the modulo p counting principle. In *Proceedings of the 26th ACM STOC*, pages 402–411, 1994.
4. N. Alon and R. Boppana. The monotone circuit complexity of Boolean functions. *Combinatorica*, 7(1):1–22, 1987.
5. P. Beame, R. Impagliazzo, J. Krajíček, T. Pitassi, and P. Pudlák. Lower bounds on Hilbert's Nullstellensatz and propositional proofs. To appear in *Proc. of the London Math. Soc.*, 1994.
6. P. Beame and T. Pitassi. Exponential separation between the matching principles and the pigeonhole principle. Submitted to *Annals of Pure and Applied Logic*, 1993.
7. S. Bellantoni, T. Pitassi, and A. Urquhart. Approximation of small depth Frege proofs. *SIAM Journal on Computing*, 21(6):1161–1179, 1992.
8. A. Blake. *Canonical expressions in Boolean algebra*. PhD thesis, University of Chicago, 1937.
9. M. Bonet, T. Pitassi, and R. Raz. Lower bounds for cutting planes proofs with small coefficients. In *Proceedings of the 27th ACM STOC*, pages 575–584, 1995.
10. S. R. Buss. *Bounded Arithmetic*. Bibliopolis, Napoli, 1986.
11. S. Buss, R. Impagliazzo, J. Krajíček, P. Pudlák, A. Razborov, and J. Sgall. Proof complexity in algebraic systems and bounded depth Frege systems with modular counting. Submitted to *Computational Complexity*, 1996.
12. V. Chvátal and E. Szemerédi. Many hard examples for resolution. *Journal of the ACM*, 35(4):759–768, 1988.

13. S. A. Cook. Feasibly constructive proofs and the propositional calculus. In *Proceedings of the 7th Annual ACM Symposium on the Theory of Computing*, pages 83–97, 1975.

14. S. A. Cook and A. R. Reckhow. The relative efficiency of propositional proof systems. *Journal of Symbolic Logic*, 44(1):36–50, 1979.

15. W. Cook, C. R. Coullard, and G. Turán. On the complexity of cutting plane proofs. *Discrete Applied Mathematics*, 18:25–38, 1987.

16. M. Davis and H. Putnam. A computing procedure for quantification theory. *Journal of the ACM*, 7(3):210–215, 1960.

17. P. Hájek and P. Pudlák. *Metamathematics of First-Order Arithmetic*. Springer-Verlag, 1993.

18. A. Haken. The intractability or fesolution. *Theoretical Computer Science*, 39:297–308, 1985.

19. A. Haken. Counting bottlenecks to show monotone $\mathbf{P} \neq \mathbf{NP}$. In *Proceedings of the 36th IEEE FOCS*, 1995.

20. J. Håstad. *Computational limitations on Small Depth Circuits*. PhD thesis, Massachusetts Institute of Technology, 1986.

21. J. Krajíček. Lower bounds to the size of constant-depth propositional proofs. *Journal of Symbolic Logic*, 59(1):73–86, 1994.

22. J. Krajíček. *Bounded arithmetic, propositional logic and complexity theory*. Cambridge University Press, 1994.

23. J. Krajíček. Interpolation theorems, lower bounds for proof systems and independence results for bounded arithmetic. To appear in *Journal of Symbolic Logic*, 1994.

24. J. Krajíček and P. Pudlák. Some consequences of cryptographical conjectures for S_2^1 and EF. To appear in the Proceedings of the meeting *Logic and Computational Complexity*, Ed. D. Leivant, 1995.

25. J. Krajíček, P. Pudlák, and A. R. Woods. Exponential lower bounds to the size of bounded depth frege proofs of the pigeonhole principle. *Random Structures and Algorithms*, 7(1):15–39, 1995.

26. T. Pitassi, P. Beame, and R. Impagliazzo. Exponential lower bounds for the pigeonhole principle. *Computational Complexity*, 3:97–140, 1993.

27. P. Pudlák. Lower bounds for resolution and cutting planes proofs and monotone computations. Submitted to *Journal of Symbolic Logic*, 1995.

28. А. А. Разборов. Нижние оценки размера схем ограниченной глубины в полном базисе, содержащем функцию логического сложения. *Матем. Зам.*, 41(4):598–607, 1987. A. A. Razborov, Lower bounds on the size of bounded-depth networks over a complete basis with logical addition, *Mathem. Notes of the Academy of Sci. of the USSR*, 41(4):333–338, 1987.

29. A. Razborov. An equivalence between second order bounded domain bounded arithmetic and first order bounded arithmetic. In P. Clote and J. Krajíček, editors, *Arithmetic, Proof Theory and Computational Complexity*, pages 247–277. Oxford University Press, 1992.

30. A. Razborov. On small depth threshold circuits. In *Proceedings of the SWAT 92, Lecture Notes in Computer Science*, 621, pages 42–52, New York/Berlin, 1992. Springer-Verlag.

31. A. Razborov. Bounded Arithmetic and lower bounds in Boolean complexity. In P. Clote and J. Remmel, editors, *Feasible Mathematics II. Progress in Computer Science and Applied Logic*, vol. 13, pages 344–386. Birkhaüser, 1995.

32. A. Razborov. Unprovability of lower bounds on circuit size in certain fragments of Bounded Arithmetic. *Изв. АН СССР, сер. матем.* (*Izvestiya of the RAN*), 59(1):201–222, 1995. See also *Izvestiya: Mathematics 59:1, 205-227*.

33. A. Razborov. On provably disjoint NP-pairs. Technical Report RS-94-36, Basic Research in Computer Science Center, Aarhus, Denmark, 1994.

34. A. Razborov and S. Rudich. Natural proofs. To appear in *Journal of Computer and System Sciences* (for the preliminary version see *Proceedings of the 26th ACM Symposium on Theory of Computing*, pp. 204-213), 1994.

35. R. A. Reckhow. On the lengths of proofs in the propositional calculus. Technical Report 87, University of Toronto, 1976.

36. S. Riis. *Independence in Bounded Arithmetic*. PhD thesis, Oxford University, 1993.

37. S. Riis. Count(q) does not imply Count(p). Technical Report RS-94-21, Basic Research in Computer Science Center, Aarhus, Denmark, 1994.

38. J. A. Robinson. A machine-oriented logic based on the resolution principle. *Journal of the ACM*, 12(1):23–41, 1965.

39. A. L. Selman. Complexity issues in cryptography. *Proceedings of Symposia in Applied Mathematics*, 38:92–107, 1989.

40. R. Smolensky. Algebraic methods in the theory of lower bounds for Boolean circuit complexity. In *Proceedings of the 19th ACM Symposium on Theory of Computing*, pages 77–82, 1987.

41. G. Takeuti. *RSUV* isomorphisms. In P. Clote and J. Krajíček, editors, *Arithmetic, Proof Theory and Computational Complexity*, pages 364–386. Oxford University Press, 1992.

42. Г. С. Цейтин. О сложности вывода в исчислении высказываний. In А. О. Слисенко, editor, *Исследования по конструктивной математике и математической логике*, II; *Записки научных семинаров ЛОМИ, т. 8*, pages 234–259. Наука, Ленинград, 1968. Engl. translation: G. C. Tseitin, On the complexity of derivations in propositional calculus, in: *Studies in mathematics and mathematical logic, Part II*, ed. A. O. Slissenko, pp. 115-125.

43. A. Urquhart. Hard examples for resolution. *Journal of the ACM*, 34(1):209–219, 1987.

44. A. Urquhart. The complexity of propositional proofs. *Bulletin of Symbolic Logic*, 1:425–467, 1995.

Algebraic Characterizations of Decorated Trace Equivalences over Tree-Like Structures (Extended Abstract)*

Xiao Jun Chen[1] and Rocco De Nicola[2]

[1] Dipartimento di Scienze dell'Informazione, Università di Roma "La Sapienza"
E-mail: chen@dsi.uniroma1.it
[2] Dipartimento di Sistemi ed Informatica, Università di Firenze
E-mail: denicola@dsi2.ing.unifi.it

Abstract. A possible approach to studying behavioural equivalences in labelled transition systems is that of characterizing them in terms of homomorphic transformations. This characterization permits relying on algebraic techniques for proving systems properties and reduces equivalence checking of two systems to studying the relationships among the elements of their structures. Different algebraic characterizations of bisimulation-based equivalences in terms of particular transition systems homomorphisms have been proposed in the literature. Here we show, by an example, that trace-based equivalences are not locally characterizable and thus that the above results cannot be extended to these equivalences. However, similar results can be obtained if we confine ourselves to restricted classes of transition systems. Here, the algebraic characterizations of three well known decorated-trace equivalences (*ready trace*, *ready* and *failure equivalence*) for tree-like structures are presented.

1 Introduction

A possible approach to studying behavioural equivalences in labelled transition systems is that of characterizing them in terms of homomorphic transformations. This characterization permits relying on algebraic techniques for proving systems properties and reduces equivalence checking of two systems to studying the relationships among the elements of their structures.

Given a behavioural equivalence, one may provide adequate conditions on transformations in order to preserve the equivalence or the modalities of the logics adequate for it. In [5], it has been shown that the *abstraction homomorphisms*, introduced in [6] to simplify labelled event structures, preserve, and actually fully characterize, the strong and weak bisimulation equivalences of [10]. *Saturating homomorphisms* for a given logic are used in [1, 2], and are applied to characterize a number of logically defined equivalences, such as strong

---

* This work has been partially founded by EEC, HCM Project EXPRESS, and by CNR: Progetto "Specifica ad Alto Livello e Verifica di Sistemi Digitali"

bisimulation equivalence [10], the generalized transition system bisimulation induced by Future Perfect logic [8] and the branching bisimulation characterized by Hennessy-Milner logic with "until" operators [13].

All the above mentioned approaches have focused on bisimulation-based equivalences, or on their corresponding modal logics. Decorated trace equivalences are a large family of equivalences weaker than the bisimulation-based ones, that can be obtained via effective testing in the style of [12]. Among decorated trace equivalences, we are interested in those equivalences that rely on analyzing the set of the actions that can be performed from the states reached via each trace. More specifically, we will consider *ready trace equivalence* [3, 16], *ready equivalence* [14], *failure/testing equivalence* [4, 11]. Readers are referred to [17, 18] for an exhaustive overview of these equivalences and for a discussion of experimental settings that lead to them.

When looking for algebraic characterizations of decorated trace equivalences similar to those based on bisimulation, one should be aware of the facts that:

1. these equivalences are not inductively defined;
2. the equivalence of two elements depends both on their future capabilities and on their past traces.

These two observations give an indication that decorated trace equivalences cannot be fully characterized in general: we provide a simple example to show this; see Figure 2 of Section 3. We will consider a restricted classes of transition systems, namely, transition trees and confluent transition systems. Transition trees are transition systems with *unique access paths* for each node, while confluent transition systems are directed acyclic graphs with *unique access trace* for each node. Over these classes, we define ready homomorphism, and failure morphism to characterize ready trace, ready and failure equivalence. More precisely, we will show that:

1. two transition trees are ready trace equivalent if and only if they have a common image under ready homomorphism;
2. two confluent transition systems are ready equivalent if and only if they have a common image under ready homomorphism;
3. two confluent transition systems are failure equivalent if and only if they have a common image under failure morphism.

In this paper, we restrict attention to the strong variants of the equivalences, i.e. we assume that all labels represent visible actions. The generalization to richer sets of actions is however straightforward.

The rest of the paper is organized as follows: Section 2 introduces the necessary notational background on labelled transition systems, and on the considered decorated trace equivalences. Section 3 presents ready homomorphism that, applied to transition trees, fully characterizes ready trace equivalence. Section 4 extends our approach to confluent transition systems: Within this class ready homomorphism fully characterizes ready equivalence. Section 5 presents failure morphisms over confluent transition systems and shows that they induce failure equivalence. The last section is dedicated to a few concluding remarks.

2 Background and Notations

In this section, we introduce the basic definitions for labelled transition systems and homomorphism, as well as the notations for the decorated trace equivalences discussed in the paper, namely *ready trace equivalence* [3, 16], *ready equivalence* [14], and *failure equivalence* [4, 11]. As already mentioned in the Introduction, we will restrict attention to systems without silent moves.

Definition 1 *Labelled Transition Systems.*
A *labelled transition system* is a quadruple $\langle S, A, \rightarrow, s_0 \rangle$ where S is a countable set of states, A is a countable set of elementary actions, $\rightarrow \subseteq S \times A \times S$ is a set of transitions, and $s_0 \in S$ is the initial state.

A labelled transition system $\langle S, A, \rightarrow, s_0 \rangle$ is *finitely branching*, if $\forall s \in S$, $\{(a, s') \mid (s, a, s') \in \rightarrow\}$ is finite. In this paper, we consider only labelled transition systems that are finitely branching.

In the following, we will use $a, b, a_1, \ldots, a_n, b_1, \ldots, b_n, \ldots$ to range over A. A^* will be used to denote the set of strings over A and we will use σ to range over it. $P(A)$ will be used to denote the powerset of A. Moreover, a transition $(s, a, s') \in \rightarrow$ will be denoted by $s \xrightarrow{a} s'$ and labelled transition systems will be sometimes called transition systems for short. Furthermore, we will use the following conventions:

- $s \xrightarrow{a}$ will stand for $\exists s'$ such that $s \xrightarrow{a} s'$;
- $s \xrightarrow{\sigma} s_n$ where $\sigma = a_1 \ldots a_n$, will stand for
$$\exists s_1, \ldots, s_{n-1} \text{ such that } s \xrightarrow{a_1} s_1 \ldots \xrightarrow{a_n} s_n;$$
- $s \xrightarrow{\sigma}$ will stand for $\exists s'$ such that $s \xrightarrow{\sigma} s'$;
- $s \not\xrightarrow{\sigma}$ stands for *not* $s \xrightarrow{\sigma}$;
- A sequence of successive transitions will be called a *path*;
- $I(s)$ will be used to denote the set of initial actions from state s:
$$I(s) = \{a \in A \mid s \xrightarrow{a}\}.$$

Definition 2 *Transition System Homomorphisms.*
Let $T = \langle S, A, \rightarrow, s_0 \rangle$ and $T' = \langle S', A, \rightarrow', s_0' \rangle$ be two transition systems. $h : S \longrightarrow S'$ is a *transition system homomorphism* if

$$h(\rightarrow) \subseteq \rightarrow' \text{ where } h(\rightarrow) = \{h(s) \xrightarrow{a}' h(s') \mid s \xrightarrow{a} s'\}.$$

A homomorphism h is *surjective* if $h(S) = S'$.

Definition 3 *Ready Trace Equivalence* [3, 16].

(i) A sequence $X_0 a_1 X_1 a_2 \ldots a_n X_n \in P(A) \times (A \times P(A))^*$ is a *ready trace* of state s, if there exist states $s_1 \ldots s_n$ such that
$$s \xrightarrow{a_1} s_1 \xrightarrow{a_2} \ldots \xrightarrow{a_n} s_n, \ I(s) = X_0 \text{ and } I(s_i) = X_i \text{ for } i = 1, \ldots, n.$$

(ii) If $RT(s)$ denotes the set of ready traces of state s, two transition systems $T = \langle S, A, \rightarrow, s_0 \rangle$ and $T' = \langle S', A, \rightarrow', s_0' \rangle$, are *ready trace equivalent* if $RT(s_0) = RT(s_0')$.

Definition 4 *Ready Equivalence* [14].

(i) A pair $\langle \sigma, X \rangle \in A^* \times P(A)$ is a *ready pair* of state s, if there exists a state s' such that $s \xrightarrow{\sigma} s'$ and $I(s') = X$.

(ii) If $R(s)$ denotes the set of ready pairs of state s, two transition systems $T = \langle S, A, \rightarrow, s_0 \rangle$ and $T' = \langle S', A, \rightarrow', s_0' \rangle$, are *ready equivalent* if $R(s_0) = R(s_0')$.

The *failure semantics* is introduced in [4] and used in the construction of a model for the process algebra named CSP [9]. The impact of such an equivalence on labelled transition systems and its characterization as a testing equivalence are studied in [11].

Definition 5 *Failure Equivalence* [4].

(i) $\langle \sigma, X \rangle \in A^* \times P(A)$ is a *failure pair* of a state s, if there is a state s' such that $s \xrightarrow{\sigma} s'$ and $I(s') \cap X = \emptyset$.

(ii) If $F(s)$ is used to denote the set of failure pairs of s, two transition systems $T = \langle S, A, \rightarrow, s_0 \rangle$ and $T' = \langle S', A, \rightarrow', s_0' \rangle$, are *failure equivalent* if $F(s_0) = F(s_0')$.

Definition 6 *Strong Bisimulation* [15].

(i) A *bisimulation* between two transition systems $T = \langle S, A, \rightarrow, s_0 \rangle$ and $T' = \langle S', A, \rightarrow', s_0' \rangle$, is a binary relation $U \subseteq S \times S'$ such that
1. $\forall s \in S, \exists s' \in S' : sUs'$ and $\forall s' \in S', \exists s \in S : sUs'$,
2. if sUt then, $\forall a \in A$:
    (a) $s \xrightarrow{a} s'$, then $\exists t' : t \xrightarrow{a}' t'$ and $s'Ut'$;
    (b) $t \xrightarrow{a}' t'$, then $\exists s' : s \xrightarrow{a} s'$ and $s'Ut'$.

(ii) We say T and T' are *bisimilar* if there is a bisimulation between them.

Definition 7 *Abstraction Homomorphisms* [5].
Let $T = \langle S, A, \rightarrow, s_0 \rangle$ and $T' = \langle S', A, \rightarrow', s_0' \rangle$ be two transition systems. $h : S \rightarrow S'$ is called an *abstraction homomorphism* if it is a surjective transition system homomorphism satisfying:

$$\forall s_1 \in S, a \in A, s_2' \in S'. \; h(s_1) \xrightarrow{a}' s_2' \text{ implies}$$

$$\exists s_2 \in S. \; s_1 \xrightarrow{a} s_2 \text{ and } h(s_2) = s_2'.$$

The following theorem [5] establishes a clear relation between bisimulation and abstraction homomorphism, formulated as in [2, 7].

Theorem 8. *Two transition systems T and T' are bisimilar if and only if they have a common image under abstraction homomorphism.*

Definition 9 *Transition Trees and Confluent Transition Systems.*

1. A *transition tree* is a transition system where every state is reachable from
the initial state s_0 and we have $a_i = b_i$, $r_i = t_i$ $(i = 1, \ldots, n)$ whenever
$$s_0 \xrightarrow{a_1} r_1 \ldots \xrightarrow{a_n} r_n = s_f, \quad s_0 \xrightarrow{b_1} t_1 \ldots \xrightarrow{b_n} t_n = s_f$$

2. A *confluent transition system* is a transition system where every state is
reachable from the initial state s_0 and
$$s_0 \xrightarrow{\sigma_1} s_f, \quad s_0 \xrightarrow{\sigma_2} s_f \text{ implies } \sigma_1 = \sigma_2.$$

When only transition trees or confluent transition systems are considered,
each state has a unique access trace, and we use $AT(s)$ to denote the access
trace of s:

$$AT(s) = \sigma \text{ iff } s_0 \xrightarrow{\sigma} s. \qquad\qquad (s_0 \text{ is the initial state})$$

3 Ready Trace Equivalence of Transition Trees

As we know, abstraction homomorphism preserves strong equivalence, and thus
it preserves all the equivalences weaker than strong equivalence as well. Naturally, for weaker equivalences, one would expect weaker conditions than those
for abstraction homomorphism. For example, the homomorphism of Figure 1
preserves ready trace equivalence, failure equivalence, etc., but it does not sat-

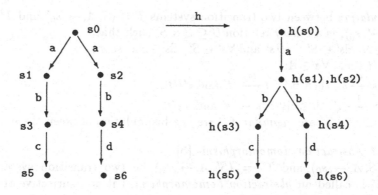

Fig. 1. A ready homomorphism for decorated trace equivalences

isfy the condition for abstraction homomorphism. Indeed, for $h(s_2) \xrightarrow{b} h(s_3)$,
there does not exist t such that $s_2 \xrightarrow{b} t$ and $h(t) = h(s_3)$.

We say that a homomorphism is *locally definable* if it can be defined only
in terms of single actions or action sequences of bound length. According to
this definition, we have that abstraction homomorphisms are locally definable.

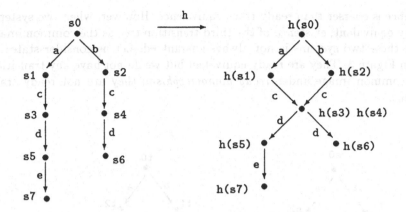

Fig. 2. Processes that are not trace equivalent but are locally homomorphic

On the other hand, we have that the decorated trace equivalences we consider cannot be characterized in terms of locally definable homomorphisms. This is mainly due to the fact that they heavily rely on trace equivalence, and this is not locally characterizable on general transition systems. This can be appreciated by considering the transition systems of Figure 2. The two systems are not trace equivalent, yet it is hard to imagine local conditions that permit differentiating s_3 and s_4 without considering their whole history/future.

If we restrict ourself to considering a subset of transition systems, namely transition trees, it can be shown that the *ready homomorphism* defined below fully characterizes ready trace equivalence. We have to remark that an important aspect of our homomorphisms is that they put requirement only on traces from the initial states, and thus we have a ready homomorphism between the two trees of Figure 1 in spite of the fact that

$$\{b\}b\{d\} \in RT(h(s_1)) \text{ but } \{b\}b\{d\} \notin RT(s_1).$$

Definition 10 *Ready Homomorphisms.*
Let $T = \langle S, A, \rightarrow, s_0 \rangle$ and $T' = \langle S', A, \rightarrow', s'_0 \rangle$ be two transition systems. A surjective transition system homomorphism $h : S \rightarrow S'$ is a *ready homomorphism* if it satisfies:

1. $h(s_0) = s'_0$;
2. $\forall s \in S, a \in A.\ h(s) \xrightarrow{a}' \text{ implies } s \xrightarrow{a}.$

Theorem 11. *Two transition trees are ready trace equivalent if and only if they have a common image under ready homomorphism.*

4 Ready Equivalence of Confluent Transition Systems

If there exists a *ready homomorphism* between two transition trees, then from the previous section we know that they are also ready equivalent. Indeed, ready equivalence is coarser than ready trace equivalence. However, when two systems are ready equivalent, existence of the third transition tree as the common image between these two systems is not always guaranteed. Let us consider states s_0 and t_0 in Figure 3. They are ready equivalent but we do not have any transition tree as common image under *ready homomorphism*; they are not ready trace equivalent.

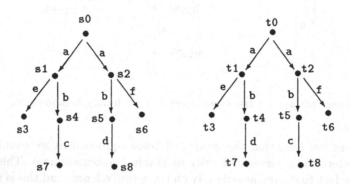

Fig. 3. Processes that are ready equivalent but not ready trace equivalent

The main difference between the two equivalences lies in the different placing of the requirements on the decorations: although both of them use $I(s)$ as decorations on s, in ready trace equivalence the requirement are put after each step of the trace while for ready equivalence, there are requirements only at the end of the trace. Thus, s_0 and t_0 in Figure 3 are not ready trace equivalent, because

$$\{a\}a\{b,e\}b\{c\} \in RT(s_0), \text{ while we have } \{a\}a\{b,e\}b\{c\} \notin RT(t_0),$$

yet for ready pair, we do not require that the state between actions a and b should have action set $\{b,e\}$, i.e. we only consider $\langle ab, \{c\}\rangle$, and we have $\langle ab, \{c\}\rangle \in R(s_0)$ and $\langle ab, \{c\}\rangle \in R(t_0)$.

Taking this main difference into account, it is natural to look for standard representatives of ready equivalence that have a richer structure, and permit flattening some of the information along the traces. More specifically, we will have to permit states that can be accessed from the initial state, via different paths with different decorations, but still labelled by the same trace.

As an example, consider Figure 4: it can be seen as derived from s_0 or t_0 in Figure 3, by adding two new transitions: $s_1 \overset{b}{\longrightarrow} s_5$, $s_2 \overset{b}{\longrightarrow} s_4$ (or correspondingly, $t_1 \overset{b}{\longrightarrow} t_5$, $t_2 \overset{b}{\longrightarrow} t_4$).

This change does not preserve ready trace equivalence, but it does preserve ready equivalence. This way, we can move from r_0 to r_1 with initial actions set $\{b,e\}$, and then from r_1 to r_5 with initial actions set $\{d\}$, etc.:

$$\{a\}a\{b,e\}b\{c\} \in RT(r_0), \quad \{a\}a\{b,e\}b\{d\} \in RT(r_0),$$

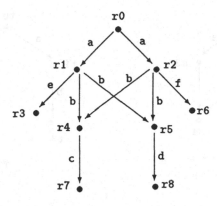

Fig. 4. Standard Representative for Ready Equivalence

$\{a\}a\{b,f\}b\{c\} \in RT(r_0)$, $\{a\}a\{b,f\}b\{d\} \in RT(r_0)$.

This is the key idea of the construction of a homomorphism between s_0 and r_0, and between t_0 and r_0:

$h_1(s_i) = r_i$ for $i = 1, \ldots, 8$,

$h_2(t_4) = r_5$, $h_2(t_5) = r_4$, $h_2(t_i) = r_i$ for $i = 1, 2, 3, 6, 7, 8$.

If one wants to consider also systems with states reachable via different paths then he has to abandon transition trees; a more generous class is needed. A result similar to that of the previous section can be proved for confluent transition systems (r_0 is indeed *confluent*) and ready equivalence.

Theorem 12. *Two confluent transition systems are ready equivalent if and only if they have a common image under ready homomorphism.*

5 Failure Equivalence of Confluent Transition Systems

We proceed now, by examining failure equivalence. In this case, transition system homomorphisms are too demanding. We need to consider "unstructured" morphism between the states. Figure 5 is an evidence of this: As a confluent transition system, $T2$ has minimal set of states, and any non-trivial addition of transitions (without the addition of new states) will not preserve failure equivalence. Thus $T2$ should be used as the common image of $T1$ and $T2$. However, a suitable morphism h from $T1$ to $T2$ should have $h(s_5) = t_1$ or $h(s_5) = t_2$, but such an h cannot be a homomorphism: $h(s_5) \xrightarrow{b} h(s_3)$ and $h(s_5) \xrightarrow{c} h(s_4)$ cannot be presented at the same time.

In the rest of this section, we discuss a class of failure morphisms which fully characterizes failure equivalence.

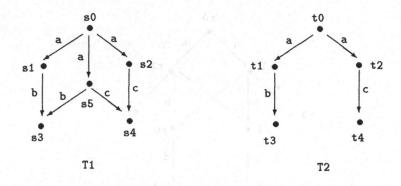

Fig. 5. Failure equivalent processes that are not homomorphic

The definition of failure morphism relies of the following predicate $mini(\_)$, that singles out those states s with the property that there exists no other state with the same access trace AT as s but has a smaller set of next actions.

Definition 13. $mini(s)$ iff $\not\exists t$ s.t. $AT(t) = AT(s) \wedge I(t) \subset I(s)$.

Example 1. In Figure 6, we have $mini(s_i)$ for $i = 0, 1, 3, \ldots, 11$, but $\neg mini(s_2)$.

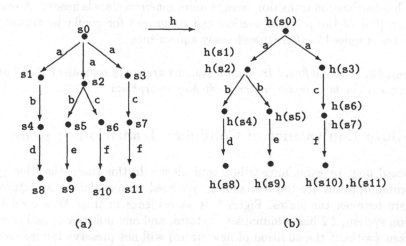

Fig. 6. Standardization for failure equivalence: reducing states

Definition 14 *failure morphism.*
Let $T1 = \langle S, A, \rightarrow, s_0 \rangle$ and $T2 = \langle S', A, \rightarrow', s_0' \rangle$ be two *confluent* transition

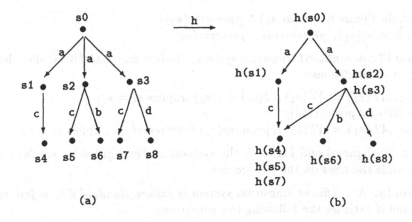

Fig. 7. Standardization for failure equivalence: reducing states

systems. $h : S \to S'$ is called *failure morphism* if it satisfies the following three conditions:

1. $\forall s \in S.\ AT(s) = AT(h(s))$;
2. $\forall s \in S.\ mini(s)$ implies $(s \xrightarrow{a}$ iff $h(s) \xrightarrow{a}')$;
3. $\forall s' \in S'.\ mini(s')$ implies $\exists s \in S.\ h(s) = s' \land (s \xrightarrow{a}$ iff $h(s) \xrightarrow{a}')$.

Example 2. Figure 6 and 7 provide two *failure morphisms*.

Theorem 15. *Let $T1 = \langle S, A, \to, s_0 \rangle$ and $T2 = \langle S', A, \to', s_0' \rangle$ be two confluent transition systems. If there exists a failure morphism $h : S \to S'$, then $F(s_0) = F(s_0')$.*

To demonstrate that any two failure equivalent systems have a common image under *failure morphism*, we introduce *failure standard* system to determine existence of common image. *Failure standard* system provides a standard representation for the class of failure equivalence over confluent transition systems. An important fact that we would like to mention is that our construction of *failure standard* system does not require any auxiliary invisible action τ, as in [12].

To define failure standard systems, we use another auxiliary predicate *preserve*. It captures those states s that, although does not satisfy *mini*, but have a next action which cannot be performed by any other state accessible via the same trace of s.

Definition 16. *preserve(s)* iff

$$\neg mini(s) \land \exists a.\ s \xrightarrow{a} \land \forall t \neq s.\ s.t.\ AT(t) = AT(s) : t \not\xrightarrow{a} .$$

Example 3. In Figure 6: $\neg mini(s_2) \wedge \neg preserve(s_2)$.
In Figure 7: $mini(s_1)$, $preserve(s_2)$, $preserve(s_3)$.

Definition 17. A *confluent* transition system is *failure minimal* if it satisfies the following three conditions:

P1 $\forall s_1, s_2.\ \mathcal{AT}(s_1) = \mathcal{AT}(s_2) \wedge I(s_1) = I(s_2)$ implies $s_1 = s_2$;
P2 $\forall s.\ mini(s) \vee preserve(s)$;
P3 $\forall s_1, s_2.\ \mathcal{AT}(s_1) = \mathcal{AT}(s_2) \wedge preserve(s_1) \wedge preserve(s_2)$ implies $s_1 = s_2$.

Example 4. In Figure 6 and Figure 7, the systems on the right sides are *failure minimal*, while the ones on the left are not.

Definition 18. A *confluent* transition system is *failure standard* if it is *failure minimal* and it satisfies the following two conditions:

P4 $\forall s.\ preserve(s)$ implies $\forall r, a, r'.\ s.t.\ \mathcal{AT}(r) = \mathcal{AT}(s) \wedge r \xrightarrow{a} r' : s \xrightarrow{a} r'$;
P5 $\forall s, a.\ s \xrightarrow{a}$ implies $\forall r, t.\ s.t.\ \mathcal{AT}(r) = \mathcal{AT}(s) \wedge r \xrightarrow{a} t : s \xrightarrow{a} t$.

Example 5. Figure 4 is a *failure standard* one while Figure 3 are not.

Theorem 19. *Any two failure equivalent confluent transition systems have a common image under failure morphism.*

6 Conclusion and Open Problem

We have investigated the algebraic characterizations in terms of transition system homomorphisms of three behavioral equivalences over subclasses of Labelled Transition Systems (LTS). The considered equivalences are all trace-based and all weaker than the bisimulation based ones, for which similar results existed. Indeed, these results have been the inspiration of our work.

We have also shown that the results obtained for bisimulation-based equivalences cannot be extended to general LTS when considering trace-based equivalences. We have argued that these equivalences are not locally characterizable for the general class of LTS. Eventhough, the present paper constitutes a step forward for important subclasses of LTS, namely, tree-like structures, that are widely used to model nondeterministic computations.

Among the problems left open by our contribution, we would like to single out that of extending our results to failure trace equivalence over transition trees or over confluent transition systems. We conclude by reporting the definition of this equivalence that undoubtely has an experimental appeal.

Two transition systems are *failure trace equivalent* if they have the same set of failure traces from their initial states.

A sequence $X_0 a_1 X_1 a_2 \ldots a_n X_n \in P(A) \times (A \times P(A))^*$ is a *failure trace* of state s, if there exist states $s_1 \ldots s_n$ such that

$$s \xrightarrow{a_1} s_1 \xrightarrow{a_2} \ldots \xrightarrow{a_n} s_n,\ I(s) \cap X_0 = \emptyset \text{ and } I(s_i) \cap X_i = \emptyset \text{ for } i = 1, \ldots, n.$$

Acknowledgement The authors gratefully acknowledge the anonymous referees for useful comments on the submitted version of this paper.

References

1. A. Arnold and A. Dicky. An algebraic characterization of transition system equivalences. *Information and Computation*, 82:198–229, 1989.
2. A. Arnold and A. Dicky. Equivalences and preorders of transition systems. *MFCS'93*, LNCS 711:20–31, 1993.
3. J.C.M. Baeten, J.A. Bergstra, and J.W. Klop. Ready-trace semantics for concrete process algebra with the priority operator. *The Computer Journal*, 30(6):498–506, 1987.
4. S.D. Brookes, C.A.R. Hoare, and A.W. Roscoe. A theory of communicating sequential processes. *JACM*, 31(3):560–599, 1984.
5. I. Castellani. Bisimulations and abstraction homomorphisms. *J. Comput. System Sci.*, 34:210–235, 1987.
6. I. Castellani, P. Franceschi, and U. Montanari. Labelled event structures: a model for observable concurrency. In D. Børner, editor, *Formal Description of Programming Concepts II*, pages 383–400, North Holland, 1983.
7. P. Degano, R. De Nicola, and U. Montanari. A distributed operational semantics for CCS based on condition/event systems. *Acta Informatica*, 26:59–91, 1988.
8. M. Hennessy and C. Stirling. The power of the future perfect in program logics. *Information and Control*, 67:23–52, 1985.
9. C.A.R. Hoare. *Communicating Sequential Processes*. Prentice Hall, 1985.
10. R. Milner. *Communication and Concurrency*. Prentice-Hall, London, 1989.
11. R. De Nicola. Extentional equivalences for transition system. *Acta Informatica*, 24:211–237, 1987.
12. R. De Nicola and M.C.B. Hennessy. Testing equivalences for processes. *Theoretical Computer Science*, 34:83–133, 1984.
13. R. De Nicola and F. Vaandrager. Three logics for branching bisimulation. *J. ACM*, 42(2):458–487, 1995.
14. E.R. Olderog and C.A.R. Hoare. Specification-oriented semantics for communicating processes. *Acta Informatica*, 23:9–66, 1986.
15. D. Park. Concurrency and automata on infinite sequences. *Proc. GI*, LNCS 104:167–183, 1981.
16. A. Pnueli. Linear and branching structures in the semantics and logics of reactive systems. *ICALP'85*, LNCS 194:15–32, 1985.
17. R.J. van Glabbeek. The linear time - branching time spectrum. *CONCUR'90*, LNCS 458:278–297, 1990.
18. R.J. van Glabbeek. The linear time - branching time spectrum ii (the semantics of sequential systems with silent moves). *CONCUR'93*, LNCS 715:66–81, 1993.

Fast Asynchronous Systems in Dense Time

Lars Jenner, Walter Vogler *

Inst. f. Mathematik, Univ. Augsburg, D-86135 Augsburg, Germany
e-mail: {vogler|jenner}@informatik.uni-augsburg.de

Abstract

A testing scenario in the sense of De Nicola and Hennessy is developed to measure the worst-case efficiency of asynchronous systems using dense time. For all three variants considered, it is shown that one can equivalently use discrete time; in the discrete versions, one variant coincides with an approach based on discrete time in [Vog95b], and thus we can clarify the assumptions behind this approach. The resulting testing-preorders are characterized with some kind of refusal traces and shown to satisfy some properties that make them attractive as faster-than relations. The three testing-preorders are incomparable in general, but for some interesting classes of systems implications are shown.

1 Introduction

In the testing approach of [DNH84], reactive systems are compared by embedding them – with a parallel composition operator $\|$ – in test environments, which are also arbitrary reactive systems having a special action ω. One variant of testing (must-testing) considers the worst-case behaviour: a system N performs successfully in an environment O if *every* run of $N \| O$ reaches success, signalled by ω. If some system N_1 performs successfully whenever a second system N_2 does, then N_1 is called an implementation of the specification N_2; of course, an implementation may be successful in more environments than specified. This approach only takes into account the functionality of systems, i.e. which actions can be performed. To take also into account the efficiency of systems, we can add a time bound D to our tests and require that every run reaches success within time D [Vog95a]. In this efficiency testing approach, an implementation cannot only be successful in more environments than the specification, it can also be successful faster; i.e. the implementation (or testing) preorder can serve as a faster-than relation.

To apply efficiency testing, we have to measure the duration of a run. This is no problem, if the parallel system $N \| O$ is synchronous, i.e. if all components perform their actions according to a common global time scale (see [Vog95a]). In asynchronous systems, the components work with indeterminate relative speeds. Usually, this is interpreted as: components may idle unnecessarily or actions

---

*This work was partially supported by the DFG-project 'Halbordnungstesten'.

may take more time than necessary; under this interpretation, the worst-case behaviour is to idle until time D is up and, thus, no test at all is satisfied.

Nevertheless, [Vog95b] develops a scenario of efficiency testing for asynchronous systems and studies the corresponding faster-than relation, using labelled safe Petri nets as system models. This is based on a different interpretation of asynchronous systems: it is assumed that the components are guaranteed to perform each enabled action within at most one unit of time; thus, a component does not delay arbitrarily; instead, all other components may work very fast in comparison. Hence, the relative speeds of the components are still arbitrary, i.e. the system is really asynchronous; this idea goes back to at least [LF81].

The approach of [Vog95b] describes time with natural numbers; since 1 is the upper time bound for an action, the faster actions take time 0. With this choice, it is not immediately convincing that the technical development of [Vog95b] really meets the intuition given above. To get a better understanding, we develop here the same intuitive idea using *dense* time. In our approach, an action may start some time after activation and it may end some time later, provided the end occurs at most one unit of time after activation. We also consider two variants: in the a-variant, all the time is spent on the action, i.e. an enabled action starts without delay and ends within time 1. On the contrary, actions are instantaneous in the i-variant and all the time is taken in the activation phase, which again takes at most time 1. For the three variants, we define satisfaction of an efficiency test and the corresponding testing preorders.

Our first main result shows that, in all three cases, we can replace the modelling of runs with dense time by an equivalent model using discrete time. This makes the testing approach much easier to work with since it gives a finite state space for a finite system. With our result, we also clarify the ideas behind the approach of [Vog95b], since we recover that model as the discrete version of the a-variant. Furthermore, we treat the following problem: using discrete time, it is unavoidable that arbitrary (finite) sequences of (possibly causally related) actions may occur within zero time; this might be regarded as counterintuitive. We show that our approach does not depend on this phenomenon: if we require in the dense time approach that different actions are separated by some nonzero time, we still get the same testing preorder.

The testing preorder in the discrete a-variant is characterized in [Vog95b] with some kind of refusal traces. As a second main result, we give characterizations in the same style for the other two variants, which imply the decidability of the testing preorders. In the basic variant, the characterization turns out to be somewhat involved; to understand it, it is useful to view refusal sets as sets of correctly treated actions, a view that might be helpful also in other settings. Using the characterizations, we can show that the three testing preorders are precongruences for parallel composition, hiding, restriction and relabelling.

To demonstrate that the testing preorder in the a-variant is really a sensible faster-than relation, three constructions of a system N' from a system N are introduced in [Vog95b], where it is intuitively clear that N and N' should be functionally equivalent, but that N should be faster. As a third main result, we show in Section 5 that this is indeed the case also in the other two variants.

Although the variants are intensionally closely related, it turns out that the three testing preorders are in general incomparable. As further main results, we exhibit some classes of systems where from a faster-than result in one variant we can deduce a faster-than result in another. These results can be used in the case of two implementations of a bounded buffer. Due to lack of space, technical developments are only sketched and proofs are only indicated at most.

2 Basic notions

We assume some basic knowledge of Petri nets like presets, the firing rule for transitions t or steps μ or the language of a (labelled) net; see e.g. [Vog95b] for our notation. We use finite *safe* Petri nets (place/transition-nets) (S, T, W, l, M_N), without isolated transitions whose transitions are labelled by l with actions from some infinite alphabet Σ' or with the empty word λ, indicating an internal, unobservable action. We can extend the labelling to steps and sequences of steps or transitions as usual, whereby deleting internal actions. Σ' contains a special action ω and we put $\Sigma = \Sigma' - \{\omega\}$.

For each set A of transitions or actions, A^+ and A^- denote disjoint copies of A whose elements are called *transition* or *action parts* and denoted a^+ resp. a^-, $a \in A$; a^+ will stand for the start of a, which only empties the corresponding preset, while a^- indicates the end of a, producing the tokens for the corresponding postset. Any labelling function l is extended to transition parts by $l(t^+) = l(t)^+$ and $l(t^-) = l(t)^-$ if $l(t) \neq \lambda$ and $l(t^+) = l(t^-) = \lambda$ if $l(t) = \lambda$.

We use parallel composition $\|_A$ with synchronization, i.e. if we combine nets N_1 and N_2 with $\|_A$, then they run in parallel and have to synchronize on actions from $A \subseteq \Sigma'$. To construct the composed net, we have to combine each a-labelled transition t_1 of N_1 with each a-labelled transition t_2 from N_2 if $a \in A$. Parallel composition is an important operator for the modular construction of nets.

We also consider shortly *relabelling* with $f : \Sigma' \to \Sigma'$, which changes the observable actions according to the function f, *hiding* some $a \in \Sigma'$, which changes all labels a to λ, and *restricting* $a \in \Sigma'$, which deletes all a-labelled transitions.

3 Timed behaviour of asynchronous systems

Our first definition describes the asynchronous behaviour of a parallel system. We assume that the components are guaranteed to perform each enabled action within at most one unit of time; this upper time bound allows the relative speeds of the components to vary arbitrarily, since we have no lower time bound. Thus, the behaviour we define is truly asynchronous.

Technically speaking, we require that each enabled transition starts and ends firing within time 1 – unless it is disabled within this time. For this purpose, we keep track of the remaining time an enabled or firing transition has using a function ρ; $\rho(t)$ is initialized to 1, when t gets enabled. As dense time domain we choose the reals, hence we will speak of continuous firing; $\mathbb{R}^+$ is the set of positive real numbers. We denote a constant function by this constant, possibly indexed by the function's domain.

Definition 3.1: A *continuous instantaneous description CID* of a net N is a quadrupel (M, A, C, ρ) consisting of a marking M of N, two sets $A \subseteq T$ and $C \subseteq T$ of *activated* and *current(ly firing) transitions* and a function $\rho : A \cup C \to [0, 1]$ describing the *residual activation* resp. *firing time* of an activated resp. current transition. The *initial CID* is $CID_N = (M_N, A_N, \emptyset, \rho_N)$ with $A_N = \{t \mid M_N[t\rangle\}$ and $\rho_N = 1_{A_N}$. We write $(M, A, C, \rho)[\varepsilon\rangle_c (M', A', C', \rho')$ if $A' = \{t' \mid M'[t'\rangle\}$ and one of the following cases applies:

1. $\varepsilon = t^+$, $t \in A$, $M' = M - {}^\bullet t$, $C' = C \cup \{t\}$, $\rho' = \rho|_{(A' \cup C')}$
2. $\varepsilon = t^-$, $t \in C$, $M' = M + t^\bullet$, $C' = C - \{t\}$, $\rho' = \rho|_{(A \cup C')} \cup 1_{(A'-A)}$
3. $\varepsilon = (r)$, $r \in \mathbb{R}^+$, $r \le \min \rho(A \cup C)$, $M' = M$, $A' = A$, $C' = C$, $\rho' = \rho - r$.

The set $CFS(N) = \{w \mid CID_N[w\rangle_c CID\}$ is the set of *continuous firing sequences* of N, the set $CL(N) = \{l(w) \mid w \in CFS(N)\}$ is the *continuous language* of N. We let l preserve time steps, i.e. $l((r)) = (r)$. ∎ 3.1

Part 3 above ensures that every transition that is enabled for one unit of time fires and ends firing within that unit, but by 1 and 2 it may also act faster.

Definition 3.2: For w in $CL(N)$ resp. $CFS(N)$, $\alpha(w)$ is the sequence of action resp. transition parts in w, and $\zeta(w)$ is the *duration*, i.e. the sum of time steps in w. ∎ 3.2

To see whether a system N performs successfully in a testing environment O, we have to check that in each run of $N\|_\Sigma O$ the success action ω is performed at some given time R at the latest. To be sure that we have seen everything that occurs up to time R, we only look at runs w with $\zeta(w) > R$.

Definition 3.3: A net is *testable* if none of its transitions is labelled with ω. A *continuously timed test* is a pair (O, R), where O is a net (the *test net*) and $R \in \mathbb{R}_0^+$. A testable net N *c-satisfies* such a test, if each $w \in CL(N\|_\Sigma O)$ with $\zeta(w) > R$ contains some ω^+. For testable nets N_1 and N_2, we call N_1 a *continuously faster implementation* of N_2, $N_1 \sqsupseteq_c N_2$, if N_1 c-satisfies all continuously timed tests that N_2 satisfies. ∎ 3.3

Our first aim is to find out about the slowest firing sequences, for these will decide satisfaction of a timed test; we will conclude that we can restrict attention to the discrete sublanguage DCL of the continuous language.

Definition 3.4: For a net N is $DCL(N) = \{v \in CL(N) \mid$ for all time steps (r) in v: $r = 1\}$. $DCL(N)$ is also generated by the suitably defined *d-continuous firing sequences* $DCFS(N)$. A *discretely timed test* is a pair (O, D), where O is a net and $D \in \mathbb{N}_0$. A testable net N *d-satisfies* such a test (O, D), N *$must_d$* (O, D), if each $v \in DCL(N\|_\Sigma O)$ with $\zeta(v) > D$ contains some ω^+; $N_1 \sqsupseteq_d N_2 \Leftrightarrow$ $(\forall (O, D) : N_2 \ must_d \ (O, D) \Rightarrow N_1 \ must_d \ (O, D))$. ∎ 3.4

The first part of 3.5 states that for every $w \in CFS$ we can find a $v \in DCFS$ that has the same action sequence but is slower; v is constructed from w by letting one time unit pass in v whenever the cumulated time in w exceeds the next natural number. Hence, if w is a run where ω does not occur in time, then v is such a run, too. From this, one can derive Theorem 3.6.

Lemma 3.5: i) For a net N there is for each $w \in CFS(N)$ a $v \in DCFS(N)$ with $\alpha(v) = \alpha(w)$ and $\zeta(v) \ge \zeta(w)$. Furthermore, transition parts in v are separated by (1) only if the corresponding transition parts in w are separated by some (r).

ii) For each $v \in DCFS(N)$ there is a $v' \in DCFS(N)$ that starts with a (1)-time-step and satisfies $\alpha(v') = \alpha(v)$ and $\zeta(v') \geq \zeta(v)$. ∎ 3.5

Theorem 3.6: The relations $\sqsupseteq_c$ and $\sqsupseteq_d$ coincide. ∎ 3.6

The construction of a DCL-sequence from a CL-sequence makes it very obvious that several events can occur at the same moment, i.e. without any time passing inbetween. In particular, a long sequence of events where one event causes the next could occur in zero-time. This could be regarded as unrealistic by some readers. In contrast, we could require that between any two events a positive amount of time has to pass. Before we continue our normalization of the continuous language, we demonstrate that this 'non-zero' requirement does not change the testing preorder.

Definition 3.7: A $w \in CFS(N)$ is a *non-zero continuous firing sequence* ($w \in NZCFS(N)$ and $l(w) \in NZCL(N)$), if in w transition parts and time steps (r) alternate. Analogously to the above, we define *nz-satisfaction* of continuously timed tests and the corresponding testing preorder $\sqsupseteq_{nz}$. ∎ 3.7

Theorem 3.8: The relations $\sqsupseteq_c$ and $\sqsupseteq_{nz}$ coincide. ∎ 3.8

Our aim is now to normalize the d-continuous language DCL to a discrete language DL. Starting from DL, it will be easier to find a characterization for the testing preorder $\sqsupseteq_c$. We will write the time steps (1) as σ and assume, using Lemma 3.5 ii), that all sequences start with a σ; the initial σ is left implicit, i.e. it will actually be omitted. The behaviour inbetween two σ's is called a *round*.

To define DL, we contract each t^+ and t^- appearing in the same round of a DCL-sequence to t. We collect the remaining t^+ in a round to a step μ, which starts at the end of this round and ends after the σ; hence, we can leave the corresponding t^- in the next round implicit. This way, we can omit the set C from the CID's. Finally, the residual time ρ has only values in $\{0, 1\}$; we replace it by a set U of urgent transitions containing those transitions with $\rho(t) = 0$.

Definition 3.9: A *discrete instantaneous description DID* of a net N is a tuple (M, U) consisting of a marking M of N and a set U of *urgent transitions*. The initial DID ist $DID_N = (M_N, U_N)$ with $U_N = \{t \mid M_N[t\rangle\}$.

We write $(M, U)[\varepsilon\rangle(M', U')$ if one of the following cases applies:

1. $\varepsilon = t \in T$, $M[t\rangle M'$, $U' = U \setminus (^\bullet t)^\bullet$
2. $\varepsilon = \mu\sigma$, $\mu \subseteq T$, $M[\mu\rangle M'$, $\mu \cap U = \emptyset = U \setminus (^\bullet\mu)^\bullet$, $U' = \{t \mid (M -^\bullet \mu)[t\rangle\}$

The set $DFS(N) = \{w \mid DID_N[w\rangle DID\}$ is the set of *discrete firing sequences* of N, the set $DL(N) = \{l(w) \mid w \in DFS(N)\}$ is the *discrete language* of N, where $l(\sigma) = \sigma$; $\zeta(w)$ is the number of σ's in w.

A testable net N *satisfies* a *discretely timed test* (O, D), if each $w \in DL(N\|_\Sigma O)$ with $\zeta(w) \geq D$ (!) contains some ω. N_1 is a *faster implementation* than N_2, $N_1 \sqsupseteq N_2$, if it satisfies all discretely timed tests that N_2 satisfies. ∎ 3.9

The initial set U_N contains all initially activated transitions as we assume an ('invisible') (1)-time-step in the beginning. When defining satisfaction of a test, we consider sequences w with $\zeta(w) \geq D$, because due to the invisible (1)-time-step these are the sequences with $\zeta(w) > D$ from the DCL-point of view. The condition $\mu \cap U = \emptyset$ requires that no urgent transition in μ is delayed over the following time step, and $U \setminus ({}^\bullet\mu)^\bullet = \emptyset$ ensures that all remaining urgent transitions are deactivated by the step. Time passes between the start and the end of the step; hence, transitions that are enabled after the start, i.e. under $M - {}^\bullet\mu$, are urgent after the end.

Theorem 3.10: The relations $\sqsupseteq_c$ and $\sqsupseteq$ coincide.   ■ 3.10

Our aim is now to characterize the test-preorder $\sqsupseteq$. In the classical case [DNH84], this is done by the failure semantics which contains pairs (w, X) where w is an executable action sequence and X is a set of actions that can be refused by the system in the state reached after w. To understand our characterization of $\sqsupseteq$, an unusual view of failure semantics seems appropriate: if (w, X) is a failure pair, w is a partial run of the system, i.e. the system is (possibly) stopped prematurely; but the actions in X are treated correctly when the system is stopped, since they are not possible at this stage. What we need to characterize $\sqsupseteq$ is a kind of refusal trace semantics (see [Phi87]) which gives information on correctly treated actions and transitions.

Instead of σ, we use a substep ν of μ of *correctly started transitions* and a set of actions to indicate a time-step. The transitions in ν are started within μ and not urgent, i.e. they are treated correctly w.r.t. the requirement $\mu \cap U = \emptyset$. The set X contains actions that are not urgent when the time-step occurs, i.e. are treated properly concerning the condition $U \setminus ({}^\bullet\mu)^\bullet = \emptyset$. In both parts, internal actions have to be treated properly.

Definition 3.11: For discrete instantaneous descriptions (M, U) and (M', U') we write $(M, U)[\varepsilon\rangle_r (M', U')$ if one of the following cases applies:
1. $\varepsilon = t \in T$, $M[t\rangle M'$, $U' = U \setminus ({}^\bullet t)^\bullet$
2. $\varepsilon = \mu\nu X$, $\mu, \nu \subseteq T$, $M[\mu\rangle M'$, $X \subseteq \Sigma'$, $U' - \{t \mid (M - {}^\bullet\mu)[t\rangle\}$,
   $\nu \subseteq \mu$, $\nu \cap U = \emptyset$, $\forall t \in \mu - \nu: l(t) \neq \lambda$, $\forall t \in U \setminus ({}^\bullet\mu)^\bullet : l(t) \notin X \cup \{\lambda\}$

The corresponding sequences are called *discrete refusal firing sequences*, their set is denoted by $DRFS(N)$. $DRT(N) = \{l(w) \mid w \in DRFS(N)\}$ is the set of *discrete refusal traces* where $l(\mu\nu X) = l(\mu)l(\nu)X$.   ■ 3.11

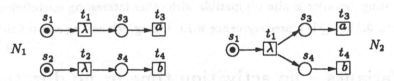

Figure 1: Two DRT-equivalent nets

In both nets of Figure 1, initially the λ-labelled transitions are activated and urgent. When the first time-step $\mu\nu X$ occurs, these internal transitions are not allowed in $U \setminus ({}^\bullet\mu)^\bullet$; since they are furthermore not in conflict with any other transition, they must occur before or within the first μ. If they were in the first μ, they would also be in the first ν since $\forall t \in \mu - \nu: l(t) \neq \lambda$; this contradicts

$\nu \cap U = \emptyset$. Thus, essentially, each *DRFS*-sequence has to start with the λ-transitions firing instantaneously (using part 1 of Definition 3.11), leaving both nets in the same situation; this implies $DRT(N_1) = DRT(N_2)$.

With some care, one can see that we can obtain $DL(N)$ from $DRT(N)$: take those sequences in which all parts $\mu\nu X$ satisfy $\mu = \nu$ and $X = \Sigma'$ and replace $\mu\mu\Sigma'$ by $\mu\sigma$.

Proposition 3.12: $DRT(N_1) \subseteq DRT(N_2)$ implies $DL(N_1) \subseteq DL(N_2)$. ∎ 3.12

The next aim is to show that the *DRT*-semantics induces a congruence for parallel composition; for this, one has to define $\|_A$ for discrete refusal traces. This is more or less as usual, except: a combined transition (t_1, t_2) of some $N_1\|_A N_2$ is urgent, if t_1 is urgent in N_1 and t_2 is urgent in N_2. Thus, correctly started actions can be combined with incorrect ones to give correctly started actions; hence, in the resulting steps the number of correctly started actions can be up to the sum of correctly started actions from the steps we combine; we omit the details. But with a long detailed case-analysis, one shows how to determine $DRT(N_1\|_A N_2)$ from $DRT(N_1)$ and $DRT(N_2)$. Using this result, one can show the characterization of our first testing preorder; for the proof, one has to construct for each $w \in DRT(N)$ a tricky test net that reveals w.

Theorem 3.13: For testable nets, $N_1 \sqsupseteq N_2$ iff $DRT(N_1) \subseteq DRT(N_2)$. ∎ 3.13

Theorem 3.14: The *DRT*-semantics is fully abstract w.r.t. *DL* and parallel composition $\|_A$ of nets, i.e. it gives the coarsest congruence for $\|_A$ that respects *DL*-equivalence. $\sqsupseteq$ is a precongruence for parallel composition. ∎ 3.14

Theorem 3.13 essentially reduces $\sqsupseteq$ to an inclusion of regular languages; the only small problem is that the refusal sets X can be arbitrarily large, but when comparing N_1 and N_2 it is obviously sufficient to draw these sets from the finite set $l_1(T_1) \cup l_2(T_2)$. Thus, $\sqsupseteq$ is in particular decidable, which is not obvious from the start, where we have an infinite (even uncountable) state space according to Definition 3.1. Similarly, $\sqsupseteq_a$ and $\sqsupseteq_i$ below will turn out to be decidable. In the literature, similar results exist that reduce an infinite state space arising from the use of dense time to a finite one, starting with [AD94]; but as far as we know, they are not applicable to our setting.

The testing preorder is also compatible with other interesting operations:

Theorem 3.15: $\sqsupseteq$ is a precongruence w.r.t. hiding, relabelling and restriction.
∎ 3.15

4 Variants – no activation time or no duration

In our asynchronous firing rule of Definition 3.1, occurrence of a transition has two phases: the activation phase lasts from the enabling moment to the start of firing, the firing phase from there to the end of firing. According to Definition 3.1, both phases together last at most time 1. Two variants also seem plausible: we could assume that the activation phase is instantaneous and that all the time the transition takes is spent on firing. Or we could assume – as it is often done –

that the transition has no duration, i.e. the firing phase is instantaneous, while the activation phase may take up to one unit of time.

Now, both these variants will be studied and the corresponding testing preorders will be translated to testing preorders based on discrete behaviour. The firing rules we define below are special cases of Definition 3.1, i.e. the firing sequences according to the definitions to come will be continuous firing sequences.

We start with the case where the activation phase takes no time; for historical reasons which will become clear, we speak of continuous asynchronous firing.

Definition 4.1: Continuous asynchronous firing, denoted by $[\rangle_{ca}$, is defined as $[\rangle_c$, except that in part 3 of Definition 3.1 we require $A = A' = \emptyset$. The set $CAFS(N) = \{w \mid CID_N[w\rangle_{ca} CID\}$ is the set of *ca-firing sequences* of N; the *continuous asynchronous language* of N is $CAL(N) = \{l(w) \mid w \in CAFS(N)\}$.

We define *ca-satisfaction* analogously to c-satisfaction in Definition 3.3 and denote the corresponding testing preorder by $\sqsupseteq_{ca}$.   ■ 4.1

To obtain a discrete version of $\sqsupseteq_{ca}$, we can again consider a discretized version as an intermediate step; the additional separation property of 3.5 i) guarantees that this lemma can also be applied in the present variant. We finally arrive at a simple discrete version. Since an enabled transition has to start firing immediately, and firing ends automatically after the next time step σ at the latest, we do not have to keep track of urgent transitions and can simply use markings to describe system states. The following definition coincides essentially with the asynchronous firing rule given in [Vog95b]. Note that we do not assume an implicit initial time step in this variant.

Definition 4.2: For markings M, M' we write $M[\varepsilon\rangle_a M'$ in one of the following cases:

1. $\varepsilon = t \in T$, $M[t\rangle M'$
2. $\varepsilon = \mu\sigma$, $\mu \subseteq T$, $M[\mu\rangle M'$ and μ is a maximal step

Based on this, we define *asynchronous firing sequences*, the *asynchronous language* $AL(N)$, *a-satisfaction* of discretely timed tests (requiring $\zeta(w) > D$!) and the *a-faster-implementation* preorder $\sqsupseteq_a$.   ■ 4.2

As in the general case, ca- and a-testing give the same relation on nets.

Theorem 4.3: The relations $\sqsupseteq_{ca}$ and $\sqsupseteq_a$ coincide.   ■ 4.3

As mentioned above, $\sqsupseteq_a$ has already been studied in [Vog95b]. For completeness, we simply state the characterization obtained there in a form that is slightly modified for uniformity. While the occurrence of $\mu\sigma$ in some asynchronous firing sequence requires that μ is a maximal step, we allow the occurrence of μX in the following definition, where X lists actions that are treated correctly w.r.t. maximality.

Definition 4.4: For markings M, M' we write $M[\varepsilon\rangle_{ra} M'$ if one of the following cases applies:

1. $\varepsilon = t \in T$, $M[t\rangle M'$
2. $\varepsilon = \mu X$, $\mu \subseteq T$, $X \subseteq \Sigma'$, $\forall t \in T: (M - {}^\bullet\mu)[t\rangle \Rightarrow l(t) \notin X \cup \{\lambda\}$, $M[\mu\rangle M'$

This defines the sequences $ARFS(N)$ and the *a-refusal traces* $ART(N)$.   ■ 4.4

Theorem 4.5: For testable nets, $N_1 \sqsupseteq_a N_2$ iff $ART(N_1) \subseteq ART(N_2)$. ■ 4.5

Similarly as above, $\sqsupseteq_a$ is a precongruence w.r.t. parallel composition, hiding, relabelling and restriction; ART-equivalence is fully abstract w.r.t. asynchronous equivalence and parallel composition of nets, see [Vog95b].

Now we come to the second variant, which we call the i-variant since here all transitions are without duration, i.e. instantaneous.

Definition 4.6: Continuous instantaneous firing, denoted by $[\rangle_{ci}$, is defined as $[\rangle_c$, except that in part 3 of Definition 3.1 we require $C = C' = \emptyset$. The set $CIFS(N) = \{w \mid CID_N[w\rangle_{ci} CID\}$ is the set of *ci-firing sequences*; the *continuous instantaneous language* of N is $CIL(N) = \{l(w) \mid w \in CIFS(N)\}$.

We define *ci-satisfaction* analogously to c-satisfaction in Definition 3.3 and denote the corresponding testing preorder by $\sqsupseteq_{ci}$. ■ 4.6

Once again, we can first consider a discretized version as an intermediate step; the additional separation property of Lemma 3.5 i) gives applicability in this variant, too. As described before Definition 3.9, we can again rearrange the transition parts. This time, each round simply is transformed to a firing sequence, and the behaviour is as follows: implicitly, 1 unit of time elapses and all enabled transitions become urgent, i.e. have ρ-value 0; some transitions fire such that each urgent transition fires or is disabled, i.e. all enabled transitions have ρ-value 1 now; these transitions become urgent with the next (1) and so on. We arrive at the following i-firing rule, which gives rise to a testing preorder coinciding with $\sqsupseteq_{ci}$.

Definition 4.7: For DID's (M, U) and (M', U') we write $(M, U)[\varepsilon\rangle_i (M', U')$ if one of the following cases applies:
1. $\varepsilon = t \in T$, $M[t\rangle M'$, $U' = U \setminus ({}^\bullet t)^\bullet$
2. $\varepsilon = \sigma$, $M = M'$, $U = \emptyset$, $U' = \{t \mid M[t\rangle\}$

$IFS(N) = \{w \in (T \cup \{\sigma\})^* \mid DID_N[w\rangle\}$ is the set of *i-firing sequences* and gives rise to the *i-language* $IL(N)$ of N. A testable net N *i-satisfies* a discretely timed test (O, D) if each $w \in IL(N\|_\Sigma O)$ with $\zeta(w) \geq D$ contains some ω; the *i-faster-implementation* preorder $\sqsupseteq_i$ is defined as usual. ■ 4.7

Theorem 4.8: The relations $\sqsupseteq_{ci}$ and $\sqsupseteq_i$ coincide. ■ 4.8

The characterization of the i-variant is – apart from the use of DID's – very simple. We define i-refusal traces, where the requirement for σ to occur, namely that no transition is urgent, is weakened and σ is replaced by a set X of non-urgent actions.

Definition 4.9: For DID's (M, U) and (M', U'), we write $(M, U)[\varepsilon\rangle_{ri} (M', U')$ if one of the following cases applies:
1. $\varepsilon = t \in T$, $M[t\rangle M'$, $U' = U \setminus ({}^\bullet t)^\bullet$
2. $\varepsilon = X$, $X \subseteq \Sigma'$, $\forall t \in U : l(t) \notin X \cup \{\lambda\}$, $M = M'$, $U' = \{t \mid M[t\rangle\}$

This defines the sequences $IRFS(N)$ and the *i-refusal traces* $IRT(N)$. ■ 4.9

Definition 4.9 is the special case of Definition 3.11 where in 2 the steps μ and ν are empty. Similarly, we can specialize the proofs of Theorem 3.13 and 3.14:

Theorem 4.10: For testable nets, $N_1 \sqsupseteq_i N_2$ iff $IRT(N_1) \subseteq IRT(N_2)$. ■ 4.10

Theorem 4.11: The IRT-semantics is fully abstract w.r.t. IL and parallel composition of nets. $\sqsupseteq_i$ is a precongruence w.r.t. parallel composition. ■ 4.11

Theorem 4.12: $\sqsupseteq_i$ is a precongruence w.r.t. hiding, relabelling and restriction. ■ 4.12

5 Properties of the testing preorders

To check our testing preorders, it is often helpful to use forward simulations; see e.g. [LV95] for a survey on the use of simulations. As an application, one can show some properties of $\sqsupseteq$, $\sqsupseteq_a$ and $\sqsupseteq_i$ one might intuitively expect from a faster-than relation.

Definition 5.1: Let N be a net. The τ-prefix $\tau.N$ of N is obtained by removing all tokens, adding a new marked place s and a new λ-labelled transition t with ${}^{\bullet}t = \{s\}$ and $t^{\bullet} = M_N$.

N' is an *elongation* of N, if it is obtained from N by choosing a transition t, adding a new unmarked place s and a new λ-labelled transition t' with ${}^{\bullet}t' = \{s\}$ and $t'^{\bullet} = t^{\bullet}$ and, finally, redefining $t^{\bullet}$ by $t^{\bullet} := \{s\}$.

Call a transition t of N *persistent*, if no reachable marking M with $M[t\rangle$ enables a transition t' with ${}^{\bullet}t \cap {}^{\bullet}t' \neq \emptyset$. A net is persistent, if each transition is persistent, it is *internally persistent* (*ip*), if each internal transition is persistent.

N' is a *sequentialisation* of N, if it is obtained from N by choosing two transitions t and t' and adding a new marked place s to the pre- and postsets of t and t'; N' is an *ip-sequentialisation* if t' is internal and persistent. ■ 5.1

One would expect intuitively, that N and $\tau.N$ exhibit the same behaviour except that $\tau.N$ might take a bit more time for the additional initialisation; i.e. N should be faster than $\tau.N$ and similarly also than any elongation or sequentialisation. This is shown for $\sqsupseteq_a$ in [Vog95b] and can be shown for $\sqsupseteq$ and $\sqsupseteq_i$ as well. The result in the $\sqsupseteq_i$-case holds for arbitrary sequentialisations, while in the $\sqsupseteq$- and $\sqsupseteq_a$-cases it only holds for *ip*-sequentialisations; i.e. only in the *i*-variant the 'truly concurrent' execution of two actions is always faster than their arbitrary interleaving. It was already argued in [Vog95b] why the parallel execution of two actions may sometimes waste time, namely if the two actions block the two copies of a resource which is needed for some other time-critical activity; in this case, the resource is not available for the duration of the two actions – an effect that cannot occur if the actions are durationless.

Theorem 5.2: N is a $(a$-, i-) faster implementation of $\tau.N$, of each elongation N' and of each *ip*-sequentialisation N''. Furthermore, N is an *i*-faster implementation of each sequentialisation N''. ■ 5.2

It can be shown that no implications hold between our three testing preorders in general; in fact, not even equivalence in one variant is enough to imply the preorder in another. We only give one of the necessary examples. The nets in Figure 1 are DRT- and IRT-equivalent, but not ART-equivalent, since $l_1(t_1 \binom{t_2}{t_3}\{a, b\}) = (a)\{a, b\} \in ART(N_1) \setminus ART(N_2)$.

Although, in general, the three testing preorders $\sqsupseteq$, $\sqsupseteq_a$ and $\sqsupseteq_i$ are incomparable, we can show some implications for special classes of nets.

Theorem 5.3: Let N_1 and N_2 be nets, where N_2 is internally persistent; then $N_1 \sqsupseteq N_2$ implies $N_1 \sqsupseteq_i N_2$. ∎ 5.3

From the extremely simple example in Figure 1, it seems that there is not much hope to get an implication from $\sqsupseteq$ to $\sqsupseteq_a$, if there are internal transitions around. But we have the following theorem:

Theorem 5.4: Let N_1 and N_2 be nets without internal transitions. Then $N_1 \sqsupseteq N_2$ implies $N_1 \sqsupseteq_i N_2$ and $N_1 \sqsupseteq_a N_2$. ∎ 5.4

Quite surprisingly, we can identify an interesting class of nets where $\sqsupseteq_a$ implies $\sqsupseteq$. This class is slightly more general than the class of persistent nets without autoconcurrency, and this generalization will be needed in an application below; the class contains all *ip*-nets where the visible transitions are labelled injectively.

Definition 5.5: An *ip*-net N is *nearly persistent* (*np*), if it is without autoconcurrency and satisfies the following: if for some reachable marking M of N, we have $M[t\rangle$ and $M[t'\rangle$ for $t \neq t'$, but t and t' are in conflict, then t is the only transition labelled $l(t)$ and similarly for t'. (Note that $l(t) \neq \lambda \neq l(t')$ since N is an *ip*-net.) The *conflict region* of $a \in \Sigma$ induced by N is $c(a) = \{a\} \cup \{l(t') \mid \exists t : l(t) = a$ and t and t' are conflicting $\}$. ∎ 5.5

The proof of the following theorem is based on the surprising and not so easy observation that the *d*-refusal traces of an *np*-net can be reconstructed from those *a*-refusal traces with no visible behaviour before the first refusal set.

Theorem 5.6: Let N_1 and N_2 be two nearly persistent nets that induce the same conflict regions. Then $N_1 \sqsupseteq_a N_2$ implies $N_1 \sqsupseteq N_2$. ∎ 5.6

As an application, we have considered two implementations of a bounded buffer for values 0 and 1 – *PIPE* and *BUFFD*. *PIPE* is the usual sequence of 1-buffers: there are two possibly conflicting transitions that perform *in0* or *in1*, if the first cell is empty; then the values are pushed through the pipe by persistent internal transitions and finally output by persistent transitions with *out0* or *out1*. *BUFFD* has an input and an output cell as *PIPE*, but the other values are stored in an array in a cyclic fashion. A long proof in [Vog95b] shows that *BUFFD* is – in the *a*-variant – faster than *PIPE*; since these nets are nearly persistent, our results above allow to deduce that the first implementation is also faster than the second one in the other two variants. Also the results of [Vog95b] regarding a third implementation *BUFFC* carry over from the *a*- to the other two variants.

6 Related literature

The bounded-buffer example above was originally taken from [AKH92]. The approach there is based on some bisimulation-type preorder, hence sometimes more discriminating than ours. Visible actions are regarded as instantaneous and costs are measured as the number of internal actions; as an interleaving approach, it disregards the parallel execution of actions and, hence, is sometimes less discriminating, too.

Faster-than relations are also presented in [HR90, CZ91, MT91, CGR95]. In [MT91], also a bisimulation-type preorder is defined and actions are regarded as instantaneous; a unit-time-delay operator with a special treatment is introduced, which makes the comparison to our approach very difficult. Such an operator is also used in [HR90], where a testing scenario is developed based on the maximal progress assumption, which is suitable for synchronous systems. Time-consuming actions are considered in [CZ91, CGR95]; in the testing approach of [CZ91], transition systems are used as models and, hence, parallel execution is a priori excluded. Finally, [CGR95] takes a bisimulation-like approach where local time-stamps are attached to actions, but actions do not necessarily occur in the order given by these time-stamps – again this is a very different idea and no relation to our approach is obvious. Neither of the two testing approaches uses time bounds in the tests.

References

[AD94]   R. Alur and D. Dill. A theory of timed automata. *Theoret. Comput. Sci.*, 126:183–235, 1994.

[AKH92] S. Arun-Kumar and M. Hennessy. An efficiency preorder for processes. *Acta Informatica*, 29:737–760, 1992.

[CGR95] F. Corradini, R. Gorrieri, and M. Roccetti. Performance preorder and competitive equivalence. unpublished manuscript, 1995.

[CZ91]   R. Cleaveland and A. Zwarico. A theory of testing for real-time. In *Proc. 6th Symp. on Logic in Computer Science*, pages 110–119. IEEE Computer Society Press, 1991.

[DNH84] R. De Nicola and M.C.B. Hennessy. Testing equivalence for processes. *Theoret. Comput. Sci.*, 34:83–133, 1984.

[HR90]   M. Hennessy and T. Regan. A temporal process algebra. Technical Report 2/90, Dept. Comp. Sci. Univ. of Sussex, Brighton, 1990.

[LF81]   N. Lynch and M. Fischer. On describing the behaviour and implementation of distributed systems. *Theoret. Comput. Sci.*, 13:17–43, 1981.

[LV95]   N. Lynch and F. Vaandrager. Forward and backward simulations I: Untimed systems. *Information and Computation*, 121:214–233, 1995.

[MT91]   F. Moller and C. Tofts. Relating processes with respect to speed. In J. Baeten and J. Groote, editors, *CONCUR '91*, Lect. Notes Comp. Sci. 527, 424–438. Springer, 1991.

[Phi87] I. Phillips. Refusal testing. *Theoret. Comput. Sci.*, 50:241–284, 1987.

[Vog95a] W. Vogler. Timed testing of concurrent systems. *Information and Computation*, 121:149–171, 1995.

[Vog95b] W. Vogler. Faster asynchronous systems. In I. Lee and S. Smolka, editors, *CONCUR 95*, Lect. Notes Comp. Sci. 962, 299–312. Springer, 1995.

A Hierarchy Theorem for the μ-Calculus

Giacomo Lenzi

Scuola Normale Superiore, Piazza dei Cavalieri 7, 56100 PISA, Italy

Abstract. We consider the *positive mu-calculus with successors* $P\mu S$, namely a variant of Kozen's modal mu-calculus L_μ [9] where negation is suppressed and where the basic modalities are a sequence of successor operators $l_1, \ldots, l_n, \ldots$. In particular we are interested in the sublanguages of $P\mu S$ determined by the value of the *Emerson-Lei alternation depth* [6]. For every $n \in \mathbb{N}$ we exhibit a formula ϕ_n whose expression in $P\mu S$ requires at least alternation depth n. In particular our result gives a new proof of the strict hierarchy theorem for $P\mu S$ which follows from [1].

1 Introduction

In this paper we are concerned with the *positive mu-calculus with successors* $P\mu S$, a variant of the propositional mu-calculus L_μ by Pratt [13] and Kozen [9]. The mu-calculus is an extension of Modal or Temporal logic with two fixpoint operators, the *least fixpoint operator*, μ, and the *greatest fixpoint operator*, ν. Intuitively, if X is a fixpoint variable ranging over sets and $\phi(X)$ is a mu-calculus formula containing X, then $\phi(X)$ defines a monotone function from sets to sets, and the formula $\mu X.\phi(X)$ expresses the least fixpoint of this function, and $\nu X.\phi(X)$ expresses the greatest fixpoint of this function.

The mu-calculus unifies many usual logics of programs, such as Propositional Dynamic Logic PDL [8], the Computation Tree Logic CTL [2], and the Full Branching Time Temporal Logic CTL* [4]. This makes the mu-calculus a suitable logic for reasoning about concurrent systems (the idea of applying temporal logic to concurrency dates back to Pnueli [12]. For this kind of applications a system is represented as a graph, and the usual correctness properties (safety, liveness, fairness, etc.) correspond to mu-calculus formulas. So, in order to reason on systems, we need a way to decide whether a given graph satisfies a given formula; this is the *model checking problem*. When the model is finite, we have the finite model checking problem. It is shown in [5] that the mu-calculus (finite) model checking problem is in $NP \cap co-NP$; however a polynomial time model-checker has still to be found.

The model-checking algorithm of [6] is polynomial in the length of the formula for formulas with fixed *alternation depth*, where intuitively the alternation depth is the maximal number of nested, alternated μ's and ν's in the formula; thus, in view of looking for a polynomial model-checker for the whole mu-calculus, it is interesting to investigate the relative expressiveness of the fragments with fixed a.d.

In fact if a translation of the mu-calculus into low alternation depth existed, we could hope to make our translation efficient and so recover an efficient model-checker from [6]. However such a translation does not exist by the hierarchy theorem of [1], which says that the alternation depth hierarchy of the mu-calculus is strict. So the challenging problem of finding an efficient model checker, or proving that it does not exist, remains open. However it may be interesting to say something more about the a.d. hierarchy itself, and this is the aim of the present paper. In particular we will consider the positive mu-calculus with successors $P\mu S$, and we will exhibit a sequence of "hard" formulas ϕ_n, such that expressing ϕ_n in $P\mu S$ requires at least a.d. n.

The rest of the paper is organized as follows. In section 2 we present the language $P\mu S$. In section 3 we introduce the notion of Emerson-Lei alternation depth, taken from [6]. In section 4 we state and comment the main result of this paper. In section 5 we introduce an ad hoc technical notion, called "collecting" a mu-calculus sentence; in section 6 we recall the standard notion of premodel [7] and we derive from it the ad hoc notion of "collected premodel" of a mu-calculus sentence; in section 7 we give some more new definitions; finally in section 8 we prove the main result.

2 The positive mu-calculus with successors

In this section we prepare the result of section 4 by defining the syntax and semantics of the mu-calculus we will work with. . Precisely we will introduce the positive mu-calculus with successors $P\mu S$. This mu-calculus is obtained from standard mu-calculus L_μ in two steps:

1) remove the negation;

2) replace the modalities $[l], < l >$ with a sequence of successor operators $l_1, l_2, \ldots$.

Intuitively a formula of $P\mu S$ whose successor operators are among $l_1, \ldots, l_n$ should be interpreted over the complete n-ary tree where each node has n distinguished children, and $l_i \phi$ should mean that ϕ holds at thew i-th child of the current node.

So the syntax of a $P\mu S$ formula ϕ is:

$$\phi ::= P | \phi_1 \wedge \phi_2 | \phi_1 \vee \phi_2 | l_1 \phi | l_2 \phi | \ldots | X | \mu X.\phi | \nu X.\phi, \tag{1}$$

where P ranges over a set $Lett$ of propositional letters, $l_1, l_2, \ldots$ ranges over a set Lab of labels and X ranges over a set Var of propositional variables.

We define the sublanguage P_n of $P\mu S$ as the set of formulas whose successor operators are among $l_1, \ldots, l_n$.

Although the negation is not present in $P\mu S$, we will use negated formulas from time to time.

A useful operator on formulas is the duality: given a formula ϕ, we define its dual ϕ^d as the result of exchanging the pairs $\wedge$ and $\vee$, μ and ν.

The precise meaning of a formula can be given by specifying the semantics of the formula. Let us define the ordinary Kripke semantics for $P\mu S$.

First we need the notion of transition system for P_n. A transition system for P_n is a tuple

$$\mathcal{T} = (T, f_1, f_2, \ldots, f_n), \tag{2}$$

where T is a nonempty set and $f_1, f_2, \ldots, f_n$ is a sequence of functions from T to T.

A valuation $\mathcal{V}$ on a transition system $\mathcal{T}$ for P_n is a function from $Lett \cup Var$ to the powerset of T: this means that letters and variables are interpreted by subsets of T.

A model for P_n is a pair $\mathcal{M} = (\mathcal{T}, \mathcal{V})$. Given a model $\mathcal{M}$ for P_n and a formula ϕ of P_n, the semantics of ϕ in $\mathcal{M}$ is a subset $||\phi||\mathcal{M}$ of T defined inductively in this way:

$$||P||\mathcal{M} = \mathcal{V}(P); \tag{3}$$

$$||X||\mathcal{M} = \mathcal{V}(X); \tag{4}$$

$$||\phi_1 \wedge \pi_2||\mathcal{M} = ||\phi_1||\mathcal{M} \cap ||\phi_2||\mathcal{M}; \tag{5}$$

$$||\phi_1 \vee \phi_2||\mathcal{M} = ||\phi_1||\mathcal{M} \cup ||\phi_2||\mathcal{M}; \tag{6}$$

$$||l_i\phi||\mathcal{M} = \{x \in T | f_i(x) \in ||\phi||\mathcal{M}\}; \tag{7}$$

$$||\mu X.\phi||\mathcal{M} = \bigcap\{E \subseteq T| \; ||\phi||\mathcal{M}[X \backslash E] \subseteq E\}; \tag{8}$$

$$||\nu X.\phi||\mathcal{M} = \bigcup\{E \subseteq T| \; ||\phi||\mathcal{M}[X \backslash E] \supseteq E\}, \tag{9}$$

where $\mathcal{M}[X \backslash E]$ is the model obtained from $\mathcal{M}$ by assigning the variable X to E and leaving all other values of $\mathcal{M}$ unchanged.

By Tarski's Theorem [14], the last two clauses set $||\mu X.\phi||$ equal to the least fixpoint of the function f sending $E \subseteq T$ to $||\phi||\mathcal{M}[X \backslash E]$, and set $||\nu X.\phi||$ equal to the greatest fixpoint of f.

We define for each n the canonical transition system for P_n

$$\mathcal{T}_n = (\{l_1, \ldots, l_n\}^*, f_1, \ldots, f_n) \tag{10}$$

where $\{l_1, \ldots, l_n\}^* = T_n$ is the set of all the finite words on the alphabet $\{l_1, \ldots, l_n\}$, and for $x \in T_n$ we let $f_i(x) = xl_i$. We let Λ be the empty word.

We say that two P_n formulas ϕ, ϕ' are semantically equivalent, written $\phi \Leftrightarrow \phi'$, if for all models $\mathcal{M}$ for P_n we have $||\phi||\mathcal{M} = ||\phi'||\mathcal{M}$; likewise we say that ϕ implies ϕ', written $\phi \Rightarrow \phi'$, if for any model $\mathcal{M}$ for P_n we have $||\phi||\mathcal{M} \subseteq ||\phi'||\mathcal{M}$.

If $x \in T_n$, $\mathcal{M}$ is a model and ϕ is a PL_n formula, we say that ϕ is true in $\mathcal{M}$ at x, written $\mathcal{M}, x \models \phi$, if $x \in ||\phi||\mathcal{M}$; moreover we say that $\mathcal{M}$ is a model of ϕ, written $\mathcal{M} \models \phi$, if we have $\mathcal{M}, \Lambda \models \phi$.

Given a model $\mathcal{M}$ on $\mathcal{T}_n$ and a point $x \in T_n$, we define $\mathcal{M}(x)$ the model $\mathcal{M}$ shifted at x; namely the model $\mathcal{M}(x)$ is defined by

$$\mathcal{M}(x)(A) = \{y \in T_n | xy \in \mathcal{M}(A)\}, \tag{11}$$

for each $A \in Lett \cup Var$.

In particular we note that we have $\mathcal{M}, x \models \phi$ if and only if $\mathcal{M}(x) \models \phi$.

3 Alternation depth

In this section we consider one of the various notions of alternation depth present in the literature, namely the Emerson-Lei alternation depth. For other definitions see [10].

Intuitively the alternation depth of a mu-calculus formula is the maximum number of nested alternated μ's and ν's. However it turns out that there are several ways of formalizing this idea. The most popular one is Emerson-Lei alternation depth, defined by means of the following hierarchies Σ_n, Π_n: :

Definition 1. $\Sigma_0 = \Pi_0$ is the set of $P\mu S$ formulas having no fixpoint operators. For all $n \in \mathbf{N}$, Σ_{n+1} is the smallest set of $P\mu S$ formulas such that:

$$\Sigma_n \cup \Pi_n \subseteq \Sigma_{n+1}; \tag{12}$$

$$if \ \phi \in \Sigma_{n+1}, \ then \ \mu X.\phi \in \Sigma_{n+1}; \tag{13}$$

$$if \ \phi_1, \phi_2 \in \Sigma_{n+1} \ then \ \phi_1 \wedge \phi_2, \phi_1 \vee \phi_2, l_i\phi \in \Sigma_{n+1}. \tag{14}$$

if $\phi, \phi_1, \ldots, \phi_m \in \Sigma_{n+1}$, $P_1, \ldots, P_m$ are propositional letters and $\phi_1, \ldots, \phi_m$ are closed, then

$$\phi[P_1 \backslash \phi_1, \ldots, P_m \backslash \phi_m] \in \Sigma_{n+1}. \tag{15}$$

Π_{n+1} is defined dually by exchanging μ and ν in condition 13.

The Emerson-Lei alternation depth of a formula ϕ is the least n such that $\phi \in \Sigma_n \cup \Pi_n$.

We remark that the Emerson-Lei definition occurs in the measure of the efficiency of the Emerson-Lei model checking algorithm of [6]: in fact, such an algorithm runs in polynomial time in the length of the formula for formulas with fixed EL-alternation depth.

As we said in the Introduction, the Emerson-Lei model checker is a motivation to study the problem whether the alternation depth collapses or not, namely whether there is an $n \in \mathbf{N}$ such that every formula is semantically equivalent to a formula with a.d. n. This problem has been recently settled by Bradfield in [1]; in fact:

Theorem 2. [1] The Emerson-Lei hierarchy does not collapse.

4 Statement of the main result

In this section we state and comment the main result of this paper, namely:

Theorem 3. Define the following sequence ϕ_n of sentences of P_n:

$$\phi_0(P) = P; \tag{16}$$

$$\phi_n(P) = \nu X_n.P \wedge \phi_{n-1}^d(l_n X_n), \ for \ n > 0. \tag{17}$$

Then every sentence $\psi \in P_n$ semantically equivalent to ϕ_n has alternation depth at least n.

The proof of this theorem will be prepared in sections 5 to 7 and will be completely carried out in section 8. Here let us just make some comments on it.

We note that this theorem gives explicit examples to the strictness of the a.d. hierarchy in the mu-calculus $P\mu S$; such counterexamples are not given in [1] which gives a rather indirect proof. In this sense, our theorem is not a mere repetition of Bradfield's results.

We note also that we can obtain a "modal" version of the theorem by replacing $P\mu S$ with the positive part of L_μ and, say, l_n with $< l_n >$.

Of course several extensions of theorem 3 are desirable. First of all one can conjecture that theorem 3 holds not only for the positive mu-calculus but also for the whole mu-calculus L_μ with negation, but this is not known to the author. Moreover it would be interesting to understand better the nature of the alternation depth levels, for example by finding sufficient conditions for a formula to be hard. In another direction, it would be interesting to find applications of the formulas with high alternation depth or even with alternation depth greater than two.

Summing up, we can say that the main value of theorem 3 is not that of being the conclusion of something but rather the beginning of some (hopefully not too hard) work.

5 Collecting a mu-calculus sentence

In this and the following two sections we prepare the proof of theorem 3 by a series of definitions. The first definition is that of collection of a P_n sentence with respect to label l_n. The idea is that every sentence $\psi \in P_n$ can be rewritten as $\chi(\Theta)$, where χ is a sentence without l_n and Θ is a tuple of sentences beginning with l_n. It may be checked that every P_n sentence ψ can be collected just by unrolling fixpoints, and moreover the result of collecting is uniquely determined, so that we call it *collection* of ψ.

Example 1. Let

$$\psi_1 = \mu X.l_3(\mu Y.l_1 X \vee l_2 Y); \tag{18}$$

the collection of ψ_1 is

$$\psi_c = l_3(\mu Y.l_1\psi_1 \vee l_2 Y). \tag{19}$$

Now we may wish to iterate the collection of ψ as far as possible. This means that, after writing $\psi = \chi(\Theta)$ and letting $\Theta = l_n\Theta'$, we can collect every sentence in Θ' (if any), and so on. After k collection steps we naturally obtain a finite tree of sentences of depth k, which we call the k-th partial collection tree of ψ. At the end we obtain a possibly infinite tree, which we call the (total) collection tree of ψ.

Now we recall from [7] the standard notion of the Fischer-Ladner closure of a sentence θ, denoted $FL(\theta)$, namely the smallest set of sentences containing θ and closed under negation, subsentence and fixpoint unrolling. For example the Fischer-Ladner closure of $\mu X.P \vee aX$ contains eight sentences: $\mu X.P \vee aX$,

$P \vee a(\mu X.P \vee aX$, P, $a(\mu X.P \vee aX)$, $\nu X.\neg P \wedge aX$, $\neg P \wedge a(\nu X.\neg P \wedge aX)$, $\neg P$, $a(\ nuX.\neg P \wedge aX)$.

It results that $FL(\theta)$ is a finite set for any mu-calculus sentence θ, and its size is linear in the length of θ. Moreover all the sentences at the leaves of the k-th collection tree of a sentence θ belong to $FL(\theta)$.

6 Collecting premodels of mu-calculus sentences

In this section we want to relate the standard notion of premodel of a mu-calculus sentence [7] with our ad hoc notion of collection; the result will be the definition of a collected premodel.

Let us begin by recalling what a premodel is in the particular case of the mu-calculus P_n.

Definition 4. Let ψ a sentence of P_n and let $\mathcal{M}$ be a model of ψ. A premodel of ψ compatible with $\mathcal{M}$ is a relation p between words $w \in \{l_1, \ldots, l_n\}^*$ and sentences $\theta \in P_n$ such that:

$$(\Lambda, \psi) \in p; \tag{20}$$

$$if\ (w, P) \in p\ and\ P\ is\ a\ propositional\ letter\ then\ \mathcal{M}, w \models P; \tag{21}$$

$$if\ (w, \theta_1 \vee \theta_2) \in p\ then\ (w, \theta_1) \in P\ or\ (w, \theta_2) \in P; \tag{22}$$

$$if\ (w, \theta_1 \wedge \theta_2) \in p\ then\ (w, \theta_1) \in P\ and\ (w, \theta_2) \in P; \tag{23}$$

$$if\ (w, l\theta) \in p\ and\ l\ is\ a\ label\ then\ (lw, \theta) \in p; \tag{24}$$

$$if\ (w, \mu X.\theta(X)) \in p\ then\ (w, \theta(\mu X.\theta(X))) \in p; \tag{25}$$

$$if\ (w, \nu X.\theta(X)) \in p\ then\ (w, \theta(\nu X.\theta(X))) \in p. \tag{26}$$

We note that, at each pair $(w, \theta_1 \vee \theta_2)$, the premodel must make a choice; this choice may be important for successful (i.e. terminating) evaluation of the least fixpoints. So in order to lead the premodel to successful fixpoint evaluation, a premodel can be supplied with a choice function f, which associates to any pair $(w, \theta_1 \vee \theta_2)$ one of the disjoints (w, θ_1) or (w, θ_2).

A premodel p and a choice function f determine a derivation relation $\rightarrow = \overset{p,f}{\rightarrow}$ such that $x \rightarrow y$ iff one of the following cases holds:

$$x = (w, \theta_1 \vee \theta_2)\ and\ y = f(x); \tag{27}$$

$$x = (w, \theta_1 \wedge \theta_2)\ and\ y = (w, \theta_1)\ or\ y = (w, \theta_2); \tag{28}$$

$$x = (w, l\theta)\ and\ y = (wl, \theta); \tag{29}$$

$$x = (w, \mu X.\theta(X))\ and\ y = (w, \theta(\mu X.\theta(X))); \tag{30}$$

$$x = (w, \nu X.\theta(X))\ and\ y = (w, \theta(\nu X.\theta(X))). \tag{31}$$

Termination of least fixpoint evaluation can be formalized as absence of "re-generations" of mu-sentences. Precisely, let $\mu X.\gamma$ be a mu-sentence and let ρ be an infinite derivation path:

$$\rho = (w_0, \theta_0) \rightarrow (w_1, \theta_1) \rightarrow (w_2, \theta_2) \rightarrow \cdots \tag{32}$$

We say that ρ regenerates $\mu X.\gamma$ if:

$$\mu X.\gamma = \theta_i \ for \ infinitely \ many \ i; \tag{33}$$

$$\mu X.\gamma \ is \ a \ subsentence \ of \ \theta_i \ for \ all \ i \ except \ for \ a \ finite \ number. \tag{34}$$

Definition 5. We say that the premodel p is wellfounded if it has a choice function whose associated derivation relation has no regeneration paths.

From [7] we deduce immediately the following theorem:

Theorem 6. *A sentence ψ is true in a model $\mathcal{M}$ if and only if there is a wellfounded premodel p of ψ compatible with $\mathcal{M}$.*

So far we have recalled the standard notion of premodel. Now we propose a slight variation of this notion which is naturally associated to the idea of collecting. Precisely let ψ be a sentence of P_n, let $\mathcal{M}$ be a model of ψ and let p be a premodel of ψ compatible with $\mathcal{M}$. The collection of p, denoted $c(p)$, is the subset of p given by the pairs $w, \theta \in p$ such that either $w = \Lambda$ or w is a word ending with l_n.

If $c(p)$ is a collected premodel and f is a choice function on p, we define also the collected derivation relation $\to$ on $c \times c$ by: $x \to y$ iff $x = (w, \theta)$ and $y = (wvl_n, \theta')$, where:

$$v \in \{l_1, \ldots, l_n\}^*; \tag{35}$$

$$\theta \ collects \ as \ \chi(l_n\theta', l_n\Theta'); \tag{36}$$

$$x(\overset{p,f}{\to})^+ y. \tag{37}$$

Here $\overset{p,f}{\to}^+$ denotes the transitive closure of the relation $\overset{p,f}{\to}$.

7 Some more definitions

Before proving theorem 3 we give some more new definitions which will be used in the proof. The new definitions are those of finiteness up to l_n, l_n topology and hard and relativized hierarchies.

We say that a model on T_n is finite if it is the unwinding of a finite model in the usual sense (see for example [3]). Finite models are important for us because of the finite model property of the mu-calculus [7]: a sentence has a model if and only if it has a finite model.

Now we consider an ad hoc approximation of finiteness, which we call finiteness up to l_n. Given $k \in \mathbf{N}$, the set S_k of the models of size k up to l_n is defined as follows:

$$S_{k,0} \ is \ the \ set \ of \ all \ models \ on \ T_n; \tag{38}$$

$$S_{k,\alpha+1} \ is \ the \ set \ of \ the \ models$$
$$obtained \ by \ taking \ a \ size \ k \ model \ \mathcal{N} \ on \ T_{n-1}$$
$$and \ attaching \ a \ model \ of \ S_{k,\alpha} \ to \ each \ point \ of \ \mathcal{N}; \tag{39}$$

$$S_{k,\lambda} = \bigcap_{\alpha < \lambda} S_{k,\alpha} \ for \ \lambda \ limit \ ordinal; \tag{40}$$

$$S_k = \bigcap_\alpha S_{k,\alpha}. \tag{41}$$

Next we consider a natural topology on the set of the models on T_n, and we call it l_n-topology.

First, given two models M, M' on T_n, we say that M and M' are equal up to level k, written $M =_k M'$, if for every point $w \in T_n$ with at most k l_n's and for every propositional letter P, we have $M, w \models P$ iff $M', w \models P$. The l_n topology is the topology generated by the basic open sets

$$(M)_k = \{M' | M =_k M'\}. \tag{42}$$

We have also a natural notion of l_n-convergence. Let (M_h) be a sequence of models and let M_∞ be a model. We say that M_h converges to M_∞ if, for any k, $M_h =_k M_\infty$ for all h sufficiently large.

It may be checked that, for any $k \in \mathbf{N}$, the set S_k is "l_n-compact": any sequence (M_h) in S_k has a converging subsequence (M_{h_i}).

Finally we define the hard and relativized hierarchies. Let us begin with the hard hierarchies:

Definition 7. The hierarchies $\Sigma_n - hard$ and $\Pi_n - hard$ are defined by:

$$\Sigma_0 - hard = \Pi_0 - hard = P\mu S; \tag{43}$$

$$\Sigma_{n+1} - hard = P\mu S - (\Sigma_n \cup \Pi_{n+1}); \tag{44}$$

$$\Pi_{n+1} - hard = P\mu S - (\Pi_n \cup \Sigma_{n+1}). \tag{45}$$

Now let us pass to the hierarchies relativized to a propositional letter P:

Definition 8. A $P\mu S$ formula ϕ is Σ_n relative to a letter P, briefly $\Sigma_n(P)$, if the formula ϕ' is Σ_n, where ϕ' is ϕ with all the maximal subsentences without P, say $\psi_1, \ldots, \psi_k$, replaced by fresh propositional letters $Q_1, \ldots, Q_k$. The definition of $\Pi_n(P)$ is dual.

The above definition can be combined into a definition of the hard, relativized hierarchies:

Definition 9. The hierarchies $\Sigma_n(P) - hard$ and $\Pi_n(P) - hard$ are defined by:

$$\Sigma_0(P) - hard = \Pi_0(P) - hard = P\mu S; \tag{46}$$

$$\Sigma_{n+1}(P) - hard = P\mu S - (\Sigma_n(P) \cup \Pi_{n+1}(P)); \tag{47}$$

$$\Pi_{n+1}(P) - hard = P\mu S - (\Pi_n(P) \cup \Sigma_{n+1}(P)). \tag{48}$$

We conclude this section with some examples.

Example 2. $\mu X.P \vee aX$ is $\Sigma_1(P) - hard$ but is $\Sigma_0(Q) \cap \Pi_0(Q)$ for any letter $Q \neq P$.

Example 3. $\mu X.(\nu Y.P \wedge aY) \vee Q \vee bX$ is $\Sigma_2(P)$ and $\Pi_2(P)$ and is $\Sigma_1(Q)$.

Example 4. $P \vee \nu X.\mu Y.a((Q \wedge X) \vee Y)$ is $\Pi_2(Q)$ but is $\Sigma_0(P)$.

8 The main argument

In this section we complete the proof of theorem 3 by proving the following lemma.

Lemma 10. *Let ψ a sentence of PL_n.*
1) *If $\psi \Leftrightarrow \phi_n$ then $\psi \in \Pi_n(P) - hard;$*
2) *If $\psi \Leftrightarrow \phi_n^d$, then $\psi \in \Sigma_n(P) - hard.$*

Proof. We prove 1) and 2) by simultaneous induction on n. Note that the case $n = 0$ is trivial, and for each n 2) is dual to 1). So it is enough to prove 1) for $n > 0$, assuming 1) and 2) for $n - 1$. We can assume that every subsentence of ψ contains a P up to eliminating trivial subsentences.

The rest of the proof is divided into ten steps.

Step 1 We will use the following definitions.

Definition 11. A sequence of k sentences $\theta_1, \ldots, \theta_k$ is a *collection branch* if it is a path through the k-th collection tree of θ_1, namely if each θ_i $(1 \leq i < k)$ collects as $\chi_i(l_n\theta_{i+1}, l_n\Theta_{i+1})$.

Definition 12. A collection branch $\theta_1, \ldots, \theta_k$ is *well-guarded* if:
1) every sentence θ_i contains an occurrence of P;
2) for every $i < k$ the sentence θ_i collects as $\chi_i(Q_{i+1} \setminus l_n\theta_{i+1}, R_{i+1} \setminus l_n\Theta_{i+1})$, where Q_{i+1} is a fresh propositional letter, R_{i+1} is a tuple of fresh propositional letters and the sentence $\chi_i(Q_{i+1}, R_{i+1})$ is $\Sigma_{n-1}(Q_{i+1})$-hard.

Likewise are defined infinite collection branches and infinite well-guarded collection branches.

Step 2) Assume $\psi \Leftrightarrow \phi_n$. For each $h \in \mathbf{N}$ collect ϕ_n h times, thus obtaining

$$\phi_n \Leftrightarrow f_h(\phi_n), \tag{49}$$

where the sentence $f_h(Q)$ is obtained by applying h times $P \wedge \phi_{n-1}^d$ and l_n to Q with the appropriate parentheses.

Step 3 For each $h \in \mathbf{N}$ collect ψ h times, thus obtaining

$$\psi \Leftrightarrow c_h(D_h), \tag{50}$$

where c_h is a suitable sentence and D_h is a tuple of sentences. Let E_h be the tuple of the sentences $\eta \in D_h$ such that:
1) each model of η has some P;
2) the length $h + 1$ collection branch $\theta_0 = \psi, \theta_1, \ldots, \theta_h = \eta$ is well-guarded.

Let F_h the tuple D_h minus E_h. Then we can also perform a "selective" collection of ψ by collecting at step h only the sentences in E_h and not the sentences in F_h. Thus we obtain

$$\psi \Leftrightarrow d_h(E_h, F_h, F_{h-1}, \ldots, F_1). \tag{51}$$

Finally let F_h' be the tuple of the elements $f \in F_h$ such that f is not identically true and not identically false.

Step 4 By the finite model property of the mu-calculus there is a k such that every satisfiable conjunction of elements of $FL(\psi)$ has a model of size k.

Step 5 Define P^* to be the sentence

$$\mu X.P \vee l_1 X \vee l_2 X \vee \ldots \vee l_n X. \tag{52}$$

For each $h \in \mathbf{N}$ define the sentence

$$\psi_h = d_h(E_h \backslash false, F'_h \backslash P^*, F'_{h-1} \backslash P^*, \ldots, F'_1 \backslash P^*). \tag{53}$$

Assume that, for some h, $\phi_n \cap S_k \Rightarrow \psi_h$. Then by definition of k we have $\phi_n \Rightarrow \psi_h$, and by interpolation we obtain

$$\phi_n \Leftrightarrow f_h(\phi_n) \Leftrightarrow d_h(E_h \backslash false, F'_h \backslash \phi_n, F'_{h-1} \backslash \phi_n, \ldots, F'_1 \ldots \phi_n). \tag{54}$$

This is the point where we need the hypothesis that $\psi \in P\mu S$ rather than just $\psi \in L_n$: in fact it is not clear at all how to generalize the argument to the whole L_μ.

Step 6 From the previous equivalence we obtain an absurdity by applying h times the inductive hypothesis that $P \wedge \phi^d_{n-1}(Q)$ requires a $\Sigma_{n-1}(Q)$-hard formula. So for any h there is a model $M_h \in S_k \wedge \phi_n \wedge \neg \psi_h$.

Step 7 Take a subsequence M_{h_k} l_n-converging to a model $M_\infty \in S_k$. Then M_∞ verifies ϕ_n as ϕ_n is closed in S_k, and hence $M_\infty \models \psi$. Moreover, as ψ_h is open in S_k and $\psi_h \Rightarrow \psi_{h+1}$, we conclude that M_∞ does not verify ψ_h for any h.

Step 8 Take a wellfounded premodel p, f of M_∞, ψ. Then for any h, since M_∞ verifies $\neg \psi_h$, there is a sequence of h words of T_n $w_{h0} = \Lambda, w_{h1}, \ldots, w_{hh}$ and a sequence of sentences $\eta_{h0} = \psi, \eta_{h1}, \ldots, \eta_{hh}$ such that for any i:
1) the word w_{hi} has exactly i l_n's and (if $i > 0$) ends with l_n;
2) $\eta_{hi} \in E_i$;
3) $(w_{hi}, \eta_{hi}) \in p$.

Step 9 Since M_∞ has size k up to l_n, we can choose x_{hi} in the previous step such that the pairs (w_{hi}, η_{hi}) form a finitely branching subtree of the collected derivation tree of p, f. Hence by König lemma we have an infinite collected derivation sequence in p, f:

$$(w_0, \eta_0) = (\Lambda, \psi) \rightarrow (w_1, \eta_1) \rightarrow (w_2, \eta_2) \rightarrow \ldots \tag{55}$$

whose second components $\eta_0, \eta_1, \eta_2 \ldots$ form an infinite well-guarded collection branch and each η_h is in E_h. In particular each η_h has at least one occurrence of P.

Step 10 By definition of well-guardedness, ψ must be Π_n-hard, otherwise the infinite derivation sequence of the previous step would give rise to a regeneration of a μ-sentence, contrary to the wellfoundedness of p, f. Moreover ψ must be $\Pi_n(P)$-hard just because every subsentence of ψ contains a P.

References

1. Bradfield, J. C.: The modal mu-calculus alternation hierarchy is strict. Manuscript, 1995.
2. Clarke, E. M., Emerson, E. A. : Design and synthesis of synchronization skeletons using Branching Time Temporal Logic. Proc. Workshop on Logics of Programs, Lecture Notes in Computer Science 131, 1981, pp. 52-71.
3. Emerson, E. A.: Temporal and Modal Logic. In Handbook of Theoretical Computer Science vol. B (J. van Leeuwen, ed.), North-Holland, 1990, pp. 995-1072.
4. Emerson, E. A., Halpern, J.: "Sometimes" and "Not never" revisited: on Branching versus Linear Time Temporal Logic. Journal of the ACM 33, 1986, pp. 151-178.
5. Emerson, E. A., Jutla, C.: The complexity of tree automata and logics of programs. Extended version from FOCS '88, 1988.
6. Emerson, E. A. , Lei, C. L.: Efficient model checking in fragments of the propositional mu-calculus. Proc. First IEEE Symp. on Logic in Computer Science, 1986, pp. 267-278.
7. Emerson, E. A., Streett, R.: An automata-theoretic decision procedure for the propositional mu-calculus. Information and Computation 81, 3, 1989, pp. 249-264.
8. Fischer, M, Ladner, R.: Propositional Dynamic Logic of regular programs. JCSS 18, 1979, pp. 194-211.
9. Kozen, D. : Results on the propositional mu-calculus. Theoretical Computer Science 27, 1983, pp. 333-354.
10. Kaivola, R.: On modal mu-calculus and Büchi tree automata. Inf. Proc. Letters 54, 1995, pp. 17-22.
11. Niwiński, D.: On fixed point clones. Proc. 13th ICALP, Lecture Notes in Computer Science 226, 1986, pp. 464-473.
12. Pnueli, A.: The temporal logic of programs. 18th IEEE Symp. on Foundations of Computer Science, 1977, pp. 46-57.
13. Pratt, V.: A decidable mu-calculus. 22nd IEEE Symp. on Foundations of Computer Science, 1981, pp. 421-427.
14. Tarski, A.: A lattice-theoretical fixpoint theorem and its applications. Pacific Journal of Mathematics 55, 1955, pp. 285-309.

An Effective Tableau System
for the Linear Time μ-Calculus

Julian Bradfield†, Javier Esparza‡, Angelika Mader‡

Abstract: We present a tableau system for the model checking problem
of the linear time μ-calculus. It improves the system of Stirling and
Walker by simplifying the success condition for a tableau. In our system
success for a leaf is determined by the path leading to it, whereas Stirling
and Walker's method requires the examination of a potentially infinite
number of paths extending over the whole tableau.

Keywords: temporal logic, linear-time μ-calculus, local model-checking,
tableau systems

1 Introduction.

Tableau techniques have been used for more than twenty years in order to establish validity of modal logics [HC68,Fit83]. A tableau system for a logic has
three parts: deduction rules, termination conditions, and success conditions.
The rules are goal-directed; they tell, given a sequent formalizing the statement
we want to prove true, how to obtain subgoals. The termination conditions tell
when to stop the construction of the proof tree, which we then call a tableau.
Finally, the success conditions indicate when a tableau succeeds in establishing
the truth of the root sequent.

Stirling has advocated the use of tableau techniques for local model-checking
problems [Sti87]. Local model-checking asks whether a particular state has a
temporal property, rather than, what is the set of states that satisfy it. Since
tableau techniques check the properties of a given state by reference to properties
of adjacent states, local model-checking may avoid having to compute all the
states of the system.

Tableau techniques are particularly suitable for computer-*assisted* verification. They give very good insight into why a property holds. Also, they allow
the verifier to apply her knowledge of the system to select the most promising
course of action, by deciding which rule to apply or which branch of a proof tree
to explore first. (Compare the standard automata-theoretic techniques: they are
efficient and easy to automate, but require some expertise to use and understand
'by hand'. For verification where human input is expected, we believe it better
to use only the formula and the model, with as few auxiliary constructions as
are necessary, which should be immediately related to the logic in question.)

† LFCS, University of Edinburgh, King's Buildings, Edinburgh, United Kingdom, EH9 3JZ;
jcb@dcs.ed.ac.uk

‡ Institut für Informatik, Technische Universität München, Arcisstr. 21, 80333 München,
Germany; {esparza,mader}@informatik.tu-muenchen.de

Stirling and Walker have proposed tableau systems for μ-calculi, a group of fixpoint logics very popular in the formal verification community. A system for the modal μ-calculus (a branching time logic) is presented in [SW91], and a system for the linear-time μ-calculus (the linear-time counterpart of the modal μ-calculus), can be found in [SW90].

While the system of [SW91] is very simple and satisfactory, the one presented in [SW90] has very complicated success conditions. In the case of the modal μ-calculus, deciding whether a leaf of the tableau is successful can be done by examining the path of the tableau leading to it. On the contrary, the success condition of [SW90] requires the examination of a potentially infinite number of so-called extended paths, which are structures that may extend all over the tableau. The decidability of the success condition is difficult to prove, and in fact this point is not addressed in [SW90].

Stirling and Walker were aware of this problem, and they wrote ([SW90], p. 176) that "it may be possible to find a simpler definition of successful termination". This is precisely the contribution of this paper. We provide a simple, alternative tableau system, in which the success condition of a terminal only depends on the path leading to it. Our approach uses some ideas of [Kai95], where Kaivola addresses the satisfiability problem for the linear-time μ-calculus.

The paper is structured as follows. In section 2 we give basic definitions and results about the linear-time μ-calculus. In section 3 we present the tableau system, while section 4 illustrates it on an example. The proofs of soundness and completeness are in section 5, while section 6 discusses complexity issues.

This work has been partially supported by a British–German Academic Collaboration Grant from the DAAD and the British Council (all authors) and by Project A3-SAM of the Sonderforschungsbereich 342 (Esparza and Mader). We thank the anonymous referees for improvements to the paper.

2 The Linear Time μ-Calculus.

We now define linear time μ-calculus syntax and semantics, and some notation for later use. The language is built from propositions, variables, boolean connectives, the minimal and maximal fixpoint operators μ and ν, and two temporal operators, the *strong nexttime* $\bigcirc$ and the *weak nexttime* $\odot$. Intuitively, $\bigcirc\phi$ means 'there is a next moment in time and ϕ is true at that moment', whereas $\odot\phi$ means 'if there is a next moment in time, then ϕ is true at that moment'.

Definition 1. Fix two disjoint countable sets, $\mathcal{Z}_C$, the set of *propositions*, and $\mathcal{Z}_V$, the set of *variables*, and define $\mathcal{Z} = \mathcal{Z}_C \cup \mathcal{Z}_V$. The formulae of νTL are defined by the abstract syntax:

$$\phi ::= Q \mid Z \mid \phi_1 \wedge \phi_2 \mid \phi_1 \vee \phi_2 \mid \bigcirc\phi \mid \odot\phi \mid \mu Z.\phi \mid \nu Z.\phi$$

where Q varies over $\mathcal{Z}_C$ and Z over $\mathcal{Z}_V$. The symbol σ is used in formulae to mean either ν or μ. An occurrence of a variable Z in ϕ is *bound* iff it is within a subformula $\sigma Z.\phi'$ of ϕ and *free* otherwise. If Z is a variable, $\phi[\phi'/Z]$ is the result of simultaneously substituting ϕ' for all free occurrences of Z in ϕ. ◁

Furthermore, in any νTL-formula ϕ, we assume that all the bound variables are distinct, and that all occurrences of bound variables are guarded, i.e. that each occurrence of variable Z in $\sigma Z.\phi$ is in a subformula of the type $\bigcirc\phi'$ or $\odot\phi'$. Any formula can be effectively transformed into an equivalent one fulfilling these restrictions. Note that negation is not in the logic; however, any formula can be negated using the De Morgan dualites.

Definition 2. A *transition system* is a set S of states together with a binary relation $\to$ on states. A *run* of a transition system is a maximal sequence $s_0 \to s_1 \to \cdots$ (which may be finite, if a state has no successor, or infinite). Let $\mathcal{R}$ be the set of runs. We shall let σ (outside formulae) range over runs, and if $\sigma = s_0 \to s_1 \to \cdots$ then we write $\sigma(i)$ for s_i, and σ^i for $s_i \to s_{i+1} \to \cdots$. We write $\Sigma(s)$ for the set $\{\sigma \mid \sigma(0) = s\}$ of runs starting at s.

A mu-calculus *model* is a transition system $T = (S, \to)$ together with a valuation $\mathcal{V} \colon Z_V \to 2^{\mathcal{R}}$, and a valuation $\mathcal{W} \colon Z_C \to 2^S$. The denotation $\|\phi\|_{\mathcal{V},\mathcal{W}}^T$ of a mu-formula ϕ in the model $(T, \mathcal{V}, \mathcal{W})$ is given by the following rules (omitting the superscript T and the subscript $\mathcal{W}$, which do not change):

$$\|Z\|_\mathcal{V} = \mathcal{V}(Z) \qquad\qquad \|Q\|_\mathcal{V} = \{\sigma \mid \sigma(0) \in \mathcal{W}(Q)\}$$

$$\|\phi_1 \wedge \phi_2\|_\mathcal{V} = \|\phi_1\|_\mathcal{V} \cap \|\phi_2\|_\mathcal{V} \qquad \|\phi_1 \vee \phi_2\|_\mathcal{V} = \|\phi_1\|_\mathcal{V} \cup \|\phi_2\|_\mathcal{V}$$

$$\|\bigcirc\phi\|_\mathcal{V} = \{\sigma \mid \sigma^1 \in \|\phi\|_\mathcal{V}\} \qquad \|\odot\phi\|_\mathcal{V} = \{\sigma \mid \sigma^1 \in \|\phi\|_\mathcal{V}\} \cup \{\sigma \mid \sigma(0) \not\to\}$$

$$\|\nu Z.\phi\|_\mathcal{V} = \bigcup\{R \subseteq \mathcal{R} \mid \|\phi\|_{\mathcal{V}[Z:=R]} \supseteq R\}$$

$$\|\mu Z.\phi\|_\mathcal{V} = \bigcap\{R \subseteq \mathcal{R} \mid \|\phi\|_{\mathcal{V}[Z:=R]} \subseteq R\}$$

where $\mathcal{V}[Z := R]$ is the valuation $\mathcal{V}'$ which agrees with $\mathcal{V}$ save that $\mathcal{V}'(Z) = R$.

A run σ satisfies ϕ, $\sigma \vDash \phi$, iff $\sigma \in \|\phi\|_\mathcal{V}$, and $s \vDash \phi$ for a state s iff all runs starting at s satisfy ϕ. $\lhd$

These preliminary definitions apply to finite or infinite systems. However, as we are interested in decidability, we shall from now on be concerned only with finite transition systems.

3 The Tableau System.

3.1 The sequents.

In the tableau system for the modal μ-calculus described in [SW91], the sequents have the form $s \vdash \phi$, where s is a state and ϕ is a formula. For the linear-time μ-calculus the sequents have to be a bit more complicated (as was already observed in [SW90]). The reason has to do with disjunction. In the modal case, $s \vDash A \vee B$ means $s \in \|A\| \cup \|B\|$, and therefore implies either $s \vDash A$ or $s \vDash B$. In consequence, the two rules

$$\frac{s \vdash A \vee B}{s \vdash A} \qquad \frac{s \vdash A \vee B}{s \vdash B}$$

are complete. In the linear-time case, $s \vDash A \vee B$ means $\Sigma(s) \subseteq \|A\| \cup \|B\|$. Since $\Sigma(s)$ is a set of runs, we can no longer infer $s \vDash \|A\|$ or $s \vDash \|B\|$: some runs in $\Sigma(s)$ may satisfy A but not B, and others B but not A. The solution is to allow sets of formulae in the right hand side of a sequent, which are interpreted disjunctively. This way, the rule

$$\frac{s \vdash A \vee B}{s \vdash A, B}$$

is sound and complete.

3.2 The rules.

If we have sets of formulae on the right of a sequent, there would usually be a choice of rules to apply, since several of the formulae may be eligible to have a rule applied. For technical reasons, it is helpful to have a unique rule to apply to a given sequent, so our system makes use of concurrent rule applications: a rule application in the tableau is actually the concurrent application of a maximal number of the basic rules; for example,

$$\frac{s \vdash A \vee B, C \wedge D}{s \vdash A, B, C \qquad s \vdash A, B, D}$$

is the concurrent application of the $\vee$ and $\wedge$ basic rules.

The basic rules are as one expects, namely

$\wedge \quad \dfrac{s \vdash \Gamma, A \wedge B}{s \vdash \Gamma, A \qquad s \vdash \Gamma, B}$ $\qquad\qquad \vee \quad \dfrac{s \vdash \Gamma, A \vee B}{s \vdash \Gamma, A, B}$

$Q \quad \dfrac{s \vdash \Gamma, Q}{s \vdash \Gamma} \quad$ where s fails Q $\qquad \sigma Z \quad \dfrac{s \vdash \Gamma, \sigma Z.A}{s \vdash \Gamma, A[\sigma Z.A/Z]}$

$\bigcirc \quad \dfrac{s \vdash \bigcirc\Gamma, \bigcirc\Delta}{s_1 \vdash \Gamma, \Delta \quad \ldots \quad s_n \vdash \Gamma, \Delta} \quad$ where $\{s_1, \ldots, s_n\} = \{\, s' \mid s \rightarrow s' \,\}$

A rule is then the concurrent application of a maximal set of basic rules. We have the following lemma trivially from the definitions:

Lemma 3. The antecedent of a rule is true if and only if all its consequents are true. $\qquad\qquad\qquad\qquad\qquad\qquad\qquad\qquad\qquad\qquad\qquad\qquad\qquad\qquad\qquad\Box$

Notice that the result of the application of a rule to a sequent is completely determined by the sequent. In other words, the children are completely determined by the parent. Notice also that the $\bigcirc$ rule cannot be applied concurrently with any other rule, since it requires that all formulae on the right start with a next operator.

3.3 Paths, internal paths, and terminals.

A *proof tree* is a tree of sequents constructed by the iterated application of these rules, starting with a root $s_0 \vdash \phi_0$. We write $\mathbf{n} : s \vdash \Gamma$ to mean that $\mathbf{n}$ is labelled with the sequent $s \vdash \Gamma$; we write $\mathbf{n} \simeq \mathbf{n}'$ to mean that $\mathbf{n}$ and $\mathbf{n}'$ are labelled with the same sequent. We extend these notations to sequences of nodes in the obvious way. (We shall use $\mathbf{n}, \mathbf{m}$ to range over nodes in a proof tree.)

Associated with a path π of the proof tree is a sequence $s \to s_1 \to \cdots \to s_m$ of transitions arising from the applications of the O-rule on π. We call this sequence $trans(\pi)$.

The price to pay for allowing sets of formulae in the right hand side of a sequent is that a path of a proof tree is an object with a rather complicated internal structure: a set of *internal paths* describing the dependencies between formulae at different nodes. The path

$$\frac{\mathbf{n}_1 : s \vdash (OA \wedge B) \vee OB}{\dfrac{\mathbf{n}_2 : s \vdash OA \wedge B, OB}{\dfrac{\mathbf{n}_3 : s \vdash OA, OB}{\mathbf{n}_4 : s' \vdash A, B}}}$$

has the following internal paths:

$$
\begin{array}{cc}
\mathbf{n}_1, (OA \wedge B) \vee OB & \mathbf{n}_1, (OA \wedge B) \vee OB \\
\downarrow & \downarrow \\
\mathbf{n}_2, OA \wedge B & \mathbf{n}_2, OB \\
\downarrow & \downarrow \\
\mathbf{n}_3, OA & \mathbf{n}_3, OB \\
\downarrow & \downarrow \\
\mathbf{n}_4, A & \mathbf{n}_4, B
\end{array}
$$

It is intuitively clear that the truth of a sequent depends on the structure of the internal paths starting at it: in particular, it is important which μ or ν-variables are unfolded in those paths. Since the terminals of our tableau system will use these notions, we define them formally:

Definition 4. Let $\pi = \mathbf{n}_1 \mathbf{n}_2 \ldots$ be a path of a proof tree. An *internal path* of π is a finite or infinite sequence of pairs $(\mathbf{n}_1, \phi_1)(\mathbf{n}_2, \phi_2) \ldots$ such that ϕ_i appears in $\mathbf{n}_i$, and for any two consecutive pairs $(\mathbf{n}_i, \phi_i)(\mathbf{n}_{i+1}, \phi_{i+1})$, one of the following cases holds:

- $\mathbf{n}_{i+1}$ is a child of $\mathbf{n}_i$, no basic rule is applied to ϕ_i, and $\phi_{i+1} = \phi_i$, or
- $\mathbf{n}_{i+1}$ is a child of $\mathbf{n}_i$, some basic rule different from Q is applied to ϕ_i, and ϕ_{i+1} is the formula given by the basic rule application.

An *internal circuit* of a finite path $\pi = \mathbf{n}_1 \mathbf{n}_2 \ldots \mathbf{n}_k$ such that $\mathbf{n}_1 \simeq \mathbf{n}_k$, is a finite sequence of internal paths of π

$$((\mathbf{n}_1, \phi_1) \ldots (\mathbf{n}_k, \phi_k)) \quad ((\mathbf{n}_1, \phi_{k+1}) \ldots (\mathbf{n}_k, \phi_{2k})) \ldots$$

$$\ldots ((\mathbf{n}_1, \phi_{jk+1}) \ldots (\mathbf{n}_k, \phi_{(j+1)k})) \qquad \text{for } j \in \mathbb{N}$$

such that $\phi_{ik+1} = \phi_{ik}$ and $\phi_1 = \phi_{(j+1)k}$ and there are no two identical pairs in the circuit.

The *characteristic* of a finite internal path is either the highest variable that is unfolded (i.e. has the σZ rule applied to its fix-point) in it, or the symbol $\perp$ if no variable is unfolded; the characteristic of an infinite internal path is the highest variable that is unfolded infinitely often. If the characteristic of an internal path is a μ-variable (ν-variable), then we also say that the path has μ-characteristic (ν-characteristic).

Let $\mathbf{n} \dots \mathbf{n}'$ be a path. The relation $Int(\mathbf{n}, \mathbf{n}')$ is defined as the set of triples (ϕ, ϕ', Z) such that there exists an internal path $(\mathbf{n}, \phi) \dots (\mathbf{n}', \phi')$ with characteristic Z. Sometimes we denote the triple (ϕ, ϕ', Z) as $(\mathbf{n}, \phi) \xrightarrow{Z} (\mathbf{n}', \phi')$.

◁

It is easy to see that if the formula at the root of a proof tree is guarded, then the characteristic of any internal circuit is always different from $\perp$.

We can now define the terminal nodes, which are rather similar to those given by Kaivola in [Kai95] for the satisfiability problem.

Definition 5. A node $\mathbf{n} : s \vdash \Gamma$ is a *terminal* if
  (i) $\Gamma = \varnothing$; or
 (ii) $\Gamma = \odot\Gamma', \odot\Gamma''$ (where Γ'' is non-empty) and s has no successor; or
(iii) $\Gamma = \odot\Gamma'$ and s has no successor; or
 (iv) $Q \in \Gamma$ and s satisfies Q; or
  (v) $\mathbf{n}$ has a predecessor $\mathbf{n}' \simeq \mathbf{n}$ and every internal circuit of the path $\mathbf{n}' \dots \mathbf{n}$ has μ-characteristic; we call $\mathbf{n}'$ the *companion* of $\mathbf{n}$; or
 (vi) $\mathbf{n}$ is not a terminal of type (v), and it has two predecessors $\mathbf{n}'' \simeq \mathbf{n}' \simeq \mathbf{n}$, with $\mathbf{n}''$ above $\mathbf{n}'$, such that $Int(\mathbf{n}'', \mathbf{n}') = Int(\mathbf{n}'', \mathbf{n})$; the nodes $\mathbf{n}''$ and $\mathbf{n}'$ are called the *second* and *first companions* of $\mathbf{n}$.
Terminals of types (i), (iii) and (v) are *unsuccessful*, and terminals of types (ii), (iv) and (vi) are *successful*.

A *tableau* is a finite proof tree whose leaves (and no other node) are terminals. A tableau is *successful* if all its terminals are successful. ◁

We have the following result

Proposition 6. There is a unique tableau with a given root.
Proof. The children of a nonterminal node are determined solely by the node, and the termination conditions are deterministic. □

We briefly explain the intuition behind the definition of the terminals. Each path of a tableau can be seen as an attempt to construct a *false* run of the system, i.e., a run which does not satisfy the formula at the root. The terminals identify the points at which we have gathered enough information, either to construct such a run (unsuccessful terminal), or to give up searching the continuations of the path (successful terminal). Let π be a path of the tableau ending in a terminal $\mathbf{n}$, and let $\sigma = trans(\pi)$.

• If $\mathbf{n}$ is of type (i), then its parent is of the form $s \vdash Q$, and no run starting at s satisfies Q. So every run of the form $\sigma\sigma'$ is false.
• if $\mathbf{n}$ is of type (ii), then σ is a true run, and there are no continuations of σ.
• if $\mathbf{n}$ is of type (iii), then σ is already a false run.

- If **n** is of type (iv), then all runs of the form $\sigma\sigma'$ are true; therefore, we can give up the search.
- If **n** is of type (v), then the run $\sigma_1(\sigma_2)^\omega$ is false, where $\sigma_2 = trans(\mathbf{n}' \ldots \mathbf{n})$ and $\sigma_1\sigma_2 = \sigma$; loosely speaking, the reason is that in any chain of dependencies corresponding to this run some μ-variable is unfolded infinitely often.
- If n is of type (vi), then we cannot say that all runs of the form $\sigma\sigma'$ are true. However, if such a run is false, then there exists also a false run of the form $\sigma''\sigma'$, where $\sigma'' = trans(\mathbf{n_0} \ldots \mathbf{n}'' \ldots \mathbf{n}')$. Moreover, the path showing the falsity of $\sigma''\sigma'$ will be shorter than the path for $\sigma\sigma'$ (loosely speaking, it will not contain the part $\mathbf{n}' \ldots \mathbf{n}$). So we can give up the search.

This concludes the presentation of the tableau system. A comparison with the systems of [SW90] and [Kai95] can be found in the conclusions.

4 Example.

For an example, we shall check the formula

$$\nu Z.\mu X.O((Z \wedge P) \vee X)$$

'(on every path) infinitely often P' on the state s of the following system, where P holds at the state t.

This formula fails at s, so the constructed tableau will contain an unsuccessful terminal. We shall only show the first branch containing an unsuccessful terminal, since that is sufficient to fail.

For readability, we use the variables Z, X to represent the associated fix-point formulae; and ϕ abbreviates $((Z \wedge P) \vee X)$. The terminals marked with $\#$ are the unsuccessful terminals (all of type (v)); both of them fail due to an internal path of the form

$$s, O\phi \rightarrow s, (Z \wedge P) \vee X \rightarrow s, X \rightarrow s, O\phi$$

which is the only internal path from the companion $s \vdash O\phi$ to the terminal $s \vdash O\phi$.

Tableaux can be quite large—we have only given one unsuccessful branch for this one. If the property being checked is true, one must construct the entire tableau, but as noted above, in the case of failure it suffices to find one unsuccessful branch. It is this feature which allows local model-checking: the user may construct the tableau depth-first, and decide at every node with several children which one to examine first.

The second tableau is the tableau for the same formula on the system with the self-loop from s to s deleted. In this case, the formula is true, and accordingly the tableau is successful. The successful terminals (marked with $\checkmark$) are some of

$$\frac{s \vdash \nu Z.\mu X.\mathrm{O}\phi}{s \vdash \mu X.\mathrm{O}\phi}$$

$$\frac{}{s \vdash \mathrm{O}\phi}$$

$$\frac{s \vdash (Z \wedge P) \vee X}{s \vdash Z \wedge P, X} \qquad \frac{t \vdash (Z \wedge P) \vee X}{\cdots}$$

$$\frac{s \vdash P, \mathrm{O}\phi}{\# \ s \vdash \mathrm{O}\phi} \qquad \frac{s \vdash Z, \mathrm{O}\phi}{\cdots}$$

First example tableau

type (iv) and some of type (vi). One of the type (vi) leaves is labelled **n**, and its companions are labelled $\mathbf{n}''$ and $\mathbf{n}'$; the others are all similar. The relations $Int(\mathbf{n}'', \mathbf{n}')$ and $Int(\mathbf{n}'', \mathbf{n})$ are both

$$\{(\mathrm{O}\phi, \mathrm{O}\phi, Z), (\mathrm{O}\phi, \mathrm{O}\phi, X)\}$$

This example also shows that tableaux can be very redundant. For instance, the two subtableaux rooted at $\mathbf{n}'$ and the similar node $\mathbf{m}'$ to its right (labelled $s \vdash \mathrm{O}\phi$). This redundancy is typical of tableau methods, and there exist several techniques that palliate it, which are out of the scope of this paper (see [Mad92,Mad95]).

5 Soundness and Completeness.

In this section we state the necessary soundness and completeness theorems for the tableau system. The proofs of these theorems are somewhat technical and intricate, especially the soundness proof, and regrettably the space constraints for these Proceedings have necessitated their removal. We therefore only give a few lines sketching the strategy, and refer the reader to the technical report version of this paper, available on the Web at
http://www.dcs.ed.ac.uk/home/jcb/Research/papers.html#lintab
(in the UK), or
http://papa.informatik.tu-muenchen.de/forschung/sfb342_a3/refs.html
(in Germany).

To prove these theorems, the standard notions of approximants for fixpoints, and of signatures, are defined.

Definition 7. For all ordinals $\alpha \in \mathrm{Ord}$, the *fixpoint approximants* $\mu^\alpha Z.\phi$ and $\nu^\alpha Z.\phi$ are defined by: $\mu^0 Z.\phi = \mathrm{ff}$, $\nu^0 Z.\phi = \mathrm{tt}$, $\sigma^{\alpha+1} Z.\phi = \phi[\sigma^\alpha Z.\phi/Z]$, $\mu^\lambda Z.\phi = \bigvee_{\alpha<\lambda} \mu^\alpha Z.\phi$ and $\nu^\lambda Z.\phi = \bigwedge_{\alpha<\lambda} \nu^\alpha Z.\phi$, where λ is a limit ordinal. ◁

Proposition 8. In a model $T = (S, \rightarrow)$, $\mu Z.\phi = \bigvee_{\alpha<\kappa} \mu^\alpha Z.\phi$ for some $\kappa \leq 2^{|S|}$, and $\nu Z.\phi = \bigwedge_{\alpha<\kappa} \nu^\alpha Z.\phi$ for some $\kappa \leq 2^{|S|}$. □

$$s \vdash \nu Z.\mu X.\bigcirc\phi$$
$$s \vdash \mu X.\bigcirc\phi$$
$$\mathbf{n''}: s \vdash \bigcirc\phi$$
$$t \vdash (Z \wedge P) \vee X$$
$$t \vdash Z \wedge P, X$$

$$\checkmark t \vdash P, \bigcirc\phi \qquad\qquad t \vdash Z, \bigcirc\phi$$
$$t \vdash X, \bigcirc\phi$$
$$t \vdash \bigcirc\phi$$
$$s \vdash (Z \wedge P) \vee X$$
$$s \vdash Z \wedge P, X$$

$$s \vdash P, \bigcirc\phi \qquad\qquad\qquad\qquad s \vdash Z, \bigcirc\phi$$
$$\mathbf{n'}: s \vdash \bigcirc\phi \qquad\qquad\qquad\qquad s \vdash X, \bigcirc\phi$$
$$t \vdash (Z \wedge P) \vee X \qquad\qquad \mathbf{m'}: s \vdash \bigcirc\phi$$
$$t \vdash Z \wedge P, X \qquad\qquad\qquad t \vdash (Z \wedge P) \vee X$$

$$t \vdash Z, \bigcirc\phi \quad \checkmark t \vdash P, \bigcirc\phi \qquad\qquad t \vdash Z \wedge P, X$$
$$t \vdash X, \bigcirc\phi \qquad\qquad\qquad\qquad t \vdash Z, \bigcirc\phi \quad \checkmark t \vdash P, \bigcirc\phi$$
$$t \vdash \bigcirc\phi \qquad\qquad\qquad\qquad\qquad t \vdash X, \bigcirc\phi$$
$$s \vdash (Z \wedge P) \vee X \qquad\qquad\qquad t \vdash \bigcirc\phi$$
$$s \vdash Z \wedge P, X \qquad\qquad\qquad s \vdash (Z \wedge P) \vee X$$

$$s \vdash P, \bigcirc\phi \qquad s \vdash Z, \bigcirc\phi \qquad\qquad s \vdash Z \wedge P, X$$
$$\checkmark s \vdash \bigcirc\phi \qquad s \vdash X, \bigcirc\phi$$
$$\checkmark \ \mathbf{n}: s \vdash \bigcirc\phi \qquad\qquad s \vdash P, \bigcirc\phi \qquad s \vdash Z, \bigcirc\phi$$
$$\checkmark s \vdash \bigcirc\phi \qquad s \vdash X, \bigcirc\phi$$
$$\checkmark s \vdash \bigcirc\phi$$

Second example tableau

(These syntactic approximants are infinitary formulae which have the obvious semantics; alternatively, one can define semantic approximants directly.)

Definition 9. The μ-*signature* μ-$sig(\sigma, \phi)$ of a run σ at a formula ϕ (where $\sigma \models \phi$) is the lexicographically least sequence $\zeta_1, \zeta_2, \ldots, \zeta_k$ such that $\sigma \models \phi[\mu^{\zeta_i} Z_i.\phi_i / \mu Z_i.\phi_i]$ where $\mu Z_i.\phi_i$ are the μ subformulae of ϕ in order of depth (i.e. in some (fixed) order such that subformulae appear after any containing subformulae).

Dually, the ν-*signature* of $\sigma \not\models \phi$ is defined as the least sequence such that $\sigma \not\models \phi[\nu^{\zeta_i} Z_i.\phi_i / \nu Z_i.\phi_i]$. ◁

Theorem 10. The (unique) tableau for a given root is finite.

Proof. Let τ be the tableau with root $s_0 \vdash \phi_0$, let n be the number of states of the transition system, and let m be the number of symbols of ϕ_0. It is easy to see that the size of the closure (i.e. the subformulae, modulo unfolding) of ϕ_0 is bounded by m. Therefore, τ contains at most $n \cdot 2^m$ different sequents, and there are at most 2^{m^3} different *Int* relations. By the definition of the terminals of type (vi), every path of τ has at most length $O(n \cdot 2^{m^3})$. Since a node has finitely many children, a tableau contains finitely many nodes. □

Theorem 11. If $s_0 \models \phi_0$, then there exists a successful tableau.

Proof. (Sketch) Starting with root $s_0 \vdash \phi_0$, apply the rules until a tableau is constructed. The construction terminates by Theorem 10. Assume that the tableau so constructed contains an unsuccessful terminal $\mathbf{n} : s \vdash \Gamma$. We prove that $\mathbf{n}$ or its parent is a false node, which contradicts Lemma 3. The interesting case is when $\mathbf{n}$ is an unsuccessful terminal of type (v). In that case, we take the run generated by the internal path from $\mathbf{n}$'s companion to $\mathbf{n}$, and show that this run does not satisfy any formula in Γ. The technique for this is a familiar one in proofs for fixpoint tableau systems: follow the run round the internal path, and use the backward soundness of the tableau rules to find either a contradiction after finite time, or an infinite descending chain of μ-signatures, which is also impossible. □

Theorem 12. If there exists a successful tableau, then $s_0 \models \phi_0$.

Proof. (Sketch) Let τ be the tableau for $s_0 \vdash \phi_0$. Suppose that τ is successful, but that $s_0 \not\models \phi_0$; we shall derive a contradiction.

An extended path of τ is a finite or infinite sequence of nodes $\mathbf{n}_1 \mathbf{n}_2 \ldots$ such that for every two consecutive nodes $\mathbf{n}_i$, $\mathbf{n}_{i+1}$ either $\mathbf{n}_i$ is a σ-terminal and $\mathbf{n}_{i+1}$ a child of its (first) companion, or $\mathbf{n}_i$ is a nonterminal and $\mathbf{n}_{i+1}$ is one of its children. The internal paths of an extended path are defined as those of normal paths.

Assuming that $s_0 \not\models \phi_0$ there must be a run σ_0 which does not satisfy ϕ_0. We shall use this run to show the existence of an unsuccessful terminal, contradicting the success of τ. Again, we use the run to construct a suitable extended path so that every node of the path is false. Since the ν-signatures must decrease, we obtain a path along which some least fixpoint is infinitely often unfolded. We then apply a delicate, but terminating, procedure to the path which demonstrates the existence of an unsuccessful terminal of type (v), contradicting the hypothesis. (This procedure essentially 'cuts out' portions of the path between companions and terminals.) □

6 Complexity Issues.

Let us first consider the problem of deciding if, given a state s and a formula ϕ, some path starting at s does *not* satisfy ϕ (this is the complement of the model-checking problem as defined in this paper).

We may nondeterministically guess a path of the tableau with root $s \vdash \phi$ leading to an unsuccessful terminal. The length of a path is $O(n \cdot 2^{m^3})$, where n is the number of states of the transition system and m the number of symbols of the formula. So we have a nondeterministic algorithm with linear space complexity in the size of the system and exponential in the length of the formula.

It is possible to do better if we observe that we do not have to store all the path. We may nondeterministically guess its length k, and the position k' of the companion. Then, it suffices to store the current node n_i, and, from the moment we reach the $k+1$-th node on, the relation $Int(n_k, n_i)$, because the characteristics of the internal paths can be obtained from there. An element of the Int relation can be stored in $O(m)$ space, and there are at most $O(m^3)$ of them. So we need $O(m^4 + \log n)$ space (including the space to store k and k'). It is still possible to improve things a bit by observing that we do not really need to store every element of $Int(n_k, n_i)$, because we are only interested in the characteristics of the internal circuits. In consequence, for every pair of formulae (ϕ, ψ), we only need to store the element $(\phi, \psi, Z) \in Int(n_k, n_i)$ having the *highest* Z. This reduces the space to $O(m^3 + \log n)$. Some other tricks (see [Kai95]) reduce it further to $O(m^2 \log m + \log n)$. Using standard results of complexity theory, it is then possible to obtain deterministic algorithms with $2^{O(m^2 \log m + \log n)}$ space and time, or $O(f)$ space and $O(2^f)$ time, where $f \in O((m^2 \log m + \log n)^2)$. These deterministic results also apply, of course, to the model-checking problem.

7 Conclusions.

We have presented a tableau system for the linear-time μ-calculus with simpler success conditions than those of Stirling and Walker's system [SW90]. The success condition of [SW90] uses the notion of *extended path*. Loosely speaking, an extended path is a sequence of nodes in which the successor of an element is either one of its children or, in case the element is a fixpoint terminal, its companion (in [SW90] fixpoint terminals only have one companion). The success condition for a terminal $\mathbf{n}$ requires to examine the internal circuits of *all* the extended paths leading from the companion $\mathbf{n'}$ of $\mathbf{n}$ to $\mathbf{n}$. Such extended paths may visit many different terminals of the tableau, because the subtree with root $\mathbf{n'}$ may have many different leaves. This makes the condition very difficult to verify. In fact, it is not easy to prove that the condition is decidable (this point is not addressed in [SW90]). Our success condition requires to examine only the internal circuits of the *unique* path leading from the companion to the terminal, and makes the complexity analysis very simple.

The price to pay for the simpler success condition is a larger tableau. In [SW90], the construction of a branch of the tableau terminates whenever we hit a sequent we have seen before (in the branch). In our system, we may have to continue further. So, in some sense, our tableau can be seen as an unfolding of the tableau of [SW90], in which the success conditions are more transparent and easier to check.

Kaivola also sketches a tableau system for the satisfiability problem at the end of [Kai95]. The system checks if some run satisfies a formula. A tableau consists of one path and a *recurrence point*, a node that must be guessed by the user. These points increase very much the number of possible tableaux for a given root. Our system improves Kaivola's by getting rid of recurrence points. In this way, we obtain a 'conventional' tableau system, closer to the system for the modal μ-calculus of [SW91].

The complexity of our system (discussed in the previous section) coincides with the complexity of model-checking techniques based on automata theory (see, for instance, [Var88]). Our aim is not to compete with these techniques, which are probably better for automatic verification, but to provide a logical system which may be used to solve small examples by hand, explains why a formula holds, and allows us to use knowledge about the system to speed up the verification process.

8 References.

[Fit83] M. Fitting, *Proof Methods for Modal and Intuitionistic Logics* (Reidel, 1983).

[HC68] G.E. Hughes and M.J. Creswell, *An Introduction to Modal Logic*, (Methuen and Co., 1968)

[Kai95] R. Kaivola, A simple decision method for the linear time mu-calculus, *Proc. Int. Workshop on Structures in Concurrency Theory* (J. Desel, ed.) (1995)

[Mad92] A. Mader, Tableau recycling, *Proc. CAV '92*, LNCS **663** (1992).

[Mad95] A. Mader, Modal μ-calculus, model checking and Gauß elimination, *Proc. TACAS'95*, to appear in LNCS. (1995)

[Sti87] C. P. Stirling, Modal logics for communicating systems. *Theoret. Comput. Sci.* **49** 311–347 (1987).

[SW90] C. Stirling and D. Walker, CCS, liveness, and local model checking in the linear time mu-calculus, *Proc. First International Workshop on Automatic Verification Methods for Finite State Systems*, LNCS **407** 166–178. (1990).

[SW91] C. Stirling and D. Walker, Local model checking in the modal mu-calculus, *Theor. Comput. Sci.* **89**, 161–177. (1991).

[Var88] M. Vardi, A temporal fixpoint calculus, *Proc. 15th PoPL*, 250–259. (1988)

Characterizing Fairness Implementability for Multiparty Interaction

Yuh-Jzer Joung *

Dept. of Info. Management, National Taiwan University, Taipei, Taiwan
joung@ccms.ntu.edu.tw

We present a semantic criterion for fairness notions for multiparty interaction. We show that if a fairness notion violates the criterion, then no deterministic algorithm for multiparty interaction scheduling can satisfy the fairness requirement. Conversely, the implementation is possible if the criterion is obeyed. Thus, the criterion is sufficient and necessary to guarantee fairness implementability. To our knowledge, this is the first such criterion to appear in the literature.

We then use this criterion to examine several important fairness notions, including *strong interaction fairness*, *strong process fairness*, *weak process fairness*, *U-fairness*, and *hyperfairness*. All, except weak process fairness, fail to pass the criterion.

1 Introduction

In recent developments of languages for distributed computing (e.g., [7, 4, 6][2]), a new family of structures, generally referred to as *multiparty interactions*, has evolved from CSP's biparty synchronous communications [8]. They allow an arbitrary number of processes to synchronize, exchange data, and perform some joint computations. Nondeterminism is also supported by permitting multiparty interactions to serve as guards in alternative and repetitive commands. Multiparty interactions provide a higher level of abstraction and encourage modular programming and design.

We focus here on the implementation of multiparty interactions, which is concerned with synchronizing asynchronous processes to participate in interactions so that no conflicting interactions (i.e., interactions involving a common member of process) execute simultaneously. The problem is typically associated with some notions of fairness to prevent "unfair" computations that favor a particular process or interaction. In general, a fairness notion excludes from all possible computations some that would otherwise be legal. Consequently, any subset of computations could be considered as a fairness notion for the system. Not many of them, however, are useful, and so criteria have been proposed for determining their appropriateness, including [1]:

Feasibility: Every partial computation can be extended to a fair one.

Equivalence-robustness: Two equivalent computations should both be fair or both be unfair. Computations are *equivalent* if they are identical up to the order of independent actions.

---

* This research was supported by the National Science Council, Taipei, Taiwan, under Grant NSC 85-2213-E-002-059.

[2] A taxonomy of programming languages offering linguistic support for multiparty interaction is presented in [11].

Feasibility is often demonstrated by an explicit scheduler, which has access to the variables of all processes in the system, and can even control the progress of them (e.g., [1, 13]). In each step the scheduler determines for the system the next action to execute, which is either a local action of some process or an interaction. A fairness notion then is said to be *feasible* iff there exists a scheduler to generate only the fair computations. Clearly, without such a scheduler the fairness notion could not possibly be implemented. So it is necessary for a fairness notion to be implementable.

Note that an explicit scheduler does not directly correspond to a real implementation. This is because in practice a process's local actions are independent of any other process's local actions and interactions. So an implementation should not control when and how a process will execute its own local actions, but should only determine when and which enabled interaction to execute. Scheduling independent local actions and interactions in a purely interleaving mode turns the system's concurrency behavior into a sequential one and thus is highly undesirable. Moreover, if the time when a process will be ready for interaction is not determined a priori (which is usually the case in real applications), then deferring the scheduling of enabled interactions until some other process has finished its local action could result in a deadlock.

It turns out that, if a scheduler can only determine which interaction to execute, then feasibility alone does not necessarily guarantee implementability. To illustrate, the notion of *strong interaction fairness (SIF)*, which requires that an interaction that is infinitely often enabled be executed infinitely often, is feasible [1], but it's implementation has been proven impossible [9, 18].

On the other hand, it is observed that most equivalence-robust fairness notions are implementable. (See [1] for a list of such fairness notions.) This holds even if only interactions can be scheduled.[3] However, equivalence-robustness is not necessary for every implementable fairness notion. For example, consider the notion of *weak process fairness* (WPF), which requires a process continually ready for an enabled interaction (not necessarily the same one, though) be able to execute some interaction eventually. WPF is not equivalence-robust [1], but it can be implemented in a system consisting of only biparty interactions [9, 17].

Summary of Technical Results

In this paper we propose a new criterion for appraising fairness notions. The criterion requires that a fairness notion be realized by an *abstract scheduling function* such that all computations produced by this function are fair, and all other computations *indistinguishable* from the produced computations are also fair. Intuitively, the abstract function captures the scheduling policy adopted by a concrete scheduling program, while the indistinctness relation expresses properties of computations that cannot be distinguished by any asynchronous distributed system.

---

[3] As we shall see in Section 4, the observation is not coincidental because, under a strong notion of feasibility, equivalence-robustness suffices to guarantee implementability.

Assume the following in the underlying model of computation:

A1. One process's readiness for multiparty interaction can be known by another only through communications, and the time it takes two processes to communicate is nonnegligible.

A2. A process decides autonomously when it will attempt an interaction, and at a time that cannot be predicted in advance.

We show that if a fairness notion violates the criterion, then no deterministic algorithm for multiparty interaction scheduling can satisfy the fairness requirement. For fairness notions that satisfy the criterion, we also present a general algorithm to implement them in an asynchronous system where processes communicate exclusively by biparty message passing. Thus, the criterion is sufficient and necessary to guarantee fairness implementability.

Using this criterion, we obtain several possibility and impossibility results. For example, we show that WPF can be implemented successfully even if interactions are multipartied. The following important fairness notions, however, are all impossible to implement: SIF, *strong process fairness* (SPF) [1], *U-fairness* [2], and *hyperfairness* [3].

The rest of the paper is organized as follows. Section 2 presents an abstract model for process interaction, and an implementation model for interaction scheduling. Section 3 presents our semantic criterion and shows that it is necessary and sufficient to determine fairness implementability. Section 4 exploits fairness properties derived from the criterion. In Section 5 we use the criterion to examine several fairness notions that are commonly associated with multiparty interactions. Section 6 concludes.

2 Preliminaries

2.1 An Abstract Model for Process Interaction

An *interaction system* is a pair $M = (P_M, I_M)$, where P_M is a finite set of processes, and I_M is a finite set of interactions. Each interaction x involves a fixed set $P_x \subseteq P_M$ of participants. A process is either in an *idle* state or in a *ready* state. Initially, all processes are idle. An idle process may autonomously become ready, where it wishes to execute one interaction, from a set of potential interactions of which it is a member. An interaction can be executed only if all of its participants are ready, and the execution drives them back to an idle state. Moreover, a process can execute only one interaction at a time.

A state s of M consists of the state of each process at some instant of time. By $[s]_p$ we mean the state of p in s, and by $[s]_{p.\,aim}$ we mean the set of potential interactions p is ready to execute. An interaction x is *enabled* in s iff every process $p \in P_x$ is ready for x. Let S_M denote the set of all possible states of M. State transitions are written as $s \xrightarrow{a} s'$, where $s, s' \in S_M$, and a is the action resulting in the transition. State transitions are either one of the following two forms:

ready: $s \xrightarrow{p.I} s'$ iff $[s]_p = idle \land [s']_p = ready \land [s']_{p.aim} = I$,
    and $\forall q \in P_M, p \neq q \Rightarrow ([s]_q = [s']_q \land [s]_{q.aim} = [s']_{q.aim})$.

interaction: $s \xrightarrow{x} s'$ iff $\forall p \in P_x ([s]_p = ready \land x \in [s]_{p.aim} \land [s']_p = idle \land [s']_{p.aim} = \emptyset)$, and $\forall p \notin P_x([s]_p = [s']_p \land [s]_{p.aim} = [s']_{p.aim})$.

A *run* π is a sequence of the form

$$s_0 \xrightarrow{a_1} s_1 \xrightarrow{a_2} s_2 \dots$$

where s_0 is the initial state (i.e., $\forall p \in P_M, [s_0]_p = idle \land [s_0]_{p.aim} = \emptyset$), and each $s_i \xrightarrow{a_{i+1}} s_{i+1}$ is a state transition of the system. In particular, π is *complete* if it is infinite or it ends up in a state in which all processes are ready but no interaction is enabled; otherwise, π is *partial*. Since π is uniquely determined by the sequence of actions executed in the run, we often write it as $a_1 a_2 \dots$. Set $run(M)$ denotes all possible complete runs of M.

Definition 1. A *fairness notion* F is a function which, given an interaction system M, returns a set of runs $F(M) \subseteq run(M)$. We say that π is *F-fair* (or simply *fair* when the context is clear) if $\pi \in F(M)$.

2.2   An Implementation Model for Interaction Scheduling

We now consider the implementation of multiparty interaction. By this we mean augmenting each process in an interaction system with variables and actions, and possibly introducing auxiliary processes so that each ready process knows when and which interaction to execute.

Formally, a *scheduling program* for an interaction system $M = (P_M, I_M)$ is a quintuple

$$\Sigma_M = (P_M, I_M, Aux, \bigcup_{p \in P_M \cup Aux} V_p, \bigcup_{p \in P_M \cup Aux} A_p),$$

where P_M is the set of *primary* processes, I_M is the set of interactions allowed in M, Aux is a set of *auxiliary* processes (possibly empty) that are added to assist the coordination, V_p is the set of variables local to process p, and A_p is the set of actions executed by p. We assume that processes communicate by reliable, biparty asynchronous message passing, although our results in this paper hold as well if communication is by accessing shared variables.

Moreover, we assume that for each primary process p, V_p contains a variable $p.state$, which designates whether p is idle or ready, a variable $p.aim$ which designates the set of potential interactions for which p is ready, and a variable $p.commit$ which designates the interaction p has committed to execute. Variable $p.commit$ is set only once in each ready state, and is undefined if p is idle. If some process has committed to x, then all other participants of x will eventually commit to x, and these commitments should not depend on the state of any other process not involved in x.

For every process p, if p is auxiliary, then actions in A_p are of the form:

$$b_\sigma(V_p); \quad message\text{-}receptions \longrightarrow f_\sigma(V_p - \{p.state, p.aim\}), \qquad (1)$$
$$message\text{-}sendings$$

where $b_\sigma(V_p)$ is a Boolean condition on the variables in V_p, $f_\sigma(V)$ represents the effect of the execution to the variables in V, *message-receptions* express the messages to be received, and *message-sendings* describe the messages to be sent out when the action is executed. The action is executed only if it is *enabled*, i.e., $b_\sigma(V_p)$ evaluates to true and the messages specified in the reception-list have arrived.

If p is primary, then in addition to actions of form (1), A_p also contains actions of the following form to capture the ready transitions:

$$p.state = idle \longrightarrow p.state := ready, \tag{2}$$
$$p.aim := \text{an arbitrary subset of } \{x \in I_M \,|\, p \in P_x\},$$
$$f_\sigma(V_p - \{p.state, p.aim\}),$$
$$message\text{-}sendings$$

Moreover, to represent interactions and state transitions from ready to idle (i.e., the termination of interactions), for each $x \in I_M$ and $p \in P_x$, A_p contains an action of the following form executed jointly by the participants of x:

$$\forall p \in P_x, p.commit = x \longrightarrow p.state := idle, \tag{3}$$
$$p.aim := \emptyset,$$
$$p.commit := \perp,$$
$$f_\sigma(V_p - \{p.state, p.commit, p.aim\}),$$
$$message\text{-}sendings$$

As usual, a *computation* Π of Σ_M is a maximal sequence of the form

$$\alpha_0 \xrightarrow{\sigma_1} \alpha_1 \xrightarrow{\sigma_2} \alpha_2 \ldots$$

where α_0 is an initial state, and each $\alpha_{i-1} \xrightarrow{\sigma_i} \alpha_i$ represents a state transition of the program. Like runs, we often write Π as $\sigma_1\sigma_2\ldots$ (assuming some fixed initial state). Note that a finite computation must end up in a state in which no action is enabled for execution. Since an idle process can autonomously become ready, in the final state all primary processes must be ready for interaction. In particular, if there is an enabled interaction, then the finite computation is *deadlocked*. We shall consider only scheduling programs that produce no deadlocked computation.

Also noteworthy is the fact that under the *minimal progress assumption* [14]— any process with an enabled action will eventually execute some action—actions of form (2) do not fully respect Assumption A2. That is, a scheduling program may simply wait until all processes become ready, and then decide on an interaction for execution. To avoid this, we consider only *wait-free* scheduling programs where the establishment of interactions do not depend on idle processes to become ready.

Definition 2. A scheduling program Σ_M is *wait-free* iff for every partial computation Π, if there is an interaction enabled in Π, then Π has a continuation such that some interaction will be executed and no process makes a ready transition in the interim.

To abstract runs from computations, we introduce the following definitions. Let $\Pi = \sigma_1\sigma_2\ldots\sigma_n\ldots$ be a computation of Σ_M. Suppose that σ_n denotes the execution of some instance of x. Then, prior to σ_n all participants of x must have committed to x. Let σ_j be the first commitment. Then we say that σ_j *establishes* the instance of x. The run corresponding to Π, denoted by $[\Pi]_M$, is $[\sigma_1]_M[\sigma_2]_M \ldots [\sigma_n]_M \ldots$, where $[\sigma_i]_M$ is defined as follows:

1. $[\sigma_i]_M = p.aim$ if σ_i results in p's transition into a ready state willing to execute the set aim of interactions,
2. $[\sigma_i]_M = x$ if σ_i establishes an instance of x, and
3. $[\sigma_i]_M = \epsilon$ otherwise, where ϵ denotes the empty sequence such that for all finite π, $\pi\epsilon = \epsilon\pi = \pi$.

Definition 3. A fairness notion F is *implementable* for M iff there exists a wait-free scheduling program Σ_M such that for every computation Π of Σ_M, $[\Pi]_M \in F(M)$.

3   The Criterion

The fairness implementability criterion depends on a strong notion of feasibility and an indistinctness relation between runs. Strong-feasibility is realized by the following two functions, one to capture the processes' autonomy in making their ready transitions, and the other for interaction scheduling.

Definition 4.
1. An *adversary* A of M is a function which given a run π of M returns either an empty sequence ϵ or a sequence of actions $p_1.I_1 \ldots p_k.I_k$ as the continuation of π, where process p_i, $1 \leq i \leq k$, must be idle at the end of π, and $I_i \subseteq \{x \in I_M \mid p_i \in P_x\}$. Moreover, $A(\pi) = \epsilon$ only if π is complete or contains some enabled interaction.
2. A *nonpreemptive scheduler* S of M is a function which given a run π of M returns either ϵ or an interaction x enabled in π as the continuation of π. Moreover, $S(\pi) = \epsilon$ only if no interaction is enabled in π.
3. The run generated by S versus A is defined by $r(S,A) = r^{i \to \infty}(S,A)$, where

$$r^i(S,A) = \begin{cases} \epsilon & : \ i = 0 \\ r^{i-1}(S,A) \cdot A(r^{i-1}(S,A)) & : \ i = 2n-1, n \in N \\ r^{i-1}(S,A) \cdot S(r^{i-1}(S,A)) & : \ i = 2n, n \in N \end{cases}$$

Note that if we do not restrict $A(\pi)$ to generate a nonempty sequence when π is partial and contains no enabled interaction, then no scheduler S versus A could possibly generate a complete run. Similarly, if $S(\pi) = \epsilon$ when some interaction is enabled in π, then A in response could return ϵ, and thus $r(S,A)$ would not be complete.

Definition 5. A fairness notion F is *strongly-feasible* for M iff there exists a nonpreemptive scheduler S such that for every adversary A, $r(S,A) \in F(M)$.

We now introduce indistinctness. Let

$$\pi = p_{1,1}.I_{1,1} \ldots p_{1,k_1}.I_{1,k_1} x_1 \ p_{2,1}.I_{2,1} \ldots p_{2,k_2}.I_{2,k_2} x_2 \ldots$$

where $x_1, x_2, \ldots$ are interactions executed in π. We say that ρ is obtained from π by an *interprocess permutation* if

$$\rho = q_{1,1}.J_{1,1}\ldots q_{1,k_1}.J_{1,k_1}x_1 \; q_{2,1}.J_{2,1}\ldots q_{2,k_2}.J_{2,k_2}x_2\ldots$$

such that, for each $i > 0$, $q_{i,1}.J_{i,1},\ldots,q_{i,k_i}.J_{i,k_i}$ is a permutation of $p_{i,1}.I_{i,1}$, $\ldots, p_{i,k_i}.I_{i,k_i}$. Furthermore, ψ is obtained from π by a *retraction* if it is obtained from π by moving, for each $i > 0$, some initial segment (possibly empty) $p_{i,1}.I_{i,1},\ldots,p_{i,h_i}.I_{i,h_i}$ of $p_{i,1}.I_{i,1}\ldots p_{i,k_i}.I_{i,k_i}$ *forward* just before x_{i-1}, where $p_{i,1},\ldots,p_{i,h_i} \notin P_{x_{i-1}}$. Run ρ is *indistinguishable* from π, denoted by $\rho \rightsquigarrow \pi$, iff ρ can be obtained from π by an interprocess permutation optionally followed by a retraction. The set of runs that are indistinguishable from π is denoted by $obeq(\pi)$. Note that $\pi \in obeq(\pi)$.

To illustrate, consider the following three runs

$$\pi_1 = (p_1 p_2 x p_3 p_4 y)^\omega$$
$$\pi_2 = (p_1 p_2 p_3 p_4 x y)^\omega$$
$$\pi_3 = (p_3 p_4 p_1 p_2 x y)^\omega$$

where $P_x = \{p_1, p_2\}$ and $P_y = \{p_3, p_4\}$. For notational simplicity we use p_i to denote the ready transition $p_i.I$ where $I = \{x \in I_M \,|\, p_i \in P_x\}$, i.e., p_i readies for all interactions of which it is a member. In this example, $\pi_2 \rightsquigarrow \pi_1$ because π_2 can be obtained from π_1 by moving each occurrence of $p_3 p_4$ ahead of x. Also, $\pi_3 \rightsquigarrow \pi_2$ because π_3 differs from π_2 only in the permutation of $p_1 p_2 p_3 p_4$. Note that $\pi_1 \not\rightsquigarrow \pi_2$ and $\pi_3 \not\rightsquigarrow \pi_1$. So indistinctness relation is neither symmetric nor transitive.

Moreover, if $\rho \rightsquigarrow \pi$ then the two runs must be equivalent. The converse may not necessarily hold, however. This can be illustrated by the above example where π_1 and π_3 are equivalent but they are not indistinguishable from each other. Thus, indistinctness is strictly stronger than the equivalence relation defined earlier by permuting independent actions.

Intuitively, $\rho \rightsquigarrow \pi$ means that any process or scheduler in an interaction system which locally observes π as the system's computation (run) cannot tell whether π or ρ is the actual computation of the system (i.e., a computation observed externally). As a result, if π is fair, then so is ρ. Otherwise, fairness of the system cannot be guaranteed. Thus, the fairness implementability criterion requires that for every run π generated by a nonpreemptive scheduler S, all runs in $obeq(\pi)$ be fair. This is stated formally in the following theorem.

Theorem 6 (Fairness Implementability Criterion). *A fairness notion F is implementable for M iff there exists a nonpreemptive scheduler S such that for every adversary A, $obeq(r(S,A)) \subseteq F(M)$.*

Proof sketch. The only-if direction can be proved by defining a nonpreemptive scheduler S to extract the interaction scheduling of Σ_M that implements F. The adversaries correspond to all possible interleavings of the process's state transitions from idle to ready.

For the if-direction of the proof, suppose there exists a nonpreemptive scheduler S such that for every adversary A, $obeq(r(S,A)) \subseteq F(M)$. We present a wait-free scheduling program $Simulate(S)$ which employs a coordinator to simulate the behavior of S (see Fig. 1). Like S, the coordinator proceeds in rounds. In each round, it first waits for idle processes to inform it of their readiness. Each process p is required to send a message $ready(p, I)$ to the coordinator when it is ready for the set I of interactions.

When the coordinator learns that some interaction has been enabled, it initiates a querying procedure, attempting to confirm if the other processes which have not yet informed the coordinator of their readiness are indeed idle. To do so, the coordinator sends a query message to each of them and waits for the response. The querying procedure terminates if every queried process replies an *idle* message to the query indicating that the process was idle when it received the query. If some process responds with $ready(p, I)$, then the coordinator has to re-initiate a querying procedure. Note that the number of querying procedures re-initiated is no greater than the total number of processes in the system.

When the coordinator has finished its querying procedures, it determines an interaction for execution by simulating the scheduling of S. Let $ready(p_{i,1}, I_{i,1})$, ..., $ready(p_{i,k}, I_{i,k_i})$ be the sequence of ready messages the coordinator receives in this round. Then, in the simulation the coordinator assumes that the adversary A is giving the sequence of ready transitions $p_{i,1}.I_{i,1}, \ldots, p_{i,k_i}.I_{i,k_i}$ to S. Let x_i be the interaction chosen by S. Then the coordinator finishes this round by sending a message $commit(x_i)$ to inform each process in P_{x_i} to execute x_i. The commit messages are acknowledged by the receivers. Note that if some interaction is enabled, then S (and thus the coordinator) must schedule an interaction for execution. Hence, $Simulate(S)$ is wait-free.

To show that for every computation Π of $Simulate(S)$, $[\Pi]_M \in F(M)$, first consider an extreme case in which the coordinator observes the exact ordering of the idle processes' ready transitions. Let Ψ denote the computation rendered by this scenario. Clearly, $[\Psi]_M = r(S, A)$. So $[\Psi]_M \in F(M)$. Next, by a careful analysis of all possible interleavings of independent actions of Ψ, we can show that for every computation Π that can actually occur in the system (w.r.t. an external observer), $[\Pi]_M \in obeq([\Psi]_M)$. So $[\Pi]_M \in F(M)$. The if-direction of the theorem is thus established. □

4 Properties of Implementable Fairness Notions

Recall that if $\rho \rightsquigarrow \pi$, then ρ and π must be equivalent. If equivalent runs are either all fair or all unfair, then $obeq(\pi)$ contains either all fair runs or all unfair runs. Therefore, if the fairness notion in consideration is also strongly-feasible, then by Theorem 6 it must be implementable. We thus have the following corollary.

Corollary 7. *If F is strongly-feasible and equivalence-robust for M, then F is implementable for M.*

Clearly, the above corollary does not exclude the possibility of a non-equivalence-robust (but strongly-feasible) fairness notion being implementable. Similarly, a fairness notion F may still be implementable even if $F(M)$ contains some π such that $obeq(\pi) \not\subseteq F(M)$. The trick is to find a nonpreemptive scheduler S that can avoid generating "odd" runs like π whose $obeq(\pi)$ contains an unfair run. This technique is used in the paper to prove the possibility result for WPF.

The code of each process p :

```
{    make a ready transition p.I  ⟶
         send a ready(p, I) message to the coordinator
□    receive a query message from the coordinator  ⟶
         if state is idle, reply to the query with a message idle;
         otherwise, ignore this query (because the process has already sent a ready
         message to the coordinator when it entered its ready state.)
□    receive commit(x) from the coordinator  ⟶
         acknowledge the message and then participate in x
}
```

The code of the coordinator :

variables :
new_ready: a queue of elements $p.I$ indicating that the coordinator
      has received message $ready(p, I)$ in current round;
yet_handled: a set of elements $p.I$ indicating that the coordinator has received
      message $ready(p, I)$ but has not yet scheduled any interaction for process p;
active: a flag indicating if the coordinator has learned that an interaction is enabled;
i: the number of rounds that have proceeded so far;

```
{    not active, receive message ready(p, I)  ⟶
         add p.I to new_ready and yet_handled;
         if some interaction has been enabled, then set active to true.
□    when active  ⟶
         while true do {
             let P = {p ∈ P_M | p.I ∉ yet_handled};
             send query messages to each process in P and wait for the responses;
             if all the queries are answered idle, then exit this while-loop;
                 otherwise, for each response that is of type ready(p, I)
                 add p.I to new_ready and yet_handled }
         assume that the adversary assigns new_ready as the ready transitions in round i;
         let x_i be the interaction chosen by S; then for each p ∈ P_{x_i} do {
             send commit(x_i) to p;
             delete the entry p.I from yet_handled; }
         wait for acknowledgments to the commit messages;
         new_ready := ∅; active := false; i := i + 1;
}
```

Fig. 1. The scheduling program $Simulate(S)$.

However, there are runs that cannot be avoided by any nonpreemptive scheduler. So if these runs happen to be "odd", then the fairness in question is not possible. Examples of runs that must be generated by every nonpreemptive scheduler are *singular* runs, which are defined below.

Definition 8. A run π is *singular* iff in every state of the run at most one interaction is enabled.

If a nonpreemptive scheduler faces a situation in which only one interaction is enabled, then by definition the scheduler must select it for execution. So every nonpreemptive scheduler for M must generate all singular runs of M.

Lemma 9. *If $\pi \in run(M)$ is singular, then for every nonpreemptive scheduler S of M, there exists an adversary A such that $\pi = r(S, A)$.*

Therefore, if F treats some run indistinguishable from a singular run as unfair, then by Theorem 6 F must not be implementable. This is stated in the following lemma, and, as we shall see in the following section, is very useful in proving fairness impossibilities.

Lemma 10. *If there exists a singular run $\pi \in run(M)$ such that $obeq(\pi) \not\subseteq F(M)$, then F is not implementable for M.*

5 Applications of the Criterion

In this section we use the proposed criterion to examine the implementability of existing fairness notions, including SIF, SPF, WPF, *U-fairness* [2] and *hyperfairness* [3]. Due to the space limitation, we present here only the impossibility result for SIF. The impossibility result for SPF can be established similarly. U-fairness and hyperfairness are not implementable because they are not even strongly-feasible. Conversely, WPF is implementable because we can show that if a run π satisfies SIF (π then must also satisfy WPF), then all runs in $obeq(\pi)$ satisfy WPF. Since SIF is strongly-feasible, by the criterion, WPF is implementable.

5.1 Strong Interaction Fairness

Using Lemma 10, we can establish an impossibility result for SIF. For intuition, observe that if a run π satisfies SIF but some run indistinguishable from π does not, then π must contain infinitely many segments of the form

$$y q_1 q_2 \cdots q_k$$

where $k \geq 1$, $q_1, q_2, \ldots q_k \notin P_y$, such that moving $q_1 q_2 \ldots q_k$ forward ahead of y (i.e., deferring y until q_k) causes some interaction x enabled immediately after q_k. This implies that $P_y \cap P_x \neq \emptyset$ and $\{q_1, q_2, \ldots, q_k\} \cap P_x \neq \emptyset$. Furthermore, x must be enabled only a finite number of times in π.

According to the segment $y q_1 q_2 \ldots q_k$, some process in $\{q_1, q_2, \ldots, q_k\} \cap P_x$ is idle before the execution of y, and becomes ready when y is executed. That is, some process in $\{q_1, q_2, \ldots, q_k\} \cap P_x$ must have executed some interaction z and then remained idle before the execution of y. This implies that $P_x \cap P_z \not\subseteq P_y$, and π is of the form

$$\cdots \cdots y q_1 q_2 \cdots q_k \cdots z \cdots y q_1 q_2 \cdots q_k \cdots z \cdots$$

Moreover, there must exist a process in $P_x \cap P_y$ which, after participating in y, remains idle before the next execution of z. Otherwise, the processes in $P_x \cap P_y$ and $\{q_1, q_2, \ldots, q_k\}$ would all be ready before the next execution of z, and so x would be enabled infinitely often. So, $P_x \cap P_y \not\subseteq P_z$. Hence, if the structure of

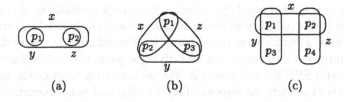

Fig. 2. Structures of interaction systems for which SIF is not possible.

the underlying interaction system can satisfy the above conditions and a run like π is inevitable (e.g., π is singular) to every nonpreemptive scheduler, then SIF would be impossible for the system.

Such a setting is indeed possible, as can be illustrated by the system depicted in Fig. 2(c). Let $\pi = (p_1 p_3 y p_2 p_4 z)^\omega$, and $\rho = (p_1 p_3 p_2 y p_4 z)^\omega$. Then, π satisfies SIF and is singular, and $\rho \rightsquigarrow \pi$. However, ρ does not satisfy SIF because x is enabled each time p_2 is ready but x is never executed. So by Lemma 10 SIF is not possible for the system. The following theorem characterizes the system structure for which SIF is not possible.

Theorem 11. *Let* $M = (P_M, I_M)$. *Assume* $\exists x, y, z$, $P_x \cap P_y \not\subseteq P_z$ *and* $P_x \cap P_z \not\subseteq P_y$. *Then SIF cannot be implemented for* M.

According to the above theorem, if interactions can involve only a single process, then Fig. 2(a) is the smallest system for which SIF is not possible. Otherwise, the one shown in Fig. 2(b) would be the smallest. In either case, SIF is not possible for both biparty and multiparty interactions.

Conversely, SIF can be implemented for systems where either no two interactions y and z conflict with a third interaction x, or if they do then it must be the case that $P_y \cap P_x \subseteq P_z$ or $P_z \cap P_x \subseteq P_y$. This can be proved by presenting a nonpreemptive scheduler S such that for every π generated by S the runs in $obeq(\pi)$ are fair.

Theorem 12. *Let* $M = (P_M, I_M)$. *Assume (1)* $|I_M| \leq 2$, *or (2)* $\forall x, y, z \in I_M$, *if* $P_x \cap P_y \neq \emptyset$ *and* $P_x \cap P_z \neq \emptyset$, *then either* $P_x \cap P_y \subseteq P_z$ *or* $P_x \cap P_z \subseteq P_y$. *Then SIF can be implemented for* M.

6 Concluding Remarks

Numerous algorithms have been proposed for scheduling multiparty interactions, e.g., [16, 5, 15, 12, 17, 10]. From our results, it is not surprising to see that only weak interaction fairness[4] has been widely implemented in these algorithms.

Our impossibility results for SIF and SPF improve upon previous results by Tsay and Bagrodia [18] and by Joung [9] in three ways: Firstly, our results do

---

[4] *Weak interaction fairness* (WIF) means that if an interaction is continually enabled, then some of its participants will eventually engage in an interaction. As is well-known, WIF is much weaker than most fairness notions.

not depend on any system topology underlying the implementation. By contrast, each process in [18] is paired with a coordinating process to schedule interactions, while [9] assumes a centralized coordinator for the scheduling. Secondly, they establish the impossibility results by identifying a particular set of interactions for which SIF and SPF are not possible. We are able to determine the structure of interactions that renders the impossibility. Finally, and most importantly, they observe the impossibility phenomena in a specific implementation model. We, however, have generalized the model and lifted its properties to the semantic level. Thus, our criterion is applicable to *all* possible fairness notions for multiparty interaction.

Acknowledgments. We thank Reino Kurki-Suonio for comments on an earlier version of this paper, and the referees for comments on the presentation.

References

1. K.R. Apt, N. Francez, and S. Katz. Appraising fairness in languages for distributed programming. *Distributed Computing*, 2(4):226–241, 1988.
2. P.C. Attie, I.R. Forman, and E. Levy. On fairness as an abstraction for the design of distributed systems. In *Proc. of the 10th ICDCS*, pp. 150–157, 1990.
3. P.C. Attie, N. Francez, and O. Grumberg. Fairness and hyperfairness in multiparty interactions. *Distributed Computing*, 6:245–254, 1993.
4. R.J.R. Back and R. Kurki-Suonio. Distributed cooperation with action systems. *ACM TOPLAS*, 10(4):513–554, Oct. 1988.
5. R. Bagrodia. Process synchronization: Design and performance evaluation of distributed algorithms. *IEEE TSE*, SE-15(9):1053–1065, 1989.
6. N. Francez and I.R. Forman. *Interacting Processes: A Multiparty Approach to Coordinated Distributed Programming*. Addison Wesley, 1995.
7. N. Francez, B. Hailpern, and G. Taubenfeld. Script: A communication abstraction mechanism. *Science of Computer Programming*, 6(1):35–88, Jan. 1986.
8. C.A.R. Hoare. Communicating sequential processes. *Communications of the ACM*, 21(8):666–677, Aug. 1978.
9. Y.-J. Joung. *On the Design and Implementation of Multiparty Interaction*. PhD thesis, Dept. of Computer Science, SUNY at Stony Brook, May 1992.
10. Y.-J. Joung and S.A. Smolka. Coordinating first-order multiparty interactions. *ACM TOPLAS*, 16(3), May 1994.
11. Y.-J. Joung and S.A. Smolka. A comprehensive study of the complexity of multiparty interaction. *Journal of the ACM*. To appear.
12. D. Kumar. An implementation of N-party synchronization using tokens. In *Proc. of the 10th ICDCS*, pp. 320–327, 1990.
13. E.R. Olderog and K.R. Apt. Fairness in parallel programs: The transformational approach. *ACM TOPLAS*, 10(3):420–455, July 1988.
14. S. Owicki and L. Lamport. Proving liveness properties of concurrent programs. *ACM TOPLAS*, 4(3):455–495, 1982.
15. M.H. Park and M. Kim. A distributed synchronization scheme for fair multiprocess handshakes. *Information Processing Letters*, 34:131–138, Apr. 1990.
16. S. Ramesh. A new and efficient implementation of multiprocess synchronization. In *Proc. Conf. on PARLE, LNCS 259*, pp. 387–401, 1987.
17. Y.-K. Tsay and R.L. Bagrodia. A real-time algorithm for fair interprocess synchronization. In *Proc. of the 12th ICDCS*, pp. 716–723, 1992.
18. Y.-K. Tsay and R.L. Bagrodia. Some impossibility results in interprocess synchronization. *Distributed Computing*, 6(4):221–231, 1993.

Termination of Context-Sensitive Rewriting by Rewriting*

Salvador Lucas

Departamento de Sistemas Informáticos y Computación
Universidad Politécnica de Valencia
Camino de Vera s/n, E-46071 Valencia, Spain.
e.mail: slucas@dsic.upv.es

Abstract. Context-sensitive rewriting allows us to deal with certain context-replacing restrictions in performing reduction steps. This is useful in avoiding undesirable reductions. In this paper, we study termination of context-sensitive rewriting. We give some conditions under which it is possible to analyze the termination of the context-sensitive rewrite relation induced by a term rewriting system by means of any method to prove termination of rewriting. This allows us to apply all known (and forthcoming) results on termination to context-sensitive rewriting.

Keywords: term rewriting, functional programming, termination.

1 Introduction

Term Rewriting is traditionally considered the operational semantics of functional programming languages [3, 16]. A functional program is a set of equations that are interpreted as left-to-right rewrite rules [5, 15]. Execution of the program for any input data consists of the evaluation of the term using the rewrite rules until it cannot be further reduced (i.e. until a *normal form* is reached). In a terminating system, no infinite computations are possible.

In context-sensitive rewriting [12, 13], for each symbol in the signature Σ we fix the set of *replacing* arguments by means of a mapping $\mu : \Sigma \to \wp(I\!\!N)$ (the *replacement map*) which specifies the set of positions which can eventually be reduced. This is useful to express some requirements on the 'intended meaning' of a function. For example, consider the standard functional definition of the *if-then-else* operation:

$if(true, x, y) \to x$
$if(false, x, y) \to y$

Given an input $if(cond, t, s)$, we are not interested in reductions on the second and third arguments of the operation if until the condition $cond$ has been evaluated. By defining a replacement map such as $\mu(if) = \{1\}$, the undesirable reductions on the second and third immediate subterms of the term $if(cond, t, s)$ can be avoided with no extra control.

* This work has been partially supported by CICYT under grant TIC 95-0433-C03-03.

Context-sensitive rewriting has some strong connections with the lazy strategies of functional programming languages. Actually, the mechanism which allows us to prevent the evaluation of some expressions can also be used to declare that these reductions need not to be done. Although giving support to lazy reduction techniques is not the main motivation of our work, context-sensitive rewriting does allow us to express fixed, meaningful and easily implementable restrictions on rewriting, similarly to the way that lazy reduction techniques approximate 'needed' reductions à la Huet and Lévy [10]. An approach to these questions which is related with our work can be found in [9].

In this paper, we address the problem of the termination of context-sensitive rewriting by using any method to prove that a TRS terminates. Termination of rewriting is normally proved by defining a noetherian ordering $>$ on terms and checking whether $t > s$ whenever $t \to s$ (Manna and Ness' theorem, [14]). This is equivalent to proving that each instance of every rule $l \to r$ in the TRS satisfies $l > r$ in a monotonic noetherian ordering $>$ (Kamin and Lévy, [8]). As suggested by Manna and Ness, '*it is often convenient to separate reduction orderings into a homomorphism from terms to an algebra with a noetherian ordering*'. This can be achieved by means of *termination functions* [2, 4]. A termination function is a *homomorphism* from terms to a Σ-algebra (where Σ is the underlying signature). It can also be considered as an *extraction function* from terms to multisets of selected immediate subterms (see [4]). We show that we can often prove termination of context-sensitive rewriting by first transforming the original TRS into a *contracted* TRS on a new signature by means of a *contracting* function. This function can be thought of as a *pseudo*-termination function in that it *projects* the original set of terms onto a new set of *contracted* terms on which we can use the standard methods.

The paper is organized as follows. In Section 2, we briefly recall the technical concepts used in the remainder of the paper. In Section 3, we formulate some basic properties of context-sensitive rewriting. Section 4 provides two basic criteria to ensure termination of context-sensitive rewriting. Section 5 concludes.

2   Preliminaries

Let us first introduce the main notations used in the paper from [5, 11]. V denotes a countable infinite set of variables and Σ denotes a set of function symbols {f, g, ...}, each with a fixed arity given by a function $ar : \Sigma \to I\!N$. By $T(\Sigma, V)$ we denote the set of terms. $Var(t)$ is the set of variable symbols in the term t.

Terms are viewed as labelled trees in the usual way. Occurrences $u, v, \ldots$ are represented by chains of positive natural numbers used to address subterms of t. Occurrences are ordered by the standard prefix ordering: $u \leq v$ iff there is a chain v' such that $v = u.v'$. $O(t)$ denotes the set of occurrences of a term t. $t|_u$ is the subterm at occurrence u of t. $t[s]_u$ is the term t with the subterm at the occurrence u replaced with s. We refer to any term C that is the same as t everywhere except below u, i.e. there exists a term s such that $C[s]_u = t$, as the *context* within the replacement occurs.

A rewrite rule is an ordered pair (l, r), written $l \rightarrow r$, with $l, r \in T(\Sigma, V)$, $l \notin V$ and $Var(r) \subseteq Var(l)$. l is said to be the left-hand side (*lhs*) of the rule and r is the right-hand side (*rhs*). A TRS is a pair $\mathcal{R} = (\Sigma, R)$ where R is a set of rewrite rules. For a given TRS $\mathcal{R} = (\Sigma, R)$, a term t rewrites to a term s (at the occurrence u), written $t \rightarrow_{\mathcal{R}} s$, if $t|_u = \sigma(l)$ and $s = t[\sigma(r)]_u$, for some rule $l \rightarrow r$ in R, occurrence u in t and substitution σ. $\rightarrow_{\mathcal{R}}$ is the one-step rewrite relation for $\mathcal{R}$.

$\mathbb{N}_k^+$ is an initial segment $\{1, 2, \ldots k\}$ of the set of positive natural numbers $\mathbb{N}^+$, where $\mathbb{N}_0^+ = \emptyset$. We let R denote a binary relation on A, i.e., $R \subseteq A \times A$. We will often write $x R y$ for $(x, y) \in R$. We let R^* denote the reflexive-transitive closure of R. A relation R is *noetherian* (or *terminating*) iff there is no infinite sequence $a_1 R a_2 \cdots a_n R a_{n+1} \cdots$.

Given a set A, we denote a k-tuple $x_1, \ldots, x_k$ of elements as $\tilde{x}$, i.e. $\tilde{x} \in A^k$. By $\tilde{x}[y]_i$ for any $i, 1 \leq i \leq k$ we denote a k-tuple where x_i is replaced by y in $\tilde{x}$. Given a k-tuple $\tilde{x}$ and a set of positive natural numbers $J = \{j_1, \ldots, j_q\} \subseteq \mathbb{N}_k^+$, denote as $\tilde{x}_J$ the q-ary tuple of elements in $\tilde{x}$ *extracted* by J, i.e. $\tilde{x}_J = x_{j_1}, \ldots, x_{j_q}$. Note that the i-th component x_i of $\tilde{x}$ is the $|\mathbb{N}_i^+ \cap J|$-th component of $\tilde{x}_J$. Given a set B, a function $\psi : A \rightarrow B$ and a k-ary tuple $\tilde{x} \in A^k$, we denote as $\psi(\tilde{x})$ the k-ary tuple of transformed-by-ψ elements, i.e. $\psi(x_1), \ldots, \psi(x_k)$.

3 Context-sensitive rewriting

3.1 Basic definitions of context-sensitive rewriting

The idea behind the concept of context-sensitive rewriting [12, 13] is to impose a syntactic replacement condition which prevents some reductions. This is achieved by using a map on the signature we call replacement map.

Definition 3.1 (Replacement map) *Let Σ be a signature. A mapping μ : $\Sigma \rightarrow \wp(\mathbb{N})$ is a replacement map (or Σ-map) for the signature Σ iff for all $f \in \Sigma$. $\mu(f) \subseteq \mathcal{N}_{ar(f)}^+$.*

The ordering $\subseteq$ on $\wp(\mathbb{N})$ extends to an ordering $\sqsubseteq$ on M_Σ, the set of all Σ-maps: $\mu \sqsubseteq \mu'$ iff $\mu(f) \subseteq \mu'(f)$ for all $f \in \Sigma$. It is immediate to see that $(M_\Sigma, \sqsubseteq, \mu_\perp, \mu_\top)$ is a complete lattice where $\mu_\perp(f) = \emptyset$ and $\mu_\top(f) = \mathbb{N}_{ar(f)}^+$ for all $f \in \Sigma$.

The replacement map determines the argument positions which can be reduced for each symbol on the signature. The replacement condition indicates the occurrences which can be rewritten.

Definition 3.2 (Replacement condition) *Let Σ be a signature and μ be a Σ-map. Let $t \in T(\Sigma, V)$ be a term. The replacement condition is a relation $\gamma_{\mu,t}$ defined on the set of occurrences $O(t)$ as follows:*

$\gamma_{\mu,t}(\varepsilon)$.

$\gamma_{\mu, f(\tilde{t}[s]_i)}(i.u) \Leftrightarrow (i \in \mu(f)) \wedge \gamma_{\mu, s}(u)$.

We say that the occurrence u of a term t satisfies the replacement condition $\gamma_{\mu,t}$ iff $\gamma_{\mu,t}(u)$. We also say that u is a replacing occurrence of t or that the subterm $t|_u$ is a replacing subterm.

We write $\gamma_t(u)$ when the replacement map is clear from the context. Some new notation entailed by the concept of replacement condition: $O^\mu(t)$ is the set of *replacing* occurrences of a term t: $u \in O^\mu(t) \Leftrightarrow u \in O(t) \wedge \gamma_t(u)$. $O_s^\mu(t)$ is the (possibly empty) set of replacing occurrences of a subterm s of t: $u \in O_s^\mu(t) \Leftrightarrow u \in O^\mu(t) \wedge t|_u = s$. $Var^\mu(t)$ is the set of *replacing variables* of t: $x \in Var^\mu(t) \Leftrightarrow O_x^\mu(t) \neq \emptyset$.

Definition 3.3 (One-step context-sensitive rewrite relation) *Let* $\mathcal{R} = (\Sigma, R)$ *be a TRS and* μ *be a* Σ*-map. A term* t μ*-rewrites to a term* s*, written* $t \hookrightarrow_{\mathcal{R}(\mu)} s$*, if* $t \rightarrow_\mathcal{R} s$ *at the replacing occurrence* $u \in O^\mu(t)$*.* $\hookrightarrow_{\mathcal{R}(\mu)}$ *is the one-step context-sensitive rewrite relation of* $\mathcal{R}$ *wrt* μ*.* $\hookrightarrow_{\mathcal{R}(\mu)}^*$ *is the context-sensitive rewrite relation of* $\mathcal{R}$ *wrt* μ*.*

In the sequel, we assume that the TRS $\mathcal{R} = (\Sigma, R)$ and Σ-map μ are given, and we drop references to them by just writing $\hookrightarrow$ instead of $\hookrightarrow_{\mathcal{R}(\mu)}$ to denote the one-step context-sensitive rewrite relation.

Example 3.4 *Let* $\rho : f(x) \rightarrow x$ *be a rule and* $t = g(f(a), f(b))$ *be a term.*
a) If $\mu(g) = \emptyset$*, then* ρ *cannot be applied to reduce* t*, because neither the occurrence* $u = 1$ *nor* $u = 2$ *satisfy the replacement condition wrt* μ*.*
b) If $\mu(g) = \{1\}$*, then* ρ *can be applied to reduce* $f(a)$*, but it cannot be applied to reduce* $f(b)$*.*
c) If $\mu(g) = \{2\}$*, then* ρ *can be applied to reduce* $f(b)$*, but it cannot be applied to reduce* $f(a)$*.*
d) If $\mu(g) = \{1, 2\}$*, then* ρ *can be applied to reduce both* $f(a)$ *and* $f(b)$*.*

By analogy to standard rewriting, context-sensitive is closed under *replacing* context application, i.e., $t \hookrightarrow s \Rightarrow C[t]_u \hookrightarrow C[s]_u$ for every $u \in O^\mu(C)$.

Remark 3.5 *Note that with the top* Σ*-map* μ_T *the replacement condition is satisfied for every term* t *and occurrence* $u \in O(t)$*. This is to say that, given a TRS* $\mathcal{R}$*, the one-step context-sensitive rewrite relation for* μ_T *coincides with the standard one-step rewrite relation, i.e.* $\hookrightarrow_{\mathcal{R}(\mu_T)} = \rightarrow_\mathcal{R}$*.*

Given a TRS, the following proposition relates the one-step context-sensitive rewrite relations induced by comparable replacement maps.

Proposition 3.6 (Monotonicity of $\hookrightarrow$ **with respect to** $\sqsubseteq$**)** *Let* $\mathcal{R} = (\Sigma, R)$ *be a TRS and* μ, μ' *be* Σ*-maps. Then* $\mu \sqsubseteq \mu' \Rightarrow \hookrightarrow_{\mathcal{R}(\mu)} \subseteq \hookrightarrow_{\mathcal{R}(\mu')}$*.*

3.2 Examples of application

Improving the evaluation of expressions.

The rules which define the *short-cut* boolean operators *and*, *or* of Lisp are:
$$and(true, x) \rightarrow x \qquad or(true, x) \rightarrow true$$
$$and(false, x) \rightarrow false \qquad or(false, x) \rightarrow x$$

It is sensible to evaluate up to the normal form the first argument of an input term $and(t, s)$ $(or(t, s))$, since the outcome is necessary to determine which rule has to be applied to the root. According to the rules, if the first argument reduces to *false* (*true*), then any reduction on the second argument is useless, since it does not contribute to computing the value of the resulting expression.

A replacement map $\mu(and) = \mu(or) = \{1\}$, has the effect of delaying the evaluation of redexes which are different from the first argument of the term.

In manipulating lists, another case of study arises. Let us consider the *head* and *tail* functions, whose behavior is as follows:

$head(x :: L) \rightarrow x$

$tail(x :: L) \rightarrow L$

head only requires the evaluation of the head of the list and *tail* only evaluates the rest of the list. This points are only of interest if *cons* (i.e. '::') does not evaluate its arguments systematically. Programming Languages have been conceived where *cons* does not evaluate its arguments (or, at least it does not evaluate its second argument) (see [6, 11]). This behavior can easily be achieved in context-sensitive rewriting, by defining $\mu(::) = \emptyset$ (or $\mu(::) = \{1\}$).

Mechanizing Inference Systems.

Context-sensitive rewriting can be formulated as an Inference System by considering each rewrite rule $l \rightarrow r$ as a (scheme of) axiom: $\frac{}{l \rightarrow r}$, and introducing as many context-passing rules [11] as replacing indices $i \in \mu(f)$ for each k-ary function symbol 'f' in the signature. That is, for each *replacing* argument position i, we define the (scheme of) rule $\frac{t \rightarrow s}{f(t_1,...,t_{i-1},t,...,t_k) \rightarrow f(t_1,...,t_{i-1},s,...,t_k)}$.

Different computational systems have been formalized by using Inference Systems whose axioms and rules match these patterns, as they are weak β-reduction, call-by-name reduction strategies of λ-calculus, and π-calculus. Therefore, we can use context-sensitive rewriting as a suitable mechanization of the execution of processes of these systems. In this way, we can exploit existing rewriting devices (reduction machines, graph reduction, etc.) for efficient implementations [7, 15].

4 μ-termination

We would like to compare the termination behavior of the standard rewriting in a given TRS $\mathcal{R}$ and the context-sensitive rewriting in $\mathcal{R}$ given a replacement map. To distinguish them, we introduce the following preparatory definition.

Definition 4.1 (μ-termination) *Let $\mathcal{R} = (\Sigma, R)$ be a TRS and μ be a Σ-map. We say that $\mathcal{R}$ μ-terminates (or $\mathcal{R}$ is μ-terminating) iff $\hookrightarrow_{\mathcal{R}(\mu)}$ is noetherian.*

Proposition 4.2 *Let $\mathcal{R} = (\Sigma, R)$ be a TRS. Let μ, μ' be two Σ-maps such that $\mu \sqsubseteq \mu'$. If $\mathcal{R}$ μ'-terminates, then $\mathcal{R}$ μ-terminates.*

PROOF. By contradiction, using Proposition 3.6. □

Now we consider the Remark 3.5 to show that any criteria establishing the termination of a TRS $\mathcal{R}$ also establishes the μ-termination of $\mathcal{R}$.

Theorem 4.3 *Let $\mathcal{R} = (\Sigma, R)$ be a TRS and μ be a Σ-map. If $\mathcal{R}$ terminates, then $\mathcal{R}$ μ-terminates.*

PROOF. Immediate from Proposition 4.2, since $\to_{\mathcal{R}} = \hookrightarrow_{\mathcal{R}(\mu_T)}$. $\qquad\qquad\square$

From Theorem 4.3, we can directly use any ordering $>$ on $\mathcal{T}(\Sigma, V)$ (for example the recursive path ordering $>^*$ in [1, 2]) to check μ-termination of the context-sensitive rewriting by just considering the rules in the TRS (without taking into account the replacement map). Actually we only need to prove that $\sigma(l) > \sigma(r)$ for all rules $l \to r$ in the TRS and substitutions σ. If this comparison succeeds, then Theorem 4.3 ensures that any context-sensitive rewriting will terminate. We refer to this method as the $\mathcal{C}_{\mathcal{R}}$-*criterion* to check μ-termination. Nevertheless, we can consider many μ-terminating TRSs despite the fact that the $\mathcal{C}_{\mathcal{R}}$-test does not succeed.

Example 4.4 *Let us consider the nonterminating TRS $\mathcal{R} = (\{f, g\}, \{f(g(x), y) \to f(x, f(g(x), y))\})$. If we take μ such that $\mu(f) = \mu(g) = \{1\}$, then it is easy to verify that any μ-rewriting on $\mathcal{R}$ will eventually terminate.*

The example shows that the $\mathcal{C}_{\mathcal{R}}$-criterion is not complete. It also illustrates that, when comparing the *lhs* and *rhs* of a rewrite rule $l \to r$ in order to determine μ-termination of the TRS, we only need to take into account the *replacing* arguments occurring in l and r. Roughly speaking, since $\mu(f) = \{1\}$ we would find $l > r$, i.e. $f(g(x), y) > f(x, f(g(x), y))$ if we forget the second immediate argument both in $f(g(x), y)$ and $f(x, f(g(x), y))$. This is similar to checking whether $f(g(x), \square) > f(x, \square)$ (where $\square$ expresses the 'unnecessary' arguments, that will not be taken into account in order to prove $l > r$). This idea will be formalized in the following sections.

4.1 The contractive transformation

In the following section, we are able to state an interesting result: we can investigate μ-termination of a TRS by studying termination of a contracted TRS built using a new signature we call μ-*contracted signature* which we define as follows.

Definition 4.5 (μ-contracted signature) *Let Σ be a signature and μ be a Σ-map. The μ-contracted signature Σ^μ is as follows: $f_\mu \in \Sigma^\mu \wedge ar(f_\mu) = |\mu(f)| \Leftrightarrow f \in \Sigma$.*

We connect terms from both signatures Σ and Σ^μ by means a function.

Definition 4.6 (μ-contracting function) *Let Σ be a signature and μ be a Σ-map. The μ-contracting function $\tau_\mu : \mathcal{T}(\Sigma, V) \to \mathcal{T}(\Sigma^\mu, V)$ is as follows:*

$- \tau_\mu(x) = x$ *if $x \in V$*

$$- \tau_\mu(f(\tilde{t})) = f_\mu(\tau_\mu(\tilde{t}_{\mu(f)}))$$

Note that this function *drops* the non-replacing immediate subterms of a term t and constructs a 'μ-contracted' term by joining the (also transformed) replacing subterms below the corresponding operator of the μ-contracted signature. See below for an illustrative example.

We can observe that this is not a termination function (see [4]) since it is neither a *homomorphism* (the operations in the contracted signature could have a lesser arity than the corresponding operation in the original one) nor an *extraction function* (because (i) the extracted-by-μ subterms are joined by a new function symbol of the contracted signature and (ii) the subterms are also transformed by the contracting function). Nevertheless, it deals with a task similar to termination functions, i.e., translating the original set of terms $T(\Sigma, V)$ into a set (in this case $T(\Sigma^\mu, V)$) on which we can demonstrate the terminating behavior of context-sensitive rewriting.

Example 4.7 *(Continued from Example 4.4) We can transform the TRS under consideration into a 'contracted' TRS $\mathcal{R}^\mu$ by applying τ_μ. This gives: $\mathcal{R}^\mu = (\{f_\mu, g_\mu\}, \{f_\mu(g_\mu(x)) \rightarrow f_\mu(x)\})$.*

The next proposition shows how the μ-contracting function changes the location of the immediate subterms.

Proposition 4.8 *Let Σ be a signature and μ be a Σ-map. If $i \in \mu(f)$, then we have $\tau_\mu(f(\tilde{t}[s]_i)) = f_\mu(\tau_\mu(\tilde{t}_{\mu(f)}[s]_j))$ with $j = |\mathcal{N}_i^+ \cap \mu(f)|$.*

PROOF. Immediate from definition of τ_μ. □

From this proposition, we know that the i-th immediate (replacing) subterm of $t = f(\tilde{t})$ becomes the $|\mathcal{N}_i^+ \cap \mu(f)|$-th immediate subterm of $\tau_\mu(t)$.

Proposition 4.9 (Surjectivity of τ_μ) *The μ-contracting function τ_μ is surjective. That is, for each $s \in T(\Sigma^\mu, V)$, there exists $t \in T(\Sigma, V)$ such that $s = \tau_\mu(t)$.*

PROOF. By structural induction on $s \in T(\Sigma^\mu, V)$.

1. Base case. If $s = x \in V$, define $t = x$. If s is a constant f_μ, let f be the corresponding symbol of Σ. Then we consider the following two cases:
   (a) $ar(f) = 0$. Define $t = f$.
   (b) $ar(f) > 0$. By definition of τ_μ, this means that $\mu(f) = \emptyset$. Then, we let $t = f(\tilde{t})$ for any $\tilde{t} \in T(\Sigma, V)^{ar(f)}$.
2. Let $s = f_\mu(\tilde{s})$, with $\tilde{s} \in T(\Sigma^\mu, V)^{ar(f_\mu)}$. By I.H., there exists $\tilde{t}' \in T(\Sigma, V)^{ar(f_\mu)}$ s.t. $\tilde{s} = \tau_\mu(\tilde{t}')$. Define $t = f(\tilde{t})$, for some $\tilde{t} \in T(\Sigma, V)^{ar(f)}$ s.t. $\tilde{t}_{\mu(f)} = \tilde{t}'$.

□

This proposition allows us to uniquely consider any term $s \in T(\Sigma^\mu, V)$ as the image by τ_μ of some term $t \in T(\Sigma, V)$. Therefore, we often write $\tau_\mu(t)$ instead of s when $s = \tau_\mu(t)$.

We need to establish some properties of the contracting function which will be used later. In particular, we are interested in relating the structure of a term in $T(\Sigma, V)$ and the corresponding term (by τ_μ) in $T(\Sigma^\mu, V)$. We introduce the following definition.

Definition 4.10 *Let Σ be a term and μ be a Σ-map. Let $t \in T(\Sigma, V)$. The function $\nu_{\mu,t} : O(t) \rightarrow (\mathbb{N} - \{0\})^*$ is as follows:*

- $\nu_{\mu,t}(\epsilon) = \epsilon$
- $\nu_{\mu,f(\tilde{t}[s]_i)}(i.u) = \epsilon$ *if $i \notin \mu(f)$*
- $\nu_{\mu,f(\tilde{t}[s]_i)}(i.u) = |\mathcal{N}_i^+ \cap \mu(f)|.\nu_{\mu,s}(u)$ *if $i \in \mu(f)$*

We will write just ν_t instead of $\nu_{\mu,t}$ when the replacement map μ is clear from the context. This function allows us to obtain the set of occurrences of a contracted term from the occurrences of the original term.

Proposition 4.11 *Let Σ be a signature, μ be a Σ-map and τ_μ the μ-contracting function. Let $t \in T(\Sigma, V)$ and $u \in O(t)$. Then we have $O(\tau_\mu(t)) = \nu_t(O(t))$.*

An interesting fact concerning the contracting function is related to the preservation of the replacing subterms.

Proposition 4.12 (τ_μ preserves the replacing subterms) *Let Σ be a signature, μ be a Σ-map and τ_μ the μ-contracting function. Let $t \in T(\Sigma, V)$ and $u \in O^\mu(t)$. Assume that $z = t|_u$. Then $\tau_\mu(z) = \tau_\mu(t)|_v$, where $v \in O(\tau_\mu(t))$ satisfies that $v = \nu_t(u)$.*

PROOF. Structural induction. The base case is immediate. For the induction step, assume $t = f(\tilde{t}[z]_{i.u'})$, where $u = i.u'$. Then, we can apply the induction hypothesis to the i-th immediate subterm s of t, since $z = s|_{u'}$. Therefore, $\tau_\mu(z) = \tau_\mu(s)|_{v'}$, where $v' = \nu_s(u')$. Since $i.u' \in O^\mu(t)$, $i \in \mu(f)$ and by Proposition 4.8, we get $\tau_\mu(t) = \tau_\mu(f(\tilde{t}[s]_i)) = f_\mu(\tau_\mu(\tilde{t}_{\mu(f)}[s]_j))$, with $j = |\mathcal{N}_i^+ \cap \mu(f)|$. Therefore, the subterm $\tau_\mu(z)$ occurs at position $j.v'$ in $\tau_\mu(t)$. By Definition 4.10, we have $\nu_t(i.u') = j.\nu_s(u') = j.v' = v$, and the conclusion follows. □

Corollary 4.13 (τ_μ preserves the replacing variables) *Let Σ be a signature, μ be a Σ-map and τ_μ the μ-contracting function. Let $t \in T(\Sigma, V)$. If $x \in Var^\mu(t)$, then $x \in Var(\tau_\mu(t))$.*

Proposition 4.14 *Let Σ be a signature, μ be a Σ-map and τ_μ the μ-contracting function. Let $t, s \in T(\Sigma, V)$. If $Var^\mu(t) \subseteq Var^\mu(s)$, then $Var(\tau_\mu(t)) \subseteq Var(\tau_\mu(s))$.*

PROOF. Let us take $x \in Var^\mu(t)$. By Corollary 4.13, $x \in Var(\tau_\mu(t))$. Since $Var^\mu(t) \subseteq Var^\mu(s)$, we also have $x \in Var(\tau_\mu(s))$, which proves the claim. □

In the following proposition, we prove that the substitution application commutes with the μ-contracting function.

Proposition 4.15 *Let Σ be a signature, μ be a Σ-map and τ_μ the μ-contracting function. Let $t \in T(\Sigma, V)$ and $\sigma : V \to T(\Sigma, V)$ a substitution. Let $\sigma_\mu : V \to T(\Sigma^\mu, V)$ such that $\sigma_\mu(x) = \tau_\mu(\sigma(x))$ for all $x \in V$. Then, we have $\tau_\mu(\sigma(t)) = \sigma_\mu(\tau_\mu(t))$.*

PROOF. By structural induction. If $t = x \in V$, then the result is immediate, since $\tau_\mu(x) = x$. For the induction step, we take $t = f(\tilde{t})$ for some $\tilde{t} \in T(\Sigma, V)^{ar(f)}$. By I.H., $\tau_\mu(\sigma(t)) = \tau_\mu(\sigma(f(\tilde{t}))) = \tau_\mu(f(\sigma(\tilde{t}))) = f_\mu(\tau_\mu(\sigma(\tilde{t}_{\mu(f)}))) = f_\mu(\sigma_\mu(\tau_\mu(\tilde{t}_{\mu(f)}))) = \sigma_\mu(f_\mu(\tau_\mu(\tilde{t}_{\mu(f)}))) = \sigma_\mu(\tau_\mu(t))$. □

4.2 μ-termination and contracted signature

The μ-contracting function allows us to prove μ-termination in context-sensitive rewriting by proving termination of the 'contracted' TRS using any method. However, there is one complication. The μ-contraction of the rewrite rules $l \to r$ of a TRS could lead to pairs $(\tau_\mu(l), \tau_\mu(r))$ which can not be considered as rewrite rules. For example: the rule $f(g(x), y) \to f(y, f(g(x), y))$ with $\mu(f) = \mu(g) = \{1\}$ yields (by μ-contracting) the 'rule' $f_\mu(g_\mu(x)) \to f_\mu(y)$, which does not satisfy the standard non-extra variable condition of rewriting rules.

We overcome this problem by imposing a syntactical restriction on the TRSs on which we want to analyse μ-termination.

Definition 4.16 (TRS with conservative replacing variables) *Let Σ be a signature and μ be a Σ-map. A TRS $R = (\Sigma, R)$ has conservative replacing variables if every replacing variable in the rhs of a rule in R is a replacing variable in the lhs of this rule. Formally: $Var^\mu(r) \subseteq Var^\mu(l)$ for every rule $l \to r \in R$.*

Definition 4.17 (μ-contracted TRS) *Let $R = (\Sigma, R)$ be a TRS and μ be a Σ-map. If R is a TRS with conservative replacing variables, we define the μ-contraction $R^\mu = (\Sigma^\mu, R^\mu)$ of R as follows:*

- Σ^μ *is the μ-contracted signature in Definition 4.5.*
- $R^\mu = \{\tau_\mu(l) \to \tau_\mu(r) \mid l \to r \in R\}$.

The following proposition establishes the correctness of Definition 4.17.

Proposition 4.18 *Let $R = (\Sigma, R)$ be a TRS and μ be a Σ-map. If R is a TRS with conservative replacing variables, then the μ-contraction $R^\mu = (\Sigma^\mu, R^\mu)$ of of R is a TRS.*

PROOF. For each rule $\tau_\mu(l) \to \tau_\mu(r) \in R^\mu$, we need to check, whether it verifies $\tau_\mu(l) \notin V$ and $Var(\tau_\mu(r)) \subseteq Var(\tau_\mu(l))$. The first condition is immediate from Definition 4.6 if R is a TRS. The second condition follows immediately from Definition 4.16 and Proposition 4.14. □

In order to connect μ-rewriting for a TRS and a replacement map and rewriting sequences using the contracted TRS, we introduce the following proposition.

Proposition 4.19 (μ-rewriting in $\mathcal{R}$ and rewriting in $\mathcal{R}^\mu$) *Let $\mathcal{R} = (\Sigma, R)$ be a TRS with conservative replacing variables and μ be a Σ-map. Let $\mathcal{R}^\mu = (\Sigma^\mu, R^\mu)$ be the contraction of $\mathcal{R}$. If $t \hookrightarrow_{\mathcal{R}(\mu)} s$, then $\tau_\mu(t) \to_{\mathcal{R}^\mu} \tau_\mu(s)$.*

PROOF. If $t \hookrightarrow_{\mathcal{R}(\mu)} s$, then there exists $u \in O^\mu(t)$, $l \to r \in R$ and substitution $\sigma : V \to \mathcal{T}(\Sigma, V)$ such that $t|_u = \sigma(l)$ and $s = t[\sigma(r)]_u$. Assume $v = \nu_t(u)$, and $\sigma_\mu : V \to \mathcal{T}(\Sigma^\mu, V)$ a substitution such that $\sigma_\mu(x) = \tau_\mu(\sigma(x))$ for all $x \in V$. By Proposition 4.12 and Proposition 4.15, we have that $\tau_\mu(t)|_v = \tau_\mu(\sigma(l)) = \sigma_\mu(\tau_\mu(l))$. By Definition 4.17, if $l \to r \in R$, then $\tau_\mu(l) \to \tau_\mu(r) \in R^\mu$. Therefore the term $\tau_\mu(t)$ can be rewritten at occurrence v by using rule $\tau_\mu(l) \to \tau_\mu(r) \in R^\mu$ and matching σ_μ. The rewritten term is $s' = \tau_\mu(t)[\sigma_\mu(\tau_\mu(r))]_v = \tau_\mu(t)[\tau_\mu(\sigma(r))]_v = \tau_\mu(t[\sigma(r)]_u) = \tau_\mu(s)$. $\square$

The next example shows that the converse statement does not hold, i.e. $\tau_\mu(t) \to_{\mathcal{R}^\mu} \tau_\mu(s)$ does not imply $t \hookrightarrow_{\mathcal{R}(\mu)} s$.

Example 4.20 *Let $\mathcal{R} = (\{a, b, f\}, \{f(a, b) \to a\})$ and consider a replacement map μ such that $\mu(f) = \{1\}$. Since $\mathcal{R}^\mu = (\{a_\mu, b_\mu, f_\mu\}, \{f_\mu(a_\mu) \to a_\mu\})$ and $\tau_\mu(f(a, a)) = f_\mu(a_\mu)$, we get $\tau_\mu(f(a, a)) \to_{\mathcal{R}^\mu} \tau_\mu(a)$. However, $f(a, a) \not\hookrightarrow_{\mathcal{R}(\mu)} a$.*

The following theorem establishes the μ-termination of a TRS and a replacement map when the contracted TRS terminates.

Theorem 4.21 (μ-termination of $\mathcal{R}$ when $\mathcal{R}^\mu$ terminates) *Let $\mathcal{R} = (\Sigma, R)$ be a TRS and μ be a Σ-map. Let $\mathcal{R}^\mu = (\Sigma^\mu, R^\mu)$ be the contracted TRS of $\mathcal{R}$. If $\mathcal{R}$ is a TRS with conservative replacing variables, then $\mathcal{R}$ μ-terminates if $\mathcal{R}^\mu$ terminates.*

PROOF. By contradiction. Assume that $\mathcal{R}$ does not μ-terminate. Then there exists an infinite μ-derivation $t_1 \hookrightarrow t_2 \hookrightarrow \cdots \hookrightarrow t_n \hookrightarrow \cdots$ in $\mathcal{R}$. By Proposition 4.19, we get an infinite derivation $\tau_\mu(t_1) \to \tau_\mu(t_2) \to \cdots \to \tau_\mu(t_n) \to \cdots$ in $\mathcal{R}^\mu$, which contradicts the initial hypothesis. $\square$

This theorem gives us a new criterion to detect μ-termination of a TRS. Actually, we can use any standard method to check termination of the μ-contracted TRS. We refer to this new method as the $\mathcal{C}_{\mathcal{R}}^\mu$-*criterion* to check μ-termination.

Example 4.22 *We can apply the $\mathcal{C}_{\mathcal{R}}^\mu$-criterion to demonstrate the μ-termination of the TRS $\mathcal{R}$ in the Example 4.4, using the μ-contracted TRS $\mathcal{R}^\mu$ obtained in Example 4.7. By using a recursive path ordering $>^*$ on $\mathcal{T}(\Sigma^\mu, V)$ based on any precedence on $\Sigma^\mu = \{f_\mu, g_\mu\}$, we get $f_\mu(g_\mu(x)) >^* f_\mu(x)$, i.e. $\mathcal{R}^\mu$ terminates, and therefore $\mathcal{R}$ μ-terminates.*

Sometimes we can not use directly the $\mathcal{C}_{\mathcal{R}}^\mu$-criterion, but it turns still useful in combination with Proposition 4.2.

Example 4.23 *Let us consider the TRS $\mathcal{R}$ and the replacement map μ of Example 4.4. Let μ' be given as follows: $\mu'(f) = \{1\}$ and $\mu'(g) = \emptyset$. Then, $\mathcal{R}$ does not satisfies the conservative replacing variables property w.r.t μ'. Hence we can not apply the $\mathcal{C}_{\mathcal{R}}^{\mu'}$-criterion. However, since $\mu' \sqsubseteq \mu$ and $\mathcal{R}$ is μ-terminating (see Example 4.22), by Proposition 4.2, $\mathcal{R}$ is μ'-terminating.*

4.3 Checking μ-termination

We use the above results to establish a suitable criterion to check μ-termination. First, an example reveals the *uncompleteness* of the $C_{\mathcal{R}}^{\mu}$-criterion.

Example 4.24 *Let us consider the TRS* $\mathcal{R} = (\{f, g\}, \{f(g(x)) \rightarrow f(x)\})$, *and a replacement map* μ *given by* $\mu(f) = \emptyset$. *For this replacement map,* $\mathcal{R}$ *is a TRS with conservative replacing variables. Then, it is easy to demonstrate that* $\mathcal{R}$ *is a* μ-terminating *TRS (it suffices to consider the* $C_{\mathcal{R}}$-criterion *and apply a recursive path ordering, for example, to* $\mathcal{R}$), *but the* μ-contracted *TRS* $\mathcal{R}^{\mu} = (\{f_{\mu}, g_{\mu}\}, \{f_{\mu} \rightarrow f_{\mu}\})$ *is not terminating.*

Since we could establish the μ-termination of $\mathcal{R}$ by means of the $C_{\mathcal{R}}$-criterion, the example points out the fact that the $C_{\mathcal{R}}^{\mu}$-criterion does not include the $C_{\mathcal{R}}$-criterion. On the other hand, from Example 4.4 we conclude that the $C_{\mathcal{R}}$-criterion does not include the $C_{\mathcal{R}}^{\mu}$-criterion either.

The example 4.24 also shows that Theorem 4.21 has not a symmetric counterpart, that is, we can not ensure termination of a μ-contracted TRS from the μ-termination of the original.

Now we can formulate a mixed criterion involving both $C_{\mathcal{R}}$ and $C_{\mathcal{R}}^{\mu}$ criteria to test μ-termination.

Theorem 4.25 (μ-termination of a TRS) *Let* $\mathcal{R} = (\Sigma, R)$ *be a TRS with conservative replacing variables given a* Σ-map μ. *Let* $\mathcal{R}^{\mu} = (\Sigma^{\mu}, R^{\mu})$ *be the contracted TRS of* $\mathcal{R}$. *Then* $\mathcal{R}$ μ-terminates *if* $\mathcal{R}$ *or* $\mathcal{R}^{\mu}$ *terminates.*

PROOF. Immediate from Theorem 4.3 and Theorem 4.21. □

Note that this theorem only applies to TRSs with conservative replacing variables. Fortunately, for general TRSs, we still have the Theorem 4.3 ($C_{\mathcal{R}}$-criterion) to prove μ-termination.

5 Conclusions

Context-sensitive rewriting allows us to deal with context-replacing restrictions in performing reduction steps, thus avoiding some undesirable reductions. Termination is always preserved or improved, as we have shown in this paper.

The restrictions imposed by context-sensitive rewriting are easily implementable. However, in order to get complete benefits from context-sensitive rewriting, more research is needed. In particular, it is interesting to study conditions to ensure that, by defining a concrete replacement map for a given TRS, computations are restricted but not *too much*, i.e., they will be able to reduce sufficiently a term by context-sensitive rewriting. This would allow us, implicitly, to define the adequate replacement map which is powerful enough to perform computations with context-sensitive rewriting. This task will be addressed in a further work.

References

1. N. Dershowitz. Orderings for term rewriting systems. *Theoretical Computer Science* 17:279-301, 1982.
2. N. Dershowitz. Termination of rewriting. *Journal on Symbolic Computation*, 3:69-115, 1987.
3. N. Dershowitz. A Taste of Rewrite Systems. In P.E. Lauer, editor, *Proc. of Functional Programming, Concurrency, Simulation and Automated Reasoning*, LNCS 693:199-228, Springer-Verlag, Berlin, 1993.
4. N. Dershowitz and C. Hoot. Natural Termination. *Theoretical Computer Science* 142(2):179-207, 1995.
5. N. Dershowitz and J.P. Jouannaud. Rewrite Systems. In J. van Leeuwen, editor, *Handbook of Theoretical Computer Science*, volume B: Formal Models and Semantics, pages 243-320. Elsevier, Amsterdam and The MIT Press, Cambridge, MA, 1990.
6. D.P. Friedman and D.S. Wise. CONS should not evaluate its arguments. In S. Michaelson and R. Milner, editors, *Automata, Languages and Programming*, pages 257-284, Edinburgh University Press, 1976.
7. J. Goguen, C. Kirchner and J. Meseguer. Concurrent Term Rewriting as a Model of Computation. In J.H. Fasel, R.M. Keller, editors, *Graph Reduction. Proceedings. of a Workshop.* LNCS 279:53-93. Springer-Verlag, Berlin, 1987.
8. S. Kamin and J.J. Lévy. Two generalizations of the recursive path ordering. University of Illinois, 1980.
9. J.F.Th. Kamperman and H.R. Walters. Lazy Rewriting and Eager Machinery. In J. Hsiang, editor, *Proc. of the 6th International Conference on Rewriting Techniques and Applications, RTA'95*, LNCS 914:147-162, Springer-Verlag, Berlin, 1995.
10. G. Huet and J.J. Lévy. Computations in orthogonal term rewriting systems. In J.L. Lassez and G. Plotkin, editors, *Computational logic: essays in honour of J. Alan Robinson*. MIT Press, Cambridge, MA, 1991.
11. R. Lalement. Computation as Logic. Masson-Prentice Hall International, 1993.
12. S. Lucas. Computational properties in context-sensitive rewriting. In M. Alpuente and M.I. Sessa, editors, *Proc. of 1995 Joint Conference on Declarative Programming, GULP-PRODE'95*, pages 435-446, 1995.
13. S. Lucas. Fundamentals of context-sensitive rewriting. In M. Bartōsek, J. Staudek and J. Wiedermann, editors, *Proc. of XXII Seminar on Current Trends in Theory and Practice of Informatics, SOFSEM'95*, LNCS 1012:405-412, Springer-Verlag, Berlin, 1995.
14. Z. Manna and S Ness. On the termination of Markov algorithms. In *Proc. of the Third Hawaii International Conference on System Science*, pages 789-792, 1970.
15. S. L. Peyton-Jones. The Implementation of Functional Programming Languages. Prentice-Hall International, London, 1987.
16. C. Reade. Elements of Functional Programming. Addison-Wesley Publishing Company, 1987.

A Complete Gentzen-Style Axiomatization for Set Constraints

Allan Cheng[*] and Dexter Kozen

Computer Science Department
Cornell University
Ithaca, New York 14853, USA
e-mail:{acheng,kozen}@cs.cornell.edu

Abstract. Set constraints are inclusion relations between expressions denoting sets of ground terms over a ranked alphabet. They are the main ingredient in set-based program analysis. In this paper we provide a Gentzen-style axiomatization for sequents $\Phi \vdash \Psi$, where Φ and Ψ are finite sets of set constraints, based on the axioms of termset algebra. Sequents of the restricted form $\Phi \vdash \bot$ correspond to positive set constraints, and those of the more general form $\Phi \vdash \Psi$ correspond to systems of mixed positive and negative set constraints. We show that the deductive system is (i) complete for the restricted sequents $\Phi \vdash \bot$ over standard models, (ii) incomplete for general sequents $\Phi \vdash \Psi$ over standard models, but (iii) complete for general sequents over set-theoretic termset algebras.

1 Introduction

Set constraints are inclusions between expressions denoting sets of ground terms. They have been used extensively in program analysis and type inference for many years [AM91a, AM91b, Hei93, HJ90b, JM79, Mis84, MR85, Rey69, YO88]. Considerable recent effort has focussed on the complexity of the satisfiability problem [AKVW93, AKW95, AW92, BGW93, CP94a, CP94b, GTT93a, GTT93b, HJ90a, Ste94]. Set constraints have also recently been used to define a constraint logic programming language over sets of ground terms that generalizes ordinary logic programming over an Herbrand domain [Koz94].

Set constraints exhibit a rich mathematical structure. There are strong connections to automata theory [GTT93a, GTT93b], type theory [KPS93, KPS94], first-order monadic logic [BGW93, CP94a], Boolean algebras with operators [JT51, JT52], and modal logic [Koz93]. There are algebraic and topological formulations, corresponding roughly to "soft" and "hard" typing respectively, which are related by Stone duality [Koz93, Koz95].

An axiomatization of the main properties of set constraints was proposed in [Koz93]. General models of these axioms are called *termset algebras*. In [Koz93],

---

[*] Visiting from Aarhus, **BRICS**, Basic Research in Computer Science, Center of the Danish National Research Foundation. e-mail:acheng@daimi.aau.dk

a representation theorem was proved showing that every termset algebra is isomorphic to a set-theoretic termset algebra. These models include the standard models in which set expressions are interpreted as sets of ground terms, as well as nonstandard models in which set expressions are interpreted as sets of states of *term automata* [KPS92].

In this paper we propose a Gentzen-style axiomatization involving sequents of the form $\Phi \vdash \Psi$, where Φ and Ψ are finite sets of set constraints. The intended interpretation of the sequent $\Phi \vdash \Psi$ is that if all the constraints in Φ hold of some model, then at least one of the constraints Ψ holds in that model.

This axiomatization can be thought of as a deductive system for refuting unsatisfiable systems of mixed positive and negative constraints. Deriving the sequent $\Phi \vdash \Psi$ is tantamount to refuting the mixed system $\Phi \cup \{s \neq t \mid s = t \in \Psi\}$. Systems of the restricted form $\Phi \vdash \bot$ correspond to systems of positive set constraints alone.

For this deductive system, we prove

(i) completeness over standard models for satisfiability of positive set constraints alone (if Φ is unsatisfiable, then Φ is refutable, *i.e.*, $\Phi \vdash \bot$ is derivable);

(ii) incompleteness over standard models for satisfiability of mixed positive and negative constraints (*i.e.*, not all valid sequents $\Phi \vdash \Psi$ are derivable);

(iii) completeness over nonstandard models (all set-theoretic termset algebras) for satisfiability of mixed positive and negative constraints (*i.e.*, all valid sequents $\Phi \vdash \Psi$ are derivable).

We feel that these results are of both theoretical and practical interest. Theoretically, they shed light on the distinction between exclusively positive and mixed positive and negative constraints. Although several interesting results involving the decidability and complexity of negative constraints have appeared [CP94b, GTT93b, AKW95, Ste94], the distinction between the two cases is still far from clear from a deductive standpoint.

Practically, we were interested in recasting the axioms of [Koz93] in a Gentzen style so as to take advantage of one of a number of automated deduction systems to implement a constraint solving package [Gri87]. We foresee this as being a useful alternative approach to building a set constraint solver for use in program analysis or constraint logic programming over set constraints.

This paper is organized as follows. In §2–§5, we briefly review the basic definitions and known results we will need regarding set constraints, termset algebras, term automata, and normal forms. These are included here for the sake of self-containment. In §6 we present our main results. Finally, in §7 we draw conclusions and discuss future work.

2 Set Expressions and Set Constraints

Let Σ be a finite ranked alphabet consisting of symbols f, each with an associated arity. Symbols in Σ of arity 0, 1, 2, and n are called *nullary, unary, binary,* and *n-ary*, respectively. Nullary elements are denoted by $a, b, \ldots$ and are called *constants*. The set of elements of Σ of arity n is denoted Σ_n. In the sequel, the use of expressions of the form $f(t_1, \ldots, t_n)$ carries the implicit assumption that f is of arity n.

The set of *ground terms* over Σ is denoted T_Σ. It is the least set such that if $t_1, \ldots, t_n \in T_\Sigma$ and $f \in \Sigma_n$, then $f(t_1, \ldots, t_n) \in T_\Sigma$. If $X = \{x, y, \ldots\}$ is a set of variables, then $T_\Sigma(X)$ denotes the set of terms over Σ and X, considering variables in X as symbols of arity 0.

Let B $= (\cup, \cap, \sim, 0, 1)$ denote the usual signature of Boolean algebra. Let $\Sigma +$B denote the signature consisting of the disjoint union of Σ and B. Boolean operators such as $-$ (set difference) and $\oplus$ (symmetric difference) are defined from these as usual. A *set expression* over X is an element of $T_{\Sigma+\text{B}}(X)$. We use $s, t, \ldots$ to denote set expressions. A typical set expression could be:

$$f(g(x \cup y), \sim (g(a) \cap b))$$

where g, f are symbols of arity 1 and 2, respectively, a, b are constants, and $x, y \in X$. A *Boolean expression* over X is an element of $T_\text{B}(X)$.

A *positive set constraint* is a formal inclusion $s \subseteq t$, where s and t are set expressions. For notational convenience we allow equational constraints $s = t$, although inclusions and equations are interdefinable: $s \subseteq t$ is equivalent to $s \cup t = t$, and $s = t$ to $s \oplus t \subseteq 0$. A *negative set constraint* is the negation of a positive set constraint: $s \not\subseteq t$ or $s \neq t$. We use $\varphi, \psi, \ldots$ to denote set constraints and $\Phi, \Psi, \ldots$ to denote finite sets of set constraints.

3 Axioms of Termset Algebra

In [Koz93], the following axiomatization of the algebra of sets of ground terms was introduced:

$$f(\ldots, x \cup y, \ldots) = f(\ldots, x, \ldots) \cup f(\ldots, y, \ldots) \tag{1}$$

$$f(\ldots, x - y, \ldots) = f(\ldots, x, \ldots) - f(\ldots, y, \ldots) \tag{2}$$

$$\bigcup_{f \in \Sigma} f(1, \ldots, 1) = 1 \tag{3}$$

$$f(1, \ldots, 1) \cap g(1, \ldots, 1) = 0, \ f \neq g \tag{4}$$

$$f(x_1, \ldots, x_n) = 0 \Rightarrow \bigvee_{i=1}^{n} (x_i = 0) \tag{5}$$

$$\text{axioms of Boolean algebra} \tag{6}$$

The ellipses in (1) and (2) indicate that the explicitly given arguments occur in corresponding places, and that the implicit arguments in corresponding places agree. Models of the axioms are called *termset algebras*. The standard interpretation 2^{T_Σ}, where the Boolean operators have their usual set-theoretic interpretations and elements $f \in \Sigma_n$ are interpreted as

$$f : (2^{T_\Sigma})^n \to 2^{T_\Sigma}$$
$$f(A_1,\ldots,A_n) = \{f(t_1,\ldots,t_n) \,|\, t_i \in A_i,\ 1 \leq i \leq n\}\,,$$

forms a model of these axioms.

Some immediate consequences of these axioms are

$$f(\ldots,0,\ldots) = 0 \tag{7}$$
$$f(\ldots,\sim x,\ldots) = f(\ldots,1,\ldots) - f(\ldots,x,\ldots) \tag{8}$$
$$f(\ldots,x \oplus y,\ldots) = f(\ldots,x,\ldots) \oplus f(\ldots,y,\ldots) \tag{9}$$
$$f(\ldots,x \cap y,\ldots) = f(\ldots,x,\ldots) \cap f(\ldots,y,\ldots) \tag{10}$$
$$x \subseteq y \Rightarrow f(\ldots,x,\ldots) \subseteq f(\ldots,y,\ldots)\,. \tag{11}$$

Also, a *generalized DeMorgan law* can be derived:

$$\sim f(x_1,\ldots,x_n) = \bigcup_{g \neq f} g(1,\ldots,1) \cup \bigcup_{i=1}^{n} f(\underbrace{1,\ldots,1}_{i-1}, \sim x_i, \underbrace{1,\ldots,1}_{n-i}) \tag{12}$$

The law intuitively says that a ground term *not* having head symbol f and i^{th} subterm satisfying x_i either has head symbol different from f or has head symbol f but one of its ith subterms does not satisfy x_i. The law is useful for pushing occurrences of the negation operator $\sim$ inward.

4   Term Automata and Models

Following Courcelle [Cou83], we define *(Σ-)terms*.

Definition 1. Let ω denote the set of natural numbers and let Σ be a finite ranked alphabet. A *(Σ-)term* is a partial function $t : \omega^* \to \Sigma$ whose domain is nonempty, prefix-closed, and respects arities in the sense that if $t(\gamma)$ is defined then

$$\{i \mid t(\gamma i) \text{ is defined}\} = \{1, 2, \ldots, \text{arity}(t(\gamma))\}\,.$$

If α is in the domain of t, the subterm of t rooted at α is the the term $\lambda\beta.t(\alpha\beta)$. A term is (in)finite if its domain is (in)finite, and is *regular* if it has only finitely many subterms. The set of finite Σ-terms is also denoted T_Σ.

Example 1. The finite term $f(g(a), f(a, g(b)))$ is formally a partial map t with domain $\{\epsilon, 1, 2, 11, 21, 22, 221\}$ such that $t(\epsilon) = t(2) = f$, $t(1) = t(22) = g$, $t(11) = t(21) = a$, and $t(221) = b$. The infinite term $f(a, f(a, f(a, \ldots)))$ is formally a map s whose domain is the infinite set described by the regular expression $2^* + 2^*1$ such that $s(\alpha) = f$ for $\alpha \in 2^*$ and $s(\alpha) = a$ for $\alpha \in 2^*1$. The infinite term s is regular since it has only two subterms, namely s and a.

4.1   Term Automata

It is well known that an infinite regular term can be represented by a finite labeled graph such that the infinite term is obtained by "unwinding" the graph (see [Cou83, Col82]). We use the automata-theoretic formulation introduced in [KPS92] of this idea.

Definition 2. A *term automaton* over Σ is a tuple $\mathcal{M} = (Q, \Sigma, \ell, \delta)$ where:

- Q is a set of *states* (not necessarily finite)
- Σ is a ranked alphabet
- $\ell : Q \to \Sigma$ is a *labeling*
- $\delta : Q \times \omega \to Q$ is a partial function such that for all $q \in Q$,

$$\{i \mid \delta(q, i) \text{ is defined}\} = \{1, 2, \ldots, \text{arity}(\ell(q))\} \ .$$

The function δ extends uniquely to a partial function $\widehat{\delta} : Q \times \omega^* \to Q$ according to the inductive definition

$$\widehat{\delta}(q, \epsilon) = q$$
$$\widehat{\delta}(q, \gamma i) = \delta(\widehat{\delta}(q, \gamma), i) \ ,$$

with the understanding that δ is strict (undefined if one of its arguments is undefined). For each $q \in Q$, the partial function

$$t_q = \lambda \gamma . \ell(\widehat{\delta}(q, \gamma))$$

is a Σ-term in the sense of Definition 1. Notice that t_p may equal t_q even though $p \neq q$.

A term automaton $\mathcal{M}$ is *closed* if for any $f \in \Sigma_n$ and $q_1, \ldots, q_n \in Q$ there exists a $q \in Q$ such that

$$\ell(q) = f \text{ and } \delta(q, i) = q_i \ , \ 1 \leq i \leq n \ . \tag{13}$$

A *model* is a closed term automaton $\mathcal{M}$. We refer to the states of $\mathcal{M}$—rather then their associated partial functions t_q—as the *terms* of $\mathcal{M}$, and use the notation $\mathbf{t} \in \mathcal{M}$ to indicate $\mathbf{t} \in Q$. A term $\mathbf{t}'$ of $\mathcal{M}$ is a *subterm* of $\mathbf{t}$ at depth k if there exists a $\gamma \in \omega^k$ such that $\widehat{\delta}(\mathbf{t}, \gamma) = \mathbf{t}'$. A term $\mathbf{t}$ of $\mathcal{M}$ is (in)finite if $t_{\mathbf{t}}$ is (in)finite, and said to be labeled by t' if $t_{\mathbf{t}} = t'$. The model is *standard* if the function $q \mapsto t_q : Q \to T_\Sigma$ is a bijection. We denote a standard model by T_Σ.

Remark. For any term automaton $\mathcal{M} = (Q, \Sigma, \ell, \delta)$ there is a closed term automaton $\mathcal{M}' = (Q', \Sigma, \ell', \delta')$ such that $Q \subseteq Q'$, ℓ' and δ' coincide with ℓ and δ on states from Q, and Q' is a minimal set of states—with respect to subset inclusion—with these properties; $\mathcal{M}'$ is said to be a *minimal closure* of $\mathcal{M}$. $\mathcal{M}'$ can be obtained as follows: Let $\mathcal{M}_0 = \mathcal{M}$ and let $\mathcal{M}_{i+1}$ be obtained from $\mathcal{M}_i$ by adding exactly one new term $\mathbf{t}$ to Q_i for every $f \in \Sigma_n$ and $\mathbf{t}_1, \ldots, \mathbf{t}_n \in Q_i$ for which (13) doesn't hold. ℓ_{i+1} is the extension of ℓ_i that maps $\mathbf{t}$ to f and δ_{i+1} is the extension of δ_i that maps $(\mathbf{t}, i)$ to $\mathbf{t}_i$, $1 \leq i \leq n$. Define $\mathcal{M}'$ as the ω-limit of these term automata.

4.2 Term Automata and Set-Theoretic Termset Algebras

Let $\mathcal{M}$ be the term automaton $(Q, \Sigma, \ell, \delta)$. For $f \in \Sigma_n$, define the partial function $R_f^{\mathcal{M}} : Q \to Q^n$ and the set-theoretic function $f^{\mathcal{M}} : (2^Q)^n \to 2^Q$ by

$$R_f^{\mathcal{M}}(q) = \begin{cases} (\delta(q,1), \ldots, \delta(q,n)) \text{, if } \ell(q) = f \\ \text{undefined ,} \qquad\qquad \text{otherwise.} \end{cases} \tag{14}$$

$$f^{\mathcal{M}}(A_1, \ldots, A_n) = \{q \in Q \mid \ell(q) = f \text{ and } \delta(q,i) \in A_i, \ 1 \le i \le n\}$$
$$= (R_f^{\mathcal{M}})^{-1}(A_1 \times \cdots \times A_n) . \tag{15}$$

Set expressions are interpreted over 2^Q, the powerset of Q, which forms an algebra of signature $\Sigma+B$, where the Boolean operators have their usual set-theoretic interpretations and elements $f \in \Sigma$ are interpreted as $f^{\mathcal{M}}$. If $\mathcal{M}$ is closed, one can show that this gives a termset algebra. Such an algebra, or a subalgebra of such an algebra, is called a *set-theoretic termset algebra*.

Let $\mathcal{M}$ be a model. A *set valuation* over $\mathcal{M}$ is a map

$$\sigma : X \to 2^Q$$

assigning a subset of terms of $\mathcal{M}$ to each variable in X. We can extend any set valuation σ uniquely to a $(\Sigma+B)$-homomorphism

$$\sigma : T_{\Sigma+B}(X) \to 2^Q$$

by induction on the structure of set expressions in the usual way. A set valuation σ over $\mathcal{M}$ satisfies the positive set constraint $s \subseteq t$ if $\sigma(s) \subseteq \sigma(t)$, and satisfies the negative set constraint $s \not\subseteq t$ if $\sigma(s) \not\subseteq \sigma(t)$. We write $\sigma \models_{\mathcal{M}} \Phi$ if σ satisfies all set constraints in Φ; Φ is said to be *satisfiable* in $\mathcal{M}$ and σ a *solution* to Φ. The set Φ is satisfiable if it is satisfiable over some model. We write $\Phi \models_{\mathcal{M}} \Psi$ if $\sigma \models_{\mathcal{M}} \Phi$ implies $\sigma \models_{\mathcal{M}} \psi$ for some $\psi \in \Psi$. When no confusion is possible, we suppress the subscript $\mathcal{M}$.

5 Systems in Normal Form and Solutions

Let $X' \subseteq X$. *Positive (negative) literals* from X' are expressions x ($\sim x$) for $x \in X'$. A maximal conjunction of literals from X' is a conjunction of positive and negative literals from X', where each variable in X' occurs exactly once.

A triple (t_B, Φ, Δ) is a system of set constraints in *normal form* (or just a system in normal form) if there is a finite set $X' \subseteq X$ such that (i) $t_B \in T_B(X')$ is of the form $\cup_{\alpha \in U} \alpha$, for some set U of maximal conjunctions of literals from X', (ii) for each $f \in \Sigma_n$ and $\alpha_1, \ldots, \alpha_n \in U$ there is exactly one set constraint in Φ of the form $f(\alpha_1, \ldots, \alpha_n) \subseteq \bigcup_{\alpha \in E_{f(\alpha_1,\ldots,\alpha_n)}} \alpha$, where $E_{f(\alpha_1,\ldots,\alpha_n)} \subseteq U$, and (iii) Δ is a finite set of Boolean expressions $\{\bigcup_{\alpha \in I_1} \alpha, \ldots, \bigcup_{\alpha \in I_m} \alpha\}$, where $I_k \subseteq U$ for $1 \le k \le m$. The set U is referred to as the set of atoms[2] specified by t_B.

---

[2] The elements of U are the atoms of the free Boolean algebra on generators X' modulo t_B.

The triple (t_B, Φ, Δ) corresponds to the set of set constraints $\{t_B = 1\} \cup \Phi \cup \{\bigcup_{\alpha \in I_1} \alpha \neq 0, \ldots, \bigcup_{\alpha \in I_m} \alpha \neq 0\}$ and is said to be (un)satisfiable if the latter is. A set valuation satisfies (t_B, Φ, Δ) if it satisfies the corresponding set constraints. If Δ is empty, we denote the system in normal form by (t_B, Φ) and call it a system of positive set constraints in normal form (or just a positive system in normal form). Every system of mixed positive and negative set constraints is equivalent to a system in normal form [AKW95].

Each positive system in normal form (t_B, Φ) has an associated hypergraph; the nodes are the elements of U and the hyperedges are specified by the sets $E_{f(\alpha_1,\ldots,\alpha_n)}$. Let $\mathcal{M}$ be a model. A *run* over $\mathcal{M}$ through the hypergraph is a function $\theta : Q \to U$ such that

$$\theta(\mathbf{t}) \in E_{f(\theta(\mathbf{t}_1),\ldots,\theta(\mathbf{t}_n))} \, ,$$

where $\ell(\mathbf{t}) = f \in \Sigma_n$ and $\delta(\mathbf{t}, i) = \mathbf{t}_i$, for $1 \leq i \leq n$. Each subset $U' \subseteq U$ induces a subhypergraph by restricting the nodes and hyperedges to U'. The subhypergraph induced by U' is *closed* if for each $f \in \Sigma_n$ and $\alpha_1, \ldots, \alpha_n \in U'$ the set $E_{f(\alpha_1,\ldots,\alpha_n)} \cap U'$ is nonempty. It can be proved that (t_B, Φ) is satisfiable over a standard model if and only if there is a nonempty $U' \subseteq U$ that induces a closed subhypergraph in the hypergraph associated with (t_B, Φ). Intuitively, from a run θ one can obtain a set valuation σ_θ over a standard model satisfying (t_B, Φ), and—vice versa—from a set valuation σ satisfying (t_B, Φ) one can obtain a run θ_σ over a standard model through the hypergraph associated with (t_B, Φ). For details see [AKVW93, Koz93, Koz95].

6 Completeness and Incompleteness

In this section we give a Gentzen-style axiomatization for sequents $\Phi \vdash \Psi$, based on the axioms of termset algebra. The intended interpretation of the sequent $\Phi \vdash \Psi$ is that if all the constraints in Φ hold of some model, then at least one of the constraints Ψ holds in that model. We prove (i) completeness over standard models for satisfiability of positive set constraints (if Φ is unsatisfiable, then Φ is refutable, *i.e.*, $\Phi \vdash \bot$ is derivable), (ii) incompleteness over standard models for satisfiability of mixed positive and negative set constraints (*i.e.*, not all true sequents $\Phi \vdash \Psi$ are derivable), and (iii) completeness over nonstandard models.

Any set constraint can be represented as an inclusion $s \subseteq t$, or an equation $u = 0$, or an equation $v = 1$. In the following, any set expression s occurring in a context expecting a set constraint denotes the set constraint $s = 1$. An inclusion $s \subseteq t$ can then be represented as the term $\sim s \cup t$, denoting the set constraint $\sim s \cup t = 1$, and an equation $s = t$ as the term $(\sim s \cup t) \cap (\sim t \cup s)$. A set Φ denotes the conjunction or disjunction of its elements, depending on whether it occurs on the left or right side of a $\vdash$, respectively. A comma denotes conjunction or disjunction, depending on whether it occurs on the left or right side of a $\vdash$, respectively. We use $\bot$ for the empty disjunction on the right side of $\vdash$; $\bot$ can be read as 0. The rules are:

$$\Phi \vdash \Phi \ \text{(ident)} \qquad \frac{\Phi \vdash \Psi}{\Phi', \Phi \vdash \Psi, \Psi'} \ \text{(weakening)}$$

$$\frac{\Phi, \sim t_i \vdash \Psi, 1 \leq i \leq n}{\Phi, \sim f(t_1, \ldots, t_n) \vdash \Psi} \ (f\text{-intro} \vdash)$$

$$\frac{\Phi, s, t \vdash \Psi}{\Phi, s \cap t \vdash \Psi} \ (\cap\text{-intro} \vdash) \qquad \frac{\Phi, s \cap t \vdash \Psi}{\Phi, s, t \vdash \Psi} \ (\cap\text{-elim} \vdash)$$

$$\frac{\Phi, \varphi[t \leftarrow t'], t = t' \vdash \Psi}{\Phi, \varphi, t = t' \vdash \Psi} \ \text{(substitution} \vdash)$$

$$\frac{\Phi, t = t' \vdash \psi[t \leftarrow t'], \Psi}{\Phi, t = t' \vdash \psi, \Psi} \ (\vdash \text{substitution})$$

For x not in Φ, t:

$$\frac{\Phi, x = t \vdash \Psi}{\Phi \vdash \Psi} \ (x\text{-elim} \vdash)$$

For any instance $s = t$ of the termset algebra axioms:

$$\frac{\Phi, s \vdash \Psi}{\Phi, t \vdash \Psi} \ \text{(termset} \vdash) \qquad \frac{\Phi \vdash \Psi, s}{\Phi \vdash \Psi, t} \ (\vdash \text{termset})$$

The sequents above and under a bar are referred to as the premises and conclusion of the rule, respectively. $\varphi[t \leftarrow t']$ denotes the substitution of all occurrences of the expression t in φ by the expression t'.

Derivation trees are inductively defined finite trees whose nodes are labeled with sequents $\Phi \vdash \Psi$. A single node labeled with any sequent $\Phi \vdash \Psi$ is a derivation tree, and if there exist derivation trees $\mathcal{T}_1, \ldots, \mathcal{T}_n$ whose roots are labeled with sequents matching the premises of a rule, then the tree whose root is labeled with the conclusion of that rule and has $\mathcal{T}_1, \ldots, \mathcal{T}_n$ as immediate subtrees is itself a derivation tree. A sequent $\Phi \vdash \Psi$ is *derivable from a set S* of sequents if and only if there is a derivation tree all of whose leaves are labeled by sequents in S and whose root is labeled $\Phi \vdash \Psi$. If S only contains sequents of the form $\Delta \vdash \Delta$ or $\Delta, \sim c \vdash \Gamma$ (corresponding to the rules (ident) and (f-intro $\vdash$) for $n = 0$, respectively), then the derivation tree is called a *tableau* and $\Phi \vdash \Psi$ is said to be *derivable*.

Lemma 3. *All rules are sound.*

The following theorem shows that the deductive system is complete over standard models for satisfiability of positive set constraints.

Theorem 4. *If a finite set of positive set constraints Φ is unsatisfiable in any standard model, then $\Phi \vdash \perp$ is derivable.*

Proof sketch. First, one derives $\Phi \vdash \bot$ from $t_B, \Phi' \vdash \bot$, where (t_B, Φ') is a positive system in normal form. Then, using the rule (f-intro $\vdash$) one splits the derivation of $t_B, \Phi' \vdash \bot$ into several subgoals, each of which is derivable from sequents of the form $t'_B, \Phi'' \vdash \bot$, where (t'_B, Φ'') is a positive system in normal form. Applying the procedure recursively yields a derivation of $t_B, \Phi' \vdash \bot$. The full proof can be found in [CK95]. □

Now suppose we are given a set of mixed positive and negative set constraints $\Phi = \{s_1 = t_1, \ldots, s_n = t_n\} \cup \{s'_1 \neq t'_1, \ldots, s'_m \neq t'_m\}$. Observe that Φ is unsatisfiable if and only if $\{s_1 = t_1, \ldots, s_n = t_n\} \models \{s'_1 = t'_1, \ldots, s'_m = t'_m\}$. The following theorem shows that the deductive system is incomplete over standard models for satisfiability of mixed positive and negative set constraints.

Theorem 5. *The axiomatization is incomplete for systems of mixed positive and negative set constraints over standard models.*

Proof. The sequent $x = f(x) \models x = 0$ certainly holds in all standard models. However, $x = f(x) \vdash x = 0$ cannot be derived, since the rules are sound for nonstandard models as well, and if infinite terms are allowed then $x = f(x) \models x = 0$ is no longer valid: in any model containing an infinite term labeled $f(f(f(\ldots)))$, the set of terms labeled $f(f(f(\ldots)))$ is a nontrivial solution to the set constraint $x = f(x)$. □

We continue by considering nonstandard models.

Lemma 6. *A system of set constraints in normal form (t_B, Φ, Δ), where $\Delta = \{\bigcup_{\alpha \in I_1} \alpha, \ldots, \bigcup_{\alpha \in I_m} \alpha\}$, is satisfiable if and only if there exists a set $U' \subseteq U$ such that*

$$\forall f \in \Sigma_n. \forall \alpha_1, \ldots, \alpha_n \in U'. E_{f(\alpha_1, \ldots, \alpha_n)} \cap U' \neq \varnothing , \quad (16)$$

$$\forall \alpha \in U'. \exists f \in \Sigma_n. \exists \alpha_1, \ldots, \alpha_n \in U'. \alpha \in E_{f(\alpha_1, \ldots, \alpha_n)} , \text{ and} \quad (17)$$

$$\forall 1 \leq k \leq m. I_k \cap U' \neq \varnothing , \quad (18)$$

where U are the atoms corresponding to t_B.

Proof sketch. The nontrivial part of the proof is the "if" direction. First, one considers a finite number of disjoint standard models. They represent all the finite terms of a model over which (t_B, Φ, Δ) is satisfiable. Then, one takes the minimal closure, adds infinite regular terms, and finally takes the minimal closure again. In [CK95] we give the details of this construction including a set valuation over the obtained nonstandard model satisfying (t_B, Φ, Δ). □

The last theorem shows that our deductive system is complete for satisfiability of mixed positive and negative set constraints.

Theorem 7. *If a finite set of mixed positive and negative set constraints*

$$\{s_1 = t_1, \ldots, s_n = t_n\} \cup \{s_1' \neq t_1', \ldots, s_m' \neq t_m'\}$$

is unsatisfiable, then

$$s_1 = t_1, \ldots, s_n = t_n \vdash s_1' = t_1', \ldots, s_m' = t_m'$$

is derivable.

Proof sketch. First, one derives $s_1 = t_1, \ldots, s_n = t_n \vdash s_1' = t_1', \ldots, s_m' = t_m'$ from

$$t_B, \Phi \vdash \bigcup_{\alpha \in I_1} \alpha = 0, \ldots, \bigcup_{\alpha \in I_m} \alpha = 0, \tag{19}$$

where $(t_B, \Phi, \{\bigcup_{\alpha \in I_1} \alpha, \ldots, \bigcup_{\alpha \in I_m} \alpha\})$ is a system in normal form, corresponding to an equivalent set of mixed positive and negative set constraints. Then, using Lemma 6 and a technique similar to that from the proof of Theorem 4 yields a derivation of $s_1 = t_1, \ldots, s_n = t_n \vdash s_1' = t_1', \ldots, s_m' = t_m'$. The full proof can be found in [CK95]. □

7 Conclusion

In this paper we have introduced and investigated a deductive system for deriving sequents $\Phi \vdash \Psi$, where Φ and Ψ are finite sets of set constraints. Using standard and nonstandard models involving set-theoretic termset algebras as introduced in [Koz93], we have shown that the deductive system is (i) complete for restricted sequents of the form $\Phi \vdash \perp$ over standard models, (ii) incomplete for general sequents $\Phi \vdash \Psi$ over standard models, but (iii) complete for general sequents over nonstandard models.

Having chosen term automata as the basis for our models, we naturally get models that allow "multiple copies" of a term t, *i.e.* we may have $t_p = t_q$ for different states p and q of the term automaton. One natural and interesting question that remains is whether the system is complete for general sequents over models that forbid such "multiple copies" but allow infinite terms.

Acknowledgments

The support of the Danish Research Academy and Danish Research Council under contract SNF-journal number 11-0773 grant 5100.7314, the National Science Foundation under grant CCR-9317320, and the U.S. Army Research Office through the ACSyAM branch of the Mathematical Sciences Institute of Cornell University under contract DAAL03-91-C-0027 is gratefully acknowledged.

References

[AKVW93] Alexander Aiken, Dexter Kozen, Moshe Vardi, and Edward Wimmers. The complexity of set constraints. In E. Börger, Y. Gurevich, and K. Meinke, editors, *Proc. 1993 Conf. Computer Science Logic (CSL'93)*, volume 832 of *Lect. Notes in Comput. Sci.*, pages 1–17. Eur. Assoc. Comput. Sci. Logic, Springer, September 1993.

[AKW95] Alexander Aiken, Dexter Kozen, and Edward Wimmers. Decidability of systems of set constraints with negative constraints. *Infor. and Comput.*, 122(1):30–44, October 1995.

[AM91a] A. Aiken and B. Murphy. Implementing regular tree expressions. In *Proc. 1991 Conf. Functional Programming Languages and Computer Architecture*, pages 427–447, August 1991.

[AM91b] A. Aiken and B. Murphy. Static type inference in a dynamically typed language. In *Proc. 18th Symp. Principles of Programming Languages*, pages 279–290. ACM, January 1991.

[AW92] A. Aiken and E. Wimmers. Solving systems of set constraints. In *Proc. 7th Symp. Logic in Computer Science*, pages 329–340. IEEE, June 1992.

[BGW93] L. Bachmair, H. Ganzinger, and U. Waldmann. Set constraints are the monadic class. In *Proc. 8th Symp. Logic in Computer Science*, pages 75–83. IEEE, June 1993.

[CK95] A. Cheng and D. Kozen. A Complete Gentzen-style Axiomatization for Set Constraints. Tech. Rep. 95-1518, Computer Science Department, Cornell University, May 1995.

[Col82] A. Colmerauer. PROLOG and infinite trees. In S.-Å. Tärnlund and K. L. Clark, editors, *Logic Programming*, pages 231–251. Academic Press, January 1982.

[Cou83] Bruno Courcelle. Fundamental properties of infinite trees. *Theor. Comput. Sci.*, 25:95–169, 1983.

[CP94a] W. Charatonik and L. Pacholski. Negative set constraints with equality. In *Proc. 9th Symp. Logic in Computer Science*, pages 128–136. IEEE, July 1994.

[CP94b] W. Charatonik and L. Pacholski. Set constraints with projections are in *NEXPTIME*. In *Proc. 35th Symp. Foundations of Computer Science*, pages 642–653. IEEE, November 1994.

[Gri87] Timothy G. Griffin. An environment for formal systems. Technical Report TR87-846, Cornell University, June 1987.

[GTT93a] R. Gilleron, S. Tison, and M. Tommasi. Solving systems of set constraints using tree automata. In *Proc. Symp. Theor. Aspects of Comput. Sci.*, volume 665, pages 505–514. Springer-Verlag Lect. Notes in Comput. Sci., February 1993.

[GTT93b] R. Gilleron, S. Tison, and M. Tommasi. Solving systems of set constraints with negated subset relationships. In *Proc. 34th Symp. Foundations of Comput. Sci.*, pages 372–380. IEEE, November 1993.

[Hei93] Nevin Heintze. *Set Based Program Analysis*. PhD thesis, Carnegie Mellon University, 1993.

[HJ90a] N. Heintze and J. Jaffar. A decision procedure for a class of set constraints. In *Proc. 5th Symp. Logic in Computer Science*, pages 42–51. IEEE, June 1990.

[HJ90b]   N. Heintze and J. Jaffar. A finite presentation theorem for approximating logic programs. In *Proc. 17th Symp. Principles of Programming Languages*, pages 197–209. ACM, January 1990.

[JM79]   N. D. Jones and S. S. Muchnick. Flow analysis and optimization of LISP-like structures. In *Proc. 6th Symp. Principles of Programming Languages*, pages 244–256. ACM, January 1979.

[JT51]   B. Jónsson and A. Tarski. Boolean algebras with operators. *Amer. J. Math.*, 73:891–939, 1951.

[JT52]   B. Jónsson and A. Tarski. Boolean algebras with operators. *Amer. J. Math.*, 74:127–162, 1952.

[Koz93]   Dexter Kozen. Logical aspects of set constraints. In E. Börger, Y. Gurevich, and K. Meinke, editors, *Proc. 1993 Conf. Computer Science Logic (CSL'93)*, volume 832 of *Lect. Notes in Comput. Sci.*, pages 175–188. Eur. Assoc. Comput. Sci. Logic, Springer, September 1993.

[Koz94]   Dexter Kozen. Set constraints and logic programming (abstract). In J.-P. Jouannaud, editor, *Proc. First Conf. Constraints in Computational Logics (CCL'94)*, volume 845 of *Lect. Notes in Comput. Sci.*, pages 302–303. ESPRIT, Springer, September 1994.

[Koz95]   Dexter Kozen. Rational spaces and set constraints. In Peter D. Mosses, Mogens Nielsen, and Michael I. Schwartzbach, editors, *Proc. Sixth Int. Joint Conf. Theory and Practice of Software Develop. (TAPSOFT'95)*, volume 915 of *Lect. Notes in Comput. Sci.*, pages 42–61. Springer, May 1995.

[KPS92]   Dexter Kozen, Jens Palsberg, and Michael I. Schwartzbach. Efficient inference of partial types. In *Proc. 33rd Symp. Found. Comput. Sci.*, pages 363–371. IEEE, October 1992.

[KPS93]   Dexter Kozen, Jens Palsberg, and Michael I. Schwartzbach. Efficient recursive subtyping. In *Proc. 20th Symp. Princip. Programming Lang.*, pages 419–428. ACM, January 1993.

[KPS94]   Dexter Kozen, Jens Palsberg, and Michael I. Schwartzbach. Efficient inference of partial types. *J. Comput. Syst. Sci.*, 49(2):306–324, October 1994.

[Mis84]   P. Mishra. Towards a theory of types in PROLOG. In *Proc. 1st Symp. Logic Programming*, pages 289–298. IEEE, 1984.

[MR85]   P. Mishra and U. Reddy. Declaration-free type checking. In *Proc. 12th Symp. Principles of Programming Languages*, pages 7–21. ACM, 1985.

[Rey69]   J. C. Reynolds. Automatic computation of data set definitions. In *Information Processing 68*, pages 456–461. North-Holland, 1969.

[Ste94]   K. Stefánsson. Systems of set constraints with negative constraints are NEXPTIME-complete. In *Proc. 9th Symp. Logic in Computer Science*, pages 137–141. IEEE, June 1994.

[YO88]   J. Young and P. O'Keefe. Experience with a type evaluator. In D. Bjørner, A. P. Ershov, and N. D. Jones, editors, *Partial Evaluation and Mixed Computation*, pages 573–581. North-Holland, 1988.

Fatal Errors in Conditional Expressions
(Extended Abstract)

Michel Billaud

Laboratoire Bordelais de Recherche en Informatique, Université Bordeaux 1,
(UA 1304 du C.N.R.S) - 351 Cours de la Libération, 33405 Talence Cedex - France
billaud@labri.u-bordeaux.fr

Abstract. We propose a semantic model based on multivalued logic for the study of conditional expressions in programming languages. This model allows one to distinguish between different types of observable errors, whereas the coarser 3-valued model considers only one type of *undefined result*.

We investigate the equational properties of this multivalued model. We present a finite equational basis for *if-then-else*, for the *short-circuit* operators $+, \cdot, \neg$, and also for the *fully-evaluated* operators $\oplus, \odot$.

Finally, we present formulas for the number of distinct n-ary functions that can be expressed as combinations of short-circuit or fully-evaluated operators.

1 Conditional Expressions

1.1 Conditional Expressions and Three-valued Logic

Conditional expressions are aimed at representing boolean expressions in programming languages, but they have different properties because they are subject to evaluation errors.

The pioneering paper [3] by J. McCarthy presents a mathematical study of conditional expressions, where *undefined results* are represented by an extra value $\bot$ added to the set $B = \{\mathbf{t}, \mathbf{f}\}$ of ordinary truth values. The properties of the *if-then-else* operator "$\rightarrow$" defined by

$$(\mathbf{t} \rightarrow x, y) = x \qquad (\mathbf{f} \rightarrow x, y) = y \qquad (\bot \rightarrow x, y) = \bot \qquad \forall x, y \in \{\mathbf{t}, \mathbf{f}, \bot\}$$

are investigated from the algebraic point of view: the main result of the paper shows that the equational system $\mathcal{MC}$ (table 1) is an *equational basis* for the identities in this 3-valued algebra, that is every valid identity can be deduced from $\mathcal{MC}_1 - \mathcal{MC}_9$.

This operator plays a central role, as other conditional operators[1] can be derived from "$\rightarrow$": negation $\neg x = (x \rightarrow \mathbf{f}, \mathbf{t})$, *short-circuit or* $x + y = (x \rightarrow \mathbf{t}, y)$, *short-circuit and* $x.y = (x \rightarrow y, \mathbf{f})$, *fully-evaluated or* $x \oplus y = (x \rightarrow (y \rightarrow \mathbf{t}, \mathbf{t}), y)$, *fully-evaluated and* $x \odot y = (x \rightarrow y, (y \rightarrow \mathbf{f}, \mathbf{f}))$.

The 3-valued logic approach is now a classical topic [6], but it is still a live subject [5, 7, 12] with ongoing research and publications in the recent past [8, 9, 10, 11].

---

[1] shown here in their left-to-right version.

Table 1. MacCarthy's axioms for "→" in 3-valued logic

$$\mathcal{MC}_1 \qquad\qquad (\mathbf{t} \to A, B) \approx A$$
$$\mathcal{MC}_2 \qquad\qquad (\mathbf{f} \to A, B) \approx B$$
$$\mathcal{MC}_3 \qquad\qquad (P \to \mathbf{t}, \mathbf{f}) \approx P$$
$$\mathcal{MC}_4 \qquad ((P \to Q, R) \to A, B) \approx (P \to (Q \to A, B), (R \to A, B))$$
$$\mathcal{MC}_5 \qquad\qquad (P \to P, B) \approx (P \to \mathbf{t}, B)$$
$$\mathcal{MC}_6 \qquad\qquad (P \to A, P) \approx (P \to A, \mathbf{f})$$
$$\mathcal{MC}_7 \qquad (P \to (P \to A, B), C) \approx (P \to A, C)$$
$$\mathcal{MC}_8 \qquad (P \to A, (P \to B, C)) \approx (P \to A, C)$$
$$\mathcal{MC}_9 \; (P \to (Q \to A, B), (Q \to C, D)) \approx (Q \to (P \to A, C), (P \to B, D))$$

1.2 Dealing with Fatal Errors

In this paper we propose a more detailed study of *fatal errors* in conditional expressions. Informally, a *fatal error* is an event which happens during a computation and terminates it. At this stage, the major difference with the "undefined result" approach is that we allow for *several* types of errors. This approach aims to be more realistic, as we are often able to tell an "array bounds error" from a "division by zero" or an "endless loop" when an expression like t[i]==0)||(x/y<3)) goes wrong.

Fatal errors are easy to give a mathematical model based on multivalued logic: let Ω be a set of fatal errors with at least two distinct elements $\bot_1, \bot_2$, and $M = B \cup \Omega$. The *if-then-else* operator is defined by

$$(\mathbf{t} \to x, y) = x \quad (\mathbf{f} \to x, y) = y \quad (\omega \to x, y) = \omega \qquad \forall x, y \in \Omega, \forall x, y \in M$$

and the operators $\neg, ., +, \oplus, \odot$ are derived from $(\_ \to \_, \_)$ as in 1.1.

The equational properties of the multivalued model are clearly different from the 3-valued case; for example the fully-evaluated operator $\oplus$ is not commutative, since $\bot_1 \oplus \bot_2 = \bot_1$ and $\bot_2 \oplus \bot_1 = \bot_2$.

1.3 Paper Overview

This paper starts with definitions from equational theory. In section 3 we provide abstract definitions for the notion of sequentiality, and for fully-evaluated functions. The class FE of fully-evaluated functions, is shown to be contained in each subclone of SE (the class of all sequentialy evaluated functions) which contains a sequential extension for each boolean function. In section 4 we present a finite equational basis for the signature $\{\mathbf{t}, \mathbf{f}, \to\}$ and also for the (more familiar) operators $\{\mathbf{t}, \mathbf{f}, +, ., \neg\}$. The next section focuses on FE. We provide a finite basis for $\{\mathbf{t}, \mathbf{f}, \oplus, \odot\}$. Finally (section 6) we give recursive formulas for the *free spectra* of SE and FE, that is the number of distinct functions of n variables that can be expressed in each class.

2 Definitions and Notations

We use the following classical definitions from universal algebra [2, 4], with slightly modified notations for Id and D:

A *signature* $\mathcal{F}$ is a set of symbols f with a fixed arity $ar(f) \geq 0$. An *algebra* $\mathbf{A} = \langle A; F \rangle$ of type $\mathcal{F}$ is a pair where A is a non-empty set and $F = \{f^{\mathbf{A}} : A^{ar(f)} \to A \mid f \in \mathcal{F}\}$ is a family of operations indexed by $\mathcal{F}$. A mapping $h : A \to B$ is an *homomorphism* from $\mathbf{A}$ to $\mathbf{B}$ (algebras of the same type) if $h(f^{\mathbf{A}}(a_1, \ldots, _n)) = f^{\mathbf{B}}(h(a_1), \ldots, h(a_n))$ for each n-ary f and every sequence $a_1, \ldots, a_n$ from A.

Let X be a set of *variables*. $T(\mathcal{F}, X)$ is the set of finite terms built over $\mathcal{F}$ and X. The algebra of terms $\mathbf{T}(\mathcal{F}, X)$ is equipped with the natural *term constructor* functions associated to $\mathcal{F}$. An *identity* is an unordered pair of terms, written $p \approx q$. It is *valid* with respect to an algebra $\mathbf{A}$ (notation: $\mathbf{A} \models p \approx q$) if for every homomorphism $h : \mathbf{T}(\mathcal{F}, X) \to \mathbf{A}$ one has $h(p) = h(q)$. We note $Id(\mathbf{A}, \mathcal{F})$ the set of all such valid identities. Let Σ be a set of identities. One writes $\Sigma \vdash p \approx q$ if $p \approx q$ belongs to the *deductive closure* $D(\Sigma, \mathcal{F})$, that is the set of identities which can be deduced from Σ using the classical rules of equational logic: reflexivity, symmetry and transitivity of $\approx$, closure under replacement and substitution by terms from $T(\mathcal{F}, X)$. We call Σ an *equational basis* for the identities of $\mathbf{A}$ on the signature $\mathcal{F}$ if $Id(\mathbf{A}, \mathcal{F}) = D(\Sigma, \mathcal{F})$.

Let C be a class of functions over an algebra. C is a *clone* if it is closed by composition and contains all projections. A clone contained in C is called a *subclone* of C. Let F be a set of functions; $cl(F)$ is the least clone which contains F (it is the clone generated by F).

Let $\mathcal{F} = \{\mathbf{t}, \mathbf{f}, +, ., \oplus, \odot, \neg, \to\}$, $\mathcal{F}_1 = \{\mathbf{t}, \mathbf{f}, \to\}$, $\mathcal{F}_2 = \{\mathbf{t}, \mathbf{f}, +, ., \neg\}$ and $\mathcal{F}_f = \{\mathbf{t}, \mathbf{f}, \oplus, \odot, \neg\}$. $\mathbf{M}$ denotes the algebra $\langle M; F \rangle$ where F is the set of operations corresponding to $\mathcal{F}$ defined on $M = \{\mathbf{t}, \mathbf{f}\} \cup \Omega$ as defined in 1.1. $\mathbf{B}$ is the 2-valued subalgebra of $\mathbf{M}$ with operators restricted to $\{\mathbf{t}, \mathbf{f}\}$.

In a compound term $(P \to C, A)$ we call the subterm P the *premise*, C the *conclusion*, and A the *alternative*.

3 Sequential Extensions with Errors

Not all functions over M represent conditional expressions that can be computed sequentially, so we first define the classes of functions we are dealing with: SE the class of all *sequential extensions with errors* of functions over B, and FE the subclass of *fully-evaluated* functions.

These definitions do not rely directly on some arbitrarily chosen evaluation mechanism, nor on an *a priori* set of primitive operators, but on the expression of constraints on the propagation of errors. Representation theorems then show that SE is generated by $\mathcal{F}_1$, (and the equivalent signature $\mathcal{F}_2$) and FE by $\mathcal{F}_f$.

We show that every clone of sequential functions which contains extensions of all 2-valued function contains FE as a subclone. In other words, in every system of primitive sequential operators whose restrictions on B generate all boolean functions, it is possible to build representations for all fully-evaluated functions.

3.1 Sequentiality

Unlike Vuillemin's [14], our definition of sequentiality is not related to domain theory, but it is directly based on the notion of fatal errors.

Informally, the definitions below express the idea that, when a non-constant expression is evaluated in a sequential way, one of the variables has to be designated as the first candidate for evaluation. If its evaluation returns an error, then the whole computation breaks. If not, the control is transferred to a subexpression (which is also sequentially evaluated) where the value of the key is known.

Definition 1. Let $f : B^n \to B$. We call $f' : M^n \to M$ an *extension with errors* of f if

- $f'(b_1 \ldots b_n) = f(b_1 \ldots b_n)$ for all $b_i \in B$
- for all $x_1 \ldots x_n \in M$, if $f'(x_1 \ldots x_n) \in \Omega$, then there is a k such that $\forall \omega \in \Omega$:

$$f'(x_1 \ldots x_{k-1}, \omega, x_{k+1} \ldots x_n) = \omega$$

Remark. The latter point expresses the fact that an error in the result in a given context can be attributed to a faulty key parameter which transmits *all* sorts of errors in the same context $x_1 \ldots x_{k-1}, -, x_{k+1} \ldots x_n$.

Remark. If the function f' is constant then its value is in B.

Definition 2. Let $g : M^n \to M$ be an extension with errors of some function over B. The function g is *sequential* if either g is constant over M^n, or there exists a k such that

- $g(x_1 \ldots x_{k-1}, \omega, x_{k+1} \ldots x_n) = \omega \; \forall \omega \in \Omega, x_1 \ldots x_{k-1}, x_{k+1} \ldots x_n \in M$
- the functions g_t^k, g_f^k (with arity $n-1$) are sequential, where $g_b^k(x_1 \ldots x_{n-1}) = g(x_1 \ldots x_{k-1}, b, x_k \ldots x_{n-1})$

SE is the class of all sequential extensions of functions on B.

Remark. The key is unique because Ω has at least 2 distinct elements.

Example 1. The negation operator has a unique sequential extension: it is not a constant function, thus its only parameter is the key. Consequently all extensions n of the negation satisfy the equations $n(\mathbf{t}) = \mathbf{f}$, $n(\mathbf{f}) = \mathbf{t}$, $n(\omega) = \omega \; \forall \omega \in \Omega$, and then n is "$\neg$".

Example 2. The binary "or" operator has 4 sequential extensions. Let $f(x, y)$ such an extension. As f is not constant, we first suppose that x is the key. Then: $f(\omega, y) = \omega, \; \forall \omega \in \Omega, \forall y \in M$. The functions $\lambda y.f(\mathbf{f}, y)$ and $\lambda y.f(\mathbf{t}, y)$ are also required to be sequential. Since $\lambda y.f(\mathbf{f}, y)$ is unary and not constant, one has $f(\mathbf{f}, \omega) = \omega \; \forall \omega \in \Omega$.

If we suppose that $\lambda(y).f(\mathbf{t}, y)$ is constant, then $f(\mathbf{t}, y) = y \forall y$, so f is "$+$". If is not constant, then $f(\mathbf{t}, \omega) = \omega \forall \omega \in \Omega$, so f actually is "$\oplus$".

Taking y as the key of $f(x, y)$ yields the right-to-left versions of "$+$" and "$\oplus$".

Theorem 3 Representation theorem for SE. $SE = cl(\mathcal{F}_1) = cl(\mathcal{F}_2)$

Proof. Obviously SE is a clone, and that $\to, +, \cdot, \neg$ are in SE. Thus $cl(\mathcal{F}_1) \subseteq SE$ and $cl(\mathcal{F}_2) \subseteq SE$. Conversely, let f be in SE. We built a *concrete representation* of f as a term $r(f) \in T(\mathcal{F}_1, \{x_1 \ldots x_n\})$ (resp. $T(\mathcal{F}_2, \{x_1 \ldots x_n\})$) by induction: if f is a constant, it has its value in **B** thus $r(f) = \mathbf{t}$ or $r(f) = \mathbf{f}$. If f is not a constant, it has a key x_k and then we take $r(f) = (x_k \to r(f_t^k), r(f_f^k))$ (resp. $r(f) = x_k . r(f_t^k) + \neg x_k . r(f_f^k)$)

3.2 Fully-evaluated Functions

Definition 4. A function f depends on its k-th parameter if there are values $x_1, \ldots, x_n$ and $x_k' \neq x_k$ such that $f(x_1 \ldots x_k \ldots x_n) \neq f(x_1 \ldots x_k' \ldots x_n)$

Definition 5. Let $f : M^n \to M$ be a function in SE. f is fully-evaluated if, for each k such that f depends on its k-th parameter, one has

$$f(x_1 \ldots x_{k-1}, \omega, x_{k+1} \ldots x_n) \in \Omega$$

where all x_i range over M, and $\omega \in \Omega$. FE is the class of all fully-evaluated functions.

Theorem 6 Representation theorem for FE. $FE = cl(\mathcal{F}_f)$

Proof. Similar to 3, taking $r(f) = (x_k \odot r(f_t^k)) \oplus (\neg x_k \odot r(f_f^k))$ as a representation for the nonconstant function f with key x_k.

Theorem 7. *Let $K \subseteq SE$ be a clone such that every $f : B^n \to B$ has at least an extension in K. Then $FE \subseteq K$.*

Proof. Such a K obviously contains "$\neg$" and at least an extension of "or". Suppose that the "$+$" operator is in K. One has $(x \to y, z) = xy + (\neg x)z = \neg(\neg x + \neg y) + \neg(\neg\neg x + \neg z)$, thus the "if-then-else" operator belongs to K and $K = SE$. On the other hand, if "$+$" is not in K, K contains $\oplus$, and also $\odot$ as $x \odot y = \neg(\neg x \oplus \neg y)$. Consequently $FE \subseteq K$.

Remark. There are other strict subclones of SE which contain extensions of every 2-valued function, for example $cl(\{\alpha, \oplus, \odot, \neg, \mathbf{t}, \mathbf{f}\})$, where $\alpha(x, y) = x + y.\mathbf{f}$.

However, a systematic study of the subclones of SE is far beyond the scope of this paper. See [13] for the 2-valued case, and [1] for 3-valued logic.

4 Equational Bases for Conditional Expressions

4.1 A Finite Basis for $\mathcal{F}_1$

Proposition 8. *The system $\mathcal{A}$ below is sound, that is $D(\mathcal{A}, \mathcal{F}_1) \subseteq Id(\mathbf{M}, \mathcal{F}_1)$*

$$
\begin{array}{ll}
\mathcal{A}_1 & (\mathbf{t} \to A, B) \approx A \\
\mathcal{A}_2 & (\mathbf{f} \to A, B) \approx B \\
\mathcal{A}_3 & (P \to \mathbf{t}, \mathbf{f}) \approx P \\
\mathcal{A}_4 & ((P \to Q, R) \to A, B) \approx (P \to (Q \to A, B), (R \to A, B)) \\
\mathcal{A}_5 & (P \to P, B) \approx (P \to \mathbf{t}, B) \\
\mathcal{A}_6 & (P \to A, P) \approx (P \to A, \mathbf{f}) \\
\mathcal{A}_7 & (P \to (Q \to A, B), C) \approx (P \to ((P \to Q, Q') \to (P \to A, A'), \\
& \qquad\qquad\qquad\qquad\qquad\qquad\qquad (P \to B, B')), C) \\
\mathcal{A}_8 & (P \to A, (Q \to B, C)) \approx (P \to A, ((P \to Q', Q) \to (P \to B', B), \\
& \qquad\qquad\qquad\qquad\qquad\qquad\qquad\qquad (P \to C', C)))
\end{array}
$$

Definition 9. A term $t \in T(\mathcal{F}_1, X)$ is a *binary decision diagram* (BDD for short) if each compound subterm $(p \to a, b)$ of t is such that $p \in X$, $a, b \notin X$. A BDD t is a *canonical form* (CF) if no compound subterm $(v \to a, b)$ of t has an occurrence of v in a or b.

Lemma 10. *For each term E there is a BDD E' such that $M \models E \approx E'$, and one has $\mathcal{A}_1 - \mathcal{A}_4 \vdash E \approx E'$.*

Proof. Constructively: unfold compound premises by $\mathcal{A}_4$; replace each variable v that appears as the conclusion or the alternative of a compound term by $(v \to \mathbf{t}, \mathbf{f})$ using $\mathcal{A}_3$; eliminate constant premises by $\mathcal{A}_1$ and $\mathcal{A}_2$.

Lemma 11. *Let $v \in X$. For all terms a, b one has:*

$$\mathcal{A}_5 - \mathcal{A}_8 \vdash (v \to a, b) \approx (v \to a[\mathbf{t}/v], b[\mathbf{f}/v])$$

Proof. Let $E = (v \to a, b)$. One has $\mathcal{A}_5, \mathcal{A}_7 \vdash E \approx (v \to a[\mathbf{t}/v], b)$ because:

- if $a \in \{\mathbf{t}, \mathbf{f}\} \cup X \setminus \{v\}$, clearly $a[\mathbf{t}/v] = a$
- if $a = v$, $\mathcal{A}_5 \vdash E = (v \to v, b) \approx (v \to \mathbf{t}, b)$
- if $a = (q \to x, y)$, use $\mathcal{A}_7$, then apply the induction hypothesis to the subterms q, x, y of a, and use $\mathcal{A}_7$ backwards.

Similarly, $\mathcal{A}_6, \mathcal{A}_8 \vdash E \approx (v \to a, b[\mathbf{f}/v])$. This proves the lemma.

Lemma 12. *For each term E there is a unique term $cf(E) \in CF$ such that $\mathbf{M} \models E \approx cf(E)$, and one has $\mathcal{A} \vdash E \approx cf(E)$.*

Proof. **Existence** of $cf(E)$: By lemma 10 there is a BDD E' such that $\mathcal{A}_1 - \mathcal{A}_4 \vdash E \approx E'$. Applying lemma 11 to E' to remove repeated variables, and then simplifying constant premises by $\mathcal{A}_1, \mathcal{A}_2$ clearly yields a term in CF equivalent to E. **Uniqueness:** we show that $\forall E \in CF$, then for every $E' \in CF$ such that $\mathbf{M} \models E \approx E'$ one has $E = E'$. By structural induction on E:

- If $E = \mathbf{t}$ then E' is different from $\mathbf{f}$. Also E' cannot be a compound term $(v \to a, b)$, because in this case $E[\perp/v] = \mathbf{t}$ and $E'[\perp/v]) = \perp$. Therefore $E' = \mathbf{t}$.

- For the same reasons, if $E = \mathbf{f}$ then $E' = \mathbf{f}$.
- Also, if $E = (v \to a, b)$, then E' is a compound term $(v' \to a', b')$. Obviously $v = v'$ (otherwise the substitution $\sigma = [\bot/v, \omega/v']$ yields distinct values for $E\sigma = \bot$ and $E'\sigma = \omega$). As $E \in CF$, v has no occurrence in a, so $E[\mathbf{t}/v] = (\mathbf{t} \to a, b[\mathbf{t}/v]) \approx a$ (by $\mathcal{A}_1$). Likewise $(\mathbf{t} \to a', b'[\mathbf{t}/v]) \approx a'$, thus $a \approx a'$ and (by induction hypothesis) $a = a'$. Similarly $b = b'$.

Theorem 13. *$\mathcal{A}$ is an equational basis for $Id(\mathbf{M}, \mathcal{F}_1)$.*

Proof. It follows from lemma 8 that $D(\mathcal{A}, \mathcal{F}_1) \subseteq Id(\mathbf{M}, \mathcal{F}_1)$. Conversely, let p, q be two terms from $T(\mathcal{F}_1, X)$ such that $\mathbf{M} \models p \approx q$. By lemma 12 their canonical forms are the same, and $\mathcal{A} \vdash p \approx cf(p) = cf(q) \approx q$, thus $p \approx q \in D(\mathcal{A}, \mathcal{F}_1)$.

4.2 A Finite Basis for $\mathcal{F}_2$

As we can express $+, \cdot, \neg$ using $\to$ and vice-versa, the question of the *existence* of a finite basis for the subsignature $\mathcal{F}_2 = \{\mathbf{t}, \mathbf{f}, +, ., \neg\}$ is closed in principle: a basis is obtained by a mere translation of the equations of $\mathcal{A}_1 - \mathcal{A}_8$ using the function ψ below. But such a translation is not really satisfactory, as the resulting equations are very large and do not provide much insight. A more direct approach is required.

Proposition 14. $\phi : T(\mathcal{F}_2, X) \to T(\mathcal{F}_1, X)$, $\psi : T(\mathcal{F}_1, X) \to T(\mathcal{F}_2, X)$ *below* $\mathbf{M} \models p \approx \phi(p)$ *and* $\mathbf{M} \models q \approx \psi(q)$ $\forall p \in T(\mathcal{F}_2, X), q \in T(\mathcal{F}_1, X)$

- $\phi(\mathbf{t}) = \mathbf{t}$, $\phi(\mathbf{f}) = \mathbf{f}$, $\phi(v) = v$, $\phi(\neg a) = (\phi(a) \to \mathbf{f}, \mathbf{t})$, $\phi(a + b) = (\phi(a) \to \mathbf{t}, \phi(b))$, $\phi(a.b) = (\phi(a) \to \phi(b), \mathbf{f})$, $\forall v \in X, \forall a, b \in T(\mathcal{F}_2, X)$.
- $\psi(\mathbf{t}) = \mathbf{t}$, $\psi(\mathbf{f}) = \mathbf{f}$, $\psi(v) = v$, $\psi(p \to q, r) = \psi(p).\psi(q) + (\neg\psi(p)).\psi(r)$ $\forall v \in X, \forall p, q, r \in T(\mathcal{F}_1, X)$.

Theorem 15. *The system $\mathcal{B}$ is a basis for $Id(\mathbf{M}, \mathcal{F}_2)$*

$$
\begin{array}{lc}
B_1 & \neg\mathbf{t} \approx \mathbf{f} \\
B_2 & \neg\neg A \approx A \\
B_3 & \neg(A.B) \approx \neg A + \neg B \\
B_4 & (A + B) + C \approx A + (B + C) \\
B_5 & A.(B + C) \approx A.B + A.C \\
B_6 & (A + B).C \approx A.C + \neg A.(B.C) \\
B_7 & \mathbf{t}.A \approx A \\
B_8 & \mathbf{f}.A \approx \mathbf{f} \\
B_9 & A.\mathbf{t} \approx A \\
B_{10} & A.(B.A) \approx A.B \\
B_{11} & A.(B.\neg A) \approx (A.B).\mathbf{f} \\
B_{12} & A.B + \neg A.C \approx \neg A.C + A.B \\
B_{13} & A.B + C \approx (A.B + \neg A.C) + A.C
\end{array}
$$

Proof. Follows from lemmas 16, 20, and 21 below.

Lemma 16. $D(\mathcal{B}, \mathcal{F}_2)) \subseteq Id(\mathbf{M}, \mathcal{F}_2)$

Definition 17. We define the *dual* $t^\sharp$ of a term t inductively by $v^\sharp = v \; \forall v \in X$, $\mathbf{t}^\sharp = \mathbf{f}$, $\mathbf{f}^\sharp = \mathbf{t}$, $(p+q)^\sharp = p^\sharp.q^\sharp$, $(p.q)^\sharp = p^\sharp + q^\sharp$, $(\neg p)^\sharp = \neg(p^\sharp)$. The dual of an identity $p \approx q$ is $p^\sharp \approx q^\sharp$.

Lemma 18. $\mathcal{B} \vdash p \approx q \iff \mathcal{B} \vdash p^\sharp \approx q^\sharp$.

Proof. Follows from the property: let t be a term with variables $x_1, \ldots x_n$. Then
$$\mathcal{B}_1 - \mathcal{B}_3 \vdash t^\sharp[X_1/x_1 \ldots X_n/x_n] \approx \neg t[\neg X_1/x_1 \ldots \neg X_n/x_n]$$

In the sequel we will use the associativity of "." ($\mathcal{B}_4$) without explicit mention.

Definition 19. The set $NF \subseteq T(\mathcal{F}_2, X)$ of terms in *normal form* is defined inductively as the least set containing $\mathbf{t}, \mathbf{f}$ and such that $v.A + \neg v.B \in NF$ for all $v \in X$ and $A, B \in NF$ without occurrences of v.

Lemma 20. *For each $t \in T(\mathcal{F}_2, X)$ there is a unique term $nf(t) \in NF$ such that $\mathbf{M} \models t \approx nf(t)$.*

Proof. We show that $nf(t) = \psi(cf(\phi(t)))$ has the required properties: a structural induction reveals that $p \in CF$ implies $\psi(p) \in NF$, thus $nf(t) \in NF$. $\mathbf{M} \models t \approx nf(t)$ follows from lemmas 12 and 14. For the proof of uniqueness the technique is the same as in lemma 12.

Lemma 21. $\mathcal{B} \vdash t \approx nf(t)$

Proof. Auxiliary definitions for normal forms are required:

A *normalized product* is a term $f_1.f_2.\ldots.f_n$, where each factor $f_i (i < n)$ is either a variable or the negation of a variable, $f_n \in \{\mathbf{t}, \mathbf{f}\}$, and no variable appears in two distinct factors.

A *normalized sum* is a term $p_1 + p_2 + \ldots + p_n$ such that each $p_i (i < n)$ is a normalized product different from $\mathbf{t}$ or $\mathbf{f}$, and the last product p_n is $\mathbf{t}$ or $\mathbf{f}$.

First we show that, for each term t there is a normalized sum $ns(t)$ such that $\mathcal{B}_1 - \mathcal{B}_{11} \vdash t \approx ns(t)$: $\mathcal{B}_1 - \mathcal{B}_6$ (and the dual identities) allow us to develop any term under the form of a sum of products of variables, negations of variables and constants. Each of these products can be normalized by $\mathcal{B}_{10} - \mathcal{B}_{11}$ (deleting multiple occurrences of the same variable) $\mathcal{B}_7 - \mathcal{B}_8$ (deleting useless factors), and introducing a last factor $\mathbf{t}$ by $\mathcal{B}_9$ if needed. The resulting sum is then normalized by the duals of $\mathcal{B}_7 - \mathcal{B}_9$.

The lemma follows by an induction on the number n of variables in $ns(t)$: if $n = 0$, clearly $ns(t) \in \{\mathbf{t}, \mathbf{f}\} \subset NF$. If $n > 0$, $ns(t)$ is the sum of at least two normalized products $ns(t) = f_1 + f_2 + \ldots + f_n$. The product f_1 is not a constant, and thus it starts with a v or $\neg v$ for some $v \in X$. By $\mathcal{B}_{13}$, $ns(t)$ turns into $f_1 + \neg v.f_2 + v.f_2 + \ldots + \neg v.f_n + v.f_n$. Each product is then normalized in order to remove extra occurrences of v. By $\mathcal{B}_{12}$ we split that expression into a sum of factors starting with v plus a sum of factors starting with $\neg v$. A factorization by $\mathcal{B}_5$ puts that sum under the form $v.A + \neg v.B$. As there is no occurrence of v in A and B, the induction hypothesis can be applied to A and B, and this leads to $nf(t) = v.nf(A) + \neg v.nf(B)$

5 A Basis for Fully-Evaluated Operators

The following lemma reveals a fundamental link between 2-valued and multivalued interpretations of fully-evaluated operators:

Lemma 22. $\mathbf{M} \models p \approx q \Leftrightarrow$ *iff* $\mathbf{M} \models p \odot \mathbf{f} \approx q \odot \mathbf{f}$ *and* $\mathbf{B} \models p \approx q$

We will see later (Lemma 29) that the condition "$\mathbf{M} \models p \odot \mathbf{f} \approx q \odot \mathbf{f}$" can be tested by a simple traversal of p and q.

Proof. ($\Rightarrow$) Because $\mathbf{B}$ is a subalgebra of $\mathbf{M}$. ($\Leftarrow$) Let $h : T(\mathcal{F}_f, X) \to \mathbf{M}$ be an homomorphism. Let us suppose that $h(p) \neq h(q)$. There are three cases (up to the permutation of p and q):

1. $h(p) = \mathbf{t}$ and $h(q) = \mathbf{f}$: then we define $h' : T(\mathcal{F}_f, X) \to \mathbf{M}$ by $h'(v) = h(v)$ for the variables of p and q, and $h'(v) = \mathbf{t}$ for other variables. Clearly $h'(p) \neq h'(q)$, thus $\mathbf{B} \not\models p \approx q$.
2. $h(p) \in \{\mathbf{t}, \mathbf{f}\}$ and $h(q) \in \Omega$. Then $h(p \odot \mathbf{f}) = \mathbf{f} \neq h(q \odot \mathbf{f}) = h(q) \in \Omega$.
3. $h(p) \in \Omega$ and $h(q) \in \Omega$: then $h(p \odot \mathbf{f}) = h(p) \neq h(q) = h(q \odot \mathbf{f})$.

A corollary is that the equivalence of terms in the 2-valued model amounts to the equivalence in the multivalued model up to some "residual terms":

Lemma 23. $\mathbf{B} \models A \approx B \Longleftrightarrow \mathbf{M} \models A \oplus (B \odot \mathbf{f}) \approx (A \odot \mathbf{f}) \oplus B$

Proof. From the previous lemma, using the property

$$\mathbf{M} \models (A \oplus (B \odot \mathbf{f})) \odot \mathbf{f} \approx ((A \odot \mathbf{f}) \oplus B) \odot \mathbf{f}$$

Lemma 24. *The following system C is sound.*

C_1	$\neg \mathbf{t} \approx \mathbf{f}$
C_2	$\neg\neg A \approx A$
C_3	$\neg(A \odot B) \approx \neg A \oplus \neg B$
C_4	$(A \oplus B) \oplus C \approx A \oplus (B \oplus C)$
C_5	$A \odot (B \oplus C) \approx (A \odot B) \oplus (A \odot C)$
C_6	$(A \oplus B) \odot C \approx (((A \odot \mathbf{f}) \oplus (B \odot \mathbf{f})) \oplus (A \odot C)) \oplus (B \odot C)$
C_7	$A \odot \mathbf{t} \approx A$
C_8	$\mathbf{t} \odot A \approx A$
C_9	$A \odot \mathbf{f} \approx \mathbf{f} \odot A$
C_{10}	$\neg A \odot \mathbf{f} \approx A \odot \mathbf{f}$
C_{11}	$A \odot A \approx A$
C_{12}	$A \odot \neg A \approx A \odot \mathbf{f}$
C_{13}	$A \oplus B \approx (A \odot \mathbf{f}) \oplus (B \oplus A)$

Lemma 25. $\oplus$ *and* $\odot$ *are dual operators.* $\mathbf{M} \models p \approx q \Leftrightarrow \mathbf{M} \models p^\mathrm{l} \approx q^\mathrm{l}$.

Proposition 26. *The following identities are equational consequences of $\mathcal{C}$:*

$$\mathcal{C}_{14} \qquad A \oplus B \oplus A \approx A \oplus B$$
$$\mathcal{C}_{15} \qquad A \odot B \odot \neg A \approx A \odot B \odot \mathbf{f}$$
$$\mathcal{C}_{16} \qquad (p \oplus q) \odot \mathbf{f} \approx (p \odot \mathbf{f}) \oplus (q \odot \mathbf{f})$$
$$\mathcal{C}_{17} \qquad (p \odot q) \odot \mathbf{f} \approx (p \odot \mathbf{f}) \oplus (q \odot \mathbf{f})$$
$$\mathcal{C}_{18} \qquad A \approx (A \odot \mathbf{f}) \oplus A$$
$$\mathcal{C}_{19} \qquad A \odot B \odot C \approx (A \odot \mathbf{f}) \oplus (B \odot A \odot C)$$
$$\mathcal{C}_{20} \qquad A \oplus \neg A \approx A \odot \mathbf{f}$$
$$\mathcal{C}_{21} \ (A \odot B) \oplus (\neg A \odot B) \approx A \odot \mathbf{f} \oplus B$$

Definition 27. Let $\Lambda(t)$ be the sequence of variables of a term t in their left-to-right order of first appearance. It is defined inductively by $\Lambda(\mathbf{t}) = \Lambda(\mathbf{f}) = \langle\rangle$, $\Lambda(x) = \langle v \rangle \ \forall v \in X$, $\Lambda(\neg t) = \Lambda(t)$, $\Lambda(p \oplus q) = \Lambda(p \odot q) = \Lambda(p) \diamond \Lambda(q)$, where "$\diamond$" is the concatenation without repetition of sequences (keeping only the leftmost occurrence of each variable).

Definition 28. The binary operations $\oplus, \odot$ are generalized over sequences of terms:

$$\bigoplus_{t \in \langle t_1 \ldots t_n \rangle} f(t) = f(t_1) \oplus \ldots \oplus f(t_n) \qquad\qquad \bigoplus_{t \in \langle\rangle} f(t) = \mathbf{f}$$
$$\bigodot_{t \in \langle t_1 \ldots t_n \rangle} f(t) = f(t_1) \odot \ldots \odot f(t_n) \qquad\qquad \bigodot_{t \in \langle\rangle} f(t) = \mathbf{t}$$

Lemma 29. $\mathcal{C} \vdash p \odot \mathbf{f} \approx \bigoplus_{v \in \Lambda(p)} v \odot \mathbf{f}$

Proof. Induction on p, using $\mathcal{C}_8, \mathcal{C}_{11}, \mathcal{C}_{10}, \mathcal{C}_{16}, \mathcal{C}_{17}$ and trivial properties of "$\diamond$".

The definition of normal forms requires arbitrary total ordering relations on X and on the set of products of variables (and negations of variables):

Definition 30. A *complete product* over a finite subset Y of X is a term $t = \bigodot_{v \in \mu} f(v)$ where μ is the finite ordered sequence (without repetitions) of all variables of Y, and $f(v) \in \{v, \neg v\}$

An *expanded sum* over Y is a term $t = \bigoplus_{t_i \in S} t_i$, where S is an ordered sequence (without repetitions) of complete products over Y.

A term t is a *normal form* (NF) if $t \in \{\mathbf{t}, \mathbf{f}\}$ or $t = G \oplus E$ where $G = \bigoplus_{v \in V} v \odot \mathbf{f}$ is the "guard" of t, V is a finite sequence of variables without repetitions, and E is an expanded sum over V.

Lemma 31. *If* $\mathbf{M} \models p \approx q$ *with* $p, q \in NF$, *then* $p = q$.

Proof. From lemma 22 and uniqueness of expanded sums for 2-valued terms.

Lemma 32. *For each term t there is a term $t' \in NF$ such that $\mathcal{C} \vdash t \approx t'$.*

Proof. If t has no variables, it can be reduced to $\mathbf{t}$ or $\mathbf{f}$ using $\mathcal{C}_1, \mathcal{C}_2, \mathcal{C}_7 - \mathcal{C}_{12}$ and their duals. Otherwise we first build the "guard" of the normal form, remarking $(\mathcal{C}_{18}$ and lemma 29) that $t \approx (t \odot \mathbf{f}) \oplus t \approx (\bigoplus_{v \in \Lambda(t)} v \odot \mathbf{f}) \oplus t$. The normal form of t is obtained by a sequence of transformations of the expansion part:

1. Turn t into a sum of products of constants, variables and negations of variables (using $C_1 - C_6$).
2. If a variable appears two or more times in the same product, replace its useless occurrences by t or f according to $C_{14}^{\sharp}$ and C_{15}.
3. Remove the constant t from the products $(C_7 - C_8)$.
4. Remove the products that contain f: they are simply "residual terms" and thus they can be deleted by application of C_9, C_{14} and lemma 29.
5. Turn the products into complete products by the following algorithm:

    for each variable v appearing in t (in decreasing order) do:

    for each product p in the sum do:

    (a) if $p = \alpha \odot v \odot \beta$ replace p by $v \odot \alpha \odot \beta$
    (b) if $p = \alpha \odot \neg v \odot \beta$ replace p by $\neg v \odot \alpha \odot \beta$
    (c) if v has no occurrence in p, replace p by $(v \odot p) \oplus (\neg v \odot p)$

    Each step is a combination of identities of C: (a) and (b) stem from C_{19} and (c) from C_{21}.
6. Reorder the sum of complete products by C_{13}, and remove duplicates by $C_{11}^{\sharp}$.

As a consequence of the soundness lemma 24 and completeness lemma 32:

Theorem 33. C *is a basis for* $Id(M, \mathcal{F}_f)$.

6 Concluding Remarks

6.1 Counting Functions

It is well-known that there are exactly 2^{2^n} distincts functions of n variables in 2-valued logic. What about the number of functions in our multivalued logic?

Sequential functions: as all n-ary sequential functions from SE are represented by terms built over the signature $\mathcal{F}_1$ and variables from $X_n = \{v_1 \ldots v_n\}$, counting n-ary functions then amounts (lemma 12) to find the number $F(n)$ of the elements of $K(X_n)$, the set of canonical forms of $T(\mathcal{F}_1, X_n)$. This set satisfies

$$K(X_n) = \{t, f\} \cup \{(v_i \to p, q) \mid 1 \leq i \leq n \land p, q \in K(X_n \setminus \{v_i\})\}$$

and thus one has $F(n) = nF(n-1)^2 + 2$ for $n > 0$. Obviously $F(0) = 2$.

Fully-evaluated functions: the representative term (in normal form) of a n-ary function in FE is determined by its guard (a sequence of k distinct variables from X_n) and its expansion over the variables of its guard.

There are $\frac{n!}{(n-k)!}$ distinct guards of k variables from X_n, and for each such guard there is 2^{2^k} distinct expansions, so the number $G(n)$ of distinct fully-evaluated functions with n variables is: $G(n) = \sum_{k=0}^{k=n} \frac{n!}{(n-k)!} 2^{2^k}$.

The first values of $F(n)$ and $G(n)$ are

n	0	1	2	3	4
$F(n)$	2	6	74	16,430	1,079,779,602
$G(n)$	2	6	42	1,646	1,579,218

6.2 Directions for Further Research

Other "if-then-else" Operators. In Lisp dialects conditional operators return other values than t and f. The logical value "false" is represented by an element nil. It there a finite axiomatization for the conditional operator defined by $(x \rightarrow y, z) = x$ if x is an error, z if x=nil, and y otherwise?

Axiomatic Semantics of Exceptions. Exceptions (errors that can be recovered) are easily modeled using multivalued logics.

The algebraic properties of the programming language features $raise(error)$ and $on(error, e_1, e_2)$ (meaning: evaluate e_1, if $error$ is raised during the evaluation then return the value of e_2 instead) should be investigated.

References

1. J. BERMAN, *Free spectra of 3-element algebras*, in: R. Freese and O. García, eds. Universal Algebra and Lattice Theory, Lecture Notes in Mathematics 1004, (Springer, Berlin, 1983), pp. 10-53.
2. S. BURRIS and H.P. SANKAPPANAVAR, *A Course in Universal Algebra*, Graduate Texts in Mathematics 78 (Springer-Verlag, 1978).
3. John McCARTHY *A basis for a mathematical theory of computation*, Computer Programming and Formal Systems, P. Braffort D. Hirschberg editors North Holland Publishing Company, 1963, pages 33-70
4. M. P. COHN, Universal Algebra, Harper and Row, 1965.
5. Stephen L. BLOOM and Ralph TINDELL *Varieties of "if-then-else"*, SIAM J. Comput. Vol. 12, No. 4, pp. 677-707 November 1983
6. D. GRIES, *The Science of Programming*, Springer, New York, 1981.
7. Irène GUESSARIAN and José MESEGUER *On the axiomatization of "if-then-else"* SIAM J. Comput Vol 16, No. 2, pp. 332-357 April 1987
8. Fernando GUZMÁN and Craig C. SQUIER; *The algebra of conditional logic*, Algebra Universalis, 27 (1990), pp. 88-110.
9. Fernando GUZMÁN; *Three-valued logics in the semantics of programming languages*, Unpublished manuscript, March 1992.
10. Ernest G. MANES *Equations for "if-then-else"*, Proceedings Math. Found. of Programming Semantics, 7th international conference Pittsburgh, march 25-28, 1991 LNCS 598 pp. 446-456 Springer-Verlag, 1992
11. Ernest G. MANES *Adas and the equational theory of if-then-else*, Algebra Universalis, 30 (1993) pp. 272-394.
12. Alan H. MEKLER and Evelyn M. NELSON; *Equational bases for if-then-else*, SIAM J. Comput. 16 (1987) pp. 465-485.
13. Emil L. POST, *The two-valued iterative systems of mathematical logic*, Annals of Math. Studies (5), Princeton University Press (1941).
14. J. VUILLEMIN, *Correct and optimal implementation of recursion in a simple programming language*, J. Comp. Systems Sci. 9 (1974), 332-354.

Different Types of Arrow Between Logical Frameworks

Till Mossakowski

University of Bremen, Dept. of Computer Science, P.O.Box 33 04 40, D-28334 Bremen
Phone: +49-421-218-2935, E-mail: till@informatik.uni-bremen.de

1 Introduction

There is a variety of specification languages for the formal specification and development of correct software systems, for example, see [26, 14]. They differ in purpose, expressiveness, level of abstraction (requirement, design, implementation), notation, available tools etc. It has been argued that one should not construct one universal all-purpose specification language (which would be very clumsy), but rather relate different languages based on different logical frameworks. Thus there is a need for a meta-notion of logical framework and a notion of map between logical frameworks in order to relate them. This leads to a category of logical frameworks.

In the literature, there are described not only one, but many categories of logical frameworks: institutions [13] with maps of institutions [16], specification frames [10], institutions with simulations [2], pre-institutions with transformations [21], institutional frames [27], π- and τ-institutions [12, 23], institutions with contexts [19], etc. Some of these were related by Maura Cerioli in her thesis [6]. Thus, there are quite different types of logical frameworks, and each type of logical framework (together with a type of arrow between logical frameworks) leads to a new category of logical frameworks.

The purpose of the morphisms in the above mentioned categories of logical frameworks is that of encoding, or representing one logical framework into another one. [1] With this, one can

- write heterogeneous specifications with components written in different logical frameworks [13, 25, 3],
- switch between different types of logical framework by borrowing missing logical structure along maps [7], e.g. endow an institution with an entailment relation and a proof calculus,
- compare expressiveness of different logical frameworks (*within* one type of logical framework, of course) [17],
- use logical framework independent specification language constructs [22, 6] being preserved by maps.

But if we do not follow some neat structuring principle in the meta theory (i. e. types of logical framework), there is the danger to end in the same Babylonian realm of

---

[1] There also are categories of logical frameworks with the purpose of building one framework above another, see [13, 25, 24, 18].

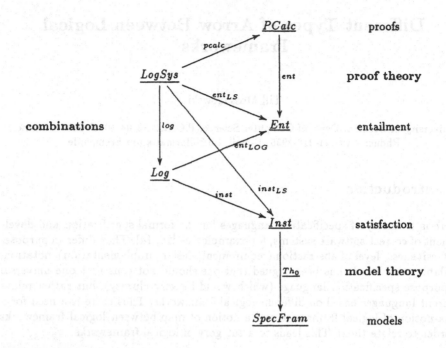

PCalc		proofs
LogSys		proof theory
combinations	*Ent*	entailment
Log		
	Inst	satisfaction
	Tho	model theory
	SpecFram	models

Fig. 1. Different types of logical framework and their relations

different languages as in the object theories (i. e. different logical frameworks). After all, the meta theory should help us to structure and compare the object theories, and not to produce the same diversity of notions!

The main argument for having several types of logical framework is the distinction between proof theory and model theory. Except for the case of specification frames, this was examined by Cerioli and Meseguer [7], see Fig. 1.

A second argument is the observation that different degrees of more or less good representation of one logical framework in another are possible. Many examples require some intermediate notion between map of institutions and map of specification frames. This has to do with the distinction of signatures and sentences in an institution. This distinction is necessary (using just specification frames abandons the notion of sentence and of theorem proving), but there are many choices for what should be included into the signature part, and what into the sentences. Therefore the need of types of arrow which may "mix up" this distinction. In this paper we argue that these different types of arrow can be generated by one basic type of arrow and monadic constructions on categories of logical frameworks, with the effect of automatically having functors relating the new categories of logical frameworks with the old ones.

The paper is organized as follows: in Sect. 2, some types of logical framework and some categorical notions are recalled. Section 3 then introduces, using monads and adjunctions, one well-known and three new notions of maps between institutions, which vary in the strictness of keeping the signature–sentence distinction. In each case, we briefly show the application to different logical frameworks. Section 4 concludes the paper. Due to lack of space, we omit proofs, which will appear elsewhere.

2 Preliminaries

2.1 Specification Frames

Specification frames by Ehrig, Pepper and Orejas [10] formalize abstract theories and models of theories, while there are no notions of sentence and satisfaction:

A *specification frame* $F = (\underline{Th}^F, \underline{Mod}^F)$ consists of

1. a category $\underline{Th}^F$ of theories
2. a functor $\underline{Mod}^F : (\underline{Th}^F)^{op} \longrightarrow \underline{CAT}$ giving the category of *models* of a theory

We omit the index F when it is clear from the context and write $M'|_\sigma$ (the σ-*reduct* of M' under σ) for $\underline{Mod}(\sigma)(M')$. M' is called an *expansion* of $M'|_\sigma$.

A map of specification frames $\mu: F \longrightarrow F'$ consists of

1. a functor $\Phi: \underline{Th}^F \longrightarrow \underline{Th}^{F'}$ and
2. a natural transformation $\beta: \underline{Mod}^{F'} \circ \Phi^{op} \longrightarrow \underline{Mod}^F$

Composition of maps $F \xrightarrow{\mu''} F'' = F \xrightarrow{\mu} F' \xrightarrow{\mu'} F''$ is defined by $\Phi'' = \Phi' \circ \Phi$ and $\beta'' = \beta \circ (\beta'_\Phi)$. This gives us a (quasi-)category $\underline{SpecFram}$ of specification frames.

2.2 Institutions

Institutions introduced by Goguen and Burstall [13] split theories into signatures and sentences, thus the area of logic starts here:

An *institution* $I = (\underline{Sign}^I, sen^I, \underline{Mod}^I, \models^I)$ consists of

1. a specification frame $(\underline{Sign}^I, \underline{Mod}^I)$, where $\underline{Sign}^I$ is the category of *signatures*,
2. a functor $sen^I : \underline{Sign}^I \longrightarrow \underline{Set}$ giving the set of *sentences* over a given signature,
3. a satisfaction relation $\models^I_\Sigma \subseteq |\underline{Mod}^I(\Sigma)| \times sen^I(\Sigma)$ for each $\Sigma \in \underline{Sign}^I$

such that for each morphism $\sigma: \Sigma \longrightarrow \Sigma'$ in $\underline{Sign}^I$ the *Satisfaction Condition*

$$M' \models^I_{\Sigma'} sen^I(\sigma)(\varphi) \iff M'|_\sigma \models^I_\Sigma \varphi$$

holds for each model $M' \in |\underline{Mod}^I(\Sigma')|$ and each sentence $\varphi \in sen^I(\Sigma)$.

Given institutions I and J, a *plain map of institutions* [16] $\mu = (\Phi, \alpha, \beta): I \longrightarrow J$ consists of

- a map of specification frames $(\Phi, \beta): (\underline{Sign}^I, \underline{Mod}^I) \longrightarrow (\underline{Sign}^J, \underline{Mod}^J)$
- a natural transformation $\alpha: sen^I \longrightarrow sen^J \circ \Phi$ and

such that the following property is satisfied for $M' \in |\underline{Mod}^J(\Phi(\Sigma))|$ and $\varphi \in sen^I(\Sigma)$:

$$M' \models^J_{\Phi(\Sigma)} \alpha_\Sigma(\varphi) \iff \beta_\Sigma(M') \models^I_\Sigma \varphi$$

2.3 Transporting Logical Structure Along Maps

We state the fundamental definition of Cerioli and Meseguer [7]: Let $\underline{C}$ and $\underline{D}$ be categories and let be $(\eta, \epsilon): U \dashv R: \underline{C} \longrightarrow \underline{D}$ an adjoint situation. The idea is that $\underline{C}$ and $\underline{D}$ are categories of logical frameworks, $\underline{D}$ having richer structure and U forgetting this richer structure (though U is a left adjoint!). If for each morphism $c: C \longrightarrow U(D)$ in $\underline{C}$ the pullback

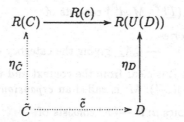

exists, we say that C *admits extension under R and U*, and $\tilde{c}: \tilde{C} \longrightarrow D$ is called the *extension of c by R and U*. Roughly speaking, $\tilde{C}$ is constructed out of C by enriching it with the features of D translated by c. Indeed, Cerioli and Meseguer show that this situation is given for all arrows in Fig. 1, except $Th_0: \underline{Inst} \longrightarrow \underline{SpecFram}$.

2.4 Monads

We recall some facts about monads and Kleisli categories from [5].

A *monad* on a category $\underline{X}$ is a triple $\mathbf{T} = (T, \eta, \xi)$ where $T: \underline{X} \longrightarrow \underline{X}$ is a functor and $\eta: 1_{\underline{X}} \longrightarrow T$ and $\xi: T \circ T \longrightarrow T$ are natural transformations satisfying the commutativity conditions $\xi \circ (\eta_T) = 1_T = \xi \circ (T\eta)$, $\xi \circ (\xi_T) = \xi \circ (T\xi)$.

A monad $\mathbf{T}$ induces the Kleisli category $\underline{X}_{\mathbf{T}}$. Our intuition behind it is that its morphisms go into some "enriched" target and thus can be constructed more flexibly than morphisms of $\underline{X}$. $\underline{X}_{\mathbf{T}}$ is defined by:

- the objects of $\underline{X}_{\mathbf{T}}$ are those of $\underline{X}$
- a morphism $f: X \longrightarrow Y$ in $\underline{X}_{\mathbf{T}}$ is a morphism $f: X \longrightarrow T(Y)$ in $\underline{X}$
- the composite of $f: X \longrightarrow Y$ and $g: Y \longrightarrow Z$ is given in $\underline{X}$ by the composite

$$X \xrightarrow{f} T(Y) \xrightarrow{T(g)} T(T(Z)) \xrightarrow{\xi_Z} T(Z)$$

- the identity on an object X of $\underline{X}_{\mathbf{T}}$ is just $\eta_X: X \longrightarrow T(X)$ in $\underline{X}$

Moreover, there is a pair of adjoint functors between $\underline{X}$ and $\underline{X}_{\mathbf{T}}$. $U_{\mathbf{T}}(X \xrightarrow{F} T(Y))$ is $T(X) \xrightarrow{\xi_Y \circ T(f)} T(Y)$, and $F_{\mathbf{T}}(X \xrightarrow{f} Y)$ is $X \xrightarrow{\eta_Y \circ f} T(Y)$. The unit of the adjunction is just η.

Given any adjoint situation $(\zeta, \epsilon): F \dashv U: \underline{A} \longrightarrow \underline{X}$, putting $T = U \circ F$, $\xi = U\zeta_F$ yields a monad (T, ζ, ξ) on $\underline{X}$. Moreover, there is a full and faithful comparison functor $K_T: \underline{X}_T \longrightarrow A$, comparing the Kleisli adjunction of the monad with the given adjunction. It is given by $K_T(X \xrightarrow{f} T(Y)) = F(X) \xrightarrow{F(f)} F(U(F(Y))) \xrightarrow{\epsilon_{F(Y)}} F(Y) = F(X) \xrightarrow{f^\#} F(Y)$.

2.5 Multiple Pushouts, Multiple Pullbacks and Amalgamation

Let $(A, (A \xrightarrow{g_i} B_i)_{i \in I})$ be a source (that is, a class of arrows with common codomain. Note that the codomain must be given separately because I may be empty). A colimit of the diagram consisting of all objects and arrows of the source is called a *multiple pushout* of the source. A category with *canonical* multiple pushouts is a category endowed with a distinguished multiple pushout for each source, such that the result of pasting together distinguished pushouts again is a distinguished pushout. The dual notion is *multiple pullback* (see [1, 11L]).

3 Refinements of the Notion of Plain Map of Institution

To work just with abstract theories and having no notion of axiom, as specification frames do, is a very semantic point of view not amenable to theorem proving. In many contexts, we need institutions and their distinction between the notion of signature and of sentence. But there are many choices for what should be included into the signature part, and what into the sentences. For example, in type theories [15], typing information is put into the sentences, while in HEP-theories [20], not only typing, but also definedness axioms are put into the signatures.

The problem now arises when considering maps. Plain maps (of institutions or entailment systems), which keep signatures and sentences separated, occur in many contexts, but typically only as trivial sub-logical frameworks. But there are many representations, which should even be called embeddings, which are not plain.

Meseguer's notion of simple map (of institutions or entailment systems) [16] is a first solution: signatures are mapped to theories, while sentences are mapped to sentences. But there are contexts, when this does not suffice. Thus there is a need for notions of maps located between simple maps of institutions and maps of specification frames. In the rest of this section, different types of maps are introduced using Kleisli categories of monads and adjunctions.

3.1 The Theory Monad and Simple Maps of Institutions

We introduce Meseguer's simple maps of institutions here not from scratch, but rather in a structured way using a monadic construction to enrich the target institution.

An institution $I = (\underline{Sign}, sen, \underline{Mod}, \models)$ induces the category of *theories with axiom-preserving theory morphisms* $Th_0(I)$.

Objects are theories $T = (\Sigma, \Gamma)$, where $\Sigma \in |\underline{Sign}|$ and $\Gamma \subseteq sen(\Sigma)$ (with Γ not necessarily closed under consequence). We set $sign(T) = \Sigma$ and $ax(T) = \Gamma$. Morphisms $\sigma: (\Sigma, \Gamma) \longrightarrow (\Sigma', \Gamma')$ in $Th_0(I)$ are signature morphisms $\sigma: \Sigma \longrightarrow \Sigma'$ such that $\sigma(\Gamma) \subseteq \Gamma'$. sen, $\underline{Mod}$ and $\models$ can be easily extended to theories as well. Thus we get a new institution $\underline{Th_0}(I) = (Th_0(I), sen^{Th_0}, \underline{Mod}^{Th_0}, \models^{Th_0})$.

Given a plain map $\mu = (\Phi, \alpha, \beta): I \longrightarrow J$, let Φ^α, the α-extension to theories of Φ, map (Σ, Γ) to $(sign(\Phi(\Sigma)), ax(\Phi(\Sigma)) \cup \alpha_\Sigma(\Gamma))$. Likewise, α and β can easily be extended to $\underline{Th_0}(I)$, so we get a plain map of institution $\underline{Th_0}(\mu) = (\Phi^\alpha, \alpha^{Th_0}, \beta^{Th_0}): \underline{Th_0}(I) \longrightarrow \underline{Th_0}(J)$, and $\underline{Th_0}: \underline{PlainInst} \longrightarrow \underline{PlainInst}$ becomes a functor. There is an obvious inclusion $\eta_I: I \longrightarrow \underline{Th_0}(I)$, and a projection $\xi_T: \underline{Th_0}(\underline{Th_0}(I)) \longrightarrow \underline{Th_0}(I)$ mapping $((\Sigma, \Gamma), \Gamma')$ to $(\Sigma, \Gamma \cup \Gamma')$ for signatures and keeping sentences and models.

Definition 1. The monad $\underline{Th_0} = (\underline{Th_0}, \eta, \xi)$ over $\underline{PlainInst}$ induces the Kleisli category $\underline{Inst} = \underline{PlainInst}_{\underline{Th_0}}$: institutions and simple maps of institutions.

Recall that objects of the Kleisli category are the same as those of $\underline{PlainInst}$, but morphisms $\mu: I \longrightarrow J$ are maps $\mu: I \longrightarrow \underline{Th_0}(J)$ in $\underline{PlainInst}$, thus we are allowed to map a signature not just to a signature, but to a theory, which gives us more flexibility when setting up maps of institutions.

A typical example is the simple map of institutions from partial first-order logic to total first-order logic where partial functions are mapped to total functions plus a partial congruence relation, sentences are translated by replacing existential equality by the partial congruence and models are translated by factoring modulo the congruence. Here, $\Phi(\Sigma)$ has to contain the congruence axioms, for example. This translation allows to re-use the proof theory via borrowing, see [8]. There are a plenty of other simple maps, see [6, 7, 2, 16].

3.2 The Conjunctive Monad and Conjunctive Maps of Institutions

Now simple maps of institutions cover more, but still not all desirable representations of institutions. A second step is to allow sentences being mapped not to single sentences, but to (finite) sets of sentences.

An institution $I = (\underline{Sign}, sen, \underline{Mod}, \models)$ can be enriched to the institution $\bigwedge(I) = (\underline{Sign}, sen^\wedge, \underline{Mod}, \models^\wedge)$, where $sen^\wedge = \mathcal{P}_{fin} \circ sen$, the composition of sen with the functor $\mathcal{P}_{fin}: \underline{Set} \longrightarrow \underline{Set}$ giving the set of finite subsets, and $M \models^\wedge_\Sigma S$ iff for all $\varphi \in S$, $M \models_\Sigma \varphi$. Again, this construction can easily be turned into a monad $\mathbf{Conj} = (\bigwedge, \eta, \xi)$ (this time acting on $\underline{Inst}$).

Definition 2. The morphisms of the Kleisli category $\bigwedge \underline{Inst} = \underline{Inst}_{\mathbf{Conj}}$ are called conjunctive simple maps of institutions.

Conjunctive maps are not needed, of course, if the target institution already has conjunction available. But consider the map from partial algebras with strong equations to partial algebras with conditional existence equations: Here, a strong equation

$t_1 \overset{s}{=} t_2$ is mapped to $(t_1 \overset{e}{=} t_1 \longrightarrow t_1 \overset{e}{=} t_2) \wedge (t_2 \overset{e}{=} t_2 \longrightarrow t_1 \overset{e}{=} t_2)$.

A more complex example is the conjunctive simple map of institutions from the institution of set-valued left exact sketches (see [4]) to the institution of limit theories (an extension of Horn theories with unique-existential quanfitication in the conclusion, see [9]) which is described in [17]. A diagram over a sketch is translated to a set of equations, while a cone over a sketch is translated to the statement that the cone is a limiting cone, which can be expressed with the unique-existential quantifiers of limit theories, but only with a set of limit sentences.

3.3 The Theory Extension Monad and Weak Maps of Institutions

Still, there is the need for even more complex representations of institutions. In many examples, it is only possible to map theories to theories. Now the theory functor $\underline{Th_0}$ from subsection 3.1 can be viewed as a functor $\underline{Inst} \longrightarrow \underline{SpecFram}$: it takes an institution $I = (\underline{Sign}, sen, \underline{Mod}, \models)$ to the specification frame $(Th_0(I), \underline{Mod})$. By abuse of language, we also denote it by $\underline{Th_0}$.

Definition 3. A weak map of institution $\mu: I \longrightarrow J$ is a map of specification frames $\mu: Th_0(I) \longrightarrow Th_0(J)$. Composition is that of $\underline{SpecFram}$. This gives us a category $\underline{WeakInst}$ of institutions and weak maps of institutions.

In some cases, a weak map can be shown to be modular. That means, the mapping of theories to theories can be split into two maps: one mapping signatures to theories, and the other mapping sentences to extensions of these theories. Now this construction is not just a monad as in the previous subsections, but comes very naturally via an adjunction between specification frames and institutions. Unfortunately, there seems to be no adjunction between specification frames and institutions in general, so we have to restrict both categories.

Let $\underline{SpecFram}^{amal}$ (resp. $\underline{Inst}^{amal}$) be the restriction of $\underline{SpecFram}$ (resp. $\underline{Inst}$) to those objects having *amalgamation* and those maps preserving amalgamation. A specification frame resp. institution has amalgamation, if $\underline{Th}$ (resp. $\underline{Sign}$) has canonical multiple pushouts and $\underline{Mod}$ maps multiple pushouts to multiple pullbacks in $\underline{CAT}$. A map of specification frames preserves amalgamation, iff (1) Φ preserves multiple pushouts and (2) β_T is a natural equivalence. A simple map of institutions preserves amalgamation, iff (1) Φ preserves multiple pushouts and (2) β_Σ is a natural equivalence with $M \models_\Sigma \varphi \iff \beta'_\Sigma(M) \models_{\Phi(\Sigma)} \alpha_\Sigma(\varphi)$ where β'_Σ is an inverse-up-to-isomorphism of β_Σ.

Now $\underline{Th_0}$ can be restricted to $\underline{Th_0}: \underline{Inst}^{amal} \longrightarrow \underline{SpecFram}^{amal}$. This theory functor has a right adjoint, the theory extension functor $\exists!$. For a specification frame $F = (\underline{Th}, \underline{Mod})$, we set $\exists!(F) = (\underline{Th}, sen, \underline{Mod}, \models)$, where $sen(T)$ is the set of all theory extensions, i. e. theory morphisms $\sigma: T \longrightarrow T'$ in $\underline{Th}$ such that $M'_1 |_\sigma \cong M'_2 |_\sigma \Rightarrow M'_1 \cong M'_2$. Sentences are translated by pushing out. Now $M \models \sigma: T \longrightarrow T'$ iff "$M \in \underline{Mod}(T')$", i. e. iff there is some $M' \in \underline{Mod}(T')$ with $M' |_\sigma \cong M$. Note that M' has to be unique up to isomorphism. For morphisms, $\exists!(\Phi, \beta) = (\Phi, \Phi, \beta)$.

The unit $\eta_I: I \longrightarrow \exists!(\underline{Th_0}(I))$ of the adjunction is given by mapping a signature Σ to the theory $(\Sigma, \emptyset)$, mapping a Σ-sentence φ to the theory extension $(\Sigma, \emptyset) \hookrightarrow (\Sigma, \{\varphi\})$ and leaving models unchanged.

The counit $\epsilon_F: \underline{Th_0}(\exists!(F)) \longrightarrow F$ of the adjunction maps a theory $TT = (T, \{T \xrightarrow{\sigma_i} T_i \mid i \in I\})$ to T', where $(T', (\theta_i: T_i \longrightarrow T')_{i \in I}) = Colim(T, (T \xrightarrow{\sigma_i} T_i)_{i \in I})$. Let $\sigma: T \longrightarrow T'$ denote $\theta_i \circ \sigma_i$ (note that, by $(T', (\theta_i: T_i \longrightarrow T')_{i \in I})$ being a cocone, this is independent of i). Then β_{TT} is $\underline{Mod}(\sigma)$, being a natural equivalence.

Theorem 4. $(\eta, \epsilon): \underline{Th_0} \dashv \exists!: \underline{SpecFram}^{amal} \longrightarrow \underline{Inst}^{amal}$ is an adjoint situation, inducing a monad Ext on $\underline{Inst}^{amal}$ with Kleisli category $\underline{Inst}^{amal}_{Ext}$.

Proposition 5. The functor $K: \underline{Inst}^{amal}_{Ext} \longrightarrow \underline{WeakInst}^{amal}$ being the identity on objects and K_{Ext} on morphisms is an isomorphism.

Thus a weak map of institutions $\mu: \underline{Th_0}(I) \longrightarrow \underline{Th_0}(J) \in \underline{WeakInst}^{amal}$ can be modularized as a simple map of institutions $K^{-1}(\mu): I \longrightarrow \exists!(\underline{Th_0}(J))$. Note that μ applied to a theory just is the multiple pushout of all theory extensions resulting from the application of $K^{-1}(\mu)$ to each sentence of the theory.

It is easy to see that each conjunctive map also is a weak map of institutions, this gives us an embedding $weak_\wedge: \bigwedge \underline{Inst}^{amal} \longrightarrow \underline{WeakInst}$.

Typical examples for weak maps of institutions are the following constructions:

- explicit definition of λ-abstraction: There is a weak map from higher-order logic to higher-order logic without λ-abstraction. A HOL-formula is translated by recursively replacing each subterm $\lambda x_\alpha.t_\beta$ in a context $x_{1\alpha_1}, \ldots, x_{n\alpha_n}$ with $f\, x_{1\alpha_1} \ldots x_{n\alpha_n}$, where $f: \alpha_1 \rightarrow \ldots \rightarrow \alpha_n \rightarrow \alpha \rightarrow \beta$ is a new function symbol with an equation $f\, x_{1\alpha_1} \ldots x_{n\alpha_n}\, x_\alpha = t_\beta$ in the theory extension.
- explicit definition of description operators: There is a weak map from partial function higher-order logic with definite description operators [11] to partial function higher-order logic *without* description operators. A PHOL-formula is translated to a theory extension containing the formula itself plus the axioms Δ_α from [11] for each description operator $d_{(\alpha \rightarrow *) \rightarrow \alpha}$ contained in the formula.
- replacing relations by projections (for example, a binary relation (graph) is represented by a new sort (edges) and two projections (source, target). There is a weak map from limit theories to partial algebras with conditional existence equations [17]. Each conditional axiom is translated to the stipulation of the existence of a partial function between the new sorts representing the relations.

By the way: there is no weak map from limit theories to Horn Clause theories, that is, the meta-level $\exists!$ cannot simulate the object-level $\exists!$ in this case.

We also can use the adjunction between $\underline{SpecFram}^{amal}$ and $\underline{Inst}^{amal}$ for borrowing logical structure. A specification frame $F = (\underline{Th}, \underline{Mod})$ may borrow sentences from an institution J along a map of specification frames $\mu = (\Phi, \beta): F \longrightarrow \underline{Th_0}(J)$. Given such a map, let $I = (\underline{Th}^F, sen, \underline{Mod}^F, \models)$ with $sen(T) = \{(T \xrightarrow{\sigma} T', \varphi) \mid \Phi(\sigma) = (\Sigma, \Gamma) \longrightarrow (\Sigma, \Gamma \cup \{\varphi\}) \in sen^{\exists!(\underline{Th_0}(J))}(\Sigma)\}$ and $M \models_\Sigma (T \xrightarrow{\sigma} T', \varphi)$ iff $M \models_\Sigma^{\exists!(F)} T \xrightarrow{\sigma} T'$.

Theorem 6. *Let $\Psi(T) = (T, \emptyset)$. The following diagram is a pullback*

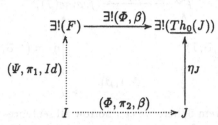

That is, F can borrow those sentences from J which do not lead to new theories. For example, if we take F to be Lawvere theories or FP-sketches over *Set* [4] and map them to their equational theories (therefore we have to equip equational theories with some notion of derived signature morphism), we can consider usual equations as new axioms.

3.4 The Model Class Monad and Semi-Maps of Institutions

In the previous subsection, we generated sentences from the theory extensions of a specification frame. Now we explore the other possibility: to generate sentences from model classes.

Let $frame(\underline{Sign}, sen, \underline{Mod}, \models)$ be just $(\underline{Sign}, \underline{Mod})$ (a very poor specification frame), and $frame(\Phi, \alpha, \beta) = (\Phi, \beta)$. $(\underline{Th}, \underline{Mod})^{\heartsuit} = (\underline{Th}, sen, \underline{Mod}, \models)$ with $sen(T) = \{ M \mid M \subseteq \underline{Mod}(T) \mid\}$ and $sen(T \xrightarrow{\sigma} T')(M) = (\underline{Mod}(\sigma))^{-1}(M)$. Satisfaction is just the element relation. $(\Phi, \beta)^{\heartsuit} = (\Phi, \beta^{-1}, \beta)$.

The unit of the adjunction $\eta_I: I \longrightarrow (frame(I))^{\heartsuit}$ is given by $\eta_I = (1, \overline{\alpha}, 1)$ with $\overline{\alpha}_{\Sigma}(\varphi) = \{ M \in \underline{Mod}(\Sigma) \mid M \models_{\Sigma} \varphi \}$. The counit $\epsilon_F: frame(F^{\heartsuit}) \longrightarrow F$ is just the identity.

Theorem 7. $(\eta, \epsilon): frame \dashv (\_)^{\heartsuit}: \underline{SpecFram} \longrightarrow \underline{PlainInst}$ *is an adjoint situation, inducing monad* **Mod** *on* $\underline{PlainInst}$.

Definition 8. *The morphisms of the Kleisli category* $\underline{SemiInst} = \underline{PlainInst}_{\mathbf{Mod}}$ *are called semi-maps of institutions.*

Since $K_{\mathbf{Mod}}$ is full and faithful, a semi-map $\mu: I \longrightarrow J$ is (in one-one correspondence with) a map of specification frames $K_{\mathbf{Mod}}(\mu): frame(I) \longrightarrow frame(J)$, that is, just a plain map without a component mapping sentences. Such semi-maps are related to institution semi-morphisms [25], which are defined similarly except that signatures and models are mapped covariantly. For an example, see [25].

Again, we have the possibility of borrowing sentences from an institution along a map of specification frames $\mu = (\Phi, \beta): F \longrightarrow frame(J)$. Given such a map, let $F = (\underline{Th}, \underline{Mod})$. We put $I := (\underline{Th}, sen^J \circ \Phi, \underline{Mod}, \models)$, where $M \models_T \varphi$ iff $M' \models^J_{sign(\Phi(T))} \varphi$ for any $M' \in \underline{Mod}(\Phi(T))$ with $\beta_T(M)' = M$. (This definition is independent of the choice of M'.)

Theorem 9. *The following diagram is a pullback*

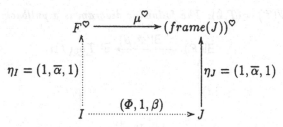

Thus, in contrast to theorem 6, here we can borrow all sentences from an institution. This borrowing of sentences from an institution much resembles the introduction of duplex institutions in [13], with the main difference that we here only require a map of specification frames (where [13] require an institution morphism).

Since the specification frame $frame(I)$ is very poor, in applications, it may be more useful to consider $frame(\underline{Th_0}(I))$. However, the composite of $frame$ and $\underline{Th_0}$ probably is not adjoint (note that $frame$ is left adjoint, while $\underline{Th_0}\colon \underline{Inst} \longrightarrow \underline{PlainInst}$ is right adjoint).

For example, if F are Lawvere theories or sketches, we now can borrow also conditional equations and thus strictly enlarging the power of F.

4 Conclusion

We argue that there are problems with the distinction between signatures and sentences within the notion of institution. While the distinction is useful and necessary, maps between institutions ofter have to go beyond this distinction. Therefore, three new notions of map between institutions, conjunctive maps, weak maps and semi-maps, have been introduced. These allow to relax the distinction between signatures and sentences while still being sentence-structured, opposed to mere maps of specification frames.

When introducing these types of arrow we have followed some guidelines which allow to relate different categories of logical framework by pair of adjoint functors. Along two of those pairs, there is a useful borrowing of sentences from an institution, extendeding the work of Cerioli and Meseguer on borrowing logical structure along maps [7] to the relation between specification frames and institutions.

Thus, there is not one common type of logical framework, but there is the chance to introduce new types of logical framework (driven by a systematic study of examples) in a manner that different types of logical framework are well-related via nice general mathematical tools and theorems.

The technical results of the paper are summarized in lower half of Fig. 2, where the arrows $\exists!$, $U_{\mathbf{Ext}} \circ K^{-1}$ and $K \circ F_{\mathbf{Ext}}$ only exist when everything is restricted to the case with amalgamation.

Acknowlegdements

I wish to thank Jo Goguen and Rod Burstall for inventing institutions, Maura Cerioli

168

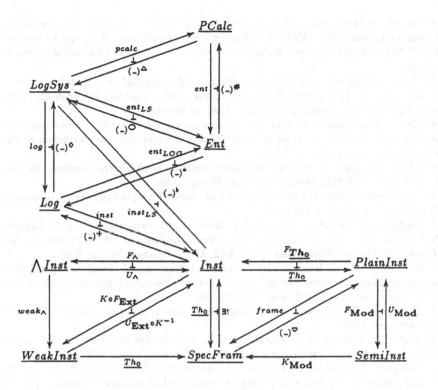

Fig. 2. Functors and adjunctions between different categories of logical frameworks

and José Meseguer for setting up the diagram of categories of logical frameworks allowing borrowing, Andrzej Tarlecki for the idea to work with constructions on the target, Uwe Wolter for pointing out the problems with the signature–sentence distinction, and all of them for (more or less intensive) discussions.

References

[1] J. Adámek, H. Herrlich, G. Strecker. *Abstract and Concrete Categories.* Wiley, New York, 1990.

[2] E. Astesiano, M. Cerioli. Relationships between logical frameworks. In M. Bidoit, C. Choppy, eds., *Proc. 8th ADT workshop, LNCS* 655, 126–143. Springer Verlag, 1992.

[3] E. Astesiano, M. Cerioli. Multiparadigm specification languages: a first attempt at foundations. In J.F. Groote C.M.D.J. Andrews, ed., *Semantics of Specification Languages (SoSl 93)*, Workshops in Computing, 168–185. Springer Verlag, 1994.

[4] M. Barr, C. Wells. *Toposes, Triples and Theories, Grundlehren der mathematischen Wissenschaften* 278. Springer Verlag, 1985.

[5] F. Borceux. *Handbook of Categorical Algebra I – III.* Cambridge University Press, 1994.

[6] M. Cerioli. *Relationships between Logical Formalisms.* PhD thesis, TD-4/93, Università di Pisa-Genova-Udine, 1993.

[7] M. Cerioli, J. Meseguer. May I borrow your logic? In A.M. Borzyszkowski, S.Sokolowski, eds., *Proc. MFCS'93 (Mathematical Foundations of Computer Science)*, *LNCS* **711**, 342–351. Springer Verlag, Berlin, 1993. To appear in Theoretical Computer Science.

[8] M. Cerioli, T. Mossakowski, H. Reichel. From total equational to partial first order. In E. Astesiano, H.-J. Kreowski, B. Krieg-Brückner, eds., *Algebraic Foundations of Systems Specifications*. Chapman and Hall, 1996. to appear.

[9] M. Coste. Localisation, spectra and sheaf representation. In M.P. Fourman, C.J. Mulvey, D.S. Scott, eds., *Application of Sheaves, Lecture Notes in Mathematics* **753**, 212–238. Springer Verlag, 1979.

[10] H. Ehrig, P. Pepper, F. Orejas. On recent trends in algebraic specification. In *Proc. ICALP'89, LNCS* **372**, 263–288. Springer Verlag, 1989.

[11] W. A. Farmer. A partial functions version of Church's simple type theory. *Journal of Symbolic Logic* **55**, 1269–1291, 1991.

[12] J. Fiadeiro, A. Sernadas. Structuring theories on consequence. In D. Sannella, A. Tarlecki, eds., *Recent Trends in Data Type Specification, 5th Workshop on Specification of Abstract Data Types, LNCS* **332**, 44–72. Springer Verlag, 1988.

[13] J. A. Goguen, R. M. Burstall. Institutions: Abstract model theory for specification and programming. *Journal of the Association for Computing Machinery* **39**, 95–146, 1992. Predecessor in: LNCS 164(1984):221–256.

[14] J. A. Goguen, T. Winkler. Introducing OBJ3. Research report SRI-CSL-88-9, SRI International, 1988.

[15] R. Harper, F. Honsell, G. D. Plotkin. A framework for defining logics. In *Proceedings of the second Symposium on Logic in Computer Science (Ithaca, NY)*. IEEE, 1987.

[16] J. Meseguer. General logics. In *Logic Colloquium 87*, 275–329. North Holland, 1989.

[17] T. Mossakowski. Equivalences among various logical frameworks of partial algebras. In H. Kleine Büning et al., ed., *Computer Science Logic. 9th Workshop, CSL'95. Paderborn, Germany, September 1995*. Springer LNCS, 1996. To appear.

[18] T. Mossakowski. Using colimits of parchments to systematically construct institutions of partial algebras. In M. Haveraaen et al., ed., *Recent Trends in Data Type Specifications. 11th Workshop on Specification of Abstract Data Types*, LNCS. To appear. Springer Verlag, 1996.

[19] W. Pawlowski. Institutions with contexts. In M. Haveraaen et al., ed., *Recent Trends in Data Type Specifications. 11th Workshop on Specification of Abstract Data Types*, LNCS. To appear. Springer Verlag, 1996.

[20] H. Reichel. *Initial Computability, Algebraic Specifications and Partial Algebras*. Oxford Science Publications, 1987.

[21] A. Salibra, G. Scollo. A soft stairway to institutions. In M. Bidoit, C. Choppy, eds., *Proc. 8th ADT workshop, LNCS* **655**, 310–329. Springer Verlag, 1992.

[22] D. Sannella, A. Tarlecki. Specifications in an arbitrary institution. *Information and Computation* **76**, 165–210, 1988.

[23] A. Sernadas, C. Sernadas. Theory spaces. Research report, DMIST, 1096 Lisboa, Portugal, 1995. Presented at ISCORE'95 and ADT/COMPASS'95.

[24] A. Tarlecki. Working with multiple logical systems. Unpublished manuscript.

[25] A. Tarlecki. Moving between logical systems. In M. Haveraaen et al., ed., *Recent Trends in Data Type Specifications. 11th Workshop on Specification of Abstract Data Types*, LNCS. To appear. Springer Verlag, 1996.

[26] M. Wirsing. Structured algebraic specifications: A kernel language. *Theoretical Computer Science* **42**, 123–249, 1986.

[27] Uwe Wolter. Institutional frames. In *Recent Trends in Data Type Specification. Proceedings, LNCS* **906**, 469–482. Springer Verlag, London, 1995.

Effective Models of Polymorphism, Subtyping and Recursion (extended abstract)

John Mitchell * and Ramesh Viswanathan **

Abstract. We explore a class of models of polymorphism, subtyping and recursion based on a combination of traditional recursion theory and simple domain theory. A significant property of our primary model is that types are coded by natural numbers using any index of their supremum operator. This leads to a distinctive view of polymorphic functions that has many of the usual parametricity properties. It also gives a distinctive but entirely coherent interpretation of subtyping. An alternate construction points out some peculiarities of computability theory based on natural number codings. Specifically, the polymorphic fixed point is computable by a single algorithm at all types when we construct the model over untyped call-by-value lambda terms, but not when we use Gödel numbers for computable functions. This is consistent with trends away from natural numbers in the field of abstract recursion theory. Although our development and analysis of each structure is completely elementary, both structures may be obtained as the result of interpreting standard domain constructions in effective models of constructive logic.

1 Introduction

The semantic models for polymorphism, subtyping and recursion studied in this paper are based on a simple, concrete connection between mathematical semantics and compilation. Our driving intuition is that the compiler for a language such as ML compiles each expression into a bit string that could be regarded as the Gödel code for some function, pair, number, or other datum. For each type, we therefore have a set of Gödel numbers representing the elements of the type. Since semantic models are intended to provide some basis for equational reasoning, we must also define an equivalence relation on the set of Gödel codes for each type. We may therefore define a membership predicate and equivalence relation, which together give us a so-called partial equivalence relation (per), for each type and take the resulting quotients of subsets of the natural numbers as the "domains" of our model. While this idea has proven useful in the semantics of polymorphism, the use of per models for recursive types and subtyping has not been as thoroughly studied. The main goal of this paper is to identify a

* Supported in part by NSF Grant CCR-9303099 and a grant from the TRW Foundation. Address: Department of Computer Science, Stanford University, Stanford, CA 94305. Email: mitchell@cs.stanford.edu

** Supported by NSF Grant CCR-9303099. Address: Bell Laboratories, 101 Crawfords Corner Road, Holmdel, NJ 07733. Email: rv@research.att.com

class of pers in which we can interpret polymorphism, subtyping and recursion, and to explore the properties of the resulting models. Three objectives in our development are (1) the class of pers identified should be as generous as possible, with precise technical justifications for any choices made, (2) the presentation of the models should be concrete so as to make transparent their properties using simple recursion-theoretic arguments, and (3) the models should be faithful to the intuitive correspondence with compilation.

A representative sample of recent studies using partial equivalence relations were presented at the 1990 IEEE Logic in Computer Science Conference. The first of three papers describes a special class of partial equivalence relations over the natural numbers called *extensional pers* [FRMS92], the second uses partial equivalence relations over partially-ordered domains instead of natural numbers [AP90], and the third works in the constructive setting of the *effective topos*, which boils down to partial equivalence relations over the natural numbers [Pho90]. Since our goal is to develop a semantic model that can be viewed as a quotient of the result of compilation, the use of pers over special partially-ordered domains does not meet our objectives. While extensional pers do give us quotients of sets of bit-strings, the standard and expected representations of basic datatypes such as natural numbers and booleans, as well as type constructors such as products, function spaces and lifting, require modification if they are to be "extensional" (in the technical sense of extensional pers). Therefore, this model too does not have the basic properties we would like. Finally, [Pho90] uses categorical logic to show that there exists a subcategory of the effective topos that admits an interpretation of recursion and polymorphism. Our work draws upon the ideas outlined in [Pho90], although our methodology of developing models from computational intuition and our attempts to characterize concrete properties that might not be apparent from the categorical interpretation of intuitionistic logic set our efforts apart.

We begin with the category **Per** whose objects (interpretation of types) are partial equivalence relations over the natural numbers and whose maps (interpretation of terms) are computable functions, in the usual sense of computability theory. While this category is a well-known model of polymorphism and subtyping, it is not useful for interpreting recursion. We demonstrate this in Sect. 3 by defining a per A of partial functions and a total recursive f on A that has no fixed point. This counter example to a generalization of the recursion theorem to arbitrary pers is based on intuition from domain theory: if we order A in the natural way, it is not "complete". This suggests that the appropriate way to extend the recursion theorem to subquotients is by borrowing ideas from domain theory. In the resulting combination of recursion-theoretic and domain-theoretic techniques, it turns out that we need far fewer ideas from domain theory than [AP90, BM92], for example.

Using essentially the same "intrinsic order" as [AP90, Pho90], we identify a class of "cpo-like" pers, each such per R having a computable function sup_R that gives the least upper bound of any constructible chain in R. We call these *effective cpo's;* although the name is similar, these are completely different from the effectively presented cpo's described in [GS90], for example. By a generalization of the Myhill-Shepherdson theorem from recursion theory, we show that all computable functions between effective cpo's are continuous. If an effective cpo has a least element, then the usual domain theoretic arguments give us an effective least fixed point operator. Serendipitously, almost all of our basic desiderata are satisfied by this construction. The natural numbers (discretely ordered) and other basic data types are easily seen to be effective cpo's. The usual product, function space and other constructions preserve the property of being an effective cpo. Moreover, the intrinsic orderings turn out to be the usual "pointwise orderings" of domain theory. Since it is easy to use the supremum operator to compute a least fixed point, the full subcategory **ECpo** of effective cpo's and computable functions gives us an elementary semantics of first-order type operators and recursion.

In the model using Gödel numbers of partial recursive functions, the interpretation of polymorphic types involves coding types as natural numbers. The traditional approach, when we are not concerned with recursion, is to make the coding relation trivial, leading to an interpretation of polymorphic types as intersection of infinite indexed families. However, the intersection of a constructive family of effective cpo's is not necessarily an effective cpo. We therefore represent a type by any Gödel code of its supremum operator, producing an interpretation of a polymorphic type $\forall t.A(t)$ as the collection of recursive functions mapping any supremum operator sup_R to an element of the per $A(R)$. Surprisingly, this interpretation of polymorphism still satisfies the expected parametricity principles. For example, we show that the per $\forall t.t$ is empty and that the only element of the per $\forall t.t \rightarrow t$ is the identity function. We can also define a $\forall^p$ quantifier where we interpret the domain of quantification to be pointed effective cpos. The difference from the $\forall$ quantifier arises because the code for a pointed effective cpo naturally gives both its least element and its supremum operator. Continuing our examination of parametricity properties, we show that the least fixed point operator is the only fixed point operator of type $\forall^p t.(t \rightarrow t) \rightarrow t$. Due to least elements, the type $\forall^p t.t$ becomes nonempty and $\forall^p t.(t \rightarrow t)$ contains more than the identity function. In short, the $\forall$ quantifier supports the parametricity principles of the system **F** while the $\forall^p$ quantifier supports the parametricity principles of system **F** extended with fixed points.

A standard method for obtaining recursive types, described in [SP82], involves showing that suitable functors on **Cpo**-enriched categories have initial F-algebras. To obtain analogous results, we work with **ECpo**-enriched categories whose homsets each form an effective cpo. In addition, and in a departure from applications of the Smyth-Plotkin method we have seen before, we ask for functors that are effective on objects in addition to morphisms. This means that a functor F on **ECpo** is *effective* if the supremum function on $F(R)$ is uniformly computable from the supremum function on R. If each homset is also pointed, we have a **PECpo**-enriched category. As for morphisms in **Per**, we can show that any effective functor is automatically locally continuous (in the sense of [SP82]) and thus any effective endofunctor on *effectively complete* **PECpo**-

enriched categories has an initial F-algebra. Our categories **ECpo** and **PECpo** can be shown to be effectively complete, as well as **ECpo**-enriched and **PECpo**-enriched, respectively. This gives us initial F-algebras for all effective functors. All of this feels relatively routine; once we have the idea of realizers for effective cpos, we are simply working out effective versions of the usual arguments.

In other per models, the interpretation of subtyping has always been the subset relation on pers, namely, R is a subtype of S iff the membership predicate and equality relations of R are subsets of those for S. Intuitively, this means that a supertype may have more elements than a subtype, but also distinct elements of a subtype may become identified at a supertype. In our model, we cannot simply follow this approach since we need coherence between the supremum operators at subtype and supertype, and between least elements if both are pointed. More specifically, we say an effective cpo R is a subtype of effective cpo S if in addition to being a subset, the least upper bound of any increasing chain in R is also a least upper bound in S. Moreover, if R is pointed then any index of the least element of R must also be one for S. All the usual examples of subtyping hold with our modified definition of subtyping.

These ideas can be combined to yield a sound and coherent model of a calculus with higher-order polymorphism, recursion and subtyping — however, due to space considerations, we omit the details of the construction in this paper.

We began with the intent of forming an extensional model by taking quotients of sets of bit strings (or Gödel numbers of functions). One reason for doing this was to obtain a close correspondence with compilation, since the denotation of an expression could be viewed as the equivalence class of the bit string produced by compiling this expression. However, a compiler for the kind of language we are interested in may generate code without using type information. A reasonable compiler could parse expressions, annotate the parse trees during type checking, but then discard or ignore the type annotations during code generation. A consequence is that we expected to find a uniform, type-independent algorithm computing the fixed point operator at each type. It is a surprising aspect of our effort that we were unable to accomplish this. Specifically, in Sect. 7, we use computability arguments to show that there is no uniform supremum operator and that no untyped fixed-point operator can be the index of the typed fixed-point operator at each type. The proof uses algorithms that apply natural number operations such as equality test to the Gödel codes of function arguments. This is allowed in per structures as long as the resulting functional has the right extensional behavior.

Because of the non-uniformity associated with natural number codes, we next investigate a per model based on untyped call-by-value lambda terms. The result, described in Sect. 8, is a very similar category of effective cpo's, but with a single untyped term computing all fixed-points. This uniformity suggests that compilation is more profitably viewed as a translation to untyped lambda terms. While we initially regarded bit strings as the result of compilation, we realize upon reflection that the result of compiling a typed program is always a "typed algorithm." In other words, although machine language allows us to do bit string operations on function code, for example, no such manipulation need occur in the object code produced by compiling a typed program. Since natural numbers are distinct from function expressions in the untyped call-by-value calculus, there is just enough of an inherent type distinction to provide a uniform fixed-point algorithm. While this may be a slight surprise to computer scientists, our eventual preference for call-by-value terms appears consistent with the move away from natural number coding in the field of recursion theory (e.g., [Bar75]).

2 Preliminaries

Let $\varphi_0, \varphi_1, \ldots$ be any enumeration of the partial recursive functions. We use $n \cdot m$ to denote $\varphi_n(m)$. We use pr for a computable pairing function on the natural numbers and $(n)_1$ and $(n)_2$ to denote the first and second projections of n. Let $\mathcal{N}$ denote the set of natural numbers. A partial equivalence relation (per) $R \subseteq \mathcal{N} \times \mathcal{N}$ is a symmetric transitive relation. For any per R, $n \colon R$ denotes that $n \, R \, n$; for any $n \colon R$, $[n]_R$ denotes the equivalence class $\{m \mid n \, R \, m\}$ and the set represented by R is defined to be $[R] = \{[n]_R \mid nRn\}$. We say that a natural number n realizes an element $a \in [R]$, $n \vdash_R a$, if $n \in a$. A function f (set-theoretic) from $[R]$ to $[S]$ is realized by a natural number n, denoted by $n \vdash_{R,S} f$ if for all $a \in [R]$, $m \vdash_R a$ we have that $n \cdot m \vdash_S f(a)$ and we call a function $f \colon [R] \to [S]$ effective or computable if it has a realizer. We often omit the subscripts R, S from $n \vdash_{R,S} f$ when they are clear from the context. We use $[n]^{R,S}$ to denote the unique map from $[R]$ to $[S]$ that n realizes, if it realizes one. The category **Per** has as objects pers, and as morphisms from R to S the effective maps from $[R]$ to $[S]$. For any set $B \subseteq \mathcal{N}$, we define the per $D_B = \{\langle n, n \rangle \mid n \in B\}$, the identity relation on B. We use N to denote the per $D_{\mathcal{N}}$ and 1 to denote the per $D_{\{0\}}$. If R, S are pers, we define the pers $R \times S, R \to S, R \rightharpoonup S$ by $n \, R \times S \, m$ iff $(n)_1 \, R \, (m)_1$ and $(n)_2 \, S \, (m)_2$, $n \, R \to S \, m$ iff $\forall x, y \in \mathcal{N} . \, x \, R \, y \Rightarrow (n \cdot x) \, S \, (m \cdot y)$, $n \, R \rightharpoonup S \, m$ iff $\forall x, y \in \mathcal{N} . \, x \, R \, y \Rightarrow (n \cdot x) \downarrow$ or $(m \cdot y) \downarrow \Rightarrow (n \cdot x) \, S \, (m \cdot y)$ where we use $(n \cdot x) \, S \, (m \cdot y)$ to mean that $(n \cdot x) \downarrow$ and $(m \cdot y) \downarrow$ and they are related in S. We can define the lifting of a per R as $R_\perp = 1 \rightharpoonup R$.

Let A be any set and $\vdash_A \subseteq \mathcal{N} \times A$ be a binary relation which we often call a realization relation. Recall that $\vdash_A$ is *onto* if for all $a \in A$, there is an n with $n \vdash_A a$ and a *function* if whenever $n \vdash_A a, n \vdash_A b$ we have that $a = b$. Every onto function $\vdash_A$ on a set A gives rise to a per, namely, $\{\langle n, m \rangle \mid n, m \vdash_A a, a \in A\}$. Suppose $\vdash_A$ is some fixed onto realization relation on a set A and $\vdash_B$ is another fixed onto realization relation on a set B. We call a function (set-theoretic) $f \colon A \to B$ *effective* if there is an n such that for all $a \in A, m \vdash_A a$ we have that $n \cdot m \vdash_B f(a)$; we take n to be the realizer of f. We also say that $f(a)$ is *effectively realizable* from a to mean that the function f is effective.

3 Fixed Points and Effective Cpos

We begin by showing that the full category **Per** is not suitable for interpreting fixed points.

Example 1 (Failure of Fixed Points). Take R to be the per **1** and S to be the per of partial functions that are defined exactly on some finite prefix of the natural numbers. More formally, S is defined by $m \, S \, n$ iff $\exists u \in \mathcal{N} \; \forall x < u \; (m \cdot x = n \cdot x)$ and $\forall x \geq u \; (m \cdot x \uparrow$ and $n \cdot x \uparrow)$. Let F be an index of the function, which on input $f : (R \rightharpoonup S)$, returns an index of the function, which on input $0 : \mathbf{1}$ returns an index of the function which on input n computes $g = f \cdot 0$ and returns 0 if $n = 0$ and returns $g \cdot (n - 1)$ otherwise. It can be verified that $F : (R \rightharpoonup S) \to (R \rightharpoonup S)$ and that $[F]^{(R \rightharpoonup S),(R \rightharpoonup S)}$ has no fixed point.

An important idea in the above example is that if we order the partial functions that are the elements of the per S by the pointwise ordering, as we would in domain theory, then S is not a cpo. Therefore, it seems natural to circumvent the problem illustrated by the per S by selecting a class of cpo-like pers. The first step is to associate an order with each per. To define an ordering on the elements of a per R, we focus attention on effective partial functions from R to **1**. Since, **1** is a one-element per, the only significant behavior of any $f : R \rightharpoonup \mathbf{1}$ on an $x : R$ is its convergence. We can thus think of $f : R \rightharpoonup \mathbf{1}$ as a partial decision algorithm computing with R-inputs or a *computable test* on R, with convergence announcing success. We then consider an $x : R$ to be distinguishable from $y : R$ if there is a computable test converging on x and diverging on y. Thus, an $x : R$ has computationally no more information than $y : R$ if x is not computationally distinguishable from y. Thus, for any per R and $x, y : R$, we define

$$x \leq_R y \quad \text{iff} \quad \forall f : R \rightharpoonup \mathbf{1}. \; f \cdot x \downarrow \text{ implies } f \cdot y \downarrow$$

For $[x]_R, [y]_R \in [R]$, we define $[x]_R \leq_{[R]} [y]_R$ iff $x \leq_R y$; $\leq_{[R]}$ is easily seen to be a preorder.

Taking this to be our *intrinsic ordering*, we consider the class of pers whose intrinsic orderings are complete partial orders in the internal language of the topos **Per**. For any per R, define the per $Seq(R)$ of *effective increasing sequences* as the largest subset of the per $N \to R$ such that $s : Seq(R)$ iff for all $n \in \mathcal{N}$, $s \cdot n \leq_R s \cdot (n + 1)$. We call a function (possibly partial) from $[Seq(R)]$ to $[R]$ a *supremum* function if it maps an element $[s]$ of $[Seq(R)]$ to a least upper bound of $\{[s \cdot n]_R \mid n \in \mathcal{N}\}$ with respect to the preorder $\leq_{[R]}$ if it exists and undefined otherwise. If the preorder $\leq_{[R]}$ on $[R]$ is antisymmetric, then there is a unique supremum function; in this case, we denote it by sup_R. We define a per R to be an *effective cpo* if

Antisymmetry For $a, b \in [R]$, if $a \leq_{[R]} b$ and $b \leq_{[R]} a$ then $a = b$.

Computable Completeness The function $sup_R : [Seq(R)] \to [R]$ is a morphism in **Per**, *i.e.*, it is *total* and *effective*.

The following lemma can be seen as a generalization of the Myhill-Shepherdson theorem.

Lemma 1 (Continuity). *Suppose that R, S are effective cpos. Then any effective $f : [R] \to [S]$ is continuous, i.e., if $s : Seq(R)$ then $f(sup_R([s]_{Seq(R)})) = sup_S(f \circ [s]^{N,R})$.*

By analogy with ordinary cpos, we say that an effective cpo R is *pointed* if $[R]$ contains a least element with respect to the intrinsic order $\leq_{[R]}$; we denote its least equivalence class by $\perp_R$.

Lemma 2 (Fixed Points). *Suppose that R is a pointed effective cpo. Every effective $f : [R] \to [R]$ has a least fixed point with respect to the ordering $\leq_{[R]}$. The function $fix_R : [R \to R] \to [R]$ mapping realizers for morphisms to their least fixed point is effective and is effectively realizable from sup_R and $\perp_R$.*

While our assumptions on the properties of an effective cpos were important to our proof of the existence of a computable fixed point operator, we can also show that they are the weakest conditions possible. We can construct a per R whose intrinsic order is not antisymmetric, and a per S whose intrinsic order is antisymmetric and for which sup_S is total but not effective; both R and S can be shown not to admit any effective fixed point operator. It can also be seen that a per that is not pointed cannot admit a fixed point operator.

The basic constructions on pers all yield effective cpos. For any $B \subseteq \mathcal{N}$, the intrinsic order on the per D_B is easily seen to be discrete and hence an effective cpo trivially and it is pointed iff $|B| = 1$; we now state the closure under the other constructions.

Proposition 3 (Product Spaces). *If R, S are pers then $x \leq_{R \times S} y$ iff $(x)_1 \leq_R (y)_1$ and $(x)_2 \leq_S (y)_2$. Thus, if R and S are effective cpos then $R \times S$ is an effective cpo, with $sup_{R \times S}$ effectively realizable from sup_R and sup_S. If R, S are pointed then $R \times S$ is pointed with $\perp_{R \times S}$ effectively realizable from $\perp_R, \perp_S$.*

Proposition 4 (Total Function Spaces). *If R, S are pers and S is an effective cpo, then $f \leq_{R \to S} g$ iff $\forall x : A.f \cdot x \leq_S g \cdot x$. Thus, if R and S are effective cpos then $R \to S$ is an effective cpo, with $sup_{R \to S}$ effectively realizable from sup_S. If S is pointed then $R \to S$ is pointed with $\perp_{R \to S}$ effectively realizable from $\perp_S$.*

Proposition 5 (Partial Function Spaces). *If R, S are pers and S is an effective cpo, then $f \leq_{R \rightharpoonup S} g$ iff $\forall x : A.f \cdot x \downarrow \Rightarrow f \cdot x \leq_S g \cdot x$ Thus, if R and S are effective cpos then $R \rightharpoonup S$ is an effective cpo, with $sup_{R \rightharpoonup S}$ computable from sup_S. The per $R \rightharpoonup S$ is always pointed with $\perp_{R \rightharpoonup S}$ realized by the index of the function "$\lambda x \in \mathcal{N}.diverge$" independent of R, S.*

While the proof of the "pointwise" intrinsic ordering for products is standard (c.f. [AP90]) and is true for arbitrary pers, the intrinsic ordering for the pers $R \to S, R \rightharpoonup S$ is pointwise only when the per S is an effective cpo. An important point to note about the closure of effective cpos under these operations, as given by the above propositions, is that it is constructive, i.e., the supremum operator for the product or function space pers of R, S can be computed from the supremum operator for R and S. Since lifting is defined using partial functions, it follows that effective cpos are closed under lifting as well.

4 Polymorphism

Let ECpo and PECpo be the set of all effective cpos and pointed effective cpos respectively. Define an *effective set* to be a set K equipped with an *onto* binary relation $\vdash_K \subseteq \mathcal{N} \times K$. For any function $C: K \to \text{ECpo}$, we can define the per $\forall^K(C)$ as follows:

$$f \, \forall^K(C) \, g \quad \text{iff} \quad \forall a \in K, x, y \in \mathcal{N} \quad x, y \vdash_K a \Rightarrow (f \cdot x) \, C(a) \, (g \cdot y)$$

It is easy to see that $\forall^K(C)$ is symmetric and transitive, *i.e.*, that it is a per. We take ECpo to be an effective set with $n \vdash_{\text{ECpo}} R$ iff $n \vdash_{Seq(R),R} sup_R$. And, we consider PECpo an effective set with $n \vdash_{\text{PECpo}} R$ iff $(n)_1 \vdash_{\text{ECpo}} R$ and $(n)_2 \vdash_R \perp_R$. Essentially the realizers for ECpo and PECpo are proofs, in the internal language of the effective topos, of the membership of a per in the particular subset. The following lemma shows that if C is an effective map (as defined in Sect. 2) from K to ECpo then $\forall^K(C)$ is an effective cpo. As usual, we also give the conditions under which it is pointed.

Lemma 6 (Higher-Order Polymorphism). *Let $C: K \to \text{ECpo}$ be an effective map. Then $f \leq_{\forall^K(C)} g$ iff $\forall a \in K, x \in \mathcal{N} \quad x \vdash_K a \Rightarrow (f \cdot x) \leq_{C(a)} (g \cdot x)$. Thus, $\forall^K(C)$ is an effective cpo with $sup_{\forall^K(C)}$ effectively realizable from C. If C induces an effective map from K to PECpo then $\forall^K(C)$ is pointed with $\perp_{\forall^K(C)}$ effectively realizable from the effective map $C: K \to \text{PECpo}$.*

As particular instances of Lemma 6, we can define constructors $\forall^{\text{ECpo}}$ which we denote $\forall$, and $\forall^{\text{PECpo}}$ which we denote $\forall^p$, yielding effective cpos. The following proposition shows that in this semantics of polymorphism as well, the type $\forall t.t$ is empty and $\forall t.(t \to t)$ only contains the identity function. It thus suggests that all the usual parametricity principles are validated even by this interpretation of polymorphism that is less "uniform" than intersection. Essentially, parametricity in our context arises from the fact that the realization relations $\vdash_{\text{ECpo}}, \vdash_{\text{PECpo}}$ are *not functions* — thus, any element of the polymorphic per must behave uniformly on all pers that admit common realizers for their supremum functions.

Proposition 7 (Parametricity). *For $\mathcal{I}: \text{ECpo} \to \text{ECpo}$ defined by $\mathcal{I}(R) = R$, there is no n with $n: \forall(\mathcal{I})$. For $\mathcal{F}: \text{ECpo} \to \text{ECpo}$ defined by $\mathcal{F}(R) = R \to R$, if $n: \forall(\mathcal{F})$ then for any effective cpo R, natural number x, if $x \vdash sup_R$ then $n \cdot x \vdash_R id_R$.*

Turning next to the $\forall^p$ quantifier, consider the map $C: \text{PECpo} \to \text{ECpo}$ given by $C(R) = (R \to R) \to R$ which is effective by Prop. 4. By Lemma 2 we have a realizer $fix: \forall^p(C)$, which computes the least fixed point at every pointed type. The following proposition shows that this is the *only* fixed point operator of this polymorphic type:

Proposition 8 (Polymorphic Fixed Point). *Consider $C: \text{PECpo} \to \text{ECpo}$ defined by $C(R) = (R \to R) \to R$. Then we have a realizer $fix: \forall^p(C)$ computing the*

*least fixed point at every pointed type. Further, suppose that we have an $f: \forall^p(C)$
such that for any pointed effective cpo R, and $x \vdash sup_R, b \vdash \perp_R$ we have that
$(f \cdot pr(x, b)) = n$ such that for any $g: R \to R$, $n \cdot g R g \cdot (n \cdot g)$, i.e., f is a fixed
point operator. Then $f \forall^p(C)$ fix.*

Just like Prop. 7, we can prove that the type $\forall^p t.t$ has exactly one element and
the type $\forall^p t.(t \to t)$ has exactly two elements; the polymorphic operator $\forall^p$ thus
supports the parametricity principles of system **F** extended with recursion on
values.

5 Recursive Types

A standard method for obtaining recursive types is that given by [SP82] which
shows that suitable functors on **Cpo**-enriched categories admit recursive solutions. While these results are not directly applicable to the category of effective
cpos, we obtain analogous results by considering everything "effectively" and
reformulating the framework and results of [SP82] for a suitable notion of effec-
tivity. The main departure from earlier work in this regard (*e.g.*, [AP90]) is that
effectiveness of a functor is a condition on its behavior on objects in addition to
arrows.

Let **C** be a category. We call it *effective* if it is equipped with a realization
relation, $\vdash_C$, on objects and morphisms that is *onto on objects* and an *onto
function on each homset*, with identity morphisms effectively realizable from
the objects and composition effective; define effective functors between effective
categories in the obvious way. By the condition on $\vdash_C$, we can define a per
corresponding to each homset as in Sect. 2. We say that **C** is an **ECpo**-enriched
category if the per corresponding to each homset $C(A, B)$ is an effective cpo
with its supremum function effectively realizable from the objects A, B; this is of
course just the internal proof that every homset is a cpo. Define the category C^{ep}
whose objects are those of **C** and whose morphisms are embedding-projection
pairs (while the embeddings and projections determine each other uniquely, they
may not necessarily do so *effectively*; it therefore does not suffice to consider the
subcategory of only embeddings or projections). As for morphisms in **Per**, any
effective functor between two **Ecpo**-enriched categories is automatically locally
continuous in the sense of [SP82]. Call an effective category **C** a **PECpo**-enriched
category if it is an **ECpo**-enriched category and the per corresponding to each
homset $C(A, B)$ is pointed with its least element, $\perp_{A,B}$, effectively realizable
from the objects A, B.

Lemma 9 (Recursive Solutions). *Suppose that **C** is PECpo-enriched, effectively ω^{op}-complete, has terminal object $\perp$, and that composition in **C** is
left-strict, i.e., for any $A \xrightarrow{f} B$ we have $\perp_{B,C} \circ f = \perp_{A,C}$. Then for any effective functor $F: C^{ep} \to C^{ep}$, we have an object $\mu(F)$ in **C** and morphisms
$fold^F: F(\mu(F)) \to \mu(F)$, $unfold^F: \mu(F) \to F(\mu(F))$ with $fold^F$, $unfold^F$ constituting an isomorphism pair in **C**. Moreover, the object $\mu(F)$ and the morphisms
$fold^F$, $unfold^F$ are effectively realizable from the effective functor F.*

Define **ECpo** to be the full subcategory of **Per**, of the effective cpos. The realization relation on objects of **ECpo** is as in Sect. 4 for ECpo and on morphisms as in Sect. 2. Take **PECpo** to be the full subcategory of **ECpo** of pointed effective cpos with the realization relation on objects as in Sect. 4 for PECpo.

Lemma 10. *The categories* **ECPo** *and* **PECpo** *are effectively* ω^{op}*-complete.*

By Prop. 4, **ECpo** is an **ECpo**-enriched category and **PECpo** is a **PECpo**-enriched category; it can also be seen that **PECpo** satisfies the other conditions of Lemma 9. Hence, any effective functor $F: \textbf{ECpo}^{\text{ep}} \to \textbf{ECpo}^{\text{ep}}$ that induces an effective functor from **PECpo**^{ep} to **PECpo**^{ep} has a recursive solution. This condition on the preservation of pointedness is not an accident of our method of constructing recursive solutions; there are functors that do not map pointed pers to pointed pers and provably cannot have any recursive solutions.

6 Subtyping

Suppose we have pers R, S with $R \subseteq S$ as relations. Since the index of $\lambda x \in \mathcal{N}.x$ realizes an effective map from R to S, we have by Lemma 1, that if $x \leq_R y$ then $x \leq_S y$. Thus, if $R \subseteq S$, then $Seq(R) \subseteq Seq(S)$. For effective cpos R, S, we define that R is a subtype of S, denoted $R <: S$, iff $R \subseteq S$, if $n \vdash sup_R$ and $s: Seq(R)$ then $n \cdot s \vdash sup_S(s)$, and if R is pointed and $n \vdash_R \perp_R$ then S is pointed with $n \vdash_S \perp_S$. Essentially then, R is a subtype of S, if the proof of any "interesting property" in R is also a proof in S; the interesting properties in our context being membership in a per, equality in a per, and membership in the subcategories **ECpo**, **PECpo**. This notion of subtyping is reflexive and transitive and one obtains all the usual subtyping relations between the various type constructors. We can define the per $Top = \mathcal{N} \times \mathcal{N}$ as the supertype of all types.

7 Uniform Fixed Point Realizer

Using the definition of the fixed point operator in terms of the supremum operator, we can obtain a uniform algorithm for computing the fixed point at all types if there is a single algorithm for computing the supremum operator. However, the following lemma shows that this is impossible even for $N_\perp, (N \to N)_\perp$.

Lemma 11 (Failure of Uniform Suprema). *There is no natural number f such that $f \vdash sup_N$ and $f \vdash sup_{N \to N}$, i.e., such that if $s: Seq(N)$ then $f \cdot s \vdash_N sup_N(s)$ and if $s: Seq(N \to N)$ then $f \cdot s \vdash_{N \to N} sup_{N \to N}(s)$.*

Corollary 12. *There is no natural number f such that $f \vdash sup_{N_\perp}$ and $f \vdash sup_{(N \to N)_\perp}$.*

We now show that for any reasonable class of pers over natural numbers, there cannot be a uniform fixed point operator. Consider the more restricted problem of uniformly computing fixed points of $f\colon (R \to S) \to (R \to S)$ for some class of pers R, S. A natural candidate for a uniform fixed point operator is the index, Y, of the untyped call-by-value fixed point operator (also given by the first recursion theorem *e.g.*, see [Cut80]). Of course, untyped lambda calculus abounds with other fixed point operators (*c.f.* [Bar84]); the following definition captures the essential property of any untyped fixed point operator.

Definition 13. A natural number f is an *untyped fixed point operator* if for any $n\colon N \to N$ we have $f \cdot n \; (N \to N) \; n \cdot (f \cdot n)$.

Lemma 14 (Failure of Untyped Fixed Points). *Let f be any untyped fixed point operator. Suppose that R, S are pers with $\neg(k(R \to S)k)$, for some $k \in \mathcal{N}$. Then there is an $e\colon (R \to S) \to (R \to S)$ with $\neg(f \cdot e(R \to S)f \cdot e)$. Thus, f does not even realize a morphism from $((R \to S) \to (R \to S))$ to $(R \to S)$ and hence cannot realize a fixed point operator for the per $R \to S$.*

In particular, the index Y cannot be a realizer for the fixed point operator on all $R \to S$. The definition of e, in the proof of Lemma 14, uses an equality test on its argument y which essentially depends on the fact that realizers for functions ($y\colon R \to S$ in this case) are accessible as realizers of natural numbers and thus susceptible to equality tests. In Sect. 8 we show that this is the only reason for the failure of Y as a fixed point operator.

8 Effective Cpos over Lambda Terms

We consider an untyped call-by-value λ-calculus that has term constants corresponding to numerals and some basic operations on natural numbers. It thus corresponds closely to the partial combinatory algebra of turing machines that we have considered so far, except for introducing a distinction between codes of natural numbers (numerals) and codes of functions (λ-abstractions). We take the partial combinatory algebra whose elements are observational congruence classes of the terms of the calculus and consider effective cpos over them.

The untyped term $Y = \lambda f. (\lambda x. f (\lambda z. xxz)) (\lambda x. f (\lambda z. xxz))$ can be shown to be a realizer for the least fixed point operator $fix^{R \to S}$ for any effective cpos R, S, using the observational preorder $\sqsubseteq$ on terms. The partial combinatory algebra, equipped with this ordering, is not a cpo; however, if $a_n, n \in \mathcal{N}$ are such that $a_n \sqsubseteq a_{n+1}$, we denote its least upperbound by $\bigsqcup_n a_n$ when it exists. For any effective cpo R, we can prove the remarkable property that if $s\colon Seq(R)$ is an increasing sequence such that $s \cdot n \sqsubseteq s \cdot (n+1)$ then $\bigsqcup_n (s \cdot n) \vdash_R sup_R(s)$, *i.e.*, if an increasing sequence is also an increasing sequence with respect to $\sqsubseteq$ then its supremum in the effective cpo must be the equivalence class of the least upper bound with respect to the global ordering. Now, consider the elements F_n defined by $F_0 = \lambda f. \Omega$, where Ω is any divergent term and $F_{n+1} = \lambda f. f (\lambda z. F_n f z)$. Because of the distinction between numerals and λ-abstractions, $F_n \sqsubseteq F_{n+1}$ and $Y = \bigsqcup_n F_n$. We also have that $F_n\colon ((R \to S) \to (R \to S)) \to (R \to S)$ and that $fix^{R \to S} = sup_{((R \to S) \to (R \to S)) \to (R \to S)}(s)$ where s is the sequence with $s \cdot n = F_n$. From all this it follows that $Y \vdash fix^{R \to S}$.

9 Conclusion

In this paper, we have developed and analysed per models for recursive types, polymorphism, and subtyping. In developing our models, the emphasis has been on simplicity and on understanding the necessity of any conditions we impose on our structures. While similar structures have been given using the interpretation of constructive logic in a topos, the presentation of the models here is more direct and the analysis of their properties uses only simple recursion-theoretic arguments. Our main technical results fall into two parts. The first set concern the properties of the model using partial equivalence relations over natural numbers, where types are coded by realizers of their supremum operators. We have established parametricity properties for an interpetation of polymorphism which is less uniform than "intersection", and coherence of the semantics of subtyping inspite of a non-uniform fixed point operator. Our second set of results concern the comparison of uniformity of fixed-point operators in Gödel number and lambda term models. We have shown the surprising property that no untyped fixed point operator can be a fixed point operator for any per that is a strict subquotient of the natural numbers, and established the correctness of an untyped fixed point operator when we consider pers over lambda terms instead.

References

[AP90]   M Abadi and G.D. Plotkin. A PER model of polymorphism and recursive types. In *Proc. IEEE Symp. on Logic in Computer Science*, pages 355–365, 1990.

[Bar75] J. Barwise. *Admissible sets and structures*. Springer-Verlag, Berlin, 1975.

[Bar84] H.P. Barendregt. *The Lambda Calculus: Its Syntax and Semantics*. North-Holland, Amsterdam, 1984. Second edition.

[BM92]   K. Bruce and J.C. Mitchell. PER models of subtyping, recursive types and higher-order polymorphism. In *Proc. 19th ACM Symp. on Principles of Programming Languages*, pages 316–327, January 1992.

[Cut80] N.J. Cutland. *Computability: An introduction to recursive function theory*. Cambridge Univ. Press, Cambridge, 1980.

[FRMS92] P. Freyd, G. Rosolini, P. Mulry, and D.S. Scott. Extensional PER's. *Information and Computation*, 98(2):211–227, 1992. Preliminary version appeared in *Proc. IEEE Symp. on Logic in Computer Science*, IEEE, 1990, 346–354.

[GS90]   C.A. Gunter and D.S. Scott. Semantic domains. In J. van Leeuwen, editor, *Handbook of Theoretical Computer Science, Volume B*, pages 633–674. North-Holland, Amsterdam, 1990.

[Pho90] W. Phoa. Effective domains and intrinsic structure. In *Proc. IEEE Symp. on Logic in Computer Science*, pages 366–377, 1990.

[SP82]   M. Smyth and G.D. Plotkin. The category-theoretic solution of recursive domain equations. *SIAM J. Computing*, 11:761–783, 1982.

Regularity for a Large Class of Context-Free Processes is Decidable

D.J.B. Bosscher[1] *   W.O.D. Griffioen[1,2] **

[1] CWI
P.O. Box 94079, 1090 GB Amsterdam, The Netherlands
{doeko,griffioe}@cwi.nl
[2] Computing Science Institute, University of Nijmegen
P.O. Box 9010, 6500 GL Nijmegen, The Netherlands

Abstract. Regularity of context-free processes has been proved to be decidable for BPA systems by [MM94] and normed context-free processes by [Kru95]. In this paper the decidable class of regular context-free processes is enlarged to that of context-free processes over so-called NRD specifications (definition in the paper). Furthermore an upper bound is given for the number of states modulo bisimulation.

1 Introduction

A classical result from formal language theory is that language equivalence is undecidable for context-free grammars. In [BH64] the correspondence Lemma of Post [Pos46] is used to prove this and the undecidability of regularity of context-free languages. The picture changes if these grammars are studied as process specifications modulo a given equivalence for process graphs.

A well-known equivalence for process theory, bisimulation equivalence is decidable for context-free processes [BBK93, CHS92]. This leads one to believe bisimulation is a sufficiently strong equivalence to allow decidability of regularity of context-free process graphs. In this paper decidability of regularity for a large class of context-free process graphs is proved.

A process graph is regular iff it is (strongly) bisimilar to a process graph with finitely many states. Context-free process graphs are denoted by "guarded recursive specifications" over Basic Process Algebra (BPA) [BW90]. Below three examples of BPA specifications for a process name X are given.

$$(1) \begin{aligned} X &= a \cdot Y + c \cdot Z \\ Y &= d \cdot X + e \cdot Y \\ Z &= c \end{aligned} \qquad (2) \begin{aligned} X &= a \cdot X \cdot Y + c \\ Y &= b \cdot Y \end{aligned} \qquad (3) \begin{aligned} X &= a \cdot Y \cdot Z \\ Y &= a \cdot Y \cdot B + c \\ Z &= b \cdot Z \\ B &= b \end{aligned}$$

---

* Research supported by Esprit BRA 7166 CONCUR 2.
** Research supported by the Netherlands Organization for Scientific Research (NWO) under contract SION 612-316-125.

All three specifications define a regular process graph for X. The process graph for X is regular with respect to first specification because the specification is linear. It is folklore that the class of regular processes is the same as the class of processes which can be denoted by a linear specification. In fact one of the reasons for using regular processes is that these can be described precisely by such specifications, which allow easy implementation and checking of modal and temporal properties [Hol89].

The second specification defines the process graph for X also as a regular process, but to see this is already more difficult, since the specification is not linear (X has summand $a \cdot X \cdot Y$ in the defining equation). In the paper by [MM94] it is proved that specifications which do not define "normed stackings" define only process names with regular process graphs. The same result is proved for normed process graphs by [Kru95]. Both papers give a method to generate the linear specification for a process graph, provided it is regular.

In this paper we extend the class of specifications further by allowing specifications such as 3. The specification is not linear and allows a normed stacking for Y. However X, which depends on Y defines a regular process graph: the idea is that the "context-free behavior" of process name Y is somehow neutralized by Z. Furthermore we give an explicit upper bound for the number of states of the process graph modulo bisimulation and a method to generate the linear specification.

As was pointed out by Didier Caucal our method does not work for every context-free process graph. We must restrict ourselves to specifications which have only so-called weakly deterministic process names in normed repeats (see Section 3 for the definition), so-called process graphs over NRD specifications (process **N**ames occurring in a weakly **N**ormed **R**epeat are weakly **D**eterministic).

The plan of the paper is as follows. In the second section we introduce (context-free) processes and process graphs formally and give some preliminary definitions. In the third section we prove the decidability result. We conclude with a remark on generating a linear specification for a regular process graph.

2   Context-free Processes and Regularity

As usual we refer to context-free processes as processes over specifications in Basic Process Algebra. Therefore, in this section we define the basic notions of BPA. We start by giving the syntax and semantics of BPA. Next we define process graphs and bisimulation equivalence as used throughout this paper. For a detailed description of the relation between language and process theory we refer the interested reader to [HM96].

The abstract syntax of BPA is given by

$$p ::= a \mid X \mid p + p \mid p \cdot p$$

where a ranges over a *finite* set Act of atomic actions and X over *countable infinite* set *Names* of process names. The $+$ is the usual process algebraic notation for choice and $\cdot$ for sequential composition. We look at *recursive* processes, processes where the meaning of process names is given by a (finite) set of equations of the form

$$\Delta = \{X_i = p_i | 1 \leq i \leq k\}$$

where $i, k \in \mathbb{N}$, the X_i are distinct process names and the process names used in p_i are those defined in $\{X_1, ..., X_k\}$. In the sequel we will only look at guarded recursive equations, i.e. every summand of the processes p_i in the equations $X_i = p_i$ starts with an action.

The operational semantics of a BPA expression, given a specification Δ, is the smallest transition relation $\rightarrow_\Delta \subset$ BPA $\times$ *Act* $\times$(BPA $\cup \{\epsilon\}$) containing the transitions provable by the following rules:

$$\frac{p \xrightarrow{a} p'}{p + q \xrightarrow{a} p'} \qquad \frac{q \xrightarrow{a} q'}{p + q \xrightarrow{a} q'}$$

$$\frac{p \xrightarrow{a} p'}{p \cdot q \xrightarrow{a} p' \cdot q} \qquad \frac{}{a \xrightarrow{a} \epsilon} \qquad \frac{p \xrightarrow{a} p'}{X \xrightarrow{a} p'}$$

where $a \in Act$, $X = p \in \Delta$ and ϵ is a special state not in BPA which denotes the empty or *terminated* state. For convenience we identify the states $\epsilon \cdot p$ and $p \cdot \epsilon$ with p.

We will use V_Δ for the set of process names defined in Δ. We use ν, ρ, σ to range over V_Δ^* and $|\nu|$ as the length of a sequence ν of process names. We use ν^k as an abbreviation for a sequence $\nu \cdot ... \cdot \nu$ of length k. Notice that we thus identify string concatenation with sequential composition and use that concatenation is associative. We speak of the head and the tail of a sequence in the obvious way. As a convention we use that ν^0 is equal to ϵ.

Most of the time we omit actions in transitions. We use the Greek letter ϕ to range over transition sequences starting with a process name in V_Δ. $\phi(0)$ denotes the first state in transition sequence, $\phi(i-1)$ the i-th state in ϕ. The Greek letters χ and ψ are used to range over transition *sub*sequences, i.e. choppings off of transitions sequences. We mean by $|\phi|$ the length of a transition sequence given by the number of transitions. We say that a process-expression p is weakly normed, also denoted by $p \downarrow$, if there is a transition sequence $p \rightarrow_\Delta ... \rightarrow_\Delta \epsilon$ [3]. A process-expression p is perpetual, also denoted by $p \uparrow$ if it is not weakly normed. By $|V_\Delta|$ we mean the size of the set V_Δ. If no confusion can arise we use $|\nu|, |\phi|, |V_\Delta|$ without stating which kind of length or size is used.

---

[3] Notice that a weakly normed process can arrive after one or more transitions in a state which is *not* weakly normed, i.e. it is not *normed* in the usual meaning.

In rest of this paper we mostly look at process graphs of process-expressions defined in Δ in the context of bisimulation equivalence. As usual we define process graphs as rooted, labeled transition systems.

Definition 2.1 Let *Act* be a given set of actions.

1. A *labeled transition system* over *Act* is a pair $(S, \rightarrow)$ where S is a set of states and $\rightarrow \subseteq S \times Act \times S$ is the transition relation,
2. A *process graph* over *Act* is a triple $g = (r, S, \rightarrow)$ with $(S, \rightarrow)$ a labeled transition system over *Act* and $r \in S$ is the root state.

In this paper the equivalence between process graphs is strong bisimulation equivalence.

Definition 2.2 Let $g = (r, S, \rightarrow)$ and $g' = (r', S', \rightarrow)$ be process graphs.

1. $R \subseteq S \times S'$ is a *bisimulation* iff for all $(p, q) \in R$ it holds that
   (a) If $p \xrightarrow{a} p'$, then there is a q' such that $q \xrightarrow{a} q'$ and $(p', q') \in R$,
   (b) If $q \xrightarrow{a} q'$, then there is a p' such that $p \xrightarrow{a} p'$ and $(p', q') \in R$.
2. g and g' are *bisimilar* iff there is a bisimulation relating the roots of g and g'.

Now we are ready to define what we mean by regularity.

Definition 2.3 A process p is *regular* iff the process graph $(p, \text{BPA} \cup \{\epsilon\}, \rightarrow_\Delta)$ is bisimilar to a process graph with a finite number of states.

3 Decidability of Regularity

In this section we prove that regularity for a large class of context-free processes is decidable.

For the proof we can restrict ourselves to BPA systems in restricted Greibach Normal Form (rGNF), i.e. systems of the form
$\Delta = \{X_i = a_{i1} \cdot \sigma_{i1} + \ldots + a_{in_i} \cdot \sigma_{in_i} | 1 \leq i \leq m\}$, where σ_{ij} is a process name sequence containing *at most two* process names. It is folklore that any guarded BPA specification specification can effectively be represented in restricted Greibach Normal Form (rGNF)[4] maintaining bisimulation equivalence of the process graphs of the defined processes. We have the following useful properties.

Proposition 3.1 Let Δ be a BPA system in rGNF and $X \in V_\Delta$.

1. Let $\phi \equiv X \rightarrow_\Delta \ldots$ be a transition sequence. For every $i \in \mathbb{N}$, $\phi(i)$ is a sequence of process names in V_Δ,

---

[4] For context-free processes see e.g. [BBK87, BP95]. Similar result for context-free languages and language equality [HU79].

2. If $\sigma \to_\Delta \rho$, then $|\rho| \le |\sigma| + 1$.

Remark 3.2 Notice that Proposition 3.1 gives that the states reachable from a process name are *sequences* of process names.

Definition 3.3 Let $X = p \in \Delta$. An action is *a unique in X*, iff p has only one summand starting with a.

Definition 3.4 A transition (sub)sequence χ is *weakly deterministic* iff for every transition $\chi(i) \xrightarrow{a} \chi(i+1)$, a is unique in the head of $\chi(i)$.

Definition 3.5 Let $W_0, W_1, \cdots \subseteq V_\Delta$ be sets inductively defined as follows

1. $W_0 = \emptyset$,
2. $X \in W_{i+1}$ iff either $X \in W_i$ or $X = p \in \Delta$ and there is an action a such that a is unique in p, and the summand starting with a has all process names in W_i.

Let $W_i = W_{i+1}$ for some $i \in \mathsf{N}$. Then W_i is the smallest set of *weakly deterministic* process names for Δ, denoted as W_Δ.

In Lemma 3.17 we prove that elements of W_Δ all start a weakly deterministic and terminating transition (sub)sequence.

Remark 3.6 Notice that W_Δ can be effectively computed (as $W_{|V_\Delta|}$).

We define *weakly normed repeat invariance* as the criterion for regularity. Informally it states that looping through a specification while stacking does not result in infinitely many non bisimilar states.

Definition 3.7 Let Δ be a specification in rGNF. Let ϕ be a transition sequence $\phi(0) \xrightarrow{a_0}_\Delta \phi(1) \xrightarrow{a_1}_\Delta \cdots$.

1. The transition sequence $\phi(0) \to_\Delta \cdots \to_\Delta \phi(i)$ is a *repeat* iff $\phi(i) \equiv \phi(0) \cdot \sigma$.
   (a) which is *cyclic* iff $\phi(i) \equiv \phi(0)$,
   (b) which is *perpetual* iff $\phi(i) \uparrow$ and $\phi(i) \not\equiv \phi(0)$,
   (c) which is *weakly normed* iff $\phi(i) \downarrow$ and $\phi(i) \not\equiv \phi(0)$.
2. The transition subsequence $\chi(0) \to_\Delta \cdots \to_\Delta \chi(i)$ *corresponds* to the repeat $\phi(0) \to_\Delta \cdots \to_\Delta \phi(i)$ iff $\chi(0) \equiv \phi(0) \cdot \sigma, ..., \chi(i) \equiv \phi(i) \cdot \sigma$,
3. A transition sequence *uses* a repeat when it has a transition subsequence that corresponds to a repeat,
4. A repeat $\phi(0) \to_\Delta \cdots \to_\Delta \phi(i)$ is *lonely* iff $\phi(0) \to_\Delta \cdots \to_\Delta \phi(i)$ uses no other repeats,
5. A transition sequence ϕ is a possible entry to a weakly normed repeat, *entry* for short, iff
   (a) ϕ starts in the root r,
   (b) uses no weakly normed repeats and,
   (c) passes only through pair-wise non bisimilar states,

6. An entry ϕ is *weakly normed repeat invariant* for a (subsequence corresponding to a) lonely weakly normed repeat χ iff
   (a) the last state of ϕ is equal to the first state of χ,
   (b) the first and last state of χ are bisimilar.
7. r is *weakly normed repeat invariant* iff every entry to a (subsequence corresponding to a) lonely weakly normed repeat is weakly normed repeat invariant.

Definition 3.8 Let Δ be a specification in rGNF. A process name Y *occurs in a weakly normed repeat* $\phi(0) \to \dots \to \phi(i)$, iff $\phi(i) \equiv \phi(0) \cdot \sigma$ and Y occurs in σ.

In the sequel we assume a fixed but arbitrary system of equations Δ in restricted Greibach Normal Form with the root process name r in V_Δ in a process graph $(r, \text{BPA} \cup \{\epsilon\}, \to_\Delta)$, all as defined in the previous section.

Finally we are able to define the class of process for which the decidability proof holds.

Definition 3.9 We say that the specification Δ is NRD (weakly Normed Repeat variables are weakly Deterministic) iff the set of process names of V_Δ occurring in a weakly normed repeat is a subset of W_Δ.

Remark 3.10 Note that it is decidable if a specification is a NRD specification.

Example 1. Let Δ be defined as

$$\Delta = \{X = a \cdot Y,$$
$$Y = a \cdot Y \cdot B + d,$$
$$B = b\}$$

where X is the root process name. The transition sequence $\chi \equiv Y \xrightarrow{a} Y \cdot B$ is a weakly normed repeat, because $Y \cdot B \downarrow$ and $Y \not\equiv Y \cdot B$. Furthermore χ is a lonely repeat. Contrast this with $\chi' \equiv Y \xrightarrow{a} Y \cdot B \xrightarrow{a} Y \cdot B \cdot B$ which is a weakly normed but not a lonely repeat, because it uses χ.

The transition sequence $\phi \equiv X \xrightarrow{a} Y$ is an entry, because it uses no weakly normed repeat and $X \not\equiv Y$. ϕ is not weakly normed repeat invariant for χ because the last state Y of ϕ, is not bisimilar to the last state $Y \cdot B$ of χ.

The root process name X is not weakly normed repeat invariant, because the entry ϕ is not weakly normed repeat invariant for the lonely, weakly normed repeat χ. The reader can verify easily that the specification is NRD. It allows only process names from W_Δ in a weakly normed repeat: the process name B occurring in the weakly normed repeat $Y \xrightarrow{a} Y \cdot B$, is in W_1.

The first part of the proof establishes that a weakly normed repeat invariant root cannot start an *infinite* transition sequence of pair-wise non bisimilar states. This implies regularity with the following well-known fact.

Proposition 3.11 If r is not regular, then there exists an *infinite* transition sequence $r \to_\Delta \dots$ where all states are pair-wise non bisimilar.

Lemma 3.12 The length of a transition subsequence $\chi \equiv \sigma \rightarrow_\Delta \ldots$ which uses *no* repeats is maximally $|\sigma|.(2^{|V_\Delta|} - 1)$.

Proof. By a well-founded simultaneous induction on the number of different process names h that occur left-most in states of χ and the length l of $|\sigma|$. Now assume the Induction Hypothesis holds for all tuples which are lexicographically smaller than tuples (h, l). Distinguish the following cases in the Induction Step.

1. If $l = 1$, then $\chi \equiv X \rightarrow_\Delta \rho \rightarrow_\Delta \ldots$ for some $X \in V_\Delta$. We only prove it for the case that $\rho \not\equiv \epsilon$, for $\rho \equiv \epsilon$ the result is immediate. The Induction Hypothesis holds for $\psi \equiv \rho \rightarrow_\Delta \ldots$ because X cannot occur left-most in states of ψ any more, otherwise ψ uses a repeat starting with X. By Proposition 3.1 $|\rho| \leq 2$ and so $|\chi| = 1 + |\psi| \leq 1 + 2.(2^{|V_\Delta - \{X\}|} - 1) = |\sigma|.(2^{|V_\Delta|} - 1)$,
2. If $l > 1$, then $\chi \equiv X \cdot \rho \rightarrow_\Delta \ldots$, for some $X \in V_\Delta, \rho \in V_\Delta^+$. By the Induction Hypothesis any transition sequence without using repeats $X \rightarrow_\Delta \ldots \epsilon$ has a length not exceeding $2^{|V_\Delta|} - 1$ and so does the associated transition (sub)sequence $X \cdot \rho \rightarrow_\Delta \ldots \rightarrow_\Delta \rho$, which does not use the presence of the tail ρ. Also by the Induction Hypothesis the transition (sub)sequence $\rho \rightarrow_\Delta \ldots$ has a maximal length of $|\rho|.(2^{|V_\Delta|} - 1)$, which gives a total maximal length of $|\sigma|.(2^{|V_\Delta|} - 1)$.

We use that a perpetual repeat passes through a state with a perpetual process name at the first or second position.

Lemma 3.13 If ϕ is a perpetual repeat, then there is a state ρ in ϕ so that $\rho \equiv P \cdot \sigma$ or $\rho \equiv Y \cdot P \cdot \sigma$, where $Y, P \in V_\Delta$, $Y \downarrow$ and $P \uparrow$.

Proof. Let N_Δ and P_Δ be the subsets of weakly normed and perpetual process names of V_Δ respectively. Suppose $\phi(i)$ is the last state of the perpetual repeat ϕ. By definition $\phi(i)$ is of the form $X \cdot \rho$ with $X \cdot \rho$ perpetual. It is easily verified that either X is perpetual, in which case we are finished, or there is a $P \in P_\Delta$ so that $\rho \equiv \nu \cdot P \cdot \sigma$ with $\nu \in N_\Delta^+$. Let $\phi(j)$ be the first state so that P is present in the process name sequence. Suppose the preceding state $\phi(j-1)$ has the process name $Y \in N_\Delta$ at the first position in the process name sequence. Using the definition of Y in Δ, P is introduced. Because Δ is in rGNF, this implies that the perpetual process name P is introduced at the first or second position.

Lemma 3.14 The length of a transition sequence, which uses no weakly normed repeat and passes only through pair-wise non bisimilar states is maximally $|V_\Delta|^2.2^{|V_\Delta|}$.

Proof. Let N_Δ and P_Δ be as in the previous proof. Suppose $\phi \equiv X \rightarrow_\Delta \ldots$ is a transition sequence which uses no weakly normed repeat. With Lemma 3.12 we know that after $2^{|V_\Delta|}$ transitions ϕ has used at least one cyclic or perpetual repeat. This repeat cannot be cyclic, because ϕ has then at least two bisimilar states. So ϕ has used a perpetual repeat in the first $2^{|V_\Delta|}$ transitions. By Lemma

3.13 it has either passed through a state ρ of the form (1) $\rho \equiv P \cdot \sigma$ or (2) $\rho \equiv X \cdot P \cdot \sigma$, where $X \in N_\Delta$ and $P \in P_\Delta$.

In case (1) in the state ρ' following ρ in ϕ, there is a perpetual process name at the first or second position. Because P is perpetual the tail σ cannot shift left-most in the transition from ρ to ρ'. Δ is in rGNF and therefore ρ' is of the form $\rho' \equiv \nu \cdot \sigma$ and $1 \leq |\nu| \leq 2$. Every transition from P is to another perpetual state, so ν has one perpetual process name. But then ρ' has a perpetual process name at the first or second position.

In case (2) ϕ passes through a state with a perpetual process name at the first or second position in $2^{|V_\Delta|}$ transitions. If ϕ does not use the presence of the tail of ρ, then this implies with the previous argument that in $2^{|V_\Delta|}$ steps from ρ, ϕ passes through a state with a perpetual process name at the first or second position. If ϕ does use the presence of the tail of ρ, then by Lemma 3.12 in $2^{|V_\Delta|}$ transitions X and its reducts have disappeared and is P the left-most process name. So in the gaps between perpetual states with a perpetual process name at the first or second position in ϕ there are at most $2^{|V_\Delta|}$ connecting transitions.

Notice that there are maximally $|P_\Delta| + |N_\Delta|.|P_\Delta|$ of such non bisimilar perpetual states in ϕ, because $P \cdot \nu \leftrightarrows P \cdot \sigma$ and $X \cdot P \cdot \nu \leftrightarrows X \cdot P \cdot \sigma$ for every $P \in P_\Delta$, ν and σ. If $|N_\Delta| = 0$ we have a maximal length of $|V_\Delta| - 1$, see Remark below. If $|N_\Delta| > 0$ simple arithmetic gives us that $|P_\Delta| + |N_\Delta|.|P_\Delta| \leq |V_\Delta|^2 - 1$. Hence the maximal length is the number of different perpetual states with a perpetual process name at the first or second position plus one times the maximal length of a gap, i.e. $|V_\Delta|^2.2^{|V_\Delta|}$.

Remark 3.15 Actually we conjecture the maximal length to have an upper bound of $2^{|V_\Delta|} - 1$. The reader can verify this in the two simple cases that either *no* process names are perpetual (Lemma 3.12) or all process names in V_Δ are perpetual. In the last case the maximum number of states modulo bisimulation is even $|V_\Delta|$ and hence the maximal length $|V_\Delta| - 1$.

We show the soundness of the criterion by proving that for *all* context-free processes, not only over NRD specifications, it disallows infinite transition sequences passing through pair-wise non bisimilar states and hence implies regularity.

Theorem 3.16 If r is weakly normed repeat invariant, then r is regular.

Proof. Assume that r is weakly normed repeat invariant, and suppose r is not regular. The non regularity implies with Proposition 3.11 that there is a transition sequence ϕ starting in r passing through infinitely many pair-wise non bisimilar states.

First suppose ϕ uses no weakly normed repeat. By Lemma 3.14 we conclude that ϕ has a finite length. Contradiction. Therefore ϕ uses a weakly normed repeat.

Let χ be the first (subsequence corresponding to a) weakly normed repeat used in ϕ. χ cannot use a cyclic repeat, because then it would have two bisimilar states. Also it cannot use a perpetual repeat, because then the last state of χ would be perpetual. Therefore χ is a (subsequence corresponding to a) *lonely, weakly normed repeat.*

Let ϕ' be the part of ϕ until the first state of the first lonely weakly normed repeat χ. ϕ (and therefore ϕ') starts in the root and passes only through pairwise non bisimilar states. So ϕ' is a entry for χ and by assumption ϕ' is weakly normed repeat invariant for χ as well. But then χ has two bisimilar states, i.e. its first and last state and hence ϕ too. Contradiction.

Unlike Theorem 3.16 we do not prove for every process that is regular, that it is weakly normed repeat invariant. Here we do need that the specifications allow only process names in W_Δ occurring in weakly normed repeats. Without this restriction Lemma 3.18 does not hold. In [BCS95] an ingenious counter example to that extent is given. Based on it, Bernhard Steffen supplied us a specification of a process name which is not weakly normed repeat invariant, i.e. specification $\{S = a \cdot X \cdot Z, X = b + c \cdot X \cdot Y, Y = d + d \cdot Y + d \cdot Z, Z = e \cdot Z\}$. As the reader can verify S has a regular process graph, but $X \cdot Y \cdot Z \not\leftrightarrow X \cdot Z$, i.e. S is not weakly normed repeat invariant and the cancellation property does not hold: $Y \cdot Z \leftrightarrow Y \cdot Y \cdot Z \not\Longrightarrow Y \cdot Z \leftrightarrow Z$.

Now we begin with a proof of the completeness of the criterion.

Lemma 3.17 If $\nu \in W_\Delta{}^+$, then there is a weakly deterministic transition subsequence $\nu \xrightarrow{a_0}_\Delta \ldots \xrightarrow{a_n}_\Delta \epsilon$.

Proof. Assume without loss of generality that $\nu \equiv X \cdot \rho$. First we prove that a weakly deterministic subsequence from ν to ρ exits. Then by induction on the number of names in ν we are done.

With each variable we associate the smallest i such that it is a member of the set W_i in Definition 3.5.

Now the sequence is constructed as follows: For each transition we pick the a summand $a \cdot X \cdot Y$ (a and $a \cdot X$ likewise) such that the initial action is unique and the process names Y, Z are in "lower" W_i.

Lemma 3.18 Let $\nu, \rho, \sigma \in V_\Delta{}^*$ and ν weakly deterministic. If $\nu \cdot \rho \leftrightarrow \nu \cdot \sigma$, then $\rho \leftrightarrow \sigma$.

Proof. Because ν is weakly deterministic, from Lemma 3.17 follows that a weakly deterministic transition subsequence from $\nu \cdot \rho$ to ρ exists. A corresponding transition subsequence starts in $\nu \cdot \sigma$ and ends in σ, because each transition is labeled with an unique action (unique with respect to the head process name in the state). By Definition of bisimulation we conclude that $\rho \leftrightarrow \sigma$.

In the proof we need the Approximation Induction Principle (AIP), which states that if the unfolded graphs of processes are bisimilar down to an arbitrary depth, then the processes are bisimilar. The principle and its proof are described in [BBK87].

Proposition 3.19 Let $\pi_i : \text{BPA} \to \text{BPA}$, $i \in \mathbb{N}^*$ be the projection operators defined as follows. Let $a \in Act$, $p, p_1, p_2 \in BPA$,

$$
\begin{aligned}
\pi_1(a \cdot p) &= a \\
\pi_i(a) &= a \\
\pi_{i+1}(a \cdot p) &= \pi_i(p) \\
\pi_i(p_1 + p_2) &= \pi_i(p_1) + \pi_i(p_2).
\end{aligned}
$$

If $\pi_m(p_1) \leftrightarrow \pi_m(p_2)$ for all $m \in \mathbb{N}^*$, then $p_1 \leftrightarrow p_2$.

The completeness of our criterion is proved by showing for a regular process that the first and last state of a (subsequence corresponding to) a lonely, weakly normed repeat have to be bisimilar.

Theorem 3.20 If r is regular, then r is weakly normed repeat invariant.

Proof. Suppose ϕ is an arbitrary entry and χ a connecting transition subsequence corresponding to a lonely, weakly normed repeat. Suppose that the last state $X \cdot \rho$ of ϕ is equal to the first state of χ and the last state of χ is $X \cdot \nu \cdot \rho$. It is easily verified that χ can be extended with a transition subsequence $X \cdot \nu \cdot \rho \to_\Delta \ldots \to_\Delta X \cdot \nu^2 \cdot \rho \ldots$ etc.. Because X is weakly normed, r can reach states $\nu^n \cdot \rho$ for every $n \in \mathbb{N}$. By assumption $\nu \in W_\Delta^+$ and hence $\nu^m \in W_\Delta^+$ for $m > 0$. Because r is regular, the pigeon hole principle gives that there are smallest $k, l \in \mathbb{N}$, $k > l$ so that $\nu^k \cdot \rho \leftrightarrow \nu^l \cdot \rho$. A (repeated) application of Lemma 3.18 gives that $\nu^{k-l} \cdot \rho \leftrightarrow \rho$. This implies that for every $m \in \mathbb{N}$, (1) $X \cdot \nu \cdot \nu^{(k-l) \cdot m} \cdot \rho \leftrightarrow X \cdot \nu \cdot \rho$ and (2) $X \cdot \nu^{(k-l) \cdot (m+1)} \cdot \rho \leftrightarrow X \cdot \rho$. Because ν is weakly normed, by Lemma 3.17 there exists a terminating transition (sub)sequence $\chi \equiv \nu^m \to_\Delta \ldots \to_\Delta \epsilon$. Because $\nu \not\equiv \epsilon$, χ has *at least* length m and so (3) for all $m, k \in \mathbb{N}^*$, $\pi_m(\nu^{m+1} \cdot \rho) \leftrightarrow \pi_m(\nu^{m+1+k} \cdot \rho)$. Therefore for all $m \in \mathbb{N}^*$,

$$
\pi_m(X \cdot \nu \cdot \rho) \overset{(1)}{\leftrightarrow} \pi_m(X \cdot \nu \cdot \nu^{(k-l) \cdot m} \cdot \rho) \overset{(3)}{\leftrightarrow} \pi_m(X \cdot \nu^{(k-l) \cdot (m+1)} \cdot \rho) \overset{(2)}{\leftrightarrow} \pi_m(X \cdot \rho).
$$

An application of AIP (Proposition 3.19) now gives that the last states of ϕ and χ are bisimilar. So ϕ is a weakly normed repeat invariant entry for χ.

It remains to be proved that the criterion is decidable.

Theorem 3.21 It is decidable if r is weakly normed repeat invariant.

Proof. BPA is finitely branching and the maximal length of an entry is bounded (Lemma 3.14). It is easily verified that the length of a lonely, weakly normed repeat is bounded too. Therefore there are only finitely many possible connecting combinations and they are of bounded length too. Verifying that such a transition sequence is weakly normed repeat invariant for a (transition subsequence corresponding to a) lonely weakly normed repeat can be done effectively, using that bisimulation equivalence is decidable for all context-free processes [CHS92].

Now we can finish with the decidability result.

Corollary 3.22 The regularity of context-free processes over NRD specifications is decidable.

Proof. Immediate from Theorems 3.16, 3.20 and 3.21.

Lemma 3.23 The number of non bisimilar states of a regular process graph given by a process name over a NRD specification is bounded by $\sum_{i=0}^{|V_\Delta|^2 . 2^{|V_\Delta|}} bf^i$, where bf is the branching factor [5].

Proof. Let R be such a process name. Each state can be reached from R passing only through pair-wise non bisimilar states with some transition sequence ϕ. Lemma 3.14 gives that the maximal length of ϕ, if it uses no weakly normed repeats is $|V_\Delta|^2 . 2^{|V_\Delta|}$. If ϕ is one longer, ϕ has to use a weakly normed repeat. By Theorem 3.20 and the regularity of the process graph, R is weakly normed repeat invariant. Therefore there are *at least* two bisimilar states in ϕ. This gives rise to a tree with branching factor bf and a maximal depth bounded by $|V_\Delta|^2 . 2^{|V_\Delta|}$.

4   Conclusions

Difference with the previously mentioned papers [MM94, Kru95] [6] is that we distinguish a root process name and allow the process to have perpetual "sub processes". Whereas the criterion for regularity for the classes in [MM94, Kru95] is relatively cheap to check, our criterion has a substantial computational complexity. If we want to use our criterion of weakly normed repeat invariance, this involves checking bisimulation equivalence between *perpetual* context-free processes [7]. Checking bisimulation equivalence of *normed* processes is known to have a polynomial time complexity [HM95], whereas the complexity of checking bisimulation of arbitrary context-free processes is (doubly) exponential [BCS95].

In principle our proof also gives a method to generate the equivalent linear specification for a given specification satisfying our constraints. The naïve way is simply to "list" all non bisimilar states. Start with the root process name and generate new non bisimilar states using the specification and the algorithm in [BCS95]. Lemma 3.23 tells us that this could be very expensive.

Acknowledgements. We thank Jan Friso Groote, Alban Ponse and Frits Vaandrager for proof-reading. Olaf Burkhart, Didier Caucal, Bernhard Steffen, Faron Moller, Sjouke Mauw and Colin Stirling are thanked for advice.

---

[5] i.e. the maximum number of summands in a definition for a process name in the specification.

[6] We refer to the results in BPA, in the last paper is also dealt with parallelism, but this is not treated in this paper.

[7] Note that in our criterion we have to check if entries are repeat normed invariant for lonely normed repeats. This comes down to checking if $X \cdot \rho$ the last state of the entry and $X \cdot \sigma \cdot \rho$, the last state of the lonely normed repeat are bisimilar. This implies that $X \cdot \rho \uparrow$ and $X \cdot \sigma \cdot \rho \uparrow$.

193

References

[BBK87] J.C.M. Baeten, J.A. Bergstra, and J.W. Klop. On the consistency of Koomen's fair abstraction rule. *Theoretical Computer Science*, 51(1/2):129–176, 1987.

[BBK93] J.C.M. Baeten, J.A. Bergstra, and J.W. Klop. Decidability of Bisimulation Equivalence for Processes generating Context-free Languages. *Journal of the ACM*, 40(3):653–682, 1993.

[BCS95] O. Burkart, D. Caucal, and B. Steffen. An Elementary Bisimulation Decision Procedure for Arbitrary Context-Free Processes. In *MFCS '95*, volume 969 of *Lecture Notes in Computer Science*, pages 423–433. Springer-Verlag, 1995.

[BH64] Y. Bar-Hillel. *Language and Information*. Series in Logic. Addison-Wesley, 1964.

[BP95] D.J.B. Bosscher and A. Ponse. Translating a Process Algebra with Symbolic Data Values to Linear Format. In Uffe H. Engberg, Kim G. Larsen, and Arne Skou, editors, *Proceedings of the Workshop on Tools and Algorithms for the Construction and the Analysis of Systems*, volume NS-95-2 of *BRICS Notes Series*, pages 119–130, 1995.

[BW90] J.C.M. Baeten and W.P. Weijland. *Process Algebra*. Cambridge Tracts in Theoretical Computer Science 18. Cambridge University Press, 1990.

[CHS92] S. Christensen, H. Hüttel, and C. Stirling. Bisimulation is Decidable for all Context-free Processes. In W.R. Cleaveland, editor, *Proceedings of CONCUR 92*, volume 630 of *Lecture Notes in Computer Science*, pages 138–147. Springer-Verlag, 1992.

[HM95] Y. Hirshfeld and F. Moller. Deciding Equivalences in Simple Process Algebras. In A. Ponse, M. de Rijke, and Y. Venema, editors, *Modal Logic and Process Algebra*, volume 53 of *CSLI Lecture Notes*, pages 151–169. CSLI Publications, Stanford, 1995.

[HM96] Y. Hirshfeld and F. Moller. Decidability Results in Automata and Process Theory. In *Logics for Concurrency: Automata vs Structure*, Springer Lecture Notes in Computer Science, 1996. To appear. Previously presented as lecture notes at the VIII-th Banff Higher Order Workshop *"Theories of Concurrency: Structure vs Automata "* in 1994.

[Hol89] Uno Holmer. Translating Static CCS Agents into Regular Form. PMG report 51, Department of Computer Science, Chalmers University of Technology and the University of Göteborg, 1989.

[HU79] J.E. Hopcroft and J.D. Ullman. *Introduction to Automata Theory, Languages and Computation*. Addison-Wesley, 1979.

[Kru95] A. Kručera. Deciding Regularity in Process Algebras. Technical Report RS-95-52, BRICS (Basic Research in Computer Science, Centre of the Danish National Research Foundation), 1995.

[MM94] S. Mauw and H. Mulder. Regularity of BPA-Systems is Decidable. In Bengt Jonsson and Joachim Parrow, editors, *CONCUR'94: Concurrency Theory*, volume 836 of *Lecture Notes in Computer Science*, pages 34–47. Springer-Verlag, 1994.

[Pos46] E.L. Post. A variant of a recursively unsolvable problem. *Bulletin of the American Mathematical Society*, 52:264–268, 1946.

On Infinite Transition Graphs
Having a Decidable Monadic Theory

Didier Caucal

IRISA, Campus de Beaulieu, 35042 Rennes, France
E-mail: caucal@irisa.fr

Abstract. We define a family of graphs whose the monadic theory is linearly reducible to the monadic theory S2S of the complete deterministic binary tree. This family contains strictly the context-free graphs investigated by Muller and Schupp, and also the equational graphs defined by Courcelle. Using words for vertices, we give a complete set of representatives by prefix rewriting of rational languages. This subset is a boolean algebra preserved by transitive closure of arcs and by rational restriction on vertices.

1 Introduction

We consider the verification of properties by structures. The properties are the monadic second-order sentences, and the structures are the labelled directed graphs. Rabin has shown that the complete deterministic tree Λ on two labels has a decidable monadic theory [Ra 69] : we can decide whether a given property expressed by a monadic sentence, is satisfied by the tree Λ. Later Muller and Schupp have extended this decision result to the context-free graphs [MS 85] : a context-free graph is a rooted graph of finite degree which has a finite number of non isomorphic connected components by decomposition by distance from a (any) vertex. These context-free graphs are also the transition graphs of pushdown automata [MS 85]. Finally Courcelle has shown that the monadic theory remains decidable for the equational graphs [Co 90] : an equational graph is a graph generated by a deterministic graph grammar. For rooted graphs of finite degree, these equational graphs are the context-free graphs [Ca 90]. These decision results of [MS 85] and [Co 90] are extensions of the definability method used by Rabin.

Another approach is to find transformations on graphs which preserve the decision of the monadic theory, and to apply these transformations to graphs having a decidable monadic theory (see for instance [Th 91]). A first transformation has been given by Shelah [Sh 75] and proved by Stupp [St 75]: if a graph has a decidable monadic theory then its "tree-graph" has a decidable monadic theory.

A way to find a transformation f on graphs that preserves the decision of the monadic theory, is to translate f into an "equivalent" transformation f^* on monadic formulas: for any graph G, $f(G)$ satisfies a sentence φ if and only if G

satisfies $f^*(\varphi)$. This method has been applied for instance in [Co 94], [CW 95], and especially in [Sem 84], [W 96] for an extension of the tree-graph transformation. We give here two transformations on graphs which have direct equivalent transformations on monadic formulas: they are based on the fact that the existence of a path labelled by a rational language can be expressed by an equivalent monadic formula. By closure of Λ under these two operations, we get a family F of graphs which have a decidable monadic theory as a corollary of Rabin's theorem. We show that this family F is a strict extension of the equational graphs. By taking words as vertices, we extract a complete subset F_0 of representatives up to isomorphism, such that F_0 remains closed under the two operations defining F, and is a boolean algebra.

2 A family of graphs with a decidable monadic theory

We define two transformations on graphs which can be translated on formulas in such a way that the decision of the monadic theory is simply preserved. The first transformation is the rational restriction on labels and the second transformation is the inverse rational mapping on labels. We start with the complete and deterministic tree on two labels which has a decidable monadic theory [Ra 69]. By applying to this tree the second transformation followed by the first one, we obtain a family of graphs with a decidable monadic theory, and which is closed by these two transformations.

We take an alphabet (finite set of symbols) T of *terminals* containing at least two symbols a, b. Here a graph is a set of arcs labelled on T. This means that a *graph* G is a subset of $V \times T \times V$ where V is an arbitrary set. Any (s, a, t) of G is a *labelled arc* of *source* s, of *target* t, with *label* a, and is identified with the labelled transition $s \xrightarrow{a}_{G} t$ or directly $s \xrightarrow{a} t$ if G is understood. We denote by

$$V_G := \{\, s \mid \exists\, a\, \exists\, t,\ s \xrightarrow{a} t \ \vee\ t \xrightarrow{a} s \,\}$$

the set of *vertices* of G. A graph is *deterministic* if distinct arcs with the same source have distinct labels: if $r \xrightarrow{a} s$ and $r \xrightarrow{a} t$ then $s = t$. And a graph is *complete* if for every label a, every vertex is source of an arc labelled by a: $\forall\, a \in T,\ \forall\, s \in V_G,\ \exists\, t,\ s \xrightarrow{a} t$.
The existence of a *path* in G from vertex s to vertex t and labelled by a word $w \in T^*$ is denoted by $s \xRightarrow{w}_{G} t$ or directly by $s \xRightarrow{w} t$ if G is understood: $s \xRightarrow{\varepsilon} s$ and $s \xRightarrow{aw} t$ if there is some vertex r such that $s \xrightarrow{a} r$ and $r \xRightarrow{w} t$.
We say that a vertex r is a *root* of G if every vertex is accessible from r: $\forall\, s \in V_G,\ \exists\, w \in T^*,\ r \xRightarrow{w} s$. And a graph is a *tree* if it has a root r which is target of no arc, and every vertex $s \neq r$ is target of a unique arc.

To construct monadic second-order formulas, we take two disjoint denumerable sets: a set of *vertex variables* and a set of *vertex set variables*. *Atomic formulas* have one of the following two forms:

$$x \in X \quad \text{or} \quad x \xrightarrow{a} y$$

where X is a vertex set variable, x and y are vertex variables, and $a \in T$.

From the atomic formulas, we construct as usual the *monadic second-order formulas* with the propositional connectives $\neg, \wedge$ and the existential quantifier $\exists$ acting on these two kind of variables. A *sentence* is a formula without free variable. The set of monadic second-order sentences $MTh(G)$ satisfied by a graph G forms the *monadic theory* of G.

Note that two isomorphic graphs satisfy the same sentences. Instead of renaming vertices, we consider the *restriction* $G_{|U}$ of G to an arbitrary set U as follows:

$$G_{|U} := G \cap (U \times T \times U) = \{ s \xrightarrow[G]{a} t \mid s, t \in U \}.$$

Analogously we consider the restriction φ_U (resp. $\varphi_{|U}$) of any sentence φ to a set U by imposing that vertex variables (resp. and vertex set variables) are interpreted only by vertices in U (resp. by subsets of vertices in U) i.e. by induction on the structure of any formula:

$$(x \in X)_U = x \in X \qquad\qquad (x \in X)_{|U} = x \in X$$
$$(x \xrightarrow{a} y)_U = x \xrightarrow{a} y \qquad\quad (x \xrightarrow{a} y)_{|U} = x \xrightarrow{a} y$$
$$(\neg \varphi)_U = \neg(\varphi_U) \qquad\qquad (\neg \varphi)_{|U} = \neg(\varphi_{|U})$$
$$(\varphi \wedge \psi)_U = \varphi_U \wedge \psi_U \qquad (\varphi \wedge \psi)_{|U} = \varphi_{|U} \wedge \psi_{|U}$$
$$(\exists x \ \varphi)_U = \exists x \ (x \in U \wedge \varphi_U) \qquad (\exists x \ \varphi)_{|U} = \exists x \ (x \in U \wedge \varphi_{|U})$$
$$(\exists X \ \varphi)_U = \exists X \ \varphi_U \qquad (\exists X \ \varphi)_{|U} = \exists X \ (X \subseteq U \wedge \varphi_{|U})$$

These restrictions of graphs and sentences are dual.

Lemma 2.1 *Given a graph G, a set U and a monadic sentence φ, we have*

$$G_{|U} \models \varphi \iff G \models \varphi_U \iff G \models \varphi_{|U}.$$

Note that φ_U and $\varphi_{|U}$ are not monadic formulas. But for any monadic formula φ, we give a general condition on U to transform φ_U (or $\varphi_{|U}$) into an equivalent monadic formula i.e. to transform $x \in U$ into a monadic formula.

Basic transformations are well-known to express for instance the propositional connectives $\vee$, $\Longrightarrow$ and the universal quantifier $\forall$:

$$\varphi \vee \psi : \neg (\neg \varphi \wedge \neg \psi) \qquad \varphi \Rightarrow \psi : \neg (\varphi \wedge \neg \psi) \qquad \forall X \ \varphi : \neg (\exists X \ \neg \varphi)$$

Similarly the existence of a path $s \overset{L}{\Longrightarrow} t$ from s to t and labelled by a word in $L \subseteq T^*$ can be expressed by a monadic formula when $L \in Rat(T^*)$ is a rational language; this is done by induction on the rational structure of L:

$$x \overset{\emptyset}{\Longrightarrow} y : \exists X \ (x \in X \wedge \neg (x \in X)) \qquad \text{i.e. a false formula}$$

$$x \overset{\{a\}}{\Longrightarrow} y : x \xrightarrow{a} y$$

$$x \overset{L+M}{\Longrightarrow} y : x \overset{L}{\Longrightarrow} y \vee x \overset{M}{\Longrightarrow} y$$

$$x \overset{L.M}{\Longrightarrow} y : \exists z \ (x \overset{L}{\Longrightarrow} z \wedge z \overset{M}{\Longrightarrow} y)$$

$$x \overset{L^*}{\Longrightarrow} y : \forall X \ ((x \in X \wedge \forall p \forall q((p \in X \wedge p \overset{L}{\Longrightarrow} q) \Rightarrow q \in X)) \Rightarrow y \in X)$$

where the transformation for $\overset{L^*}{\Longrightarrow}$ is the reflexive and transitive closure $(\overset{L}{\Longrightarrow})^*$ of $\overset{L}{\Longrightarrow} : x \overset{L^*}{\Longrightarrow} y$ if and only if every vertex set X containing x and closed by $\overset{L}{\Longrightarrow}$ contains y.

So we consider the restriction $G_{\|r,L}$ of a graph G to the vertices accessible from a vertex r by a path labelled in $L \subseteq T^*$:

$$G_{\|r,L} := G_{|\{s \mid r \overset{L}{\Longrightarrow} s\}} = \{ s \overset{a}{\underset{G}{\to}} t \mid r \overset{L}{\Longrightarrow} s \wedge r \overset{L}{\Longrightarrow} t \}.$$

For instance taking the following deterministic complete tree Λ on $\{a, b\}$ and its root r i.e. its vertex satisfying the formula $\neg (\exists y \ y \overset{a}{\to} x)$, then its restriction $\Lambda_{\|r,b^*a^*}$ is the following graph:

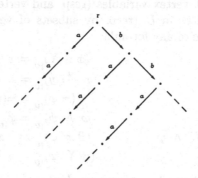

By Lemma 2.1, the restriction of a graph preserves the decision of the monadic theory if we restrict to the vertices accessible from a given vertex by a path labelled in a given rational language.

Proposition 2.2 *Given a graph G, a rational language L over T, and a monadic formula $\varphi(x)$ satisfied by a unique vertex r, we have*
$$MTh(G) \text{ decidable} \implies MTh(G_{\|r,L}) \text{ decidable}.$$

To move by inverse arcs, we introduce a new alphabet $\overline{T} := \{ \overline{a} \mid a \in T \}$ in bijection with T. Any transition $u \overset{\overline{a}}{\to} v$ means that $v \overset{a}{\to} u$ is an arc of G.

We extend by composition the existence of a path $\overset{w}{\Longrightarrow}$ labelled by a word w in $(T \cup \overline{T})^*$.

A *label-mapping* is a mapping from T into the family $2^{(T \cup \overline{T})^*}$ of languages over $T \cup \overline{T}$.

The *inverse* $h^{-1}(G)$ of a graph G according to a label-mapping h is the following graph:

$$h^{-1}(G) := \{ s \overset{a}{\to} t \mid \exists w \in h(a), \ s \overset{w}{\underset{G}{\Longrightarrow}} t \}.$$

For instance for $h(a) = \{b\}$, $h(b) = \{b\overline{b}a\}$, $h(c) = \{\overline{a}ba\}$, $h(d) = \emptyset$ for every other d in T, the inverse $h^{-1}(\Lambda_{\|r,b^*a^*})$ of the previous graph by h is the following graph:

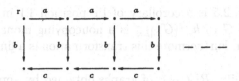

Similarly to the restriction, we define the transformation φ_h by h of any formula φ by replacing each atomic formula $x \xrightarrow{a} y$ by the existence of a path $x \xRightarrow{h(a)} y$ labelled in $h(a)$ i.e. by induction on the structure of any formula:

$$(x \in X)_h = x \in X \qquad (x \xrightarrow{a} y)_h = x \xRightarrow{h(a)} y$$
$$(\neg \varphi)_h = \neg(\varphi_h) \qquad (\varphi \wedge \psi)_h = \varphi_h \wedge \psi_h$$
$$(\exists x \; \varphi)_h = \exists x \; \varphi_h \qquad (\exists X \; \varphi)_h = \exists X \; \varphi_h$$

Note that $(\varphi_U)_h = (\varphi_h)_U$ which is denoted by $\varphi_{U,h}$ and $(\varphi_{|U})_h = (\varphi_h)_{|U}$ which is denoted by $\varphi_{|U,h}$.

Lemma 2.3 *Given a graph G, a label-mapping h and a monadic sentence φ, we have*
$$h^{-1}(G) \models \varphi \iff G \models \varphi_{U,h} \iff G \models \varphi_{|U,h}$$
where $U = V_{h^{-1}(G)}$ is the set of vertices of $h^{-1}(G)$.

However $\varphi_{U,h}$ and $\varphi_{|U,h}$ are not monadic formulas. But we can transform
$$x \in V_{h^{-1}(G)} \quad \text{by} \quad \exists y \, (x \xRightarrow{L} y \vee y \xRightarrow{L} x) \quad \text{for} \quad L = \bigcup_{a \in T} h(a) \, ,$$
and we have seen that $x \xRightarrow{h(a)} y$ can be transformed into a monadic formula if $h(a)$ is rational (and with $x \xRightarrow{\{\overline{a}\}} y$ transformed into $y \xrightarrow{a} x$).

So the inverse according to a label-mapping h preserves the decision of the monadic theory when h is rational: $h(a) \in Rat((T \cup \overline{T})^*)$ for any $a \in T$.

Proposition 2.4 *Given a graph G and a rational label-mapping h, we have*
$$MTh(G) \text{ decidable} \implies MTh(h^{-1}(G)) \text{ decidable.}$$

Let us compose Proposition 2.2 with Proposition 2.4.

Proposition 2.5 *Given a graph G with a unique root r, a rational label-mapping h, and a rational label language $L \in Rat((T \cup \overline{T})^*)$, we have*
$$MTh(G) \text{ decidable} \implies MTh(h^{-1}(G)_{|L_G}) \text{ decidable}$$
with $L_G := \{ s \mid r \xRightarrow[G]{L} s \}$ the set of vertices accessible in G from r by a path in L.

Note that Proposition 2.5 is a corollary of Proposition 3.1 in [Co 94] because the transformation of G to $h^{-1}(G)_{\mid L_G}$ is a noncopying monadic second-order definable transduction. Furthermore this transformation is a linear reduction for the monadic theory.

We will study the family REC_{Rat} of graphs obtained by applying to the tree Λ an inverse rational label-mapping followed by a rational restriction:

$$REC_{Rat} := \left\{ h^{-1}(\Lambda)_{\mid L_\Lambda} \ \middle| \ \begin{array}{l} \Lambda \text{ is a complete deterministic tree on } \{a,b\} \\ h : T \longrightarrow Rat(\{a,b,\overline{a},\overline{b}\}^*) \\ L \in Rat(\{a,b\}^*) \end{array} \right\}$$

We consider also the sub-family REC_{Fin} by using only finite label-mappings:

$$REC_{Fin} := \left\{ h^{-1}(\Lambda)_{\mid L_\Lambda} \ \middle| \ \begin{array}{l} \Lambda \text{ is a complete deterministic tree on } \{a,b\} \\ h : T \longrightarrow Fin(\{a,b,\overline{a},\overline{b}\}^*) \\ L \in Rat(\{a,b\}^*) \end{array} \right\}$$

where $Fin(V)$ is the set of finite subsets of a set V.

We start with any complete and deterministic tree on two labels because Rabin has shown that we can decide any monadic sentence on it.

Theorem 2.6 [Ra 69] *Any complete and deterministic tree on two labels has a decidable monadic theory.*

Let us apply Proposition 2.5 to this result of Rabin.

Corollary 2.7 *Any graph in REC_{Rat} has a decidable monadic theory.*

In fact REC_{Rat} is the closure of the complete and deterministic trees on $\{a,b\}$ by the operations of Proposition 2.2 and Proposition 2.4.

Proposition 2.8 *The family REC_{Rat} is closed by rational restriction and by inverse rational label-mapping.*

Proposition 2.8 is proved using a complete set of representatives. We will obtain other closure properties and particularly we will deduce that this family contains strictly the graphs generated by deterministic graph grammars.

3   Complete sets of representatives

We show that the rational restrictions on vertices of prefix transition graphs of labelled word rewriting systems is a complete set of representatives of REC_{Fin}

(Corollary 3.4). This set of representatives contains the context-free graphs of [MS 85]. In fact REC_{Fin} is exactly the class of regular graphs of finite degree (Theorem 3.9) where a regular graph (or equational graph) is a graph generated by a deterministic graph grammar.

We show that the rational restrictions on vertices of prefix transition graphs of labelled recognizable relations is a complete set of representatives of REC_{Rat} (Corollary 3.4). It follows that REC_{Rat} contains strictly the class of regular graphs (Proposition 3.10).

Finally we extend these sets of representatives to the rationally controlled prefix transition graphs of labelled recognizable relations. This set is also a complete set of representatives of REC_{Rat} (Proposition 3.13). But it is a boolean algebra preserved by inverse rational label-mapping and by rational restriction on vertices (Theorems 3.14).

We take an alphabet $N \subseteq T$ containing the symbols a, b. A representant of the complete deterministic trees labelled on N is the tree Δ_N in $N^* \times N \times N^*$ defined by

$$\Delta_N := \{ u \xrightarrow{a} au \mid a \in N \land u \in N^* \} .$$

The *right closure* $G.N^*$ of a graph G in $N^* \times T \times N^*$ is

$$G.N^* := \{ uw \xrightarrow{a} vw \mid u \xrightarrow{a} v \in G \land w \in N^* \}$$

the set of prefix transitions of G. For instance

$$\Delta_N = \{ \epsilon \xrightarrow{a} a \mid a \in N \}.N^* .$$

Note that a finite graph G in $N^* \times T \times N^*$ is a labelled rewriting system i.e. a finite set of rules over N and labelled in T; and the right closure of G is the prefix rewriting according to G. For instance the *identity graph* $\{ u \xrightarrow{a} u \mid a \in T \land u \in N^* \}$ is the right closure of the finite graph $\{ \epsilon \xrightarrow{a} \epsilon \mid a \in T \}$.

We denote by $U \xrightarrow{a} V := \{ u \xrightarrow{a} v \mid u \in U \land v \in V \}$ the graph of the transitions from $U \subseteq N^*$ to $V \subseteq N^*$ and labelled by $a \in T$. A *recognizable graph* is a finite union of such graphs $U \xrightarrow{a} V$ where $U, V \in Rat(N^*)$. We denote by $Rec(N^* \times T \times N^*)$ the family of recognizable graphs.

Note that the unlabelled recognizable graphs $\{ (u, v) \mid \exists a \in T, u \xrightarrow{a} v \in G \}$ are the *recognizable relations* in $N^* \times N^*$ (by Mezei's theorem) [Be 79]. For instance, the *full graph* $\{ u \xrightarrow{a} v \mid a \in T \land u, v \in N^* \}$ is the recognizable graph $\bigcup \{ N^* \xrightarrow{a} N^* \mid a \in T \}$ and is equal to its right closure.

The right closures of recognizable graphs in $N^* \times T \times N^*$ are exactly the inverse rational mappings $h^{-1}(\Delta_N)$ of Δ_N i.e. for $h : T \longrightarrow Rat((N \cup \overline{N})^*)$.

Theorem 3.1 *The inverse rational (resp. finite) label-mappings $h^{-1}(\Delta_N)$ of Δ_N are effectively the right closures $G.N^*$ of the recognizable graphs (resp. finite graphs) G :*

$$h^{-1}(\Delta_N) = G.N^*$$

with h rational (resp. finite) $\iff$ G recognizable (resp. finite).

This theorem implies some direct generalizations of known results. A first consequence follows from the closure by composition of (extended) rational label-mappings.

Corollary 3.2 *The right closures of recognizable graphs is a class closed effectively by inverse rational label-mapping.*

For instance take the right closure $G.N^*$ of $G = \{x \xrightarrow{a} \epsilon, x^2 \xrightarrow{b} x^3\}$ with $N = \{x\}$. Its inverse $h^{-1}(G.N^*)$ by h defined by $h(a) = \{b\}$ and $h(b) = \{baa\}$, is the following graph:

which is the right closure of $\{x^2 \xrightarrow{a} x^3, x^2 \xrightarrow{b} x\}$.

A consequence of Corollary 3.2 is that the unlabelled right closures of recognizable graphs are preserved by reflexive and transitive closure. More precisely, the *prefix rewriting* $\underset{R}{\longmapsto}$ of any binary relation R on N^* is the unlabelled graph $R.N^*$, i.e.

$$\underset{R}{\longmapsto} := \{ (uw, vw) \mid u \, R \, v \wedge w \in N^* \}$$

and its reflexive and transitive closure $\underset{R}{\overset{*}{\longmapsto}}$ is the *prefix derivation* of R.

Corollary 3.3 *The prefix derivation of any recognizable relation is effectively the prefix rewriting of a recognizable relation.*

In particular for any finite relation, its prefix derivation is a rational transduction [BN 84], and this remains true for any recognizable relation. Thus for the right closure of any recognizable graph, the set of vertices accessible from any rational set is rational; this extend the rationality of words accessible from a given word by prefix derivation of a finite relation [Bü 64].

Another consequence of Theorem 3.1 is that the following family:

$$REC_{\mid Rat} := \{ (G.N^*)_{\mid U} \mid G \in Rec(N^* \times T \times N^*) \wedge U \in Rat(N^*) \}$$

is a complete set of representatives of REC_{Rat}.

Corollary 3.4 *The set $REC_{\mid Rat}$ of the rational restrictions on vertices of the right closures of recognizable graphs (resp. finite graphs) is a complete set of representatives of REC_{Rat} (resp. REC_{Fin}).*

Note that Corollary 3.4 is true in particular for $N = \{a, b\}$. By Corollary 2.7, any rational restriction on vertices of the right closure of any recognizable graph has a decidable monadic theory.

Corollary 3.5 $MTh((G.N^*)_{|U})$ *is decidable for any* $G \in Rec(N^* \times T \times N^*)$ *and for any* $U \in Rat(N^*)$.

A particular case are the pushdown transition graphs (called also context-free graphs) considered in [MS 85]. A *pushdown transition graph* is the graph $R.N^*_{|U}$ of the right closure of a pushdown automaton R in $Q.P \times T \times Q.P^*$ with $N = P \cup Q$, and restricted to the set $U = \{ s \mid r \overset{*}{\underset{R}{\longmapsto}} s \}$ of vertices accessible from a given axiom $r \in Q.P^*$. By Corollary 3.3, we deduce the well-known fact that U is rational, and it remains to apply Corollary 3.5 to get Theorem 4.4 of [MS 85].

Corollary 3.6 [MS 85] *Any pushdown transition graph has a decidable monadic theory.*

But the pushdown automata define up to isomorphism the same accessible prefix transition graphs than the labelled rewriting systems.

Proposition 3.7 [Ca 90] *The pushdown transition graphs form effectively a complete set of representatives of the rooted right closures of finite graphs.*

Instead of labelled (word) rewriting systems, we can also use a subclass of term context-free grammars [Ca 92]. We will now show that REC_{Rat} contains also the graphs generated by deterministic graph grammars. These graphs are the equational graphs of [Co 90], and are called here *regular graphs*. Several basic properties of regular graphs are given in [Ca 95].
The regular graphs generalize the pushdown transition graphs. In fact the pushdown transition graphs are the rooted graphs of finite degree which can be decomposed by distance from any vertex [MS 85]. As the decomposition is dual to the generation, this implies that any pushdown transition graph is a rooted regular graph of finite degree, and this is a result due to Muller and Schupp. Furthermore the inverse inclusion remains true and this correspondance is effective.

Proposition 3.8 [Ca 90] *The pushdown transition graphs form effectively a complete set of representatives of the rooted regular graphs of finite degree.*

For instance every deterministic graph grammar generating a rooted graph G of finite degree, is mapped effectively into a pushdown automaton with an axiom such that its accessible prefix transition graph is isomorphic to G, and the reverse transformation is also effective.
To generalize Proposition 3.8 to all the regular graphs of finite degree, it suffices to take the class of graphs of all the prefix transitions of pushdown automata

(or labelled rewriting systems) and to extend this class by restriction to rational vertex sets instead to the rational set of vertices accessible from an axiom.

Theorem 3.9 [Ca 95] REC_{Fin} *is effectively the family of regular graphs of finite degree.*

We get the regular graphs of infinite degree with inverse rational mappings.

Proposition 3.10 $REC_{Rat} \supset$ *effectively the regular graphs.*

For the strict containment, we consider the rational label-mapping h defined by $h(a) = a$ and $h(b) = \overline{a}^+$, and the rational language $L = a^*$.
Then $h^{-1}(\Delta_{\{a,b\}}) \mid_L = h^{-1}(\Delta_{\{a\}}) \mid_L$ is the following graph:

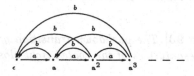

i.e. the right closure $G.\{a\}^*$ of the recognizable graph $G = \{\epsilon \xrightarrow{a} a \, , \, a^+ \xrightarrow{b} \epsilon\}$. By definition this graph is in REC_{Rat} but it is not regular because it has infinitely many vertex out-degrees (a^n is of out-degree $n + 1$). It remains to apply Corollary 2.7 to get Theorem 7.11 of [Co 90].

Corollary 3.11 [Co 90] *Any regular graph has a decidable monadic theory.*

Thus Corollary 3.6 and Corollary 3.11 have been obtained by using the following complete set $REC_{|Rat}$ of representatives of REC_{Rat} (see Corollary 3.4). Although $REC_{|Rat}$ is obviously closed by rational restriction on vertices, it is not closed for instance by inverse finite label-mapping, nor by union. A simple extension is the family $\bigcup_f REC_{|Rat}$ of finite unions of graphs in $REC_{|Rat}$. But $\bigcup_f REC_{|Rat}$ is not closed by complement. So we introduce another complete set of representatives of REC_{Rat}. Following [Ch 82], we extend the right closures of recognizable graphs to rational right closures.

Definition 3.12 A *rational right closure of a recognizable graph* is a finite union of graphs
$$(U \xrightarrow{a} V).W := \{ uw \xrightarrow{a} vw \mid u \in U \wedge v \in V \wedge w \in W \}$$
where $U, V, W \in Rat(N^*)$.

This extension remains in REC_{Rat} .

Proposition 3.13 *The rational right closures of recognizable graphs form a complete set of representatives of REC_{Rat} .*

Contrary to $REC_{|Rat}$ the rational right closures of recognizable graphs are preserved by boolean operations, by composition and by transitive closure, hence by inverse rational label-mapping.

Theorem 3.14 *The rational right closures of recognizable graphs form an effective boolean algebra effectively preserved by inverse rational label-mapping and by rational restriction on vertices.*

Theorem 3.14 with Proposition 3.13 give Proposition 2.8.

Let us conclude with a remark. We have seen that the decision of the monadic theory is preserved by inverse rational label-mapping. But this is false by inverse context-free label-mapping. For instance with $N = \{x, y\}$, the following (linear) context-free mapping:

$$h(a) = \{\overline{x}xx\,,\,\overline{y}yx\} \text{ and } h(b) = \{\,(\overline{x}\overline{x}x + \overline{x}yy)^n\overline{y}yy(\overline{x}xx + \overline{y}yx)^n \mid n \geq 0\,\}$$

defines by inverse of Δ_N followed by the restriction to the rational language $L = x^*y^+$ a graph $h^{-1}(\Delta_N)\mid_L$ which is the usual grid having an undecidable monadic theory (see among others [Se 91]).

Acknowledgements

Let me thank Wolfgang Thomas for a remark at the origin of Theorem 3.1, hence of this paper. This paper has been partly supported by ESPRIT BRA 6317 (ASMICS). Due to a lack of space, this paper has been seriously reduced. The full version (with proofs) can be accessed by ftp irisa (anonymous) in the file: (/site/irisa/ftp) local/caucal/monadic.ps.gz or by Web ftp://ftp.irisa.fr/local/caucal/monadic.ps.gz

References

[Be 79] J. BERSTEL *Transductions and context-free languages*, Ed. Teubner, pp. 1–278, 1979.

[BN 84] L. BOASSON and M. NIVAT *Centers of context-free languages*, Internal Report LITP 84–44, 1984.

[Bü 64] R. BÜCHI *Regular canonical systems*, Archiv für Mathematische Logik und Grundlagenforschung 6, pp. 91–111, 1964 or in *The collected works of J. Richard Büchi*, edited by S. Mac Lane and D. Siefkes, Springer-Verlag, New York, pp. 317–337, 1990.

[Ca 90] D. CAUCAL *On the regular structure of prefix rewriting*, CAAP 90, LNCS 431, pp. 87–102, 1990, extended in TCS 106, pp. 61–86, 1992.

[Ca 92] D. CAUCAL *Monadic theory of term rewritings*, 7th IEEE Symp., LICS 92, pp. 266–273, 1992.

[Ca 95] D. CAUCAL *Bisimulation of context-free grammars and of pushdown automata*, CSLI volume 53 "Modal logic and process algebra", pp. 85–106, Stanford, 1995. The full version is accessible by ftp (see acknowledgements) in the file bisimulation.ps.gz

[Ch 82] L. CHOTTIN *Langages algébriques et systèmes de réécriture rationnels*, RAIRO-TIA 16-2, pp. 93–112, 1982.

[Co 90] B. COURCELLE *Graph rewriting: an algebraic and logic approach*, Handbook of TCS, Vol. B, Elsevier, pp. 193–242, 1990.

[Co 94] B. COURCELLE *Monadic second-order definable graph transductions: a survey*, TCS 126, pp. 53–75, 1994.

[CW 95] B. COURCELLE and I. WALUKIEWICZ *Monadic second-order logic, graphs and unfolding of transition systems*, CSL 95, 1995.

[MS 85] D. MULLER and P. SCHUPP *The theory of ends, pushdown automata, and second-order logic*, TCS 37, pp. 51–75, 1985.

[Ra 69] M. RABIN *Decidability of second-order theories and automata on infinite trees*, Trans. Amer. Math. Soc. 141, pp. 1–35, 1969.

[Sa 79] J. SAKAROVITCH *Syntaxe des langages de Chomsky; essai sur le déterminisme*, Thèse de doctorat d'état Paris VII, pp. 1–175, 1979.

[Se 91] D. SEESE *The structure of the models of decidable monadic theories of graphs*, Annals of Pure and Applied Logic 53, pp. 169–195, 1991.

[Sem 84] A. SEMENOV *Decidability of monadic theories*, MFCS 84, LNCS 176, pp. 162–175, 1984.

[Sé 93] G. SÉNIZERGUES *Formal languages and word-rewriting*, Ecole de Printemps 93.

[Sh 75] S. SHELAH *The monadic theory of order*, Ann. Math. 102, pp. 379–419, 1975.

[St 75] J. STUPP *The lattice-model is recursive in the original model*, Manuscript, The Hebrew University, 1975.

[Th 90] W. THOMAS *Automata on infinite objects*, Handbook of TCS, Vol. B, Elsevier, pp. 135–191, 1990.

[Th 91] W. THOMAS *On logics, tilings, and automata*, ICALP 91, LNCS 510, pp. 191–231, 1991.

[W 96] I. WALUKIEWICZ *Monadic second order logic on tree-like structures*, STACS 96, LNCS 1046, pp. 401–413, 1996.

Semi-Groups Acting on Context-Free Graphs

Géraud Sénizergues

LaBRI
Université de Bordeaux I
351, Cours de la Libération 33405 Talence, France **

Abstract. Let Γ be a context-free graph. We give sufficient conditions on a semi-group of bisimulations H to ensure that the quotient $H\backslash\Gamma$ is context-free. Using these sufficient conditions we show that the quotient $\mathrm{Aut}(\Gamma)\backslash\Gamma$ of Γ by its full group of automorphisms is always context-free.

Keywords: infinite graphs; ends; pushdown automata; automorphisms; bisimulations; groups; semi-groups.

1 Introduction

The automorphisms of a given *graph* Γ can be studied from several points of view: one can study the general structure of *every automorphism* and classify them with respect to this structure (see [18] for example), one can study the structure of the *group of automorphisms* (see [21] in the case of finite graphs) and more generally the structure of any *group acting* on Γ (see [4] for trees, [15] for graphs). Finally, one can study the structure of the *quotient* $H\backslash\Gamma$ of Γ by a group H acting on Γ.

The notion of a *context-free* graph (which is closely related to the notion of *pushdown automaton*) appears in several fields of theoretical computer science: originally in the frontier between formal language theory and combinatorial group theory ([1, 22, 23], see [7, p.95-100] for a survey) and later also in algebraic graph theory ([5, 6], see [13] for a survey), in model theory ([12, 14], see [13] for a survey) and in semantics of processes ([3],[8],[11],[20], see [10] for a survey).

The aim of this work is to study the structure of $H\backslash\Gamma$ where Γ is a context-free graph and H a group acting on Γ. The study of H itself is done in [25] (while, in some sense, the fundamental article [22] treats the case where Γ is exactly the Cayley graph of H).

The technical tools we use are of two kinds: we use *geometrical* arguments about the graph Γ (leaning on results of [23, 28, 9, 25]) and also some classical *language-theoretic* arguments which may be applied via the characterization of context-free graphs as the computation-graphs of pda's ([23]). Surprizingly, we do not need any substantial results of group theory to establish the structure of $H\backslash\Gamma$: the notion of *semi-group* acting on Γ seems the easiest way of treating the action of a group on Γ (see theorem 3.5, corollary 3.11 and the proof of theorem 3.15).

** mailing adress:LaBRI and UFR Math-info, Université Bordeaux1, 351 Cours de la libération -33405- Talence Cedex.
email:ges@labri.u-bordeaux.fr; fax: 56-84-66-69

2 Preliminaries

2.1 Graphs

Let X be a finite alphabet and let $\bar{X}$ be a disjoint alphabet given with a fixed bijection $\alpha : X \to \bar{X}$. We call $X \cup \bar{X}$ a *paired* alphabet and for every x in X (resp. in $\bar{X}$) we denote by $\bar{x}$ the letter $\alpha(x)$ (resp. $\alpha^{-1}(x)$). We call a *graph over X* any pair $\Gamma = (V_\Gamma, E_\Gamma)$ where V_Γ is a set and E_Γ is a subset of $V_\Gamma \times (X \cup \bar{X}) \times V_\Gamma$ fufilling the following *symetry* condition:

$$\forall v, v' \in V_\Gamma, \forall x \in X \cup \bar{X}, (v, x, v') \in E_\Gamma \Leftrightarrow (v', \bar{x}, v) \in E_\Gamma. \tag{1}$$

In the following we fix X and by *graph* we always mean *graph over X*. Let $\sim$ be some equivalence relation over the set V_Γ and let $\Pi_\sim$ denote the canonical projection from V_Γ onto $\sim \backslash V_\Gamma$. The *quotient graph* $\sim \backslash \Gamma$ is defined as:

$$\sim \backslash \Gamma = (\sim \backslash V_\Gamma, \sim \backslash E_\Gamma), \text{ where } \sim \backslash E_\Gamma = \{\Pi_\sim(v), x, \Pi_\sim(v') \mid (v, x, v') \in E_\Gamma\}. \tag{2}$$

Connectivity

Let Γ be some graph. By $d(v, v')$ we denote the *distance* between two vertices v, v' in Γ : it is the minimal length of a path from v to v' in Γ. Given non-empty subsets $V, V' \subseteq V_\Gamma$, the distance between V and V' is defined as $d(V, V') = \min\{d(v, v') \mid v \in V, v' \in V'\}$. Let κ be an integer. We call κ-*path* in Γ any sequence $(v_0, \ldots, v_i, \ldots, v_l)$ of vertices such that,

$$\forall i \in [0, l-1], d(v_i, v_{i+1}) \le \kappa.$$

A subset $V' \subseteq V_\Gamma$ will be said κ-*connected* iff, $\forall v, v' \in V'$, there exists a κ-path in Γ from v to v', whose vertices are all in V'. Given a subset $V' \subseteq V_\Gamma$, the *frontier* $\Delta V'$ of V' is defined by :

$$\Delta V' = \{v \in V' \mid \exists v' \in V_\Gamma - V', \exists x \in X \cup \overline{X}, (v', x, v) \in E_\Gamma\},$$

and the *cofrontier* $\partial V'$ of V' is defined by :

$$\partial V' = \{(v, x, v') \in E_\Gamma \mid v \in V', x \in X \cup \overline{X}, v' \in V_\Gamma - V'\}.$$

Ends

By $B(v, n)$ (resp. $S(v, n)$) we denote the ball $\{v' \in V_\Gamma \mid d(v, v') < n\}$ (resp. the sphere $\{v' \in V_\Gamma \mid d(v, v') = n\}$). Given some vertex $v \in V_\Gamma$, we call v-*end* of Γ (relative to the ball $B(v, n)$) any subset of V_Γ which is a connected component of $\Gamma - B(v, n)$ for some radius n (terminology of [23]). More generally , given a subset $P \subseteq V_\Gamma$, we call P-end any connected component of $\Gamma - B(P, n)$ for some radius n. Let E, E' be two P-ends. A map $\varphi : E \to E'$ is a *end-isomorphism* from E to E' iff φ is a graph isomorphism (for the induced subgraphs) such that $\varphi(\Delta E) = \Delta E'$.

We recall that a graph Γ is said *context-free* iff it is connected and there exists some vertex $v_0 \in V_\Gamma$ such that Γ has only finitely many classes of v_0-ends modulo end-isomorphism ([23]).

Morphisms, simulations

Definition 1 [24]. Let $\Gamma = (V_\Gamma, E_\Gamma), \Gamma' = (V_{\Gamma'}, E_{\Gamma'})$ be two graphs. Let ρ be some binary relation $\rho \subseteq V_\Gamma \times V_{\Gamma'}$.
1- ρ is a *simulation* iff $\mathrm{dom}(\rho) = V_\Gamma$ and for all $v, w \in V_\Gamma, v' \in V_{\Gamma'}, \sigma \in X \cup \bar{X}$, such that $(v, \sigma, w) \in E_\Gamma$ and $v \sim v'$, there exists $w' \in V_{\Gamma'}$ such that $(v', \sigma, w') \in E_{\Gamma'}$ and $w \sim w'$.
2- ρ is a *bisimulation* iff ρ and ρ^{-1} are simulations.
A map $\varphi : V_\Gamma \to V_{\Gamma'}$ is a graph *homomorphism* from Γ to Γ' iff it is a bisimulation . [3]

Remark 2.1
1-An equivalence relation $\sim$ over V_Γ is a bisimulation over Γ iff the canonical projection $\Pi_\sim$ is an homomorphism.
2-The fact that we only consider graphs fulfilling the symetry condition(1) implies that , if ρ is a bisimulation from Γ to Γ' it is also a bisimulation from Γ^{-1} to Γ'^{-1}, which is not the case for general labelled graphs.

A binary relation $\rho \subseteq V_\Gamma \times V_{\Gamma'}$ is said (v, κ)-*connected* iff, for every $v' \in \rho v$ (i.e. v' such that $(v, v') \in \rho$), there exists some κ-path in Γ from v to v' : $(v_0, \ldots, v_i, \ldots, v_l)$ such that, $\forall i \in [0, l-1], (v_i, v_{i+1}) \in \rho$. The two following lemmas are easy.

Lemma 2.2 *Let m_1, m_2 be (v, κ)-connected bisimulations over some graph Γ. Then $m_1 \circ m_2$ is a (v, κ)-connected bisimulation.*

Lemma 2.3 *Let $\mathcal{M}$ be a set of (v, κ)-connected bisimulations over some graph Γ. Then $\bigcup_{m \in \mathcal{M}} m$ is a (v, κ)-connected bisimulation too.*

2.2 Automata

Finite automata
A n-tape finite automaton (abbreviated n-f.a. in the sequel) is a 5-tuple:

$$\mathcal{A} =< X, Q, \delta, r, F >$$

where, as usual, X is the input-alphabet, Q is the set of states, r is the initial state, $F \subseteq Q$ is the set of final states and δ, the set of transitions, is a subset of $Q \times (X \cup \{\epsilon\})^n \times Q$. $\mathcal{A}$ is said *letter-to-letter* when $\delta \subseteq Q \times X^n \times Q$. We denote by $\mathrm{L}(\mathcal{A})$ the language recognized by $\mathcal{A}$. A subset $L \subseteq (X^*)^n$ is said *rational* iff there exists some n-f.a. $\mathcal{A}$ such that $L = \mathrm{L}(\mathcal{A})$. A subset $L \subseteq (X^*)^n$ (equivalently , a n-ary relation over X^*) is called *length-preserving* iff, for every $(u_1, \ldots, u_i, \ldots, u_n) \in L, |u_1| = \ldots = |u_i| = \ldots = |u_n|$.

Theorem 2.4 ([16], [17]) *Let $L \subseteq (X^*)^n$ be length-preserving. Then L is rational iff L is recognized by some letter-to-letter n-f.a.*

We recall that the *recognizable* subsets of X^{*n} are exactly the subsets L of the form

$$L = \bigcup_{1 \leq i \leq p} R_{i,1} \times \ldots \times R_{i,n}$$

where $p \in \mathbb{N}$ and all the $R_{i,j}$ are rational subsets of X^*. If L is rational and R is recognizable, then $L \cap R$ is rational too. (We refer to [17] for more details about n-f.a.).

---

[3] This definition is a straightforward adaptation to *graphs* over X of the definition given in [2, p.212] for *transition systems*.

Pushdown automata

A *pushdown automaton* on the alphabet X is a 6-tuple $\mathcal{A} = <X, Y, Q, \delta, q_0, y_0>$ where Y is the finite stack-alphabet, Q is the finite set of states, $q_0 \in Q$ is the initial state, y_0 is the initial stack-symbol and δ, the transition function, is a mapping $\delta : YQ \times (X \cup \{\epsilon\}) \to \mathcal{P}(Y^*Q)$. For every $\omega q, \omega' q' \in Y^*Q$ and $f \in X^*$, we note $\omega q \xrightarrow{f}_{\mathcal{A}} \omega' q'$ iff there exists some computation of $\mathcal{A}$, starting on ωq, reading f on the input tape and ending on $\omega' q'$. $\mathcal{A}$ is said *real-time* iff, for every $yq \in YQ$, $\mathrm{Card}(\delta(yq, \epsilon)) = 0$. (We refer to [19] for more details about p.d.a.).

2.3 Pushdown-structures

We call *transition-graph* of the pda $\mathcal{A}$, denoted $T(\mathcal{A})$, the graph:
$T(\mathcal{A}) = (V_{T(\mathcal{A})}, E_{T(\mathcal{A})})$ where $V_{T(\mathcal{A})} = Y^*Q$ and

$$E_{T(\mathcal{A})} = \{(\omega q, x, \omega' q') \in Y^*Q \times X \times Y^*Q \mid \omega q \xrightarrow{x}_{\mathcal{A}} \omega' q'\}$$
$$\cup \{(\omega' q', \bar{x}, \omega q) \in Y^*Q \times \bar{X} \times Y^*Q \mid \omega q \xrightarrow{x}_{\mathcal{A}} \omega' q'\}.$$

We call *computation-graph* of $\mathcal{A}$, denoted $\mathcal{C}(\mathcal{A})$, the *connected component* in $T(\mathcal{A})$ of $y_0 q_0$.

We call *pushdown-structure* over a graph Γ any pair $(\mathcal{A}, \varphi)$ where $\mathcal{A}$ is a real-time pushdown-automaton over the alphabet X and $\varphi : \mathcal{C}(\mathcal{A}) \longrightarrow \Gamma$ is an isomorphism. We call *initial vertex* of the p-structure the vertex $v_0 = \varphi(y_0 q_0)$, where $y_0 q_0$ is the initial configuration of $\mathcal{A}$.

For every pair of graphs endowed with a p-structure $(\Gamma_1, \mathcal{A}_1, \varphi_1), (\Gamma_2, \mathcal{A}_2, \varphi_2)$, a binary relation $\rho \subseteq V_{\Gamma_1} \times V_{\Gamma_2}$ will be said *rational* iff $\varphi_2^{-1} \circ \rho \circ \varphi_1$ is a rational subset of $(Y_1 \cup Q_1 \cup Y_2 \cup Q_2)^* \times (Y_1 \cup Q_1 \cup Y_2 \cup Q_2)^*$ (see §2.2). The following lemma will then be useful

Lemma 2.5 *For every two real-time pushdown automata* $(\mathcal{A}_1, \mathcal{A}_2)$, *the computation graphs* $\mathcal{C}(\mathcal{A}_1), \mathcal{C}(\mathcal{A}_2)$ *are isomorphic iff there exists some rational isomorphism* ψ *from* $\mathcal{C}(\mathcal{A}_1)$ *to* $\mathcal{C}(\mathcal{A}_2)$.

This result was exposed at [28]. A full proof is given in [26]. $\square$
Within this terminology, the fundamental theorem of Muller and Schupp ([23]) can be rephrased as follows.

Theorem 2.6 *Let* Γ *be a graph over* X. Γ *is context-free if and only if* Γ *has some pushdown structure.*

Let us say that two p-structures $(\mathcal{A}_1, \varphi_1), (\mathcal{A}_2, \varphi_2)$ over the same graph Γ are *equivalent* iff $\mathrm{Id} : (\Gamma, \mathcal{A}_1, \varphi_1) \to (\Gamma, \mathcal{A}_2, \varphi_2)$ is rational.

Canonical p-structures

The notion of *canonical automaton* associated with a context-free graph Γ has been defined in [23]. We use here a slight adaptation of this notion that we call a *P-canonical automaton*. Given a finite subset P of V_Γ, the P-canonical automaton describes the P-ends of Γ in the same way as the ordinary canonical (called here the v_0-canonical) automaton was describing the v_0-ends. (It is proved in [9] that Γ must have only finitely P-ends up to end-isomorphism). Instead of a completely precise definition (which would be somewhat long and technical), we just state here the key property of such a P-canonical automaton $\mathcal{A} = <X, Y, Q, \delta, q_0, y_0>$: if $(\mathcal{A}, \varphi)$ is a p-structure over Γ where $\mathcal{A}$ is P-canonical then

$$\forall \omega q \in V_{C(\mathcal{A})}, \mathrm{d}(P, \varphi(\omega q)) = | \omega | - 1 \qquad (3)$$

i.e. the distance from the subset P to some vertex v is equal to the stack-height (minus 1) of the configuration representing v.

Let us remark that by (3), $\varphi^{-1}(P)$ must be of the form $y_0 Q_0$ for some $Q_0 \subseteq Q$.

Proposition 2.7 *Let Γ be a context-free graph, P a non-empty finite set of vertices of Γ and $(\mathcal{A}, \varphi)$ a p-structure over Γ. Then there exists some P-canonical p-structure $(\mathcal{A}', \varphi')$ over Γ which is equivalent to $(\mathcal{A}, \varphi)$.*

Connectivity and p-structures

Lemma 2.8 *For every rational subset K of a context-free graph Γ endowed with a p-structure $(\mathcal{A}, \varphi)$, there exists some integer κ such that K is κ-connected.*

A key-result for our investigations is the following

Theorem 2.9 ([25]) *Let Γ be a context-free graph and let $\sim$ be the bisimulation induced by its group of automorphisms $\mathrm{Aut}(\Gamma)$. Then, for every $v \in V_\Gamma$, there exists some integer κ (which may be computed from some p-structure $(\mathcal{A}, \varphi)$ over Γ) such that $\sim$ is (v, κ)-connected.*

3 Quotient of a context-free graph by a semi-group of bisimulations

We fix some context-free graph Γ throughout this section, as well as some p-structure $(\mathcal{A}, \varphi)$ over Γ.

3.1 Quotient by a rational bisimulation of finite order

Let us say that an equivalence relation $\sim$ over a set V has *finite order* iff every class modulo $\sim$ is finite.

Theorem 3.1 *Let $\sim$ be an equivalence relation on Γ such that:*

1. *$\sim$ is a bisimulation*
2. *$\sim$ is rational*
3. *$\sim$ has finite order*

Then the graph $\sim \backslash \Gamma$ is context-free.

Let us consider the set $P = [v_0]_\sim$. This set is finite by hypothesis (3). Let us denote by $\Pi : \Gamma \to \sim \backslash \Gamma$ the canonical projection.

Lemma 3.2 $\forall v \in V_\Gamma, \mathrm{d}(P, v) = \mathrm{d}(\Pi(v_0), \Pi(v))$.

Proof: The proof is not difficult, see [27]. □

Lemma 3.3 *Let E be some P-end of Γ relative to the ball $\mathrm{B}(P, n)$ for some integer n. Then $\Pi(E)$ is a $\Pi(v_0)$-end of $\sim \backslash \Gamma$ relative to the ball $\mathrm{B}(\Pi(v_0), n)$. Moreover every $\Pi(v_0)$-end of $\sim \backslash \Gamma$ is of this form.*

Proof: 1- Let E be a P-end of Γ relative to $\mathrm{B}(P, n)$. As E is connected and Π is a morphism, $\Pi(E)$ is connected. By lemma 3.2 $\Pi(E)$ is included in $\Gamma - \mathrm{B}(\Pi(v_0), n)$.
Let us consider some

$$(\Pi(e), \sigma, w) \in \partial \Pi(E). \tag{4}$$

As Π is a bisimulation there exists some $v \in \Pi^{-1}(w)$ such that $(e, \sigma, v) \in E_\Gamma$. If $\mathrm{d}(P, v) \geq n$ then v must belong to E, hence $\Pi(v) \in \Pi(E)$, contradicting the hypothesis (4). Hence $\mathrm{d}(P, v) = n - 1$. We have shown that

$$\Delta \Pi(E) \subseteq \mathrm{S}(\Pi(v_0), n).$$

As $\Pi(E)$ is a connected subset of $\sim \backslash \Gamma - \mathrm{B}(\Pi(v_0), n)$ whose frontier is included in $\mathrm{S}(\Pi(v_0), n)$, it must be a connected component of $\Gamma - \mathrm{B}(\Pi(v_0), n)$, i.e. a $\Pi(v_0)$-end.
2- Let F be some $\Pi(v_0)$-end of $\sim \backslash \Gamma$ relative to $\mathrm{B}(\Pi(v_0), n)$. Let E be some connected component of $\Pi^{-1}(F)$. By lemma 3.2 $E \subseteq \Gamma - \mathrm{B}(P, n)$. Let us consider some

$$(e, \sigma, v) \in \partial E. \tag{5}$$

If $\mathrm{d}(P, v) \geq n$ then $\mathrm{d}(\Pi(v_0), \Pi(v)) \geq n$, hence the edge $(\Pi(e), \sigma, \Pi(v))$ is in $\Gamma - \mathrm{B}(\Pi(v_0), n)$, which implies that $\Pi(v) \in F$. So the edge (e, σ, v) connects two points of $\Pi^{-1}(F)$, and v must belong to E, contradicting hypothesis (5). Hence $\mathrm{d}(P, v) = n - 1$. We have shown that

$$\Delta E \subseteq \mathrm{S}(P, n).$$

As E is a connected subset of $\Gamma - \mathrm{B}(P, n)$ whose frontier is included in $\mathrm{S}(P, n)$, it must be a connected component of $\Gamma - \mathrm{B}(P, n)$, i.e. a P-end.
$\square$

Let $\psi : E \to E'$ be a map from a P-end E to a P-end E'. We say that ψ is a end-isomorphism *preserving* $\sim$ (or is a $\sim$-*end-isomorphism*) iff ψ is a end-isomorphism and

$$\forall v, w \in E, v \sim w \Rightarrow \psi(v) \sim \psi(w).$$

Lemma 3.4 Γ *has only finitely many P-ends up to $\sim$-end-isomorphisms.*

Proof: Let us consider a P-*canonical* structure $(\mathcal{A}', \varphi')$ over Γ which is rationaly equivalent to the initial structure $(\mathcal{A}, \varphi)$ (see proposition 2.7). Let us use the notation $\mathcal{A}' = \langle X, Y', Q', \delta', Q'_0, y'_0 \rangle$ and $Z' = Y' \cup Q'$. $\sim$ is a rational relation over Z'^* which *preserves the length*. By theorem 2.4 $\sim$ is recognized by some *letter-to-letter* finite 2-automaton $\mathcal{S} = \langle X, Q_\mathcal{S}, \delta_\mathcal{S}, r_\mathcal{S}, F_\mathcal{S} \rangle$. This automaton $\mathcal{S}$ can be choosen *deterministic* because, it can be considered as a 1-automaton over $Z' \times Z'$. Let E, E' be two P-ends and let ω (resp. ω') be the word over Z' associated with E (resp. E'). Let us set $E \equiv E'$ iff ω, ω' have the same rightmost letter and $\delta_\mathcal{S}(r_\mathcal{S}, \omega) = \delta_\mathcal{S}(r_\mathcal{S}, \omega')$. One can check that $\equiv$ defines only finitely many classes and that for every P-ends E, E', $E \equiv E' \Rightarrow E, E'$ are $\sim$-end-isomorphic . $\square$

Theorem 3.1 follows from lemma 3.3 and lemma 3.4. $\square$

3.2 Quotient by a semi-group of bisimulations

Theorem 3.5 (Main theorem) *Let H be some semi-group of bisimulations of Γ fulfilling the following conditions:*

(0) $\forall h \in H, h^{-1} \in H$
(1) H is finitely generated
(2) $\forall h \in H$, h is rational
(3) $\forall v, v' \in V_\Gamma$, there exists some finite subset $F(v,v') \subseteq H$ such that

$$\bigcup_{h \in H, (v,v') \in h} h \subseteq \bigcup_{h \in F(v,v')} h.$$

Then $H \backslash \Gamma$ is context-free.

Let us notice that, under hypothesis (0), the binary relation $\sim_H = \bigcup_{h \in H} h$ is an equivalence relation over V_Γ which is a bisimulation. By $H \backslash \Gamma$ we mean the quotient graph $\sim_H \backslash \Gamma$ as it is defined by definition (2) of §2.1.

Construction of a set of vertices R. Let $\mathcal{G}$ be some finite set of generators of H. Hypothesis (2) of the theorem implies that each set hv_0, for $h \in \mathcal{G}$, is rational. By lemma 2.8, there exists some integer κ_0 such that, for every $h \in \mathcal{G}$, hv_0 is κ_0-connected. By lemma 2.2 every $h \in H$ is then (v_0, κ_0)-connected. Fom now on, we fix such an integer κ_0 and we let κ_1 be the maximum diameter of the frontiers of the ends of Γ. We set

$$\kappa_2 = \kappa_0 + \kappa_1.$$

For every integer n we denote by $\mathcal{E}_n$ the set of ends of Γ relative to the ball $B(v_0, n)$. We define a subset R of vertices by

$$R = B(v_0, \kappa_2) \cup \bigcup_{E \in \mathcal{E}_{\kappa_2}, Hv_0 \cap E = \emptyset} E \tag{6}$$

Lemma 3.6 ΔR *is finite.*

Proof: ΔR is included in the sphere of center v_0 and radius κ_2. $\square$

Lemma 3.7 $\forall v \in V_\Gamma, [v]_{\sim_H} \cap R \neq \emptyset.$

Proof: Let $v \in V_\Gamma$. If $v \in R$, clearly $[v]_{\sim_H} \cap R \neq \emptyset$.
Let us suppose now that $v \notin R$.
By definition (6), there exists some end $E \in \mathcal{E}_{\kappa_2}$ such that

$$v \in E \text{ and } Hv_0 \cap E \neq \emptyset.$$

As Hv_0 is κ_0-connected and $Hv_0 \cap E \neq \emptyset$, there exists some vertex w_0 such that

$$w_0 \in Hv_0 \cap E \text{ and } d(w_0, \Delta E) < \kappa_0. \tag{7}$$

Let us consider a vertex $v' \in \Delta E$ which is the intersection of some minimal path from v_0 to v with ΔE :

$$d(v_0, v) = d(v_0, v') + d(v', v). \tag{8}$$

From (7) and the definition of κ_1 we get

$$d(v', w_0) < \kappa_1 + \kappa_0 = \kappa_2 \tag{9}$$

Using (8),(9) we obtain

$$d(w_0, v) \leq d(w_0, v') + d(v', v) < \kappa_2 + d(v', v) = d(v_0, v). \qquad (10)$$

As $w_0 \sim_H v_0$ and $\sim_H$ is a bisimulation, there exists some $w \sim_H v$ with $d(w_0, v) = d(v_0, w)$ (see figure 1).
Using (10) we obtain that, for every $v \in V_\Gamma - R$, there exists some $w \sim_H v$ such that

$$d(v_0, w) < d(v_0, v).$$

(See figure1). By induction on $d(v_0, v)$ it follows that every class $\pmod{\sim_H}$ has at least one representative in R. $\square$

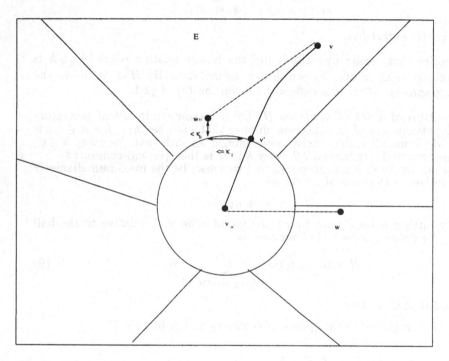

Fig. 1. R is a set of representatives

Lemma 3.8 *There exists some finite subset $K \subseteq H$ such that $\sim_H |R =\sim_K |R$.*

Proof: Let us introduce two finite subsets K_1, K of H by :

$$K_1 = \bigcup_{w, w' \in B(v_0, 2 \cdot \kappa_2)} F(w, w') \quad , \quad K = K_1 \circ K_1^{-1}.$$

We show below that, if $v_1, w_1 \in R$ are such that

$$d(v_1, Hv_0 \cap R) = d(Hw_1 \cap R, Hv_0 \cap R) \text{ and } Hv_1 \cap R = Hw_1 \cap R \qquad (11)$$

then $w_1 \in K_1 v_1$. But the integer $d(Hw_1 \cap R, Hv_0 \cap R)$ must be equal to some $d(v_1, Hv_0 \cap R)$, where $v_1 \in Hw_1 \cap R$. Hence each class Hw_1 contains at least one vertex v_1 fulfilling equation (11) and it will follow that

$$\forall w_1, w_1' \in R, w_1 \sim_H w_1' \Rightarrow w_1 \in (K_1 \circ K_1^{-1})w_1'.$$

Let v_1, w_1 fulfilling (11). Let $w_0 \in Hv_0 \cap R$ such that

$$d(v_1, w_0) = d(Hv_1 \cap R, Hv_0 \cap R). \tag{12}$$

Let $h_1 \in H$ such that $w_1 \in h_1 v_1$.
As h_1 is a bisimulation, there exists some $w_0' \in h_1 w_0$ such that

$$d(v_1, w_0) = d(w_1, w_0') \tag{13}$$

By minimality of $d(v_1, w_0)$(12) and equality (13)we have

$$d(w_1, w_0') \leq d(w_1, v_0) \tag{14}$$

case 1: $d(w_1, v_0) < \kappa_2$.
Then by (14), $d(w_1, w_0') < \kappa_2$, hence $d(v_0, w_0') < 2 \cdot \kappa_2$.
case 2: $d(w_1, v_0) \geq \kappa_2$ and $d(w_0', v_0) < \kappa_2$.
It is clear that $d(v_0, w_0') < 2 \cdot \kappa_2$.
case 3: $d(w_1, v_0) \geq \kappa_2$ and $d(w_0', v_0) \geq \kappa_2$.
By definition of R (see equation (6)), $w_0' \notin R$. Hence there are two distinct
ends $E_1, E_0' \in \mathcal{E}_{\kappa_2}$ such that $w_1 \in E_1, w_0' \in E_0'$ (see figure 2).
As E_1, E_0' must be disjoint we have:

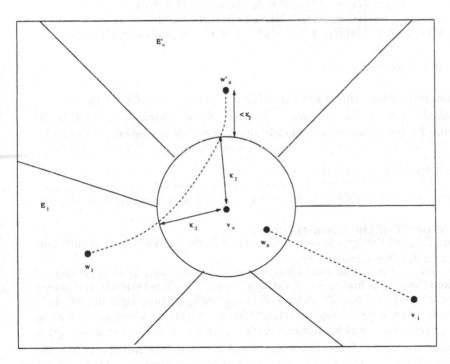

Fig. 2. Construction of K finite, case 3

$$d(w_1, \Delta E_1) + 1 + d(\Delta E_0', w_0') \le d(w_1, w_0') \le d(w_1, v_0) \le d(w_1, \Delta E_1) + \kappa_2.$$

Hence $1 + d(\Delta E_0', w_0') \le \kappa_2$, which leads to

$$d(v_0, w_0') \le \kappa_2 + d(\Delta E_0', w_0') < 2 \cdot \kappa_2.$$

In every case $d(v_0, w_0') < 2 \cdot \kappa_2$, hence

$$h_1 \subseteq \bigcup_{h \in H, (w_0, w_0') \in h} h \subseteq \bigcup_{h \in F(w_0, w_0')} h \subseteq \bigcup_{h \in K_1} h.$$

But $w_1 \in h_1 v_1$, so that $w_1 \in K_1 v_1$ (as required). $\square$

Construction of a graph $\hat{R}$

Let us denote by Π_H (resp. Π_K) the canonical projection from Γ onto $H\backslash\Gamma$ (resp. from $\hat{R}$ onto $K\backslash\hat{R}$). Let us consider some map $\tau :\sim_H \backslash V_\Gamma \to R$ fufilling

$$\forall v \in V_\Gamma, \tau(\Pi_H(v)) \in \Pi_H(v).$$

We construct now a graph $\hat{R} = (V_{\hat{R}}, E_{\hat{R}})$ where $V_{\hat{R}} = R$ and $E_{\hat{R}}$ is defined by:

$$E_{\hat{R}} = \{(v, \sigma, v') \mid v, v' \in R \text{ and } (v, \sigma, v') \in E_\Gamma\}$$
$$\cup \{(v, \sigma, \tau(\Pi_H(v'))) \mid v \in R, v' \notin R \text{ and }$$
$$(v, \sigma, v') \in E_\Gamma\} \cup \{(\tau(\Pi_H(v)), \sigma, v') \mid v \notin R, v' \in R \text{ and } (v, \sigma, v') \in E_\Gamma\}.$$

Lemma 3.9 $\hat{R}$ *is context-free.*

Proof: Let us consider the graph $\Gamma_R = (R, E_\Gamma \cap (R \times (X \cup \bar{X}) \times R))$. Almost all the ends of Γ_R are ends of Γ, hence Γ_R is context-free. As $\hat{R}$ is obtained from Γ_R by adjunction of finitely many edges, $\hat{R}$ is context-free too. $\square$

Lemma 3.10 $H\backslash\Gamma \approx K\backslash\hat{R}$.

Proof: One can check that $\bar{\tau} = \Pi_K \circ \tau$ is a graph isomorphism from $H\backslash\Gamma$ to $K\backslash\hat{R}$. $\square$

Sketch of proof of the main theorem

Let H and Γ fulfill the hypotheses of theorem 3.5. By lemma 3.10 it is sufficient to prove that $K\backslash\hat{R}$ is context-free.

As each element of K is a bisimulation, $\sim_K$ is a bisimulation. As each element of K is rational and K is finite, $\sim_K$ is rational over Γ. As R is rational, the subset $\varphi^{-1}(R) \times \varphi^{-1}(R)$ of $Z^* \times Z^*$ (where Z is the configuration alphabet of $\mathcal{A}$) is recognizable . Hence $(\varphi^{-1} \circ \sim_K \circ \varphi) \cap (\varphi^{-1}(R) \times \varphi^{-1}(R))$ is a rational subset of $Z^* \times Z^*$, hence $\sim_K |R$ is a rational relation over Γ_R. It follows that $\sim_K |R$ is a rational relation over $\hat{R}$ too (for some suitable p-structure inherited from the p-structure of Γ). As $Kv_0 \cap R$ is finite and $\hat{R}$ is connected, $\sim_K |R$ has finite order over $\hat{R}$.

The three hypotheses of theorem 3.1 are met by $K, \hat{R}$, hence $K\backslash\hat{R}$ is context-free. $\square$

Case of groups

Corollary 3.11 *Let H be a group of automorphisms of Γ fulfilling the following conditions:*

(1) H is finitely generated
(2) $\forall h \in H$, h is rational
(3) $\mathrm{Stab}_H(v_0)$ is finite

Then $H \backslash \Gamma$ is context-free.

Proof: Let us notice that, as Γ is connected , $\mathrm{Stab}_H(v_0)$ is finite iff all the stabilizers $\mathrm{Stab}_H(v)$ for $v \in V_\Gamma$ are finite. Hence the corollary follows immediately from theorem 3.5. □

Remark 3.12 *It is noteworthy that*
1- In the particular case where $H = \mathrm{Aut}(\Gamma)$ the hypotheses of theorem 3.11 are equivalent one to each other: (1) $\Leftrightarrow$ (3) is stated in [25, Theorem 22] while (1) $\Leftrightarrow$ (2) can be easily deduced from the proof of [25, Theorem 22].
2- In the general case, the conclusion of corollary 3.11 may fail as soon as one of the three hypotheses is not met (see [27, part 4]).

3.3   Quotient by the group of automorphisms

We denote here by $\sim$ the equivalence relation over V_Γ induced by the action of $\mathrm{Aut}(\Gamma)$ (i.e. $\sim_{\mathrm{Aut}(\Gamma)}$). For every subset $T \subseteq V_\Gamma$ we denote by $\mathrm{Aut}(\Gamma, T)$ the subgroup of elements of $\mathrm{Aut}(\Gamma)$ which preserve the set T:

$$\mathrm{Aut}(\Gamma, T) = \{h \in \mathrm{Aut}(\Gamma) \mid h(T) = T\}.$$

Lemma 3.13 $\sim_{\mathrm{Aut}(\Gamma, v_0)}$ *is a rational relation.*

Proof: The proof is technical (see [27]). Assuming that the p-structure $(\mathcal{A}, \varphi)$ is v_0-canonical (by proposition 2.7), we exhibit a 2-f.a. recognizing $\varphi^{-1} \circ \sim_{\mathrm{Aut}(\Gamma, v_0)} \circ \varphi$. □

Lemma 3.14 *There exists some subgroup $K < \mathrm{Aut}(\Gamma)$ which is finitely generated, acts rationaly and transitively on $[v_0]_\sim$.*

Proof: Let κ be some integer satisfying the conclusion of theorem 2.9 for the vertex v_0. For every $v \in B(v_0, \kappa) \cap [v_0]_\sim$ there exists some $\varphi_v \in \mathrm{Aut}(\Gamma)$ which is rational and maps v_0 to v (lemma 2.5). Let us consider

$$\mathcal{K} = \{\varphi_v \mid v \in B(v_0, \kappa) \cap [v_0]_\sim\} \text{ and } K = < \mathcal{K} >$$

(i.e. K is the subgroup generated by $\mathcal{K}$).
One can prove by induction on the integer n that : for every κ-path $(v_0, v_1, \ldots, v_n)$ where $v_i \in [v_0]_\sim$,

$$\text{there exists } k \in \mathcal{K}^n \text{ such that } kv_0 = v_n. \tag{15}$$

By theorem 2.9, $[v_0]_\sim$ is κ-connected, hence, by equation (15),

$$\forall v \in [v_0]_\sim, \exists k \in K, kv_0 = v.$$

□

Theorem 3.15 $Aut(\Gamma)\backslash\Gamma$ *is context-free.*

Proof: Let us consider the finite set of bisimulations: $\mathcal{H} = \{\sim_{Aut(\Gamma,v_0)}\} \cup \mathcal{K}$.
Let H be the semi-group of relations generated by $\mathcal{H}$. From lemma 3.14 one can
deduce that $Aut(\Gamma) = K \circ Aut(\Gamma,v_0)$, hence $\sim_{Aut(\Gamma)}=\sim_H$.
As $\mathcal{H}$ is finite and closed under inverse, H fulfills hypotheses (0)(1) of theorem
3.5. As each element of $\mathcal{H}$ is rational (by lemma 3.13 and by construction of $\mathcal{K}$),
H fufills hypothesis (2) of theorem 3.5.
Let $v, v' \in V_\Gamma$. Let us consider $n = d(v_0,v), n' = d(v,v')$.

$$\bigcup_{\substack{h\in H \\ (v,v')\in h}} h = \bigcup_{\substack{h\in Aut(\Gamma,v_0) \\ hv=v'}} h \subseteq \bigcup_{\substack{h\in Aut(\Gamma,v_0) \\ hv_0\in B(v_0,2\cdot n+n')}} h \subseteq \bigcup_{\substack{f\in \mathcal{K}^{2\cdot n+n'}\circ \sim_{Aut(\Gamma,v_0)}}} f.$$

Hence the finite set $F(v,v') = \mathcal{K}^{2\cdot n+n'}\circ \sim_{Aut(\Gamma,v_0)}$ fulfills equation (3) of theorem 3.5. Applying theorem 3.5 we obtain that $H\backslash\Gamma$ is context-free, and by the
equality $\sim_{Aut(\Gamma)}=\sim_H$ we conclude that $Aut(\Gamma)\backslash\Gamma$ is context-free. $\square$

One can easily generalize this theorem as follows

Theorem 3.16 *Let Γ be a context-free graph and T be any rational subset of
V_Γ (with respect to some p-structure on Γ). Then $Aut(\Gamma,T)\backslash\Gamma$ is context-free.*

Remark: When Γ is defined as the computation graph of some pda $\mathcal{A}$, and T
is defined as the set of *terminal* configurations of $\mathcal{A}$, $Aut(\Gamma,T)\backslash\Gamma$ can be viewed
as the computation graph of a new pda $\bar{A}$ with set of terminal configurations
$\Pi(T)$, which recognizes the *same language* as $\mathcal{A}$. Moreover this automaton $\bar{A}$ is
automorphism-free in the sense that the group $Aut(\mathcal{C}(\bar{A}), \Pi(T))$ is trivial.

4 Discussion

One natural research perspective arising from this work is to treat the case of a
group H acting on a connected graph Γ of bounded degree and finite tree-width.
Considering the results of [4, 22, 25], it seems natural to expect the group H to
be virtually- free (provided the stabilizer of some vertex is finite). Considering
our corollary 3.11 it seems natural to expect the graph $H\backslash\Gamma$ to have still finite
tree-width (provided H is finitely generated).
Another natural question is whether, under the hypotheses of theorem 3.15, the
canonical projection Π is rational (equivalently, whether $\sim_{Aut(\Gamma)}$ is rational, by
a corollary of theorem 3.5)? Example 4 of [27] just shows that this property must
be specific of the full group $Aut(\Gamma)$ and we expect this property to be true.
Aknowledgements: We thank an anonymous referee for pertinent criticism
and useful suggestions.

References

1. A.V. Anisimov. Group languages. *Kibernetica 4*, pages 18–24, 1971.
2. A. Arnold and A. Dicky. Transition systems equivalence. *Information and Computation 82*, pages 198–229, 1989.
3. J. Baeten, J. Bergstra, and J. Klop. Decidability of bisimulation equivalence for
   processes generating context-free languages. In *Proceedings of PARLE 87*, pages
   94–111. LNCS 259, 1987.
4. H. Bass and J.P. Serre. *Arbres,amalgames, SL2*. SMF, collection astérisque, nr
   46, 1983.

5. M. Bauderon. Infinite hypergraph I, basic properties (fundamental study). *TCS 82*, pages 177–214, 1991.
6. M. Bauderon. Infinite hypergraph II, systems of recursive equations. *TCS 103*, pages 165–190, 1992.
7. J. Berstel and L. Boasson. Context-free languages. In *Handbook of theoretical computer science, vol.B, Chapter 2*, pages 59–102. Elsevier, 1991.
8. D. Caucal. Graphes canoniques de graphes algébriques. *RAIRO TIA, nr 24-4*, pages 339–352, 1990.
9. D. Caucal. On the regular structure of prefix rewritings. *TCS 106*, pages 61–86, 1992.
10. D. Caucal. Bisimulation of context-free grammars and of pushdown automata. *To appear in CSLI, Modal Logic and process algebra, vol. 53, Stanford*, pages 1–20, 1995.
11. S. Christensen, H. Hüttel, and C. Stirling. Bisimulation equivalence is decidable for all context-free processes. In *Proceedings of CONCUR 92*, pages 138–147. LNCS 630, 1992.
12. B. Courcelle. The monadic second-order logic of graphs ii: infinite graphs of bounded width. *Math. Systems Theory 21*, pages 187–221, 1989.
13. B. Courcelle. Graph rewriting: and algebraic and logic approach. *In Jan van Leeuwen, editor, Handbook of Theoretical Computer Science, volume B*, pages 193–242, 1990.
14. B. Courcelle. The monadic second-order logic of graphs IV: definability properties of equational graphs. *Annals of Pure and Applied Logic*, pages 183–255, 1990.
15. W. Dicks and M.J. Dunwoody. *Groups acting on graphs*. Cambridge University Press, 1990.
16. S. Eilenberg. *Automata, Languages and Machines, vol. A*. Academic Press, 1974.
17. C. Frougny and J. Sakarovitch. Synchronized rational relations of finite and infinite words. *TCS 108*, pages 45–82, 1993.
18. R. Halin. Automorphisms and endomorphisms of infinite locally finite graphs. *Abh.Math.Sem.Univ.Hamburg 39*, pages 251–283, 1973.
19. M.A. Harrison. *Introduction To Formal Language Theory*. Addison-Wesley, 1978.
20. Y. Hirshfeld, M. Jerrum, and F. Moller. A polynomial algorithm for deciding bisimilarity of normed context-free processes. In *Proceedings of Dagstuhl seminar on Algorithms in Automata theory*. To appear in TCS, 1994.
21. C.M. Hoffmann. *Group-theoretic algorithms and graph-isomorphism*. LNCS 136, 1982.
22. D.E. Muller and P.E. Schupp. Groups, the theory of ends and context-free languages. *JCSS vol 26, no 3*, pages 295–310, 1983.
23. D.E. Muller and P.E. Schupp. The theory of ends, pushdown automata and second-order logic. *TCS 37*, pages 51–75, 1985.
24. D. Park. Concurrency and automata on infinite sequences. *LNCS 104*, pages 167–183, 1981.
25. L. Pelecq. Automorphism groups of context-free graphs. *Accepted for publication in TCS*, pages 1–17, 1995.
26. G. Sénizergues. Graphes context-free. *Notes de cours de DEA de l'université de Bordeaux 1*, pages 1–98, 1989.
27. G. Sénizergues. Semi-groups acting on context-free graphs. *LaBRI internal report nr 1094-95, can be accessed at the URL, http://www.labri.u-bordeaux.fr/*, 1995.
28. G. Sénizergues. The isomorphism problem for the transition-graphs of pushdown automata. *Talk given at FAIT 89, Wuerzburg, Germany*, December 1989.

Hard Sets Method and Semilinear Reservoir Method with Applications

Leonid P. Lisovik
Department of Cybernetics
Kiev State University, Kiev, 252017, Ukraine
Email: Lis@cyber.univ.kiev.ua

Abstract

Here is proposed a variant of the lecture devoted to mentioned two methods with applications in automata theory, formal language theory, schematology and with solution of the equivalence problem for finite substitutions on regular language.

1.Introduction

In paper [1] Karel Culik II discussed three recently developed techniques for proving the decidability of equivalence problems in the formal language theory. We cannot help saying also about the long series of papers intended to the equivalence problem for deterministic push-down automata (DPDAs). These papers based on well known Valiant's methods of "alternating stacking" and "parallel stacking". The last achievement in this way was the decidability of the equivalence problem for real-time deterministic push-down automata [V.Romanovsky (1986) and M.Oyamaguchi (1987)]. The other approach was introduced earlier by A. Korenjak and J. Hopcroft (1966). They have obtained the decidability of equivalence problem for simple DPDAs by means of using the systems of equality over languages. The latter approach was developed by V.Mejtus in two papers (see Kibernetika, 1989, N 5, p.14 - 27 and Kibernetika, 1992, N 5, p. 20 - 45). These two papers have represented the attempt to decide in general the equivalence problem for DPDAs, however on account of its incompleteness these papers must be very thoroughly investigated for further elaboration. New original approach was found recently by T.Harju and J.Karhumäki in order to decide the long-standing equivalence problem for multitape finite automata [2].

The methods mentioned above in the title of the paper aren't discussed in [1] or in other papers in English, so it is necessary to consider these methods on the whole. What are means and notions involved for studying the decidability problems by hard sets method? There are the trace technique and quasi-identities and series of quasi-identities in semigroup. What are main problems which have been decided by means of these two method? There are the following three problems.

A. The equivalence problem for finite-valued finite transducers [3].

B. The equivalence problem for linear unary recursive schemes with individual constants [4].

C. The equivalence problem for metalinear unary recursive schemes with individual constants [5].

Besides there was obtained the applications to combinatorial group theory, mathematical logic and formal language theory. Hard sets method was applied to the problems A and B. Semilinear reservoir method was applied to the problem C. The last problem is a generalization of the problem B, which was a problem from paper of S.Garland and D.Luckham [6]. The problem C is the most difficult problem. It is necessary to remark that even the emptiness problem for metalinear unary recursive schemes with individual constants was open and very difficult.

What are means and notions involved in studying the decidability problems by semilinear reservoir method? There are the semilinear sets (cf. [7]), so-called ΣTC-transducers (see [8] and definitions below), the non deterministic jumps technique and the specific technique for representation of the computational processes by means of infinite labelled trees (see [9]).

2. Definitions

A subset H of a semigroup S is called a hard subset [10] if the following condition holds in the semigroup S. For any its elements v, γ, δ, ν if elements vv, $v\gamma\nu$ and $v\delta\nu$ are belonging to set H then element $v\gamma\delta\nu$ belongs to H also. Let $N=\{0, 1, ...\}$ and $N^n=\{(x_1, ..., x_n)| x_i \in N, 1 \leq i \leq n\}$ for every integer n. A set $L \subseteq N^n$ is said to be semilinear set if it is a finite union of linear sets (see [7]). Every linear set L, where $L \subseteq N^n$, has a form $L=L(c,P)$, where $c \in N^n$, P $\subseteq N^n$, P is the finite set, $P=\{ p_j \mid 1 \leq j \leq m \}$ and

$$L(c,P) = \{x|x = c + \sum_{j=1}^{m} k_j p_j, \text{ for } k_j \in N, 1 \leq j \leq m\}$$

Theorem 1. Let H be a hard subset of a free commutative monoid $S=<N^n, +>$. Then H is a semilinear set [11].

The semigroup S is said to be a H-semigroup if the set $D=\{(a,a)|a \in S\}$ is a hard subset of semigroup $S \times S$, which is a Cartesian product of semigroup S with itself. The semigroup S is H-semigroup iff the quasi-identity

$$vv = v_1\nu_1 \wedge v\gamma\nu = v_1\gamma_1\nu_1 \wedge v\delta\nu = v_1\delta_1\nu_1 \longrightarrow v\gamma\delta\nu = v_1\gamma_1\delta_1\nu_1$$

holds in semigroup S. Every commutative semigroup and every semigroup embedded in group are H-semigroup. If S is Cartesian product of H semigroups S_1 and S_2 then $S =S_1 X S_2$ is also H-semigroup. The coverings of semigroups and regular events (or sets) by finite unions of hard sets in semigroups were considered [10]. This question is directly connected with the problem A, when coverings of regular events by finite unions of hard sets are considered in a finite Cartesian products with factors, which are the free finitely generated monoids.

Theorem 2. Let S be finitely presented semigroup and H_i, $1 \leq i \leq m$, be decidable hard subsets of semigroup S. Given regular event R in semigroup S, it is decidable whether the inclusion $R \subseteq \bigcup_{i=1}^{m} H_i$ is valid [10].

The notion of ΣTC-transducer have been defined in [8]. The ΣTC-transducer is nondeterministic Turing machine over input labeled full infinite n-ary trees with finite number of counters and output tape and satisfying some conditions. It has bounded number of returns in counters and bounded number of movement across any rib.

Theorem 3. The emptiness problem for ΣTC-transducers is decidable [8]

Theorem 4. The equivalence problem for deterministic ΣTC-transducers is decidable [8].

The semilinear reservoir method involved also the special technique for disposition on labeled tree the long words which may be appeared in computation process being simulated. There remains the necessity for more widened explanation of this technique. It is concerned also the nondeterministic jumps technique. In first approximation it was presented in [8]. Later the semilinear reservoir method was presented thoroughly and in its generality in three papers (1983, 1985, 1986). Then in paper (1993) was showed the possibility of this method for resolving the equivalence problem and the inclusion problem in various classes of languages. Besides one paper must be appeared. On the whole the semilinear reservoir method is difficult for explanation.

3. From regular sets in semigroups to finite transducers

Regular sets in a semigroup $S(\triangle)$, where $\triangle$ is its finite set of generators, defined by regular expressions in the alphabet $\triangle$. At the end of 1974 was obtained the series of result on the decidability of equivalence problem for regular sets in semigroups. This problem was studied in a free finitely generated group, commutative finitely presented semigroup (two papers in 1975), free products of groups (paper in 1977). From 1975 we studied this problem in Cartesian product of free group and abelian group. Simultaneously the equivalence problem for deterministic finite transducers was investigated. Initially it was our exercise for students. It turned out to be difficult question for student, so we put one's hand to this question and then to the equivalence problem for single-valued finite transducers. As a result three different solution of this question was obtained in 1975-1976 and two papers was published in 1978 (see Kibernetika, 1978, N3, p.1-7). Later it became known for us that the same question was an open problem of Greibach (see problem 3.1.29 in [12]). These questions for deterministic and single-valued finite transducers have been decided also by M.Schützenberger (see[13]) and M.Blattner and T.Head (see J.Comp. Syst. Sci. 15 (1977), 310-327 and 19 (1979), 45-49).

The following our question which was raised and resolved in 1976 was question about the decidability of the equivalence problem for finite-valued finite

222

transducers (see [3]). It is astonishingly that a simple solution of the equivalence problem for single-valued finite transducer we extracted from the complex solution for finite-valued case. The method applied in [3] was presented in still more general form as a method of hard sets in [10]. On the whole the research program involved the equivalence problems and the inclusion problems for the following classes of transducers and program schemes.

1. Finite transducers
   1.1. Deterministic finite transducers
   1.2. Single-valued finite transducers
   1.3. Finite system of single-valued finite transducers
   1.4. Finite-valued finite transducers
   1.5. Finite transducers with one letter output alphabet
2. Deterministic finite transducers with additional finite-turned counters
3. Two-way transducers
   3.1. Two-way deterministic finite transducers
   3.2. Two-way deterministic finite transducers with additional finite-turned counters
4. Program schemata
   4.1. Linear unary recursive schemes with individual constants
   4.2. Metalinear unary recursive schemes
   4.3. Linear unary recursive schemes with reversible operators
   4.4. Metalinear unary recursive schemes with individual constants
   4.5. Ultralinear unary recursive schemes with individual constants
5. Transducers over infinite labeled trees
   5.1. LC-schemes
   5.2. ML-schemes
   5.3. ΣTC-transducers
   5.4. MLC-schemes
   5.5. ULC-schemes

Remark that for $i=1, 2, 4, 5$ the transducers of class 5. i correspond to the schemes of class 4 $i.$, when they are considered over free interpretations. The equivalence problem in class 1.5 is undecidable (O.Ibarra (1976) and L.Lisovik (1977) independently, cf. [14, 15]). The equivalence problem for class 4.3 is still open. All other equivalence problems mentioned above have been solved up till now. Concerning the class 4.3, we achieved some partial progress in deciding of the equivalence problem for linear unary recursive scheme with reversible operators (see [16,17]. Our approach to this problem is connected with trace technique and series of quasi-identities in free monoid and in free finite generated group. Simultaneously we consider that attempts to solve this problem as a whole will lead to further development of hard sets method.

In the general setting the equivalence problem was researched in subclasses of following generalized classes.

6. Transducers with output tape
7. Transducers with output semigroup $S(\Delta)$

8. Regular sets in semigroups of form $\Sigma^+ \times S(\Delta)$

9. Regular sets in semigroups

For example, in [16] was researched the equivalence problem for two-way deterministic finite transducers with output semigroup $S(\Delta)$, which is a free group. Note that other setting for the equivalence problem in class 4.3 is one in the following class.

10. Linear unary recursive schemes over free group

It is typical for many author's papers to decide the equivalence problem in concrete classes of transducers with output tape in generalized form for transducers with output semigroup. Such approach in many cases have admitted to find simple and simultaneously general forms of proofs. As variants of output semigroups were used free monoids, free groups, finite presented commutative semigroups, every group with decidable word problem, semigroups embedded in groups with decidable word problem, any H-semigroup with decidable word problem, free or Cartesian product of semigroups (or group) and so on.

We give below some examples of propositions. Let $A = (K, \Sigma, \Delta, H, q_0, F)$ be a (nondeterministic normalized) finite transducer where K, Σ and Δ denote nonempty finite sets of states, input symbols and output simbols respectively, $q_0 \in K$, $F \subseteq K$ and $H \subseteq K \times (\Sigma \cup \{\varepsilon\}) \times \Delta^* \times K$. Then it defines the mapping $O(A) \subseteq \Sigma^+ \times \Delta^+$, where $O(A) = \{(v,w) \mid v \in \Sigma^+, w \in \Delta^+, w \in A(v)\}$. Let Δ be a set of generators of some fixed semigroup $S(\Delta)$. Let $\varphi : \Delta^+ \longrightarrow S(\Delta)$ be canonical homomorphism. Then we denote by $O_\varphi(A)$ the mapping $O_\varphi(A) = \{(v, \varphi(w)) \mid (v,w) \in O(A)\}$.

Theorem 5. Let $S(\Delta)$ be a H-semigroup with decidable word problem. Given any integer m and n, and single-valued finite transducers A_i, $1 \leq i \leq m+n$, with input alphabet Σ and output alphabet Δ, it is decidable whether the equality

$$\bigcup_{j=1}^{m} O_\varphi(A_j) = \bigcup_{j=m+1}^{m+n} O_\varphi(A_j)$$

is valid [3].

The mapping $O_\varphi(A)$ is called k-valued if for every $v \in \Sigma^+$ the set $W(A,\varphi,v) = \{\varphi(w) \mid (v,w) \in O_\varphi(A)\}$ contains at most k elements, where k is fixed integer, $k \geq 1$. The mapping $O_\varphi(A)$ is called finite-valued if there exists some integer k such that $O_\varphi(A)$ is k-valued mapping. The transducer A is called finite-valued if the mapping $O(A)$ is finite-valued.

Theorem 6. Let $S(\Delta)$ be a semigroup which is embedded in a group with decidable word problem. Given two mapping $O_\varphi(A)$, i=1, 2, defined correspondently by finite-valued finite transducers A_i, $i=1, 2$, it is decidable whether the equality $O_\varphi(A_1) = O_\varphi(A_2)$ is valid [3].

The proof of theorem 6 in [3] used the following decomposition lemma.

Lemma 1. For every finite-valued finite transducer A there exist integer k and single-valued finite transducers A_i, $1 \leq i \leq k$, such that the equality

$$O(A) = \bigcup_{i=1}^{k} O(A_i)$$

holds.

The complete proof for effective version of the decomposition lemma was given later in 1988 by A.Weber (see [18]). Note that this lemma in its noneffective version was stated in [13] without complete proof. A.Weber [42] proved also the following strong form of Lemma 1. For every k-valued finite transducer A there exist exactly k single-valued finite transducers $A_i, 1 \leq i \leq k$, such that the equality $O(A) = \overset{k}{\underset{i=1}{\cup}} O(A_i)$ holds. In [18] A.Weber in spirit of our approach have obtained the proof of theorem 6 in case when semigroup $S(\triangle)$ is a free monoid. The same result was obtained also by K.Culik II and J.Karhumäki (cf. [19]). They have obtained this result in some more general form for finite-valued finite-transducer on input HDTOL-language. In their solution was proposed another method connected with the Ehrenfeucht conjecture and G.Makanin's result (1977) about the decidability of the existence solution problem for finite systems of equations in free semigroup. The Ehrenfeucht conjecture have been affirmated by M.Albert and J.Lawrence [20] and V.Guba in 1985 (see [45]).

Now we have proof for such general decomposition theorem.

Theorem 7. Let $S(\triangle)$ be a semigroup which is embedded in group G. Given a finite transducer A, where $O_\varphi(A)$ is finite-valued mapping, there exist integer k and finite transducers $A_i, 1 \leq i \leq k$, such that every $O_\varphi(A_i)$ is single-valued mapping and the equality

$$O_\varphi(A) = \overset{k}{\underset{i=1}{\cup}} O(A_i)$$

holds. Besides if the word problem in group G is decidable then integer k and finite transducers $A_i, 1 \leq i \leq k$, may be constructed effectively (obtained in 1990, published in [21]).

Theorem 8. Let $S(\triangle)$ be a semigroup embedded in group G, which has decidable word problem. Given any finite transducer A and B, where $O_\varphi(B)$ is finite-valued mapping, it is effectively checked the inclusion $O_\varphi(A) \subseteq O_\varphi(B)$.

By construction from [3], we have the proof of Theorem 8 based on Theorem 7. Recently we have obtained a short proof of Theorem 8 only by means of hard sets method and without usage the decomposition theorem. Such proof have been obtained in 1990 (see [21]). A.Weber proved [41] the decidability of finite-valuedness problem for finite transducers (see also [21]).

4. Connections with program schemata and formal languages

The main problems for program schemata are the equivalence problem, inclusion problem, weak equivalence problem, emptiness problem, totality problem and freedom problem. All these problems excluding the freedom problem are decidable for the following classes of program schemata and classes of transducers.

11. Metalinear unary recursive schemes (1978).

12. Metalinear unary recursive schemes with additional finite number of finite-turned counters (1983).

13. Metalinear unary recursive schemes with individual constants (1985).

14. Ultralinear unary recursive schemes with individual constants (1993, see [22]).

15. Semilinear macrotransducers (1986). It is the most general class of transducers investigated by means of semilinear reservoir method.

16. Finite systems of semilinear macrotransducers (1986).

17. Janov's schemes with procedure which is linear unary recursive schemes (1985).

The decidability of equivalence problem in the last class is corollary from the decidability of equivalence problem for real-strict DPDAs (cf. [23]) and E.Friedman's result about the connection between the equivalence problem for unary recursive schemes and the equivalence problem for DPDAs (see J. Comp. Syst. Sci. 14(1977) 344-359). It is concerning also of the equivalence problem for metalinear unary recursive schemes.

Note that now we can easy extract the decidability of equivalence problem for deterministic single-valued finite transducers from the linear unary recursive schemes. The last was established in 1970 (see [6]).

The equivalence problem and inclusion problem are decidable for

18. Languages recognizable by two-way deterministic metalinear push-down automata [22].

The inclusion problem is decidable for

19. Deterministic metalinear context-free languages [22].

The decidabilities were established by means of semilinear reservoir method for classes 12-16, 18, 19 and by means of hard sets method for classes 11, 17 (see [24]).

5. Open problems, other applications and additional results

20. Let F be a finite language and H(F) be a hard set generated by set F. Is it true that the set H(F) is always a regular language?

21. Let $\varphi:\Delta^+ \to S(\Delta)$ is a canonical homomorphish into a free group $S(\Delta)$. Is it decidable the finiteness problem for sets of form $\varphi(L(A))$, where $L(A)$ is a language recognizable by push-down automaton A?

22. The set $L \subseteq \Delta^+$ is called a ΣTC-generatable language if it can be generated by some ΣTC-transducer A. The equivalence problem for transducers from a class $\mathfrak{S}$ on input ΣTC-generatable language W(A) is decidable for many classes $\mathfrak{S}$. Now we continue the investigation of the class of ΣTC-generatable languages.

23. (In combinatorial group theory)

Theorem 9.1. Let $S(\Delta)$ be a group. A subset $H \subseteq S(\Delta)$ is a hard set iff it is a left coset [10].

Theorem 9.2. Let $S(\Delta)$ be a group, L is a regular language, $L \subseteq \Delta^+$ and G is a subgroup with generating set $\varphi(L)$. Then G is finitely generated subgroup [10].

Corollary 9.1. Let $S(\triangle)$ be a group. A hard set $H \subseteq S(\triangle)$, generated by a regular set $\varphi(L)$, is finitely generated hard set.

Let $\wp$ be the class of finitely generated groups G satisfying the following condition. Given finitely generated subgroup G_1, G_2 and elements a_1, a_2 of group G, it is decidable whether the intersection $a_1 G_1 \cap a_2 G_2$ is empty.

Theorem 9.3. The class $\wp$ is closed under operation of free product [25].

For any integer $k \geq 2$ the subset H of semigroup S is called a k-hard subset if the following condition holds in the semigroup S. For any its elements v, ν, γ_i, where $1 \leq i \leq k$, the implication

$$\forall(\varepsilon_1, ..., \varepsilon_k) \in \{0, 1\}^k \ (\sum_{i=1}^{k} \varepsilon_i < k \to v(\gamma_1)^{\varepsilon_1}... (\gamma_k)^{\varepsilon_k}\nu \in H) \to v\gamma_1...\gamma_k\nu \in H$$

is valid. Here it means that $(\gamma_i)^1 = \gamma_i$ and $(\gamma_i)^0$ is empty word.

Theorem 9.4. Let G be a group and $T = \{(a,b,c)| \ a,b \in G, \ c = ab\}$ is its Kelley's table. Then T is a 3-hard set in a group G^3 [26].

24. (In the theory of finitely presented commutative semigroups)

Theorem 10.1. Let $S(\triangle)$ be a finitely presented commutative semigroup. If a subset $H \subseteq S(\triangle)$ is a hard set then it is a regular set in semigroup $S(\triangle)$ (see [11]).

Theorem 10.2. Let $S(\triangle)$ be a finitely presented commutative semigroup and $\Re$ is class of regular sets in semigroup $S(\triangle)$. The class $\Re$ is effectively closed under Boolean operations [11].

Corollary 10.1. It is decidable the first order theory of any finitely presented commutative semigroup $S(\triangle)$.

The last result have been obtained in 1966 by M.Taitslin. The proof of M.Taitslin was very difficult. It was fulfilled by means of model completeness method.

25. Bounded context-free languages and permutation letters of words

Let n be any fixed integer, $n \geq 1$, and $\triangle = \{a_i | 1 \leq i \leq n\}$ is an alphabet. For every set $M \subseteq N^n$ denote by $\pi_n^{-1}(M)$ the bounded language $\{w | w = a_1^{k_1}...a_n^{k_n}$ where $(k_1, ..., k_n) \in M\}$. For every language $L \subseteq \triangle^*$ denote by comm(L) the language $\{w |$ there exists word $v \in L$ such that w may be obtained from v by permutation letters of word v$\}$. The language comm(L) is called the commutation closure of language L.

There are two open problems.

25.1. Is it decidable whether $\pi_n^{-1}(M)$ is a context-free language for given semilinear set $M \subseteq N^n$?

25.2. To find the necessary and sufficient (and effectively checkable) condition which means that the commutative closure of a context-free language L is also a context-free language.

The useful technique for (partial) deciding these two problems was developed in [28, 29]. In paper [29] have obtained the following result.

Theorem 11. Let M be a semilinear set, $M \subseteq N^n$, $n \geq 4$, where

$$M = \bigcup_{i=1}^{m} L(c_i, P)$$

for some $c_i \in N^n$, $1 \leq i \leq m$, $P \subseteq N^n$, P is a common system of periods, and

$\pi_n^{-1}(M)$ is a context-free language. The language $comm(\pi_n^{-1}(M))$ is a context-free language iff for any sequence $< \alpha_1, \alpha_2, \alpha_3, \alpha_4 >$ where $1 \le \alpha_i \le n$, i=1,2,3,4, the language $comm(\pi_n^{-1}(M) \cap (b_1{}^*b_2{}^*b_3{}^*b_4{}^*$ is a context-free language, where $b_i = a_{\alpha_i}$

The proof of theorem 11 was published also in [30]. Earlier was obtained such result.

25.3. If $\triangle = \{a_1, a_2, a_3\}$ is alphabet and L is a bounded context-free language, where $L \subseteq a_1{}^*a_2{}^*a_3{}^*$, then comm(L) also is a context-free language.

It gave a solution of a problem of Ginsburg [7]. The result 25.3 was obtained by J.-F. Perrot [31] and independently by H.Maurer [32] and also in 1971 by L.Lisovik (see also [33]). Note that theorem 11 was proved later by J.Beauquier, M.Blattner and M.Latteux in [34]. The proof of theorem 11 is difficult.

By the technique of paper [28] the problem 25.1 was decided in case when there exists finite number of mutually disjoint lattices H_i, $1 \le i \le$ s, such that

$$M = \bigcup_{i=1}^{k} M_i$$

where $M_i = M \cap H_i$ and

$$M_i = \bigcup_{i=1}^{m_i} L(c_{ij}, P_i)$$

for all integer i, $1 \le i \le$ s.

We presuppose that the technique of paper [28] is sufficient for deciding on the whole the problem 25.1. We presuppose also that theorem 11 holds for every semilinear set M such that $\pi_n^{-1}(M)$ is a context-free language. Similar conjecture was announced in paper [30]. Analogous hypothesis was announced (in personal communication) by M.Fliess in 1970 (as we known from [34]). More correctly M.Fliess' hypothesis has the following form.

25.4. The language comm(L) is context-free language if comm(L) $\cap$ $a_{\sigma(1)}{}^*$ ... $a_{\sigma(n)}{}^*$ is context-free language for every sequence $< \sigma(1),...,\sigma(n) >$ where $\{\sigma(i)| \ 1 \le i \le n \ \} = \{1,...,n\}$.

26. On Rabin's automata and its generalizations

It is well known that Buchi's automata over infinite words and Rabin's automata over infinite labeled trees has interesting applications in mathematical logic (see [35, 36]).

26.1. It is decidable the emptiness problem for Rabin's automata augmented with one finite-turned counter [37].

26.2. It is decidable the singular second order theory of two successors with additional unary predicate P(x) [38]. There are many appropriate variants of predicate P(x), for example

$P(x) \leftrightarrow \exists n \ (|x| = n^2)$,
$P(x) \leftrightarrow \exists n \ (|x| = n^k)$, $k \in N$,
$P(x) \leftrightarrow \exists n \ (|x| = 2^n)$,
$P(x) \leftrightarrow \exists n \ (|x| = n!)$

and other predicates constructed in the same way as in paper of C.Elgot and M.Rabin (see [39]).

27. The equivalence problem for drawers

Definition 27.1. A drawers is a ten-tuple $A=(K, \Sigma, \Delta, n, \delta, \lambda, \xi, q_0, F, Z_0)$, where K is a finite set of states, Σ is a finite set (input alphabet), Δ is a finite set (output alphabet), n is integer, $n \geq 1$, δ, λ and ξ are mappings

$\delta: K \times \Sigma \to K$ (function of transitions),

$\lambda: K \times \Sigma \times \Delta \to \Delta \backslash \{Z_0\}$ (function of exits),

$\xi: K \times \Sigma \to \{-1, 0, 1\}^n$ (function of shifts),

q_0(K is an initial state, $F \subseteq K$ is a set of final states, $Z_0 \in \Delta$ is a fixed symbol (a marker).

On the containing a drawer is a finite automaton augmented with output n-ary lattice which is partitioned on cells with coordinates $(x_1,...,x_n)$, where $x_i \in Z$, $1 \leq i \leq n$, $Z = \{0, \pm 1, ...\}$. For example in case n=2 a drawer has output plane. A drawer A is reading an input word w from the left to the right and simultaneously it moves its control head on the output plane and changes its state and output symbols in cells. Initially it has output lattice with symbol Z_0 in all cells and its control head is setting on a cell θ coordinated by $(0,...,0)$. The actions is fulfilled in accordance with the mapping δ, λ and ξ. When the input word w have been read and the drawer A have gone into a final state, then the mapping $\mu: \{-1, 0, 1\}^n \to \Delta$, which have been obtained is considered as a result of the drawer A on the input word w.

27.1. The equivalence problem for drawers is decidable [40].

28. It is known the open problem about decidability of the equivalence problem for finite substitutions on regular languagee [46]. We give a solution of this problem. It is undecidable [43]. As a corollary it was shown the undecidability of the equivalence problem for nondeterministic generalized sequential machines (NGSM) with two states [43]. These results are based on paper [44] where it have been shown that one special property for markoff chains is undecidable.

29. The inclusion problem for finite substitutions on regular language is undecidable (see [47, 48]).

30. Superhard sets.

A subset H of a semigroup S is called a superhard subset [26] if for any elements u, γ, δ, v , γ_1, δ_1, w of semigroup S if elements uvw, $u\gamma v \gamma_1 w$ and $u\delta v \delta_1 w$ are belonging to set H then element $u\gamma\delta v\delta_1\gamma_1 w$ belongs to H also.

30.1. Let H_i , $1 \leq i \leq m$, be decidable superhard subsets of a free monoid Δ^*. Given context-free language L it is decidable whether the inclusion $L \subseteq \bigcup_{i=1}^{m} H_i$ is valid [26].

30.2. In [26] the finitely generated groups defined by automata (finite, pushdown, nested stack and linear bounded automata respectively) were investigeted. We say that a group G is defined by automaton A if it has representation $G = \langle \Delta; w=1 \ (w \in L) \rangle$ where L is a language recognizable by automaton A. It is open the question about existence of nested stack group which isn't context-free group (some possible example of such group see in [26]).

31. The finite ω - transducer A has the same definition as the finite transducer A but it defines the mapping $O_1(A) \subseteq \Sigma^\omega \times \Delta^\omega$ with infinite input and

output words. Given finite $\omega-$ transducer A it is decidable whether or not the mapping $O_1(A)$ is finite valued [21].

References

[1] K.Culik II, New technique for proving the decidability of equivalence problems, Theoret. Comput. Sci. 71 (1990) 29-45.

[2] T.Harju and J.Karhumäki, The equivalence problem of multitape finite automata, Theoret. Comput. Sci. 78 (1991) 347-355.

[3] L.Lisovik, The equivalence problem for finite-valued automata over semigroups, Kibernetika 4(1979) 24-27.

[4] L.Lisovik, The equivalence problem for transducers over labeled trees, Dokladu of Ukrainian Acad. Sci. 6(1980) 77-79.

[5] L.Lisovik, Metalinear schemes with individual constants, Programmirovanie 2 (1985) 29-38.

[6] S.Garland and D.Luckham, Program schemes, recursion schemes and formal languages (Report, California Univ., Los Angeles, 1971).

[7] S.Ginsburg, The mathematical theory of context-free languages (McGraw-Hill, New York, 1966).

[8] L.Lisovik, To the equivalence problem for transducers over Σ-trees with finite-turned counters, Kibernetika, 5(1984) 19-24.

[9] L.Lisovik, Metalinear recursive schemes over labeled trees, Programmirovanie 5(1983) 13-22.

[10] L.Lisovik, Hard sets and finite coverings of semigroup, Kibernetika, 1(1980) 12-16.

[11] L.Lisovik, The sets definable in the theories of commutative semigroup, Dokladu of Ukrainian Acad. Sci. 8(1981) 80-83.

[12] A.Aho and J.Ullman, The theory of parsing, translation and compiling (Prentice-Hall, Englewood Cliffs, N.Y. 1972).

[13] M.Schützenberger, Sur les relations rationnelles entre monoides libres, Theoret. Comput. Sci. 3(1976) 243-259.

[14] O.Ibarra, The unsolvability of the equivalence problem for ε-free NGSM's with unary input (output) alphabet and applications, SIAM J.Comput. 4(1978) 524-532.

[15] L.Lisovik, The identity problems of regular events over Cartesian product of free and cyclic semigroup, Dokladu of Ukrainian Acad. Sci. 6(1979) 410-413.

[16] L.Lisovik, On finite transducers over labeled trees and quasi-identities in free semigroup, Problems of cybernetics 40(1983) 19-42.

[17] L.Lisovik, Quasi-correlations in free group and equivalence problems for transducers, Algebra and Logic 25(1986) 51-86.

[18] A.Weber, A decomposition theorem for finite-valued transducers and an application to the equivalence problem (Interner bericht 1/88, Fächbereich Informatik, Universitat Frankfurt, 1988) 29 p.

[19] K.Culik II and J.Karhumäki, The equivalence problem for finite-valued transducers (on HDTOL languages) is decidable, Theoret. Comput. Sci. 47 (1986) 71-84.

[20] M.Albert and J.Lawrence, A proof of Ehrenfeucht conjecture, Theoret. Comput. Sci. 41(1985) 121-123.

[21] L.Lisovik, The problems of inclusion and ambiguity for regular events in semigroups, Discrete Mathematics 5(1993) 54-74.

[22] L.Lisovik, The problem of inclusion and equivalence for program schemes and formal languages, Kibernetika and Systems Analysis (former Kibernetika) 2(1993) 62-73.

[23] M.Oyamaguchi, N.Honda and Y.Inagaki, The equivalence problem for real-time strict deterministic languages, Inform. and Control 45(1980) 90-115.

[24] L.Lisovik, On regular algebra of functionals over labeled trees, Kibernetika, 5(1985) 25-30.

[25] L.Lisovik, On the problem of occurrence in to regular subsets in the free product of subgroup, Siberian Mathematical Journal 20(1979) No 4.

[26] L.Lisovik and V.Red'ko, Regular events in semigroups, Problems of cybernetics, 37(1980) 155-184.

[27] M.Taitslin, On elementary theories of commutative semigroups, Algebra and Logic, 5(1966) 55-89.

[28] L.Lisovik, The lattice structure of semilinear sets and the context-freeness problem in class of bounded languages, The theory of languages and processors (collection), Inst. of Cybernetics, Ukrainian Acad. Sci., Kiev, 1974, 89-99.

[29] L.Lisovik, About preserving the context-freeness under permutation letters of words. The theory of languages and processors (collection), Inst. of Cybernetics, Ukrainian Acad. Sci., Kiev, 1974, 78-88.

[30] L.Lisovik, Commutative closure of context-free languages, Kibernetika, 4(1978) 23-28.

[31] J.F.Perrot, Sur la fermeture commutative des c-langages, C.r. Acad. Aci. 265(1967) No 20.

[32] H.Maurer, The solution of a problem by Ginsburg, Information Processing Lett. 1(1971) No 1.

[33] T.Oshiba, On permuting letters of words in context-free languages, Inform. and Control 20(1972) No 5.

[34] J.Beauquier, M.Blattner and M.Latteux, On commutative context-free languages, J. Comput. System Sci. 35(1987) 311-320.

[35] M. Rabin, Definability of second-order theories and automata on infinite trees, Trans. Amer. Math. Soc. 141(1969) 1-35.

[36] W. Thomas, Automata on infinite objects, Handbook of Theoretical Computer Science, Vol. B, Elsevier Science Publishers B.V. (1990).

[37] L.Lisovik, Σ-automata with finite-turned counters, Kibernetika 3(1988) 19-22.

[38] L.Lisovik, On construction of singular theories of two successors with additional predicate, Algebra and Logic 23 (1984) 266-277.

[39] C.Elgot and M.Rabin, Decidability and undecidability of extensions of second (first) order theory of (generalized) successor, J. Symbol. Logic 31 (1966) 169-181.

[40] L.Lisovik, Semilinear sets and decidable problems, Mathematical questions of cybernetics 1 (1988) 201-222.

[41] A.Weber, On the valuedness of finite transducers, Acta Informatica 27(1990) 749-780.

[42] A.Weber, Decomposing finite - valued transducers and deciding their equivalence, SIAM J. Comput. 22(1993)175-202.

[43] L.Lisovik, About equivalence of transducers on regular languages, Kibernetika and Systems Analysis 6(1995)158-159.

[44] L.Lisovik, An undecidable problem for countable Markoff chains, Kibernetika 2(1991) 1-6.

[45] A. Salomaa, Jewels of formal language theory (Russian translation, with. afterword by A.A.Muchnik and A.L.Semenov), Mir Publishers, Moskow, 1986.

[46] J. Karhumäki, Problems and solutions, P97. Bulletin of the EATCS 25(1985)184-185.

[47] Y. Maon, On the equivalence of some transductions involving letter to letter morphisms on regular languages, Acta Informatica 23(1986)585-596.

[48] P.Turakainen, On some transducer equivalence problems for families of languages, Intern. J. Computer Math. 23(1988) 99-124

Random Polynomials and Polynomial Factorization

Philippe Flajolet,[1] Xavier Gourdon,[1] and Daniel Panario[2]

[1] Algorithms Project, INRIA Rocquencourt, F-78153 Le Chesnay, France.
[2] Department of Computer Science, University of Toronto, Toronto, Canada M5S-1A4.
E-mails: Philippe.Flajolet@inria.fr, Xavier.Gourdon@inria.fr, daniel@cs.toronto.edu.

Abstract. We give a precise average-case analysis of a complete polynomial factorization chain over finite fields by methods based on generating functions and singularity analysis.

1. Introduction

Polynomial factorization is basic to many areas of computer algebra [12], algebraic coding theory [1], computational number theory and cryptography [2, 6, 18, 20]. Its implications include finding complete partial fraction decompositions (a problem itself useful for symbolic integration), designing cyclic redundancy codes, computing the number of points on elliptic curves and building arithmetic public key cryptosystems.

Polynomial factorization may be carried out over any field, but the efficient algorithms are essentially *probabilistic* and they eventually rely on factoring over a *finite field* $\mathbb{F}_q$ where q is a prime or the power of a prime, see [16] for an excellent introduction. This paper derives basic properties of random polynomials over finite fields that are of interest in the study of factoring algorithms. We show that the most important characteristics can be treated systematically by methods of "analytic combinatorics" based on generating functions and singularity analysis.

We have elected here to consider a classical factorization chain over finite fields that is at the same time simple, fairly efficient, and *complete*. It is close to what is used internally in the Maple computer algebra system [12] and to what is likely to be required of a general purpose computer algebra system that mostly deals with polynomials of intermediate "size". Our factorization chain may not be the fastest at the moment, compare for instance with Shoup's technique [24]. However the discipline of completely analyzing such algorithms, which is in the line of Knuth's works [16], reveals parameters that are of intrinsic interest for polynomial factoring in general. To the best of our knowledge, such a task has not been undertaken systematically beyond rough (mostly worst-case) bounds.

Our reference factorization chain comprises the following three classical steps:

ERF: *Elimination of repeated factors* replaces a polynomial by a square-free form that contains all the irreducible factors of the original polynomial with exponents reduced to 1.

DDF: *Distinct-degree factorization* splits a squarefree polynomial into a product of polynomials whose irreducible factors all have the same degree.

EDF: *Equal-degree factorization* factors a polynomial the irreducible factors of which all have the same degree.

The top-level code of our factorization chain (in pseudo-Maple) is given below.

```
procedure factor(f : polynomial);
    1: a := ERF(f);
    2: b := DDF(a);
       F := 1;
    3: for k from 1 to n do
           F := F . EDF(b[k],k);
       od;
    4: return(F . factor(f/a));
end;
```

Computational model. All average-case analyses are expressed as asymptotic forms in n, the degree of the polynomial to be factored. We *fix a finite field* $\mathbb{F}_q$ with $q = p^m$ (p prime) and consider the polynomial ring $\mathbb{F}_q[x]$, see [12, 16, 19]. For simplicity of exposition, we assume here that the characteristic p is odd, but the algorithms and their analyses can be easily adapted to the otherwise important cases of $\mathbb{F}_2$ and $\mathbb{F}_{2^m}$. Our model assumes that a basic field operation has cost $\mathcal{O}(1)$; then the cost of a sum is $\mathcal{O}(n)$ and the cost of a product, a division or a gcd is $\mathcal{O}(n^2)$, when applied to polynomials of degree $\leq n$. For *dominant asymptotics*, we can freely restrict attention to polynomial products and gcd's whose costs can be taken under the standard form

$$\text{product: } \tau_1 n^2, \qquad \text{gcd: } \tau_2 n^2.$$

2. Summary of results

It is well-known [1, 16] that a random polynomial of degree n is irreducible with probability tending to 0 and has close to $\log n$ factors on average and with a high probability [4, 10]. Thus, the factorization of a random polynomial over a finite field is almost surely nontrivial.

The first phase *ERF* of our factorization chain classically starts with the elimination of repeated factors, a simplified form of squarefree factorization described in Section 4. Theorem 1 quantifies this process and shows that up to smaller order terms, the expected cost is dominated by a single gcd of the polynomial f to be factored and its derivative f', so that it is $\mathcal{O}(n^2)$ on average. In a precise technical sense, most of the factorization cost results from the subsequent phases since the non-squarefree part has average degree $\mathcal{O}(1)$.

The second phase *DDF* that is described in Section 5 splits the squarefree part a of the polynomial to be factored into a product $a = b_1 \cdot b_2 \cdots b_n$, where b_k is itself the product of the irreducible factors of a that have degree k. This phase is based on elementary properties of finite fields and is the one with the highest computational cost, namely $\mathcal{O}(n^3)$ on average. Theorems 3,4,5 provide a precise comparison of three strategies: the naïve rule, the "half-degree" rule and the "early abort" rule whose costs are found to be in the approximate proportion $1 :: \frac{3}{4} :: \frac{2}{3}$. Thus a savings of about one third results from controlling the DDF phase by the early abort strategy. At the end of this phase, the factorization is complete with a probability ranging asymptotically between 0.56 and 0.67, see Theorem 6.

The third phase *EDF* can be exactly analysed and it is found that its expected cost is comparatively small, being $\mathcal{O}(n^2)$, see Theorems 7,8 for precise statements. For each nontrivial factor b_k, it involves a recursive refinement process again based on properties of finite fields. The analysis is close to that of digital trees known as "tries" [15] but under a biased probability model.

Precise statements are given in the next few pages with an explicit dependency on the field cardinality q, and some of them involve number-theoretic functions that can be both evaluated and estimated easily. Therefore, the results obtained allow us to quantify precisely what goes on. A simplified picture is as follows. The ERF phase involves with high probability little more than a single polynomial gcd. The DDF phase of cost $\mathcal{O}(n^3)$ is the one that is most intensive computationally, where control by the "early-abort" strategy is expected to bring gains close to 36% at no extra cost. The last phase of EDF is executed less than 50% of the time and its cost is again small compared to that of DDF.

3. Basic methodology

This paper relies heavily on a symbolic use of *generating functions* (GF's). These are used to express enumerative properties of random polynomials and also to derive direct asymptotic results from singularities. General references are Chapter 3 of Berlekamp's book [1], the exercise section 4.6.2 of Knuth's book [16], and the paper by Flajolet and Odlyzko [9] for asymptotic methods.

3.1. Generating functions. We specialize our discussion to polynomials over a finite field $\mathbb{F}_q$. Let $\mathcal{I}$ be the collection of monic irreducible polynomials. The two expressions

$$(1) \qquad \mathcal{Q} = \prod_{\omega \in \mathcal{I}}(1+\omega), \qquad \text{and} \qquad \mathcal{P} = \prod_{\omega \in \mathcal{I}}(1-\omega)^{-1}.$$

when expanded by distributivity "generate" formally the family $\mathcal{Q}$ of monic squarefree polynomials and $\mathcal{P}$ of all monic polynomials. In this context, $\mathcal{I}$ may itself be identified with the formal sum $\mathcal{I} = \sum_{\omega \in \mathcal{I}} \omega$.

Let z be a formal variable. The substitution $\omega \mapsto z^{|\omega|}$ with $|\omega|$ the degree of $\omega \in \mathcal{I}$ produces generating functions by a well-known process For instance, $I(z) = \sum_{\omega \in \mathcal{I}} z^{|\omega|} = \sum_n I_n z^n$, where I_n is the number of polynomials in $\mathcal{I}$ having degree n. The same substitution applied to $\mathcal{P}$ and $\mathcal{Q}$ yields two series, $P(z)$ and $Q(z)$, that are found to satisfy

$$(2) \qquad Q(z) = \prod_{n=1}^{\infty}(1+z^n)^{I_n}, \qquad P(z) = \prod_{n=1}^{\infty}(1-z^n)^{-I_n}.$$

Then, the coefficients $Q_n = [z^n]Q(z)$ and $P_n = [z^n]P(z)$ represent the number of polynomials of degree n in $\mathcal{Q}$ and $\mathcal{P}$ respectively.

Since P_n has value q^n, we have $P(z) = (1 - qz)^{-1}$, and the second relation of (2) implicitly determines I_n by a well-known process based on Moebius inversion [1]

$$(3) \qquad I_n = \frac{1}{n} \sum_{k|n} \mu(k) q^{n/k}, \qquad \text{so that} \qquad I_n = \frac{q^n}{n} + O\left(\frac{q^{n/2}}{n}\right),$$

$$(4) \qquad I(z) = \sum_{k=1}^{\infty} \frac{\mu(k)}{k} \log \frac{1}{1 - qz^k}.$$

Thus a fraction extremely close to $1/n$ of all polynomials of degree n are irreducible. This result was first proven by Gauss [11] for prime fields (see also [8]).

As regards Q_n, the formula $1 + z = (1 - z^2)/(1 - z)$ applied to the infinite products for $P(z), Q(z)$ entails

$$(5) \qquad Q(z) = \frac{P(z)}{P(z^2)} = \frac{1 - qz^2}{1 - qz}, \qquad \text{and} \qquad Q_n = q^{n-1}(q - 1) \quad (n \geq 2),$$

with $Q_0 = 1$, $Q_1 = q$. Apparently, this result was given for the first time in [5].
Parameters. We need extensions of this symbolic method in order to take care of characteristic parameters of polynomial factorization. Let Φ be a class of monic polynomials, χ some integer–valued parameter on Φ. The sum

$$\Phi(z, u) = \sum_{\omega \in \Phi} z^{|\omega|} u^{\chi(\omega)}$$

is such that the coefficient $[z^n u^k] \Phi(z, u)$ represents the number of polynomials of degree n and χ–parameter equal to k. For additive parameters χ, the product decompositions above generalize, provided one uses the translation rule $\omega \mapsto z^{|\omega|} u^{\chi(\omega)}$. The technique of rearranging logarithms of infinite products is useful in simplifying such expressions.

Averages and standard deviations are obtained by taking successive derivatives of bivariate generating functions with respect to u, then setting $u = 1$.

3.2. Asymptotic analysis. Generating functions (GFs) encode exact informations on their coefficients. Furthermore, their behaviour near their dominant positive singularity is an important source of coefficient asymptotics.

Most of the generating functions $f(z)$ to be studied in this paper are singular at $z = 1/q$ with an isolated singularity of the algebraic-logarithmic type. In that case, an expansion near $z = 1/q$ of the form

$$(6) \qquad f(z) = \frac{1}{(1 - qz)^{\alpha}} \left(\log \frac{1}{1 - qz}\right)^k (1 + o(1)).$$

is translated to coefficients by the method known as singularity analysis [9, 21]

$$(7) \qquad [z^n] f(z) = q^n \frac{n^{\alpha-1}}{\Gamma(\alpha)} (\log n)^k (1 + o(1)),$$

whenever $\alpha \neq 0, -1, -2, \ldots$. This requires analytic continuation (isolated singularity), a condition for instance satisfied by the GF's of Theorems 1,3,4.

The same translation can be effected under a variety of alternative conditions corresponding either to Darboux's method [7] or to Tauberian theorems of the

Hardy-Littlewood-Karamata type [13, 21, 22]. Such alternative conditions are needed in Theorems 5 and 6 where the GF's have a natural boundary.

The permutation model. The following property is well-known. The joint distribution of degrees in the prime decomposition of a random polynomial over $\mathbb{F}_q$ having degree n admits as a limit, when the cardinality q of the base field tends to infinity (n staying fixed!), the joint distribution of cycle lengths in random permutations of size n. Accordingly GF's of random polynomials at z/q converge to GF's of corresponding permutation families when $q \to +\infty$.

This gives rise to a useful heuristic for large field cardinalities. An instance is mentioned in [13] in connection with the probability that a random polynomial admits factors of distinct degrees which, for large q and large n is found to approach $e^{-\gamma}$. Our Theorem 6 illustrates an instance of this situation.

4. Elimination of repeated factors (ERF)

The first step in the factorization chain of a polynomial is the *elimination of repeated factors* (ERF). In characteristic 0, this is achieved by the gcd of f and its derivative f'. In finite characteristics, additional control is needed in order to deal with pth powers whose derivatives are 0, see [12, 16]. The auxiliary computation of pth roots, $g^{1/p}$, is performed in the classical way described in [12, p. 344] for example.

```
procedure ERF(f : polynomial);
        g := gcd(f,f'); h := f/g; k := gcd(g,h);
        while k<>1 do  g := g/k; k := gcd(g,h)  od;
        if  g <> 1 then h := h*ERF(g^(1/p)) fi;
        return(h);
end;
```

Theorem 1. *(i) A random polynomial of degree $n \geq 2$ in $\mathbb{F}_q[x]$ has a probability $1 - 1/q$ to be squarefree.*
(ii) The degree of the non-squarefree part of a random polynomial has expected value asymptotic to

$$C_q = \sum_{n \geq 1} \frac{n I_n}{q^{2n} - q^n},$$

and a geometrically decaying probability tail. We have $C_q \sim 1/q$ as $q \to \infty$.

PROOF. Part (i) is classical and is the consequence of Eq. (5). As for (ii), the bivariate generating function of the degree of the non-squarefree part of monic polynomials in $\mathbb{F}_q[x]$ is, by the symbolic methods of Section 3,

$$P(z, u) = \prod_{n \geq 1} \left(1 + \frac{z^n}{1 - u^n z^n} \right)^{I_n}.$$

The mean degree of the non-squarefree part is obtained from the derivative $P_u(z, 1)$ by singularity analysis. The generating function $P(z, 3/2)$ is dominated by $P(z)$ near its dominant singularity, so that the geometrically decaying probability tail holds. Finally, the asymptotic value of C_q as $q \to \infty$ is obtained by means of the expansion $n I_n = q^n + \mathcal{O}(q^{n/2})$. □

Theorem 1 has important consequences for the recursive structure of the
factor procedure. First, the overall cost of the recursive calls (Step 4 in the
top-level procedure) remains $\mathcal{O}(1)$ on average. Next, alternative strategies giving
the full squarefree factorization [12, p.345] have asymptotically equivalent costs.
Finally, the ERF phase has a cost dominated by its first gcd.

Theorem 2. *The expected cost of the ERF phase applied to a random polynomial of degree n is asymptotically that of a single gcd,*

$$\overline{\tau ERF}_n \sim \tau_2 n^2.$$

5. Distinct-degree factorization (DDF)

The second stage of our reference algorithm requires finding the *distinct-degree
factorization* (DDF) of the squarefree polynomial a. This means expressing a in
the form $b_1 \cdot b_2 \cdots b_n$ where b_k is the product of irreducible factors of degree k.
The principle is that the polynomial $x^{q^k} - x \in \mathbb{F}_q[x]$ is the product of all monic
irreducible polynomials in $\mathbb{F}_q[x]$ whose degree divides k (see [19], p. 91).

```
procedure DDF(a : polynomial);  [a is assumed squarefree]
         n := deg(a); g := a; h := x;
         for k := 1 to n do
1.          h := h^q mod g;
2.          b[k] := gcd(h-x,g);
3.          g := g/b[k]; [a without irred factors of deg<=k]
4.          if b[k] <> 1 then h := h mod g fi;
         od;
         return(b[1].b[2]...b[n]);
end;
```

The computation in step 1 is done by means of the classical *binary powering*
method [16, p. 441-442]. With $\nu(q)$ the number of ones in the binary representation of q, the number of products needed to compute h^q (mod g) is

$$(8) \qquad \lambda(q) = \lfloor \log_2 q \rfloor + \nu(q) - 1.$$

By the exponential tail result of Thm 1, we need only consider the cost of DDF
applied to the squarefree part a of the input polynomial f and our subsequent
analyses are all relative to the statistics induced by a random input f of degree n.

Theorem 3. *The expected cost of the basic DDF phase satisfies*

$$\overline{\tau DDF}_n \sim \frac{5}{12} \left(\lambda(q)\tau_1 + \tau_2 \right) n^3 \quad where \quad \lambda(q) = \lfloor \log_2 q \rfloor + \nu(q) - 1.$$

PROOF. The cost of the basic DDF is $C_1 + C_2 + C_3 + C_4$, where C_j denotes
the cost of line number j. We let $\overline{C_j}$ be the expectation of C_j. Since the mean
number of factors of f is $\mathcal{O}(\log n)$, we find $\overline{C_3} + \overline{C_4} = \mathcal{O}(n^2 \log n)$.

Let d_k denote the degree of polynomial g when the kth iteration of the main
loop starts; the parameter d_k is also the sum of the degrees of the distinct factors

of f with degree $\geq k$. The quantity $C_1 + C_2$ is equal to $(\lambda(q)\tau_1 + \tau_2)\sum_{k \geq 1} d_k^2$. The bivariate generating function associated with d_k is, by the basic decompositions,

$$P_k(z, u) = \prod_{j < k}\left(\frac{1}{1 - z^j}\right)^{I_j} \prod_{j \geq k}\left(1 + u^j \frac{z^j}{1 - z^j}\right)^{I_j}.$$

The expected value of $C = \sum_{k \geq 1} d_k^2$ is then given by

$$\overline{C} = \frac{1}{q^n}[z^n]R(z), \qquad R(z) = \sum_{k \geq 1}\left(\frac{\partial^2 P_k}{\partial u^2}(z, u) + \frac{\partial P_k}{\partial u}(z, u)\right)_{u=1}.$$

The GF $R(z)$ involves the coefficients I_n and from the main estimate $nI_n = q^n + \mathcal{O}(q^{n/2})$, the behaviour near the dominant singularity $z = 1/q$ results: $R(z) \sim \frac{5}{2}(1 - qz)^{-4}$. Singularity analysis entails that $[z^n]R(z/q) \sim \frac{5}{12}n^3$. $\qquad\square$

5.1. The "half-degree" rule.
A natural idea is to stop the DDF loop when $k = n/2$, since at this stage the remaining factor is either 1 or it is irreducible.

Theorem 4. *The expected cost of the "half-degree rule" DDF phase satisfies*

$$\overline{\tau DDF}_n^{(HD)} \sim \frac{5}{16}(\lambda(q)\tau_1 + \tau_2)\, n^3 \qquad where \quad \lambda(q) = \lfloor \log_2 q \rfloor + \nu(q) - 1.$$

PROOF. The cost is now given by $(\lambda(q)\tau_1 + \tau_2)C^{(1)}$, where $C^{(1)} = \sum_{k \leq n/2} d_k^2$. Let D_1 be the highest degree of all irreducible factors of f. We study the difference $C^{(2)} = C - C^{(1)}$. If $D_1 \leq n/2$, we have $C^{(2)} = 0$, otherwise we have $C^{(2)} = (D_1 - \lfloor n/2 \rfloor)D_1^2$ since there can be only one factor of degree larger than $n/2$, namely D_1. Thus the mean value of $C^{(2)}$ is given by

$$(9) \qquad \overline{C^{(2)}} = \sum_{n/2 < k \leq n} \Pr(D_1 = k)\left(k - \left\lfloor\frac{n}{2}\right\rfloor\right)k^2.$$

The probability $\Pr(D_1 = k)$ is derived from the generating function $\chi_k(z)$ of polynomials whose factors have all degree $\leq k$ as

$$\Pr(D_1 = k) = \frac{1}{q^n}[z^n](\chi_k(z) - \chi_{k-1}(z)), \qquad \chi_k(z) = \prod_{j=1}^{k}\left(\frac{1}{1 - z^j}\right)^{I_j}.$$

When $k > n/2$, the n-th coefficient of $\chi_k(z) - \chi_{k-1}(z)$ is obtained from

$$\chi_k(z) - \chi_{k-1}(z) = P(z)\left(1 - (1 - z^k)^{I_k}\right)\prod_{j > k}(1 - z^j)^{I_j} = P(z)\left(I_k z^k + \mathcal{O}(z^{n+1})\right)$$

which entails $\Pr(D_1 = k) = I_k/q^k \sim 1/k$ for $n/2 < k \leq n$. Plugging this information into (9) gives $\overline{C^{(2)}} \sim \frac{5}{48}n^3$, thus $\overline{C^{(1)}} = \overline{C} - \overline{C^{(2)}} \sim \frac{15}{48}n^3$. $\qquad\square$

Thus, the half-degree rule results in a savings of 25% asymptotically.

5.2. The "early-abort" strategy. A still better strategy called "early abort" consists in stopping the main loop of DDF as soon as $2k$ exceeds the degree of the remaining factor, since then the remaining factor must be irreducible. The analysis now has some analogy to that of integer factoring given by Knuth and Trabb-Pardo [17].

Theorem 5. *The expected cost of the "early-abort rule" DDF phase satisfies*

$$\overline{\tau DDF}_n^{(EA)} \sim \delta \left(\lambda(q)\tau_1 + \tau_2\right) n^3, \qquad \text{where} \quad \delta = 0.2668903307\ldots,$$

$$\delta = \frac{5}{12} - \frac{1}{3} \int_0^\infty e^{-2x} \exp\left(-\int_x^\infty \frac{e^{-u}}{u}\,du\right) \frac{1-x^2}{x}\,dx.$$

The constant δ is a close relative of the famous Golomb constant that intervenes in the expectation of the longest cycle in a random permutation [23].

PROOF. Let D_1 and D_2 be the degrees of the two irreducible factors of f of highest degree, setting $D_2 = 0$ if a is irreducible. The iteration is now aborted at step $k = \max\{\lfloor D_1/2\rfloor, D_2\} + 1$. The cost of DDF with this stopping rule becomes $C^{(3)} = \sum_{k \leq \max\{\lfloor D_1/2\rfloor, D_2\}} d_k^2$ times the constant $(\lambda(q)\tau_1 + \tau_2)$. Consider the difference $C^{(4)} = C - C^{(3)}$. We have

$$C^{(4)} = \begin{cases} (D_1 - \lfloor D_1/2\rfloor)\, D_1^2 & \text{if } D_1/2 > D_2 \\ (D_1 - D_2)D_1^2 & \text{if } D_1/2 \leq D_2. \end{cases}$$

The generating function of polynomials for which $D_1 > 2D_2$ is given by

$$\phi_{D_1}(z) = \left[\prod_{1 \leq \ell < D_1/2} \left(\frac{1}{1-z^\ell}\right)^{I_\ell} \right] I_{D_1} \left(\frac{z^{D_1}}{1-z^{D_1}}\right),$$

and the generating function of polynomials for which $D_2 < D_1 \leq 2D_2$ is given by

$$\psi_{D_1,D_2}(z) = \left[\prod_{1 \leq \ell < D_2} \left(\frac{1}{1-z^\ell}\right)^{I_\ell} \right] \left[\left(\frac{1}{1-z^{D_2}}\right)^{I_{D_2}} - 1 \right] \left[I_{D_1} \frac{z^{D_1}}{1-z^{D_1}} \right]$$

(we do not need to take the case $D_1 = D_2$ into account since it contributes 0 to $C^{(4)}$). Hence, the GF of the cumulated values of the parameter $C^{(4)}$,

$$\Phi(z) = \sum_{D_1} \left(D_1 - \left\lfloor\frac{D_1}{2}\right\rfloor\right) D_1^2\, \phi_{D_1}(z) + \sum_{D_2 < D_1 \leq 2D_2} (D_1 - D_2)D_1^2\, \psi_{D_1,D_2}(z).$$

The analysis of this generating function near its positive dominant singularity q^{-1} is done by approximating sums with integrals (Euler-Maclaurin summation) after the change of variables $z = e^{-t}q^{-1}$. A somewhat delicate analysis shows that $\Phi(z/q) \sim c_0(1-z)^{-4}$ as $z \to 1^-$, where $c_0 = \frac{5}{2} - 6\delta$. A Tauberian argument is needed since the positive singularity is not isolated. $\square$

The global savings of the early abort rule is of 36% and the expected cost of $\mathcal{O}(\log q \cdot n^3)$ for DDF clearly dominates in the whole factorization chain.

6. The output configuration of DDF

The DDF procedure does not completely factor a polynomial that has different irreducible factors of the same degree. However, as shown by the following theorem, "most" of the factoring has been completed after DDF.

Theorem 6. *(i) The asymptotic probability of a complete DDF factorization is*

$$c_q = \prod_{n \geq 1} \left(1 + \frac{I_n}{q^n - 1}\right)(1 - q^{-n})^{I_n},$$

$c_2 \doteq 0.6656$, $c_{257} \doteq 0.5618$, $c_\infty = e^{-\gamma} \doteq 0.5614$, *where* γ *is Euler's constant.*
(ii) The expected degree of the part of the input polynomial subjected to the EDF phase is asymptotic to $\log n$.

PROOF. (i) The GF of polynomials with irreducible factors of distinct degrees

(10)
$$\prod_{n \geq 1} \left(1 + I_n \frac{z^n}{1 - z^n}\right)$$

has the equivalent form $(1 - qz)^{-1} \phi(z)$, where $\phi(z)$ is obtained by multiplying each term of the product (10) by $(1 - z^n)^{I_n}$. The function $\phi(z)$ is continuous at $1/q$ and a Tauberian-like argument applies. Finally, when q is large, the relation $nI_n = q^n + \mathcal{O}(q^{n/2})$ is used to prove that c_q tends to $\prod_{n \geq 1}(1 + 1/n)e^{-1/n} = e^{-\gamma}$.

(ii) The bivariate generating function associated to the total degree of the nontrivial part of DDF is

$$P(z, u) = \prod_{n \geq 1} \left[\left(1 + u^n \frac{z^n}{1 - z^n}\right)^{I_n} - (u^n - 1)I_n \frac{z^n}{1 - z^n}\right].$$

The corresponding mean value is $q^{-n}[z^n]R(z)$, where $R(z)$ equals $P_u(z, u)|_{u=1}$. Near $z = q^{-1}$, $R(z)$ behaves like $(1 - qz)^{-1} \log(1 - qz)^{-1}$. As before, a Tauberian-like argument is needed, giving $q^{-n}[z^n]R(z) \sim \log n$. $\square$

7. Equal-degree factorization (EDF)

From Section 5, the factorization problem is eventually reduced to factoring a collection of polynomials b_j of a special form that have all their irreducible factors of the same (known) degree j. Our reference chain uses the classical Cantor-Zassenhaus algorithm [3]. The analysis combines a recursive partioning problem akin to digital tries [15] with estimates on the degree of irreducible factors of random polynomials [14].

```
procedure EDF(b : polynomial, k : integer);
[b is a product of irreducibles of degree k]
        if degree(b) <= k then return(b) fi;
        h := randpoly(degree(b)-1);
1.      a := h^((q^k-1)/2)-1 mod b;
2.      d := gcd(a,b);
        return(EDF(d,k).EDF(b/d,k));
end;
```

7.1. EDF and digital tries. By elementary properties of finite fields, each factor of b has a probability $\alpha = \frac{q-1}{2q}$ to be a factor of d and the complementary probability $\beta = \frac{q+1}{2q}$ to divide b/d. The probability that a random choice leads to a split of b that is of type $\langle \ell, j - \ell \rangle$ is thus the Bernoulli probability $\binom{j}{\ell}\alpha^\ell \beta^{j-\ell}$.

Theorem 7. *The cost of the EDF algorithm on polynomials with j irreducible factors of degree k is $C_{j,k} =$*

$$\left(\frac{1}{2\alpha\beta}j(j-1) + j \sum_{m \geq 0} \sum_{\ell=0}^{m} \binom{m}{\ell} \alpha^{m-\ell}\beta^\ell \left(1 - (1 - \alpha^{m-\ell}\beta^\ell)^{j-1}\right) \right) (\mu_k \tau_1 + \tau_2)k^2,$$

where $\mu_k = \lambda((q^k - 1)/2) = \left\lfloor \log_2 \frac{q^k-1}{2} \right\rfloor + \nu\left(\frac{q^k-1}{2}\right) - 1$.

PROOF. A complete recursive execution of the EDF procedure is equivalent to developing a binary tree of possibilities. For a tree t with root subtrees t_0, t_1, we thus consider a general cost function of the additive type,

$$(11) \qquad C[t] = e_{|t|} + C[t_0] + C[t_1].$$

where $e_{|t|}$ is a (problem specific) "toll" function that depends on the size $|t|$ (number of nonempty external nodes) of t.

Like for tries [15], the subtree sizes obey the Bernoulli probability given above. Thus, the expectation c_j of $C[t]$ over trees of size j satisfies the recurrence

$$c_j = e_j + \sum_{\ell=0}^{j} \binom{j}{\ell} \alpha^\ell \beta^{j-\ell}(c_\ell + c_{j-\ell}) = e_j + \sum_{\ell=0}^{j} \binom{j}{\ell}(\alpha^\ell \beta^{j-\ell} + \alpha^{j-\ell}\beta^\ell)c_\ell.$$

This translates, in terms of exponential generating functions, $C(z) = \sum_j c_j z^j / j!$ and $E(z) = \sum_j e_j z^j / j!$, into the functional equation $C(z) = E(z) + e^{\beta z}C(\alpha z) + e^{\alpha z}C(\beta z)$, that iterates to give the explicit solution

$$(12) \qquad C(z) = \sum_{j \geq 0} \sum_{\ell=0}^{j} \binom{j}{\ell} E(\alpha^{j-\ell}\beta^\ell z)e^{z(1-\alpha^{j-\ell}\beta^\ell)}.$$

Here, the toll function $e_j = j^2 - \delta_{j,1}$ leads to

$$C(z) = \frac{1}{2\alpha\beta}z^2 e^z + z \sum_{m \geq 0} \sum_{\ell=0}^{m} \binom{m}{\ell} \alpha^{m-\ell}\beta^\ell \left(e^z - e^{z(1-\alpha^{m-\ell}\beta^\ell)} \right),$$

by means of Eq. (12). From there, an explicit expression for the coefficients results. The analysis is completed by finally taking into account the cost of multiplications modulo b that intervene in the computation of $h^{(q^k-1)/2} \mod b$ by the binary powering algorithm, leading to the μ_k. $\qquad\square$

7.2. Complete analysis. Completing the analysis of EDF only requires weighting the costs given by Theorem 7 by the probability $\Pr(\omega_n(k) = j)$ of finding j irreducible factors of degree k. Let $\omega_n(k)$ be the random variable counting the number of distinct irreducible factors of degree k in a random polynomial of degree n. The corresponding probability distribution can be computed by the decomposition techniques of Section 3, see [14], and one has:

$$\Pr\{\omega_n(k) = j\} = \begin{cases} \dfrac{\binom{I_k}{j}}{q^{kj}}(1 - q^{-k})^{I_k - j} \underset{q \to \infty}{\sim} e^{-1/k}\dfrac{k^{-j}}{j!} & \text{if } n \geq kI_k, \\[3mm] \dfrac{\binom{I_k}{j}}{q^{kj}} \displaystyle\sum_{\ell=0}^{\lfloor n/k \rfloor - j} (-1)^\ell \dfrac{\binom{I_k - j}{\ell}}{q^{k\ell}} & \text{if } kj \leq n < kI_k. \end{cases}$$

The distribution is essentially a negative binomial that can be approximated by a Poisson law of parameter $1/k$. Hence:

Theorem 8. *The expected cost of the EDF phase satisfies*

$$\overline{\tau EDF}_n \sim \frac{\tau_1}{\alpha\beta} \sum_{k=1}^{\lceil n/2 \rceil} \mu_k, \qquad \mu_k = \left\lceil \log_2 \frac{q^k - 1}{2} \right\rceil + \nu\left(\frac{q^k - 1}{2}\right) - 1.$$

In addition, this cost is $\mathcal{O}(n^2)$ and

$$(13) \quad \overline{\tau EDF}_n \sim \left(\frac{3}{4}\tau_1 \frac{q^2}{q^2 - 1} \log_2 q \cdot n^2\right)(1 + \xi_n + o(1)), \qquad -\frac{1}{3} \leq \xi_n \leq \frac{1}{3}.$$

PROOF. The intuition behind the proof is that the major contribution comes from situations where just 2 factors are present, the other cases having globally a very small probability of occurrence. Let $\overline{E_k}$ be the expected value of the cost of the EDF algorithm corresponding to degree k. By definition, we have $\overline{E_k} = \sum_{j>2} \Pr(\omega_n(k) = j) C_{j,k}$, where $C_{j,k}$ is given by Theorem 7.

First, the form of the distribution of the number of distinct factors implies

$$\Pr(\omega_n(k) = 2) = \frac{\binom{I_k}{2}}{q^{2k}}(1 + \mathcal{O}(1/k)) \quad \text{for } 2k \leq n, \quad \Pr(\omega_n(k) = j) = O\left(\frac{1}{j! k^j}\right).$$

Next, from Theorem 7, we deduce

$$C_{0,k} = C_{1,k} = 0, \quad C_{2,k} = \frac{2}{\alpha\beta}(\mu_k \tau_1 + \tau_2) k^2, \quad \text{and uniformly} \quad C_{j,k} = \mathcal{O}(j^2 k^3).$$

This entails that, as $k \to \infty$ with $2k \leq n$,

$$\overline{E_k} = C_{2,k}\frac{\binom{I_k}{2}}{q^{2k}}(1 + \mathcal{O}(1/k)) + \sum_{j \geq 3} O\left(\frac{k^{-j}}{j!} j^2 k^3\right) = \frac{\tau_1}{\alpha\beta}\mu_k + \mathcal{O}(1),$$

while $\overline{E_k} = 0$ for $2k > n$. Thus, the overall cost of the EDF component is $\sum_k \overline{E_k} = \frac{\tau_1}{\alpha\beta} \sum_{k=1}^{\lceil n/2 \rceil} \mu_k + \mathcal{O}(n)$. The second form is easily obtained from $k \log_2 q - 2 \le \mu_k \le 2k \log_2 q$. □

Under the unproven assumption that the binary representation of q^k behaves like that of a random integer, the arithmetic function ξ_n should be close to 0.

Acknowledgement. Work of P.F. and X. G. has been supported by the Long Term Research Project Alcom-IT (# 20244) of the European Union.

References

1. BERLEKAMP, E. R. *Algebraic Coding Theory.* Mc Graw-Hill, 1968. Revised edition, 1984.
2. BUCHMANN, J. Complexity of algorithms in number theory. In *Number Theory: Proceedings of the First Conference of the Canadian Number Theory Association* (1990), Walter de Gruyter, pp. 37–53.
3. CANTOR, D. G., AND ZASSENHAUSS, H. A new algorithm for factoring polynomials over finite fields. *Mathematics of Computation 36* (1981), 587–592.
4. CAR, M. Factorisation dans $F_q[x]$. *Comptes-Rendus de l'Académie des Sciences 294 (Ser. I)* (1982), 147–150.
5. CARLITZ, L. The arithmetic of polynomials in a Galois field. *American Journal of Mathematics 54* (1932), 39–50.
6. CHOR, B., AND RIVEST, R. A knapsack type public key cryptosystem based on on arithmetics over finite fields. *IEEE Transactions on Information Theory 34* (1988), 901–909.
7. COMTET, L. *Advanced Combinatorics.* Reidel, Dordrecht, 1974.
8. DEDEKIND, R. Abriss einer Theorie der höhern Congruenzen in Bezug auf einen reellen Primzahlmodulus. *Journal für die reine und angewandte Mathematik 54* (1857), 1–26.
9. FLAJOLET, P., AND ODLYZKO, A. M. Singularity analysis of generating functions. *SIAM Journal on Discrete Mathematics 3*, 2 (1990), 216–240.
10. FLAJOLET, P., AND SORIA, M. Gaussian limiting distributions for the number of components in combinatorial structures. *Journal of Combinatorial Theory, Series A 53* (1990), 165–182.
11. GAUSS, C. F. *Untersuchungen über höhere Mathematik.* Chelsea, New York, 1889.
12. GEDDES, K. O., CZAPOR, S. R., AND LABAHN, G. *Algorithms for Computer Algebra.* Kluwer Academic Publishers, Boston, 1992.
13. GREENE, D. H., AND KNUTH, D. E. *Mathematics for the analysis of algorithms,* second ed. Birkhauser, Boston, 1982.
14. KNOPFMACHER, J., AND KNOPFMACHER, A. Counting irreducible factors of polynomials over a finite field. *Discrete Mathematics 112* (1993), 103–118.
15. KNUTH, D. E. *The Art of Computer Programming,* vol. 3: Sorting and Searching. Addison-Wesley, 1973.
16. KNUTH, D. E. *The Art of Computer Programming,* 2nd ed., vol. 2: Seminumerical Algorithms. Addison-Wesley, 1981.
17. KNUTH, D. E., AND PARDO, L. T. Analysis of a simple factorization algorithm. *Theoretical Computer Science 3* (1976), 321–348.
18. LENSTRA, H. W. On the Chor Rivest cryptosystem. *Journal of Cryptology 3* (1991), 149–155.
19. LIDL, R., AND NIEDERREITER, H. *Finite Fields,* vol. 20 of *Encyclopedia of Mathematics and its Applications.* Addison-Wesley, 1983.
20. ODLYZKO, A. M. Discrete logarithms and their cryptographic significance. In *Advances in Cryptology* (1985), Lecture Notes in Computer Science, Springer Verlag, pp. 224–314.
21. ODLYZKO, A. M. Asymptotic enumeration methods. In *Handbook of Combinatorics,* M. G. R. Graham and L. Lovász, Eds., vol. II. Elsevier, Amsterdam, 1995, pp. 1063–1229.
22. POSTNIKOV, A. G. *Tauberian theory and its applications,* vol. 144 of *Proceedings of the Steklov Institute of Mathematics.* American Mathematical Society, 1980.
23. SHEPP, L. A., AND LLOYD, S. P. Ordered cycle lengths in a random permutation. *Transactions of the American Mathematical Society 121* (1966), 340–357.
24. SHOUP, V. A new polynomial factorization algorithm and its implementation. Preprint, 1994.

Optimal Gröbner Base Algorithms for Binomial Ideals

Ulla Koppenhagen and Ernst W. Mayr

Institut für Informatik, Technische Universität München
D-80290 München, GERMANY
e-mail: {KOPPENHA|MAYR}@INFORMATIK.TU-MUENCHEN.DE
WWW: HTTP://WWWMAYR.INFORMATIK.TU-MUENCHEN.DE/

Abstract. Little is known about upper complexity bounds for the normal form algorithms which transform a given polynomial ideal basis into a Gröbner basis. In this paper, we exhibit an optimal, exponential space algorithm for generating the reduced Gröbner basis of binomial ideals. This result is then applied to derive space optimal decision procedures for the finite enumeration and subword problems for commutative semigroups.

1 Introduction

Problems connected with polynomial ideals generated by finite sets of multivariate polynomials occur, as mathematical subproblems, in a number of different areas of computer science, like language generating and term rewriting systems, tiling problems, algebraic manifolds, motion planing, and several models for parallel systems.

However, most ideal-theoretic problems in multivariate polynomial rings are very hard to solve. B. Buchberger, in 1965, was the first to show that for polynomials over a field it is possible to construct, from an arbitrary given basis, a "detaching" basis, the so-called Gröbner basis, such that many problems of interest become easily expressible, and algorithmically solvable (see [Buc65], also [Hi64]).

Although versions of Buchberger's algorithm have been somewhat successful in practice, the complexity of the algorithm is not well understood. First steps towards an upper bound were obtained in [Bay82] and [MoMo84] where upper bounds for the degrees in a minimal Gröbner basis were derived.

However, these results did not, or only under very restrictive assumptions, imply bounds for the degree of the polynomials arising during the intermediate computations of the Gröbner basis algorithms.

Using a novel partitioning method for polynomial ideals, Dubé [Dub90] obtained the sharpened degree bound of $2 \cdot (\frac{d^2}{2} + d)^{2^{k-1}}$ (with d the maximum degree of the input basis and k the number of indeterminates) for the degree of polynomials in a reduced Gröbner basis, employing only combinatorial arguments.

Applying a result of Mayr and Meyer in [MM82] showing that commutative Thue systems can count up to double-exponentially large integers, Huynh

[Huy86] derived a doubly exponential lower bound (in the size of the problem instance) for the maximal degree of the elements of Gröbner bases as well as for the cardinality of such bases.

In this paper, we make use of the close relationship between commutative semigroups and binomial ideals (for an investigation of the algebraic structure of binomial ideals see [EiSt94]). Based on the algorithm in [MM82] for the uniform word problem in commutative semigroups we derive an exponential space algorithm for constructing the reduced Gröbner basis of a binomial ideal. By the results in [MM82] and [Huy86], this algorithm is space optimal.

This paper is organized as follows. In Sect. 2 we briefly introduce the basic notations and formalisms. In Sect. 3 we exhibit the close relationship between commutative semigroups and binomial ideals and give a characterization of the reduced Gröbner basis for binomial ideals. In the following section we describe the exponential space Gröbner basis algorithm and, in Sect. 5, show two applications of this algorithm: we derive a procedure enumerating the elements of finite congruence classes in commutative semigroups and an algorithm for the general subword problem in commutative semigroups. Both algorithms use exponential space and are space optimal.

2 Basic Concepts and Notations

Let X denote the finite set $\{x_1, \ldots, x_k\}$, and[1] $\mathbb{Q}[X]$ the (commutative) ring of polynomials with indeterminates $x_1, \ldots, x_k$ and rational coefficients. A *term* t in $x_1, \ldots, x_k$ is a product of the form $t = x_1^{e_1} \cdot x_2^{e_2} \cdots x_k^{e_k}$, with $(e_1, e_2, \ldots, e_k) \in \mathbb{N}^k$ the *degree vector* of t. By the *degree* $\deg(t)$ of a term t we shall mean the integer $e_1 + e_2 + \ldots + e_k$. Each *polynomial* $f(x_1, \ldots, x_k) \in \mathbb{Q}[X]$ is a finite sum $f(x_1, \ldots, x_k) = \sum_{i=1}^{n} a_i \cdot t_i$, with $a_i \in \mathbb{Q} - \{0\}$ the coefficient of the ith term t_i of f. The product $m_i = a_i \cdot t_i$ is called the ith *monomial* of the polynomial f. The degree of a polynomial is the maximum of the degrees of its terms. For $f_1, \ldots, f_h \in \mathbb{Q}[X]$, $\langle f_1, \ldots, f_h \rangle \subseteq \mathbb{Q}[X]$ denotes the ideal generated by $\{f_1, \ldots, f_h\}$ that is[2]

$$\langle f_1, \ldots, f_h \rangle := \left\{ \sum_{i=1}^{h} p_i f_i; \ p_i \in \mathbb{Q}[X] \text{ for } i \in I_h \right\}.$$

If $I = \langle f_1, \ldots, f_h \rangle$, $\{f_1, \ldots, f_h\}$ is called a *basis* of I.

An *admissible term ordering* $\prec$ on $\mathbb{Q}[X]$ is given by any admissible order on $\mathbb{N}^k$, i.e., any total order $<$ on $\mathbb{N}^k$ satisfying the following two conditions:

(T1) $e > (0, \ldots, 0)$ for all $e \in \mathbb{N}^k - \{(0, \ldots, 0)\}$;

(T2) $a < b \ \Rightarrow \ a + c < b + c$ for all $a, b, c \in \mathbb{N}^k$.

If $(d_1, \ldots, d_k) > (e_1, \ldots, e_k)$, we say that any monomial $a_1 \cdot x_1^{d_1} \cdots x_k^{d_k}$, $a_1 \in \mathbb{Q} - \{0\}$, is greater in the term ordering than any monomial $a_2 \cdot x_1^{e_1} \cdots x_k^{e_k}$, $a_2 \in \mathbb{Q} - \{0\}$ (written $a_1 \cdot x_1^{d_1} \cdots x_k^{d_k} \succ a_2 \cdot x_1^{e_1} \cdots x_k^{e_k}$).

---

[1] $\mathbb{Q}$ denotes the set of rationals, and $\mathbb{N}$ the set of nonnegative integers.
[2] For $n \in \mathbb{N}$, I_n denotes the set $\{1, \ldots, n\}$.

For a polynomial $f(x_1, \ldots, x_k) = \sum_{i=1}^{n} a_i \cdot t_i$ we always assume that $t_1 \succ t_2 \succ \ldots \succ t_n$. For any such nonzero polynomial $f \in \mathbb{Q}[X]$ we define the *leading term* $LT(f) := t_1$.

For the sake of constructiveness, we assume that the term ordering is given as part of the input by a $k \times k$ integer matrix T such that $a_1 \cdot x_1^{d_1} \cdots x_k^{d_k} \succ a_2 \cdot x_1^{e_1} \cdots x_k^{e_k}$ iff, for the corresponding degree vectors d and e, Td is *lexicographically greater* than Te (see [Rob85, Wei87]).

Let I be an ideal in $\mathbb{Q}[X]$, and let some admissible term ordering $\prec$ on $\mathbb{Q}[X]$ be given. A finite set $\{g_1, \ldots, g_r\}$ of polynomials from $\mathbb{Q}[X]$ is called a *Gröbner basis* of I (w.r.t. $\prec$), if

(G1) $\{g_1, \ldots, g_r\}$ is a basis of I;

(G2) $\{LT(g_1), \ldots, LT(g_r)\}$ is a basis of the *leading term ideal* of I, which is the smallest ideal containing the leading terms of all $f \in I$, or equivalently: if $f \in I$, then $LT(f) \in \langle LT(g_1), \ldots, LT(g_r) \rangle$.

A Gröbner basis is called *reduced* if no monomial in any one of its polynomials is divisible by the leading term of any other polynomial in the basis.

For a finite alphabet $X = \{x_1, \ldots, x_k\}$, let X^* denote the free commutative monoid generated by X. An element u of X^* is called a *(commutative) word*. For a word the order of the symbols is immaterial, and we shall in the sequel use an exponent notation: $u = x_1^{e_1} \ldots x_k^{e_k}$, where[3] $e_i = \Phi(u, x_i) \in \mathbb{N}$ for $i = 1, \ldots, k$. We identify any $u \in X^*$ (resp., the corresponding vector $u = (\Phi(u, x_1), \ldots, \Phi(u, x_k)) \in \mathbb{N}^k$) with the term $u = x_1^{\Phi(u, x_1)} \cdot x_2^{\Phi(u, x_2)} \cdots x_k^{\Phi(u, x_k)}$ and vice versa.

Let $\mathcal{P} = \{l_i \equiv r_i; \ i \in I_h\}$ be some (finite) commutative semigroup presentation with $l_i, r_i \in X^*$ for $i \in I_h$. We say a word $v \in X^*$ *is derived in one step* from $u \in X^*$ (written $u \to v$ ($\mathcal{P}$)) by application of the congruence $(l_i \equiv r_i) \in \mathcal{P}$ iff, for some $w \in X^*$, we have $u = wl_i$ and $v = wr_i$, or $u = wr_i$ and $v = wl_i$ (note, since '$\equiv$' is symmetric, '$\to$' is symmetric, i.e., $u \to v$ ($\mathcal{P}$) $\Leftrightarrow v \to u$ ($\mathcal{P}$)). The word u *derives* v, written $u \equiv v \mod \mathcal{P}$, iff $u \xrightarrow{*} v$ ($\mathcal{P}$), where $\xrightarrow{*}$ is the reflexive transitive closure of $\to$. More precisely we write $u \xrightarrow{+} v$ ($\mathcal{P}$), where $\xrightarrow{+}$ is the transitive closure of $\to$, if $u \xrightarrow{*} v$ ($\mathcal{P}$) and $u \neq v$. A sequence $(u_0, \ldots, u_n)$ of words $u_i \in X^*$ with $u_i \to u_{i+1}$ ($\mathcal{P}$) for $i = 0, \ldots, n-1$, is called a *derivation* (of length n) of u_n from u_0 in $\mathcal{P}$.

By $I(\mathcal{P})$ we denote the binomial $\mathbb{Q}[X]$-ideal generated by $\{l_1 - r_1, \ldots, l_h - r_h\}$, i.e.,

$$I(\mathcal{P}) := \left\{ \sum_{i=1}^{h} p_i(l_i - r_i); \ p_i \in \mathbb{Q}[X] \text{ for } i \in I_h \right\}.$$

We call such an ideal a *binomial ideal*, i.e., each polynomial in the basis is the difference of two terms. By looking at Buchberger's algorithm [Buc65] it is not hard to see that the reduced Gröbner basis of a binomial ideal still consists only of binomials.

---

[3] Let Φ be the Parikh mapping, i.e., $\Phi(u, x_i)$ (also written $(\Phi(u))_i$) indicates, for every $u \in X^*$ and $i \in \{1, \ldots, k\}$, the number of occurrences of $x_i \in X$ in u.

3 Commutative Semigroups and Binomial Ideals

3.1 The Basic Problems and Their Relationship

The *uniform word problem* for commutative semigroups is the problem of deciding for a commutative semigroup presentation $\mathcal{P}$ over some alphabet X, and two words $u, v \in X^*$ whether $u \equiv v \bmod \mathcal{P}$. The *polynomial ideal membership problem* (PIMP) is the problem of deciding for given polynomials $f, f_1, \ldots, f_h \in \mathbb{Q}[X]$ whether $f \in \langle f_1, \ldots, f_h \rangle$.

Proposition 1. [MM82] *Let* $X = \{x_1, \ldots, x_k\}$, $\mathcal{P} = \{l_i \equiv r_i;\ l_i, r_i \in X^*, i \in I_h\}$, *and* $u, v \in X^*$. *Then the following are equivalent:*

(i)   *There exist* $p_1, \ldots, p_h \in \mathbb{Q}[X]$ *such that* $\quad v - u = \sum_{i=1}^h p_i(l_i - r_i)$.

(ii) *There is a derivation* $u = \gamma_1 \to \gamma_2 \to \ldots \to \gamma_n = v\ (\mathcal{P})$ *of* v *from* u *such that for* $j \in I_n \quad \deg(\gamma_j) \leq \max\{\deg(l_i p_i),\ \deg(r_i p_i);\ i \in I_h\}$.

(iii) $u \equiv v \bmod \mathcal{P}$.

In the fundamental paper [Her26], G. Hermann gave a doubly exponential degree bound for the polynomial ideal membership problem:

Proposition 2. [Her26] *Let* $X = \{x_1, \ldots, x_k\}$, $g, g_1, \ldots, g_h \in \mathbb{Q}[X]$, *and* $d := \max\{\deg(g_i);\ i \in I_h\}$. *If* $g \in \langle g_1, \ldots, g_h \rangle$, *then there exist* $p_1, \ldots, p_h \in \mathbb{Q}[X]$ *such that*

(i) $g = \sum_{i=1}^h g_i p_i$;

(ii) $(\forall i \in I_h)\, [\deg(p_i) \leq \deg(g) + (hd)^{2^k}]$.

By $\mathrm{size}(\cdot)$ we shall denote the number of bits needed to encode the argument in some standard way (using radix representation for numbers).

Then the above two propositions yield an exponential space upper bound for the uniform word problem for commutative semigroups:

Proposition 3. [MM82] *Let* $X = \{x_1, \ldots, x_k\}$ *and* $\mathcal{P} = \{l_i \equiv r_i;\ l_i, r_i \in X^*, i \in I_h\}$. *Then there is a (deterministic) Turing machine* M *and some constant* $c > 0$ *independent of* $\mathcal{P}$, *such that* M *decides for any two words* $u, v \in X^*$ *whether* $u \equiv v \bmod \mathcal{P}$ *using at most space* $(\mathrm{size}(u, v, \mathcal{P}))^2 \cdot 2^{c \cdot k}$.

3.2 The Reduced Gröbner Basis for Binomial Ideals

Let $\mathcal{P}$ be a commutative semigroup presentation over some alphabet X.

Theorem 4. *Let* $X = \{x_1, \ldots, x_k\}$, $\mathcal{P} = \{l_i \equiv r_i;\ l_i, r_i \in X^*, i \in I_h\}$, *and* $G = \{h_1 - m_1, \ldots, h_r - m_r\}$ *the reduced Gröbner basis of the ideal* $I(\mathcal{P})$ *w.r.t. some admissible term ordering* $\prec$ $(m_i \prec h_i)$. *Then* m_i *is the minimal element (w.r.t.* $\prec$) *of the congruence class* $[h_i]_\mathcal{P}$, $i \in I_r$.

Proof. Assume that $w \neq m_i$ is the minimal element of $[h_i]_{\mathcal{P}}$ (w.r.t. $\prec$). Then $w \prec m_i$ and $m_i - w \in I(\mathcal{P})$. Since G is a Gröbner basis of $I(\mathcal{P})$, $m_i \in \langle h_1, \ldots, h_r \rangle$, i.e., there must be some $j \in I_r$ such that h_j divides m_i. But this contradicts the fact that $h_i - m_i$ is an element of the reduced Gröbner basis of $I(\mathcal{P})$. □

Theorem 5. *Let* $X = \{x_1, \ldots, x_k\}$, $\mathcal{P} = \{l_i \equiv r_i; \ l_i, r_i \in X^*, i \in I_h\}$, *and* $G = \{h_1 - m_1, \ldots, h_r - m_r\}$ *the reduced Gröbner basis of the ideal* $I(\mathcal{P})$ *w.r.t. some admissible term ordering* $\prec$ *($m_i \prec h_i$). Then* $LT(I(\mathcal{P}))$ *(the set of the leading terms of* $I(\mathcal{P})$*) is the set of all terms with nontrivial congruence class which are not the minimal element in their congruence class w.r.t.* $\prec$. $H = \{h_1, \ldots, h_r\}$ *is the set of the minimal elements of* $LT(I(\mathcal{P}))$ *w.r.t. divisibility.*

Proof. Since G is the reduced Gröbner basis of $I(\mathcal{P})$, it is clear that H is the set of minimal elements of $LT(I(\mathcal{P}))$ w.r.t. divisibility.
Since $h_i - m_i \in I(\mathcal{P})$, there is a derivation in $\mathcal{P}$ of $m_i \prec h_i$ from h_i, for all $i \in I_r$. Because G is a Gröbner basis, for any $h \in LT(I(\mathcal{P}))$ there is an $h_j \in H$ and a term t in X with $h = t \cdot h_j$. Thus, for any $h \in LT(I(\mathcal{P}))$, the congruence class $[h]_{\mathcal{P}}$ is nontrivial, and h is not the minimal element in $[h]_{\mathcal{P}}$.
Let $s \in X^*$ be a term with nontrivial congruence class. If s is not the minimal element m_s (w.r.t. $\prec$) of its congruence class $[s]_{\mathcal{P}}$, then s derives m_s, and thus, $s - m_s \in I(\mathcal{P})$, i.e., $s \in LT(I(\mathcal{P}))$. If $s = m_s$, then there is no derivation of any $t_s \prec s$ from s. Thus, no $h_j \in H$ can divide s. □

4 An Optimal Algorithm for the Reduced Gröbner Basis

In this section we give an exponential space algorithm for generating the reduced Gröbner basis of a binomial ideal. To determine the complexity of the algorithm we need the results of Sect. 3 and the following upper bound for the total degree of polynomials required in a Gröbner basis, obtained by Dubé in [Dub90].

Proposition 6. *[Dub90] Let* $F = \{f_1, \ldots, f_h\} \subset \mathbb{Q}[X]$, $I = \langle f_1, \ldots, f_h \rangle$ *the ideal generated by* F*, and let* d *be the maximum degree of any* $f \in F$*. Then for any admissible term ordering* $\prec$ *on* $\mathbb{Q}[X]$*, the degree of polynomials required in a Gröbner basis for* I *w.r.t.* $\prec$ *is bounded by* $2 \cdot \left(\frac{d^2}{2} + d\right)^{2^{k-1}}$.

Now we will generate the reduced Gröbner basis of the binomial ideal $I(\mathcal{P})$ w.r.t. some fixed admissible term ordering $\prec$, where $X = \{x_1, \ldots, x_k\}$, and $\mathcal{P} = \{l_i \equiv r_i; \ l_i, r_i \in X^*, i \in I_h\}$ (w.l.o.g. $l_i \succ r_i$). Let H denote the set $\{h_1, \ldots, h_r\}$ of the minimal elements of $LT(I(\mathcal{P}))$ w.r.t. divisibility, and m_i the minimal element of $[h_i]_{\mathcal{P}}$, $i \in I_r$, w.r.t. $\prec$. From Theorems 4 and 5 we know that the set $G = \{h_1 - m_1, \ldots, h_r - m_r\}$ is the reduced Gröbner basis of $I(\mathcal{P})$.

First we establish some technical details. For detailed proofs see [KM96].

The minimal element w.r.t. $\prec$ of a nontrivial congruence class $[u]_{\mathcal{P}}$, $u \in X^*$, is of the form $t \cdot r_i$ with $r_i \in \{r_1, \ldots, r_h\}$, $t \in X^*$. For $h = x_1^{e_1} \cdots x_k^{e_k} \in X^*$ and $i \in I_k$ such that $e_i > 0$, define $h^{(i)} := x_1^{e_1} \cdots x_i^{e_i-1} \cdots x_k^{e_k}$. Then H consists exactly of those terms $h \in X^*$ which have degree $\leq 2 \cdot \left(\frac{d^2}{2} + d\right)^{2^{k-1}}$, which are congruent to some term $t \cdot r_i \prec h$ with $r_i \in \{r_1, \ldots, r_h\}$, $t \in X^*$, and, by Proposition 6, $\deg(t \cdot r_i) \leq 2 \cdot \left(\frac{d^2}{2} + d\right)^{2^{k-1}}$, and for which, for all applicable i, $[h^{(i)}]_{\mathcal{P}}$ is trivial. By Proposition 3, the condition regarding the reducibility of h can be checked in space $(\text{size}(\mathcal{P}))^2 \cdot 2^{c \cdot k}$. Testing non-reducibility of the $h^{(i)}$ can also be done in exponential space because of Proposition 3 and

Lemma 7. [KM96] *For a term $u \in X^*$ with $\deg(u) \leq D$, $[u]_{\mathcal{P}}$ is non-trivial iff there is some $t \cdot r_i$ with $t \cdot r_i \prec u$, $r_i \in \{r_1, \ldots, r_h\}$, $t \in X^*$, and $\deg(t \cdot r_i) \leq D + 2 \cdot \left(\frac{d^2}{2} + d\right)^{2^{k-1}}$ such that $u \xrightarrow{+} t \cdot r_i$ $(\mathcal{P})$.*

From this, we derive the exponential space algorithm given in Fig. 1.

The Algorithm

Input: $\mathcal{P} = \{l_1 - r_1, \ldots, l_h - r_h\}$, admissible term ordering $\prec$
Output: the reduced Gröbner basis $G = \{h_1 - m_1, \ldots, h_r - m_r\}$ of $I(\mathcal{P})$

$d := \max\{\deg(l_i), \deg(r_i); i \in I_h\}$; $G := \emptyset$;

for each $h \in X^*$ with degree $\leq 2 \cdot \left(\frac{d^2}{2} + d\right)^{2^{k-1}}$ **do**   /* $h = x_1^{e_1} \cdots x_k^{e_k}$ */

  $D := \deg(h)$;

  $m :=$ the term $t \cdot r_i$ with $t \cdot r_i \prec h, r_i \in \{r_1, \ldots, r_h\}, t \in X^*, \deg(t \cdot r_i) \leq 2 \cdot \left(\frac{d^2}{2} + d\right)^{2^{k-1}}$

     which is $\equiv h$ and, among these terms, minimal w.r.t. $\prec$;

  if m exists **then**   /* $h \in LT(I(\mathcal{P}))$ */

    $minred := $ **true**;

    for each $i \in I_k$ with $e_i \geq 1$ **while** $minred$ **do**

      $h' := x_1^{e_1} \cdots x_i^{e_i - 1} \cdots x_k^{e_k}$;

      $m' :=$ any term $t \cdot r_i$ with $t \cdot r_i \prec h', r_i \in \{r_1, \ldots, r_h\}, t \in X^*, \deg(t \cdot r_i) \leq$
         $(D-1) + 2 \cdot \left(\frac{d^2}{2} + d\right)^{2^{k-1}}$ which is $\equiv h'$;

      if m' exists **then** $minred := $ **false** **end_if**

    end_for

    if $minred$ **then** $G := G \cup \{h - m\}$ **end_if**

  end_if

end_for

Fig. 1. Algorithm for constructing the reduced Gröbner basis of a binomial ideal

Theorem 8. *Let $X = \{x_1, \ldots, x_k\}$, $\mathcal{P} = \{l_i \equiv r_i; l_i, r_i \in X^*, i \in I_h\}$, and $\prec$ be some admissible term ordering. Then there is an algorithm which generates the reduced Gröbner basis $G = \{h_1 - m_1, \ldots, h_r - m_r\}$ of the binomial ideal $I(\mathcal{P})$ using at most space $(\text{size}(\mathcal{P}))^2 \cdot 2^{\bar{c} \cdot k} \leq 2^{c \cdot \text{size}(\mathcal{P})}$, where $\bar{c}, c > 0$ are some constants independent of $\mathcal{P}$.*

From the results in [Huy86] we know that, in the worst case, any Gröbner basis of $I(\mathcal{P})$ has maximal degree at least $2^{2^{c \cdot \text{size}(\mathcal{P})}}$ for some constant $c > 0$ independent of $\mathcal{P}$. Hence, any algorithm that computes Gröbner bases requires at least exponential space in the worst case.

5 Applications

We present space optimal decision procedures for the finite enumeration and subword problems for commutative semigroups.

5.1 The Finite Enumeration Problem for Commutative Semigroups

Let $\mathcal{P}$ be a finite commutative semigroup presentation over some alphabet X, and $u \in X^*$ a word such that the congruence class of u is bounded. Then the finite enumeration problem for commutative semigroups, or equivalently, reversible Petri nets is the problem of generating a complete list of all the elements of $[u]_{\mathcal{P}}$. We give a procedure for the solution of this problem which needs at most exponential work space.

Theorem 9. *Let* $X = \{x_1, \ldots, x_k\}$, $\mathcal{P} = \{l_i \equiv r_i;\ l_i, r_i \in X^*, i \in I_h\}$ *be a finite commutative semigroup presentation over* X, *and* $u \in X^*$ *a word such that the congruence class of* u *is bounded. Then there is an algorithm which generates the elements of* $[u]_{\mathcal{P}}$ *using at most space* $(\text{size}(u, \mathcal{P}))^2 \cdot 2^{\bar{c} \cdot k} \leq 2^{c \cdot \text{size}(u, \mathcal{P})}$, *where* $\bar{c}$, $c > 0$ *are some constants independent of* u *and* $\mathcal{P}$.

Proof. In addition to $x_1, \ldots, x_k$ we introduce $2k + 3$ new variables m, s, t, $y_1, \ldots, y_k$, and $z_1, \ldots, z_k$. Let $X' = X \cup \{m, s, t, y_1, \ldots, y_k, z_1, \ldots, z_k\}$. Given $\mathcal{P}$ and the word $u \in X^*$, we construct a new commutative semigroup presentation $\mathcal{P}'$ over X' as follows: $\mathcal{P}'$ contains the congruences

$$s \cdot x_j \equiv s \cdot y_j \cdot z_j, \quad \text{for} \quad j = 1, \ldots, k, \tag{1}$$

$$s \cdot y(u) \equiv t, \tag{2}$$

$$s \cdot u \equiv m, \tag{3}$$

and, for every congruence $l_i \equiv r_i$ in $\mathcal{P}$, the congruences

$$s \cdot y(l_i) \equiv s \cdot y(r_i), \quad \text{and} \tag{4}$$

$$t \cdot z(l_i) \equiv t \cdot z(r_i), \tag{5}$$

where y (resp., z) are the homomorphisms replacing x_j by y_j (resp., z_j).
Let $\prec$ be a lexicographic term ordering satisfying
$$m \prec a \prec s \prec b \quad \text{for all} \quad a \in \{x_1, \ldots, x_k\}, b \in \{t, y_1, \ldots, y_k, z_1, \ldots, z_k\}.$$
In the following we prove that $v \in [u]_{\mathcal{P}}$ iff $s \cdot v - m \in G$, where G is the reduced Gröbner basis of the ideal $I(\mathcal{P}')$ w.r.t. $\prec$. Then, by Theorem 8, the elements of $[u]_{\mathcal{P}}$ can be generated using at most space $(\text{size}(u, \mathcal{P}'))^2 \cdot 2^{d' \cdot k} \leq (\text{size}(u, \mathcal{P}))^2 \cdot 2^{d \cdot k}$, where d', $d > 0$ are some constants independent of u and $\mathcal{P}'$ (resp., $\mathcal{P}$).
First we establish some technical details.

Lemma 10. *Every word $w \in [s \cdot u]_{\mathcal{P}'}$ satisfies the following conditions:*

(i) $\Phi(w, s) + \Phi(w, t) + \Phi(w, m) = 1;$

(ii) if $\Phi(w, s) = 1$,

then $x_1^{\Phi(w,x_1)+\Phi(w,y_1)} \cdot x_2^{\Phi(w,x_2)+\Phi(w,y_2)} \cdots x_k^{\Phi(w,x_k)+\Phi(w,y_k)} \in [u]_{\mathcal{P}},$

$x_1^{\Phi(w,x_1)+\Phi(w,z_1)} \cdot x_2^{\Phi(w,x_2)+\Phi(w,z_2)} \cdots x_k^{\Phi(w,x_k)+\Phi(w,z_k)} \in [u]_{\mathcal{P}};$

if $\Phi(w, t) = 1$, then $\Phi(w, x_1) = \Phi(w, x_2) = \ldots = \Phi(w, x_k) = 0,$

$$\Phi(w, y_1) = \Phi(w, y_2) = \ldots = \Phi(w, y_k) = 0,$$

$$x_1^{\Phi(w,z_1)} \cdot x_2^{\Phi(w,z_2)} \cdots x_k^{\Phi(w,z_k)} \in [u]_{\mathcal{P}}.$$

Proof. Let w be any word in $[s \cdot u]_{\mathcal{P}'}$. Then there is a repetition-free derivation in $\mathcal{P}'$ leading from $s \cdot u$ to w. If $w = m$, then w is derived in one step from $s \cdot u$ by congruence (3) and w trivially satisfies conditions (i) and (ii). Note that if in a derivation starting at $s \cdot u$ congruence (3) is applied, then this derivation can only be continued by again using congruence (3), causing a repetition. If $w \neq m$, then in any repetition-free derivation starting at $s \cdot u$ leading to w only the congruences in (1) and (4) can be applied until the word $s \cdot y(u) \cdot z(u)$ is reached and changed to $t \cdot z(u)$ by congruence (2). Since $[u]_{\mathcal{P}}$ is bounded, there is no $u' \in \{y_1, \ldots, y_k\}^*$ with $s \cdot u' \cdot z(u) \in [s \cdot u]_{\mathcal{P}'}$, $u' \neq y(u)$, and $y(u)$ divides u'. Therefore, any word w occurring in this derivation of $s \cdot y(u) \cdot z(u)$ from $s \cdot u$ satisfies conditions (i) and (ii).

Then, as long as congruence (2) is not applied, by the congruences in (5) words $t \cdot z(v)$ with $v \in [u]_{\mathcal{P}}$ can be derived from $t \cdot z(u)$. Note that for all such words $t \cdot z(v)$ with $v \in [u]_{\mathcal{P}}$ $\Phi(t \cdot z(v), s) = 0$, $\Phi(t \cdot z(v), t) = 1$, $\Phi(t \cdot z(v), m) = 0$, and condition (ii) is satisfied. Congruence (2) changes $t \cdot z(v)$ to $s \cdot y(u) \cdot z(v)$ and again the congruences in (1) and (4) can be applied. As above, the words w in the resulting sub-derivation starting at $s \cdot y(u) \cdot z(v)$ satisfy (i) and (ii) with

$$x_1^{\Phi(w,x_1)+\Phi(w,z_1)} \cdot x_2^{\Phi(w,x_2)+\Phi(w,z_2)} \cdots x_k^{\Phi(w,x_k)+\Phi(w,z_k)} = v.$$

By the congruences in (4), from $s \cdot y(u) \cdot z(v)$ any word $s \cdot y(v') \cdot z(v)$ with $v' \in [u]_{\mathcal{P}}$ can be derived. Congruence (2) can only be applied to the word $s \cdot y(u) \cdot z(v)$, causing a repetition. $\quad\square$Lemma 10

Lemma 11. *Let $s \cdot v \in [s \cdot u]_{\mathcal{P}'}$ with $v \in X^*$, $v \neq u$, and let $s \cdot u = \gamma_0 \to \gamma_1 \to \ldots \to \gamma_n = s \cdot v$ be any repetition-free derivation in $\mathcal{P}'$ leading from $s \cdot u$ to $s \cdot v$. Then, there is exactly one $i \in I_{n-1}$ with $\gamma_i = s \cdot y(u) \cdot z(u)$, $\gamma_{i+1} = t \cdot z(u)$, and exactly one $j \in I_{n-1}$, $j > i$, with $\gamma_j = t \cdot z(v)$, $\gamma_{j+1} = s \cdot y(u) \cdot z(v)$.*

Lemma 12. *Let v be some word in X^*, then* $v \in [u]_{\mathcal{P}} \iff s \cdot v \in [s \cdot u]_{\mathcal{P}'}$ *.*

Proof. Follows immediately from Lemmas 10 and 11. $\quad\square$Lemma 12

Lemma 13. $[s \cdot u]_{\mathcal{P}'}$ *is bounded.*

Proof. Immediate from the definition of $\mathcal{P}'$. $\quad\square$Lemma 13

Lemma 14. *Let v be some word in X^* with $v \notin [u]_{\mathcal{P}}$, and v divides some $u' \in [u]_{\mathcal{P}}$. Then $s \cdot v$ is the minimal (w.r.t. $\prec$) element of $[s \cdot v]_{\mathcal{P}'}$.*

Proof. If $v \in X^*$ with $v \notin [u]_\mathcal{P}$, and v divides some $u' \in [u]_\mathcal{P}$, then there is some $v' \in X^* - \{\varepsilon\}$ with $u' = v \cdot v' \in [u]_\mathcal{P}$. Because of the boundedness of $[u]_\mathcal{P}$ there is no $\bar{v} \in [v]_\mathcal{P}$ with $\bar{v} = u \cdot \bar{u}$ for $\bar{u} \in X^*$, since otherwise $[u]_\mathcal{P}$ would be unbounded. Thus, in any derivation starting at $s \cdot v$ the congruences (2) and (3) can not be applied. Only the congruences in (1) and (4) can possibly be used. Since $y_i \succ x_i$ (resp., $z_i \succ x_i$) for all $i \in I_k$, $s \cdot v$ is the minimal element of $[s \cdot v]_{\mathcal{P}'}$ w.r.t. $\prec$.

$\square$Lemma 14

Note that each $v \in X^*$ is the minimal (w.r.t. $\prec$) element of $[v]_{\mathcal{P}'}$ because no congruence in $\mathcal{P}'$ is applicable.

Since $[s \cdot u]_{\mathcal{P}'}$ is bounded, it follows from Dickson's Lemma [Dic13] that each $w \in [s \cdot u]_{\mathcal{P}'}$ is minimal in $[s \cdot u]_{\mathcal{P}'}$ w.r.t. divisibility. The minimal element w.r.t. $\prec$ of $[s \cdot u]_{\mathcal{P}'}$ is m. Thus, by Lemma 14, each $s \cdot v \in [s \cdot u]_{\mathcal{P}'}$ with $v \in X^*$ is contained in the set of the minimal elements of $LT(I(\mathcal{P}'))$ w.r.t. divisibility, and hence $G \supseteq \{s \cdot v - m \mid s \cdot v \in [s \cdot u]_{\mathcal{P}'}, v \in X^*\}$ (see Theorems 4 and 5).

$\square$Theorem 9

Theorem 15. *The finite enumeration problem for commutative semigroups is exponential space complete with respect to log-lin reducibility.*

From the work in [MM82] we know that the uniform word problem for commutative semigroups is exponential space complete (the input consisting of u, v and $\mathcal{P}$). Actually, the construction in [MM82] proves the following, slightly stronger statement, from which the proof of Theorem 15 easily follows.

Proposition 16. [MM82] *Let $\mathcal{P}$ be a finite commutative semigroup presentation over X, v a word in X^*, and $u \in X^*$ a word such that $[u]_\mathcal{P}$ is bounded. Even with this restriction, the uniform word problem, i.e., the problem of deciding whether $u \equiv v \bmod \mathcal{P}$, is exponential space complete with respect to log-lin reducibility.*

5.2 The Subword Problem for Commutative Semigroups

Let $X = \{x_1, \ldots, x_k\}$ be a finite alphabet, and $\mathcal{P} = \{l_i \equiv r_i; l_i, r_i \in X^*, i \in I_h\}$ a finite commutative semigroup presentation over X. Fix some $v_1 \in X_{v_1} := \{x_1, \ldots, x_l\}$ ($l \leq k$) and, if $l < k$, let $X_{\overline{v_1}} = \{x_{l+1}, \ldots, x_k\}$ be the set of variables not occurring in v_1.
Let Y be the subset $\{x_{l_1}, \ldots, x_{l_2}\}$ of X with $l_2 \geq l$ (if $l_1 > l_2$ then $Y = \emptyset$). Similarly, Z is the subset $\{x_{l_3}, \ldots, x_k\}$ of X with $l_2 < l_3$, and $Z = \emptyset$ if $l_3 > k$. Then, for the case $l_1 < l < l_2 < l_3 < k$ we get the following picture:

$$
\overbrace{x_1, \ldots, x_{l_1-1}, \underbrace{x_{l_1}, \ldots, x_l,}_{} }^{X_{v_1}} \overbrace{x_{l+1}, \ldots, x_{l_2},}_{} \overbrace{x_{l_2+1}, \ldots, x_{l_3-1}, \underbrace{x_{l_3}, \ldots, x_k}_{Z}}^{X_{\overline{v_1}}}
$$
$$
\underbrace{\phantom{x_{l_1}, \ldots, x_l, x_{l+1}, \ldots, x_{l_2},}}_{Y}
$$

Definition. The *Subword Problem* is: Given X, $\mathcal{P}$, u, v_1, Y, and Z, decide whether there is a $v_2 \in [u]_\mathcal{P}$ such that $v_2 = v_1 \cdot x_{l_1} \cdots x_{l_2} \cdot w$ for some $w \in (Y \cup Z)^*$.

Theorem 17. *Let $X = \{x_1, \ldots, x_k\}$, and $\mathcal{P} = \{l_i \equiv r_i;\ l_i, r_i \in X^*, i \in I_h\}$ be a commutative semigroup presentation over X. Then there is an algorithm which decides for any two words u, $v_1 \in X^*$, and sets Y, $Z \subseteq X$ defined as above whether there is a $v_2 \in [u]_{\mathcal{P}}$ such that $v_2 = v_1 \cdot v \cdot w$, where $w \in (Y \cup Z)^*$, and $v = x_{l_1} \cdots x_{l_2}$ if $Y = \{x_{l_1}, \ldots, x_{l_2}\}$ resp., $v = \varepsilon$ if $Y = \emptyset$, using at most space $(\mathrm{size}(u, v_1, \mathcal{P}))^2 \cdot 2^{\bar{c} \cdot k} \le 2^{c \cdot \mathrm{size}(u, v_1, \mathcal{P})}$ for some constants $\bar{c}$, $c > 0$ independent of u, v_1 and $\mathcal{P}$.*

Proof. We show that if there is a $v_2' \in [u]_{\mathcal{P}}$ as described in the Theorem, then there is a $v_2 \in [u]_{\mathcal{P}}$ with the same properties as v_2' and v_2 can be determined in space $(\mathrm{size}(u, v_1, \mathcal{P}))^2 \cdot 2^{\bar{c} \cdot k}$.

In addition to $x_1, \ldots, x_k$ we introduce three new variables s, $\bar{s}$, and t. Let $X_t = X \cup \{s, \bar{s}, t\}$. Given $\mathcal{P}$ and the two words u, $v_1 \in X^*$, we construct a new commutative semigroup presentation $\mathcal{P}_t$ over X_t as follows: For every congruence $l_i \equiv r_i$ in $\mathcal{P}$, $\mathcal{P}_t$ contains the congruence $t \cdot l_i \equiv t \cdot r_i$. Then we add to $\mathcal{P}_t$ the congruences $s \equiv t \cdot u$, and $t \cdot v_1 \cdot v \equiv \bar{s}$.

Let $\prec$ be any lexicographic term ordering satisfying
$$s \succ t \succ x_{v_1} \succ x_{\overline{v_1}} \succ \bar{s} \succ y \succ z,$$
for all $x_{v_1} \in X_{v_1} - (Y \cup Z)$, $x_{\overline{v_1}} \in X_{\overline{v_1}} - (Y \cup Z)$, $y \in Y$ and $z \in Z$.

By Theorem 5, $s \in LT(I(\mathcal{P}_t))$, and, since s is minimal in $LT(I(\mathcal{P}_t))$ w.r.t. divisibility, $s \in H_t$, where H_t is the set of the minimal elements of $LT(I(\mathcal{P}_t))$ w.r.t. divisibility. By Theorems 4 and 5, $s - m_s \in G$, where G is the reduced Gröbner basis of $I(\mathcal{P}_t)$, and m_s is the minimal element of $[s]_{\mathcal{P}_t}$ w.r.t. $\prec$.

Because we assume that there is $v_2' \in [u]_{\mathcal{P}}$ such that $v_2' = v_1 \cdot v \cdot w'$ for some $w' \in (Y \cup Z)^*$, it follows that $t \cdot v_1 \cdot v \cdot w' \in [t \cdot u]_{\mathcal{P}_t}$. As $t \cdot v_1 \cdot v \cdot w' \equiv \bar{s} \cdot w' \bmod \mathcal{P}_t$, we have $\bar{s} \cdot w' \in [t \cdot u]_{\mathcal{P}_t}$. Since m_s is the minimal element of $[s]_{\mathcal{P}_t} = [t \cdot u]_{\mathcal{P}_t}$, we also have $m_s \prec \bar{s} \cdot w'$ or $m_s = \bar{s} \cdot w'$. In particular, the variables s, t, and the variables in $X_{v_1} - (Y \cup Z)$, $X_{\overline{v_1}} - (Y \cup Z)$ do not occur in m_s.

By induction, it can easily be seen that every word γ in a derivation in $\mathcal{P}_t$ starting at s satisfies the invariant
$$\Phi(\gamma, s) + \Phi(\gamma, \bar{s}) + \Phi(\gamma, t) = 1.$$
Also, the words γ_i in a repetition-free derivation
$$s \to t \cdot u = \gamma_0 \to \gamma_1 \to \cdots \to \gamma_{n-1} \to \gamma_n \to m_s\ (\mathcal{P}_t)$$
with $n \in \mathbb{N}$, can not contain s or $\bar{s}$. The only congruences applicable to γ_i, $i \in \{0, \ldots, n-1\}$ are the congruences $t \cdot l_i \equiv t \cdot r_i$. Thus, any repetition-free derivation in $\mathcal{P}_t$ leading from s to m_s has the form
$$s \to t \cdot u \to t \cdot \delta_1 \to \cdots \to t \cdot \delta_n = t \cdot v_1 \cdot v \cdot w \to \bar{s} \cdot w = m_s\ (\mathcal{P}_t)$$
with $n \in \mathbb{N}$, and $t \cdot \delta_i = \gamma_i$, $i \in I_n$.
We obtain the following derivation in $\mathcal{P}$ leading from u to $v_2 = v_1 \cdot v \cdot w$:
$$u \to \delta_1 \to \cdots \to \delta_n = v_1 \cdot v \cdot w = v_2\ (\mathcal{P}).$$
By Theorem 8, $m_s = \bar{s} \cdot w$ can be determined in space $(\mathrm{size}(u, v_1, \mathcal{P}_t))^2 \cdot 2^{d \cdot k}$ for some constant $d > 0$ independent of u, v_1 and $\mathcal{P}_t$, and thus, v_2 can be determined using at most space $(\mathrm{size}(u, v_1, \mathcal{P}))^2 \cdot 2^{\bar{c} \cdot k}$. $\quad\square$

Theorem 18. *The subword problem for commutative semigroups is exponential space complete with respect to log-lin reducibility.*

Proof. The claim follows since the word problem is a special case (Y and Z both empty) of the subword problem, and because of Theorem 17 we conclude the assertion. □

6 Conclusion

For the special case of binomial ideals, we have presented an algorithm for transforming any given basis into the reduced Gröbner basis using at most space $2^{c \cdot n}$, where n is the size of the problem instance, and c some constant independent of n. Since, in the worst case, any Gröbner basis can have maximal degree at least $2^{2^{c \cdot n}}$, any algorithm for computing Gröbner bases requires at least exponential space (see [MM82], [Huy86]).

As an application of our basis construction algorithm, we have presented space optimal decision procedures for the finite enumeration and subword problems for commutative semigroups. These procedures also require at most space $2^{d \cdot n}$ for some constant d independent of the size n of the problem instance. This complexity bound for the finite enumeration problem implies an analogous bound for the finite containment problem (FCP) (and the finite equality problem (FEP)) for commutative semigroups and, equivalently, for reversible Petri nets.

The finite containment problem (the finite equality problem) for general (not necessarily reversible) Petri nets is the problem of determining for any two given Petri nets with finite reachability sets whether the reachability set of the first is contained in (is equal to) that of the second. As shown in [KaMi69], FCP (FEP) for Petri nets is decidable, but the complexity of any decision procedure for FCP (FEP) for general Petri nets is non-primitive recursive [MM81], in stark contrast to the situation for reversible Petri nets investigated in this paper.

It is also interesting to note that in the case of general Petri nets, the subword problem (i.e., the submarking reachability problem) easily reduces to the word problem since, in a semi-Thue system, we can arrange irreversible transitions from one phase to another. This technique can no longer be applied in the case of Thue systems like commutative semigroups or reversible Petri nets since all transitions can, by definition, be undone.

Our investigation of the complexity of Gröbner basis algorithms for binomial ideals has benefited a lot from the close relationship to commutative semigroups. It is an interesting open problem, whether this relationship can be extended, in one way or another, to the general case of polynomial ideals over fields or (Noetherian) rings. Another open problem is finding algorithms to construct a representation of the (infinite) state space for process systems other than those defined by symmetric commutative context-free process algebras (corresponding to reversible Petri nets) and determining the complexity of various fundamental problems for process algebras in this context, like bisimulation equivalence.

References

[Bay82]   Bayer, D.: The division algorithm and the Hilbert scheme. Ph.d. thesis, Harvard University, Cambridge, MA (1982)

[Buc65]   Buchberger, B.: Ein Algorithmus zum Auffinden der Basiselemente des
            Restklassenrings nach einem nulldimensionalen Polynomideal. Ph.d. thesis,
            Department of Mathematics, University of Innsbruck (1965)

[Dic13]   Dickson, L.E.: Finiteness of the odd perfect and primitive abundant num-
            bers with n distinct prime factors. Amer. J. Math. **35** (1913) 413–422

[Dub90]   Dubé, T.W.: The structure of polynomial ideals and Gröbner bases. SIAM
            J. Comput. **19** (1990) 750–773

[EiSt94]   Eisenbud D., Sturmfels B.: Binomial Ideals. Preprint (1994)

[Her26]   Hermann, G.: Die Frage der endlich vielen Schritte in der Theorie der
            Polynomideale. Math. Ann. **95** (1926) 736–788

[Hi64]   Hironaka, H.: Resolution of singularities of an algebraic variety over a field
            of characteristic zero: I. Ann. of Math. **79(1)** (1964) 109–203

[Huy86]   Huynh, D.T.: A superexponential lower bound for Gröbner bases and
            Church-Rosser commutative Thue systems. Inf. Control **68(1-3)** (1986)
            196–206

[KaMi69]   Karp, R., Miller R.: Parallel program schemata. J. Comput. Syst. Sci. **3**
            (1969) 147–195

[KM96]   Koppenhagen, U., Mayr, E.W.: An Optimal Algorithm for Constructing
            the Reduced Gröbner Basis of Binomial Ideals. Proceedings of the Inter-
            national Symposium on Symbolic and Algebraic Computation, ISSAC '96,
            ACM Press (1996)

[MM81]   Mayr, E.W., Meyer, A.: The complexity of the finite containment problem
            for Petri nets. J. ACM **28(3)** (1981) 561–576

[MM82]   Mayr E.W., Meyer A.: The complexity of the word problems for commuta-
            tive semigroups and polynomial ideals. Adv. Math. **46(3)** (1982) 305–329

[MoMo84]   Möller, H.M., Mora F.: Upper and lower bounds for the degree of Gröbner
            bases. Proceedings of the 3rd Internat. Symp. on Symbolic and Algebraic
            Computation, EUROSAM 84, Springer Verlag, LNCS 174 (1984) 172–183

[Rob85]   Robbiano, L.: Term orderings on the polynomial ring. Proceedings of the
            10th European Conference on Computer Algebra, EUROCAL '85, Vol. 2:
            Research contributions, Springer Verlag, LNCS 204 (1985) 513–517

[Wei87]   V. Weispfenning, V.: Admissible orders and linear forms. ACM SIGSAM
            Bulletin **21(2)** (1987) 16–18

Minimum Fill-In on Circle and Circular-Arc Graphs

T. Kloks[1] * D. Kratsch[2] ** C. K. Wong[3] ***

[1] Department of Mathematics and Computing Science
Eindhoven University of Technology
P.O.Box 513, 5600 MB Eindhoven
The Netherlands
[2] Fakultät für Mathematik und Informatik
Friedrich-Schiller-Universität
07740 Jena, Germany
[3] Department of Computer Science and Engineering
The Chinese University of Hong Kong
Shatin, Hong Kong

Abstract. We present two algorithms solving the minimum fill-in problem on circle graphs and on circular-arc graphs in time $O(n^3)$.

1 Introduction

The MINIMUM FILL-IN problem is a well known and often studied problem. It stems from the optimal performance of Gaussian elimination on sparse matrices.

The knowledge on the algorithmic complexity of the MINIMUM FILL-IN problem when restricted to special graph classes is relatively small compared to that of other problems, as e.g. INDEPENDENT SET, CLIQUE, DOMINATING SET, TREEWIDTH and PATHWIDTH. Indeed, due to the lack of efficient algorithms for finding an optimal solution, in practice one usually has to work with certain heuristics for 'approximating' a minimum fill-in.

The MINIMUM FILL-IN problem 'Given a graph G and a positive integer k, decide whether there is a minimum fill-in of G with at most k edges' remains NP-complete on cobipartite graphs [17] and on bipartite graphs [16]. The only known graph classes for which the minimum fill-in can be computed by a polynomial time algorithm were for almost ten years the relatively small classes of cographs [3] and bipartite permutation graphs [14]. Now polynomial time algorithms for chordal bipartite graphs [7], multitolerance graphs [11] and d-trapezoid graphs [2] are known.

In a sense, the MINIMUM FILL-IN problem has many similarities with the TREE-WIDTH problem. Both problems ask for a chordal embedding of the graph. In the TREEWIDTH problem, one wishes to keep the maximum clique size as small as possible. In the MINIMUM FILL-IN problem one wants the total number of edges to be as

---

* This research was done while this author was with the department of computer science and engineering of the Chinese university of Hong Kong, Shatin, Hong Kong, as a research fellow.
** Email: `kratsch@minet.uni-jena.de`
*** On leave from IBM T.J. Watson Research Center, P.O.Box 218, Yorktown Heights, NY 10598, U.S.A.

small as possible. An $O(n^3)$ algorithm computing the treewidth of circle graphs is given in [6] and an $O(n^3)$ algorithm computing the treewidth of circular-arc graphs is given in [15]. We are not aware of any graph class for which the two problems TREEWIDTH and MINIMUM FILL-IN have different algorithmic complexity, although the solution for the two problems can be far apart [1] (see also [8]).

Circle graphs as well as circular-arc graphs are defined as intersection graphs of geometrical objects of a circle. This leads to a number of similarities between the two graph classes and it also allows a somewhat generic algorithm for the two different classes. To emphasize the similarities we shall present all the theoretical background results for our algorithms 'in parallel'.

We present two elegant and simple algorithms to compute the minimum fill-in of circle and circular-arc graphs. Both algorithms compute a minimum weight triangulation of a certain convex polygon and have overall running time $O(n^3)$.

This shows that circle graphs and circular-arc graphs are two more graph classes with polynomial time algorithms solving the MINIMUM FILL-IN problem. Thus the algorithmic complexity of the problems TREEWIDTH and MINIMUM FILL-IN is the same on both classes. In this way we erase two graph classes from the relatively small list of candidates for graph classes with different algorithmic complexity of TREEWIDTH and MINIMUM FILL-IN.

Since the class of permutation graphs is properly contained in the class of circle graphs, our circle graph algorithm extends an result of [2], where a $O(n^2)$ time algorithm is given that computes the minimum fill-in of trapezoid graphs, and thus of permutation graphs.

2 Preliminaries

Let $G = (V, E)$ be a graph. We denote the number of vertices of G by n. For a set $W \subseteq V$, we denote by $G[W]$ the subgraph of G induced by W.

2.1 Preliminaries on triangulations

We start with considering triangulations and minimal separators.

Definition 1. A graph is *chordal* if it does not contain a chordless cycle of length greater than three.

Definition 2. A *triangulation* of a graph G is a chordal graph H with the same vertex set as G, such that G is a subgraph of H. A triangulation H of a graph G is called a *minimal triangulation* of G, if no proper subgraph of H is a triangulation of G.

The following theorem has been shown in [12].

Theorem 3. *Let H be a triangulation of a graph G. Then H is a minimal triangulation of G if and only if each edge $e \in E(H) \setminus E(G)$ is unique chord of a cycle of length four in H.*

Definition 4. Let $G = (V, E)$ be a graph and a, b two nonadjacent vertices of G. The set $S \subseteq V$ is an a, b-*separator* if the removal of S separates a and b in distinct connected components. If no proper subset of S is an a, b-separator then S is a *minimal a, b-separator*. A *minimal separator* is a set of vertices S that is a minimal a, b-separator.

Lemma 5. *Let S be a minimal a, b-separator of the graph $G = (V, E)$ and let C_a and C_b be the connected components of $G[V \setminus S]$ containing a and b respectively. Then every vertex of S has at least one neighbor in C_a and at least one neighbor in C_b.*

We denote by $\Delta(H)$ the set of all minimal separators of a graph H. In [9] the following characterization of minimal triangulations is given.

Theorem 6. *A triangulation H of a graph G is a minimal triangulation of G if and only if the following three conditions are satisfied.*

1. *If a and b are nonadjacent vertices of H, then every minimal a, b-separator of H is also a minimal a, b-separator of G.*
2. *If S is a minimal separator of H and C a connected component of $H[V \setminus S]$, then the vertex set of C induces also a connected component in $G[V \setminus S]$.*
3. *$H = G_{\Delta(H)}$, where $G_{\Delta(H)}$ is the graph obtained from G by adding edges between every pair of vertices contained in the same set S, for any $S \in \Delta(H)$.*

Now it is convinient to define the MINIMUM FILL-IN problem as follows.

Definition 7. The MINIMUM FILL-IN problem, is the problem of finding a triangulation H of the given graph $G = (V, E)$ such that $|E(H) \setminus E(G)|$ is as small as possible. The *minimum fill-in* of the graph G, denoted by mfi(G), is the minimum number of edges which have to be added to make G chordal.

In other words, solving the MINIMUM FILL-IN problem is equivalent to finding a minimal triangulation H of the input graph G with smallest possible number of edges. Then any perfect elimination ordering of H is a minimum elimination ordering of G (see [12]).

2.2 Preliminaries on circle and circular-arc graphs

We give the necessary background material concerning the two graph classes studied in this paper. For more information on circle graphs, circular-arc graphs and related classes of graphs we refer to [5].

Definition 8. A *circle graph* $G = (V, E)$ is a graph for which one can associate with each vertex $v \in V$ a chord of a circle C such that two vertices of G are adjacent if and only if the corresponding chords have a nonempty intersection. The set of chords and the circle C are said to be a *circle model* $\mathcal{D}(G)$.

Without loss of generality we assume that no two chords of the circle model share an end point. We also assume that a circle model of the input graph is given, since there is an $O(n^2)$ time recognition algorithm for circle graphs, that also computes a circle model of the input graph, if it is a circle graph [13].

Definition 9. A *circular-arc graph* is a graph $G = (V, E)$ for which one can associate with each vertex $v \in V$ an arc on a circle C such that two vertices of G are adjacent if and only if the corresponding arcs have a nonempty intersection. The set of arcs and the circle C are said to be a *circular-arc model* $\mathcal{D}(G)$.

Without loss of generality we assume that no two arcs of a circular-arc model share an end point. If the circle C contains a point which is not contained in any arc, then the graph is an interval graph and its minimum fill-in is zero. Thus, we assume that every point of C is covered by at least one arc.

We also assume that a circular-arc model of the given graph is available. Note that, if no circular-arc model is part of the input, then we can compute one by an $O(n^2)$ time algorithm [4]. This is a recognition algorithm for circular-arc graphs that also computes a circular-arc model of the given graph, if it is a circular-arc graph.

In the following sections, we show that any minimal triangulation of a circle graph and a circular-arc graph can be represented in terms of a (planar) triangulation of a well-defined convex polygon. This is the property exploited by our algorithms. This property enables the design of simple algorithms computing the minimum fill-in. It also gives a new correctness proof for the treewidth algorithms in [6, 15].

3 Scanlines

We show how to represent the minimal separators of circle graphs and circular-arc graphs by means of scanlines.

3.1 Circle graphs

Let $G = (V, E)$ be a circle graph with circle model $\mathcal{D}(G)$.

Definition 10. Place new points on the circle C as follows. Go around the circle in clockwise order. Between every two consecutive end points of chords, place a new point. These new points are called *scanpoints* and the set of the $2n$ scanpoints of $\mathcal{D}(G)$ is denoted by Z.

Definition 11. A *scanline* of $\mathcal{D}(G)$ is a chord of the circle C, connecting two scanpoints.

Consequently, there are $\binom{2n}{2}$ different scanlines in $\mathcal{D}(G)$.

Definition 12. Two scanlines *cross* if they have a nonempty intersection but no scanpoint in common.

Definition 13. Let c_1 and c_2 be two chords of C with empty intersection. A scanline s is *between* c_1 and c_2 if every path from an end point of c_1 to an end point of c_2 along C passes through a scanpoint of s.

For any scanline s of $\mathcal{D}(G)$, we denote by $S(s)$ the set of all vertices v of G, for which the corresponding chord intersects s. For the following theorem and corollary see [6, 10].

Theorem 14. *Let a and b be nonadjacent vertices of the circle graph $G = (V, E)$. For every minimal a, b-separator S of G, there exists a scanline s of $\mathcal{D}(G)$ between the chords of a and b such that $S = S(s)$.*

Corollary 15. *A circle graph on n vertices has $O(n^2)$ minimal separators.*

3.2 Circular-arc graphs

Let $G = (V, E)$ be a circular-arc graph with circular-arc model $\mathcal{D}(G)$. For any point p of the circle C of $\mathcal{D}(G)$, we denote by $S(p)$ the set of all vertices v, for which the corresponding arc contains the point p.

Definition 16. Place new points on the circle C as follows. Go around the circle in clockwise order. Consider two consecutive end points x and y of arcs and let p be a point on the circle between x and y. If $|S(p)| < \min(|S(x)|, |S(y)|)$ then we call p a *scanpoint*. The set of scanpoints of $\mathcal{D}(G)$ is denoted by Z^*.

Also in this case, we call a chord of the circle connecting two scanpoints a scanline.

Definition 17. Let a and b be two nonadjacent vertices of G. A scanline s is *between* the arcs of a and b if every path from an end point of the arc of a to an end point of the arc of b along the circle C passes through a scanpoint of s.

For any scanline s of $\mathcal{D}(G)$, we denote by $S(s)$ the set of all vertices v of G, for which the corresponding arc contains at least one scanpoint of s. The following theorem and corollary are given in [10].

Theorem 18. *Let a and b be nonadjacent vertices of a circular-arc graph $G = (V, E)$. For every minimal a, b-separator S of G, there exists a scanline s between the arcs of a and b such that $S = S(s)$.*

Corollary 19. *A circular-arc graph on n vertices has $O(n^2)$ minimal separators.*

Characterizations of the minimal separators in terms of scanlines as in Theorem 14 and 18, and their consequences for the number of minimal separators as in Corollary 15 and 19 have been presented for various classes of intersection graphs (see e.g. [2, 8, 10]).

4 Realizers and triangulations

We introduce two similar types of convex polygons and show how certain planar triangulations of these polygons and the triangulations of the corresponding graph relate to each other.

4.1 Circle graphs

Let $G = (V, E)$ be a circle graph. Consider a circle model $\mathcal{D}(G)$ with the set Z of scanpoints.

Definition 20. Let $Y \subseteq Z$ and $|Y| \geq 3$. We denote by $\mathcal{P}(Y)$ the convex polygon with vertex set Y. The *candidate component* $G(Y)$ is the subgraph of G induced by the set of vertices corresponding to chords of $\mathcal{D}(G)$, that have a nonempty intersection with the region of $\mathcal{P}(Y)$.

Hence the edges of the polygon $\mathcal{P}(Y)$ are scanlines. Notice that $G(Z) = G$.

Definition 21. Let $Y \subseteq Z$ and $|Y| \geq 3$. For each scanline s that is an edge of the polygon $\mathcal{P}(Y)$, add an edge between any pair of nonadjacent vertices of the candidate component $G(Y)$, of which the two corresponding chords intersect the scanline s. The graph obtained in this way is called the *realizer* $R(Y)$ of the candidate component $G(Y)$.

Hence each candidate component is a subgraph of its realizer.

Suppose $|Y| = 3$. Then $\mathcal{P}(Y)$ is a triangle and each chord corresponding to a vertex of $G(Y)$ intersects exactly two edges of $\mathcal{P}(Y)$. Thus $R(Y)$ is a clique. The next two lemmata have been shown in [6].

Lemma 22. *Let $Y \subseteq Z$ and $|Y| \geq 3$. Then the realizer $R(Y)$ is a circle graph.*

Definition 23. Let $G(Y)$ be a candidate component with realizer $R(Y)$. A scanline s of $\mathcal{D}(R(Y))$ is Y-*nice*, if the scanpoints of s are elements of Y.

Lemma 24. *Let $Y \subseteq Z$ and $|Y| \geq 3$ and let S be a minimal a, b-separator in the realizer $R(Y)$. Then there is a Y-nice scanline s such that $S = S(s)$ in $\mathcal{D}(R(Y))$.*

Definition 25. Let $\mathcal{P}$ be a convex polygon with m vertices. A *(planar) triangulation* of $\mathcal{P}$ is a set of $m - 3$ non crossing diagonals in $\mathcal{P}$ that divide the interior of $\mathcal{P}$ in $m - 2$ triangles.

Definition 26. Let $Y \subseteq Z$ and $|Y| \geq 3$. Let T be a triangulation of $\mathcal{P}(Y)$. Then $H(T)$ is defined as the graph with the same vertex set as $G(Y)$ and vertices u and v of $H(T)$ are adjacent if there exists a triangle Q in T such that the two chords corresponding to u and v both intersect Q.

Note that $R(Y)$ is a spanning subgraph of $H(T)$.

Lemma 27. *The graph $H(T)$ is chordal and therefore a triangulation of $R(Y)$.*

Proof. Let $y \in Y$ be a vertex of $\mathcal{P}(Y)$ that is not incident with a diagonal of the triangulation T. Hence y is incident with exactly one triangle Q of T. If there is a chord intersecting Q but no other triangle of T, then the vertex x corresponding to this chord is a simplicial vertex of $H(T)$. Remove x from $H(T)$ and the chord from the circle model.

If there is no chord left that intersects Q but no other triangle of T, then remove y from Y. In this way, we obtain a perfect elimination ordering of the graph $H(T)$. Thus $H(T)$ is chordal. $\square$

Now we give a representation theorem for all minimal triangulations of a circle graph in terms of planar triangulations of the polygon $\mathcal{P}(Z)$.

Theorem 28. *Let $G = (V, E)$ be a circle graph. Let $\mathcal{D}(G)$ be a circle model of G and Z its set of scanpoints. Then for every minimal triangulation H of G there is a (planar) triangulation T of the polygon $\mathcal{P}(Z)$ such that $H = H(T)$.*

Proof. We claim that for any set $Y \subseteq Z$ with $|Y| \geq 3$ and any minimal triangulation H of the realizer $R(Y)$, there is a triangulation T of $\mathcal{P}(Y)$ such that $H = H(T)$. Note that the claim immediately implies the theorem, since $G = R(Z)$. We prove the claim by induction on the number of vertices in Y.

First let H be a complete graph. Then $R(Y)$ is also complete by Theorem 6. Thus $H = H(T)$ for any triangulation T of $\mathcal{P}(Y)$.

Now let S be a minimal a, b-separator of H. Since H is a minimal triangulation, S is also a minimal a, b-separator of $R(Y)$ by Theorem 6. Then there exists a Y-nice scanline s such that $S = S(s)$ in $\mathcal{D}(R(Y))$ by Lemma 24. Clearly, s divides the polygon $\mathcal{P}(Y)$ into two polygons $\mathcal{P}(Y_1)$ and $\mathcal{P}(Y_2)$. Moreover $|Y_1| < |Y|$ and $|Y_2| < |Y|$, since s is between the chords of a and b.

Consider the corresponding realizers $R(Y_1)$ and $R(Y_2)$. The subgraphs of H induced by the vertices of $R(Y_1)$ and of $R(Y_2)$ are minimal triangulations of $R(Y_1)$ and $R(Y_2)$. Now the claim follows by induction. □

The consequence of Theorem 28 and Lemma 27 is a minimum fill-in algorithm for circle graphs, that essentially computes a minimum weight triangulation of the polygon $\mathcal{P}(Z)$.

4.2 Circular-arc graphs

Let $G = (V, E)$ be a circular-arc graph with circular-arc model $\mathcal{D}(G)$. Consider the set Z^* of scanpoints of $\mathcal{D}(G)$.

Contrary to circle graphs, we only have $0 \leq |Z^*| \leq 2n$. If $|Z^*| \leq 1$, then G is complete. If $|Z^*| = 2$ then G is an interval graph having at most two maximal cliques. Thus in both cases $mfi(G) = 0$. Hence we may assume $|Z^*| \geq 3$.

Let $\mathcal{P}(Z^*)$ be the convex polygon with vertex set Z^*. Hence the edges of $\mathcal{P}(Z^*)$ are scanlines of $\mathcal{D}(G)$.

Definition 29. Let T be a (planar) triangulation of the polygon $\mathcal{P}(Z^*)$. Then the graph $H(T)$ is defined as follows. The vertex set of $H(T)$ is the same as the vertex set of G. Two vertices u and v of $H(T)$ are adjacent if either they are adjacent in G or, there exists a diagonal d in T, such that the arcs of u and v both contain a scanpoint of d.

The proof of the following lemma is omitted due to space restrictions.

Lemma 30. *$H(T)$ is chordal and therefore a triangulation of G.*

Now we give a representation theorem for the minimal triangulations of circular-arc graphs in terms of (planar) triangulations of the convex polygon $\mathcal{P}(Z^*)$.

Theorem 31. *Let $G = (V, E)$ be a circular-arc graph. Let $\mathcal{D}(G)$ be a circular-arc model of G and Z^* its set of scanpoints. Then for every minimal triangulation H of G there is a triangulation T of the polygon $\mathcal{P}(Z^*)$ such that $H = H(T)$.*

Proof. Suppose H is a complete graph. Then G is also complete. Hence $H(T)$ is a complete graph for any triangulation T of $\mathcal{P}(Z^*)$.

The proof is by a induction on the number of vertices of the graph. Assume H has two nonadjacent vertices a and b. Let S be a minimal a, b-separator in H. Since H is chordal, S is a clique in H. By Theorem 6, S is a minimal a, b-separator in G.

By Theorem 18, there exists a scanline s of $\mathcal{D}(G)$ such that $S(s) = S$ and s is between the arcs of a and b. Let α and β be the scanpoints of s. The removal of α and β from the circle $\mathcal{C}$ creates two 'halves', that we call *a-half* if it contains the arc of a and *b-half* if it contains the arc of b.

By Theorem 6, the vertex set of every connected component of $H[V \setminus S]$ induces a connected component of $G[V \setminus S]$. Let $V_1 \subseteq V$ be the set containing S and the vertex sets of all components of $G[V \setminus S]$ for which the arcs are in the a-half. Similarly, $V_2 \subseteq V$ is the set containing S and the vertex sets of all components for which the arcs are in the b-half. Clearly, $|V_1| < |V|$ and $|V_2| < |V|$.

Let $Z_1^* \subseteq Z^*$ (resp. $Z_2^* \subseteq Z^*$) be the set consisting of α, β and all scanpoints in the a-half (resp. b-half). Thus $\min(|Z_1^*|, |Z_2^*|) \geq 2$.

Case 1: $\min(|Z_1^*|, |Z_2^*|) \geq 3$.
Consider the a-half. Clearly $H[V_1]$ is chordal and thus a triangulation of $G[V_1]$. We obtain a circular-arc model of $G[V_1]$ by removing the arcs of all vertices of $V_2 \setminus S$ from $\mathcal{D}(G)$. Then replace any arc of a vertex v of S containing only one of α and β by a new arc, that has the original end point in the a-half and a new end point in the b-half, such that all these new arcs contain a fixed point of the b-half.

First assume that there is one arc of a vertex in S containing α but not β, and one arc containing β but not α. Then the new circular-arc model can be constructed in such a way that the set of scanpoints in the new model is exactly Z_1^*. Moreover, the new model is a circular-arc model of the graph $R(V_1)$ obtained from $G[V_1]$ by making S a clique. Now $H[V_1]$ is a minimal triangulation of $R(V_1)$, which can be seen best by using Corollary 3. Using induction, we obtain that there is a triangulation T_1 of $\mathcal{P}(Z_1^*)$ such that $H[V_1] = H(T_1)$.

Due to space restrictions we omit the details of the case in which the arcs of vertices in S either all contain α or all contain β.

Analogously, there is a triangulation T_2 of $\mathcal{P}(Z_2^*)$ such that $H[V_2] = H(T_2)$.

Take the scanline s and all the diagonals of T_1 and T_2. This gives a triangulation T of $\mathcal{P}(Z^*)$. Furthermore $H = H(T)$ since any edge of H, that does not have both end vertices in S, is either an edge of $H[V_1]$ or $H[V_2]$, and thus represented by some diagonal of T.

Case 2: $\min(|Z_1^*|, |Z_2^*|) = 2$.
W.l.o.g. assume $|Z_1^*| = 2$. Then consider $H[V_2]$ and the set $Z_2^* = Z^*$. $H[V_2]$ has less vertices than H. Thus, using induction there is a triangulation T_2 of $\mathcal{P}(Z^*)$ such that $H[V_2] = H(T_2)$. Consequently $H[V_2] = H(T_2)$, since all vertices of $V_1 \setminus S$ are simplicial in G (see [15]), thus $H(T) = H$. □

5 End-triangles

The concept of an end-triangle is important for obtaining efficient algorithms that compute the minimum fill-in on circle graphs and circular-arc graphs.

5.1 Circle graphs

Let $G = (V, E)$ be a circle graph with circle model $\mathcal{D}(G)$ and let $\mathcal{P}(Z)$ be the convex polygon with vertex set Z.

Definition 32. Let a and b be nonadjacent vertices of G. A triangle Q in $\mathcal{P}(Z)$, i.e., all vertices of Q are in Z, is an *end-triangle* for $\{a, b\}$, if the chords of a and b both intersect Q, but there is only one edge of Q, that is crossed by both chords.

Lemma 33. *Let T be any triangulation of $\mathcal{P}(Z)$. If the chords of two nonadjacent vertices a and b both intersect some triangle of T, then there are exactly two end-triangles for $\{a, b\}$ in T.*

Proof. Suppose the chords of a and b both intersect a triangle Q of T. Then the chords of a and b both cross at least one common edge, say r, of Q. Hence the chords of a and b also intersect a neighboring triangle Q' which shares the edge r with Q. In this way we find a path of triangles which must end with an end-triangle for $\{a, b\}$.

Notice that the set of triangles, having nonempty intersection with the chords of a and b, is exactly the path of triangles between the two end-traingles for $\{a, b\}$. This shows that there are exactly two end-triangles for $\{a, b\}$. □

Definition 34. Let Q be a triangle of $\mathcal{P}(Z)$. Then $w(Q)$, the *weight* of Q, is the number of unordered pairs $\{a, b\}$ of non-crossing chords for which Q is an end-triangle. The *weight $w(T)$ of a triangulation* T is the sum of the weights of all the triangles in T.

Corollary 35. *Let $G = (V, E)$ be a circle graph. Then*

$$mfi(G) = \frac{1}{2} \cdot \min\{w(T) \, : \, T \text{ triangulation of } \mathcal{P}(Z)\}.$$

Proof. Let T be a triangulation of $\mathcal{P}(Z)$. Consider $H(T)$. The weight of T is twice the number of edges of $H(T)$ minus the number of edges in G, since every edge in $H(T)$ which is not an edge in G is counted exactly twice, namely once for each end-triangle. □

5.2 Circular-arc graphs

Let $G = (V, E)$ be a circular-arc graph with cicular-arc model $\mathcal{D}(G)$ and let $\mathcal{P}(Z^*)$ be the convex polygon with vertex set Z^*.

Definition 36. Let a and b be two nonadjacent vertices of G. A triangle Q of $\mathcal{P}(Z^*)$ is an *end-triangle for* $\{a, b\}$, if the arcs of a and b each contain exactly one vertex of Q.

Analogously to Subsection 5.1, one can obtain the following results.

Lemma 37. *Let T be any triangulation of $\mathcal{P}(Z^*)$. If there is a diagonal s of T such that the arcs of the two nonadjacent vertices a and b both contain one scanpoint of s, then there are exactly two end-triangles for $\{a, b\}$.*

Definition 38. Let Q be a triangle of $\mathcal{P}(Z^*)$. Then $w(Q)$, the *weight of Q*, is the number of unordered pairs $\{a, b\}$, for which Q is an end-triangle.

Corollary 39. *Let $G = (V, E)$ be a circular-arc graph. Then*

$$mfi(G) = \frac{1}{2} \cdot \min\{w(T) \ : \ T \text{ triangulation of } \mathcal{P}(Z^*)\}.$$

6 The algorithms computing the minimum fill-in

In this section we describe simple polynomial time algorithms to find the minimum fill-in of circle graphs and circular-arc graphs.

Both algorithms compute the minimum weight of a triangulation of the polygon $\mathcal{P}(Z)$ and $\mathcal{P}(Z^*)$, respectively. Luckily, the weight functions are such that we can apply a classical dynamic programming algorithm for this problem, that has also been applied in [6, 15].

Let $s_1, s_2, \ldots, s_h$ be the scanpoints of the set Z (resp. Z^*) in some clockwise order. Then $h = 2n$ for circle graphs and $h \leq 2n$ for circular-arc graphs. (Note that all indices in this section are to be taken modulo h.)

Let $c(i, j, k)$ be the weight of the triangle with vertices s_i, s_j, s_k. Suppose the weights of all triangles of $\mathcal{P}(Z)$ (resp. $\mathcal{P}(Z^*)$) are given. Then there is an $O(h^3)$ algorithm that computes the minimum weight of a triangulation of $\mathcal{P}(Z)$ (resp. $\mathcal{P}(Z^*)$).

Define $w(i, t)$ as the minimum weight of a triangulation of the polygon with vertices $s_i, s_{i+1}, \ldots, s_{i+t-1}$. Then for all i, $w(i, 2) = 0$, and for all $t \in \{3, \ldots, 2n\}$

$$w(i, t) = \min_{2 \leq j < t} w(i, j) + w(i + j - 1, t - j + 1) + c(i, i + j - 1, i + t - 1).$$

By Corollary 35 and 39, the minimum fill-in of the input graph can be computed in time $O(n^3)$ plus the time for computing the weights of all the $O(n^3)$ triangles of the polygon $\mathcal{P}(Z)$ and $\mathcal{P}(Z^*)$. It is not hard to see that the weights of all triangles can be computed in time $O(n^5)$ for circle and circular-arc graphs. This would give $O(n^5)$ algorithms computing the minimum fill-in. In the remainder we show how to get faster algorithms.

6.1 Circle graphs

The weight of the triangle Q is the number of nonadjacent vertices a and b of G, for which the chords of a and b both cross Q, but there is only one edge of Q that is crossed by both chords. The weight $c(p, q, r)$ of all triangles of $\mathcal{P}(Z)$ can be computed as follows.

Let $L(i, k; j)$ (where s_j is not on the part of the circle going clockwise from s_i to s_{i+k-1}), be the number of chords that have one end point in clockwise order between s_i and s_{i+k-1} and that cross the scanline between s_i and s_j. The numbers $L(i, 2; j)$ are easy to determine, since there is a unique chord with an end point on the part of the circle going clockwise from s_i to s_{i+1}. The numbers $L(i, k + 1; j)$ can be determined as follows. Check if the chord with an end point on the part of the circle going clockwise from s_{i+k-1} to s_{i+k}, crosses with the scanline between s_i and s_j. If it does, then $L(i, k + 1; j) = L(i, k; j) + 1$ and if it does not, then $L(i, k + 1; j) = L(i, k; j)$.

Let $A(i, k; j)$ (where s_j is not in the part of the circle going clockwise from s_i to s_{i+k-1}) be the number of nonadjacent vertices a and b such that a and b have one end point on the part of the circle going clockwise from s_i to s_{i+k-1} and such that the chord of a crosses the scanline between s_i and s_j and the chord of b crosses the scanline between s_{i+k-1} and s_j. Then $A(i, 2; j) = 0$ for all i and j, since there is only one chord with an end point on the part going clockwise from s_i to s_{i+1}. Then consider the chord with an end point on the part going clockwise from s_{i+k-1} to s_{i+k}. If this chord crosses the scanline between s_{i+k} and s_j, then $A(i, k + 1; j) = A(i, k; j) + L(i, k; j)$. Otherwise $A(i, k + 1; j) = A(i, k; j)$.

Now the weight of the triangle with vertices s_p, s_q and s_r, where s_p, s_q, s_r is in clockwise order, can be computed as

$$c(p, q, r) = A(p, q - p + 1; r) + A(q, r - q + 1; p) + A(r, p - r + 1; q).$$

Consequently the weight of all triangles of $\mathcal{P}(Z)$ can be determined in $O(n^3)$ time.

Theorem 40. *There is an $O(n^3)$ time algorithm computing the minimum fill-in of a circle graph.*

6.2 Circular-arc graphs

First, check in linear time (see [5]), whether the graph is an interval graph. If so, its minimum fill-in is zero. Thus we may assume that the input graph G is not an interval graph, implying $|Z^*| \geq 3$.

We shall demonstrate that the weights of all triangles of $\mathcal{P}(Z^*)$ can be computed by solving the corresponding problem on a suitable circle model. First add scanpoints to the original set Z^* such that there is exactly one scanpoint between any two consecutive end points of arcs. Similar to the circle graph terminology we call the set of all these points Z. Clearly $Z^* \subseteq Z$. Now transform $\mathcal{D}(G)$ into a circle model $\mathcal{D}'(G)$ by replacing any arc with end points x and y with a chord connecting x and y.

Let a and b be two nonadjacent vertices of G. Then for every triangle Q of $\mathcal{P}(Z)$, the chords corresponding to a and b in $\mathcal{D}'(G)$ intersect Q, but only one edge of Q is crossed by each chord, if and only if the arcs of a and b in $\mathcal{D}(G)$ have empty intersection and contain each exactly one vertex of Q.

Consequently, the weight of all triangles of $\mathcal{P}(Z^*)$ can be computed in time $O(n^3)$ by using the algorithm for circle graphs, given in Subsection 6.1, additionally counting a pair of noncrossing chords in $\mathcal{D}'(G)$ only if the corresponding vertices are nonadjacent in the circular-arc graph G.

Theorem 41. *There is an $O(n^3)$ time algorithm computing the minimum fill-in of a circular-arc graph.*

References

1. H. Bodlaender, L. van der Gaag and T. Kloks, Some remarks on minimum edge and minimum clique triangulations, manuscript.
2. H. Bodlaender, T. Kloks, D. Kratsch and H. Müller, Treewidth and minimum fill-in on d-trapezoid graphs, Technical Report UU-CS-1995-34, Utrecht University.
3. D.G. Corneil, Y. Perl and L.K. Stewart, Cographs: recognition, applications and algorithms, *Congressus Numerantium* **43** (1984), pp. 249–258.
4. E.M. Eschen and J.P. Spinrad, An $O(n^2)$ algorithm for circular-arc graph recognition, *Proceedings of SODA '93*, pp. 128–137.
5. M.C. Golumbic, *Algorithmic Graph Theory and Perfect Graphs*, Academic Press, New York, 1980.
6. T. Kloks, Treewidth of circle graphs, *Proceedings of ISAAC'93*, Springer-Verlag, LNCS 762, 1993, pp. 108–117.
7. T. Kloks, Minimum fill-in for chordal bipartite graphs, Technical report RUU-CS-93-11, Utrecht University, 1993.
8. T. Kloks, *Treewidth–Computations and Approximations*, Springer–Verlag, Lecture Notes in Computer Science 842, 1994.
9. T. Kloks, D. Kratsch and H. Müller, Approximating the bandwidth of AT-free graphs, *Proceedings of ESA '95*, Springer-Verlag, LNCS 979, 1995, pp. 434–447.
10. D. Kratsch, The structure of graphs and the design of efficient algorithms, Habilitation (thesis), F.-Schiller-Universität Jena, 1995.
11. A. Parra, Triangulating multitolerance graphs, Technical Report 392/1994, Technische Universität Berlin.
12. D.J. Rose, R.E. Tarjan and G.S. Lueker, Algorithmic aspects of vertex elimination on graphs, *SIAM Journal on Computing* **5** (1976), pp. 266–283.
13. J. Spinrad, Recognition of circle graphs, *J. Algorithms* **16** (1994), pp. 264–282.
14. J. Spinrad, A. Brandstädt and L.K. Stewart, Bipartite permutation graphs, *Discrete Applied Mathematics* **18** (1987), pp. 279–292.
15. R. Sundaram, K. Sher Singh and C. Pandu Rangan, Treewidth of circular arc graphs, *SIAM Journal on Discrete Mathematics* **7** (1994), pp. 647–655.
16. R.E. Tarjan, Decomposition by clique separators, *Discrete Mathematics* **55** (1985), pp. 221–232.
17. M. Yannakakis, Computing the minimum fill-in is NP-complete, *SIAM Journal on Algebraic and Discrete Methods* **2** (1981), pp. 77–79.

Practical Approximation Schemes for Maximum Induced-Subgraph Problems on $K_{3,3}$-free or K_5-free Graphs

Zhi-Zhong Chen

Dept. of Math. Sci., Tokyo Denki Univ., Hatoyama, Saitama 350-03, Japan

Abstract. We show that for an integer $k \geq 2$ and an n-vertex graph G without a $K_{3,3}$ (resp., K_5) minor, we can compute k induced subgraphs of G with treewidth $\leq 3k-4$ (resp., $\leq 6k-7$) in $O(kn)$ (resp., $O(kn+n^2)$) time such that each vertex of G appears in exactly $k-1$ of these subgraphs. This leads to *practical* polynomial-time approximation schemes for various maximum induced-subgraph problems on graphs without a $K_{3,3}$ or K_5 minor. The result extends a well-known result of Baker that there are practical polynomial-time approximation schemes for various maximum induced-subgraph problems on *planar* graphs.

1 Introduction

Let π be a property on graphs. π is *hereditary* if, whenever a graph G satisfies π, every induced subgraph of G also satisfies π. Suppose π is a hereditary property. The *maximum induced subgraph problem* associated with π (MISP(π)) is the following: Given a graph $G = (V, E)$, find a maximum subset U of V that induces a subgraph satisfying π. Yannakakis showed that various natural MISP(π)'s are NP-hard even if the input graph is restricted to a planar graph [15]. Thus, it is of interest to design efficient approximation algorithms for these MISP(π)'s.

An approximation algorithm A for a maximization problem Π achieves a *performance ratio* of ρ if for every instance I of Π, the ratio of the optimal value for I to the solution value returned by A is at most ρ. A *polynomial-time approximation scheme* (PTAS) for problem Π is an approximation algorithm which given an instance I of Π and an $\epsilon > 0$, returns a solution s within time polynomial in the size of I such that the ratio of the optimal value for I to the value of s is at most $(1 + \epsilon)$.

Much work has been devoted to designing PTASs for MISP(π)'s restricted to certain special instances [1, 5, 12]. Lipton and Tarjan were the first who proved that various MISP(π)'s restricted to *planar* instances have PTASs [12]. In their approach, they applied their planar separator theorem. Unfortunately, their schemes are known to be nonpractical [6]. That is, to achieve a reasonable performance ratio (e.g., 2), the number of vertices in the input graph and/or the running time of the schemes has to be enormous ($\approx 2^{2^{400}}$). Later, Baker gave practical PTASs for the same problems using a different approach [5]. By extending Lipton & Tarjan's approach, Alon et al. [1] showed that various MISP(π)'s restricted to graphs without an excluded minor have polynomial-time approximation schemes. Like Lipton and Tarjan's schemes, Alon et al.'s schemes have the shortage of being *very* nonpractical. Very recently, Eppstein proved that if $\mathcal{F}$ is a family of graphs without an excluded minor and does not contain all apex graphs, then there is a function f such that every graph in $\mathcal{F}$ with diameter at most D has treewidth $f(D)$ [7]. Combining this result together with Baker's approach leads to PTASs for MISP(π)'s restricted to graphs in such an family $\mathcal{F}$. Unfortunately, Eppstein's proof is based on Robertson & Seymour's "planar obstruction theorem" [13] and $f(D)$ is *extremely* large (even if D is small) [7]. Consequently, the resulting PTASs are nonpractical.

Since neither Alon et al.'s schemes nor the schemes implied by Eppstein's result above are practical, it is natural to ask whether practical PTASs exist for MISP(π)'s restricted to graphs without an excluded minor. In this paper, we give an affirmative answer to this question when the minor is $K_{3,3}$ or K_5. Since neither a $K_{3,3}$ minor nor a K_5 minor can exist in a planar graph, our result extends Baker's result above. Our schemes can be viewed as a modification of Baker's schemes. Recall that Baker's schemes consist of three steps. First, decompose the input planar graph G into k ($k-1$)-outerplanar (induced) subgraphs G_1, $\cdots$, G_k such that each vertex of G appears in exactly $k-1$ of these subgraphs. Next, compute an optimal solution s_i in each G_i using dynamic programming. Finally, output the best one among $s_1, \cdots, s_k$ as a (nearly optimal) solution in the original graph G. In [5], Baker shows that the output solution has size at least $(k-1)/k$ optimal. Our schemes differ from Baker's only in the first step. This difference is essential because it is impossible to perform the first step above when G is not planar. In our schemes, the input graph G without a $K_{3,3}$ (resp., K_5) minor is decomposed into k induced subgraphs with treewidth $\leq 3k-4$ (resp., $\leq 6k-7$) in $O(kn)$ (resp., $O(kn+n^2)$) time such that each vertex of G appears in exactly $k-1$ of these subgraphs. This decomposition is based on the nice structures of graphs without a $K_{3,3}$ or K_5 minor that were developed in [2, 8, 11]. Roughly speaking, these nice structures say that a graph without a $K_{3,3}$ (resp., K_5) minor must have very special 3-connected (resp., 4-connected) components each of which can easily be decomposed into induced subgraphs of bounded treewidth. The problem is how to combine the decompositions of these components into a (single) decomposition of the original graph G. We solve this problem by organizing these components into a suitable tree. The other two steps in our schemes are the same as those in Baker's, and therefore can be done in practical polynomial (often linear) time because various MISP(π)'s restricted to graphs of bounded treewidth can be computed in practical polynomial (often linear) time by dynamic programming [4, 14]. Besides their practicality, our schemes also have the advantage of being easy to parallelize.

The paper is organized as follows. In section 2, we give basic definitions and state several elementary facts. We prove a technical lemma in section 3. This lemma shows how to combine the decompositions of certain subgraphs (e.g., 3-connected or 4-connected components) of G into a single decomposition of G. Using this lemma, we show how to decompose graphs without a $K_{3,3}$ (resp., K_5) minor in section 4 (resp., section 5). Section 6 concludes the paper.

2 Preliminaries

Throughout this paper, a graph is always connected. Unless stated explicitly, a graph is always simple, i.e., has neither multiple edges nor self-loops. Let $G = (V, E)$ be a graph. For convenience, we allow $V = \emptyset$. If $V = \emptyset$, then we call G an *empty* graph. We sometimes write $V(G)$ instead of V and $E(G)$ instead of E. The *neighborhood* of a vertex v in G is the set of vertices in G adjacent to v. For $U \subseteq V$, the *subgraph of G induced by U* is the graph (U, F) with $F = \{\{u, v\} \in \bar{E} : u, v \in U\}$ and is denoted by $G[U]$. When $U \subseteq V$, we sometimes write $G - U$ instead of $G[V - U]$.

A *contraction* of an edge $\{u,v\}$ in G is made by identifying u and v with a new vertex whose neighborhood is the union of the neighborhoods of u and v (resulting multiple edges and self-loops are deleted). A *contraction* of G is a graph obtained from G by a sequence of edge contractions. A graph H is a *minor* of G if H is the contraction of a subgraph of G. G is H-*free* if G has no minor isomorphic to H. This paper deals with $K_{3,3}$-free or K_5-free graphs.

A *tree-decomposition* of G is a pair $(\{X_i : i \in I\}, T)$, where $\{X_i : i \in I\}$ is a family of subsets of V and T is a tree with $V(T) = I$ such that (a) $\cup_{i \in I} X_i = V$, (b) for every edge $\{v, w\} \in E$, there is an X_i, $i \in I$ with $v \in X_i$ and $w \in X_i$, and (c) for all $i, j, k \in I$, if j lies on the path from i to k in T, then $X_i \cap X_k \subseteq X_j$. The *treewidth* of a tree-decomposition $(\{X_i : i \in I\}, T)$ is $\max\{|X_i| - 1 : i \in I\}$. The *treewidth* of G, denoted by $\mathrm{tw}(G)$, is the minimum treewidth of a tree-decomposition of G, taken over all possible tree-decompositions of G. The treewidth of an empty graph is defined to be 0.

Lemma 1. [Robertson & Seymour] Let $G = (V, E)$ be a graph, and R_1 and R_2 be two subsets of V such that (i) $R_1 \cap R_2 = \emptyset$ or $G[R_1 \cap R_2]$ is a clique and (ii) there is no $\{u_1, u_2\} \in E$ with $u_1 \in R_1 - R_2$ and $u_2 \in R_2 - R_1$. Then, $\mathrm{tw}(G[R_1 \cup R_2]) \leq \max\{\mathrm{tw}(G[R_1]), \mathrm{tw}(G[R_2])\}$.

A set $S \subseteq V$ is a *cutset* if $G - S$ is disconnected. A cutset S is a k-*cut* if $|S| = k$. A k-cut is *strong* if $G - S$ has at least three connected components. A graph with at least k vertices is k-*connected* if it has no $(k-1)$-cut. A *biconnected component* of G is a maximal 2-connected subgraph of G.

Let C be a cutset of G, and $G_1, ..., G_p$ be the connected components of $G - C$. For $1 \leq i \leq p$, let $G_i \cup K(C)$ be the graph obtained from $G[V(G_i) \cup C]$ by adding an edge between every pair of non-adjacent vertices in C. The graphs $G_1 \cup K(C), ..., G_p \cup K(C)$ are called the *augmented components* induced by C. Clearly, if G is k-connected and C is a k-cut of G, then all the augmented components induced by C are also k-connected.

It is well known that the biconnected components of a graph are unique. Let $\mathcal{C}^1$ be the set of all 1-cuts of G, and $\mathcal{B}$ be the set of all biconnected components of G. Consider the bipartite graph $H = (\mathcal{C}^1 \cup \mathcal{B}, F)$, where $F = \{\{C, B\} : C \in \mathcal{C}^1, B \in \mathcal{B}, \text{ and } C \subseteq V(B)\}$. It is known that H is a tree. Suppose that $\mathcal{B} = \{B_1, ..., B_q\}$. Let $I = \{1, ..., q\}$. Root the tree H at B_1 and define $T^1(G)$ to be the tree whose vertex set is I and edge set is $\{\{i, i'\} : B_i$ is the grandparent of $B_{i'}$ in the rooted tree $H\}$. (Note that $T^1(G)$ is undirected.)

Fact 1 $(\{V(B_i) : i \in I\}, T^1(G))$ is a tree-decomposition of G and can be computed from G in $O(|V|)$ time.

Suppose that G is 2-connected. Further suppose that G contains a 2-cut. Replacing G by the augmented components induced by a 2-cut is called *splitting* G. Suppose G is split, the augmented components are split, and so on, until no more splits are possible. The graphs constructed in this way are 3-connected and the set of the graphs are called a *2-decomposition* of G. Each element of a 2-decomposition of G is called a *split component* of G. It is possible for G to have two or more 2-decompositions. A split component of G must be either a triangle or a 3-connected graph with at least 4 vertices. Let $\mathcal{D}$ be a 2-decomposition of G. We use $\mathcal{C}^2(\mathcal{D})$ to denote the set of the 2-cuts used to split G into the split components in $\mathcal{D}$. Consider the bipartite graph $H = (\mathcal{C}^2(\mathcal{D}) \cup \mathcal{D}, F)$, where $F = \{\{C, D\} : C \in \mathcal{C}^2(\mathcal{D}), D \in \mathcal{D}, \text{ and } C \subseteq V(D)\}$. It is known that H is a tree. Suppose that $\mathcal{D} = \{D_1, ..., D_q\}$. Let $I = \{1, ..., q\}$. Root the tree H at D_1 and define $T^2(G, \mathcal{D})$ to be the tree whose vertex set is I and edge set is $\{\{i, i'\} : D_i$ is the grandparent of $D_{i'}$ in the rooted tree $H\}$. (Note that $T^2(G, \mathcal{D})$ is undirected.) Construct a supergraph $G^2(\mathcal{D})$ of G as follows: For each $\{u, v\} \in \mathcal{C}^2(\mathcal{D})$ with $\{u, v\} \notin E$, add the edge $\{u, v\}$ to G. Then, we have the following fact:

Fact 2 $(\{V(D_i) : i \in I\}, T^2(G, \mathcal{D}))$ is a tree-decomposition of $G^2(\mathcal{D})$.

3 A technical lemma

Let S be a set. For an integer $k \geq 2$, a *k-cover* of S is a list of k subsets of S such that each element of S is contained in exactly $k - 1$ subsets in the list.

Lemma 2. Let $G = (V, E)$ be a graph. Let k and b be two nonnegative integers with $k \geq 2$, and τ be a property on k-covers of subsets of V. Suppose that G has a tree-decomposition $(\{X_j : j \in I\}, T)$ and T has a rooted version such that the following three conditions are satisfied:
  (1) For every $j' \in I$ and every child j of j' in T, $G[X_{j'} \cap X_j]$ is a clique.
  (2) For the root $r \in I$ of T, we can compute a k-cover $\langle R_1, ..., R_k \rangle$ of X_r in $f(k, |X_r|)$ time such that
      (2a) for every $1 \leq l \leq k$, $\mathrm{tw}(G[R_l]) \leq b$ and
      (2b) for every child j'' of r in T, $\langle R_1 \cap X_{j''}, ..., R_k \cap X_{j''} \rangle$ is a k-cover of $X_r \cap X_{j''}$ satisfying τ.
  (3) For every $j' \in I$ and every child j of j' in T and every k-cover $\langle Y_1, ..., Y_k \rangle$ of $X_{j'} \cap X_j$ satisfying τ, we can compute a k-cover $\langle Z_1, ..., Z_k \rangle$ of X_j in $f(k, |X_j|)$ time such that
      (3a) for every $1 \leq l \leq k$, $Y_l = Z_l \cap X_{j'}$,
      (3b) for every $1 \leq l \leq k$, $\mathrm{tw}(G[Z_l]) \leq b$, and
      (3c) for every child j'' of j, $\langle Z_1 \cap X_{j''}, ..., Z_k \cap X_{j''} \rangle$ is a k-cover of $X_j \cap X_{j''}$ satisfying τ.
Then, we can compute a k-cover $\langle V_1, ..., V_k \rangle$ of V in $O(\sum_{j \in I} f(k, |X_j|))$ time such that for each $1 \leq l \leq k$, $\mathrm{tw}(G[V_l]) \leq b$ and $V_l \cap X_r = R_l$.

Proof. Consider the following algorithm for computing $\langle V_1, ..., V_k \rangle$:

Algorithm 1
1. Set $V_1, ..., V_k$ to be the empty set.
2. While traversing T (starting at its root r) in a breadth-first manner, perform the following steps:
    2.1. If the current vertex j is r, then compute a k-cover $\langle R_1, ..., R_k \rangle$ of X_r satisfying the two conditions (2a) and (2b) above, and further add the vertices in each R_l, $1 \leq l \leq k$, to V_l.
    2.2. If the current vertex j is not r, then find the parent j' of j in T, set $\langle Y_1, ..., Y_k \rangle = \langle V_1 \cap (X_{j'} \cap X_j), ..., V_k \cap (X_{j'} \cap X_j) \rangle$, compute a k-cover $\langle Z_1, ..., Z_k \rangle$ of X_j satisfying the conditions (3a), (3b), and (3c) above, and add the vertices in each Z_l, $1 \leq l \leq k$, to V_l.
3. Output $\langle V_1, ..., V_k \rangle$.

Next, we prove that the output $\langle V_1, ..., V_k \rangle$ of Algorithm 1 satisfies that $\mathrm{tw}(G[V_l]) \leq b$ and $V_l \cap X_r = R_l$ for each $1 \leq l \leq k$. First note that the while-loop in Algorithm 1 is executed $|I|$ times. W.l.o.g., we may assume that $I = \{1, ..., |I|\}$ and that $j + 1$ is traversed by Algorithm 1 right after j for each

$1 \leq j \leq |I| - 1$. Then, $r = 1$. For each $1 \leq j \leq |I|$ and each $1 \leq l \leq k$, let V_l^j be the content of the variable V_l right after the jth iteration of the while-loop. We claim that for each $1 \leq j \leq |I|$, $\langle V_1^j, ..., V_k^j \rangle$ is a k-cover of $\cup_{1 \leq i \leq j} X_i$ satisfying the following three conditions:

(C1) $\mathrm{tw}(G[V_l^j]) \leq b$ and $V_l^j \cap X_1 = R_l$ for each $1 \leq l \leq k$.

(C2) For each son j'' of j in T, $\langle V_1^j \cap (X_j \cap X_{j''}), ..., V_k^j \cap (X_j \cap X_{j''}) \rangle$ is a k-cover of $X_j \cap X_{j''}$ satisfying τ.

(C3) For each $1 \leq i \leq j$ and each child i'' of i in T, $\langle V_1^j \cap (X_i \cap X_{i''}), ..., V_k^j \cap (X_i \cap X_{i''}) \rangle = \langle V_1^i \cap (X_i \cap X_{i''}), ..., V_k^i \cap (X_i \cap X_{i''}) \rangle$.

The lemma implies the claim. We prove the claim by induction on j. In case $j = 1$, the claim clearly holds. Let j be some integer with $2 \leq j \leq |I|$ and assume that the claim holds for all integers i with $i \leq j - 1$. Let j' be the parent of j in T, and let $\langle Y_1, ..., Y_k \rangle = \langle V_1^{j-1} \cap (X_{j'} \cap X_j), ..., V_k^{j-1} \cap (X_{j'} \cap X_j) \rangle$. Then, since $j' \leq j - 1$, we have $\langle Y_1, ..., Y_k \rangle = \langle V_1^{j'} \cap (X_{j'} \cap X_j), ..., V_k^{j'} \cap (X_{j'} \cap X_j) \rangle$ by (C3) in the inductive hypothesis. Combining this with (C2) in the inductive hypothesis, we have that $\langle Y_1, ..., Y_k \rangle$ is a k-cover of $X_{j'} \cap X_j$ satisfying τ. Thus, in the jth execution of step 2.2, we can compute a k-cover $\langle Z_1, ..., Z_k \rangle$ of X_j satisfying the conditions (3a), (3b), and (3c) above.

Firstly, we prove that $\langle V_1^j, ..., V_k^j \rangle$ is a k-cover of $\cup_{1 \leq i \leq j} X_i$. To see this, first observe that $\langle Z_1, ..., Z_k \rangle$ is a k-cover of X_j and that $\langle V_1^j, ..., V_k^j \rangle = \langle V_1^{j-1} \cup Z_1, ..., V_k^{j-1} \cup Z_k \rangle$. Moreover, by the inductive hypothesis, $\langle V_1^{j-1}, ..., V_k^{j-1} \rangle$ is a k-cover of $\cup_{1 \leq i \leq j-1} X_i$. Thus, each $v \in \cup_{1 \leq i \leq j} X_i - ((\cup_{1 \leq i \leq j-1} X_i) \cap X_j)$ appears in exactly $k - 1$ sets in $\langle V_1^j, ..., V_k^j \rangle$. It remains to consider the vertices in $(\cup_{1 \leq i \leq j-1} X_i) \cap X_j$. Since the path from j to each i, $1 \leq i \leq j - 1$, in T must pass j', we have $(\cup_{1 \leq i \leq j-1} X_i) \cap X_j = X_{j'} \cap X_j$ by the definition of tree-decompositions. Fix a vertex $v \in X_{j'} \cap X_j$. By the inductive hypothesis, v appears in exactly $k - 1$ sets in $\langle V_1^{j-1}, ..., V_k^{j-1} \rangle$. Also, v appears in exactly $k - 1$ sets in $\langle Z_1, ..., Z_k \rangle$. Moreover, for each $1 \leq l \leq k$, $v \in V_l^{j-1}$ if and only if $v \in Z_l$ by the condition (3a) above. Thus, v appears in exactly $k - 1$ sets in $\langle V_1^j, ..., V_k^j \rangle$.

Secondly, we prove that for $1 \leq l \leq k$, $\mathrm{tw}(G[V_l^j]) \leq b$. Fix an integer l with $1 \leq l \leq k$. It suffices to prove that $\mathrm{tw}(G[V_l^j]) \leq b$. This is done by applying Lemma 1. Let us be more precise. Since $(\cup_{1 \leq i \leq j-1} X_i) \cap X_j = X_{j'} \cap X_j$, we have $V_l^{j-1} \cap Z_l \subseteq X_{j'} \cap X_j$. On the other hand, $G[X_{j'} \cap X_j]$ is a clique. Thus, $G[V_l^{j-1} \cap Z_l]$ is also a clique. Let $v_1 \in V_l^{j-1} - Z_l$ and $v_2 \in Z_l - V_l^{j-1}$. We want to show that $\{v_1, v_2\} \notin E$. Assume, on the contrary, that $\{v_1, v_2\} \in E$. Then, since the path from j to each i, $1 \leq i \leq j - 1$, in T must pass j', we have that $v_1 \in X_{j'} \cap X_j$ or $v_2 \in X_{j'} \cap X_j$ by the definition of tree-decompositions. If $v_1 \in X_{j'} \cap X_j$, then $v_1 \in (V_l^{j-1} \cap (X_{j'} \cap X_j)) - (Z_l \cap X_{j'})$; otherwise, $v_2 \in (Z_l \cap X_{j'}) - (V_l^{j-1} \cap (X_{j'} \cap X_j))$. However, this contradicts that $V_l^{j-1} \cap (X_{j'} \cap X_j) = Y_l = Z_l \cap X_{j'}$. Therefore, $\{v_1, v_2\} \notin E$. Recall that $G[V_l^{j-1} \cap Z_l]$ is a clique. Hence, if we set $R_1 = V_l^{j-1}$ and $R_2 = Z_l$, then R_1 and R_2 satisfy the conditions in Lemma 1. This implies that $\mathrm{tw}(G[V_l^j]) \leq \max\{\mathrm{tw}(G[V_l^{j-1}]), \mathrm{tw}(G[Z_l])\}$. By the inductive hypothesis, $\mathrm{tw}(G[V_l^{j-1}]) \leq b$. By the condition (3b) above, $\mathrm{tw}(G[Z_l]) \leq b$. Thus, $\mathrm{tw}(G[V_l^j]) \leq b$ by Lemma 1.

Thirdly, we prove that for each $1 \leq l \leq k$, $V_l^j \cap X_1 = R_l$. Fix an integer l with $1 \leq l \leq k$. By the inductive hypothesis, $V_l^{j-1} \cap X_1 = R_l$. Thus, to prove that $V_l^j \cap X_1 = R_l$, it suffices to prove that $Z_l \cap X_1 \subseteq V_l^{j-1} \cap X_1$. Fix a vertex $v \in Z_l \cap X_1$. Since the path from j to the root 1 in T must pass j', we have $v \in X_{j'}$ by the definition of tree-decompositions. Thus, $v \in Z_l \cap X_{j'} \cap X_1$. This together with the condition (3a) implies that $v \in Y_l \cap X_1$. Recall that $Y_l = V_l^{j-1} \cap (X_{j'} \cap X_j)$. Therefore, $v \in V_l^{j-1} \cap X_1$.

Fourthly, we prove that $\langle V_1^j, ..., V_k^j \rangle$ satisfies the condition (C2) above. Let j'' be a son of j in T. We want to show that $\langle V_1^j \cap (X_j \cap X_{j''}), ..., V_k^j \cap (X_j \cap X_{j''}) \rangle$ is a k-cover of $X_j \cap X_{j''}$ satisfying τ. Since $\langle Z_1 \cap X_{j''}, ..., Z_k \cap X_{j''} \rangle$ is a k-cover of $X_j \cap X_{j''}$ satisfying τ by the condition (3c) above, it suffices to show that for each $1 \leq l \leq k$, $V_l^j \cap X_j = Z_l$. Fix an integer l with $1 \leq l \leq k$. Since the path from j to each i, $1 \leq i \leq j - 1$, in T must pass j', we have $V_l^{j-1} \cap X_j \subseteq X_{j'}$ by the definition of tree-decompositions. Thus, $V_l^{j-1} \cap X_j \subseteq V_l^{j-1} \cap (X_{j'} \cap X_j)$. On the other hand, $V_l^{j-1} \cap (X_{j'} \cap X_j) = Z_l \cap X_{j'}$ by the condition (3a) above. Hence, $V_l^{j-1} \cap X_j \subseteq Z_l$. Noting that $V_l^j = V_l^{j-1} \cup Z_l$ and $Z_l \subseteq X_j$, we see that $V_l^j \cap X_j = Z_l$ if and only if $V_l^{j-1} \cap X_j \subseteq Z_l$. Therefore, we have $V_l^j \cap X_j = Z_l$.

Finally, we prove that $\langle V_1^j, ..., V_k^j \rangle$ satisfies the condition (C3) above. Let i be an integer with $1 \leq i \leq j$, and i'' be a child of i in T. We want to show that $\langle V_1^j \cap (X_i \cap X_{i''}), ..., V_k^j \cap (X_i \cap X_{i''}) \rangle = \langle V_1^i \cap (X_i \cap X_{i''}), ..., V_k^i \cap (X_i \cap X_{i''}) \rangle$. This clearly holds if $i = j$. So, we may assume that $i \leq j - 1$. Then, $\langle V_1^{j-1} \cap (X_i \cap X_{i''}), ..., V_k^{j-1} \cap (X_i \cap X_{i''}) \rangle = \langle V_1^i \cap (X_i \cap X_{i''}), ..., V_k^i \cap (X_i \cap X_{i''}) \rangle$ by the inductive hypothesis. By this, we only need to show that $\langle V_1^j \cap (X_i \cap X_{i''}), ..., V_k^j \cap (X_i \cap X_{i''}) \rangle = \langle V_1^{j-1} \cap (X_i \cap X_{i''}), ..., V_k^{j-1} \cap (X_i \cap X_{i''}) \rangle$. Fix an integer l with $1 \leq l \leq k$. Since the path from j to each i, $1 \leq i \leq j - 1$, in T must pass j', we have $Z_l \cap (X_i \cap X_{i''}) \subseteq X_{j'}$ by the definition of tree-decompositions. Thus, $Z_l \cap (X_i \cap X_{i''}) \subseteq Z_l \cap X_{j'} \cap (X_i \cap X_{i''})$. On the other hand, $Z_l \cap X_{j'} \subseteq V_l^{j-1}$ by the condition (3a) above. Hence, $Z_l \cap (X_i \cap X_{i''}) \subseteq V_l^{j-1} \cap (X_i \cap X_{i''})$. Noting that $V_l^j = V_l^{j-1} \cup Z_l$ and $Z_l \subseteq X_j$, we see that $V_l^j \cap (X_i \cap X_{i''}) = V_l^{j-1} \cap (X_i \cap X_{i''})$ if and only if $Z_l \cap (X_i \cap X_{i''}) \subseteq V_l^{j-1} \cap (X_i \cap X_{i''})$. Therefore, we have $V_l^j \cap (X_i \cap X_{i''}) = V_l^{j-1} \cap (X_i \cap X_{i''})$. ∎

Let $G = (V, E)$ be a graph, and $U \subseteq V$. A k-cover L of U is *completely unbalanced* if exactly one set in L is empty and the others are equal to U. A k-cover L of U is *weakly unbalanced* if there are one $u \in U$ and two sets U_1 and U_2 in L such that $U_1 = \{u\}$, $U_2 = U - \{u\}$, and all the sets in L except U_1 and U_2 are equal to U. A k-cover of U is *unbalanced* if it is either completely unbalanced or weakly unbalanced. Note that if $|U| \leq 2$, then every k-cover of U must be unbalanced. Hereafter, the property τ in Lemma 2 means "unbalanced", i.e., a k-cover L of U satisfies τ if and only if L is unbalanced.

4 Decomposing $K_{3,3}$-free graphs

We start by recalling the definition of k-outerplanar graphs introduced by Baker [5]. These graphs are defined inductively. Let G be a graph. G is 1-*outerplanar* if and only if it is outerplanar. For $k \geq 2$, G is *k-outerplanar* if and only if it has a planar embedding such that if all vertices on the exterior face (and all adjacent edges) are deleted, then the connected components of the remaining graph are all $(k-1)$-outerplanar. We utilize k-outerplanar graphs in the proof of the following lemma.

Lemma 3. Let $G = (V, E)$ be a connected planar graph, and k be an integer ≥ 2. Suppose that s_1 and s_2 are two adjacent vertices in G and $\langle Y_1, ..., Y_k \rangle$ is an unbalanced k-cover of $\{s_1, s_2\}$. Then, we can compute a k-cover $\langle Z_1, ..., Z_k \rangle$ of V in $O(k|V|)$ time such that $tw(G[Z_l]) \leq 3k - 4$ and $Z_l \cap \{s_1, s_2\} = Y_l$ for each $1 \leq l \leq k$.

Proof. We assume that $\langle Y_1, ..., Y_k \rangle$ is completely unbalanced; the other case is similar. Then, by symmetry, we may assume that $Y_1 = \emptyset$ and $Y_2 = \cdots = Y_k = \{s_1, s_2\}$. Let H be the graph obtained from G by replacing the edge $\{s_1, s_2\}$ with two edges $\{s_1, x\}$ and $\{x, s_2\}$, where x is a new vertex. It is clear that H is still planar. We perform a breadth-first-search (BFS) on H starting at x to obtain a BFS tree T. For each vertex v in H, we define the level number of v (denoted $lev(v)$) to be the length of the path from x to v in T. Note that only x has level number 0 and only s_1 and s_2 have level number 1. For each $1 \leq l \leq k$, let $V_l = \{v \in V : lev(v) \equiv l - 1 \pmod{k}\}$. Let $Z_1 = V - V_2$, $Z_2 = V - V_1$, and $Z_l = V - V_l$ for each $3 \leq l \leq k$. Obviously, $\langle Z_1, ..., Z_k \rangle$ is a k-cover of V. Moreover, the subgraph induced by each nonempty Z_l, $1 \leq l \leq k$, is $(k-1)$-outerplanar and hence has treewidth $\leq 3k - 4$ [3]. It is also clear that $Z_l \cap \{s_1, s_2\} = Y_l$ for each $1 \leq l \leq k$. ∎

Lemma 4. [2, 8]. Each split component of a 2-connected $K_{3,3}$-free graph is either isomorphic to K_5 or planar.

Lemma 5. Let $G = (V, E)$ be a 2-connected $K_{3,3}$-free graph. Then, for any $k \geq 2$, we can compute a k-cover $\langle V_1, ..., V_k \rangle$ of V in $O(k|V|)$ time such that $tw(G[V_l]) \leq 3k - 4$ for each $1 \leq l \leq k$.

Proof. Let $\mathcal{D} = \{D_1, ..., D_q\}$ be a 2-decomposition of G, and let $I = \{1, ..., q\}$. It is known that $\mathcal{D}$ can be computed in $O(|V|)$ time [9]. Moreover, $\sum_{i \in I} |V(D_i)| = O(|V|)$ [9]. W.l.o.g., we may assume that $G^2(\mathcal{D}) = G$ because a k-cover $\langle V_1, ..., V_k \rangle$ of V such that the subgraph of $G^2(\mathcal{D})$ induced by V_l has treewidth $\leq 3k - 4$ for each $1 \leq l \leq k$ is also a k-cover $\langle V_1, ..., V_k \rangle$ of V such that $tw(G[V_l]) \leq 3k - 4$ for each $1 \leq l \leq k$. Then, by Fact 2, $(\{V(D_j) : j \in I\}, T^2(G, \mathcal{D}))$ is a tree-decomposition of G. For convenience, let $T = T^2(G, \mathcal{D}))$, $b = 3k - 4$, and $X_j = V(D_j)$ and $f(k, |X_j|) = O(k|X_j|)$ for each $j \in I$. We want to apply Lemma 2 to the graph G and the tree-decomposition $(\{X_j : j \in I\}, T)$. To this end, we first (arbitrarily) choose an $r \in I$ and root T at r.

Clearly, the condition (1) in Lemma 2 is satisfied by G and $(\{X_j : j \in I\}, T)$. By Lemma 4, $G[X_r] = D_r$ is either isomorphic to K_5 or planar. Let us first suppose that $G[X_r]$ is isomorphic to K_5. Then, we set $R_1 = \emptyset$ and $R_2 = \cdots = R_k = X_r$ if $k \geq 3$; otherwise $(k = 2)$, we arbitrarily choose two vertices v_1 and v_2 in X_r and set $R_1 = \{v_1, v_2\}$ and $R_2 = X_r - R_1$. Obviously, $\langle R_1, ..., R_k \rangle$ is a k-cover of X_r satisfying the condition (2a) in Lemma 2. $\langle R_1, ..., R_k \rangle$ also satisfies the condition (2b) in Lemma 2 since $|X_r \cap X_{j''}| = 2$ for every child j'' of r in T. Next, suppose that $G[X_r]$ is a planar graph. Then, we arbitrarily choose an edge $\{s_1, s_2\}$ in $G[X_r]$, set $Y_1 = \emptyset$ and $Y_2 = \cdots = Y_k = \{s_1, s_2\}$, and use Lemma 3 to compute a k-cover $\langle R_1, ..., R_k \rangle$ of X_r in $O(k|X_r|)$ time such that $tw(G[R_l]) \leq 3k - 4$ for each $1 \leq l \leq k$. Clearly, $\langle R_1, ..., R_k \rangle$ satisfies the condition

(2a) in Lemma 2. $\langle R_1, ..., R_k \rangle$ also satisfies the condition (2b) in Lemma 2 since $|X_r \cap X_{j''}| = 2$ for every child j'' of r in T.

Fix a $j' \in I$ and a child j of j' in T. Let $\langle Y_1, ..., Y_k \rangle$ be an unbalanced k-cover of $X_{j'} \cap X_j$. W.l.o.g., we may assume that $|Y_l| \leq |Y_{l+1}|$ for each $1 \leq l \leq k - 1$. By Lemma 4, $G[X_j] = D_j$ is either isomorphic to K_5 or planar. Let us first suppose that $G[X_j]$ is isomorphic to K_5. If $k \geq 3$, then we set $Z_1 = Y_1$ and $Z_l = Y_l \cup (X_j - X_{j'})$ for each $2 \leq l \leq k$. Otherwise ($k = 2$), we arbitrarily choose a vertex $v \in X_j - X_{j'}$ and set $Z_1 = Y_1 \cup (X_j - (X_{j'} \cup \{v\}))$ and $Z_2 = Y_2 \cup \{v\}$. Then, no matter what k is, $\langle Z_1, ..., Z_k \rangle$ is a k-cover of X_j satisfying the conditions (3a), (3b), and (3c) in Lemma 2. Next, suppose that $G[X_j]$ is planar. Let $X_{j'} \cap X_j = \{s_1, s_2\}$. Note that s_1 and s_2 are adjacent in G. We use Lemma 3 to compute a k-cover $\langle Z_1, ..., Z_k \rangle$ of X_j. It should be easy to see that $\langle Z_1, ..., Z_k \rangle$ is a k-cover of X_j satisfying the conditions (3a), (3b), and (3c) in Lemma 2. ∎

Theorem 6. Let $G = (V, E)$ be a $K_{3,3}$-free graph. Then, for any $k \geq 2$, we can compute a k-cover $\langle V_1, ..., V_k \rangle$ of V in $O(k|V|)$ time such that $\text{tw}(G[V_l]) \leq 3k - 4$ for $1 \leq l \leq k$.

5 Decomposing K_5-free graphs

Suppose G is 3-connected. If G contains a *strong* 3-cut C, then replacing G by the augmented components induced by C is called *strongly splitting* G. Suppose G is strongly split, the augmented components are strongly split, and so on, until no more strong splits are possible. The set of the graphs constructed in this way are called a *strong 3-decomposition* of G.

Definition 7. We define W to be the graph obtained from a 8-cycle by adding 4 crossing edges. That is, $W = (\{1, ..., 8\}, E_1 \cup E_2)$, where $E_1 = \{\{i, i+1\} : 1 \leq i \leq 7\} \cup \{\{8, 1\}\}$ and $E_2 = \{\{i, i+4\} : 1 \leq i \leq 4\}$. A K_5-free graph G is said to be *nice* if G is 3-connected, nonplanar, and is not isomorphic to $K_{3,3}$ or W.

Fact 3 [11] Suppose that G is a nice K_5-free graph. Let C be a strong 3-cut in G. Then, the augmented components induced by C are also nice K_5-free graphs. Moreover, G has another strong 3-cut C' if and only if C' is a strong 3-cut of an augmented component of G induced by C.

Fact 4 [11] A nice K_5-free graph has a *unique* strong 3-decomposition. Moreover, each graph in the strong 3-decomposition is planar.

Suppose that $G = (V, E)$ is a nice K_5-free graph. Let $\mathcal{D}^3(G)$ be the strong 3-decomposition of G, and $\mathcal{C}^3(G)$ be the set of all strong 3-cuts in G. Define $H(G)$ to be the bipartite graph $(\mathcal{D}^3(G) \cup \mathcal{C}^3(G), F)$, where $F = \{\{D, C\} : D \in \mathcal{D}^3(G), C \in \mathcal{C}^3(G), \text{ and } C \subseteq V(D)\}$.

Lemma 8. (1) Every edge of G is contained in some graph in $\mathcal{D}^3(G)$.
    (2) If a subset S of V induces a triangle but $S \notin \mathcal{C}^3(G)$, then exactly one graph in $\mathcal{D}^3(G)$ contains the three vertices in S.
    (3) $H(G)$ is a tree. Moreover, if some vertex $u \in V$ is contained in two graphs D and D' in $\mathcal{D}^3(G)$, then u is contained in every graph on the path between D and D' in $H(G)$.

Proof. We show the lemma by induction on the number of strong 3-cuts in G. The lemma clearly holds when G has no strong 3-cut. Let $p \geq 1$, and assume that the lemma is true for every graph that has up to $p - 1$ strong 3-cuts. Consider a graph G with p strong 3-cuts. Let C be a strong 3-cut in G, and $G_1 \cup K(C)$, ..., $G_k \cup K(C)$ be the augmented components induced by C. By Fact 3, each $G_i \cup K(C)$, $1 \leq i \leq k$, is a nice K_5-free graph, $\mathcal{C}^3(G) = \cup_{1 \leq i \leq k} \mathcal{C}^3(G_i \cup K(C)) \cup \{C\}$, and $\mathcal{D}^3(G) = \cup_{1 \leq i \leq k} \mathcal{D}^3(G_i \cup K(C))$.

It is clear that every edge of G is contained in at least one of the graphs $G_1 \cup K(C)$, ..., $G_k \cup K(C)$. Moreover, by the inductive hypothesis, every edge of each $G_i \cup K(C)$, $1 \leq i \leq k$, is contained in some graph in $\mathcal{D}^3(G_i \cup K(C))$. These together with the fact that $\mathcal{D}^3(G) = \cup_{1 \leq i \leq k} \mathcal{D}^3(G_i \cup K(C))$ imply that the statement (1) in the lemma holds for G.

Suppose that $S \subseteq V$ induces a triangle but $S \notin \mathcal{C}^3(G)$. Then, there is exactly one $G_i \cup K(C)$, $1 \leq i \leq k$, containing S. Moreover, S cannot be a strong 3-cut in the graph $G_i \cup K(C)$ or else S would be a strong 3-cut in G by Fact 3. Thus, by the inductive hypothesis, exactly one graph in $\mathcal{D}^3(G_i \cup K(C))$ contains the three vertices in S. Therefore, exactly one graph in $\mathcal{D}^3(G)$ contains the three vertices in S. This implies that the statement (2) in the lemma holds for G.

By the inductive hypothesis, each $H(G_i \cup K(C))$, $1 \leq i \leq k$, is a tree. Moreover, in each graph $G_i \cup K(C)$, $1 \leq i \leq k$, C induces a triangle but is not a strong 3-cut. Thus, for each $1 \leq i \leq k$, exactly *one* graph (say, D_i) in $\mathcal{D}^3(G_i \cup K(C))$ contains the three vertices in C by the inductive hypothesis. On the other hand, $\mathcal{C}^3(G) = \cup_{1 \leq i \leq k} \mathcal{C}^3(G_i \cup K(C)) \cup \{C\}$ and $\mathcal{D}^3(G) = \cup_{1 \leq i \leq k} \mathcal{D}^3(G_i \cup K(C))$. Therefore, $H(G)$ can be obtained from C and the trees $H(G_1 \cup K(C))$, ..., $H(G_k \cup K(C))$ by adding the edges $\{C, D_1\}$, ..., $\{C, D_k\}$. This implies that $H(G)$ is a tree. Next, suppose that some vertex $u \in V$ is contained in two graphs D and D' in $\mathcal{D}^3(G)$. If the path between D and D' in $H(G)$ does not pass C, then the inductive hypothesis guarantees that u is contained in every graph on the path between D and D' in $H(G)$. So, we may assume that the path between D and D' in $H(G)$ does pass C. Then, there are two neighbors D_i and D_j, $1 \leq i, j \leq k$, of C in $H(G)$ such that D_i lies on the path between D and C and D_j lies on the path between D' and C. Moreover, u must be contained in C since every vertex shared by a pair of two graphs among $G_1 \cup K(C)$, ..., $G_k \cup K(C)$ must be contained in C. Hence, u is contained in both D_i and D_j. By the inductive hypothesis, u is contained in every graph on both the path between D and D_i in $H(G_i \cup K(C))$ and the path between D' and D_j in $H(G_j \cup K(C))$. This implies that u is contained in every graph on the path between D and D' in $H(G)$. Therefore, the statement (3) in the lemma holds for G. ∎

Suppose that $\mathcal{D}^3(G) = \{D_1, ..., D_q\}$. Let $I = \{1, ..., q\}$. Root the tree $H(G)$ at D_1 and define $T^3(G)$ to be the tree whose vertex set is I and edge set is $\{\{i, i'\}$: D_i is the grandparent of $D_{i'}$ in the rooted tree $H(G)\}$. (Note that $T^3(G)$ is undirected.) Construct a supergraph G^3 of G as follows: For each strong 3-cut C and each pair of nonadjacent vertices u and v in C, add the edge $\{u, v\}$ to G.

Corollary 9. $(\{V(D_i) : i \in I\}, T^3(G))$ is a tree-decomposition of G^3.

Lemma 10. Let $G = (V, E)$ be a connected planar graph, and $k \geq 2$. Suppose that $S \subseteq V$ induces a triangle, and $\langle Y_1, ..., Y_k \rangle$ is an unbalanced $\overline{k}$-cover of S. Then, we can compute a k-cover $\langle Z_1, ..., Z_k \rangle$ of V in $O(k|V|)$ time such that $\mathrm{tw}(G[Z_l]) \leq 6k - 7$ and $Z_l \cap S = Y_l$ for each $1 \leq l \leq k$, and $\langle Z_1 \cap S', ..., Z_k \cap S' \rangle$ is an unbalanced k-cover of S' for all subsets S' of V with $G[S']$ being a triangle.

Proof. Let $S = \{s_1, s_2, s_3\}$. We assume that $\langle Y_1, ..., Y_k \rangle$ is completely unbalanced. Then, by symmetry, we may assume that $Y_1 = \emptyset$ and $Y_2 = \cdots = Y_k = S$. Let H be the graph obtained from G by merging the three vertices in S into a new vertex $x \notin V$. Clearly, H is also planar. We perform a breadth-first-search (BFS) on H starting at x to obtain a BFS tree T. For each vertex v in H, we define $lev(v)$ to be the length of the path from x to v in T. Recall that s_1, s_2, and s_3 are not in H. We define $lev(s_1) = lev(s_2) = lev(s_3) = 0$. For each $1 \leq l \leq k$, let $Z_l = V - \{v \in V : lev(v) \equiv l - 1 \pmod k\}$. Obviously, $\langle Z_1, ..., Z_k \rangle$ is a k-cover of V, $Z_l \cap S = Y_l$ for each $1 \leq l \leq k$, and $\langle Z_1 \cap S', ..., Z_k \cap S' \rangle$ is an unbalanced k-cover of S' for all subsets S' of V with $G[S']$ being a triangle. It remains to show that $\text{tw}(G[Z_l]) \leq 6k - 7$ for each $1 \leq l \leq k$. To this end, fix an arbitrary l, $1 \leq l \leq k$. Consider a planar embedding of $\bar{G}$. In the embedding, the triangle $G[S]$ splits the plane into two regions. One of the regions is infinite and the other is finite. Let Z_l^{in} be the vertices of Z_l falling into the finite region, and $Z_l^{out} = Z_l - Z_l^{in}$. It is not difficult to see that both $G[Z_l^{in}]$ and $G[Z_l^{out}]$ are $(k-1)$-outerplanar (no matter whether $S \subseteq Z_l$ or not). From this, we observe that $G[Z_l]$ is $(2k-2)$-outerplanar. Therefore, $\text{tw}(G[Z_l]) \leq 6k - 7$ [3]. $\blacksquare$

Lemma 11. Let $G = (V, E)$ be a nice K_5-free graph, and $k \geq 2$. Suppose that s_1 and s_2 are two adjacent vertices in G and $\langle U_1, ..., U_k \rangle$ is an unbalanced k-cover of $\{s_1, s_2\}$. Then, we can compute a k-cover $\langle V_1, ..., V_k \rangle$ of V in $O(k|V| + |V|^2)$ time such that $\text{tw}(G[V_l]) \leq 6k - 7$ and $V_l \cap \{s_1, s_2\} = U_l$ for $1 \leq l \leq k$.

Proof. Let $\mathcal{D}^3(G) = \{D_1, ..., D_q\}$ be the strong 3-decomposition of G, and let $I = \{1, ..., q\}$. It is known that $\mathcal{D}^3(G)$ can be computed in $O(|V|^2)$ time [10]. W.l.o.g., we may assume that $G^3 = G$ because a k-cover $\langle V_1, ..., V_k \rangle$ of V such that the subgraph of G^3 induced by V_l has treewidth $\leq 6k - 7$ for each $1 \leq l \leq k$ is also a k-cover $\langle V_1, ..., V_k \rangle$ of V such that $\text{tw}(G[V_l]) \leq 6k - 7$ for each $1 \leq l \leq k$. Then, by Fact 9, $(\{V(D_j) : j \in I\}, \mathcal{T}^3(G))$ is a tree-decomposition of G. For convenience, let $T = \mathcal{T}^3(G)$, $b = 6k - 7$, and $X_j = V(D_j)$ and $f(k, |X_j|) = O(k|X_j|)$ for each $j \in I$. We want to apply Lemma 2 to the graph G and the tree-decomposition $(\{X_j : j \in I\}, T)$. To this end, we first choose an $r \in I$ with $\{s_1, s_2\} \subseteq X_r$ and root T at r. Such an r must exist because $\{s_1, s_2\} \in E(G)$.

Clearly, the condition (1) in Lemma 2 is satisfied by G and $(\{X_j : j \in I\}, T)$. By Fact 4, $G[X_r] = D_r$ is planar. So, by Lemma 3, we can compute a k-cover $\langle R_1, ..., R_k \rangle$ of X_r such that $\text{tw}(G[R_l]) \leq 3k - 4$ and $R_l \cap \{s_1, s_2\} = U_l$ for each $1 \leq l \leq k$. Moreover, it is clear from the proof of Lemma 3 that for every subset S of X_r with $G[S]$ being a triangle, $\langle R_1 \cap S, ..., R_k \cap S \rangle$ is an unbalanced k-cover of S. Now, it should be easy to verify that $\langle R_1, ..., R_k \rangle$ is a k-cover of X_r satisfying the conditions (2a) and (2b) in Lemma 2.

Fix a $j' \in I$ and a child j of j' in T. Let $\langle Y_1, ..., Y_k \rangle$ be an unbalanced k-cover of $X_{j'} \cap X_j$. Let $S = X_{j'} \cap X_j$. Recall that $G[S]$ is a triangle. So, we can compute a k-cover $\langle Z_1, ..., Z_k \rangle$ of X_j satisfying the conditions in Lemma 10. It should be easy to see that $\langle Z_1, ..., Z_k \rangle$ is a k-cover of X_j satisfying the conditions (3a), (3b), and (3c) in Lemma 2.

By the discussions above and Lemma 2, there is a k-cover $\langle V_1, ..., V_k \rangle$ of V such that $\text{tw}(G[V_l]) \leq 6k - 7$ and $V_l \cap X_r = R_l$ for each $1 \leq l \leq k$. Fix an l with $1 \leq l \leq k$. Recall that $R_l \cap \{s_1, s_2\} = U_l$ and that $\{s_1, s_2\} \subseteq X_r$. Thus, $V_l \cap \{s_1, s_2\} = V_l \cap (X_r \cap \{s_1, s_2\}) = R_l \cap \{s_1, s_2\} = U_l$. $\blacksquare$

Lemma 12. Let $G = (V, E)$ be a 2-connected K_5-free graph. Then, for any $k \geq 2$, we can compute a k-cover $\langle V_1, ..., V_k \rangle$ of V in $O(k|V| + |V|^2)$ time such that $\mathrm{tw}(G[V_l]) \leq 6k - 7$ for each $1 \leq l \leq k$.

Proof. Let $\mathcal{D} = \{D_1, ..., D_q\}$ be a 2-decomposition of G, and let $I = \{1, ..., q\}$. It is known that $\mathcal{D}$ can be computed in $O(|V|)$ time [9]. W.l.o.g., we may assume that $G^2(\mathcal{D}) = G$ because a k-cover $\langle V_1, ..., V_k \rangle$ of V such that the subgraph of $G^2(\mathcal{D})$ induced by V_l has treewidth $\leq 6k - 7$ for each $1 \leq l \leq k$ is also a k-cover $\langle V_1, ..., V_k \rangle$ of V such that $\mathrm{tw}(G[V_l]) \leq 6k - 7$ for each $1 \leq l \leq k$. Then, by Fact 2, $(\{V(D_j) : j \in I\}, T^2(G, \mathcal{D}))$ is a tree-decomposition of G. For convenience, let $T = T^2(G, \mathcal{D})$, $b = 6k - 7$, and $X_j = V(D_j)$ and $f(k, |X_j|) = O(k|X_j| + |X_j|^2)$ for each $j \in I$. We want to apply Lemma 2 to the graph G and the tree-decomposition $(\{X_j : j \in I\}, T)$. To this end, we first (arbitrarily) choose an $r \in I$ and root T at r.

Clearly, the condition (1) in Lemma 2 is satisfied by G and $(\{X_j : j \in I\}, T)$. To see that the condition (2) in Lemma 2 is also satisfied, we distinguish four cases as follows:

Case 1: $G[X_r]$ is planar. Then, as stated in the proof of Lemma 5, we can compute a k-cover $\langle R_1, ..., R_k \rangle$ of X_r in $O(k|X_r|)$ time satisfying the conditions (2a) and (2b) in Lemma 2.

Case 2: $G[X_r]$ is isomorphic to $K_{3,3}$. Then, we set $R_1 = \emptyset$ and $R_2 = \cdots = R_k = X_r$. Obviously, $\langle R_1, ..., R_k \rangle$ is a k-cover of X_r satisfying the conditions (2a) and (2b) in Lemma 2.

Case 3: $G[X_r]$ is isomorphic to the graph W (see Definition 7). Then, we set $R_1 = \emptyset$ and $R_2 = \cdots = R_k = X_r$ if $k \geq 3$; otherwise ($k = 2$), we (arbitrarily) choose four vertices from X_r and set R_1 to be the set of the four vertices and R_2 to be $X_r - R_1$. Obviously, $\langle R_1, ..., R_k \rangle$ is a k-cover of X_r satisfying the conditions (2a) and (2b) in Lemma 2.

Case 4: $G[X_r]$ is a nice K_5-free graph. Then, we arbitrarily choose an edge $\{s_1, s_2\}$ in $G[X_r]$ and set $U_1 = \emptyset$ and $U_2 = \cdots = U_k = \{s_1, s_2\}$. By Lemma 11, we can compute a k-cover $\langle R_1, ..., R_k \rangle$ of X_r in $O(k|X_r| + |X_r|^2)$ time such that $\mathrm{tw}(G[R_l]) \leq 6k - 7$ for each $1 \leq l \leq k$. Clearly, $\langle R_1, ..., R_k \rangle$ satisfies the condition (2a) in Lemma 2. $\langle R_1, ..., R_k \rangle$ also satisfies the condition (2b) in Lemma 2 since $|X_r \cap X_{j''}| = 2$ for every child j'' of r in T.

Since one of the above four cases must occur, the condition (2) in Lemma 2 is satisfied by G and $(\{X_j : j \in I\}, T)$. To see that the condition (3) in Lemma 2 is also satisfied, fix a $j' \in I$ and a child j of j' in T. Let $\langle Y_1, ..., Y_k \rangle$ be an unbalanced k-cover of $X_{j'} \cap X_j$, and let $X_{j'} \cap X_j = \{s_1, s_2\}$. Recall that $\{s_1, s_2\}$ is an edge in both $G[X_{j'}]$ and $G[X_j]$. Moreover, by symmetry, we may assume that $|Y_l| \leq |Y_{l+1}|$ for all $1 \leq l \leq k - 1$.

Case 1': $G[X_j]$ is planar. Then, as stated in the proof of Lemma 5, we can compute a k-cover $\langle Z_1, ..., Z_k \rangle$ of X_j in $O(k|X_j|)$ time satisfying the conditions (3a), (3b), and (3c) in Lemma 2.

Case 2': $G[X_j]$ is isomorphic to $K_{3,3}$. Then, we set $Z_1 = Y_1$ and $Z_l = Y_l \cup (X_j - X_{j'})$ for each $2 \leq l \leq k$. Clearly, $\langle Z_1, ..., Z_k \rangle$ is a k-cover of X_j satisfying the conditions (3a), (3b), and (3c) in Lemma 2.

Case 3': $G[X_j]$ is isomorphic to the graph W (see Definition 7). If $k \geq 3$, then we set $Z_1 = Y_1$ and $Z_l = Y_l \cup (X_j - X_{j'})$ for each $2 \leq l \leq k$; otherwise ($k = 2$), we (arbitrarily) choose a subset A of $X_j - X_{j'}$ with $|A| = 3$ and set $Z_1 = Y_1 \cup A$ and $Z_2 = X_j - Z_1$. Then, it is easy to verify that $\langle Z_1, ..., Z_k \rangle$ is a k-cover of X_j satisfying the conditions (3a), (3b), and (3c) in Lemma 2.

Case 4': $G[X_j]$ is a nice K_5-free graph. Then, by Lemma 11, we can compute a k-cover $\langle Z_1, ..., Z_k \rangle$ of X_j in $O(k|X_j| + |X_j|^2)$ time such that $\mathrm{tw}(G[Z_l]) \leq 6k - 7$

and $Z_l \cap \{s_1, s_2\} = Y_l$ for each $1 \leq l \leq k$. From this, it should be clear that $\langle Z_1, ..., Z_k \rangle$ satisfies the conditions (3a), (3b), and (3c) in Lemma 2.

By the discussions above and Lemma 2, we have the lemma. ∎

Theorem 13. Let $G = (V, E)$ be a K_5-free graph. Then, for any $k \geq 2$, we can compute a k-cover $\langle V_1, ..., V_k \rangle$ of V in $O(k|V| + |V|^2)$ time such that $tw(G[V_l]) \leq 6k - 7$ for each $1 \leq l \leq k$.

6 Concluding remarks

Let π be a hereditary property on graphs. Suppose that MISP(π) restricted to n-vertex graphs of treewidth $\leq k$ can be solved in $T_\pi(k, n)$ time. Then, by Theorem 6 (resp., Theorem 13), given an integer $k \geq 2$ and a $K_{3,3}$-free (resp., K_5-free) graph $G = (V, E)$, we can compute a subset U of V in $O(k|V| + T_\pi(3k - 4, |V|))$ (resp., $O(k|V| + |V|^2)$) time such that $G[U]$ satisfies π and $|U|$ is at least $(k-1)/k$ optimal. For various properties π, $T_\pi(k, n) = 2^{p(k)}q(n)$ where p and q are polynomials of low degree (often, of degree 1) [4, 14]. Hence, for such properties π, MISP(π) restricted to $K_{3,3}$-free or K_5-free graphs has a practical polynomial-time approximation scheme.

References

1. N. Alon, P. Seymour, and R. Thomas, A separator theorem for graphs without an excluded minor and its applications, *STOC'90*.
2. T. Asano, An approach to the subgraph homeomorphism problem, *TCS* **38** (1985).
3. H.L. Bodlaender, Planar graphs with bounded treewidth, Technical Report.
4. H.L. Bodlaender, Dynamic programming algorithms on graphs with bounded treewidth, *ICALP'88*, LNCS 317.
5. B.S. Baker, Approximation algorithms for NP-complete problems on planar graphs, *J. ACM* **41** (1994).
6. N. Chiba, T. Nishizeki, and N. Saito, An approximation algorithm for the maximum independent set problem on planar graphs, *SIAM-JC* **11** (1982).
7. D. Eppstein, Subgraph isomorphism in planar graphs and related problems, *SODA'95*.
8. D.W. Hall, A note on primitive skew curves, *Bull. Amer. Math. Soc.* **49** (1943).
9. J.E. Hopcroft and R.E. Tarjan, Dividing a graph into triconnected components, *SIAM-JC* **2** (1973), 135-158.
10. A. Kanevsky and V. Ramachandran, Improved algorithms for graph four-connectivity, *FOCS'87*.
11. A. Kézdy and P. McGuinness, Sequential and parallel algorithms to find a K_5 minor, *SODA'92*.
12. R.J. Lipton and R.E. Tarjan, Applications of a planar separator theorem, *SIAM-JC* **9** (1980), 615-627.
13. N. Robertson and P.D. Seymour, Graph minors V. Excluding a planar graph, *J. Combinatorial Theory Ser. B* **41** (1986).
14. J.A. Telle and A. Proskurowski, Practical algorithms on partial k-trees with an application to domination-like problems, *WADS'93*, LNCS 709.
15. M. Yannakakis, Node- and edge-deletion NP-complete problems, *STOC'78*.

Searching a Fixed Graph

Elias Koutsoupias[1], Christos Papadimitriou[2], Mihalis Yannakakis[3]

[1] CS Department, UCLA
[2] EECS Department, UC Berkeley
Research supported in part by a NSF grant
[3] Bell Laboratories, Murray Hill, NJ 07974

Abstract. We study three combinatorial optimization problems related to searching a graph that is known in advance, for an item that resides at an unknown node. The *search ratio* of a graph is the optimum *competitive ratio* (the worst-case ratio of the distance traveled before the unknown node is visited, over the distance between the node and a fixed root, minimized over all Hamiltonian walks of the graph). We also define the *randomized search ratio* (we minimize over all *distributions* of permutations). Finally, the *traveling repairman problem* seeks to minimize the expected time of visit to the unknown node, given some distribution on the nodes. All three of these novel graph-theoretic parameters are NP-complete —and MAXSNP-hard— to compute exactly; we present interesting approximation algorithms for each. We also show that the randomized search ratio and the traveling repairman problem are related via *duality* and *polyhedral separation*.

1 Introduction

Imagine that you know that an information item you need resides at *some* node of a fixed graph (say, a large network of hypertext documents), but you do not know where. You can only navigate the graph by following its edges, at unit cost (that is, we assume that there is no random access of pages). You will see the item once you arrive at the right node —and only then. What are good strategies for performing this task efficiently?

This is obviously a situation of decision-making under uncertainty, and therefore an invitation for applying the techniques of *competitive analysis* [ST85]. In fact, this is an on-line problem of a rather familiar genre: *exploration and navigation* [PY91, DP90, DKP91]. However, unlike previous formulations of such on-line problems, here we *know* the terrain[4] being explored. In other words, for each graph G and start (root) node r of G there is an optimal competitive ratio,

$$\sigma(G, r) = \min_{\pi} \max_{v \in G} \frac{d_{\pi}(r, v)}{d(r, v)}.$$

---

[4] An example of previous work on searching a known terrain is the *bridge problem*, sometimes called the *cow path problem*, [BCR88]; our work can be seen as a generalization of this problem from infinite paths to general graphs.

Here π ranges over all walks of G, starting from r, and visiting all nodes of G, while $d_\pi(r, v)$ is the distance traversed in walk π until we first visit node v. We call $\sigma(G, r)$ the *search ratio* of the graph with respect to the root. It is an interesting graph-theoretic parameter of a rather novel kind. Unfortunately, we point out (Theorem 1) that it is NP-complete to compute —in fact, our proof establishes that it is MAXSNP-complete.

We may of course want to introduce the *randomized version* of the search ratio:

$$\rho(G, r) = \min_\Delta \max_{v \in G} \frac{\mathcal{E}_\Delta[d_\pi(r, v)]}{d(r, v)},$$

where Δ ranges over *distributions of walks*. We call this parameter the *randomized search ratio*. For example, for the graph and root shown in Figure 1, the search ratio is $\frac{7}{3}$, while the randomized search ratio is 2. Computing $\rho(G, r)$ is also NP-complete; in fact, the surprising part here is that it is in NP at all —we establish this in Theorem 1 via a *linear programming formulation* of the problem. It is also MAXSNP-hard, although this too is somewhat tricky to establish.

[ht]

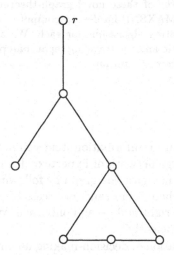

Fig. 1. A graph G with $\sigma(G, r) = \frac{7}{3}$ and $\rho(G, r) = 2$.

We present polynomial-time algorithms for approximating these parameters within a fixed factor. A simple *doubling heuristic* (repeatedly double the radius, explore the resulting graph depth-first) achieves an approximation ratio of 8 for the deterministic ratio, 16 for the randomized version.[5] We improve on this basic

---

[5] Incidentally, notice the novelty of the situation: We approximate within a bounded ratio a parameter which is *itself* a ratio of a feasible solution divided by an ideal solution! That is, we mix two well-studied compromises: one in the face of uncertainty,

algorithm in several directions: By using Christofides algorithm for traversing the graph we improve the guarantees to 6 and 12.35, respectively. By using a novel kind of randomization (and an expansion factor other than two) we improve the latter to 8.98.

Competitive analysis is supposed to be a novel alternative to the classical approach to decision-making under uncertainty: *expectation minimization*. In expectation minimization we would assume a distribution for the node sought, and optimize the expected cost. That is, we would be trying to

$$\min_{\pi} \sum_{v \in G} \mathbf{pr}(v) d_{\pi}(r, v).$$

Interestingly, this is an equivalent formulation of a rather well-studied and notoriously hard problem, the *traveling repairman problem (trp)*[6] [ACP+86, Wes95, Wil93]: Given a metric and a starting point, find the route that minimizes the sum of the arrival times at the points. It is known only how to solve on paths [ACP+86] —even the case of *trees*, even that of *caterpillars* (paths with edges sticking out), is conjectured to be hard [Wes95]. We observe that the problem can be solved in polynomial time (for any distribution) for a class of graphs slightly more general than paths: trees with a bounded number of leaves.

We also point out something rather unexpected: *The trp is the polyhedral separation problem of the dual of the randomized search ratio problem*. That is to say, if we can solve the trp for some metric, and for arbitrary distribution, then we would be able to solve the dual of the randomized search ratio problem (and thus the randomized search problem itself) for the same metric by using the ellipsoid algorithm [GLS88]; unfortunately, as we mentioned above we can only solve it in the fairly restricted case of trees with bounded number of leaves (it follows that all three problems are polynomial-time solvable in this case). In fact, the techniques in [PST91, Tar95] and [GLS88] suggest that a *polynomial-time approximation scheme* for the trp may be transferable to the randomized search ratio problem (for the same class of graphs).

The trp can be approximated within a constant factor: [BCC+94] gives an algorithm with approximation ratio 144, and [GK96] improves this to 21.55. It is possible to do better in the case of *uniform* distribution. A simple spanning tree heuristic achieves ratio 2. We give an interesting variant of Christofides' algorithm [Chr76] for the trp with *uniform* distribution, and show that it achieves approximation ratio 1.662.

---

the other in the face of computational complexity. Complexity issues are traditionally ignored in the context of competitive analysis.

[6] This problem has been studied also under the names *delivery man* [FLM93, Min89] and *minimum latency* [BCC+94, GK96]. [Wil93] calls this the *school-bus driver problem*, with this amusing explanation: A bus driver tries to deliver the children in his/her bus so as to minimize not travel time, but time weighted by the number of children (and ensuing havoc) in the bus...

2 Computing the Search Ratio

Both versions of the search ratio problem have been defined in the introduction.

Theorem 1. *Computing the search ratio and the randomized search ratio of a graph G with respect to a root node r is NP-complete and MAXSNP-hard.*

Proof. Both are easy reductions from the Hamilton path problem. Given a graph H, we define G as H plus a new node r, and edges from r to all vertices of H. It is easy to see that $\sigma(G, r)$ is h or less and $\rho(G, r)$ is $h/2$ or less, where h is the number of nodes of H, if and only if there was a Hamilton path in H. MAXSNP-hardness for the search ratio follows as well. MAXSNP-hardness for the randomized version is a little trickier, because this reduction is *not* an L-reduction. However, it can be shown that, if the smallest Hamilton walk of H is $h(1 + \epsilon)$, then the randomized search ratio of G is at least $\frac{1}{2}h(1 + \epsilon^2)$, and this suffices.

It is a little more nontrivial to argue that the randomized search ratio problem is in NP. It is not hard to verify that $\rho(G, r)$ can be reformulated as follows:

$$\min \rho$$

$$\sum_\pi x_\pi d_\pi(r, v) \le \rho \cdot d(r, v) \quad \text{(all } v) \qquad \text{(LP)}$$

$$\sum_\pi x_\pi = 1$$

$$x_\pi \ge 0$$

This is an $(n + 1) \times n!$ linear program. However, the optimum value will be a basic feasible solution having at most $n + 1$ walks with nonzero probability. Such a solution can be guessed, computed, and compared with any given bound, establishing that the problem is in NP.

2.1 Approximation

Theorem 2. *There are polynomial-time approximation algorithms for the search ratio and randomized search ratio problems with* asymptotic approximation ratio 6 *and* $\frac{3+\sqrt{10}}{2} \cdot \frac{3+2\sqrt{2}}{2} \approx 8.98$, *respectively.*

Proof. Consider the following family of heuristics:

>**for** $i := 0$ **to** m **do**
>>Let G_i be the graph G restricted to all nodes with distance x_i
>>>or less from r;   (* *comment:* x_m *is the radius of* G *)
>>
>>search G_i depth-first, and return to r.

Let S_k be the set of nodes in distance k or less from r. Then it is easy to see that the search ratio σ is at least $\frac{|S_k|-1}{k}$. In phase i, it takes $2(|S_{x_i}| - 1)$ steps to explore all nodes and return to r. If the target node is found during the n-th phase, the on-line cost is at most $2(|S_{x_1}| - 1) + 2(|S_{x_2}| - 1) + \ldots + 2(|S_{x_n}| - 1)$. Since the optimum is at least x_{n-1} the search ratio is at most

$$\frac{2(|S_{x_1}| - 1) + 2(|S_{x_2}| - 1) + \ldots + 2(|S_{x_n}| - 1)}{x_{n-1}} \leq \frac{2\sigma x_1 + 2\sigma x_2 + \ldots 2\sigma x_n}{x_{n-1}}$$

The optimal strategy is to choose $x_i = 2^i$. (Simple proof: We want to minimize $\max_n \frac{x_1 + \ldots + x_n}{x_{n-1}}$. Notice that all fractions except for the nth are nondecreasing in x_n, and the nth is decreasing; it follows that at minimax they are all equal. Call this value r. To solve for x_n we have $x_n = r(x_{n-1} - x_{n-2})$, which gives increasing x_i's only if $r \geq 4$. Adopting this minimum value gives x_is that are powers of 2, up to a constant. End of proof that powers of two are optimal.) The *doubling heuristic*, with ratio at most 8σ, results.

For the randomized search ratio, let u be a random node in S_k. Any deterministic on-line algorithm will explore on the average $(|S_k| - 1)/2$ nodes before u. Since a randomized algorithm is simply a distribution of deterministic algorithms the same holds for randomized algorithms. Therefore, for any randomized algorithm there exists a node u in S_k that is expected to be reached after $(|S_k| - 1)/2$ steps. This gives a lower bound $\frac{|S_k| - 1}{2k}$ of the randomized search ratio —half the lower bound of the deterministic search ratio. It follows that the doubling heuristic produces a search strategy which, seen as a distribution, has expected ratio at most 16 times the optimum randomized search ratio.

The above lower bound $\frac{|S_k| - 1}{k}$ (or $\frac{|S_k| - 1}{2k}$) of the optimum search ratio is too crude. A better lower bound results by improving the numerator: Any on-line algorithm needs at least p_k steps to explore all nodes in S_k, where p_k is the length of the minimum TSP path. Therefore, the optimum search ratio r is at least $\frac{p_k}{k}$. Hence, instead of exploring the nodes in S_{x_i} in a depth-first manner, we can use Christofides algorithm [Chr76]. Let χ_k be the length of the tour computed by Christofides algorithm for exploring S_k. Notice that the optimum tour that visits all nodes in S_k is at most $p_k + k$ (the result of visiting all nodes using an optimal path and then returning to the root). It follows that $\chi_k \leq 1.5(p_k + k) \leq 1.5(r + 1)k$. Hence, this simple modification of the doubling heuristic finds a ratio that is at most $6(\sigma + 1)$ (notice the additive constant, whence the "asymptotic" in the statement of the theorem).

For the randomized version, Christofides' algorithm gives an approximation ratio of $6 + 2\sqrt{10} \approx 12.35$: The lower bound is now $\frac{1}{2}|S_k|(1 + (\frac{p_k - |S_k|}{|S_k|})^2)$ (proof omitted), whereas Christofides' algorithm gives a path of length at most $\frac{3|S_k|}{2} + \frac{p_k - |S_k|}{2}$; the worst-case ratio of the two is $6 + 2\sqrt{10} \approx 12.35$.

But we can do better by a rather novel kind of randomization. Our approximation algorithm is still deterministic, but the solution it produces will in fact be a *distribution* of walks on G. In particular, suppose that the tours produced by Christofides' algorithm in the various stages are $T_1, T_2, \ldots, T_m$. The distribution we produce selects the tour $(T_1^{e_1}, \ldots, T_m^{e_m})$ with probability $\frac{1}{2^m}$, where the e_i's are either 1 or -1, denoting possible reversal of the tour. In other words, at each phase we try both the tour and its reverse, with equal probability. As a result, the target node is expected to be encountered at the middle of the last tour. Hence the ratios to be minimized now become $\frac{x_1+x_2+\cdots+x_{n-1}+\frac{1}{2}x_n}{x_{n-1}}$. The recurrence for the x_i's is now $x_n = (2r-1)x_{n-1} - 2rx_{n-2}$, which is feasible (the corresponding algebraic equation has real roots) only when $4r^2 - 12r + 1 \geq 0$, or $r \geq \frac{3}{2} + \sqrt{2}$. Hence the radius in the "doubling heuristic" now is increased by factors of $1 + \sqrt{2}$, and the approximation ratio becomes $\frac{3+\sqrt{10}}{2} \cdot \frac{3+2\sqrt{2}}{2} \approx 8.98$.

3 The Traveling Repairman Problem

The trp has been originally defined [ACP+86] on an arbitrary (non-graph) metric d on n points, where we seek to

$$\min_\pi \sum_{i=1}^{n} \sum_{j=1}^{i-1} d_{\pi(j),\pi(j+1)},$$

where $\pi(1) = r$. Here we define it on a graph G, with an arbitrary distribution **pr** on the nodes, and a root r, where we must

$$\min_\pi \sum_{v\in G} \mathbf{pr}(v)d_\pi(r,v),$$

where π now ranges over all walks of the graph. It is not hard to observe that, under very mild restrictions (not affecting, for example, approximability) these two versions are equivalent:

Theorem 3. *If the distances in the metric d are polynomially small integers, and the probabilities* **pr** *are rational numbers with small coefficients and common denominators, then the two problems are polynomially equivalent.*

Proof. In one direction we simulate distances by long paths whose intermediate nodes have zero probabilities; in the other we simulate a node with probability $\frac{A}{B}$, where B is the common denominator, by a cluster of A nodes with distance zero from one another.

We shall henceforth focus on the graph version of the trp.

Theorem 4. *The trp with the uniform distribution is NP-complete (and MAXSNP-hard).*

Proof. Another easy reduction from Hamilton path, omitted.

Can we solve the trp exactly on any interesting class of graphs? It follows from the results of [ACP+86] that it can be solved on paths. We can generalize this a little:

Theorem 5. *On trees with L leaves, the trp can be solved in $O(n^L)$ time.*

Proof. Such a tree has $O(n^L)$ subtrees; furthermore, the optimum trp solution is guaranteed to end up in a leaf, and thus dynamic programming is enabled.

It is worth mentioning that the trp with the uniform distribution is solvable for general trees; in fact, any depth-first traversal is optimal.

3.1 Polyhedral Separation

Suppose that we wish to solve the linear programming formulation LP of the randomized search ratio problem. In fact, we should solve the dual, which has manage-ably many dimensions:

$$\min \sum_v d(r,v)y_v - z$$

$$\sum_v d_\pi(r,v)y_v - z \geq 0 \quad \text{(all } \pi) \qquad \text{(DLP)}$$

$$y_v \geq 0$$

Suppose then that we wish to solve DLP by the ellipsoid algorithm [GLS88]. We are given a point $(\bar{y}, z) \in \Re^{n+1}$, and we are asking whether or not it lies within the feasible region of DLP; if not, we need a violated inequality. It is easy to see that *this is precisely the trp problem*. Hence we have:

Theorem 6. *If the trp can be solved in polynomial time for a class of graphs and any distribution, then the randomized search ratio problem can be solved exactly in polynomial time for that class of graphs.*

Corollary 7. *The search ratio problem and the randomized search ratio problem can be solved exactly in polynomial time for trees with a bounded number of leaves.*

The work of [PST91, Tar95] and [GLS88] suggest that any polynomial time *approximation scheme* for the trp can be transferred to the randomized search ratio problem in the same class of graphs.

3.2 Approximation of the Uniform trp

We have a graph G, with a fixed root r, and we wish to find a walk starting from r that visits all nodes, and minimizes the sum of the arrival times at the nodes. It is easy to see that an approximation ratio of 2 can be achieved by a simple spanning tree heuristic that traverses a spanning tree in depth-first order. Formulating and analysing the analogue of Christofides' algorithm in this setting is nontrivial. Our approximation algorithm is the following:

Find a spanning tree T of G in which r has odd degree.
Find a shortest matching M_1 of all odd-degree nodes of T.
Add M_1 to T, to obtain an Eulerian graph E_1.
Find a traversal of E_1, and its reverse, call them S_1 and S_1^r.
Find a shortest matching M_2 of all odd-degree nodes of T *except* for r
    and some other node.
Add M_2 to T to obtain an *almost* Eulerian graph E_2 with two odd-degree
    nodes.
Find a traversal of E_2, call it S_2.
Select the best among S_1, S_1^r, and S_2.

The first line is impossible if r is an articulation point belonging to an even number of components, to all as a leaf; in this case we add a new node r' adjacent only to r, and call it the root; the performance is not affected.

Theorem 8. *The algorithm above yields a solution to the trp which is at most* 1.662 *times the optimum.*

The precise ratio in the statement of the theorem is $\max_{x\in[1,2]} \frac{4x-x^2-2}{1+(x^2-5x+5)^2} \approx$ 1.6615, which is maximized when $x \approx 1.4545$ is the (unique) root between 1 and 2 of the polynomial $x^5 - 11x^4 + 44x^3 - 75x^2 + 44x + 2$.

Proof. Suppose that the length of the shortest walk in G starting from r and visiting all nodes is $n(1 + \delta) - 1$, for some δ between 0 and 1. It can be shown (proof omitted) that $\frac{1}{2}n^2(1+\delta^2) + o(n^2)$ is a *lower bound* on the trp (otherwise, we would be able to find a shorter walk from r).

Let ℓ_1 be the total length of E_1; it can be shown that the total trp cost of the smallest among S_1 and $S_1^\intercal$ is bounded from above by $\frac{\ell_1 \cdot n}{2}$ (each node is first visited on the average with delay $\frac{\ell_1}{2}$ or better in the two traversals). Next, let ℓ_2 be the total length of E_2; it is easy to see that the total trp cost of S_2 is bounded from above by $\ell_2 \cdot n - \frac{n^2}{2} + o(n^2)$.

The key observation now is that the sum of ℓ_1 and ℓ_2 is at most $(3 + \delta) \cdot n$, because the shortest walk from r contains *both* a matching of the odd-degree nodes, *and* a matching of the odd-degree nodes minus r. Hence, the performance of the algorithm is bounded above by

$$\max_{\substack{\ell_1+\ell_2\leq(3+\delta)\cdot n \\ 0\leq\delta\leq1}} \frac{\min\{\frac{\ell_1\cdot n}{2}, \ell_2\cdot n - \frac{n^2}{2}\}}{\frac{1}{2}(1+\delta^2)\cdot n^2}.$$

It turns out that this expression evaluates to $\frac{5+\sqrt{29}}{6} \approx 1.73$

To improve this to $1.6615\ldots$ we must argue that the total trp cost is bounded from above by $2\ell_2 n - \frac{1}{2}\ell_2^2 - n^2 + o(n^2)$, instead of the more pessimistic $\ell_2 \cdot n - \frac{n^2}{2} + o(n^2)$. The argument involves decomposing E_2 into a path and several Eulerian graphs, and choosing for each Eulerian graph the better of two traversals. The expression now becomes

$$\max_{\substack{\ell_1+\ell_2\leq(3+\delta)\cdot n \\ 0\leq\delta\leq1}} \frac{\min\{\frac{\ell_1\cdot n}{2}, 2\ell_2 n - \frac{1}{2}\ell_2^2 - n^2\}}{\frac{1}{2}(1+\delta^2)\cdot n^2},$$

and it evaluates to $\max_{x\in[1,2]} \frac{4x-x^2-2}{1+(x^2-5x+5)^2} \approx 1.6615$, which is maximized when $x \approx 1.4545$ is the unique root between 1 and 2 of the equation $x^5 - 11x^4 + 44x^3 - 75x^2 + 44x + 2$.

4 Open Problems

Can we achieve in polynomial time better approximations than those in Theorem 2? Naturally, we can do better for graphs for which the TSP is solvable exactly, or has a better approximation ratio than $\frac{3}{2}$, as in the case of TSP with distances 1 and 2 [PY93]. Also, our approximation of $\sigma(G, r)$ can be extended to *weighted* graphs. Can our approximation of $\rho(G, r)$ be also so extended?

Computing $\sigma(G, r)$ and $\rho(G, r)$ when G is a tree is a surprisingly tough problem. An NP-completeness proof for the tree case (not unlikely, in view of the many NP-complete mean-flow scheduling problems with a similar flavor) would establish the NP-hardness, via duality and polyhedral separation, of the trp for trees (a problem long conjectured to be NP-complete).

Can we improve the approximation ratio for the uniform trp to 1.5? Further, in the nonuniform (weighted) case the current ratio is still rather large.

Lastly, is there a polynomial-time approximation scheme for the trp for trees with general distributions? We conjecture that at least a *pseudo-polynomial* time approximation scheme exists. If so, polyhedral separation and duality could imply that the randomized search ratio problem is so approximable in trees.

References

[ACP+86] F. Afrati, S. Cosmadakis, C.H. Papadimitriou, G. Papageorgiou, and N. Papakostantinou. The complexity of the travelling repairman problem. *Informatique Théórique et Applications*, 20(1):79–87, 1986.

[BCR88]   R.A. Baeza-Yates, J.C. Culberson, and G.J.E. Rawlins. Searching with uncertainty. *SWAT 88. 1st Scandinavian Workshop on Algorithm Theory. Proceedings*, pages 176–89, 1988.

[BCC+94] A. Blum, P. Chalasani, D. Coppersmith, W. Pulleyblank, P. Raghavan, and M. Sudan. The minimum latency problem. *Proceedings 26th Annual Symposium on Theory of Computing*, pages 163–171, 1994.

[Chr76]   N. Christofides. Worst-case analysis of a new heuristic for the traveling salesman problem. *Technical report, GSIA, Carnegie-Mellon University*, 1976.

[DKP91]   X. Deng, T. Kameda, and C. Papadimitriou. How to learn an unknown environment. *Proceedings 32nd Annual Symposium on Foundations of Computer Science*, pages 298–303, 1991.

[DP90]   X. Deng and C.H. Papadimitriou. Exploring an unknown graph. *Proceedings 31st Annual Symposium on Foundations of Computer Science*, pages 355–361 vol. 1, 1990.

[FLM93]   M. Fischetti, G. Laporte, and M. Martello. The delivery man problem and cumulative matroids. *Operations Research*, vol. 41, pages 1055–1064, 1993.

[GK96]   M. Goemans and J. Kleinberg. An improved approximation ratio for the minimum latency problem. *Proceedings Annual Symposium on Discrete Algorithms*, to appear, 1996.

[GLS88]   M. Grötschel, L. Lovász, and A. Schrijver. *Geometric algorithms and combinatorial optimization*. Springer-Verlag, 1988.

[Min89]   E. Minieka. The delivery man problem on a tree network. *Annals of Operations Research*, vol. 18, pages 261–266, 1989.

[PST91]   S.A. Plotkin, D.B. Shmoys, and É. Tardos. Fast approximation algorithms
          for fractional packing and covering problems. *Proceedings 32nd Annual Sym-*
          posium on Foundations of Computer Science, pages 495–504, 1991.

[PY91]   C.H. Papadimitriou and M. Yannakakis. Shortest paths without a map.
          Theoretical Computer Science, 84(1):127–50, July 1991.

[PY93]   C.H. Papadimitriou and M. Yannakakis. The traveling salesman problem
          with distances one and two. *Mathematics of Operations Research*, 18(1):1–
          11, February 1993.

[ST85]   D.D. Sleator and R.E. Tarjan. Amortized efficiency of list update and paging
          rules. *Communications of the ACM*, 28(2):202–8, February 1985.

[Tar95]   É. Tardos. *Private communication*, 1995.

[Wes95]   D. West. *Private communication*, 1995.

[Wil93]   T. G. Will. *Extremal Results and Algorithms for Degree Sequences of*
          Graphs. PhD thesis, U. of Illinois at Urbana-Champaign, 1993.

Improved Sampling with Applications to Dynamic Graph Algorithms.

Monika Rauch Henzinger[1] * and Mikkel Thorup[2]**

[1] Digital System Research Center, 130 Lytton Ave, Palo Alto, CA
[2] Department of Computer Science, University of Copenhagen, Universitetsparken 1, 2100 Kbh. Ø, Denmark

Abstract. We state a new sampling lemma and use it to improve the running time of dynamic graph algorithms.
For the dynamic connectivity problem the previously best randomized algorithm takes expected time $O(\log^3 n)$ per update, amortized over $\Omega(m)$ updates. Using the new sampling lemma, we improve its running time to $O(\log^2 n)$. There exists a lower bound in the cell probe model for the time per operation of $\Omega(\log n/\log\log n)$ for this problem.
Similarly improved running times are achieved for 2-edge connectivity, k-weight minimum spanning tree, and bipartiteness.

1 Introduction

In this paper we present a new sampling lemma, and use it to improve the running times of various dynamic graph algorithms.

We consider the following type of problem: Let S be a set with a subset $R \subseteq S$. Membership in R may be efficiently tested. For a given parameter $r > 1$, either (i) find an element of R, or (ii) guarantee with high probability that the ratio $|R|/|S|$ is at most $1/r$, i.e. that $r|R| \leq |S|$. This problem arises in the fastest dynamic graph algorithms for various graph problems (connectivity, two-edge connectivity, k-weight minimum spanning tree, $(1 + \epsilon')$-approximate minimum spanning tree, and bipartiteness-testing) [6].

No deterministic algorithm of time less than $\Omega(|S|)$ is possible. In [6] they address the problem by sampling $O(r \log |S|)$ from S, returning any element found from R. This is hence a Monte-Carlo type algorithm, running in time $\Theta(r \log |S|)$, whose type (ii) answer is wrong with probability $1/s^{\Theta(1)}$. In this paper we give a randomized Monte-Carlo type algorithm that requires only $O(r)$ random samples of S. To be precise we will show the following lemma.

Lemma 1. Let R be a subset of a set S, and let $r, c \in \mathbb{R}_{>1}$. Set $s = |S|$. Then there is an algorithm with one of two outcomes:

---
* (*mhr@src.dec.com*) Author's Maiden Name: Monika H. Rauch. This research was done while at Cornell University, Ithaca, NY and supported by an NSF CAREER Award, Grant No. CCR-9501712.
** (*mthorup@diku.dk, http://www.diku.dk/~mthorup*).

(i) *It returns an element from R after having sampled an expected number of $O(r)$ random elements from S and having tested them for membership of R.*

(ii) *Having sampled and tested $O(s/c)$ random elements from S, it states that $|R|/|S| > 1/r$ with probability $< \exp(-s/rc)$.*

The significance of the lemma is for $s/c \gg r$. Note that the bounds in (i) and (ii) are assymptotically optimal. Trivially this is true for (i). For (ii), note that if x elements from S are sampled randomly and no element of R is found, then the probability that $|R|/|S| > 1/r$ is approximately $\exp(-x/r)$. Thus, picking $O(s/c)$ random elements is asymptotically optimal for achieving a bound of $\exp(-s/rc)$ on the probability.

1.1 Dynamic graph algorithms

Let $G = (V, E)$ be a graph with n nodes and m edges. A graph property $\mathcal{P}$ is a function that (a) maps every graph G to *true* or *false* or (b) that maps every tuple (G, u, v) to *true* or *false*, where $G = (V, E)$ is a graph and $u, v \in V$. An example for Case (a) is a function that maps every bipartite graph to *true* and every non-bipartite graph to *false*. An example for Case (b) is *connectivity* that returns *true* if u and v are connected in G and *false* otherwise.

A dynamic graph algorithm is a data structure that maintains any graph G and a graph property $\mathcal{P}$ under an arbitrary sequence of the following operations.

- *Insert(u, v)*: Add the edge (u, v) to G.
- *Delete(u, v)*: Remove the edge (u, v) from G if it exists.
- *Query(u, v)*: Return *yes* if $\mathcal{P}$ holds for u and v in G and *false* otherwise.

In this paper we improve the complexities of the following graph properties: connectivity, two-edge connectivity, k-weight minimum spanning tree, $(1 + \epsilon')$-approximate minimum spanning tree, and bipartiteness-testing.

1.2 Previous Work

Dynamic graph algorithms are compared using the (amortized or worst-case) time per operation. The best deterministic algorithms for the above graph properties take time $O(\sqrt{n})$ per update operation and $O(1)$ or $O(\log n)$ per query [3, 4]. Recently [6], Henzinger and King gave algorithms with polylogarithmic amortized time per operation using (Las-Vegas type) randomization. Their algorithms achieve the following running times:

1. $O(\log^3 n)$ to maintain a spanning tree in a graph (the connectivity problem;
2. $O(\log^4 n)$ to maintain the bridges in a graph (the 2-edge connectivity problem);
3. $O(k \log^3 n)$ to maintain a minimum spanning tree in a graph with k different weights (k-weight minimum spanning tree problem);

4. $O(\log^3 n \log U/\epsilon')$ to maintain a spanning tree whose weight is a $(1 + \epsilon')$ approximation of the weight of the minimum spanning tree, where U is the maximum weight in the graph (the $(1+\epsilon')$-approximate minimum spanning tree problem);

5. $O(\log^3 n)$ to test if the graph is bipartite (the bipartiteness-testing problem).

Fredman and Henzinger showed lower bounds of $\Omega(\log n/\log\log n)$ in the cell probe model for the first four of these problems [5] (see also [8]).

1.3 New Results

With our new sampling technique, we get the following improved running times:

1. $O(\log^2 n)$ for connectivity;
2. $O(\log^3 n)$ for 2-edge connectivity;
3. $O(k \log^2 n)$ for the k-weight minimum spanning tree problem;
4. $O(\log^2 n(\log U)/\epsilon')$ for the $(1+\epsilon')$ approximate minimum spanning tree problem, where U is the maximum weight in the graph;
5. $O(\log^2 n)$ for bipartiteness testing.

2 Improved sampling in dynamic graph algorithms

Our improvements are achieved by locally improving a certain bottleneck in the approach by Henzinger and King [6], henceforth referred to as the *HK-approach*. Rather than repeating their whole construction, we will confine ourselves to a reasonably self-contained description of this bottleneck. Our techniques for the bottleneck are of a general flavor and we expect them to be applicable in other contexts.

Let T be a spanning tree of some graph $G = (V, E)$. In the HK-approach, G is only one of many sub-graphs of the real graph. If some tree edge e is removed from T, we get two sub-trees T_1, T_2. Consider the *cut* C_e of non-tree edges with end-points in both T_1 and T_2. Any cut edge $f \in C_e$ can replace e in the sense that $T \cup \{f\} \setminus \{e\}$ is a spanning tree of G. Our general goal is to find such a cut edge f. Alternatively it is acceptable to discover that the cut C_e is sparse as defined below.

For each vertex $v \in T$, we have the set $N(v)$ of non-tree edges incident to T. Let $w(v) = |N(v)|$. For any sub-tree U of T, set $N(U) = \bigcup_{v \in V(U)} N(v)$ and $w(U) = \sum_{v \in V(U)} w(v)$. Note that $w(U)$ may be bigger than $|N(U)|$ because edges with both end-point in U are counted twice. Assume that T_1 contains no more nodes than T_2. We say that the cut C_e is *sparse* if $8 \log_2 n|C_e| < w(T_1)$. Otherwise C_e is said to be *dense*. If the cut is sparse, a cost of $O(w(T_1))$ may be attributed other operations due to an amortization in the HK-approach.

We store all edges of $N(T_1)$ in the leaves of a balanced search tree. This allows us to pick in time $O(\log n)$ a random edge from $N(T_1)$ (edges with both end-points in T_1 are picked with twice the probability of edges with one end-point in T_1) and check if its other end-point is in T_2. This is the desired approach for dense cuts. Alternatively, in time $O(w(T_1))$, we may scan all of $N(T_1)$, identifying all the edges in C_e. This is the desired approach for sparse cuts where the $O(w(T_1))$ is paid for via amortization. Unfortunately, we do not know in advance whether C_e is sparse or dense.

In the HT-approach, in time $O(\log^3 n)$, they sample $16 \log_2^2 n$ random edges from $N(T_1)$. If the sampling successfully finds an edge from C_e, this edge is returned. Otherwise, in time $O(w(T_1))$, they make a complete scan. If C_e is sparse, the scan is attributed to the amortization. The probability of C_e not being sparse is the probability of the sampling not being successful for a dense cut, which is $\leq (1 - 1/(8\log_2 n))^{16\log_2 n} < 1/n^2 = O(1/w(T_1))$. Hence the expected cost of an unduly scan (i.e. a scan even though the cut is dense) is $O(w(T_1)/w(T_1)) = O(1)$. Thus, the total expected cost is $O(\log^3 n)$. This cost remains a bottle-neck for the HK-approach as long as the time per operation is $\Omega(\log^2 n)$.

We will now apply the sampling from Lemma 1 with $R = C_e$, $S = N(T_1)$, $w(T_1)/2 \leq s \leq w(T_1) = O(n^2)$, $r = 8\log_2 n$, and $c = O(\log n)$. Moreover, the cost of sampling and testing is $O(\log n)$. Then, in case (i), we find an element from C_e in expected time $O(\log n \cdot 8\log_2) = O(\log^2 n)$. In case (ii), the cost is $O(\log n \cdot w(T_1)/\log n) = O(w(T_1))$ matching the cost of the sub-sequent scanning. If the cut turns out to be sparse this cost is attributed to the amortization. In case (ii) the probability of a dense cut is $\exp(-s/rc) = \exp(-w(T_1)/O(\log^2 n))$, so the expected contribution from unduly scanning is $O(w(T_1)\exp(-w(T_1)/O(\log^2 n))) = O(\log^2 n)$. Thus, our expected cost is $O(\log^2 n)$, as opposed to the $O(\log^3 n)$ cost achieved by the HK-approach.

The removal of a factor $O(\log n)$ explains our improvements.

3 The sampling lemma

In this section, we will prove Lemma 1. The proof gives an algorithm that uses $\log^* s$ rounds of sampling. To give an intuition for this proof and because of its ease of implementation we first (Section 3.1) show that just 2 rounds of sampling leads to a substantial reduction in the number of samples. Section 3.2 contains the proof of the lemma.

3.1 Proving a simpler lemma

In this section, we will prove the following simpler lemma, which we beleive to of practical relevance.

Lemma 2. Let R be a subset of a set S, and let $r, c \in \mathbb{R}_{>1}$. Set $s = |S|$. Then there is an algorithm with one of two outcomes:

(i) It returns an element from R after having sampled an expected number of at most $4r(\ln\ln s + 2)$ random elements from S and having tested them for membership of R.

(ii) Having sampled and tested $8r\ln s + 4r\ln\ln s$ random elements from S, it states that $|R|/|S| > 1/r$ with probability $< 1/s$.

Proof: We first give the algorithm and then prove that it fulfills the above lemma.

Algorithm A: Does the task described in Lemma 2.

A.1 Let S_1 be a random subset of S of size $4r \ln \ln s$.

A.2 $R_1 := S_1 \cap R$.

A.3 If $R_1 \neq \emptyset$, then return any $x \in R_1$.

A.4 Let S_2 be a random subset of S of size $8r \ln s$.

A.5 $R_2 := S_2 \cap R$.

A.6 If $|R_2| > 4 \ln s$, then return any $x \in R_2$

A.7 Return "$|R|/|S| > 1/r$ with probability $< 1/s$."

We show next a bound on the probability p that the algorithm returns an element from R in A.6 (Claim 2A). Afterwards we prove that the Algorithm A satisfies the conditions of Lemma 2.

CLAIM 2A $p \leq 1/\ln s$.

Proof: We divide into two cases:

Case 1: $|R|/|S| > 1/(4r)$: The algorithm did not return in A.3, so

$$p < (1 - 1/(4r))^{4r \ln \ln s} \leq e^{-4r \ln \ln s/(4r)} \leq 1/\ln s.$$

Case 2: $|R|/|S| \leq 1/(4r)$: The algorithm did return an element in Step A.6, so $|R_2| \geq 4 \ln s$. However, the expected value μ of $|R_2|$ is at most $2 \ln s$. Note that

$$p = Pr(|R_2| > (1 + \delta)\mu) \text{ with } \delta = 1.$$

Using a Chernoff bound from [2,9],

$$Pr(|R_2| > (1 + \delta)\mu) < e^{-\delta^2 \mu/3} = e^{-2 \ln s/3} \leq e^{-2(e \ln \ln s)/3} < 1/\ln s.$$

Above, it was used that $x/\ln x \geq e$ for any real $x > 0$.

$\square$

We are now ready to show that Algorithm B satisfies the conditions of Lemma 1.

(i) First we find the expected number of samples if the algorithm returns an element from R. By Claim 2A, the probability p of the algorithm returns an element from R in Step A.6 is bounded by $1/\log s$. Thus, the expected number of samples is

$$4r \ln \ln s + 8r \ln s/\ln s = 4r(\ln \ln s + 2).$$

(ii) Second we consider the case that the algorithm does not return an element from R, i.e. that the conditions in Steps A.3 and A.6 do not get satisfied.

Suppose $|R|/|S| > 1/r$. We did not return an element from R in Step A.6, so $X = |R_2| \le 2 \ln s$, but the expected value μ of $|R_2|$ is at least $4 \ln \ln s$. Note that

$$p \le Pr(|R_2| < (1-\delta)\mu) \text{ with } \delta = 1/2.$$

Using a Chernoff bound from [1],

$$Pr(|R_1| < (1-\delta)\mu) < e^{-\delta^2 \mu/2} = e^{-(1/2)^2 8 \ln s/2} = 1/s,$$

as desired. ∎

In the next section we proof the general sampling lemma.

3.2 Proving the sampling lemma

In this section, we will prove Lemma 1 constructively, presenting a concrete algorithm. First recall the statement of lemma 1:

Let R be a subset of a set S, and let $r, c \in \mathbb{R}_{>1}$. Set $s = |S|$. Then there is an algorithm with one of two outcomes:

(i) It returns an element from R after having sampled an expected number of $O(r)$ random elements from S and having tested them for membership of R.

(ii) Having sampled and tested $O(s/c)$ random elements from S, it states that $|R|/|S| > 1/r$ with probability $< \exp(-s/rc)$.

Proof: Let the increasing sequence $n_0, \ldots, n_k \cdots$ be defined such that $n_0 = 26^4$ and for $i > 0$, $n_i = \exp(n_{i-1}^{1/4})$. Let the decreasing sequence $r_0, \ldots, r_k \cdots$ be defined such that $r_0 = 2r(1 + 2n_0^{-1/4}) = 28/13 \cdot r < 3r$ and for $i > 0$, $r_i = r_{i-1}/(1 + n_{i-1}^{-1/4})$.

CLAIM 1A *For all $i \ge 0$,*

(a) $2n_i < n_{i+1}$.
(b) $2n_i^{1/4} < n_{i+1}^{1/4}$.
(c) $2r < r_i < 3r$.

Proof: Both (a) and (b) are easily verified by insertion. The r_i are decreasing, so $r_i \le r_0 < 3r$. Finally, $r_i = 2r(1 + 2n_0^{-1/4})/\prod_{j=1}^{i-1}(1 + n_j^{-1/4}) \ge 2r \exp(2n_0^{-1/4} - \sum_{j=1}^{i-1} n_j^{-1/4}) > 2r$. The last inequality uses (b). □

Algorithm B: Does the task described in Lemma 1.

B.1.$i := 0$;

B.2.While $r_i n_i < 8s/c$:

B.2.1.   Let S_i be a random subset of S of size $r_i n_i$.

B.2.2. $R_i := S_i \cap R$.

B.2.3.   If $|R_i| \geq n_i$, then return $x \in S_i \cap R$

B.2.4. $i := i + 1$;

B.3.Let S_i be a random subset of S of size $8s/c$.

B.4.$R_i := S_i \cap R$.

B.5.If $|R_i| \geq 8s/(cr_i)$, then return $x \in S_i \cap R$.

B.6.Return "$|R|/|S| > 1/r$ with probability $< \exp(-s/rc)$."

We show next a bound on the number of sampled edges (Claim 1B) and on the probability that the algorithm return an element from R in round i (Claim 1C). Afterwards we prove that the Algorithm B satisfies the conditions of Lemma 1.

Let t be the final value of i - if we return an element from R in Step B.2.3, then i is not subsequently increased.

CLAIM 1B   *For all $t \geq i \geq 0$, $\sum_{j=0}^{i} |S_j| = O(rn_i)$.*

Proof: Note that in Steps B.3–B.5, $|S_i| = 8s/c \leq r_i n_i$. Thus, for all $i \geq 0$,

$$\sum_{j=0}^{i} |S_j| \leq \sum_{j=0}^{i} r_j n_j \leq 3r \sum_{j=0}^{i} n_j = O(rn_i).$$

The last inequality uses Claim 1Aa.   □

For $i > 0$, let p_i be the probability that the algorithm returns an element from R in round i. Here the round refers to the value of i in Step B.2.3 or B.5.

CLAIM 1C   *For all $i \geq 1$, $p_i \leq n_i^{-2}$.*

Proof: We divide into two cases:

Case 1: $|R|/|S| > (1 + n_{i-1}^{-1/4}/2)/r_{i-1}$: In round $i - 1$ we did not return, so $|R_{i-1}|$ is less than $x = n_{i-1}$. However, the expected value μ of $|R_{i-1}|$ is at least $n_{i-1}(1 + n_{i-1}^{-1/4}/2)$.
Note that

$$p_i \leq Pr(|R_{i-1}| < (1 - \delta)\mu) \text{ with } \delta = (\mu - x)/\mu.$$

Using the Chernoff bound (according to [1]),

$$Pr(|R_{i-1}| < (1 - \delta)\mu) < e^{-\delta^2 \mu/2} = e^{-(\mu - x)^2/(2\mu)}.$$

For $\mu \geq n_{i-1}(1 + n_{i-1}^{-1/4}/2)$ this function is maximized for $\mu = n_{i-1}(1 + n_{i-1}^{-1/4}/2)$. Thus,

$$p_i \leq \exp(\frac{-(n_{i-1}^{3/4}/2)^2}{2n_{i-1}(1 + n_{i-1}^{-1/4}/2)}) < \exp(\frac{-n_{i-1}^{1/2}}{9}) < \exp(-2n_{i-1}^{1/4}) = n_i^{-2}.$$

The inequalities use that $n_{i-1}^{1/4} \geq n_0^{1/4} > 18 > 16$.

Case 2: $|R|/|S| \leq (1 + n_{i-1}^{-1/4}/2)/r_{i-1}$: Note that

$$\frac{1 + n_{i-1}^{-1/4}/2}{r_{i-1}} = \frac{1 + n_{i-1}^{-1/4}/2}{r_i(1 + n_{i-1}^{-1/4})} = \frac{1 - n_{i-1}^{-1/4}/2(1 + n_{i-1}^{-1/4})}{r_i} < \frac{1 - n_{i-1}^{-1/4}/2.1}{r_i}.$$

The last inequality uses that $n_{i-1}^{-1/4} \geq n_0^{1/4} > 20$. Thus we have $|R|/|S| < (1 - n_{i-1}^{-1/4}/2.1)/r_i$.

First suppose that we are returning in Step B.2.3. Then $|R_i|$ is at least $x = n_i$. However, the expected value μ of $|R_i|$ is at most $n_i(1 - n_{i-1}^{-1/4}/2.1) = n_i(1 - 1/(2.1 \ln n_i))$. Note that

$$p_i \leq Pr(|R_{i-1}| > (1 + \delta)\mu) \text{ with } \delta = (x - \mu)/\mu.$$

Using the Chernoff bound (according to [2,9]),

$$Pr(|R_{i-1}| > (1 + \delta)\mu) < e^{-\delta^2\mu/3} = e^{-(x-\mu)^2/(3\mu)}.$$

For $\mu \leq n_i(1 - 1/(2.1 \ln n_i))$ this function is maximized for $\mu = n_i(1 - 1/(2.1 \ln n_i))$. Thus,

$$p_i \leq \exp(\frac{-(n_i/(2.1 \ln n_i))^2}{3n_i(1 - 1/(2.1 \ln n_i))}) < \exp(\frac{-n_i}{13(\ln n_i)^2}) \leq n_i^{-2}.$$

For the last inequality, we use that $n_i \geq 26(\ln n_i)^3$ which follows ¿from $\ln n_i \geq n_0^{1/4} = 26$.

Next suppose that we are returning in Step B.5. Then $|R_i|$ is at least $x = 8s/(cr_i)$ and $\mu \leq (1 - 1/(2.1 \ln n_i))8s/(cr_i)$. Note that $x > n_{i-1}r_{i-1}/r_i > n_{i-1}$, since $8s/c > r_{i-1}n_{i-1}$. As above

$$p_i \leq \exp(\frac{-(x/(2.1n_{i-1}^{1/4}))^2}{3x(1 - 1/(2.1n_{i-1}^{1/4}))}) < \exp(\frac{-x}{13n_{i-1}^{1/2}})$$
$$\leq \exp(\frac{-n_{i-1}^{1/2}}{13}) \leq \exp(-2n_{i-1}^{1/4}) = n_i^{-2}.$$

For the last inequality, we actually require that $n_{i-1}^{1/4} \geq 26$.

$\square$

We are now ready to show that the Algorithm B satisfies the conditions of Lemma 1.

(i) First we find the expected number of samples if the algorithm returns an element from R. By Claim 1C, for $i > 0$, the probability p_i of the algorithm returns an element from R in round i is bounded by n_i^{-2}. Moreover, by Claim 1B, if the algorithm returns in round i, it has sampled $O(rn_i)$ edges. Finally, by Claim 1Aa, $2n_i < n_{i+1}$. The expected number of samples is thus

$$\sum_{i=0}^{\infty} p_i O(rn_i) = O(rn_0 + \sum_{i=1}^{\infty} r/n_i) = O(rn_0 + 2r/n_1) = O(r).$$

(ii) Second we consider the case that the algorithm does not return an element from R, i.e. that the conditions in Steps B.2.3 and B.5 are never satisfied. Using Claim 1B, the total sample size is $\sum_{i=0}^{t} |S_i| = O(rn_{t-1}) + 8s/c = O(s/c)$.

Suppose $|R|/|S| > 1/r$. We did not return an element from R in Step B.5, so $X = |R_t|$ is less than $x = 8s/(cr_i) < 4s/(cr)$ by Claim 1Ac. However, the expected value μ of $|R_t|$ is at least $8s/(cr)$. The probability p is now calculated as in Case 1 of the proof of Claim 1C:

$$p \le e^{-(\mu-x)^2/(2\mu)} \le \exp(\frac{-(4s/(cr))^2}{2(8s/(cr))}) \le \exp(-s/(cr)),$$

as desired.   ∎

At present, in case (i), we are making an expected number of $\le 2n_0 r_0 = 6 \cdot 26^4 r = O(r)$ samples. The constant can be reduced by adding a round -1, with $n_{-1} = 1$ (meaning that we return if we find just one representative) and $r_{-1} = 3 \cdot 14 = 42$ $(14 > \ln 26^4(1 + 2/24))$. This gives an expected number of $\le 84r$ samples, which can be further reduced by introducing more preliminary rounds.

References

1. N. Alon, J. Spencer, P. Erdös. The Probabilistic Method. *Wiley-Interscience Series*, Johan Wiley and Sons, Inc., 1992.
2. D. Angluin, L. G. Valiant. Fast probabilistic algorithms for Hamiltonian circuits and matchings. *J. Comput. System Sci.* 18 (2), 1979, 155–193.
3. D. Eppstein, Z. Galil, G. F. Italiano. Improved Sparsification. Tech. Report 93-20, Department of Information and Computer Science, University of California, Irvine, CA 92717.
4. D. Eppstein, Z. Galil, G. F. Italiano, A. Nissenzweig. Sparsification - A Technique for Speeding up Dynamic Graph Algorithms. *Proc. 33rd Symp. on Foundations of Computer Science*, 1992, 60–69.
5. M. L. Fredman and M. R. Henzinger. Lower Bounds for Fully Dynamic Connectivity Problems in Graphs. Submitted to *Algorithmica.*
6. M. R. Henzinger and V. King. Randomized Dynamic Graph Algorithms with Polylogarithmic Time per Operation. *Proc. 27th ACM Symp. on Theory of Computing*, 1995, 519–527.
7. K. Mehlhorn. Data Structures and Algorithms 1: Sorting and Searching. *EATCS Monographs on Theoretical Computer Science*, Springer-Verlag, 1984.

8. P.B. Miltersen, S. Subramanian, J.S. Vitter, and R. Tamassia. Complexity models for incremental computation. *Theoretical Computer Science*, 130, 1994, 203-236.

9. J. P. Schmidt, A. Siegel, A. Srinivasan. Chernoff-Hoeffding Bounds for Limited Independence. *SIAM J. on Discrete Mathematics* 8 (2), 1995, 223-250.

10. R.E. Tarjan and U. Vishkin. Finding biconnected components and computing tree functions in logarithmic parallel time. *SIAM J. Computing*, 14(4): 862-874, 1985.

The Expressive Power of Existential First Order Sentences of Büchi's Sequential Calculus

Jean-Eric Pin

LITP/IBP, CNRS, Université Paris VI

4 Place Jussieu, 75252 Paris Cedex 05

FRANCE

Summary. The aim of this paper is to study the first order theory of the successor, interpreted on finite words. More specifically, we complete the study of the hierarchy based on quantifier alternations (or Σ_n-hierarchy). It was known (Thomas, 1982) that this hierarchy collapses at level 2, but the expressive power of the lower levels was not characterized effectively. We give a semigroup theoretic description of the expressive power of Σ_1, the existential formulas, and $B\Sigma_1$, the boolean combinations of existential formulas. Our characterization is algebraic and makes use of the syntactic semigroup, but contrary to a number of results in this field, is not in the scope of Eilenberg's variety theorem, since $B\Sigma_1$-definable languages are not closed under residuals.

An important consequence is the following: given one of the levels of the hierarchy, there is polynomial time algorithm to decide whether the language accepted by a deterministic n-state automaton is expressible by a sentence of this level.

1. The sequential calculus

The connections between formal languages and mathematical logic were first studied by Büchi [5]. But although Büchi was primarly interested in infinite words, we will consider only finite words in this paper.

Büchi's sequential calculus is a logical formalism to specify some properties of a finite word, for instance "the factor bba occurs three times in the word, but the factor bbb does not occur". Thus, each logical sentence of this calculus defines a language, namely the set of all words that satisfy the property expressed by the formula. For instance, in our example, this language would be $A^*bbaA^*bbaA^*bbaA^* \setminus A^*bbbA^*$, where $A = \{a, b\}$ denotes the alphabet.

More formally, to each word $u \in A^+$ is associated a structure

$$\mathcal{M}_u = (\{1, 2, \ldots, |u|\}, S, (R_a)_{a \in A})$$

where S denotes the successor relation on $\{1, 2, \ldots, |u|\}$ and R_a is set of all i such that the i-th letter of u is an a. For instance, if $A = \{a, b\}$ and $u = abaab$, then $R_a = \{1, 3, 4\}$ and $R_b = \{2, 5\}$. The logical language appropriate to such models has S and the R_a's as non logical symbols, and formulas are built in the standard way by using these non-logical symbols, variables, boolean connectives, equality between elements (positions) and quantifiers. Note that the symbol $<$ is not used in this logic. We shall use the notations F_1 (resp.

F_2) for the set of first order (resp. second order) formulas with signature $\{S, (R_a)_{a \in A}\}$.

Given a sentence φ, we denote by $L(\varphi)$ the set of all words which satisfy φ, when words are considered as models. It is a well known result of Büchi that monadic second order sentences exactly define the recognizable (or regular) languages. That is, for each monadic second order sentence φ, $L(\varphi)$ is a recognizable language and, for every recognizable language L, there exists a monadic second order sentence φ such that $L(\varphi) = L$. Actually, monadic second order logic constitutes a border line in the study of the sequential calculus. Beyond that border, one enters the hard world of complexity classes [7].

2. First order

The expressive power of F_1, the set of first order formulas with signature $\{S, (R_a)_{a \in A}\}$ was first studied by Thomas [25].

2.1 The combinatorial description

Some definitions from language theory are in order to state the result of Thomas. First, we will make a distinction between *positive boolean operations* on languages, that comprise finite union and finite intersection and *boolean operations* that comprise finite union, finite intersection and complement. Given a word x and a positive integer k, it is not very difficult to express in F_1 a property like "a factor x occurs at least k times". Let us denote by $F(x, k)$ the language defined by this property. A language L of A^+ is *strongly threshold locally testable* (STLT for short) if it is a boolean combination of sets of the form $F(x, k)$ where $x \in A^+$ and $k > 0$. It is *threshold locally testable* (TLT) if it is a boolean combination of sets of the form uA^*, A^*v or $F(x, k)$ where $u, v, x \in A^+$ and $k > 0$. Note that uA^* (resp. A^*v) is the set of words having u as a prefix (resp. v as a suffix), a property that can also be expressed in F_1. The classes of *positively strongly locally threshold testable* (PSTLT) and *positively threshold locally testable* (PTLT) languages are defined similarly, by replacing "boolean combination" by "positive boolean combination" in the definition[1]. Thomas proved the following theorem.

Theorem 2.1. *A language is F_1-definable if and only if it is TLT.*

In fact, this result is a particular instance of the general fact that first order formulas can express only local properties [9, 26, 27].

Theorem 2.1 gave a combinatorial description of the F_1-definable languages but also led to the next question : given a finite deterministic automaton $\mathcal{A}$, is it decidable whether the language accepted by $\mathcal{A}$ is F_1-definable?

---

[1] The reader is referred to [3] or to [22, p. 47] for an explanation of this terminology.

2.2 The semigroup approach

This problem was solved positively by semigroup-theoretic methods. Let L be a language of A^+. The *syntactic congruence of L* is the congruence $\sim_L$ on A^+ defined by $u \sim_L v$ if and only if, for every $x, y \in A^*$,

$$xuy \in L \Longleftrightarrow xvy \in L$$

The quotient semigroup $S(L) = A^+/\sim_L$ is called the *syntactic semigroup* of L. It is also equal to the transition semigroup of the minimal automaton of A. It follows that a language is recognizable if and only if its syntactic semigroup is finite. The quotient morphism $\eta : A^+ \to S(L)$ is called the *syntactic morphism* and the subset $P = \eta(L)$ of $S(L)$ is the *syntactic image* of L. See [15] for more details.

Recall that a finite semigroup S is *aperiodic* if there exists an integer $n \geq 0$ such that, for each $s \in S$, $s^n = s^{n+1}$. Another important property was introduced by Thérien and Weiss [24]. If e and f are idempotents[2] of S, and if r, s and t are elements of S, then $erfsetf = etfserf$.

It is easier to remember this condition in terms of categories (there are also good mathematical reasons to do so). The Cauchy category of a finite semigroup S is defined as follows: the objects are the idempotents of S and, if e and f are idempotents, the arrows from e to f are the triples (e, s, f), such that $s = es = sf$. Composition of arrows is defined in the obvious way:

$$(e, s, f)(f, t, g) = (e, st, g)$$

Thus the condition above can be simply written

$$pqr = rqp \qquad\qquad (C)$$

where p and r are coterminal arrows, say, from e to f, and q is an arrow from f to e.

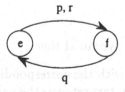

p, r

q

The condition $pqr = rqp$.

Thérien and Weiss did not explicitly mention the TLT languages in their paper but nevertheless gave the main argument of the proof of the following theorem.

---

[2] An element $e \in S$ is idempotent if $e^2 = e$. One can show that a non empty finite semigroup contains at least one idempotent.

Theorem 2.2. *A language is TLT if and only if its syntactic semigroup S is aperiodic and satisfies the condition (C).*

The link between the papers [25] and [24] was first observed in [2]. A complete proof of both results can also be found in the elegant book of Straubing on circuit complexity [22]. We complete these results by analyzing the complexity of the algorithm. More precisely, we prove the following result.

Theorem 2.3. *There is a polynomial time algorithm to decide whether the language recognized by a deterministic n-state automaton is F_1-definable.*

Proof. (Sketch) Testing for aperiodicity is PSPACE-complete [6], but it suffices to test whether the language is of "dot-depth one", which can be done in polynomial time [21]. Condition (C) can also be tested in polynomial time. It suffices to see if, for every configuration of the form represented below, in which e, f, p, q, r and y are paths in the given automaton, $q_1 \in F$ if and only if $q_2 \in F$.

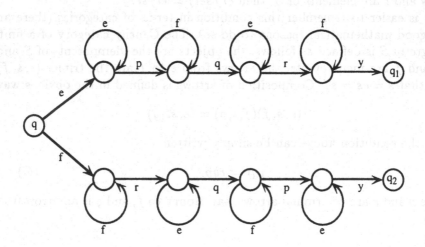

This can be easily tested in polynomial time.

Theorem 2.3 is in contrast with the corresponding result for the first order logic of the binary relation $<$, interpreted as the natural order on the integers. For this logic, McNaughton and Papert [12] gave a combinatorial description (the star-free languages) and Schützenberger [19] gave an algebraic characterization (the syntactic semigroup is aperiodic), but it was shown in [6] that the corresponding algorithm is PSPACE-complete.

3. Inside first order

The details of the landscape can be refined by considering the Σ_n-hierarchy of first order logic. It was shown by Thomas [25] that any F_1-definable language can also be defined by a Σ_2-sentence, that is, a sentence of the form

$$\exists x_1 \cdots \exists x_n \, \forall y_1 \cdots \forall y_m \; \varphi(x_1, \cdots, x_n, y_1, \cdots, y_m)$$

where φ is quantifier-free. Recall that a Σ_1-formula is of the form

$$\exists x_1 \cdots \exists x_n \; \varphi(x_1, \cdots, x_n)$$

where φ is quantifier-free. Denote by Σ_1 the set of Σ_1-formulas and by $\mathcal{B}\Sigma_1$ the set of boolean combinations[3] of Σ_1-formulas. The expressive power of Σ_1 and $\mathcal{B}\Sigma_1$ was still to be characterized. The following result was proved in [2, 3] by using Ehrenfeucht-Fraïssé games [22].

Theorem 3.1. *A language is $\mathcal{B}\Sigma_1$-definable if and only if it is STLT.*

The proof can be easily adapted to obtain a characterization of the Σ_1-definable languages

Theorem 3.2. *A language is Σ_1-definable if and only if it is PSTLT.*

These results complete the combinatorial description of the Σ_n-hierarchy, but do not solve the decidability questions: given a finite deterministic automaton $\mathcal{A}$, is it decidable whether the language accepted by $\mathcal{A}$ is $\mathcal{B}\Sigma_1$-definable (resp. Σ_1-definable)?

The main result of this paper provides a positive answer to these questions. Let S be a finite semigroup. We denote by S^1 the monoid equal to S if S has an identity, and to $S \cup \{1\}$, where 1 is a new identity, otherwise. Two elements s and t of S are said to be $\mathcal{J}$-equivalent (notation $s \, \mathcal{J} \, t$) if they generate the same ideal, that is, if there exists $x, y, u, v \in S^1$ such that $usv = t$ and $xty = s$. Let $\equiv$ be the coarsest equivalence relation on S satisfying the two following conditions

(1) for all $s, t \in S$, $s \, \mathcal{J} \, t$ implies $s \equiv t$,
(2) for all idempotents e, f of S, $esfte \equiv ftesf$

We say that a subset P of S *saturates* the $\equiv$-classes if, for all $s, t \in S$, $s \in P$ and $s \equiv t$ imply $t \in P$.

Theorem 3.3. *Let L be a recognizable language, S its syntactic semigroup and P its syntactic image. The following conditions are equivalent:*

1. *L is $\mathcal{B}\Sigma_1$-definable,*
2. *L is STLT,*
3. *S is aperiodic and satisfies (C), and P saturates the $\equiv$-classes.*

---

[3] boolean operations on formulas comprise conjunction, disjunction and negation.

For the PSTLT languages, the syntactic semigroup does not suffice, and we need the ordered syntactic semigroup, introduced in [18]. Let L be a language of A^+ and let $A^+ : \eta \to S(L)$ be its syntactic morphism. Define a relation $\preceq_L$ on A^+ by setting $u \preceq_L v$ if and only if, for every $x, y \in A^*$,

$$xvy \in L \Rightarrow xuy \in L$$

Then $\preceq_L$ is a reflexive and transitive relation such that $u \sim_L v$ if and only if $u \preceq_L v$ and $v \preceq_L u$. It follows that there is a well defined partial order on $S(L)$ defined by $\eta(u) \leq \eta(v)$ if and only if $u \preceq_L v$. This order is stable under product: if $s \leq t$ and $s' \leq t'$, then $ss' \leq tt'$. The ordered semigroup $(S(L), \leq)$ is called the *ordered syntactic semigroup* of L.

To each idempotent e is associated the subsemigroup eSe of S, defined by $eSe = \{ese \mid s \in S\}$. This is in fact a monoid, with e as an identity, called the *local submonoid* of e. Now, e is called a *local maximum* if, for every $s \in S$, $ese \leq e$. We can now formulate our characterization of the PSTLT languages.

Theorem 3.4. *Let L be a recognizable language, let S be its ordered syntactic semigroup and let P be its syntactic image. The following conditions are equivalent:*

1. *L is Σ_1-definable*
2. *L is PSTLT,*
3. *S satisfies (C), each idempotent of S is a local maximum and P saturates the equivalence $\equiv$.*

The proof of Theorem 2.3 can be adapted to the case of $B\Sigma_1$-formulas.

Corollary 3.1. *There is a polynomial time algorithm to decide whether the language recognized by a deterministic n-state automaton is $B\Sigma_1$ (resp. Σ_1-definable).*

4. Three examples

Example 4.1. Let $A = \{a, b\}$ and let $L = a^*ba^*$. Then L is recognized by the automaton shown in figure 4.1.

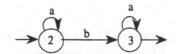

The minimal automaton of a^*ba^*.

The transitions and the relations defining the syntactic semigroup S of L are given in the following tables

	a	b	bb
1	1	2	–
2	2	–	–

$a = 1$
$b^2 = 0$

Thus $S = \{1, b, 0\}$ and $E(S) = \{1, 0\}$. The syntactic order is defined by $b \leq 0$ and $1 \leq 0$. The local semigroups are $0S0 = \{0\}$ and $1S1 = S$. The latter is not idempotent, since $b^2 \neq b$. Therefore, L is not locally testable. On the other hand, the Cauchy category of $S(L)$, represented in the figure below, satisfies the condition $pqr = rqp$.

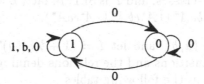

The Cauchy category of $S(L)$.

Therefore L is TLT. The syntactic image of L is $P = \{b\}$, which saturates the $\equiv$-classes. Thus L is STLT. This can be seen directly in this case, since L is the set of all words containing exactly one occurrence of b. However, L is not PSTLT since, in the local semigroup $1S1 = S$, 1 is not the top element.

Example 4.2. Let $A = \{a, b, c\}$, and let $L = c(ab)^* \cup c(ab)^*a$. Then L is recognized by the following automaton.

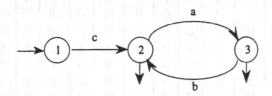

An automaton recognizing L.

The transitions and the relations defining the syntactic semigroup S of L are given in the following tables

	a	b	c	aa	ab	ba	ca
1	–	–	2	–	–	–	3
2	3	–	–	–	2	–	–
3	–	2	–	–	–	3	–

$a^2 = b^2 = c^2 = ac = bc = cb = 0$
$aba = a$
$bab = b$
$cab = c$

The $\mathcal{J}$-class structure is represented in the following diagram, where the grey box is the image of L.

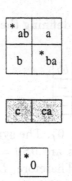

The $\mathcal{J}$-class structure.

Thus P saturates the $\mathcal{J}$-classes, and L is SLT. In fact, $L = A^*cA^* \setminus (A^*aaA^* \cup A^*acA^* \cup A^*bbA^* \cup A^*bcA^* \cup A^*cbA^* \cup A^*ccA^*)$.

Example 4.3. Let $A = \{a, b\}$, and let $L = (1 + b)a(ba)^*b^2b^*a(ba)^*(1 + b) \cup b^2b^*a(ba)^*b^2b^*$. The transitions and the relations defining the syntactic semi-group S of L are given in the following tables

Elements	1	2	3	4	5	6	7	8	9	10	11
a	6	10	0	7	10	0	0	6	7	0	6
b	11	2	3	3	0	8	4	2	9	5	9
aa	0	0	0	0	0	0	0	0	0	0	0
ab	8	5	0	4	5	0	0	8	4	0	8
ba	6	10	0	0	0	6	7	10	7	10	7
bb	9	2	3	3	0	2	3	2	9	0	9
abb	2	0	0	3	0	0	0	2	3	0	2
bab	8	5	0	0	0	8	4	5	4	5	4
bba	7	10	0	0	0	10	0	10	7	0	7
abba	10	0	0	0	0	0	0	10	0	0	10
babb	2	0	0	0	0	2	3	0	3	0	3
bbab	4	5	0	0	0	5	0	5	4	0	4
abbab	5	0	0	0	0	0	0	5	0	0	5
babba	10	0	0	0	0	10	0	0	0	0	0
bbabb	3	0	0	0	0	0	0	0	3	0	3
babbab	5	0	0	0	0	5	0	0	0	0	0

Relations :

$$aa = 0 \qquad aba = a \qquad b^3 = b^2 \qquad abbabb = 0 \qquad bbabba = 0$$

The idempotents are ab, ba, bb and 0. The $\mathcal{J}$-class structure is represented in the following diagram:

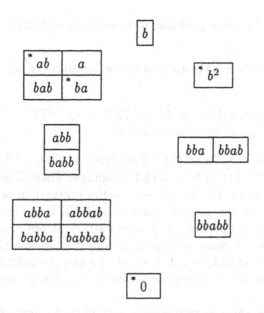

The image of the language is $P = \{bbabb, abba, abbab, babba, babbab\}$. One can verify that P saturates $\equiv$. Notice in particular that $babbab = (ba)(bb)(ba)$. Since the elements $e = ba$ and $f = bb$ are idempotent, $efe \in P$ should imply $fef \in P$, since P saturates $\equiv$. Indeed, $fef = babba \in P$. In fact, $L = (F(ab^2, 1) \cap F(b^2a, 1)) \setminus (F(aa, 1) \cup F(ab^2, 2) \cup F(b^2a, 2))$.

5. Outline of the proof of Theorem 3.3

Our proof is partly inspirated by the proof of Wilke [28], which gives a very nice characterization of the TLT languages of infinite words. However, Wilke's characterization makes use of the topology on infinite words, which is useless on finite words. We first introduce some combinatorial definitions.

Let A be a finite alphabet. If u is a word of length $\geq k$, we denote by $p_k(u)$ and $s_k(u)$, respectively, the prefix and suffix of length k of u. If u and x are two words, we denote by $\begin{bmatrix} u \\ x \end{bmatrix}$ the number of occurrences of the factor x in u. For instance $\begin{bmatrix} abababa \\ aba \end{bmatrix} = 3$, since aba occurs in three different places in $abababa$: $\underline{aba}baba$, $ab\underline{aba}ba$, $abab\underline{aba}$.

Let x and y be two integers. Then $x \equiv y$ threshold t (also denoted $x \equiv_t y$) if and only if $(x < t$ and $x = y)$ or $(x \geq t$ and $y \geq t)$. For instance the equivalence classes of $\equiv_4$ are $\{0\}, \{1\}, \{2\}, \{3\}, \{4, 5, 6, 7, \dots\}$.

For every $k, t > 0$, let $\equiv_{k,t}$ be the equivalence of finite index defined on A^+ by setting $u \equiv_{k,t} v$ if and only if, for every word x of length $\leq k$, $\begin{bmatrix} u \\ x \end{bmatrix} \equiv_t \begin{bmatrix} v \\ x \end{bmatrix}$. For instance, $abababab \equiv_{2,3} abababa$ since $abababab$ contains 4 ($\equiv 3$ threshold 3) occurrences of ab and 3 ($\equiv 3$ threshold 3) occurrences of ba, and no occurrences of aa (respectively bb).

We also define a congruence $\sim_{k,t}$ of finite index on A^+ by setting $u \sim_{k,t} v$ if

1. u and v have the same prefixes (resp. suffixes) of length $< k$,
2. $u \equiv_{k,t} v$.

The next proposition gives an alternative definition of the TLT and STLT languages.

Proposition 5.1. *A subset of A^+ is TLT (resp. STLT) if it is union of $\sim_{k,t}$-classes (resp. $\equiv_{k,t}$) for some k and t.*

The equivalence of (1) and (2) follows from Theorem 3.1. We now prove that (2) implies (3). Let L be a STLT language. Then L is union of $\equiv_{k,t}$-classes for some k and t. Let $\eta : A^+ \to S$ be the syntactic morphism of L and let P be the syntactic image of L. Since L is STLT, it is also TLT and thus, by Theorem 2.2, S is aperiodic and satisfies (C). It remains to see that P saturates the $\equiv$-classes. Since η is onto, one can fix, for each element $s \in S^1$ a word $\bar{s} \in A^*$ such that $\eta(\bar{s}) = s$. Let s and t be two $\mathcal{J}$-equivalent elements of S and suppose that $s \in P$. Then there exist $x, y, u, v \in S^1$ such that $usx = t$ and $vty = s$.

Since S is finite, there is an integer n such that, for any $s \in S$, s^n is idempotent. Assuming that $n \geq kt$, one gets $(\bar{v}\bar{u})^n \bar{s}(\bar{x}\bar{y})^n \equiv_{k,t} \bar{u}(\bar{v}\bar{u})^n \bar{s}(\bar{x}\bar{y})^n \bar{x}$. But $\eta((\bar{v}\bar{u})^n \bar{s}(\bar{x}\bar{y})^n) = s \in P$ and thus $(\bar{v}\bar{u})^n \bar{s}(\bar{x}\bar{y})^n \in L$. It follows that $\bar{u}(\bar{v}\bar{u})^n (\bar{x}\bar{y})^n \bar{x} \in L$ and thus $\eta(\bar{u}(\bar{v}\bar{u})^n (\bar{x}\bar{y})^n \bar{x}) = t \in P$.

Let now e and f be two idempotents of S and suppose that $esfte \in P$. Then, for $n \geq kt$, $\bar{e}^n \bar{s}\bar{f}^n \bar{t}\bar{e}^n \equiv_{k,t} \bar{f}^n \bar{t}\bar{e}^n \bar{s}\bar{f}^n$. But $\eta(\bar{e}^n \bar{s}\bar{f}^n \bar{t}\bar{e}^n) = esfte \in P$ and thus $\bar{e}^n \bar{s}\bar{f}^n \bar{t}\bar{e}^n \in L$. Therefore $\bar{f}^n \bar{t}\bar{e}^n \bar{s}\bar{f}^n \in L$ and thus $\eta(f^n te^n sf^n) = ftesf \in P$. Thus P saturates $\equiv$.

The direction (3) implies (2) is much more difficult. Since S is aperiodic and satisfies (C), Theorem 2.2 and Proposition 5.1 show that L is union of $\sim_{k,t}$-classes for some k and t. Unfortunately, L is not in general union of $\equiv_{k,t}$-classes, but we will show that L is union of $\equiv_{k,T}$-classes for some large T (to be precise, one takes $T = (1 + t \cdot (|A|^k)!)(1 + |A|)$). Associate with each word u a labelled graph $N(u)$ defined as follows: the vertices are the words of length $k - 1$, and if x is a word of length k, there is an edge of label $\begin{bmatrix} u \\ x \end{bmatrix}$ threshold t from the prefix length $k - 1$ of x to its suffix of length $k - 1$. The prefix (resp. suffix) of u of length $k - 1$ is called the initial (resp. final) vertex.

Thus let u and u' be two words such that $u \equiv_{k,T} u'$ and $u \in L$. Our aim is to show that $u' \in L$. If $|u| < T$ (or $|u'| < T$), then necessarily $u = u'$, thus we may assume that $|u|, |u'| \geq T$. Since $\begin{bmatrix} u \\ x \end{bmatrix} = \begin{bmatrix} u' \\ x \end{bmatrix}$ threshold T (and thus also threshold t), the labelled graphs $N(u)$ and $N(u')$ are equal, except for the initial and final vertices. We denote by i and f (resp. i' and f') the initial and final vertices of $N(u)$ (resp. $N(u')$).

In the figure below, two graphs are represented. The parameters are $k = 3$ and $t = 3$. The graph on the left hand side corresponds to the words $u = (ab)^4 (cb)^4 a$ and $u' = b(cb)^4 (ab)^4 cb$. The initial and final vertices of u (resp. u') are represented by full (resp. dotted) unlabelled arrows. The graph on the right hand side corresponds to the words $u = (ab)^4(cb)^4abcb$ and $u' = b(ab)^4(cb)^4acb$.

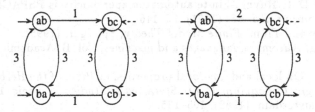

Two vertices v_1 and v_2 are in the same strongly t-component if there are two paths from v_1 to v_2 and from v_2 to v_1 using only edges of label t. For instance, in the two graphs above, ab and ba (resp. bc and cb) are in the same t-component. A non trivial combinatorial argument shows that if $u \equiv_{k,T} u'$, then two cases may arise:

1. i and i' are in the same t-component and f and f' are in the same t-component,
2. i and f are in the same t-component and i' and f' are in the same t-component.

The first and second cases are illustrated by the graphs on the right and on the left hand side, respectively. Now, one can show that in the first case, the elements $\eta(u)$ and $\eta(u')$ are $\mathcal{J}$-equivalent. Since P saturates the $\mathcal{J}$-classes, we are done in this case. In the second case, one can show that $\eta(u) \, \mathcal{J} \, \eta(v)$ and $\eta(u') \, \mathcal{J} \, \eta(v')$ for some words v and v' such that

1. $v \equiv_{k,T} v'$,
2. The initial and final vertices of $N(v)$ (resp. $N(v')$) coincide

But this last condition implies that for some idempotents e and f and for some elements p and q, $\eta(v) = epfqe$ and $\eta(v') = fqepf$. Now one can use the fact that P saturates the $\equiv$-classes.

References

1. J. Almeida, *Finite semigroups and universal algebra*, Series in Algebra Vol 3, Word Scientific, Singapore, 1994.
2. D. Beauquier and J.-E. Pin, Factors of words, in *Automata, Languages and Programming*, (G. Ausiello, M. Dezani-Ciancaglini and S. Ronchi Della Rocca, eds.), *Lecture Notes in Comput. Sci.* **372**, Springer, (1989), 63–79.
3. D. Beauquier and J.-E. Pin, Languages and scanners, *Theoret. Comput. Sci.* **84**, (1991), 3–21.
4. J.A. Brzozowski and I. Simon, Characterizations of locally testable languages, *Discrete Math.* **4**, (1973), 243-271.

5. J.R. Büchi, On a decision method in restricted second-order arithmetic, in *Proc. 1960 Int. Congr. for Logic, Methodology and Philosophy of Science*, Stanford Univ. Press, Standford, (1962), 1-11.

6. S. Cho and D.T. Huynh, Finite automaton aperiodicity is PSPACE-complete, *Theoret. Comput. Sci.* **88**, (1991), 99-116.

7. Ebbinghaus and Flum, *Finite Model Theory*, Springer, (1995).

8. S. Eilenberg, *Automata, languages and machines*, Vol. B, Academic Press, New York, 1976.

9. H. Gaifman, On local and non-local properties, in *Proc. of the Herbrandt Symposium, Logic Colloquium'81 (J. Stern, ed.)*, Studies in Logic **107**, North-Holland, Amsterdam, (1982), 105-135.

10. R. McNaughton, Algebraic decision procedures for local testability, *Math. Syst. Theor.* **8**, (1974), 60-76.

11. R. McNaughton and S. Pappert, *Counter-free Automata*, MIT Press, 1971.

12. D. Perrin, *Automata*, Chapter 1 in Handbook of Theoretical Computer Science (Van Leeuwen, J. ed.), Vol B: Formal Models and Semantics, Elsevier (1990).

13. D. Perrin and J.-E. Pin, First order logic and star-free sets, *J. Comput. System Sci.* **32**, (1986), 393-406.

14. J.-E. Pin, *Variétés de langages formels*, Masson, Paris, 1984. English translation: *Varieties of formal languages*, Plenum, New-York, 1986.

15. J.-E. Pin, Logic, Semigroups and Automata on Words, *Theor. Comp. Sc.*, to appear.

16. J.-E. Pin, Finite semigroups and recognizable languages : an introduction, in NATO Advanced Study Institute *Semigroups, Formal Languages and Groups*, J. Fountain (ed.), Kluwer academic publishers, (1995), 1-32.

17. J.-E. Pin, A variety theorem without complementation, *Izvestiya VUZ Matematika* **39** (1995) 80-90. English version, *Russian Mathem. (Iz. VUZ)* **39**, (1995), 74-83.

18. M.P. Schützenberger, On finite monoids having only trivial subgroups, *Information and Control* **8**, (1965), 190-194.

19. J. Stern, Characterization of some classes of regular events, *Theoret. Comp. Sci.* **35**, (1985), 17-42.

20. J. Stern, Complexity of some problems from the theory of automata, *Inform. and Control* **66**, (1985), 63-176.

21. H. Straubing, *Finite automata, formal logic and circuit complexity*, Birkhäuser, 1994.

22. H. Straubing, D. Thérien and W. Thomas, Regular Languages Defined with Generalized Quantifiers, in *Proc. 15th ICALP*, Springer Lecture Notes in Computer Science **317**, (1988), 561-575.

23. D. Thérien and A. Weiss, Graph congruences and wreath products, *J. Pure Applied Algebra* **35**, (1985), 205-215.

24. W. Thomas, Classifying regular events in symbolic logic, *J. Comput. Syst. Sci* **25**, (1982), 360-375.

25. W. Thomas, On logics, tilings, and automata, *Proc. 18th ICALP*, Madrid, (J. Leach Albert et al., eds.), Lect. Notes in Comp. Sci. **510**, Springer, Berlin, (1991), 441-454.

26. W. Thomas, On the Ehrenfeucht-Fraïssé Game in Theoretical Computer Science, *TAPSOFT'93*, M.C. Gaudel, J.P. Jouannaud (Eds.), Lect. Notes in Comp. Sci. **668**, Springer, Berlin, (1993), 559-568.

27. Th. Wilke, Locally threshold testable languages of infinite words, in *STACS 93*, P. Enjalbert, A. Finkel, K.W. Wagner (Eds.), Lect. Notes in Comp. Sci. **665**, Springer, Berlin, (1993), 607-616.

Fixpoints for Rabin Tree Automata Make Complementation Easy

Roope Kaivola*

Laboratory for Foundations of Computer Science
University of Edinburgh
The King's Buildings, Edinburgh EH9 3JZ, United Kingdom

Abstract Modal mu-calculus and automata on infinite trees are complementary ways of describing infinite tree languages. In this paper we describe direct fixpoint constructions for Rabin-automata, allowing us to translate modal mu-calculus inductively to Rabin-automata. As Rabin-automata can also be mapped to modal mu-calculus, the fixpoint constructions provide a new proof of the expressive equivalence of the two formalisms, and lead to a simple proof of Rabin's complementation lemma, the core of the powerful decidability result for SnS.

1 Introduction

The concept of finite automata on infinite trees was originally introduced by Rabin [14] as a technical tool for proving the decidability of SnS, the monadic second-order theory of n successors. The decidability of SnS is a powerful and fundamental result to which large classes of other decidability results can be reduced [14, 16]. Rabin's decidability proof is based on translating formulae of SnS inductively to equivalent tree automata, and determining the decidability of a formula by checking whether the corresponding tree automaton has an accepting computation. The core point is showing closure under complementation, i.e. that for any automaton recognising language L there is an automaton recognising the complement of L. This result, known as Rabin's complementation lemma, is shown in [14] by an argument that is considered notoriously complex.

Due to the central nature of Rabin's result, several researchers have worked on devising alternative, easier approaches to it. Three important developments have been the use of infinite games, alternating tree automata and fixpoint calculi.

The complementation problem has been examined in the framework of infinite games by Büchi [2, 3] and Gurevich and Harrington [7]. In short, a run of a tree automaton on an input can be viewed as a strategy in an infinite game where one of the players plays for acceptance and the other for rejection. Conversely, a finite state winning strategy in such a game can be viewed as an automaton. The issue of complementation can then be reduced to a theorem about the determinacy of these games and the existence of finite state winning strategies [7].

Alternating tree automata have been proposed by Muller and Schupp as the tool to study the decidability of SnS [11]. The concept of alternation in automata

---

* Current address: Department of Computer Science, PO Box 26 (Teollisuuskatu 23),
FIN-00014 University of Helsinki, Finland, e-mail: Roope.Kaivola@helsinki.fi

generalises the notion of nondeterminism by allowing states to be existential or universal. The advantage of alternating automata is that they are easy to complement by dualization. However, an essential property of Rabin-automata that the decidability result for SnS takes advantage of is closure under projection or existential quantification. Although trivial for normal automata, projection is hard for alternating ones and effectively requires translating them back to normal automata. The easiest way to do this is to rely on the Gurevich-Harrington theorem about determinacy and finite state strategies mentioned above.

The complementation problem has been approached by Niwiński [12, 13] and Emerson and Jutla [6] using relations between Rabin-automata and fixpoint calculi. The calculus used in [6], called the modal mu-calculus, is a language extending propositional modal logic with maximal and minimal fixpoint operators vz and μz. It has attracted considerable interest as a method for specifying properties of concurrent programs [10, 18]. Niwiński uses a slightly different calculus based on tree operators instead of modal ones. However, as the calculi are easily reducible to each other, in the following we ignore the difference and just talk about modal mu-calculus.

Like alternating automata, modal mu-calculus is trivially closed under complementation. Therefore, if we can translate mu-calculus to Rabin-automata and vice versa without invoking the complementation lemma in the process, the complementation lemma follows as an immediate corollary. A direct translation from Rabin-automata to modal mu-calculus was described by Niwiński in [13]. He also provided a reverse mapping from an essentially conjunction-free fragment of modal mu-calculus to Rabin-automata [12]. However, closure under complementation is not obvious for this fragment, so these results alone do not suffice to prove the complementation lemma.

The first translation from the full modal mu-calculus to Rabin-automata that did not rely on the complementation lemma was presented by Emerson and Jutla in [6]. They interpret mu-calculus formulae as alternating tree automata of a particular kind, called history-free automata, and reduce these to normal Rabin-automata. Although in the general case a reduction from alternating automata to normal ones requires the Gurevich-Harrington theorem, an easier construction suffices for the more restricted history-free automata. This reduction uses Safra's construction [17] for determinising automata on infinite strings, and a method for simultaneously complementing an automaton while determinising it [5].

The current paper describes a new and direct inductive translation from modal mu-calculus to Rabin-automata by providing for every operator of the logic a corresponding operation for automata. The novelty is a powerset construction for Rabin-automata corresponding to fixpoint operations. This construction extends the current author's earlier work on fixpoints for Büchi-automata [9], and is related to powerset constructions in [4, 15].

In contrast with [6], the inductive nature of the translation here allows us to deal with one operator of the logic at a time. The fixpoint constructions arise naturally from the intuition of starting a new copy of an automaton every time the fixpoint is unfolded. The correctness of the whole construction is easy to show from scratch, without recourse to the Gurevich-Harrington theorem or other such means. In line with with all the other papers on simplifying the complementation lemma, we therefore believe that the current approach yields an easier proof than any of the earlier ones.

2 Preliminaries

Definition 1. Let Σ be an arbitrary set. We use Σ^* and Σ^ω to denote the sets of finite and infinite strings of elements of Σ, respectively, and define $\Sigma^\infty = \Sigma^* \cup \Sigma^\omega$. The symbol ϵ denotes the unique empty string, $|s|$ the length of s, s_i the i-th element of s, $t \cdot s$ the concatenation of t and s, and $s \preceq t$ ($s \prec t$) means that s is a prefix (a proper prefix) of t. If $s \preceq t$, $s^{-1}t$ is the s' such that $s \cdot s' = t$. For all $k \in \mathbb{N}$, define $[k] = \{i \in \mathbb{N} \mid 1 \leq i \leq k\}$. If A and A' are sets, $A \to A'$ and $A \rightharpoonup A'$ are the sets of functions and partial functions from A to A', respectively. If f is a partial function, $\mathrm{dom}(f)$ denotes the set on which f is defined. Notation $\bar{q}$ denotes a vector, and $\bar{q}_i$ its element.

To relate modal mu-calculus and tree automata to each other, we fix some $k > 0$ and take the set of k-branching infinite trees as models of modal mu-calculus and inputs of tree automata. To overcome the technical difference that automata are normally run on trees without transition labels, and that automata can distinguish different subtrees whereas modal mu-calculus cannot, it is natural to replace the usual relativised modalities $<l>$ and $[l]$ by indexed modalities $\textcircled{1} \ldots \textcircled{k}$, one for each possible subtree, as in [8].

Definition 2. Fix a countable set $\mathcal{Z}$ of propositions. The formulae of the modal mu-calculus μK are defined by the abstract syntax:

$$\phi ::= z \mid \neg\phi \mid \phi_1 \wedge \phi_2 \mid \textcircled{i}\,\phi \mid \mu z.\phi$$

where z varies over $\mathcal{Z}$ and i over $[k]$. In $\mu z.\phi$, each occurrence of z in ϕ is required to be *positive*, i.e. in the scope of an even number of negations. The derived operators $\vee$, $\Rightarrow$, $\Leftrightarrow$ are as usual, and νz stands for $\nu z.\phi = \neg\mu z.\neg\phi[\neg z/z]$. The symbol σ refers to both the μ and ν-operators.

An occurrence of a variable z in a formula ϕ is *guarded* iff it is in a subformula of the type $\textcircled{i}\,\phi'$. A formula ϕ is guarded iff for every fixpoint subformula $\sigma z.\phi'$ of ϕ, every occurrence of z in ϕ' is guarded.

Definition 3. Let Σ be an arbitrary set. An *infinite k-branching tree labelled with Σ* is a mapping $T : [k]^* \to \Sigma$. If $p \in [k]^\omega$, p is *a path of T*, and $T[p]$ is the infinite string $a_1 a_2 a_3 \ldots \in \Sigma^\omega$ where $a_j = T(p_1 \ldots p_{j-1})$.

The most obvious class of structures serving as models of modal mu-calculus and inputs for tree automata would be the class of k-branching infinite trees labelled with subsets of $\mathcal{Z}$. However, for technical convenience we label the nodes of the trees so that a label not only specifies which atomic propositions are true in the node, but also states explicitly which atomic propositions are false.

Definition 4. Take the set $\mathcal{Z}$ of atomic propositions and define the set of negated atomic propositions $\overline{\mathcal{Z}} = \{\neg z \mid z \in \mathcal{Z}\}$. Define $\mathcal{T}$, the set of *models*, as the following set of infinite k-branching $2^{(\mathcal{Z} \cup \overline{\mathcal{Z}})}$-labelled trees:

$$\mathcal{T} = \{T \in ([k]^* \to 2^{(\mathcal{Z} \cup \overline{\mathcal{Z}})}) \mid \forall t \in [k]^* \forall z \in \mathcal{Z} : z \in T(t) \text{ iff } \neg z \notin T(t)\}$$

Definition 5. Let $T \in \mathcal{T}$ be a model. The set of nodes of T satisfying a formula ϕ, denoted $\|\phi\|_T$, is defined inductively by: $\|z\|_T = \{i \in [k]^* \mid z \in T(i)\}$, $\|\neg\phi\|_T = [k]^* \setminus \|\phi\|_T$, $\|\phi \wedge \phi'\|_T = \|\phi\|_T \cap \|\phi'\|_T$, $\|\textcircled{i}\,\phi\|_T = \{j \in [k]^* \mid j \cdot i \in \|\phi\|_T\}$, and $\|\mu z.\phi\|_T = \bigcap\{W \subseteq [k]^* \mid \|\phi\|_{T[W/z]} \subseteq W\}$, where $T[W/z]$ is defined by: $T[W/z](i) = T(i) \cup \{z\}$ if $i \in W$, $T[W/z](i) = T(i) \setminus \{z\}$ if $i \notin W$.

The *language characterised by* ϕ is $L(\phi) = \{T \in \mathcal{T} \mid \epsilon \in \|\phi\|_T\}$.

Using the derived operators, every formula of modal mu-calculus can be transformed to a form where negations are applied only to atomic propositions, i.e. to positive normal form as defined e.g. in [18]. This can be done by pushing negations inwards using DeMorgan's laws and the rules $\neg\,\textcircled{i}\,\phi = \textcircled{i}\,\neg\phi$, $\neg\mu z.\phi = \nu z.\neg\phi[\neg z/z]$ and $\neg\nu z.\phi = \mu z.\neg\phi[\neg z/z]$.

Definition 6. A formula ϕ is in *positive normal form* (abbr. *pnf*) iff it only contains atomic propositions, their negations, and the $\vee, \wedge, \textcircled{i}, \mu$ and ν-operators.

Let us then recall some definitions related to Rabin-automata.

Definition 7. A *Rabin-automaton* A for k-branching trees is a 4-tuple $A = (Q, q_0, \Delta, \Omega)$, where Q is a set of states, $q_0 \in Q$ is the initial state, $\Delta \subseteq Q \times \Sigma \times (Q \setminus \{q_0\})^k$ is the transition relation, and Ω is an acceptance condition of the form $\Omega = ((G_1, R_1), \ldots, (G_n, R_n))$, where $G_m, R_m \subseteq Q$ for every $m \in [n]$.

The notation $q \xrightarrow{Z} \bar{q}$ means $(q, Z, \bar{q}) \in \Delta$. We say that A is *guarded* with respect to a variable z iff there are no Z and $\bar{q}$ such that $z \in Z$ and $(q_0, Z, \bar{q}) \in \Delta$.

A *run of A on $T \in \mathcal{T}$* is an inf. k-branching tree π labelled with Δ such that $\pi(\epsilon) = (q_0, Z, \bar{q})$ for some Z, $\bar{q}$, and for all $i \in [k]^*$, if $\pi(i) = (q, Z, \bar{q})$, then $Z \subseteq T(i)$ and for all $j \in [k]$, $\pi(i \cdot j) = (\bar{q}_j, Z', \bar{q}')$ for some $Z', \bar{q}'$.

Define functions π^{fr}, π^{lab} and π^{to} by: for all $i \in [k]^*$, $\pi^{\mathrm{fr}}(i) = q$, $\pi^{\mathrm{lab}}(i) = Z$ and $\pi^{\mathrm{to}}(i) = \bar{q}$, where $\pi(i) = (q, Z, \bar{q})$.

A run π of A is *accepting* iff for every path $p \in [k]^\omega$ there is some index $m \in [n]$ such that
- $\pi^{\mathrm{fr}}[p]_i \in R_m$ for only finitely many $i \in \mathbb{N}$, and
- $\pi^{\mathrm{fr}}[p]_i \in G_m$ for infinitely many $i \in \mathbb{N}$.

The *language recognised by* A is $L(A) = \{T \in \mathcal{T} \mid A \text{ has an accepting run on } T\}$.

Notice that in the definition of a run we require $Z \subseteq T(i)$, not $Z = T(i)$. Intuitively, a transition label Z of the automaton specifies that every $z \in Z$ must hold in $T(i)$ for the transition to be possible, but some other propositions may hold as well.

3 Formulae and automata

The easier half of the expressive equivalence of modal mu-calculus and Rabin-automata is showing that any tree language that is Rabin-recognisable can also be characterised by a modal mu-calculus formula. Since Niwiński has already described in detail such a mapping from Rabin-automata to modal mu-calculus, we just state this half of the equivalence here without a proof. It should be pointed out that Niwiński's proof does not involve Rabin's complementation lemma.

Theorem 8. *For any Rabin-automaton A there is a modal mu-calculus formula ϕ such that $L(A) = L(\phi)$.*

Proof. See [13, Theorem 6.1].

Let us then consider the more difficult direction, showing that for any modal mu-calculus formula there is an equivalent automaton. The method to construct an automaton A_ϕ for a given formula ϕ described in the following is inductive. For each basic modal mu-calculus formula we define a corresponding automaton directly, and for conjunction, disjunction and the modal and fixpoint operators we describe a method for obtaining the required automaton from the automata corresponding to the formulae the operator is applied to. Here the fixpoint operators are the essential problem. Let us state the main result first, and discuss the required constructions after that.

Theorem 9. *For any modal mu-calculus formula ϕ, there is a Rabin-automaton A such that $L(\phi) = L(A)$.*

Proof. Take any modal mu-calculus formula ϕ. We can assume without loss of generality that ϕ is guarded (see [1, Subsect. 2.4]) and in positive normal form, as we can mechanically transform any formula into this form.

We show by induction on the structure of ϕ that for any guarded ϕ in pnf there is a Rabin-automaton A such that $L(\phi) = L(A)$, and A is guarded w.r.t. any variable z if ϕ is. The base cases, when ϕ is either z or $\neg z$ for some $z \in \mathcal{Z}$, are shown in Lemma 11, the induction steps for $\vee$, $\wedge$ and $\odot$ in Lemma 12, and for ν and μ in Propositions 20 and 26.

Corollary 10 (Complementation Lemma). *For any Rabin-automaton A there is a Rabin-automaton $\overline{A}$ such that $L(\overline{A}) = \mathcal{T} \setminus L(A)$.*

The Rabin-automata corresponding to the atomic formulae and the constructions for the boolean and modal operators are easy.

Lemma 11. *For any $z \in \mathcal{Z}$, there are Rabin-automata A_z and $A_{\neg z}$ such that $L(A_z) = L(z)$ and $L(A_{\neg z}) = L(\neg z)$.*

Proof. Define $A_z = (\{q_0, q_1\}, q_0, \Delta_z, \Omega)$ and $A_{\neg z} = (\{q_0, q_1\}, q_0, \Delta_{\neg z}, \Omega)$, where $\Omega = ((\{q_1\}, \emptyset))$, $\Delta_z = \{(q_0, \{z\}, \bar{q}), (q_1, \emptyset, \bar{q})\}$ and $\Delta_{\neg z} = \{(q_0, \{\neg z\}, \bar{q}), (q_1, \emptyset, \bar{q})\}$ where $\bar{q} = (q_1, \ldots, q_1)$.

Lemma 12. *Let ϕ_1 and ϕ_2 be modal mu-calculus formulae and A_1 and A_2 Rabin-automata such that $L(\phi_1) = L(A_1)$ and $L(\phi_2) = L(A_2)$. Then*
- *there is a Rabin-automaton A such that $L(A) = L(\phi_1 \wedge \phi_2)$,*
- *there is a Rabin-automaton A such that $L(A) = L(\phi_1 \vee \phi_2)$, and*
- *for every $i \in [k]$, there is a Rabin-automaton A such that $L(A) = L(\odot \phi_1)$.*

Proof. The claims for $\wedge$ and $\vee$ hold, since Rabin-recognisable tree languages are closed under these operations [19]. To show the claim for the modal operator $\odot$, suppose that $L(A_1) = (Q, q_0, \Delta, \Omega)$, where $\Omega = ((G_1, R_1), \ldots, (G_n, R_n))$. Define $A = (Q \cup \{q'_0, q_a\}, q'_0, \Delta', \Omega')$, where q'_0 and q_a are new elements not in Q, $\Delta' = \Delta \cup \{(q'_0, \emptyset, \bar{q}), (q_a, \emptyset, (q_a, \ldots, q_a))\}$, where $\bar{q}$ is defined by: $\bar{q}_i = q_0$ and $\bar{q}_j = q_a$ for all $j \in [k]$, $j \neq i$, and $\Omega' = ((G_1, R_1), \ldots, (G_n, R_n), (\{q_a\}, \emptyset))$. It is easy to see that $L(A) = L(\odot \phi_1)$.

4 Maximal fixpoints

Both the maximal and minimal fixpoint operations are based on a powerset construction related to similar constructions for Büchi-automata (Dam [4], Rabin [15] and the current author [9]), and the methods used by Niwiński [12] to map conjunction-free modal mu-calculus to Rabin-automata. Supposing that automaton A corresponds to ϕ, the idea can be described as follows. In order to decide whether $T \in \mathcal{T}$ is a model of $\nu z.\phi$, we start by running A down T. Each time A requires z to be true, we start a new copy of A running down that particular subtree, and repeat the process with the new copies. When two copies of A are in the same state running down the same subtree, they are joined as one. If all the runs of copies of A are accepting, T is a model of $\nu z.\phi$. This idea leads to the following definition of an intermediate automaton[1].

Definition 13. Let $A = (Q, q_0, \Delta, \Omega)$ be a Rabin-automaton and z a variable. The *intermediate automaton* $\mathrm{fix}_z A$ is a triple (Q', q'_0, Δ'), where
- $Q' = 2^Q$ is the set of states,
- $q'_0 = \{q_0\}$ the initial state,
- $\Delta' \subseteq Q' \times \Sigma \times Q'^k \times (Q \rightharpoonup \Delta)$ the transition relation,

and where $(P, Z, \overline{P}, \delta) \in \Delta'$ iff

1. The domain of δ is $P \subseteq Q$, i.e. δ is a function $\delta : P \to \Delta$. Define functions δ^{fr}, δ^{lab} and δ^{to} by: for all $q \in P$, $\delta^{\mathrm{fr}}(q) = q'$, $\delta^{\mathrm{lab}}(q) = Z'$ and $\delta^{\mathrm{to}}(q) = \bar{q}'$, where $\delta(q) = (q', Z', \bar{q}')$.
2. $\delta^{\mathrm{fr}}(q) = q$ for all $q \in P$.
3. $q_0 \in P$ iff $P = \{q_0\}$ or there is a $q \in P \setminus \{q_0\}$ such that $z \in \delta^{\mathrm{lab}}(q)$.
4. $Z = \bigcup_{q \in P} \delta^{\mathrm{lab}}(q) \setminus \{z\}$.
5. $\overline{P}_j \setminus \{q_0\} = \{\delta^{\mathrm{to}}(q)_j \mid q \in P\}$ for all $j \in [k]$,

A *run* Π of $\mathrm{fix}_z A$, and the functions Π^{fr}, Π^{lab} and Π^{to} are as in Def. 7.

In the intermediate automaton $\mathrm{fix}_z A$, the states are sets of states of A, intuitively sets of copies of A. A transition of $\mathrm{fix}_z A$ corresponds to a set of simultaneous transitions of A. Conditions 1 and 2 state that δ attaches to each $q \in P$ a transition from q, 3 states that a new copy of A is started initially and when a running copy refers to z, 4 states that a transition of $\mathrm{fix}_z A$ is possible iff all the underlying transitions of A are, and 5 says that the state of $\mathrm{fix}_z A$ after a transition is specified by the targets of the underlying transitions of A, with possibly a new copy of A started.

In order to decide whether a given a run Π of $\mathrm{fix}_z A$ corresponds to accepting runs of A in the sense outlined in the beginning of the section, we need to be able to extract structures corresponding to these individual runs of A from Π. This is captured by the following definition.

Definition 14. Let A be as in Definition 13, Π a run of $\mathrm{fix}_z A$, $t \in [k]^*$, $\Pi(t) = (P, Z, \overline{P}, \delta)$, and $q \in P$. The *trail of Π starting from state q at t* is the tree $\pi(t, q) : [k]^* \to \Delta$, such that

---

[1] The development of the theory here follows [9] very closely up to the point of intermediate automata. However, the subsequent steps in [9] are quite different, as they only need to deal with Büchi-automata, not Rabin-automata.

- $\pi(t,q)(\epsilon) = \delta(q)$, and
- for all $i \in [k]^*$ and $j \in [k]$, $\pi(t,q)(i \cdot j) = \delta'(\pi(t,q)^{\mathrm{to}}(i)_j)$, where $\Pi(t \cdot i \cdot j) = (P', Z', \overline{P}', \delta')$.

The functions $\pi(t,q)^{\mathrm{fr}}$, $\pi(t,q)^{\mathrm{lab}}$ and $\pi(t,q)^{\mathrm{to}}$ are as in Definition 7.

According to the intuition explained above, in order to decide whether a tree $T \in \mathcal{T}$ is a model of $\nu z.\phi$, we have to check that all the runs of the individual copies of A within Π are accepting ones.

Definition 15. Let $A = (Q, q_0, \Delta, \Omega)$ be a Rabin-automaton, $\Omega = ((G_1, R_1), \ldots, (G_n, R_n))$, $T \in \mathcal{T}$, and Π a run of $\mathrm{fix}_z A$ on T.

We say that Π is ν-*accepting* iff for every $t \in [k]^*$, every $q \in \Pi^{\mathrm{fr}}(t)$ and every path $p \in [k]^\omega$ of the trail $\pi(t,q)$, there is some $m \in [n]$ such that

- $\pi(t,q)^{\mathrm{fr}}[p]_i \in R_m$ for only finitely many $i \in \mathbb{N}$, and
- $\pi(t,q)^{\mathrm{fr}}[p]_i \in G_m$ for infinitely many $i \in \mathbb{N}$.

Lemma 16. *If* $L(\phi) = L(A)$, $T \in L(\nu z.\phi)$ *iff* $\mathrm{fix}_z A$ *has a* ν-*accepting run on* T.

Proof. See [9, Lemma 13]. Although formulated there for Büchi-automata, the proof is exactly the same for Rabin-automata.

Let us then express ν-acceptance as a condition making reference to the paths of the run Π of the intermediate automaton, instead of the paths of the individual trails $\pi(t,q)$ in Π.

Definition 17. Let A, Ω and Π be as in Def. 15. We say that Π is ν'-*accepting* iff for every path $p \in [k]^\omega$ of Π there is

- a point $t \prec p$,
- a set of states $P_d \subseteq \Pi^{\mathrm{fr}}(t)$,
- for every state $q \in P_d$ an index $m_q \in [n]$, and
- an infinite strictly increasing sequence of points $t = t_0 \prec t_1 \prec t_2 \prec \ldots \prec p$

such that

1. for all $q, q' \in P_d$ and all $t \preceq t' \prec p$, $\pi(t,q)^{\mathrm{fr}}(t^{-1}t') = \pi(t,q')^{\mathrm{fr}}(t^{-1}t')$ iff $q = q'$.
2. for every $i \in \mathbb{N}$ and every $q' \in \Pi^{\mathrm{fr}}(t_i)$, there is a $q \in P_d$ such that $\pi(t,q)^{\mathrm{fr}}(t^{-1}t_{i+1}) = \pi(t_i, q')^{\mathrm{fr}}(t_i^{-1}t_{i+1})$
3. for every $q \in P_d$ and every $t \preceq t' \prec p$, $\pi(t,q)^{\mathrm{fr}}(t^{-1}t') \notin R_{m_q}$, and
4. for every $i \in \mathbb{N}$ and every $q \in P_d$, there is a $t_i \preceq t' \prec t_{i+1}$ such that $\pi(t,q)^{\mathrm{fr}}(t^{-1}t') \in G_{m_q}$.

Intuitively, ν'-acceptance states that for any path p of Π, there is a finite set of 'designated' trail paths, each of these trail paths has an associated acceptance pair in Ω, and there is an infinite sequence of 'checkpoints' along p such that (1) all the designated trail paths are entirely separate, (2) any trail path from any checkpoint along p coincides with a designated trail path since the next checkpoint at the latest, (3) no designated trail path ever passes through an associated red state, and (4) every designated trail path passes through an associated green state between any two checkpoints.

Lemma 18. *A run Π of $\mathrm{fix}_z A$ is ν-accepting iff it is ν'-accepting*

Proof. It is easy to see that if Π is ν'-accepting, it is also ν-accepting. Suppose then that Π is ν-accepting, and take any path $p \in [k]^*$ of Π. Notice that if there are $t \prec p$, $P_d \subseteq \Pi^{\mathrm{fr}}(t)$ and indices m_q for every $q \in P_d$ such that conditions 1 and 3 of ν'-acceptance are fulfilled and the following conditions 2' and 4' hold, then we also have a sequence of points $t_0 \prec t_1 \prec \ldots p$ so that the conditions 2 and 4 of ν'-acceptance are fulfilled:

2' for every $t \preceq t' \prec p$ and every $q' \in \Pi^{\mathrm{fr}}(t')$, there are $q \in P_d$ and $t' \preceq t'' \prec p$ such that $\pi(t,q)^{\mathrm{fr}}(t^{-1}t'') = \pi(t',q')^{\mathrm{fr}}(t'^{-1}t'')$, and

4' for every $q \in P_d$ and infinitely many $t \preceq t' \prec p$, $\pi(t,q)^{\mathrm{fr}}(t^{-1}t') \in G_{m_q}$.

To see that we can pick t and P_d so that conditions 1 and 2' are satisfied, it suffices to notice that the sets $\Pi^{\mathrm{fr}}(t')$ are all bounded. Since Π is ν-accepting, it is easy to satisfy conditions 3 and 4', as well.

To build a Rabin-automaton on the basis of the intermediate automaton, we attach to the intermediate automaton a mechanism checking for every path p whether there exists a point $t \prec p$ and a set of states P_d fulfilling the requirements of ν'-acceptance. This mechanism consists of a table $\bar{e}$, each element $\bar{e}_i$ of which is used to check for ν'-acceptance for some particular set of designated trail paths and indices m_q. An element $\bar{e}_i$ is a 5-tuple $(P, P_d, \mathbf{m}, P_a, P_e)$, where P is the current state of the intermediate automaton, P_d specifies the states in P that are on the designated trails, $\mathbf{m}$ associates with each state in P_d an acceptance pair in Ω, P_a specifies the states in P_d for which the corresponding trail has passed through a green state in the related acceptance pair after last checkpoint, and P_e specifies the states in P on trail paths that are not in the designated trail path set, that have started before last checkpoint, and have not yet coincided with a designated trail path.

In the definition of the transition relation of the Rabin-automaton $\nu z.A$ below, the components of the table $\bar{e}$ are updated according to their intended meaning in a transition. To keep the table $\bar{e}$ bounded, duplicate entries are deleted. Furthermore, after each transition, we add to empty places in $\bar{e}$ elements checking ν'-acceptance for every possible combination of designated trail paths and indices that is not already being checked by some entry in $\bar{e}$.

Definition 19. Let A and Ω be as in Def. 15, and let $\mathrm{fix}_z A = (Q', q'_0, \Delta')$. Define

$$E = \{(P, P_d, \mathbf{m}, P_a, P_e) \in 2^Q \times 2^Q \times (Q \rightharpoonup [n]) \times 2^Q \times 2^Q \mid$$
$$P_a \subseteq P_d = \mathrm{dom}(\mathbf{m}) \subseteq P, \ P_e \subseteq P, \ P_e \cap P_d = \emptyset\}$$

For every $P' \subseteq Q$ define

$$E(P') = \{(P, P_d, \mathbf{m}, P_a, P_e) \in E \mid P = P', \ P_a = \emptyset, \ P_e = P \setminus P_d\}$$

Let $\bot$ be an empty element not in E, and define $n_\nu = 2 \cdot |E|$, $\mathcal{E} = (E \cup \{\bot\})^{n_\nu}$.
Define the Rabin-automaton $\nu z.A = (Q_\nu, q_{\nu 0}, \Delta_\nu, \Omega_\nu)$ by:
- Q_ν is the smallest subset of $Q' \times \mathcal{E}$ that contains $q_{\nu 0}$ and is closed under the transition function Δ_ν,

- $q_{\nu 0} = (\{q_0\}, \bar{e})$, where $\bar{e}$ contains one copy of every element of $E(\{q_0\})$, listed in some fixed order, and all other elements of $\bar{e}$ have the empty value $\bot$.
- If $(P, \bar{e}) \in Q_\nu$ and $(P, Z, (P^1, \ldots, P^k), \delta) \in \Delta'$ then

$$(P, \bar{e}) \xrightarrow{Z} ((P^1, \bar{e}^1), \ldots, (P^k, \bar{e}^k)) \quad \text{in } \Delta_\nu$$

where for every $j \in [k]$, $\bar{e}^j$ is as follows.
Fix $j \in [k]$, and define a vector $\bar{e}'$. For any $i \in [n_\nu]$, if $\bar{e}_i = \bot$ then $\bar{e}'_i = \bot$. Otherwise, assume that $\bar{e}_i = (P, P_d, \mathbf{m}, P_a, P_e)$. Define

- $P' = P^j$, and
- $P'_d = \{\delta^{\text{to}}(q)_j \mid q \in P_d\}$.
- For every $q' \in P'_d$, if there is a unique $q \in P_d$ such that $\delta^{\text{to}}(q)_j = q'$, then $\mathbf{m}'(q') = \mathbf{m}(q)$ for this unique q, and otherwise $\mathbf{m}'(q')$ is undefined.
- If $P_a = P_d$ and $P_e = \emptyset$, then $P'_a = \emptyset$ and $P'_e = P' \setminus P'_d$, and otherwise

$$P'_a = \{\delta^{\text{to}}(q)_j \mid q \in P_a\} \cup \{q' \in P'_d \mid q' \in G_{\mathbf{m}'(q')}\}$$
$$P'_e = \{\delta^{\text{to}}(q)_j \mid q \in P_e\} \setminus P'_d$$

If either

a there are $q_1, q_2 \in P_d$ such that $q_1 \neq q_2$ but $\delta^{\text{to}}(q_1)_j = \delta^{\text{to}}(q_2)_j$, or
b there is some $q' \in P'_d$ such that $q' \in R_{\mathbf{m}'(q')}$,

then $\bar{e}'_i = \bot$, and otherwise $\bar{e}'_i = (P', P'_d, \mathbf{m}', P'_a, P'_e)$.
The vector $\bar{e}^j$ is derived from $\bar{e}'$ by first replacing every entry $\bar{e}_i$ for which there is some $i' < i$ such that $\bar{e}_{i'} = \bar{e}_i$ by $\bot$, and then adding one copy of every value in $E(P^j)$ that does not already occur in $\bar{e}'$ to some place $\bar{e}'_i$ that was not used in $\bar{e}$ (i.e. for which $\bar{e}_i = \bot$), according to some fixed strategy.

- $\Omega_\nu = ((G_1^\nu, R_1^\nu), \ldots, (G_{n_\nu}^\nu, R_{n_\nu}^\nu))$, where for every $i \in [n_\nu]$,

$$G_i^\nu = \{(P, \bar{e}) \in Q_\nu \mid \bar{e}_i = (P, P_d, \mathbf{m}, P_a, P_e), P_a = P_d \text{ and } P_e = \emptyset\}$$
$$R_i^\nu = \{(P, \bar{e}) \in Q_\nu \mid \bar{e}_i = \bot\}$$

Proposition 20. If $L(\phi) = L(A)$, then $L(\nu z.\phi) = L(\nu z.A)$.

Proof. $T \in L(\nu z.\phi)$ iff (by Lemma 16) $\text{fix}_z A$ has a ν-accepting run Π on T iff (by Lemma 18) $\text{fix}_z A$ has a ν'-accepting run Π on T iff (easy) $T \in L(\nu Z.A)$.

5 Minimal fixpoints

The only difference between the constructions corresponding to the minimal and the maximal fixpoint operators is that in the minimal case we have to disallow infinite regeneration of the fixpoint formula. Intuitively, such regeneration occurs whenever an individual copy of A running as a part of the intermediate automaton $\text{fix}_z A$ takes a transition the label of which refers to the fixpoint variable z. The following definition expresses this intuition of one copy of A starting another copy, this another one starting a third copy etc.

Definition 21. Let A, $\mathrm{fix}_z A$ and Π be as in Def. 14, and let $t \in [k]^*$, $q \in \Pi^{\mathrm{fr}}(t)$. A sequence $d = (t_1, q_1')(t_2, q_2') \ldots \in ([k]^* \times Q)^\infty$ is a *dependency sequence* of Π from (t, q) iff

- $t \preceq t_1$, $q_1' = \pi(t, q)^{\mathrm{fr}}(t^{-1} t_1)$, and
- for all $1 \leq i < |d|$, $t_i \preceq t_{i+1}$, $z \in \pi(t_i, q_i')^{\mathrm{lab}}(\epsilon)$, and $q_{i+1}' = \pi(t_i, q_0)^{\mathrm{fr}}(t_i^{-1} t_{i+1})$

We say that d is *proper* iff $|d| > 1$, and if d is finite, that it *leads to* $(t_{|d|}, q_{|d|}')$.

We say that Π is *μ-accepting* iff Π is ν-accepting and has no infinite dependency sequences.

Lemma 22. *If $L(\phi) = L(A)$, $T \in L(\mu z.\phi)$ iff $\mathrm{fix}_z A$ has a μ-accepting run on T.*

Proof. See [9, Lemma 18].

Definition 23. Let A, Ω and Π be as in Def. 15. We say that Π is *μ'-accepting* iff for every path $p \in [k]^\omega$ of Π there is

- a point $t \prec p$,
- a set of states $P_d \subseteq \Pi^{\mathrm{fr}}(t)$,
- for every state $q \in P_d$ an index $m_q \in [n]$, and
- an infinite strictly increasing sequence of points $t = t_0 \prec t_1 \prec t_2 \prec \ldots \prec p$

such that they fulfil conditions 1-4 of Definition 17 and:

5   for every $i \in \mathbb{N}$, every $q' \in \Pi^{\mathrm{fr}}(t_i)$, and every proper dependency sequence $(t_1', q_1') \ldots (t_{h+1}', q_{h+1}')$ from (t_i, q') for which $t_h' \prec t_{i+1} \preceq t_{h+1}' \prec p$, there are $j \in [h]$ and $q \in P_d$ such that $\pi(t, q)^{\mathrm{fr}}(t^{-1} t_j') = q_j'$, and

6   for every $q \in P_d$ and every $t \preceq t' \prec p$, there is no proper dependency sequence from (t, q) to $(t', \pi(t, q)^{\mathrm{fr}}(t^{-1} t'))$.

Intuitively, condition 5 states that every proper dependency sequence from a checkpoint extending beyond the following checkpoint passes through a state on some designated trail path between the checkpoints, and condition 6 that there is no proper dependency sequence from a designated trail path back to the same trail path.

Lemma 24. *If A is guarded w.r.t. z, then a run Π of $\mathrm{fix}_z A$ is μ-accepting iff it is μ'-accepting*

Proof. Assume first that Π is μ'-accepting, hence ν'-accepting, and ν-accepting by Lemma 18. Suppose then that there is some $p \in [k]^\omega$ and an infinite dependency sequence $(t_1', q_1')(t_2', q_2') \ldots$ such that every $t_i' \prec p$. Since A is guarded w.r.t. z, $t_1' \prec t_2' \prec \ldots \prec p$. By condition 5 of μ'-acceptance and the finiteness of P_d, this means that there is some $q \in P_d$ such that $q_i = \pi(t, q)^{\mathrm{fr}}(t^{-1} t_i')$ for infinitely many i. But then there is a dependency sequence violating condition 6 for q.

Assume then that Π is μ-accepting, hence ν-accepting, and ν'-accepting by Lemma 18. Take any $p \in [k]^\omega$. As Π is ν'-accepting, there are t, P_d etc. fulfilling conditions 1-4 of μ'-acceptance. Notice that for every $q \in P_d$ there is a $t \preceq t' \prec p$ such that there is no proper dependency sequence from $(t', \pi(t, q)^{\mathrm{fr}}(t^{-1} t'))$ to $(t'', \pi(t, q)^{\mathrm{fr}}(t^{-1} t''))$ for any $t' \preceq t'' \prec p$, as otherwise we could construct an infinite dependency sequence along p in Π. Therefore, there are t, P_d, $t_0 \prec t_1 \ldots$ fulfilling conditions 1-4 and 6. To see that Π is μ'-accepting, let us show that we can pick a subsequence of $t_0 \prec t_1 \prec \ldots p$ so that condition 5 is satisfied, as well.

To see this, take any t_i and $q' \in \Pi^{\mathrm{fr}}(t_i)$. By condition 2 of ν'-acceptance, if $(t'_1, q'_1) \ldots$ is a dependency sequence from (t_i, q'), and $q'_1 \neq \pi(t, q)^{\mathrm{fr}}(t^{-1}t'_1)$ for all $q \in P_d$, then $t'_1 \prec t_{i+1}$. This means that for any $h \in \mathbb{N}$, there are only finitely many dependency sequences $(t_1, q'_1) \ldots (t'_h, q'_h)$ of length h from (t_i, q') such that $\pi(t, q)^{\mathrm{fr}}(t^{-1}t'_j) \neq q'_j$ for all j and $q \in P_d$. Since there are no infinite dependency sequences in Π, this implies by König's lemma that there are only finitely many dependency sequences $(t_1, q'_1) \ldots$ from (t_i, q') such that $\pi(t, q)^{\mathrm{fr}}(t^{-1}t'_j) \neq q'_j$ for all j and $q \in P_d$.

The construction of the Rabin-automaton $\mu z.A$ uses the same strategy as in the case of maximal fixpoints. However, the elements of the table $\bar{e}$ have a new component $\mathbf{d}$, specifying for each designated trail path the states to which there is a proper dependency sequence from some earlier point in the trail. The component P_e is also used to track dependency sequences from trails outside the designated trail path set.

Definition 25. Let A and Ω be as in Def. 15, and let $\mathrm{fix}_z A = (Q', q'_0, \Delta')$. Define

$$E = \{(P, P_d, \mathbf{m}, \mathbf{d}, P_a, P_e) \in 2^Q \times 2^Q \times (Q \rightharpoonup [n]) \times (Q \rightharpoonup 2^Q) \times 2^Q \times 2^Q \mid$$
$$P_a \subseteq P_d = \mathrm{dom}(\mathbf{m}) = \mathrm{dom}(\mathbf{d}) \subseteq P,$$
$$P_e \subseteq P, \ P_e \cap P_d = \emptyset, \ \forall q \in P_d : \mathbf{d}(q) \subseteq P\}$$

For every $P' \subseteq Q$ define

$$E(P') = \{(P, P_d, \mathbf{m}, \mathbf{d}, P_a, P_e) \in E \mid$$
$$P = P', \ P_a = \emptyset, \ P_e = P \setminus P_d, \ \forall q \in P_d : \mathbf{d}(q) = \emptyset\}$$

Define $n_\mu = 2 \cdot |E|$ and $\mathcal{E} = (E \cup \{\bot\})^{n_\mu}$. In the Rabin-automaton $\mu z.A = (Q_\mu, q_{\mu 0}, \Delta_\mu, \Omega_\mu)$ the components Q_μ, $q_{\mu 0}$ and Ω_μ are analogous to the Q_ν, $q_{\nu 0}$ and Ω_ν of Definition 19, and Δ_μ is defined as follows.

If $(P, \bar{e}) \in Q_\mu$ and $(P, Z, (P^1, \ldots, P^k), \delta) \in \Delta'$ then

$$(P, \bar{e}) \xrightarrow{Z} (P^1, \bar{e}^1), \ldots, (P^k, \bar{e}^k)) \quad \text{in } \Delta_\mu$$

where for every $j \in [k]$, $\bar{e}^j$ is derived from the following $\bar{e}'$ as in Def. 19.

For any $i \in [n_\mu]$, if $\bar{e}_i = \bot$ then $\bar{e}'_i = \bot$. Assume then $\bar{e}_i = (P, P_d, \mathbf{m}, \mathbf{d}, P_a, P_e)$. Let P', P'_d, $\mathbf{m}'$ and P'_a be as in Definition 19, and define:

- For every $q' \in P'_d$, if there is a unique $q \in P_d$ such that $\delta^{\mathrm{to}}(q)_j = q'$, then

$$\mathbf{d}'(q') = \{\delta^{\mathrm{to}}(q'')_j \mid q'' \in \mathbf{d}(q)\} \cup \{\delta^{\mathrm{to}}(q_0)_j \mid \exists q'' \in \mathbf{d}(q) \cup \{q\} : z \in \delta^{\mathrm{lab}}(q'')\},$$

  and if there is no such unique q, then $\mathbf{d}'(q')$ is undefined.

- If $P_a = P_d$ and $P_e = \emptyset$, then $P'_e = P' \setminus P'_d$, and otherwise

$$P'_e = (\{\delta^{\mathrm{to}}(q)_j \mid q \in P_e\} \cup \{\delta^{\mathrm{to}}(q_0)_j \mid \exists q'' \in P_e : z \in \delta^{\mathrm{lab}}(q'')\}) \setminus P'_d$$

If condition **a** or **b** of Definition 19 holds, or if
c there is some $q' \in P'_d$ such that $q' \in \mathbf{d}'(q')$,
then $\bar{e}'_i = \bot$, and otherwise $\bar{e}'_i = (P', P'_d, \mathbf{m}', \mathbf{d}', P'_a, P'_e)$.

Lemma 26. *If A is guarded w.r.t. z and $L(\phi) = L(A)$, then $L(\mu z.\phi) = L(\mu z.A)$.*

Proof. Straightforward from 22 and 24.

Acknowledgements

I would like to thank Colin Stirling for many useful discussions. This work has been funded by the Academy of Finland, the British Council, the CVCP Overseas Research Students Awards Scheme and the Finnish Cultural Foundation.

References

1. Banieqbal, B. & Barringer, H.: Temporal logic with fixed points, in *Temporal Logic in Specification*, LNCS vol. 398, Springer-Verlag, 1989, pp. 62-74
2. Büchi, J. R.: Using determinacy to eliminate quantifiers, in *Fundamentals of Computation Theory*, LNCS vol. 56, 1977, pp. 367-378
3. Büchi, J. R.: State-strategies for games in $F_{\sigma\delta} \cap G_{\delta\sigma}$, in *Journal of Symbolic Logic*, vol. 48, 1983, pp. 1171-1198
4. Dam, M.: Fixpoints of Büchi automata, in *Proceedings of the 12th FST & TCS*, LNCS vol. 652, Springer-Verlag, 1992, pp. 39-50
5. Emerson, E. A. & Jutla, C. S.: On simultaneously determinizing and complementing ω-automata, in *Proceedings of the 4th IEEE LICS*, 1989
6. Emerson, E. A. & Jutla, C. S.: Tree automata, mu-calculus, and determinacy, in *Proceedings of the 32nd FOCS*, 1991, pp. 368-377
7. Gurevich, Y. & Harrington, L.: Trees, Automata and Games, in *Proceedings of the 14th ACM STOC*, 1982, pp. 60-65
8. Hüttel, H.: SnS can be Modally Characterised, in *Theoretical Computer Science*, vol. 74, 1990, pp. 239-248
9. Kaivola, R.: On modal mu-calculus and Büchi tree automata, in *Information Processing Letters*, vol. 54, 1995, pp. 17-22
10. Kozen, D.: Results on the propositional μ-calculus, in *Theoretical Computer Science*, vol. 27, 1983, pp. 333-354
11. Muller, D. E. & Schupp, P. E.: Alternating automata on infinite trees, in *Theoretical Computer Science*, vol. 54, 1987, pp. 267-276
12. Niwiński, D.: On fixed point clones, in *Proc. of the 13th ICALP*, LNCS vol. 226, Springer-Verlag, 1986, pp. 402-409
13. Niwiński, D.: Fixed Points vs. Infinite Generation, in *Proceedings of the 3rd IEEE LICS* 1988, pp. 402-409
14. Rabin, M. O.: Decidability of second order theories and automata on infinite trees, in *Transactions of the Americal Mathematical Society*, vol. 141, 1969, pp. 1-35
15. Rabin, M. O.: Weakly definable relations and special automata, in Bar-Hillel, Y. (ed.): *Math. Logic and Foundations of Set Theory*, North-Holland, 1970, pp. 1-23
16. Rabin, M. O.: Decidable theories, in Barwise, J. (ed.): *Handbook of Mathematical Logic*, North-Holland, 1977, pp. 595-629
17. Safra, S.: On the complexity of ω-automata, in *Proceedings of the 29th Symposium on Foundations of Computer Science*, 1988, pp. 319-327
18. Stirling, C.: Modal and temporal logics, in Abramsky, S. & al. (eds.): *Handbook of Logic in Computer Science*, Oxford University Press, 1992, pp. 477-563
19. Thomas, W.: Automata on infinite objects, in van Leeuwen, J. (ed.): *Handbook of Theoretical Computer Science*, vol. 2, Elsevier/North-Holland, 1990, pp. 133-191

New Upper Bounds to the Limitedness of Distance Automata

Kosaburo Hashiguchi

Department of Information Technology, Faculty of Engineering, Okayama University, Tsushima, Okayama 700, Japan

Abstract. A distance automaton is a finite nondeterministic automaton with a distance function which assigns zero or one to each atomic transition and assigns a nonnegative integer to each accepted word by the plus-min principle. In this paper, we prove that the distances of all accepted words of a distance automaton is bounded by some constant if and only if they are bounded by $2^{4m^3 + m\log(m+2) + m}$, where m is the number of states of the automaton.

1 Introduction

A finite nondeterministic automaton with a distance function is called a distance automaton. The distance function assigns zero or one to each atomic transition, and assigns to each accepted word a nonnegative integer by the plus-min principle. Distance automata were first introduced in [1], and play important roles as basic computing machines for solving representation problems and the problems of determining star height, inclusion star height, relative star height, and relative inclusion star height about regular languages [2-4]. In some previous papers, distance automata were called finite automata with distance functions. The set of nonnegative integers and the infinity ∞ with the min and plus operations constitutes the tropical semiring which has been studied in [5-7]. Here the min and plus correspond to the usual addition and multiplication, respectively, in a semiring.

A distance automaton is said to be limited (in distance) if the distances of all accepted words are bounded by some nonnegative integer. The author presents the following two limitedness theorems in [1,3] : (1) a zero-deterministic distance automaton is limited if and only if the distances of all accepted words are bounded by $2^{4m^3 + 2m^2 + 5m}$; (2) an arbitrary distance automaton is limited if and only if the distances of all accepted words are bounded by $2^{2^{3m+4}}$. Here m is the number of states of the automaton. The second theorem was deduced from the first one by transforming a given distance automaton to the corresponding equivalent zero-deterministic distance automaton which has 2^m states. The following problem was left open in [3]. Problem A : dose there exist a more general argument working directly on arbitrary distance automata and producing a remarkable decrease (preforably to a simple exponential function) in the upper bound $2^{2^{3m+4}}$ above ?

H. Leung[6] presents an algorithm that runs in time $4^{O(m^2)}$ for deciding the limitedness of distance automata, but does not provide any upper bound to the limitedness of distance automata. The main result of this paper is Theorem 17 which provides the following : a distance automaton is limited if and only if the distances of all accepted words are bounded by $2^{4m^3+m\log(m+2)+m}$. Our main arguments work for arbitrary distance automata to produce directly the upper bound above, and these combinatorial arguments give an answer to Problem A above posed in [3].

This paper consists of four sections. Section 2 presents preliminaries. In Section 3, we define the notion of factor pairs of an arbitrary distance automaton $\mathcal{A}$, and define the index $I(B,y)$ to each factor pair $(B,y) \in 2^Q \times \Sigma^*$ of $\mathcal{A}$, where Q is the set of states of $\mathcal{A}$ and Σ is the input alphabet. Then, by induction on $I(B,y)$, we prove Lemma 16. Theorem 17 follows from Lemma 16 easily. Theorem 17 presents five equivalent conditions for an arbitrary distance automaton not to be limited, two of which are new types of characterizations. Theorem 26 in Section 4 presents five equivalent conditions for a zero-deterministic distance automaton not to be limited. The upper bound in Theorem 26 is almost the same as the ones already known, but we present Theorem 26 since it presents, as Theorem 17, two new types of characterizations to the unlimitedness of zero-deterministic distance automata.

2   Preliminaries

Σ is a finite nonempty alphabet. λ is the null word. For any $w \in \Sigma^*$, $|w|$ is the length of w, and for any set A, $|A|$ is the cardinality of A. ϕ is the empty set. N is the set of nonnegative integers. If a set A is a singleton, $A = \{a\}$, then A is often denoted simply by a. For any $w \in \Sigma^*$, if $w = xyz$ for $x, y, z \in \Sigma^*$, then x is a prefix of w, y is a factor of w, and z is a suffix of w. For any $w \in \Sigma^+$, a decomposition of w is a n-tuple, $1 \leq n \leq |w|$, $(x_1, \cdots, x_n)$, such that $w = x_1 \cdots x_n$ and $x_i \in \Sigma^+$ for all i.

A distance automaton $\mathcal{A}$ (over Σ) is a sixtuple, $< \Sigma, Q, \delta, S, F, d >$, where Q is a finite set of states, $\delta : Q \times \Sigma \to 2^Q$ is a transition function, $S \subset Q$ and $F \subset Q$ are the sets of initial and final states, respectively, and $d : Q \times \Sigma \times Q \to \{0, 1, \infty\}$ is a distance function. d satisfies the following : for any $(q, a, q') \in Q \times \Sigma \times Q$, $d(q, a, q') = \infty$ iff $q' \notin \delta(q, a)$. The language accepted by $\mathcal{A}$ is denoted by $L(\mathcal{A})$, and $L(\mathcal{A}) = \{w \in \Sigma^* \mid \delta(S, w) \cap F \neq \phi\}$. d is extended to $d : Q \times \Sigma^* \times Q \to N \cup \infty$ and $d : 2^Q \times \Sigma^* \times 2^Q \to N \cup \infty$ by the plus-min principle in the following way :
    For any $a \in \Sigma, w \in \Sigma^*, q, q' \in Q$ and $t, t' \subset Q$,

1. $d(q, \lambda, q') = 0$ if $q = q'$; $d(q, \lambda, q') = \infty$ otherwise ;
2. $d(q, wa, q') = min\{d(q, w, q'') + d(q'', a, q') \mid q'' \in Q\}$,
    where $i + \infty = \infty$ and $min\{i, \infty\} = i$ for any $i \in N \cup \infty$;
3. $d(t, w, t') = min\{d(q, w, q') \mid q \in t, q' \in t'\}$.

$D(\mathcal{A})$ denotes $D(\mathcal{A}) = sup\{d(S, w, F) \mid w \in L(\mathcal{A})\}$. By definition, $D(\mathcal{A}) = 0$ if $L(\mathcal{A})$ is empty. $\mathcal{A}$ is said to be limited (in distance) if $D(\mathcal{A}) \in N$.

Throughout the rest of this section, let $\mathcal{A} =< \Sigma, Q, \delta, S, F, d >$ be an arbitrary distance automaton.

Definition 1. δ_0 denotes the zero-distance function of $\mathcal{A}$, $\delta_0 : Q \times \Sigma^* \to 2^Q$, such that for any $p \in Q$ and $w \in \Sigma^*, \delta_0(p, w) = \{q \in Q \mid d(p, w, q) = 0\}$. $\mathcal{A}$ is zero-deterministic if for any $p \in Q$ and $w \in \Sigma^*, |\delta_0(p, w)| \leq 1$.

Definition 2. A factor pair of $\mathcal{A}$ is a pair $(B, y) \in 2^Q \times \Sigma^*$ such that for some $w \in L(\mathcal{A})$ and $x, z \in \Sigma^*$, it holds $w = xyz$ and $B = \delta(S, x)$. $FP(\mathcal{A})$ denotes the set of factor pairs of $\mathcal{A}$.

Definition 3. Let $(B, y) \in FP(\mathcal{A})$.

1. $\delta(B|y)$ denotes the restriction of δ to B so that for any $p \in B, \delta(B|y)(p) = \delta(p, y)$.
2. $\delta_0(B|y)$ denotes the restriction of δ_0 to B so that for any $p \in B, \delta_0(B|y)(p) = \delta_0(y)$.
3. The pair set $PS(B|y)$ of y w.r.t. B is the set of pairs $(p, q) \in B \times Q$ such that $q \in \delta(p, y)$.

Definition 4.   1. For any $A \subset Q \times Q$, $f(A)$ is the function $f(A) : Q \to 2^Q$ such that for any $p \in Q, f(A)(p) = \{q \in Q \mid (p, q) \in A\}$.
2. For any $A_1, A_2 \subset Q \times Q, A_1 A_2$ is the set, $\{(p, q) \in Q \times Q \mid for\ some\ q' \in Q, (p, q') \in A_1\ and (q', q) \in A_2\}$.
3. For any $f : Q \to 2^Q$, $A(f)$ is the set $\{(p, q) \in Q \times Q \mid q \in f(p)\}$.
4. For any $f_1, f_2 : Q \to 2^Q$, $f_2 f_1$ is the function from Q to 2^Q such that for any $p \in Q, f_2 f_1(p) = \{q \in Q \mid for\ some\ q' \in Q, q' \in f_1(p)\ and\ q \in f_2(q')\}$.
5. For any $f : Q \to 2^Q$, f^{-1} is the function from Q to 2^Q such that for any $p \in Q, f^{-1}(p) = \{q \in Q \mid p \in f(q)\}$.

Lemma 5.

1. For any $\Lambda_1, \Lambda_2, \Lambda_3 \subset Q \times Q, (A_1 A_2) A_3 = A_1 (A_2 A_3)$.
2. For any $n \geq 1, A_1, \ldots, A_n \subset Q \times Q$ and $f_1, \ldots, f_n : Q \to 2^Q$, it holds that $A(f_n \cdots f_1) = A(f_1) \cdots A(f_n)$, $f(A_1 \cdots A_n) = f(A_n) \cdots f(A_1)$, and $(f_n \cdots f_1)^{-1} = f_1^{-1} \cdots f_n^{-1}$.

Definition 6. The set $MRE(\Sigma)$ of multiplicative regular expressions over Σ is defined inductively as follows.

1. $\lambda, \phi, a \in MRE(\Sigma)$ for $a \in \Sigma$.
2. If $E_1, E_2 \in MRE(\Sigma)$, then $(E_1)^*, E_1 E_2 \in MRE(\Sigma)$.

Definition 7. Let $E \in MRE(\Sigma)$.

1. $|E|$ is the language denoted by E, where E is regarded as a regular expression.
2. For any $k \in N$, $E(k)$ is the word in Σ^* which can be obtained form E by replacing each occurrence of $*$ with k.
3. The size of E, $S(E)$, is the number of occurrences of all symbols in $\Sigma \cup \{\lambda\}$ which occur in E : we do not count the number of occurrences of $*$, (, and).

Example 1. Let $\Sigma = \{a, b\}$ and $E = a(ab^*a)^*b$. Then $S(E) = 5, E(0) = ab, E(1) = aabab, E(2) = a(ab^2a)^2b, \cdots,$ etc.

3 On Arbitrary Distance Automata

Throughout this section, let $\mathcal{A} =< \Sigma, Q, \delta, S, F, d >$ be an arbitrary distance automaton. The notations about $\mathcal{A}$ introduced in Section 2 will be used in this section with the same meaning.

Definition 8. Let $(B, y) \in FP(\mathcal{A})$. A subset C of B is a parallel subset w.r.t. (B, y) if $\delta_0(C, y) \neq \phi$ and for each prefix v of y, there exists an injection $\alpha(C, y, v) : C \to \delta_0(C, v)$ which satisfies the following (1) and (2).

1. For any prefix v of y, and any $p \in C$, it holds $\alpha(C, y, v)(p) \in \delta_0(p, v)$.
2. For any prefix v of y, any prefix u of v with $v = ux$ for $x \in \Sigma^*$ and any $p \in C$, it holds $\alpha(C, y, v)(p) \in \delta_0(\alpha(C, y, u)(p), x)$.

Definition 9. Let $(B, y) \in FP(\mathcal{A})$.

1. The index $I(B, y)$ of (B, y) is defined by :
   (a) $I(B, y) = |Q|$ if $\delta_0(B, y) = \phi$.
   (b) $I(B, y) = |Q| - max\{|C| \mid C \text{ is a parallel subset w.r.t. } (B, y)\}$ otherwise.
2. A parallel subset C w.r.t. (B, y) is maximal if it holds $I(B, y) = |Q| - |C|$.
3. Define $minI(\mathcal{A})$ and $maxI(\mathcal{A})$ by :
   (a) $minI(\mathcal{A}) = min\{I(B, y) \mid (B, y) \in FP(\mathcal{A})\}$.
   (b) $maxI(\mathcal{A}) = max\{I(B, y) \mid (B, y) \in FP(\mathcal{A})\}$.

Lemma 10. *For any $(B, y) \in FP(\mathcal{A})$, it holds $0 \leq minI(\mathcal{A}) \leq I(B, y) \leq maxI(\mathcal{A}) \leq |Q|$.*

Lemma 11. *For any $(B, \lambda) \in FP(\mathcal{A})$, it holds $I(B, \lambda) = |Q| - |B|$, and for any $p \in B$, $\delta(p, \lambda) = \delta_0(p, \lambda) = p$ and $d(p, \lambda, p) = 0$.*

Lemma 12. *For any $(B, y) \in FP(\mathcal{A})$ and $t, u, v \in \Sigma^*$ with $y = tuv$, the following hold.*

1. *$I(\delta(B, t), u) \geq I(B, y)$.*
2. *Assume that $I(\delta(B, t), u) = I(B, y)$. Let C be a maximal parallel subset w.r.t. (B, y), and $\alpha(C, y, t)$ be the injection from C into $\delta_0(C, t)$ as in Definition 8. Then $\alpha(C, y, t)(C)$ is a maximal parallel subset w.r.t. $(\delta(B, t), u)$.*

Proof. If $\delta_0(B, y) = \phi$, then the assertion is clear. Assume the contrary. (1). Let C be a maximal parallel subset w.r.t. (B, y). For each prefix x of y, let $\alpha(C, y, x)$ be the injection from C into $\delta_0(C, x)$ as in Definition 8. Then one can see easily that $\alpha(C, y, t)(C)$ is a parallel subset w.r.t. $(\delta(B, t), u)$. Thus $I(\delta(B, t), u) \leq |Q| - |\alpha(C, y, t)(C)| = |Q| - |C| = I(B, y)$. Thus (1) holds. (2) is clear since $\alpha(C, y, t)(C)$ becomes maximal if it holds $I(B, y) = I(\delta(B, t), u)$.

Lemma 13. *For any $(B, y) \in FP(\mathcal{A})$, if $I(B, y) = minI(\mathcal{A})$, then it holds (1) for any $(p, q) \in PS(B|y)$, $d(p, y, q) < |B|$, and (2) there exists $v \in \Sigma^*$ such that $|v| < 4^{|Q|^2}$, $\delta(B|v) = \delta(B|y)$, $\delta_0(B|v) = \delta_0(B|y)$ and $I(B, y) = I(B, v)$.*

Proof. Assume $I(B,y) = minI(A)$. Note that $I(B,\lambda) = I(B,y) = |Q| - |B|$. Thus B is the maximal parallel subset w.r.t. (B,y), and for any prefix u of y, it holds $|\delta(B,u)| = |\delta_0(B,u)| = |B|$. Let $(p,q) \in PS(B|y)$, and assume that $d(p,y,q) \geq |B|$. Then there exist $u_1, \cdots, u_n \in \Sigma^+$ and $p_0, p_1, \cdots, p_n \in Q$ such that (i) $n = d(p,y,q)$, (ii) $y = u_1 \cdots u_n$, (iii) $p_0 = p$ and $p_n = q$, and (iv) for each $1 \leq i \leq n$, $d(p_{i-1}, u_i, p_i) = 1$. By definition of $I(B,y)$, for each $1 \leq i \leq n$, there exists a bijection $\alpha_i : \delta(B, u_0 \cdots u_{i-1}) \to \delta(B, u_0 \cdots u_i)$ such that for each $r \in \delta(B, u_0 \cdots u_{i-1})$, it holds $\alpha_i(r) \in \delta_0(r, u_i)$, where $u_0 = \lambda$. Then for each $1 \leq i \leq n$, it holds $p_i \in \delta_0((\alpha_1 \cdots \alpha_i)^{-1}(p_i), u_1 \cdots u_i)$, where $(\alpha_1 \cdots \alpha_i)^{-1}(p_i) \in B - p_0$. Since α_i is a bijection for $1 \leq i \leq n$ and $d(p,y,q) = d(p_0, u_1, p_1) + \cdots + d(p_{n-1}, u_n, p_n)$, , it holds $(\alpha_1 \cdots \alpha_j)^{-1}(p_j) \neq (\alpha_1 \cdots \alpha_k)^{-1}(p_k)$ for all $1 \leq j < k \leq n$, which is clearly a contradiction. (2). If $|y| < 4^{|Q|^2}$, then we put $v = y$. Otherwise Let $y = a_1 \cdots a_n, n \geq 4^{|Q|^2}$, and $a_i \in \Sigma$ for all i. Since $I(B,y) = minI(A)$, it holds that for any prefix u of y, $|\delta(B,u)| = |\delta_0(B,u)| = |B|$. For each $0 \leq i \leq n$, let $f_i : B \to 2^Q \times 2^Q$ be a function such that for each $p \in B$, $f_i(p) = (\delta_0(p, a_0 a_1 \cdots a_i), \delta(p, a_0 a_1 \cdots a_i))$, where $a_0 = \lambda$. Since $|\{f | B \to 2^Q \times 2^Q\}| - |\{h | h : B \to 2^Q \times 2^Q$, and for $X = \{q \in A | (A, C) = f_i(p)$ for$/;p \in B$ and $C \subset Q\}$, it holds $|X| \neq |B|\}| \leq 4^{|Q|^2} - 2^{|Q|}$, there exist $0 \leq i < j \leq n$ such that $f_i = f_j$. We put $y' = a_0 a_1 \cdots a_i a_{j+1} \cdots a_n$. If $|y'| \geq 4^{|Q|^2}$, we repeat the same procedure. Now it is clear that (2) holds.

Definition 14. For each $0 \leq i \leq |Q|$, two integers $o_1(i)$ and $o_2(i)$ are defined inductively as follows:

1. $o_1(0) = |Q| - 1$ and $o_2(0) = 4^{|Q|^2} - 1$.
2. For $0 < i \leq |Q|$,
   (a) $o_1(i) = (|Q| - i + 2)2 \cdot 2^{4|Q|^2}(o_1(i-1) + 1)$.
   (b) $o_2(i) = 2^{7|Q|^2}(o_1(i-1) + 1)$.

Lemma 15. *Let* $m = |Q|$.

1. $o_1(m) < 2^{4m^3 + m\log(m+2)+m}$.
2. $o_2(m) < 2^{7m^3 + 2m^2 + 1}$.

Proof. (1). One can see that $o_1(m) = m(m+1)!t^m + m!t^{m-1} + (m-1)!t^{m-2} + \cdots 3!t^2 + 2!t$, where $t = 2 \cdot 2^{4m^2}$. By induction on i, $3 \leq i \leq m$, one can prove that $i!t^{i-1} + (i-1)!t^{i-2} + \cdots 2!t < 2i!t^{i-1}$. Thus $o_1(m) < m(m+1)!t^m + 2m!t^{m-1} \leq t^m((m+1)!m + (m-1)!) \leq t^m(m+2)^m = 2^{4m^3 + m\log(m+2)+m}$. (2). As above, it holds $o_2(m) = 2^{2m^2}t^m + t^{m-1} + t^{m-2} + t^{m-3} + \cdots + t = 2^{2m^2}t^m + (t^m - t)/(t-1) < 2^{2m^2}t^m \cdot 2 = 2^{7m^3 + 2m^2 + 1}$, where $t = 2^{7m^2}$.

Lemma 16. *For any* $(B,y) \in FP(A)$, *there exist* $A(B,y) \subset PS(B|y)$ *and* $E(B,y) \in MRE(\Sigma)$ *to which the following hold:*

1. *For any* $(p,q) \in A(B,y)$, $d(p,y,q) \leq o_1(I(B,y))$.

2. $S(E(B,y)) \leq o_2(I(B,y))$, and for any $k \geq 0$, it holds that $\delta(B|E(B,y)(k)) = \delta(B|y)$, $\delta_0(B|E(B,y)(k)) = \delta_0(B|y)$ and for any $(p,q) \in PS(B|y) - A(B,y)$, $d(p, E(B,y)(k), q) \geq k$.

Proof. The proof is by induction on $I(B,y)$. When $I(B,y) = minI(\mathcal{A})$, we put $A(B,y) = PS(B|y)$ and $E(B,y) = v$, where v is as in Lemma 13. The assertions are clear from the lemma. Let $I(B,y) > minI(\mathcal{A})$. If $|y| \leq o_1(I(B,y))$, then the assertions are clear as above. Assume $|y| > o_1(I(B,y))$. We consider two cases.

Case (1) : For any proper prefix u of y, it holds $I(B,y) > I(B,u)$. Let $y = ta$ for $a \in \Sigma$ and $t \in \Sigma^+$. Then $I(B,t) > I(B,y)$, and by the inductive hypothesis, there exist $A(B,t)$ and $E(B,t)$ to which the corresponding (1) and (2) hold. We put $A(B,y) = \{(p,q) \in PS(B|y) \mid$ for some $q' \in Q, (p,q') \in A(B,t)$ and $q \in \delta(q',a)\} = A(B,t)PS(\delta(B,t)|a)$ and $E(B,y) = E(B,t)a$. Now consider any $(p,q) \in A(B,y)$. There exists $q' \in \delta(p,t)$ such that $(p,q') \in A(B,t)$, $q \in \delta(q',a)$ and $d(p,y,q) \leq d(p,t,q') + d(q',a,q) \leq 1 + o_1(I(B,y) - 1) < o_1(I(B,y))$. Now consider $k \geq 0$ and any $(p,q) \in PS(B|y) - A(B,y)$. Clearly $E(B,y)(k) = E(B,t)(k)a$. Let $q' \in Q$ be such that $d(p, E(B,y)(k), q) = d(p, E(B,t)(k), q') + d(q', a, q)$. Then $(p,q') \in PS(B|t) - A(B,t)$. By the inductive hypothesis, $d(p, E(B,t)(k), q') \geq k$. Now the assertions are clear.

Case (2) : Case (1) does not hold. Let $(w_1, \cdots, w_n)$ be a decomposition of y such that (i) $n \geq 2$, (ii) $y = w_1 \cdots w_n$, (iii) either $w_n \in \Sigma$ or $|w_n| \geq 2$ and $I(\delta(B, w_1 \cdots w_{n-1}), w_n) < I(B,y)$, and (iv) for each $1 \leq i \leq n-1$, it holds that $w_i \in \Sigma^+$, $I(\delta(B, w_0 \cdots w_{i-1}), w_i) = I(B,y)$, and for $w_i = u_i a_i$ with $u_i \in \Sigma^*$ and $a_i \in \Sigma$, either $u_i = \lambda$ or $I(\delta(B, w_0 w_1 \cdots w_{i-1}), u_i) < I(B,y)$, where $w_0 = \lambda$. Let C be a maximal parallel subset w.r.t. (B,y), and for each $1 \leq i \leq n-1$, let α_i denote the injection $\alpha(C, y, w_0 w_1 \cdots w_{i-1})$ as in Definition 8. For each $1 \leq i \leq n$, we put $B_i = \delta(B, w_0 w_1 \cdots w_{i-1})$ and $C_i = \alpha(C, y, w_0 w_1 \cdots w_{i-1})(C)$. Note that $C = C_1$. By definition of $I(B,y)$, one can see that for each $1 \leq i \leq n-1$, C_i is a maximal parallel subset w.r.t. (B_i, w_i). By the inductive hypothesis and Lemma 11, for each $1 \leq i \leq n-1$, there exist $A(B_i, u_i)$ and $E(B_i, u_i)$ to which the corresponding (1) and (2) hold. We put $A(B_i, w_i) = \{(p,q) \in PS(B_i|w_i) \mid$ for some $q' \in Q, (p,q') \in A(B_i, u_i)$ and $q \in \delta(q', a_i)\} = A(B_i, u_i)PS(\delta(B_i, u_i)|a_i)$, and $E(B_i, w_i) = E(B_i, u_i)a_i$. As in case (1), one can see that the following (3) and (4) hold for each $1 \leq i \leq n-1$.

(3) For any $(p,q) \in A(B_i, w_i), d(p, w_i, q) \leq o_1(I(B,y) - 1) + 1$.

(4) $S(E(B_i, w_i)) \leq o_2(I(B,y) - 1) + 1$, and for any $k \geq 0$, it holds $\delta(B_i | E(B_i, w_i)(k)) = \delta(B_i | w_i)$, $\delta_0(B_i | E(B_i, w_i)(k)) = \delta_0(B_i | w_i)$, and for any $(p,q) \in PS(B_i | w_i) - A(B_i, w_i), d(p, E(B_i, w_i)(k), q) \geq k$.

For w_n, by the inductive hypothesis and as above, there exist $A(B_n, w_n) \subset PS(B_n | w_n)$ and $E(B_n, w_n) \in MRE(\Sigma)$ to which the corresponding (1) and (2) hold. Now we consider two subcases.

Case (2.1) : $n \leq 2^{4|Q|^2} - 2$. We put $A(B,y) = A(B_1, w_1) \cdots A(B_n, w_n)$ and $E(B,y) = E(B_1, w_1) \cdots E(B_n, w_n)$.

Claim 1 (5) For any $(p,q) \in A(B,y), d(p,y,q) \leq (2^{4|Q|^2} - 2)(o_1(I(B,y)-1)+1) < o_1(I(B,y))$. (6) $S(E(B,y)) \leq (2^{4|Q|^2} - 2)(o_2(I(B,y)-1)+1) < o_2(I(B,y))$, and for any $k \geq 0$, it holds $\delta(B|E(B, y)(k)) = \delta(B|y), \delta_0(B|E(B, y)(k)) = \delta_0(B|y)$, and for any $(p, q) \in PS(B|y) - A(B, y), d(p, E(B, y)(k), q) \geq k$.

Proof of Claim 1. (5) is clear. (6). We shall prove that for any $k \geq 1$, and $(p, q) \in PS(B|y) - A(B, y), d(p, E(B, y)(k), q) \geq k$. The other assertions are clear. Consider any $(p, q) \in PS(B|y) - A(B, y)$. Let $k \geq 1$ and $p_0, \cdots, p_n \in Q$ be such that $p_0 = p, p_n = q$ and $d(p, E(B, y)(k), q) = d(p_0, E(B_1, w_1)(k), p_1) + \cdots + d(p_{n-1}, E(B_n, w_n)(k), p_n)$. We define a set A by : $A = \{i \mid 1 \leq i \leq n$ and $p_i \notin f(A(B_1, w_1) \cdots A(B_i, w_i))(p_0)\}$. Note that $n \in A$. Let i be the minimum in A. Then $(p_{i-1}, p_i) \notin A(B_i, w_i)$. From (4) above, $d(p_{i-1}, E(B_i, w_i)(k), p_i) \geq k$. This completes the proof of Claim 1.

Now the assertions are clear for Case (2.1).

We continue the proof of Lemma 16.

Case (2.2) : $n \geq 2^{4|Q|^2} - 1$. We put $\alpha = 2^{|Q|^2} - 1$ and $c = 2^{4|Q|^2} - 2$. We first define a set $\gamma(B, y)$ as follows. For each $1 \leq i \leq c$, let g_i denote the function from B to $2^Q \times 2^Q \times 2^Q$ such that for any $p \in B, g_i(p) = (\delta(p, w_1 \cdots w_i), \delta_0(p, w_1 \cdots w_i), f(A(B_1, w_1) \cdots A(B_i, w_i))(p))$. Since $|\{h \mid h : B \to 2^Q \times 2^Q \times 2^Q\}| - |\{h \mid h : B \to 2^Q \times 2^Q \times 2^Q$, and for $h(B) = (Q_0, Q_1, Q_2)$, it holds $Q_0 = \phi$ or $Q_2 = \phi\}| \leq 8^{|B||Q|} - 2^{2|Q|}$, there exist $d_1, \cdots, d_{\alpha+1}$ such that $1 \leq d_1 < d_2 < \cdots < d_{\alpha+1} \leq c$ and $g_{d_1} = \cdots = g_{d_{\alpha+1}}$. Now for each $i \in \{d_1, \cdots, d_{\alpha+1}\}$, we define the function β_i from B_{c+1} to 2^Q such that for any $p \in B_{c+1}, \beta_i(p) = f(A(B_{i+1}, w_{i+1})A(B_{i+2}, w_{i+2}) \cdots A(B_{c+1}, w_{c+1}))^{-1}(p)$. As above, there exist $d_1 \leq i_1 < j_1 \leq d_{\alpha+1}$ such that $\beta_{i_1} = \beta_{j_1}$. If $n < 2(c + 1)$, then we put $\gamma(B, y) = \{(i_1, j_1)\}$ and our procedure ends. Otherwise we define a pair (i_2, j_2) of integers which satisfy the above conditions for $(B_{c+2}, w_{c+2} \cdots w_n)$ as above, and it holds $c + 2 \leq i_2 < j_2 < 2(c + 1)$. If $n \geq 3(c + 1)$, then we continue repreating the procedure, and finally obtain the set $\gamma(B, y) = \{(i_1, j_1), (i_2, j_2), \cdots, (i_d, j_d)\}$ with $d \geq 1, n - j_d \leq c$, and for each $1 \leq k \leq d, (c + 1)(k - 1) < i_k < j_k < k(c + 1)$. One can see easily that the following (7) holds :

(7) $j_{k+1} - j_k \leq 2(c + 1) - 2 < 2 \cdot 2^{4|Q|^2}$ for all $0 \leq k \leq d - 1$, where $j_0 = 0$.

For each $1 \leq e \leq d$, we put (i) $A_{e1} = A(B_{j_{e-1}+1}, w_{j_{e-1}+1}) \cdots A(B_{i_e}, w_{i_e})$, (ii) $F_e = \{(p, u, q', v, q) \mid (p, q) \in A(B_{i_e+1}, w_{i_e+1}) \cdots A(B_{j_e}, w_{j_e})$, and for some $h, i_e + 1 \leq h \leq j_e$ and $r, s \in \Sigma^*$ with $w_h = rs$, it holds $q' \in \delta_0(C_h, r), u = w_{i_e+1}w_{i_e+2} \cdots w_{h-1}r, v = sw_{h+1} \cdots w_{j_e}$, and $d(p, u, q') + d(q', v, q) \leq (o_1(I(B, y) - 1) + 1)(j_e - i_e)\}$, (iii) $A_{e2} = \{(p, q) \in A(B_{i_e+1}, w_{i_e+1}) \cdots A(B_{j_e}, w_{j_e}) \mid$ for some $q' \in Q$ and $u, v \in \Sigma^*, (p, u, q', v, q) \in F_e\}$, and (iv) $E_e = E(B_{j_{e-1}+1}, w_{j_{e-1}+1}) \cdots E(B_{i_e}, w_{i_e})(E(B_{i_e+1}, w_{i_e+1}) \cdots E(B_{j_e}, w_{j_e}))^*$. We define $A(B, y)$ and E by : $A(B, y) = A_{11}A_{12}A_{21}A_{22} \cdots A_{d1}A_{d2}A_{d+1}$ and $E = E_1 \cdots E_d E_{d+1}$, where $A_{d+1} = A(B_{j_d+1}, w_{j_d+1}) \cdots A(B_n, w_n)$ and $E_{d+1} = E(B_{j_d+1}, w_{j_d+1}) \cdots E(B_n, w_n)$ if $j_d < n$ and $A_{d+1} = \{(q, q) \mid q \in Q\}$ and $E_{d+1} = \lambda$ if $j_d = n$.

Claim 2 (8) For any $(p, q) \in A(B, y), d(p, y, q) \leq o_1(I(B, y))$. (9) For any $k \geq 0$, it holds that $\delta(B|E(k)) = \delta(B|y), \delta_0(B|E(k)) = \delta_0(B|y)$, and for any $(p, q) \in PS(B|y) - A(B, y), d(p, E(k), q) \geq k$.

Proof of Claim 2. (8). Consider any $(p, q) \in A(B, y)$. There exists a sequence of states, $(p_0, p_1, \cdots, p_d, p_{d+1})$ such that (i) $p_0 = p$, and (ii) for each $1 \leq e \leq d$, there exists $(r_e, u_e, p_e, v_e, q_e) \in F_e$ for some $r_e, q_e \in Q$ and $u_e, v_e \in \Sigma^*$

such that $d(p, y, q) = d(p_0, w_1 \cdots w_{i_1} u_1, p_1) + d(p_1, v_1 w_{j_1+1} \cdots w_{i_2} u_2, p_2) + \cdots + d(p_{d-1}, v_{d-1} w_{j_{d-1}+1} \cdots w_{i_d} u_d, p_d) + d(p_d, v_d w_{j_d+1} \cdots w_n, p_{d+1})$. For each $1 \leq e \leq d-1$, we put $z_e = v_e w_{j_e+1} \cdots w_{i_{e+1}} u_{e+1}$. We define an integer e_1, $1 \leq e_1 \leq d$, as follows. We put $e_1 = 1$ if for each $2 \leq e \leq d, p_e \notin \delta_0(p_1, z_1 \cdots z_{e-1})$. Otherwise let e_1 be the largest integer e, $2 \leq e \leq d$, such that $p_e \in \delta_0(p_1, z_1 \cdots z_{e-1})$. If $e_1 \neq d$, then we repeat the procedure to define e_2, $e_1 + 1 \leq e_2 \leq d$, so that we put $e_2 = e_1 + 1$ if for each $e_1 + 2 \leq e \leq d, p_e \notin \delta_0(p_{e_1}, z_{e_1+1} \cdots z_{e-1})$, and otherwise let e_2 be the largest integer e, $e_1 + 2 \leq e \leq d$ such that $p_e \in \delta_0(p_{e_1}, z_{e_1+1} \cdots z_{e-1})$. We repeat this procedure and obtain a sequence of integers, $(e_1, \cdots, e_r)$, such that $e_r = d$.

Claim 2.1 $r \leq |Q| - I(B, y)$.

Proof of Claim 2.1. Assume that $r \geq |Q| - I(B, y) + 1$. We put $v_1 = w_1 \cdots w_{i_1} u_1$, $v_2 = z_1 \cdots z_{e_1}$, $v_{r+2} = w_{j_d+1} \cdots w_n$, and for each $1 \leq b \leq r - 1$, $v_{b+2} = z_{e_b+1} \cdots z_{e_{b+1}}$. Note that $y = v_1 \cdots v_{r+2}$. Now consider the case where $\delta_0(B, y) \neq \phi$. The other case can be handled more easily. Recall that C is a maximal parallel subset w.r.t. (B, y). For each $1 \leq b \leq r + 1$, we put $D_b = \delta_0(C, v_0 v_1 \cdots v_{b-1})$, where $v_0 = \lambda$ and we put $D_1 = C$. From definition of $I(B, y)$, one can see that for each $1 \leq b \leq r$, there exists a bijection $\alpha_b : D_b \rightarrow D_{b+1}$ such that for each $p \in D_b, \alpha_b(p) \in \delta_0(p, v_b)$. Thus for each $1 \leq b \leq r, (\alpha_2 \alpha_3 \cdots \alpha_b)^{-1}(p_{e_b}) \in \alpha_1(C, v_1) - p_1$, and by definition of e_b, $1 \leq b \leq r$, it must hold for each $1 \leq b < s \leq r, (\alpha_2 \cdots \alpha_b)^{-1}(p_{e_b}) \neq (\alpha_2 \cdots \alpha_s)^{-1}(p_{e_s})$. This is clearly a contradiction since $|C| = |Q| - I(B, y)$. This completes the proof of Claim 2.1.

We shall continue the proof of Claim 2. As in the proof of (7), one can see that for each $-1 \leq b \leq r, d(p_{e_b}, v_{b+2}, p_{e_{b+1}}) \leq (2c - 1)(o_1(I(B, y) - 1) + 1) \leq 2 \cdot 2^{4|Q|^2}(o_1(I(B, y) - 1) + 1)$, where $e_{-1} = 0, e_0 = 1$ and $e_{r+1} = n$. Now (8) of Claim 2 is clear. We shall prove (9) of Claim 2. Clearly $\delta(B|E(k)) = \delta(B|y)$ and $\delta_0(B|E(k)) = \delta_0(B|y)$ for all $k \geq 0$. Consider any $(p, q) \in PS(B|y) - A(B, y)$. We shall prove that $d(p, E(k), q) \geq k$ for each $k \geq 0$. Obviously $(p, E(0), q) \geq 0$. Let $k \geq 1$, and $E(k) = E_{11}(k)(E_{12}(k))^k E_{21}(k)(E_{22}(k))^k \cdots E_{d1}(k)(E_{d2}(k))^k E_{d+1}(k)$, where for each $1 \leq b \leq d, E_{b1} = E(B_{j_{e-1}+1}, w_{j_{e-1}+1}) \cdots E(B_{i_e}, w_{i_e})$ and $E_{b2} = E(B_{i_e+1}, w_{i_e+1}) \cdots E(B_{j_e}, w_{j_e})$, where $j_0 = 0$. Let $d(p, E(k), q) = d(p, E_{01}(k), p_{10}) + d(p_{10}, E_{02}(k), p_{11}) + d(p_{11}, E_{02}(k), p_{12}) + \cdots + d(p_{1k-1}, E_{02}(k), p_{1k}) + d(p_{1k}, E_{03}(k), p_{21}) + d(p_{21}, E_{04}(k), q)$, where $E_{01} = E_{11}, E_{02} = E_{12}, E_{03} = E(B_{j_1+1}, w_{j_1+1}) \cdots E(B_{c+1}, w_{c+1})$ and $E_{04} = E(B_{c+2}, w_{c+2}) \cdots E(B_{i_2}, w_{i_2}) E_{31}(E_{32})^* \cdots E_{d1}(E_{d2})^* E_{d+1}$ so that $E = E_{01}(E_{02})^* E_{03} E_{04}$. If $(p, p_{10}) \notin A_{11}$, then, as in the proof for Case (1), one can prove that $d(p, E_{11}(k), p_{10}) \geq k$, and the assertion holds. In the same way, we may assume that for each $0 \leq b \leq k-1$, it holds $(p_{1b}, p_{1b+1}) \in A_{12}$. Moreover, if for each $0 \leq b \leq k - 1$, it holds $p_{1b+1} \notin \delta_0(p_{1b}, E_{12}(k))$, then $d(p_{1b}, E_{12}(k), p_{1b+1}) \geq 1$ and the assertion holds. Otherwise there exists $0 \leq b \leq k - 1$ such that $p_{1b+1} \in \delta_0(p_{1b}, E_{12}(k))$. Now in the state transition $p_{1b} \rightarrow p_{1b+1}$ by $E_{12}(k)$, there exist $u, v \in \Sigma^*$ and $q' \in Q$ such that $E_{12}(k) = uv, d(p_{1b}, u, q') + d(q', v, p_{1b+1}) = 0$, and $q' \in \alpha(C, y, E_{11}(k)u)(C)$ since $I(B, y) = I(\delta(B, E_{10}(k), E_{12}(k))$. By the inductive observations about relations between $E_{12}(k)$ and $w_{i_1+1} \cdots w_{j_1}$ (recall the proof of Lemma 13-(2)), these arguments imply that there exist $r, s \in \Sigma^*$ such that $w_{i_1+1} \cdots w_{j_1} = rs$, $q' \in \alpha(C, y, w_1 \cdots w_{i_1} r)(C)$ and $d(p_{1b}, r, q') + d(q', s, p_{1b+1}) \leq (o_1(I(B, y) - 1) + 1)(j_1 - i_1)$. By applying these arguments for each factor $E_{b1}(k)(E_{b2}(k))^k E_{j_b+1}(k) \cdots E_{(c+1)b}(k)$ of $E(k), 2 \leq b \leq d$, one can conclude that $(p, q) \in A(B, y)$ if it holds $d(p, E(k), q) < k$. This completes the proof of Claim 2.

Now we shall complete the proof for Case (2.2), which will complete the proof of Lemma 16. It suffices to find $E(B, y)$ with size less than $o_2(I(B, y))$. If $d < 8^{|Q|^2}$, then $n \leq (d+1)(c+1) \leq 2^{7|Q|^2}$. We put $E = E(B, y)$ and the assertion can be seen easily. Otherwise, for each $1 \leq b \leq d$, let h_b be the function from B to $2^Q \times 2^Q \times 2^Q$ such that for any $p \in Q, h_b(p) = (\delta_0(p, w_1 \cdots w_{j_b}), \delta(p, w_1 \cdots w_{j_b}), f(A_{11}A_{12}A_{21}A_{22} \cdots A_{b1}A_{b2})(p))$. As in the paragraph of the definition of (i_1, j_1) above, one can see that there exist $1 \leq b < e \leq d$ such that $h_b = h_e$. Now put $E' = E_1 \cdots E_b E_{e+1} E_{e+2} \cdots E_d E_{d+1}$. Clearly it holds $A(B, y) = A_{11} A_{12} A_{21} A_{22} \cdots A_{b1} A_{b2} A_{e+11} A_{e+12} \cdots A_{d1} A_{d2} A_{d+1}$ and for any $k \geq 0, \delta(B|E'(k)) = \delta(B|y)$ and $\delta_0(B|E'(k)) = \delta_0(B|y)$. As above, one can prove that $d(p, E'(k), q) \geq k$ for any $(p, q) \in PS(B|y) - A(B, y)$. If $d - (b - e) \geq 8^{|Q|^2}$, then we repeat the same argument. Thus one can see that there exists $E(B, y) \in MRE(\Sigma)$ to which (2) holds. This completes the proof for Case (2.2) and the proof of Lemma 16.

The following theorem is the main result of this paper, where (2) and (5) are new types of characterizations for the unlimitedness of distance automata.

Theorem 17. *For any distance automaton* $A = < \Sigma, Q, \delta, S, F, d >$, *the following five conditions are equivalent.*

1. A *is not limited.*
2. *There exists* $E \in MRE(\Sigma)$ *such that* $S(E) < o_2(A)$ *and for any integer* $k \geq 0$, $E(k) \in L(A)$ *and* $d(S, E(k), F) \geq k$.
3. *There exists* $E \in MRE(\Sigma)$ *such that for any integer* $k \geq 0$, $E(k) \in L(A)$ *and* $d(S, E(k), F) \geq k$.
4. *There exists* $w \in L(A)$ *such that* $d(S, w, F) \geq o_1(A)$.
5. *There exists* $w \in L(A)$ *such that* $|w| \leq o_3(A)$ *and* $d(S, w, F) \geq o_1(A)$.

Here $m = |Q|$, $o_1(A) = 2^{4m^3 + m \log(m+2) + m}$, $o_2(A) = 2^{7m^3 + 2m^2 + 1}$, *and* $o_3(A) = 2^{m^2(o_1(A)+1)}$.

Proof. (1)$\Rightarrow$(2). Assume A is not limited. Then there exists $w \in L(A)$ such that $d(S, w, F) \geq o_1(A)$. By Lemma 16, there exist $A(S, w) \subset PS(S|w)$ and $E(S, w) \in MRE(\Sigma)$ for which (1) and (2) in the lemma hold. If $S \times F \cap A(S, w) \neq \phi$, then $d(S, w, F) \leq o_1(I(S, w)) < o_1(A)$, a contradiction. Thus $S \times F \cap PS(S|w) \subset PS(S|w) - A(S, w)$, and for any $k \geq 0, d(S, E(S, w)(k), F)) \geq k$. (2)$\Rightarrow$(3) and (3)$\Rightarrow$(4) are obvious. (4)$\Rightarrow$(5). Assume that (4) holds. Let $v \in L(A)$ be one of the shortest words such that $d(S, v, F) \geq o_1(A)$. Assume that $|v| > o_3(A)$. Let $v = a_1 \cdots a_n$ for $n > o_3(A)$ and $a_i \in \Sigma$ for all i. For each $0 \leq i \leq n$, let f_i be the function from Q to $(2^Q)^{o_1(A)+1}$ such that for any $p \in Q, f_i(p) = (A_0, A_1, \ldots, A_{o_1(A)})$, where for each $0 \leq j < o_1(A), A_j = \{q \in$

$Q \mid d(p, a_0a_1 \cdots a_i, q) = j\}$ and $A_{o_1(\mathcal{A})} = \{q \in Q \mid d(p, a_0a_1 \cdots a_i, q) \geq o_1(\mathcal{A})\}$. Here $a_0 = \lambda$. Since $|\{f \mid f : Q \rightarrow (2^Q)^{o_1(\mathcal{A})+1}\}| = 2^{|Q|^2(o_1(\mathcal{A})+1)}$, there exist $0 \leq i < j \leq n$ such that $f_i = f_j$. But then $\delta(S|a_1 \cdots a_i a_{j+1} \cdots a_n) = \delta(B|w)$ and $d(S, a_1 \cdots a_i a_{j+1} \cdots a_n, F) \geq o_1(\mathcal{A})$, which is a contradicton to minimality of v. (5)$\Rightarrow$(1). Assume that $\mathcal{A}$ is limited. Consider any $w \in L(\mathcal{A})$. Let $A(S, w) \subset PS(S|W)$ and $E(S, w) \in MRE(\Sigma)$ be as in Lemma 16. If $S \times F \cap A(S, w) \neq \phi$, then, as above, (2) holds, a contradiction. Thus there exists $(p, q) \in S \times F \cap A(B, w)$, and by Lemma 16, $d(S, w, F) \leq d(p, w, q) \leq o_1(I(S, w)) < o_1(\mathcal{A})$. Thus (5) does not hold.

4   On Zero-Deterministic Distance Automata

By slightly changing the arguments in Section 3, one can prove the corresponding limitedness theorem about zero-deterministic distance automata. The arguments are almost the same except that in centain places, 2^Q will be replaced with $Q \cup \{\phi\}$. Let $\mathcal{A} = < \Sigma, Q, \delta, S, F, d >$ be an arbitrary zero-deterministic distance automaton. δ_0 is regarded as the function $\delta_0 : Q \times \Sigma^* \rightarrow Q \cup \{\phi\}$, and for any $w \in \Sigma^*$ and $B \subset Q$, $\delta_0(B|w)$ is the function from B to $Q \cup \{\phi\}$. Thus we write $\delta_0(p, w) = q$ and $\delta_0(B|w)(p) = q$ instead of $\delta_0(p, w) = \{q\}$ and $\delta_0(B|w)(p) = \{q\}$.

As in Definition 2, $FP(\mathcal{A})$ denotes the set of factor pairs of $\mathcal{A}$. For any $(B, y) \in FP(\mathcal{A})$, the index $I(B, y)$ of (B, y), and $minI(\mathcal{A})$ and $maxI(\mathcal{A})$ are defined as in Definitions 8, 9. In this section, we shall present only the main results omitting proofs.

Lemma 18. *For any* $(B, y) \in FP(\mathcal{A})$, $I(B, y) = |Q| - |\delta_0(B, y)|$.

Lemma 19. *For any* $(B, y) \in FP(\mathcal{A})$, *it holds* $0 \leq minI(\mathcal{A}) \leq I(B, y) \leq maxI(\mathcal{A}) \leq |Q|$.

Lemma 20. *For any* $(B, y) \in FP(\mathcal{A})$, *it holds* $I(B, \lambda) = |Q| - |B|$, *and for any* $y \in B$, $\delta(p, \lambda) = \delta_0(p, \lambda) = p$ *and* $d(p, \lambda, p) = 0$.

Lemma 21. *For any* $(B, y) \in FP(\mathcal{A})$ *and* $t, u, v \in \Sigma^*$ *with* $y = tuv$, *the following hold.*

1. $|\delta_0(\delta(B, t), u)| \geq |[delta_0(B, tu)$.
2. $I(\delta(B, t), u) \geq I(B, y)$.
3. *If* $|\delta_0(B, t)| = |\delta_0(B, tu)|$, *then there exists a bijection* $\alpha : \delta_0(B, y) \rightarrow \delta_0(B, tu)$ *such that for each* $p \in \delta_0(B, y)$, $\alpha(p) \in \delta_0(p, u)$.

Lemma 22. *For any* $(B, y) \in FP(\mathcal{A})$, *if* $I(B, y) = minI(\mathcal{A})$, *then it holds* (1) *for any* $(p, q) \in PS(B|y)$, $d(p, y, q) < Q$, *and* (2) *there exists* $v \in \Sigma^*$ *such that* $|v| < (|Q| + 1)^{|Q|2^{|Q|^2}}$, $\delta(B|v) = \delta(B|y)$, $\delta_0(B, v) = \delta_0(B|y)$ *and* $I(B, y) = I(B, v)$.

Definition 23. *For each* $0 \leq i \leq |Q|$, *two integers* $o_4(i)$ *and* $o_5(i)$ *are defined inductively as follows* :

1. $o_4(0) = |Q| - 1$ and $o_5(0) = (|Q| + 1)^{|Q|} 2^{|Q|^2} - 1$.
2. For $0 < i \leq |Q|$,
   (a) $o_4(i) = (|Q| - i + 2)(|Q| + 1)^{|Q|} 2^{3|Q|^2} (o_4(i - 1) + 1)$;
   (b) $o_5(i) = (|Q| + 1)^{2|Q|} 2^{5|Q|^2} (o_5(i - 1) + 1)$.

Lemma 24. *Let* $m = |Q|$.

1. $o_4(m) < 2^{3m^3 + (m^2 + m)\log(m+2)}$.
2. $o_5(m) < 2^{5m^3 + m^2 + (m^2 + m)\log(m+1) + 1}$.

Lemma 25. *For any* $(B, y) \in FP(\mathcal{A})$, *there exist* $A(B, y) \subset PS(B|y)$ *and* $E(B, y) \in MRE(\Sigma)$ *to which the following hold:*

1. *For any* $(p, q) \in A(B, y)$, $d(p, y, q) \leq o_4(I(B, y))$.
2. $S(E(B, y)) \leq o_5(I(B, y))$, *and for any* $k \geq 0$, *it holds that* $\delta(B|E(B, y)(k)) = \delta(B|y)$, $\delta_0(B|E(B, y)(k)) = \delta_0(B|y)$ *and for any* $(p, q) \in PS(B|y) - A(B, y)$, $d(p, E(B, y)(k), q) \geq k$.

The following theorem can be proved as Theorem 17, where (4) and (5) are new types of characterizations about the unlimitedness of zero-deterministic distance automata.

Theorem 26. *For any zero-deterministic distance automaton* $\mathcal{A} = < \Sigma, Q, \delta, S, F, d >$, *the following five conditions are equivalent.*

1. $\mathcal{A}$ *is not limited.*
2. *There exists* $E \in MRE(\Sigma)$ *such that* $S(E) < o_5(\mathcal{A})$ *and for any integer* $k \geq 0$, $E(k) \in L(\mathcal{A})$ *and* $d(S, E(k), F) \geq k$.
3. *There exists* $E \in MRE(\Sigma)$ *such that for any integer* $k \geq 0$, $E(k) \in L(\mathcal{A})$ *and* $d(S, E(k), F) \geq k$.
4. *There exists* $w \in L(\mathcal{A})$ *such that* $d(S, w, F) \geq o_4(\mathcal{A})$.
5. *There exists* $w \in L(\mathcal{A})$ *such that* $|w| \leq o_6(\mathcal{A})$ *and* $d(S, w, F) \geq o_4(\mathcal{A})$.

Here $m = |Q|$, $o_4(\mathcal{A}) = 2^{3m^3 + (m^2 + m)\log(m+2)}$, $o_5(\mathcal{A}) = 2^{5m^3 + m^2 + (2m^2 + m)\log(m+1) + 1}$, *and* $o_6(\mathcal{A}) = 2^{m^2(o_4(\mathcal{A}) + 1)}$.

References

1. K. Hashiguchi, Limitedness Theorem on finite automata with distance functions, J. Comput. System Sci. 24 (1982) 233-244.
2. K. Hashiguchi, Algorithms for determining relative star height and star height, Inform. and Control 78 (1988) 124-169.
3. K. Hashiguchi, Improved limitedness theorems on finite automata with distance functions, Theoret. Comput. Sci. 72 (1990) 27-38.
4. K. Hashiguchi, Algorithms for determining relative inclusion star height and inclusion star height, Theoret. Comput. Sci. 91 (1991) 85-100.
5. D. Krob, The equality problem for rational series with multiplicities in the tropical semiring is undecidable, In ternat, J. of Algebra and Computation 4(1994)405-425.

6. H. Leung, On the topological structure of a finitely generated semigroup of matrices, Semigroup Forum 37 (1988) 273-287.
7. I. Simon, Factorization forests of finite height, Theoret. Comput. Sci. 72 (1990) 65-94.

Recognizing Regular Expressions
by Means of Dataflow Networks

Pascal Raymond
email: Pascal.Raymond@imag.fr

VERIMAG *
Miniparc ZIRST, 38330 Montbonnot-St Martin, France

Abstract. This paper addresses the problem of building a Boolean
dataflow network (sequential circuit) recognizing the language described
by a regular expression. The main result is that both the construction
time and the size of the resulting network are linear with respect to the
size of the regular expression.

Introduction

"Grep" machine: Let Σ be a vocabulary, L be a regular language on Σ. A "grep"
machine is a machine receiving a sequence $s_0, s_1, \ldots, s_n, \ldots$ of symbols ($s_i \in \Sigma$)
and computing a sequence $b_0, b_1, \ldots, b_n, \ldots$ of Booleans, such that b_n is true if
and only if the word $s_0 s_1 \ldots s_n$ belongs to L^2.

This paper addresses the problem of building a "grep" machine for languages
described by regular expressions. This problem is rather classical [4, 11, 10, 3,
1, 2]. We propose a solution which, to our knowledge, is new: Informally, it con
sists of building, from a regular expression E, a "circuit" (or Boolean data-flow
network) exploring all the branches of a non-deterministic automaton recognizing $L(E)$. The relations between regular languages and sequential circuits are
also studied [5, 7]. But the important point here, is that the automaton is never
explicit (it is represented implicitly by a system of Boolean equations) nor deterministic. As a consequence, both the construction time and the size of the
circuit are linear with respect to the size of the expression E.

Our method is presented in a general framework based on linear systems of
equations. We first review the classical methods (deterministic finite automaton,
non-deterministic finite automaton without or with empty-labelled transitions),
by expressing them in this framework. We show that this classification coincides
with the classes of complexity (exponential, quadratic and linear).

---

* Verimag is a joint laboratory of CNRS, Institut National Polytechnique de Grenoble,
Université J. Fourier and VERILOG S.A. associated with IMAG.

[2] The "grep L" unix command is slightly different: in this case, b_n is true if and only
if it exists some i such that $s_i s_{i+1} \ldots s_n$ belongs to L, but this problem is equivalent
to the initial one with $L' = \Sigma^* L$

1 Regular expressions

1.1 Regular languages

Let Σ be a finite set of symbols; Σ^* denotes the set of all finite strings of symbols in Σ. The empty string (string of length 0) in Σ^* is denoted by ε. A language over Σ is any subset of Σ^*. The empty language is denoted by $\emptyset$. The set of regular languages over Σ is the least class of languages containing all the finite languages over Σ, and closed by union, concatenation (denoted by $\cdot$) and Kleene star operator (denoted by $*$). Any regular language can be denoted by a regular expression:

$$E ::= \emptyset \mid a \mid E \cdot E \mid E + E \mid E^*$$

The semantics of regular expressions is given by a function L which associates to each expression E the language described by E[3]:

$$L(\emptyset) = \emptyset \quad L(a) = \{a\} \quad L(E + F) = L(E) \cup L(F)$$
$$L(E \cdot F) = L(E) \cdot L(F) \quad L(E^*) = \mu\ell.\{\varepsilon\} \cup (\ell \cdot L(E))$$

We note $\simeq$ the semantic equivalence over regular expressions, i.e.:

$$E \simeq F \Leftrightarrow L(E) = L(F)$$

1.2 More "useful" regular expressions

We are interested in "non trivial" regular languages, i.e. any language $\mathcal{L}$ which is neither empty ($\mathcal{L} \neq \emptyset$) nor reduced to the empty string ($\mathcal{L} \neq \{\varepsilon\}$). We will then use a more useful syntax, which forbids the representation of trivial languages:

$$E ::= a \mid E \cdot E \mid E + E \mid E^* \mid E^\varepsilon$$

The semantic of the power-ε operator is given by: $L(E^\varepsilon) = \varepsilon \cup L(E)$.

We note Rexp the set of regular expressions, and we also define a function $\Lambda : \text{Rexp} \to \mathbb{B} = \{true, false\}$, such that $\Lambda(E)$ is true if and only if the empty string belongs to $L(E)$.

$$\Lambda(a) = false \quad \Lambda(E^\varepsilon) = true \quad \Lambda(E^*) = true$$
$$\Lambda(E + F) = \Lambda(E) \vee \Lambda(F) \quad \Lambda(E \cdot F) = \Lambda(E) \wedge \Lambda(F)$$

---

[3] $\mu\ell.f(\ell)$ denotes the least fixpoint of the equation $\ell = f(\ell)$

2 Linear systems of equations

In this section we show that the problem of building a grep machine from a regular expression can be expressed as follows: build a system of language equations of the form "$X = t_1 + \ldots + t_n$", where each t_i is a language term of a "special" form. The various known methods for building grep machines can be related to the form of the terms. This classification is largely inspired by [5, 6].

In this paper, we use the following notations: let θ be a system of equations, we note $\mathcal{V}(\theta)$ the set of language identifiers defined in θ, $\mathcal{A}(\theta) \in \mathcal{V}(\theta)$ the main language (the axiom), and for each $X \in \mathcal{V}(\theta)$, $\theta(X)$ the set of terms t_i defining X in θ. We note $\mathcal{L}(\theta)$ the solution (if it exists) of θ, i.e. the language denoted by the axiom $\mathcal{A}(\theta)$.

2.1 Suffix-linear systems

In such systems, a term is either ε, or of the form $a \cdot Y$, where a is a symbol and Y is a language identifier. More precisely, each equation is of the form:

$$X = a_1 \cdot Y_1 + \ldots + a_n \cdot Y_n \ [+\ \varepsilon]$$

Such a system θ is equivalent to a finite automaton (Q, I, F, T_Σ), where:

- $Q = \mathcal{V}(\theta)$ is the set of states,
- $I = \mathcal{A}(\theta)$ is the (unique) initial state,
- $F = \{X \in \mathcal{V}(\theta) \,|\, \varepsilon \in \theta(X)\}$ is the set of final states,
- $T_\Sigma \in Q \times \Sigma \times Q$ is the set of symbol-labelled transitions, with:
  $(X, a, Y) \in T_\Sigma \Leftrightarrow (a \cdot Y) \in \theta(X)$.

Deterministic automaton (DFA): Moreover, if in each equation $X = a_1 \cdot Y_1 + \ldots + a_n \cdot Y_n \ [+\ \varepsilon]$, all the symbols a_i are different, the automaton is deterministic, and each Y_i is said to be the a_i-derivative of X. Such a deterministic system can be built by computing derivatives of regular expressions. Brzozowski has formally defined this construction [4]. The size of the resulting deterministic automaton (and thus the cost of the construction) is, in the worst case, exponential with respect to the size of the regular expression (number of operators in the regular expression).

Non-deterministic automaton (NFA): If in each equation $X = a_1 \cdot Y_1 + \ldots + a_n \cdot Y_n \ [+\ \varepsilon]$, some a_i are equal, the resulting automaton is non-deterministic. McNaughton and Yamada defined the construction of such an automaton [10]. Berry and Sethi gave an efficient algorithm to build such an automaton from a regular expression [3]. Antimirov defined a notion of partial derivatives, which generalizes the notion of derivative for non-deterministic automata [2]. For those algorithms, the size of the resulting non-deterministic automaton (and thus the cost of the construction) is, in the worst case, quadratic with respect to the size of the regular expression. More precisely, the number of states is linear, but the number of transitions can be quadratic.

Non-deterministic automaton with ε-transitions: The previous definition can be extended with terms of the form Y. Such systems are equivalent to a non-deterministic automaton with ε-transitions: $(Q, I, F, T_\Sigma, T_\mathcal{E})$, where:

- $T_\mathcal{E} \in Q \times Q = \{(X, Y) \mid Y \in \theta(X)\}$

Thompson has defined the construction of such machines from regular expressions [11]. The main result is that both the size of the automaton and the cost of its construction are linear with respect to the size of the regular expression.

Our goal is to define a linear algorithm that builds a grep machine from a regular expression, so we choose the third kind of linear systems (i.e. non-deterministic automaton with ε-transitions).

2.2 Prefix-linear systems

A completely dual definition can be given by using terms of the form $Y \cdot a$ instead of $a \cdot Y$. Let us see the example of a non-deterministic prefix-linear system with ε-transitions. Such a system θ is equivalent to a finite automaton $(Q, I, F, T_\Sigma, T_\mathcal{E})$, where:

- $Q = \mathcal{V}(\theta)$ is the set of states,
- $I = \{X \in \mathcal{V}(\theta) \mid \varepsilon \in \theta(X)\}$ is the set of initial states,
- $F = \mathcal{A}(\theta)$ is the (unique) final state,
- $T_\Sigma \in Q \times \Sigma \times Q$ is the set of symbol-labelled transitions,
  with: $(Y, a, X) \in T_\Sigma \Leftrightarrow (Y \cdot a) \in \theta(X)$,
- $T_\mathcal{E} \in Q \times Q = \{(X, Y) \mid Y \in \theta(X)\}$ is the set of ε-labelled transitions.

In this paper, we chose to build prefix-linear systems. Indeed, the problems of building prefix- or suffix-linear systems are completely equivalent, but, as we will show in a following section, prefix-linear systems are "equivalent" to sequential circuits, and then, in some sense, directly "executable".

3 From regular expressions to prefix-linear systems

3.1 Abstract syntax for left-linear systems

We propose here a "functionnal-like" syntax which allows a system of equations to be represented by a single equation:

$$
\begin{array}{rcl}
system & ::= & X = terms \\
terms & ::= & \varepsilon \mid X \mid X \cdot a \mid terms + terms \mid \\
& & \mathsf{let}\, X = terms\, \mathsf{in}\, terms \mid \mathsf{rec}\, X = terms
\end{array}
$$

!Note that $\mathsf{rec}\, X = terms$ is a macro-notation for !$\mathsf{let}\, X = terms\, \mathsf{in}\, X$.

The semantic function $\mathcal{L}$ is naturally extended to this new syntax: $\mathcal{L}(\sigma)$ is the language denoted by the *system* σ. We also extend the semantic function for *terms*, which are, in some sense, "partially built" systems: $\mathcal{L}(\tau)$ is (if it exists) the language denoted by the system $X = \tau$, where X is any identifier not appearing in τ.

3.2 The basic algorithm

The idea is to define a function Θ which transforms a regular expression into a prefix-linear system, i.e. which verifies:

$$L(E) = \mathcal{L}(\Theta(E))$$

Moreover, we want the cost of the translation to be linear with respect to the size of the regular expression; the translation should also introduce few variables.

For that purpose, we introduce a recursive function $\Gamma(X, E)$ whose parameters are a regular expression (E) and a language identifier (X) representing the "prefixes" of E. The result of Γ is a partially built system of equations (i.e. a "*terms*" according to the previously defined abstract syntax). This will allow us to introduce variables only when needed.

The meaning of Γ is quite simple: intuitively, $\Gamma(X, E)$ "denotes" the same language as "$X \cdot E$". More formally, let L_X be the language denoted by X:

$$\mathcal{L}(\Gamma(X, E)) = L_X \cdot L(E)$$

The definition of Γ is given by induction on the structure of regular expressions:

$$\Gamma(X, a) = X \cdot a \tag{1}$$
$$\Gamma(X, E^\varepsilon) = X + \Gamma(X, E) \tag{2}$$
$$\Gamma(X, E + F) = \Gamma(X, E) + \Gamma(X, F) \tag{3}$$

Let Y be a new identifier:

$$\Gamma(X, E \cdot F) = \text{let } Y = \Gamma(X, E) \text{ in } \Gamma(Y, F) \tag{4}$$
$$\Gamma(X, E^*) = \text{rec } Y = X + \Gamma(Y, E) \tag{5}$$

The system corresponding to the regular expression E is obtained by computing $\Gamma(Z, E)$ with $Z = \varepsilon$, and then, by naming the result, let X be a new identifier:

$$\Theta(E) = X = (\text{let } Z = \varepsilon \text{ in } \Gamma(Z, E))$$

3.3 Example

Let us consider the expression $(a^* + b)^* \cdot a$; in order to simplify the notations, we only give the main result and indicate whether a new equation is generated:

$$\Gamma(Z, (a^* + b)^* \cdot a) = \text{let } Y = \Gamma(Z, (a^* + b)^*) \text{ in } \Gamma(Y, a)$$
$$= \text{let } Y = \Gamma(Z, (a^* + b)^*) \text{ in } Y \cdot a$$
$$\Gamma(Z, (a^* + b)^*) = \text{rec } T = Z + \Gamma(T, a^* + b)$$
$$\Gamma(T, a^* + b) = \Gamma(T, a^*) + \Gamma(T, b)$$
$$= (\text{rec } W = T + \Gamma(W, a)) + T \cdot b$$
$$= (\text{rec } W = T + W \cdot a) + T \cdot b$$

A classical system of equations is obtained by extracting the sub-equations and naming the axiom with X:

$$X = Y \cdot a \quad Y = T \quad T = Z + W + T \cdot b \quad W = T + W \cdot a \quad Z = \varepsilon$$

3.4 Automaton size and comparison

The size of $\Theta(E)$ (or of the equivalent NFA $(Q, I, F, T_\Sigma, T_\varepsilon)$) can be easily related to the size of E: let o be the number of symbol occurrences, d be the number of dots, e be the number of power-εoperators, and k the number of Kleene stars, we have:

$$|Q| = d + k + 2 \quad |T_\Sigma| = o \quad |T_\varepsilon| = k + e$$

The two additional states are the initial state introduced by the starting rule ("let $Z = \varepsilon$ in..."), and a final state corresponding to the axiom.

It is interesting to compare this solution where the states correspond to the dot and star operators, to the McNaughton and Yamada automata (or *normalized NFA's*) where the states correspond to the symbol occurences (in a normalized NFA, we have $|Q| = o + 1$) [10, 3].

The same notion of states (attached to dots and stars) exists in the Antimirov's work: the algorithm can then be viewed as a simple way to compute Antimirov's partial derivatives [2]. However, Antimirov's algorithm does not produce ε-transitions, so the resulting automaton is, in the worst case, quadratic with respect of the size of the regular expression.

The use of ε-transitions makes this algorithm close to the Thompson's one. But the basic Thompson NFA is, in general, bigger: between r and $2r$ states, and between r and $4r$ transitions, where r is the length of the regular expression. Our algorithm can be viewed as a way to directly produce an optimized Thompson NFA.

4 Boolean networks

We present here a simple model for sequential circuits. It consists of a Boolean operator network, together with a simple data-flow semantics. The networks are built with classical Boolean operators (true, false, not, and, or) and a delay binary operator fby, whose first argument is the initial value, and the second the Boolean value to be delayed.

4.1 Syntax

We define a syntax that allows the representation of a network (a cyclic graph in general) by a simple syntactic tree. This is done by using a "functional-like" syntax (X stands for any identifier):

$$net ::= X \mid \text{true} \mid \text{false} \mid \text{not } net \mid net \text{ and } net \mid net \text{ or } net \mid net \text{ fby } net$$
$$\mid \text{ let } X = net \text{ in } net \mid \text{rec } X = net$$

Just like for the linear systems of equations (§ 3.1), a network can also be described by a set of equations. For instance, the system of Boolean equations:

$$s = x \text{ and not } y \quad x = a \text{ or } b \quad y = \text{false fby}(x \text{ or } y)$$

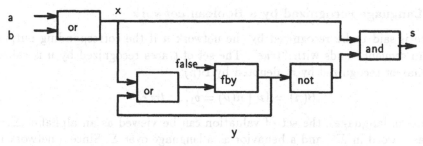

let (x = a or b) in (x and not (rec y = false fby (x or y)))

Fig. 1. A Boolean network and the corresponding expression

is equivalent to the network shown (in both syntactic and graphic form) in Fig. 1.

We only consider networks without combinational loop: recursive definitions can only appear under a fby operator. The set of correct networks is denoted by N.

4.2 Synchronous semantics

The set of free variables (inputs) of a network n is denoted by $Free(n)$. A valuation of $Free(n)$ is a function which associates a Boolean value to each variable of $Free(n)$: $\rho : Free(n) \rightarrow \mathbb{B}$. A trace over $Free(n)$ is a finite sequence of valuation: $\nu = \rho_1, ..., \rho_k$.

The synchronous semantics models the behavior of idealized sequential circuits: given an input trace ν of length k, a network n produces a sequence of k Boolean values. More precisely, each sub-network (each node in the network) produces a sequence of k Boolean values. For instance, the constant true produces a sequence of k "true". Let $\nu = \rho_1, ..., \rho_k$ be the input trace, the input x produces the sequence $\rho_1(x), ..., \rho_k(x)$. The classical operators operate pointwise on sequences; for instance, if n produces the sequence $b_1, ..., b_k$ and n' the sequence $b'_1, ..., b'_k$, then n or n' produces $b_1 \vee b'_1, ..., b_k \vee b'_k$. At last, the fby operator delays its second operand, while its first operand defines the initial value: n fby n' produces $b_1, b'_1, ..., b'_{k-1}$.

The absence of combinational loops in the network n is a sufficient property for this semantics to be operational. In other terms, if n has no combinational loop, then for any input trace ρ, the corresponding output sequence is completely determined. We then consider that any network n without combinational loop is a function from traces to Boolean sequences, and we note $n(\nu)$ the output sequence corresponding to the trace ν.

4.3 Language recognized by a Boolean network

A trace is said to be recognized by the network n if the corresponding output Boolean sequence ends with "*true*". The set of traces recognized by n is called the behavior recognized by n, denoted by $\mathcal{B}(n)$:

$$\mathcal{B}(n) = \{\nu \mid n(\nu) = b_1, ..., true\}$$

In terms of languages, the set of valuation can be viewed as an alphabet Σ, a trace as a word in Σ^*, and a behavior as a language over Σ. Since a network is clearly a finite state machine (with at most 2^m states, where m is the number of fby operators in the network), the behavior recognized by a network is a regular language over Σ.

5 Prefix-linear systems and networks

In this section, we will map a prefix-linear system of equations θ to an Boolean network $n = \Omega(\theta)$.

5.1 Encoding

A prefix-linear system describes a language over some alphabet Σ, while the networks describe behaviors over a set of Boolean variables. So the initial alphabet has to be encoded using Boolean variables. We consider here a "trivial" one-to-one encoding (i.e. we identify the alphabet to a set of Boolean variables). Indeed, the method can easily be adapted for more powerful encoding such as logarithmic encoding.

5.2 Empty trace

As they are defined, the networks cannot recognize the empty trace, so the resulting network cannot be "completely equivalent" to the source system. Let θ be a prefix-linear system, the resulting network $\Omega(\theta)$ simply verifies:

$$\mathcal{B}(\Omega(\theta)) = \mathcal{L}(\theta) \setminus \{\varepsilon\}$$

5.3 Initial states

Intuitively, each term in a system of equations will be replaced by a network recognizing its non-empty traces. In order to treat terms of the form $X \cdot a$, we need to know if the empty string belongs to X, since, if $\varepsilon \in \mathcal{L}(X)$:

$$\mathcal{L}(X \cdot a) \setminus \{\varepsilon\} = (\mathcal{L}(X) \setminus \{\varepsilon\}) \cdot a + a$$

The set of variables X such that $\varepsilon \in \mathcal{L}(X)$ is called the set of initial states (referring to the terminology of NFA's). This set is denoted by $\mathcal{I}(\theta)$ and is recursively defined by:

$$X \in \mathcal{I}(\theta) \Leftrightarrow (\varepsilon \in \theta(X)) \vee (\exists Y \in \mathcal{I}(\theta) \mid Y \in \theta(X))$$

In fact, this set can be computed during the construction of the system, using the function Λ (§ 1.2) and this computation has a linear cost (with respect to the size of the regular expression).

5.4 From prefix-linear systems to networks

The definition of the mapping function Ω is the following:

$$\Omega(\varepsilon) = \text{false} \qquad \Omega(X) = X \qquad \Omega(\tau + \tau') = \Omega(\tau) \text{ or } \Omega(\tau')$$
$$\Omega(X \cdot a) = (\text{true fby } X) \text{ and } a \quad \text{if } X \in \mathcal{I}(\theta)$$
$$\Omega(X \cdot a) = (\text{false fby } X) \text{ and } a \quad \text{if } X \notin \mathcal{I}(\theta)$$
$$\Omega(\text{let } X = \tau \text{ in } \tau') = (\text{let } X = \Omega(\tau) \text{ in } \Omega(\tau'))$$

Let us come back to the example of § 3.3 ($\theta = \Theta((a^* + b)^* \cdot a)$). We first compute $\mathcal{I}(\theta) = \{Z, T, Y, W\}$, then we obtain the network (given as a set of equations):

$$X = (\text{false fby } Y) \text{ and } a$$
$$Y = T$$
$$T = Z \text{ or } W \text{ or } (\text{true fby } T) \text{ and } b$$
$$W = T \text{ or } (\text{true fby } W) \text{ and } a$$
$$Z = \text{false}$$

5.5 Combinational loops and normalization

We have defined a method for building a sequential circuit form a regular expression: $\Omega(\Theta(E))$.

As it is computed, the intermediate system $\Theta(E)$ (and then the resulting network too) may contain combinational loops. For instance, in the example of § 3.3 ($\theta = \Theta((a^* + b)^* \cdot a)$), we have $W \in \theta(T)$ and $T \in \theta(W)$.

Such combinational loops appear in $\Theta(E)$ if and only if there exists in E a sub-expression of the form (F^*) such that $\Lambda(F)$. In order to avoid combinational loops, we define a function $Norm$ on regular expressions, which recursively replaces each sub-expression of the form (F^*) by an equivalent expression (G^*) such that $\neg \Lambda(G)$.

$$Norm(a) = a$$
$$Norm(E + F) = Norm(E) + Norm(F)$$
$$Norm(E \cdot F) = Norm(E) \cdot Norm(F)$$
$$Norm(E^\varepsilon) = Norm(E)^\varepsilon$$
$$Norm(E^*) = NormBis(E)^*$$

where:

$$NormBis(E) = Norm(E) \text{ if } \neg\Lambda(E)$$
$$NormBis(E + F) = NormBis(E) + NormBis(F) \text{ if } \Lambda(E + F)$$
$$NormBis(E \cdot F) = NormBis(E) + NormBis(F) \text{ if } \Lambda(E \cdot F)$$
$$NormBis(E^{\mathcal{E}}) = NormBis(E)$$
$$NormBis(E^*) = NormBis(E)$$

The function $Norm$ is a simple parsing which applies the "interesting" function $NormBis$ to the Kleene star operators.

Notice that this normalization does not replace the expression E^* by an expression F^* such that $L(F) = L(E) \setminus \varepsilon$. Such a transformation would have a quadratic cost since, for all regular languages ℓ and ℓ' containing the empty string: $(\ell \cdot \ell') \setminus \varepsilon = (\ell \setminus \varepsilon) \cdot \ell' + \ell \cdot (\ell' \setminus \varepsilon)$.

The normalization simply replaces the expression E^* by an expression F^* such that $L(F^*) = L(E^*)$ and $\neg\Lambda(F)$. The correctness of the algorithm is based on the following lemma: $(\varepsilon \in \ell \wedge \varepsilon \in \ell') \Rightarrow (\ell \cdot \ell')^* = (\ell + \ell')^*$.

Here are some examples of normalization:

$$Norm((a^* + b)^*) = (a + b)^* \quad Norm(((a^*)^*)^*) = a^*$$
$$Norm((a^{\mathcal{E}} \cdot b^* + c^*)^*) = (a + b + c)^*$$

6 From regular expressions to Boolean networks

Finally, let E be a regular expression over an alphabet Σ interpreted as a set of Boolean variables, we have defined a method which computes a correct "equivalent" network n, i.e. a network without combinational loop recognizing all the non-empty traces of the language $L(E)$:

$$\mathcal{B}(\Omega(\Theta(Norm(E)))) = L(E) \setminus \varepsilon$$

This method involves the computation of the function Λ on each sub-expression of E, the normalization of E, and also the computation of the initial states in the system $\Theta(Norm(E))$. But the main result is that the cost of all those treatments is linear with respect to the size of the source regular expression.

In order to outline this linear complexity, we define in this section a function Φ which directly produces a network from a regular expression. Like the function Θ, the function Φ is defined using a recursive function, Υ, which takes a special parameter representing the prefixes. This function takes two parameters to describe the prefixes:

$$\Upsilon : \mathrm{N} \times \mathrm{I\!B} \times \mathrm{Rexp} \rightarrow \mathrm{N}$$
$$(X, b, E) \mapsto n$$

The first parameter (X) is a network which is supposed to recognize the non-empty traces of the prefixes, while the second one (b) is a Boolean value indicating whether the empty trace belongs to the prefixes. Intuitively, this new parameter b allows us to directly compute the set of "initial states" (§ 5.3). The normalization of the regular expression (§ 5.5) is also performed during the construction. For that purpose, a special recursive function Υ^* is applied to the Kleene star operands.

$$\Phi(E) = \text{let } Z = \text{false in } \Upsilon(Z, true, E)$$

$$\Upsilon(X, b, a) = (b \text{ fby } X) \text{ and } a$$
$$\Upsilon(X, b, E + F) = \Upsilon(X, b, E) \text{ or } \Upsilon(X, b, F)$$
$$\Upsilon(X, b, E^\varepsilon) = X \text{ or } \Upsilon(X, b, E)$$

Let Y be a new identifier:

$$\Upsilon(X, b, E.F) = \text{let } Y = \Upsilon(X, b, E) \text{ in } \Upsilon(y, b \wedge \Lambda(E), F)$$
$$\Upsilon(X, b, E^*) = \text{rec } Y = X \text{ or } \Upsilon^*(Y, b, E)$$

$$\Upsilon^*(X, b, E) = \Upsilon(X, b, E) \quad \text{if } \neg \Lambda(E)$$
$$\Upsilon^*(X, b, E + F) = \Upsilon^*(X, b, E) \text{ or } \Upsilon^*(X, b, F) \quad \text{if } \Lambda(E + F)$$
$$\Upsilon^*(X, b, E \cdot F) = \Upsilon^*(X, b, E) \text{ or } \Upsilon^*(X, b, F) \quad \text{if } \Lambda(E \cdot F)$$
$$\Upsilon^*(X, b, E^\varepsilon) = \Upsilon^*(X, b, E)$$
$$\Upsilon^*(X, b, E^*) = \Upsilon^*(X, b, E)$$

Figure 2 shows the result of $\Upsilon(false, true, (a^* + b)^* \cdot a)$. One can see that the resulting network has the same structure as the regular expression.

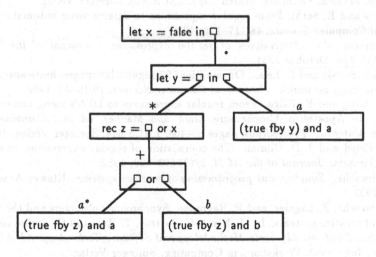

Fig. 2. The network of $(a^* + b)^* \cdot a$

7 Implementation and use

The origin of this work takes place in the domain of synchronous programming [8]. Synchronous programming offers an idealized framework for programming reactive systems. For instance, idealized sequential circuits are synchronous programs. This idealized framework also permits to perform formal verifications on programs. Critical properties are described by means of Boolean synchronous programs called *observers* [9]. Such observers can be expressed with any existing synchronous language (Lustre, Argos, Esterel), but it appeas that the classical regular constructs (sequence, iteration) are also very useful to describe safety properties.

The idea was to translate, with a minimal cost, a safety property expressed with regular constructs into a suitable synchronous program, such as a Boolean network.

A tool called reglo has been designed for this purpose. This tool translates a set of regular expressions, describing traces over a set of Boolean variables (i.e. the input alphabet is supposed to be already Boolean encoded), into an equivalent Boolean dataflow network, expressed in the language Lustre.

References

1. A. V. Aho. Algorithms for finding patterns in strings. In J. van Leeuwen, editor, *Handbook of theoretical computer science*, chapter 5, pages 257–300. Elsevier Science Publishers B. V., 1990.
2. V. Antimirov. Partial derivatives of regular expressions and finite automata constructions. In E. W. Mayr and C. Puech, editors, *Proceedings of STACS '95*, pages 455–466, Munich, Germany, March 1995. LNCS 900, Springer Verlag.
3. G. Berry and R. Sethi. From regular expressions to deterministic automata. *Theoretical Computer Science*, 48:117–126, 1986.
4. J. A. Brzozowski. Derivatives of regular expressions. *Journal of the ACM*, 11(4):481–494, October 1964.
5. J. A. Brzozowski and E. Leiss. On equations for regular languages, finite automata, and sequential networks. *Theoretical Computer Science*, 10:19–35, 1980.
6. C. H. Chang and R. Paige. From regular expressions to DFA's using compressed NFA's. In Apostolico, Crochemore, Galil, and Manber, editors, *Combinatorial Pattern Matching. Proceedings*, pages 88–108. LNCS 644, Springer Verlag, 1992.
7. R. W. Floyd and J. D. Ullman. The compilation of regular expressions into integrated circuits. *Journal of the ACM*, 29(3):603–622, 1982.
8. N. Halbwachs. *Synchronous programming of reactive systems*. Kluwer Academic Pub., 1993.
9. N. Halbwachs, F. Lagnier, and P. Raymond. Synchronous observers and the verification of reactive systems. In M. Nivat, C. Rattray, T. Rus, and G. Scollo, editors, *Third Int. Conf. on Algebraic Methodology and Software Technology, AMAST'93*, Twente, June 1993. Workshops in Computing, Springer Verlag.
10. R. McNaughton and H. Yamada. Regular expressions and state graphs for automata. *IEEE Trans. on Electronic Computers*, 9(1):39–47, 1960.
11. K. Thompson. Regular expression search algorithm. *Communications of the ACM*, 11(6):419–423, June 1968.

On the Power of Randomized Branching Programs

Farid Ablayev *   Marek Karpinski †

Abstract

We define the notion of a randomized branching program in the natural way similar to the definition of a randomized circuit. We exhibit an explicit function f_n for which we prove that:

1) f_n can be computed by polynomial size randomized read-once ordered branching program with a small one-sided error;

2) f_n cannot be computed in polynomial size by deterministic read-once branching programs;

3) f_n cannot be computed in polynomial size by deterministic read-k-times ordered branching program for $k = o(n/\log n)$ (the required deterministic size is $\exp\left(\Omega\left(\frac{n}{k}\right)\right)$).

1 Preliminaries

Different models of branching programs introduced in [13, 15], have been studied extensively in the last decade (see for example [19]). A survey of known lower bounds for different models of branching programs can be found in [17].

Developments in the field of digital design and verification have led to the introduction of restricted forms of branching programs. In particular, ordered read-once branching programs are now commonly used in the circuit verification [9], [20]. But many important functions cannot be computed by read-once branching programs of polynomial size. For more information see the survey [9] and papers [18], [16].

It is known that different models of randomized circuits with weak enough restrictions on the error of randomized computation have only polynomial advantage over nonuniform deterministic models (see [2], [4], [3], and survey

---

*Dept. of Computer Science, University of Bonn. Visiting from University of Kazan. Research partially supported by the Volkswagen-Stiftung and the Basic Research Grant 96-01-01962 Email: ablayev@ksu.ras.ru

†Dept. of Computer Science, University of Bonn, and International Computer Science Institute, Berkeley, California. Research partially supported by DFG Grant KA 673/4-1, by the ESPRIT BR Grants 7097 and EC-US 030, and by the Volkswagen-Stiftung. Email: marek@cs.uni-bonn.de

[6]). In the paper we define the notion of a randomized branching program in a natural way similar to the definition of a randomized circuit. Our goal is to show that randomized computation with a small error for read-once polynomial branching programs can be more powerful than deterministic ones. The argument that can help the intuition in this direction is that amplification method does not work for the case of restricted number of input verifications. Note that in the paper [7] it is presented an explicit function which needs exponential size for presentation by a nondeterministic read-k-times branching program for $k = o(\log n)$.

We use the variant of a definition of a branching program from the paper [7]. A deterministic branching program P for computing a function $g : \Sigma^n \to \{0,1\}$, where Σ is a finite set, is a directed acyclic multi-graph with a single source node, distinguished sink nodes labeled "accept" and "reject". For each non-sink node there is a variable x_i such that all out-edges from this node are labeled by "$x_i = \delta$" for some $\delta \in \Sigma$ and for each δ there is exactly one such labeled edge. The label "$x_i = \delta$" indicates that only inputs satisfying $x_i = \delta$ may follow this edge in the computation. We call a node v an x_i-node if all output edges of the node v are labeled by "$x_i = \delta$", $\delta \in \Sigma$.

A deterministic branching program P computes a function $g : \Sigma^n \to \{0,1\}$, in the obvious way; that is, $g(\sigma_1, \ldots, \sigma_n) = 1$ iff there is a computation on $< \sigma_1, \ldots, \sigma_n >$ starting in the source state and leading to the accepting state.

A randomized branching program is a one which has in addition to its standard (deterministic) inputs some specially designated inputs called random inputs. When these random inputs are chosen from the uniform distribution, the output of the branching program is a random variable. We call a node v of the randomized branching program a "random generator" node if output edges of the node v are labeled by random inputs.

We say a *randomized* branching program *(a,b)-computes* a function g if it outputs 1 with probability at most a for input x such that $g(x) = 0$ and outputs 1 with probability at least b for inputs x such that $g(x) = 1$. A randomized branching program computes the function g with one-sided ε-error if it $(\varepsilon, 1)$-computes the function g.

For a branching program P, we define size(P) (complexity of the branching program P) as the number of internal nodes in P.

From the definition of the complexity of a branching program it follows that the size of randomized branching program is the sum of random generator nodes and x_i-nodes.

Read-once branching programs are branching programs in which for every path, every variable is tested no more than once. A read-once ordered branching program is a read-once branching program which respects a fixed ordering π of the variables, i.e. if an edge leads from an x_i-node to an x_j-node, the condition $\pi(i) < \pi(j)$ has to be fulfilled.

A *read-k-times* branching program is a branching program with the property that no input variable x_i appears more than k times on any path in the program. A read-k-times ordered branching program is a read-k-times branching program which is partitioned into k layers such that the each layer is a read-once ordered respecting the same ordering π. In [5] it is proved that deterministic ordered read-$(k+1)$-times branching programs are more powerful than deterministic ordered read-k-times branching programs. Namely classes of functions computed by deterministic polynomial-size read-k-times ordered branching programs form proper hierarchy for $k = o(n^{1/2}/\log^2 n)$.

We exhibit an explicit function $f_n : \{0, 1, \hat{0}, \hat{1}\}^{2n} \rightarrow \{0, 1\}$, for which we prove that:

(i) Function f_n can be computed with one sided $\varepsilon(n)$-error by randomized read-once ordered branching program with the size $O\left(\frac{n^6}{\varepsilon^3(n)} \log^2 \frac{n}{\varepsilon(n)}\right)$ (Theorem 1).

(ii) Any deterministic read-once branching program that computes function f_n has the size no less than 2^n (Theorem 2).

(iii) Any deterministic read-k-times ordered branching program for computing function f_n has size no less than $2^{(n-1)/(2k-1)}$ (Theorem 3).

Function f_n can be easily defined as a boolean function. For technical reasons in the proofs we prefer to use the above notation.

Note that one can think of each internal node of a branching program as a state of a computation. This point of view is essential for the investigation of the amount of space necessary to compute functions. Restricted models of branching programs are useful for the investigation of time-space tradeoffs. We can think of read-k-time ($k \geq 1$) restrictions as a restriction on time, say time $\leq kn$ (see survey [8] for more information). This approach draws time-space tradeoff point of view to our results. Recent results on the general lower bounds on randomized space and time can be found in [1] and [11].

2 Function

Consider the finite alphabet $\Sigma = \{0, 1, \hat{0}, \hat{1}\}$. As usual Σ^* and Σ^n denote the set of all words of finite length and the length n over Σ respectively.

For $\sigma_1, \sigma_2 \in \Sigma$, $x \in \Sigma^*$ define $Proj_{\sigma_1, \sigma_2}(x)$ to be the longest subsequence x' of the sequence x that consists only of symbols σ_1 and σ_2.

Define function $f_n : \Sigma^{2n} \rightarrow \{0, 1\}$ as follows: $f(x) = 1$ iff
1) $Proj_{0,1}(x)$ and $Proj_{\hat{0},\hat{1}}(x)$ have the same length and
2) i-th symbol in $Proj_{0,1}(x)$ is σ_i iff the i-th symbol in $Proj_{\hat{0},\hat{1}}(x)$ is $\hat{\sigma}_i$ for all i.

Informally speaking inputs of f_n are words over the alphabet Σ which consists of two kinds of zeroes and two kinds of ones. $f_n(x) = 1$ iff a subsequence z of x formed by the first kind of zeroes and ones and a subsequence y of x formed by the second kind of zeroes and ones are binary notations of the same natural number.

As it is mentioned in the section above, function f_n can be easily defined as a boolean function $f'_n : \{0,1\}^{4n} \to \{0,1\}$. One can encode, for example, 0 by 00, 1 by 01, $\hat{0}$ by 10, and $\hat{1}$ by 11. Our presentation help us to make main ideas of proof methods more clear and help us to avoid several technical details in proofs.

3 Results

Theorem 1. *Function f_n can be computed with one sided $\varepsilon(n)$-error by randomized read-once ordered branching program of size*

$$O\left(\frac{n^6}{\varepsilon^3(n)} \log^2 \frac{n}{\varepsilon(n)}\right).$$

Proof: Randomized read-once ordered branching program P that computes f_n works as follows:

Phase 1. (probabilistic). Choose $d(n)$ to be some function in $O(n)$, s.t. $d(n) > 2n$. P randomly selects a prime number p from the set $Q_{d(n)} = \{p_1, p_2, \ldots, p_{d(n)}\}$ of first $d(n)$ prime numbers.

P selects a prime number p in the following way. P use $t = \lceil \log d(n) \rceil$ random variables $y_1, y_2, \ldots, y_t$, where $y_i \in \{0,1\}$ and $Prob(y_i = 1) = Prob(y_i = 0) = 1/2$. The branching program P reads its random inputs in the fixed order $y_1, y_2, \ldots, y_t$. Sequence $y = y_1 y_2 \ldots y_t$ is interpreted as binary notation of the number $N(y)$. P selects i-th prime number $p_i \in Q_{d(n)}$ iff $N(y) = i \bmod d(n)$.

Phase 2. (deterministic). Let $\sigma \in \Sigma^{2n}$ be a valuation of x. Denote $\alpha = Proj_{0,1}(\sigma)$, $\beta = Proj_{\hat{0},\hat{1}}(\sigma)$. We treat $\hat{\sigma}_i$ to be the number 0 if $\hat{\sigma}_i = \hat{0}$, and to be the number 1 if $\hat{\sigma}_i = \hat{1}$. Sequences α and β are interpreted as binary notations of numbers $N(\alpha)$ and $N(\beta)$. P reads input sequence $x = \sigma$ in the order $x_1, \ldots, x_{2n}$.

Along the computation path, P

a) verifies if $|\alpha| = |\beta|$,

b) counts modulo p the numbers $N(\alpha)$ and $N(\beta)$ ($a = N(\alpha) \bmod p$ and $b = N(\beta) \bmod p$) in the following way. In the beginning of computation $a := 0$ and $b := 0$. When P reads i-th input symbol $\sigma_i \in \{0,1\}$ of the sequence α (respectively i-th input symbol $\hat{\sigma}_i \in \{\hat{0}, \hat{1}\}$ of the sequence β) then $a := a + \sigma_i 2^i \bmod p$ (respectively $b := b + \hat{\sigma}_i 2^i \bmod p$).

Let α' and β' be first parts of the length t and k respectively of subsequences α and β that were tested during the path from the source to the internal node (state) v. For the realization of the procedure described in the *phase 2* it is sufficient to store in the state v four numbers: $t, k \in \{0, 1, \ldots, n\}$, $a = N(\alpha') \pmod p$, and $b = N(\beta') \pmod p$.

If $|\alpha| \neq |\beta|$ then P outputs 1 correct answer with probability 1.

Consider the case $|\alpha| = |\beta|$.

If $N(\alpha) = N(\beta)$ (mod p) then P outputs 1 else P outputs 0.

From the description of P it follows that if $N(\alpha) = N(\beta))$ then P with probability 1 outputs correct answer. If $N(\alpha) \neq N(\beta)$ then it can happen that $N(\alpha) = N(\beta)$ (mod p) for some $p \in Q_{d(n)}$. In these cases P make error output.

For $x = \sigma$ it holds that $|N(\alpha) - N(\beta)| \leq 2^n < p_1 p_2 \cdots p_n$ where $p_1, p_2, \ldots, p_n$ are first n prime numbers. This means that in the case when $N(\alpha) \neq N(\beta)$ the probability $\varepsilon(n)$ of the error of P on the input $x = \sigma$ is no more than $2n/d(n)$.

The size of P is no more than

$$2^{t+1} - 1 + \sum_{p \in Q_{d(n)}} \sum_{l=1}^{n} (n+1)^2 p^2.$$

It is known from the number theory that the value of the i-th prime is of order $O(i \log i)$. Therefore from the above upper bound for the $size(P)$ and from the upper bound for $\varepsilon(n)$ it follows that

$$size(P) \leq O(n^3 d^3(n) \log^2 d(n)) \leq O\left(\frac{n^6}{\varepsilon^3(n)} \log^2 \frac{n}{\varepsilon(n)}\right).$$

Theorem 2. *Any deterministic read-once branching program that computes the function f_n has the size of no less than 2^n.*

Proof: Consider an arbitrary deterministic read-once branching program P that computes function f_n. Let v be a node of the P. Let $\sigma = \sigma_1 \sigma_2 \ldots \sigma_l$ be a sequence of symbols over Σ. We will write $v = v(\sigma)$ if there is a sequence $x_{i_1}, x_{i_2}, \ldots, x_{i_l}$ of variables such that edges $x_{i_1} = \sigma_1, x_{i_2} = \sigma_2, \ldots, x_{i_l} = \sigma_l$ form a path P from the source to the node v. Denote $x(\sigma) = \{x_{i_1}, x_{i_2}, \ldots, x_{i_l}\}$.

For the node $v(\sigma)$ denote $f_{v(\sigma)}$ the function which is computed by P when the node $v(\sigma)$ is considered as a source node. $f_{v(\sigma)}$ is the sub function of f_n where we have replaced the variables read on $x(\sigma)$ by the proper constants from σ.

For proving the lower bound of the theorem it is enough to show that for any $\sigma, \sigma' \in \{0,1\}^n$, $\sigma \neq \sigma'$ it holds that $v(\sigma) \neq v(\sigma')$.

Assume that there are sequences $\sigma = \sigma_1 \sigma_2 \ldots \sigma_n \in \{0,1\}^n$ and $\sigma' = \sigma'_1 \sigma'_2 \ldots \sigma'_n \in \{0,1\}^n$ such that $\sigma \neq \sigma'$ and $v(\sigma) = v(\sigma') = v$. P is read-once. This means that $f_{v(\sigma)}$ and $f_{v(\sigma')}$ are functions over the same set of variables and $f_{v(\sigma)} = f_{v(\sigma')}$. From the definition of the function f_n we have that there exists a sequence $\hat{\sigma} \in \{\hat{0}, \hat{1}\}^n$ such that $f_{v(\sigma)}(\hat{\sigma}) = 1$ but $f_{v(\sigma')}(\hat{\sigma}) = 0$. This means that $f_{v(\sigma)} \neq f_{v(\sigma')}$.

Note that the proof of the theorem 2 can be also obtained as a corollary from theorem 2.1 [18].

Below we prove an exponential lower bound for the complexity of presentation of the function f_n by deterministic read-k-times ordered branching program. For proving it we use a method based on two-way communication game. We present this method in the lemma below for a more common notion of ordering variables for a branching program than the traditional ones.

Note that the method based on communication game is used in the paper [12] and later in [5] for proving lower bound for deterministic read-k-times ordered branching programs.

Definition 1. Call a read-once branching program a π-weak-ordered read-once branching program if it respects an ordered partition π of the variables into two parts X_1 and X_2, i.e. if an edge leads from an x_i-node to an x_j-node, where $x_i \in X_t$ and $x_j \in X_m$, then the condition $t \leq m$ has to be fulfilled.

Call a read-k-times branching program read-k-times π-weak-ordered if it is partitioned to k layers such that the each layer is a π-weak-ordered read-once respecting the same ordered partition π of variables in each layer.

A π-weak-ordering of variables of a branching program P means that if some input $x_i \in X_2$ is tested by P, then on the rest part of computation path no variables from X_1 can be tested.

We call branching program P a read-k-times weak-ordered if it is read-k-times π-weak-ordered for some ordered partition π of the set of variables of P into two sets.

From the definition it follows that if read-once (read-k-times) branching program is ordered then it is weak-ordered.

For a function $g : \Sigma^n \to \{0,1\}$ and for a partition π of the set of variables x of g into two parts X_1 and X_2, denote by $C_{k,\pi}(g)$ a k-round deterministic communication complexity of g for the communication game with two players A and B where A obtains variables from the first part X_1 of variables and B obtains variables from the second part X_2 of variables of g.

Lemma 1. *Let for a function $g : \Sigma^n \to \{0,1\}$ P be a deterministic read-k-times π-weak-ordered branching program that computes g. Then*

$$size(P) \geq 2^{(C_{2k-1,\pi}(g)-1)/(2k-1)}.$$

Proof: Consider the following communication game with two players A and B for computing function f_n. Let X_1 and X_2 be two sets determined by partition π of a set of variables x of P. Part of the input corresponding to X_1 is known to A, and part of the input corresponding to X_2 is known to B. Players A and B have the copy of P. In order to compute f_n, A and B communicate with each other in $(2k - 1)$ rounds by sending messages in each round according to the following protocol ϕ. Player A is the first one to send a message. The output is produced by B. Let $\sigma \in \Sigma^{2n}$ be a valuation of x. Denote σ_A and σ_B parts of input σ which correspond to variables from X_1 and X_2 (inputs of A and B), respectively.

For each i, $1 \leq i \leq k-1$, communication protocol ϕ simulates computation on the i-th layer of P by two communication rounds $2i - 1$ and $2i$.

First round: Player A starts simulation of P on his part σ_A of input σ from the source of P. Let v_1 be a node which is reachable by P on σ_A from the source. Player A sends node v_1 to B.

Second round: Player B on obtaining message v_1 form A starts its simulation of P on his part σ_B of input σ from the node v_1. Let v_2 be a node which is reachable by P on σ_B from the v_1. Player B sends node v_2 to A.

Last round (round $2k - 1$): Player A on obtaining message v_{2k-2} from B starts its computation from the node v_{2k-2} on his part σ_A of input σ. Let v_{2k-1} be a node which is reachable by P on σ_A from the node v_{2k-2}. Player A sends node v_{2k-1} to B. Player B on obtaining v_{2k-1} starts its part of simulation of P from the v_{2k-1} on σ_B and then outputs the result of computation.

The message that A and B has exchanged during the computation is $m = v_1 v_2 \ldots v_{2k-1}$. Call m a full message.

Denote by V_i the set of all internal nodes which can be send on the i-th round by player A to B if i is odd (by player B to A if i is even) during the computations on Σ^{2n}. Denote $d_i = |V_i|$. From our notation it follows that the number of all full messages that can be exchanged on inputs from Σ^{2n} according to protocol ϕ is no more than $\prod_{i=1}^{2k-1} d_i$.

The number of full messages used by ϕ cannot be less than $2^{C_{2k-1,\pi}(g)-1}$

$$\prod_{i=1}^{2k-1} d_i \geq 2^{C_{2k-1,\pi}(g)-1}.$$

The lower bound of the lemma follows from the inequality above (with $d = \max\{d_i : i \in \{1, 2, \ldots, 2k - 1\}\}$) for which it holds that

$$d^{2k-1} \geq \prod_{i=1}^{2k-1} d_i \geq 2^{C_{2k-1,\pi}(g)-1}$$

and hence

$$d \geq 2^{(C_{2k-1,\pi}(g)-1)/(2k-1)}.$$

∎

Theorem 3. *Any deterministic ordered read-k-times branching program that computes function f_n has the size no less than $2^{(n-1)/(2k-1)}$.*

Proof: Let P be an ordered read-k-times branching program with an ordering π of variables which computes the function f_n. Consider the following partition $\hat{\pi}$ of variables x of f_n into two parts $X_1 = \{x_{\pi(1)}, x_{\pi(2)}, \ldots, x_{\pi(n)}\}$ and $X_2 = \{x_{\pi(n+1)}, x_{\pi(n+2)}, \ldots, x_{\pi(2n)}\}$. It is obvious that P is read-k-times $\hat{\pi}$-weak-ordered.

Denote CM a communication matrix of the function f_n for the partition $\hat{\pi}$ of the variables x. Consider the $2^n \times 2^n$ sub-matrix CM' of CM which is formed by strings that correspond to the part of inputs from $\{0,1\}^n$ and columns that correspond to the part of inputs from $\{\hat{0}, \hat{1}\}^n$. Matrix CM' is the E matrix (elements of the main diagonal are 1 and all rest elements are 0). This means that

$$C_{t,\pi}(f_n) \geq n$$

for $t \geq 1$. From the lower bound for $C_{t,\pi}(f_n)$ above and the lemma 1 it follows that

$$size(P) \geq 2^{(n-1)/(2k-1)}.$$

The lower bound of the theorem follows from considering the best read-k-times ordered branching program that computes f_n. ∎

Corollary. f_n *cannot be computed by any deterministic read-k-times ordered branching programs in polynominal size for* $k = o(n/\log n)$.

4 Further Research and Open Problems

We conclude with two open problems:

1. It will be interesting to describe how to separate "hard functions" (functions for which *randomization* does not improve their branching program complexity for the restricted number of testing variables) from the functions which can be computed more efficiently using randomization. Another words it is an interesting open problem to develop new randomized lower bound techniques for branching programs.

2. What is the exact dependence of the size of randomized branching programs on the error of computation?

Acknowledgments. We thank Sasha Razborov and Roman Smolensky for a number of interesting discussions on the subject of the paper.

References

[1] F. Ablayev, Lower bounds for probabilistic space complexity: communication-automata approach, *in Proceedings of the LFCS'94, Lecture Notes in Computer Science, Springer-Verlag*, 813, (1994), 1-7.

[2] L. Adelman, Two theorems on random polynomial time, *in Proceedings of the 19-th FOCS*, (1978), 75-83.

[3] M. Ajtai and M. Ben-Or, A theorem on randomized constant depth circuits, *in Proceedings of the 16-th STOC*, (1984), 471-474.

[4] C. Bennet and J. Gill, Relative to a random oracle A, $P^A \neq NP^A \neq co - NP^A$ with probability 1, *SIAM J. Comput*, 10, (1981), 96-113.

[5] B. Bolling, M. Sauerhoff, D. Sieling, and I. Wegener, On the power of different types of restricted branching programs, *ECCC Reports 1994*, TR94-025.

[6] R. Boppana and M. Sipser, The complexity of finite functions, *in Handbook of Theoretical Computer Science*, Vol A: Algorithms and Complexity, MIT Press and Elsevier, The Netherlands, 1990, ed. J Van Leeuwen, 757-804.

[7] A. Borodin, A. Razborov, and R. Smolensky, On lower bounds for read-k-times branching programs, *Computational Complexity*, 3, (1993), 1-18.

[8] A. Borodin, Time-space tradeoffs (getting closer to barrier?), *in Proceedings of the ISAAC'93, Lecture Notes in Computer Science, Springer-Verlag*, 762, (1993), 209-220.

[9] R. Bryant, Symbolic boolean manipulation with ordered binary decision diagrams, *ACM Computing Surveys*, 24, No. 3, (1992), 293-318.

[10] R. Frevalds, Fast probabilistic algorithms, *in Proceedings of the Conference Mathematical Foundation of Computer Science 1979, Lecture Notes in Computer Science, Springer-Verlag*, 74, (1979), 57-69.

[11] R. Freivalds and M. Karpinski, Lower time bounds for randomized computation, *in Proceedings of the ICALP'95, Lecture Notes in Computer Science, Springer-Verlag*, 944, (1995), 183-195.

[12] M. Krause, Lower bounds for depth-restricted branching programs, *Information and Computation*, 91, (1991), 1-14.

[13] C. Y. Lee, Representation of switching circuits by binary-decision programs, *Bell System Technical Journal*, 38, (1959), 985-999.

[14] L. Lovasz, Communication complexity: a survey, *in "Paths, Flows and VLSI Layout", Korte, Lovasz, Proemel, Schrijver Eds., Springer-Verlag* (1990), 235-266.

[15] W. Masek, A fast algorithm for the string editing problem and decision graph complexity, M.Sc. Thesis, Massachusetts Institute of Technology, Cambridge, May 1976.

[16] S. Ponzio, A lower bound for integer multiplication with read-once branching programs, *Proceedings of the 27-th STOC*, (1995), 130-139.

[17] A. Razborov, Lower bounds for deterministic and nondeterministic branching programs, *in Proceedings of the FCT'91, Lecture Notes in Computer Science, Springer-Verlag*, 529, (1991), 47-60.

[18] J. Simon and M. Szegedy, A new lower bound theorem for read-only-once branching programs and its applications, *Advances in Computational Complexity Theory*, ed. Jin-Yi Cai, DIMACS Series, 13, AMS (1993), 183-193.

[19] I. Wegener, *The complexity of Boolean functions*. Wiley-Teubner Series in Comp. Sci., New York – Stuttgart, 1987.

[20] I. Wegener, Efficient data structures for boolean functions, *Discrete Mathematics*, 136, (1994), 347-372.

Hitting Sets Derandomize BPP

Alexander E. Andreev[1], Andrea E. F. Clementi[2], José D. P. Rolim[3]

[1] Dept. of Mathematics, University of Moscow, andreev@matis.math.msu.su
[2] Dip. di Scienze dell'Informazione, University "La Sapienza" of Rome
clementi@dsi.uniroma1.it
[3] Centre Universitaire d'Informatique, University of Geneva, CH,
rolim@cui.unige.ch

Abstract. We show that hitting sets can derandomize *any* probabilistic, two-sided error algorithm. This gives a positive answer to a fundamental open question in probabilistic algorithms. More precisely, we present a polynomial time deterministic algorithm which uses any given hitting set to approximate the fractions of 1's in the output of any boolean circuit of polynomial size. This new algorithm implies that if a quick hitting set generator with logarithmic price exists then $BPP = P$. Furthermore, we generalize this result by showing that the existence of a quick hitting set generator with price k implies that $BPTIME(t) \subseteq DTIME(2^{O(k(t^{O(1)}))})$. The existence of quick hitting set generators is thus a new weaker sufficient condition to obtain $BPP = P$; this can be considered as another strong indication that the gap between probabilistic and deterministic computational power is not large.

1 Introduction

- *Motivations and previous results.* This paper addresses the issue of the derandomization of probabilistic algorithms, i.e., the design of general methods that permit an efficient deterministic simulation of algorithms which make use of random bits. *Pseudo-Random Generators* (PSRG's) ([4, 12]) constitute the best general method to this aim. A PSRG is a function $G = \{G_n : \{0,1\}^{k(n)} \rightarrow \{0,1\}^n, n > 0\}$, denoted by $G : k(n) \rightarrow n$ that "stretches" $k(n)$ truly random bits into n pseudo-random bits; more formally G is a PSRG if for any sufficiently large n and for any boolean circuit $C : \{0,1\}^n \rightarrow \{0,1\}$ whose size is at most n we have: $|\mathbf{Pr}(C(\mathbf{y}) = 1) - \mathbf{Pr}(C(G_n(\mathbf{x})) = 1)| \leq 1/n$ (where $\mathbf{y}$ is chosen uniformly at random in $\{0,1\}^n$, and $\mathbf{x}$ in $\{0,1\}^{k(n)}$). The intuition behind this is that the output of a PSRG looks "random" to any small circuit. Nisan and Wigderson [9] showed a method to construct *quick* PSRG's (i.e. PSRG's that are computable in polynomial-time in the length of their output; notice also that if $k(n) = O(\log n)$ then the PSRG belongs to EXP) which is based on a particular hardness assumption (i.e. the existence of boolean functions in EXP having exponential *hardness* [2, 8, 9]). In particular, they proved that, using $k(n) = O(\log n)$ truly random bits, these quick PSRG's can efficiently derandomize any two-sided error, polynomial-time algorithm (i.e. any BPP-algorithm).

In this paper we show a different approach to derandomize algorithms: we give
a positive answer to the question whether *Hitting Set Generators* ([10, 7, 1]) can
achieve equivalent general performances to those obtained by PSRG's. A *Hitting
Set Generator* (HSG) is a function $H = \{H_n : \{0,1\}^{k(n)} \to \{0,1\}^n, n > 0\}$
($H : k(n) \to n$) such that, for any sufficiently large n, and for any boolean
circuit C with size at most n and such that $\mathbf{Pr}\,(C(\mathbf{y}) = 1) \geq 1/n$, it is required
to provide *just* one "example" $\mathbf{y}$ for which $C(\mathbf{y}) = 1$, i.e., there exists at least
one $\mathbf{x} \in \{0,1\}^{k(n)}$ such that $C(H_n(\mathbf{x})) = 1$.

Observe first that any PSRG is also a HSG but the *converse* is not necessarily
true [1, 7]. Informally speaking, a PSRG provides a precise approximation of the
value $\mathbf{Pr}\,(C(\mathbf{y}) = 1)$, i.e., the fraction of 1's in the output of C, for any "small"
circuit C. Thus, if C has a large fraction of 1's in its output then the PSRG must
generate an input space for which this fraction has about the same large size. On
the other hand, HSG's are not required to have this property: a HSG provides,
for any "small" circuit C having a "sufficiently large" number of $1's$ in its output,
only a witness of the fact that C is not a *null* function. Another point of view
to distinguish between PSRG's and HSG's is that Hitting Sets (i.e. the output
of HSG's) have a *monotone* property not verified by the output of PSGR's:
if $\mathbf{Im}\,(H) = \{\mathbf{Im}\,(H_n), n > 0\}$ is a Hitting Set then any other set collection
$\mathbf{Im}\,(H') = \{\mathbf{Im}\,(H'_n), n > 0\}$, such that for any $n > 0$ $\mathbf{Im}\,(H_n) \subseteq \mathbf{Im}\,(H'_n)$, is
still a Hitting Set. In general, this monotone property makes the construction of
efficient HSG's easier than the construction of PSRG's [1, 7].

The design of HSG's which use "few" random bits has been a central topic
in complexity theory over the last ten years, since these generators are often
used to reduce the required number of random bits in many known randomized
algorithms [6, 10, 5]. Moreover, in [7] and afterwards in [1], some interesting
deterministic algorithms have been introduced to construct a Hitting Set for a
restricted class of boolean functions. However, the main question left open by
these previous works is whether quick HSG's (i.e. HSG's computable in polynomial time in the length of their output, as required for PSRG's) can replace
quick PSRG's in order to efficiently derandomize *any* BPP-algorithm. In this
paper, we give a positive answer to this question.

Our Results. The main technical result of this paper can be stated in the following way:

Theorem 1. *Let $q(n)$ be any positive polynomial function. If a quick HSG $H :
k(n) \to n$ (with $k(n) = \Omega(\log n)$) exists then it is possible to construct a deter-
ministic algorithm A that, for any n and for any circuit $C(x_1, \ldots, x_n)$ of size at
most $q(n)$, computes in polynomial time in n a value $A(C)$ such that*

$$|\mathbf{Pr}\,(C = 1) - A(C)| \leq \frac{1}{q(n)}.$$

Since approximating the fraction of 1's in the output of a linear-size boolean
circuit is *BPP*-hard (a proof of this can be found in [9]), Theorem 1 directly
implies the following

Corollary 2. *Let $k(n) = O(\log n)$. If there exists a quick HSG $H : k(n) \to n$ then $BPP = P$.*

Actually, Theorem 1 gives a more general consequence: by considering the "price" $k(n)$ (with $k(n) = \Omega(\log n)$) of the HSG as a parameter, we will prove the following

Corollary 3. *If a quick HSG $H : k(n) \to n$ exists, then for any time-bound $t(n)$, we have $BPTIME(t) \subseteq DTIME(2^{O(k(t^{O(1)}))})$, where $BPTIME(t)$ is the class of languages accepted by probabilistic, two-sided error Turing Machines running in time t.*

Notice that this result is comparable to the one in [8, 9] stating that the existence of a quick PSRG $G : k(n) \to n$ implies $BPTIME(t) \subseteq DTIME(2^{O(k(t^2))})$.

Such results globally states that quick HSG's can be used as a new general method to derandomize probabilistic algorithm. Moreover, from the previous discussion on the differences between PSRG's and HSG's, our results can be considered as a new, stronger indication on the fact that the computational power of probabilistic machines is not much larger than that of deterministic machines. This conjecture was first observed in [8, 9], and furtherly supported in [2].

2 The approximation algorithm

2.1 Overall description of the algorithm

The main technical contribution of this paper is an efficient deterministic algorithm that uses a quick HSG $H : k(n) \to n$ in order to approximate the value $Pr(C = 1)$ for a boolean circuit $C : \{0,1\}^n \to \{0,1\}$ $(n > 0)$. In order to facilitate this informal descrption, we will refer to the case in which C is a boolean circuit of polynomial size, and the HSG has a logarithmic price. The approximation algorithm considers a sufficiently large *Table T* (but still polynomial in n, i.e., $|T| = h(n)$ for some polynomial $h(n)$) of inputs for C, and computes two parameters d_{min} and d_{max} where d_{min} (d_{max}) is the minimum (maximum) fraction of 1's that C generates on the set of inputs of the form $\mathbf{y} \oplus \alpha$ where $\mathbf{y} \in T$ and $\alpha \in \mathbf{Im}(H_n)$ (more precisely, the minimum (maximum) is computed with respect to α). We then prove the following inequalities $d_{min} - \epsilon(n) \leq \mathbf{Pr}(C = 1) \leq d_{max} + \epsilon(n)$, where $\epsilon(n)$ is a "small" positive function which depends on H, and that we will prove to be smaller than the inverse of any polynomial function in n. From the above inequalities, it should be clear that the algorithm provides a "good" approximation if and only if the difference $D = d_{max} - d_{min}$ is "small" (more precisely, we require this value not to be greater than $\epsilon(n)$). However, this condition is not generally satisfied and consequently a further iterative procedure must be performed. This procedure consider a polynomially bounded sequence of Tables $Y = \{T_k : k = 1, \ldots, q(n)\}$, each of them having size $h(n)$. For each T_k, it computes the parameters d_{min}^k and d_{max}^k, and checks whether the following condition is true

"THERE IS AT LEAST ONE k FOR WHICH $D^k = d_{max}^k - d_{min}^k \leq \epsilon(n)$ " (1).

If this condition is verified for some k then the procedure terminates and returns the value $(d_{max}^k + d_{min}^k)/2$ which is a "good" approximation of $\mathbf{Pr}\left(()C = 1\right)$. If Condition (1) is false then the procedure performs a *compression* phase on Y whose goal is to reduce the values D^k's. Indeed, when Condition (1) is false, the non-negligible values D^k's provide a key-information about the "behavior" of C on Y that the procedure uses to codify Y (using a convenient binary coding system) into a new binary sequence having a "large" expected number of 0's. The procedure can then efficiently compress this new binary sequence in polynomial time. The obtained string will be the new sequence of Tables on which the procedure will re-check Condition (1). The algorithm applies the compression phase until either Condition (1) is satisfied or the total length of the compressed sequence of Tables is smaller than $h(n)$. In the latter case, a "failure" answer will be returned.

A suitable construction of the input Tables, based on the HSG, will guarantee the efficiency (i.e. a polynomial bound on the maximum number of compression phases of the algorithm) and the correctness (i.e. the fact that the values D^k's actually decreases and thus the procedure will never return the "failure" answer) of the algorithm.

Due to the lack of space, this preliminary version of the paper does not contain the proofs of lemmas and theorems. Formal proofs are given in the full version (see [3]).

2.2 Preliminaries

The length of a string $x \in \{0,1\}^*$ is denoted as $l(x)$. Given any boolean circuit $C : \{0,1\}^n \to \{0,1\}$, we denote its size as $L(C)$ (note that any 2-input boolean function is here considered as one gate). The same notation is used for the circuit size complexity of any finite boolean function.

Definition 4. Let $\epsilon(n)$ be a polynomial-time computable function such that, for any $n \geq 1$, $0 < \epsilon(n) < 1$. Then, the operator $H = \{H_n : \{0,1\}^{k(n)} \to \{0,1\}^n, n > 0\}$ is a *quick $\epsilon(n)$-Hitting Set Generator* (in short, $\epsilon(n)$-HSG) if: *a).* H is computable in polynomial time in n (note that, if $k(n) = O(\log n)$ then H is computable in exponential time with respect to the length of its input), and *b)* for any boolean circuit C, such that $L(C) \leq n$ and $\mathbf{Pr}\left(C = 1\right) \geq \epsilon(n)$, there exists (at least one) $\mathbf{a} \in \{0,1\}^{k(n)}$ such that $C(H_n(\mathbf{a})) = 1$.

In which follows, we consider only *quick* HSG's, and thus we will omit the term *quick*. Furthermore, we will always assume that the price $k(n)$ is a monotone function such that, for any n, $k(n+1) - k(n) \leq 1$ and $n^\alpha \geq k(n) \geq \log n$ where $0 < \alpha < 1$.

The following properties of HSG's will be strongly used throughout the paper.

Lemma 5. *1) If $H : k(n) \to n$ is an $\epsilon_1(n)$-HSG and, for any $n > 0$ the function $\epsilon_2(n))$ is such that $\epsilon_1(n) \le \epsilon_2(n)$, then $H : k(n) \to n$ is also an $\epsilon_2(n)$-HSG.*
2) If for some constant ϵ, $0 < \epsilon < 1$, there exists an ϵ-HSG $H : k(n) \to n$ then, for any positive function $p(n)$, we can construct (using H) a $p(n)^{-1}$-HSG $H' : k'(n) \to n$, in polynomial time in $2^{k'(n)}$, where $k'(n) = O(k(n^2 p(n)))$.

Probably, the most interesting case is when $k(n) = O(\log n)$ and $p(n)$ is a polynomial since the price is still logarithmic and the construction time is polynomial in n.

2.3 A first approximation of probability

As described in Section 2.1, our goal is to derive a deterministic algorithm that, given a circuit $C : \{0,1\}^n \to \{0,1\}$, uses an ϵ-HSG $H : k(n) \to n$ and *Table* of inputs for C to approximate $\mathbf{Pr}\,(C = 1)$.

In which follows, we show how to derive a first estimation of $\mathbf{Pr}\,(C = 1)$ using a boolean sequence. Let $T = a_1 a_2 ... a_{l(T)}$ be a boolean sequence. Observe that the results shown in this section hold for any sequence T: the correct choice of T as the collection of input Tables for the approximation algorithm will be shown in Section 2.5.

We can define the string $T(n,i) \in \{0,1\}^n$ (i.e. the input for $C : \{0,1\}^n \to \{0,1\}$) as follows

$$T(n,i) = \begin{cases} a_{(i-1)n+1} a_{(i-1)n+2} ... a_{(i-1)n+n} & \text{if } i \cdot n \le l(T) \\ a_{(i-1)n+1} ... a_{l(T)} 0...0 & \text{if } i \cdot n > l(T) \end{cases}$$

where $i = 1, 2, ..., m = \lceil l(T)/n \rceil$. T represents our *input Table* for C. Given $\alpha \in \{0,1\}^n$, consider the function

$$Med(C, T, \alpha) = \frac{1}{m} \sum_{i=1}^{m} C(T(n,i) \oplus \alpha) ..$$

and its "complexity" $l(C, m) = L(C) \cdot m + c_{sym} \cdot m$ where c_{sym} is the positive constant that appears in the linear bound which holds for the circuit size of *symmetrical functions* (it is possible to prove that $c_{sym} \le 30$, see [11]). Let $H : k(n) \to n$ be an ϵ-HSG, we denote the prefix of length j of H_n as $H_{n,j}$ i.e. $H_{n,j} = (H_n^1, H_n^2, .., H_n^j)$. We can now define the following two parameters

$$d_{min}(C, T, H) = \min_{\gamma \in \{0,1\}^{k(l(C,m))}} Med(C, T, H_{l(C,m),n}(\gamma)) ,$$

$$d_{max}(C, T, H) = \max_{\gamma \in \{0,1\}^{k(l(f,m))}} Med(C, T, H_{l(C,m),n}(\gamma)).$$

Lemma 6. *If $H : k(n) \to n$ is an ϵ-HSG then*

$$d_{min}(C, T, H) - \epsilon \le \mathbf{Pr}\,(C(x_1, x_2, ..., x_n) = 1) \le d_{max}(C, T, H) + \epsilon .$$

Furthermore, the values $d_{min}(C, T, H)$ and $d_{max}(C, T, H)$ can be computed in $O(2^{k(l(C,m))} m L(C))$ time.

2.4 Compression

The quality of the approximation of $\mathbf{Pr}\,(C = 1)$ given by Lemma 6 depends on the value $D = d_{max}(C, T, H) - d_{min}(C, T, H)$. However, D can be arbitrarily large and thus a further procedure must be applied in order to reduce this value when it is not sufficiently small. The procedure is based on a suitable compression of a set of input Tables. In this section, we first describe the coding technique which permits to efficiently compress a single input Table. Then we generalize this technique in order to compress a polynomial sequence of input Tables.

Coding and decoding the Table. For the sake of simplicity, we will use the following definitions

$$d_1 = d_{min}(C, T, H) = Med(C, T, \alpha_1) \text{ and } d_2 = d_{max}(C, T, H) = Med(C, T, \alpha_2)$$

where $\alpha_1 = H_{l(C,m),n}(\gamma_1)$ and $\alpha_2 = H_{l(C,m),n}(\gamma_2)$ for some $\gamma_1, \gamma_2 \in \{0, 1\}^{k(n)}$.

The j-th component of vector $\mathbf{a}$ will be denoted as $[\mathbf{a}]^j$. Since we are considering the case in which $D > 0$, without loss of generality, we can assume that for index s we have $[\alpha_1]^s \neq [\alpha_2]^s$. Our next goal consists to show that we can codify the input Table as a boolean sequence in which the s-th component of each input string for C is always 0. Consider the operator $T^{\#} : \{1, \ldots, m\} \rightarrow \{0, 1\}^n$ such that $T^{\#}(i) = T(n, i) \oplus ([T(n, i)]^s \cdot (\alpha_1 \oplus \alpha_2))$ where the operation "$\oplus$" between two boolean vectors is performed component by component and the operation "." is the standard scalar product. The s-th component of $T^{\#}(i)$ satisfies the following equations:

$$[T^{\#}(i)]^s = [T(n, i)]^s \oplus ([T(n, i)]^s \cdot ([\alpha_1]^s \oplus [\alpha_2]^s)) = [T(n, i)]^s \oplus [T(n, i)]^s \cdot 1 = 0 . \tag{1}$$

Observe also that the set $\{T^{\#}(i) \oplus \alpha_1, T^{\#}(i) \oplus \alpha_2\}$ is equal to the set $\{T(n, i) \oplus \alpha_1 , T(n, i) \oplus \alpha_2\}$. Let

$$N(\sigma, \phi_1, \phi_2) = |\{i : [T(n, i)]^s = \sigma , \text{ and } C(T(n, i) \oplus \alpha_j) = \phi_j \ j = 1, 2\}| . \tag{2}$$

We can now introduce the function which approximates the s-th component of $T(n, i)$. Consider the function Q defined as follows:

$$Q_{N(\sigma, \phi_1, \phi_2)}(x, y) = \begin{cases} x \text{ if} & x \neq y \\ 1 \text{ if } x = y = 0 & \text{and} \quad N(1, 0, 0) \geq N(0, 0, 0) \\ 0 \text{ if } x = y = 0 & \text{and} \quad N(1, 0, 0) < N(0, 0, 0) \\ 1 \text{ if } x = y = 1 & \text{and} \quad N(1, 1, 1) \geq N(0, 1, 1) \\ 0 \text{ if } x = y = 1 & \text{and} \quad N(1, 1, 1) \geq N(0, 1, 1) \end{cases}$$

In which follows we will consider the function N as a fixed parameter, and thus we will omit the index $N(*, *, *)$ in the definition of Q. Then the approximation function is $Z(i) = Q(C(T^{\#}(i) \oplus \alpha_1), C(T^{\#}(i) \oplus \alpha_2))$, $i = 1, .., m$.

Our next goal is to estimate the number of errors generated by $Z(i)$. Let $ND(\sigma, \phi_1, \phi_2)$ be the number of indexes i such that the following conditions are satisfied:

- **i)** $[T(n,i)]^s \oplus Z(i) = 1$ (i.e. there is an error);   **ii)** $[T(n,i)]^s = \sigma$;
- **iii)** $C(T(n,i) \oplus \alpha_1) = \phi_1$;   **iv)** $C(T(n,i) \oplus \alpha_2) = \phi_2$.

The following Lemma gives an upper bound on the number of approximation errors.

Lemma 7.

$$\sum_{(\sigma,\phi_1,\phi_2)\in\{0,1\}^3} ND(\sigma,\phi_1,\phi_2) \leq m\left(\frac{1}{2} - \frac{d_2 - d_1}{2}\right).$$

In which follows, we show how to represent (i.e. codify) the input Table T using the approximation function $Z(i)$. From Lemma 7, we will prove that this representation will have a "large" number of 0's in its last component when $D = d_2 - d_1$ is not "small". The function $U(i) = [T(n,i)]^s \oplus Z(i)$, $(i = 1, 2, ..., m)$ singles out the positions in the input table in which there is an error. We thus have that $[T(n,i)]^s = U(i) \oplus Z(i) = U(i) \oplus Q(C(T^\#(i) \oplus \alpha_1), C(T^\#(i) \oplus \alpha_2))$.

The new representation of the input Table is then the following string:

$$(s, \alpha_1, \alpha_2, Q, [T^\#]^1, ..., [T^\#]^{s-1}, [T^\#]^{s+1}, ..., [T^\#]^n, U),\qquad (3)$$

where the term $[T^\#]^j$ denotes the boolean sequence consisting of all j-th components of the operator $[T^\#]$. From the string in Eq. 3, we can efficiently (i.e. in polynomial time) reconstruct the s-th bit by computing the following sequence of values.

Procedure 1

1. $T^\#(i) = ([T^\#(i)]^1, ..., [T^\#(i)]^{s-1}, 0, [T^\#(i)]^{s+1}, ..., [T^\#(i)]^n)$;
2. $[T(n,i)]^s = U(i) \oplus Q(C(T^\#(i) \oplus \alpha_1), C(T^\#(i) \oplus \alpha_2))$;
3. $T(n,i) = T^\#(i) \oplus ([T(n,i)]^s \cdot (\alpha_1 \oplus \alpha_2))$.

Observe that if we adopt the representation of T shown in Eq. 3, the string size does not decrease. However, the crucial fact for the compression phase is that, if $D = d_2 - d_1$ is not small, U contains a large number of 0's.

Compression of strings with a large number of 0's. Given any boolean string $V = V^* w$, let $l(V) = t$ be its length, $|V| = k$ be the number of 1's, and w be its last bit. We also define the following "counting" function:

$$NUM(V) = \begin{cases} 0 & \text{if } t = k \\ NUM(V^*) & \text{if } w = 0 \\ \binom{t-1}{k} + NUM(V^*) & \text{if } w = 1 \end{cases}$$

The "compressed" version of V is then the string $(l(V), |V|, NUM(V))$. Notice that we can efficiently reconstruct V from this string. The simple procedure is the following.

Procedure 2

1. Consider $V = V^* w$, if $NUM(V) = 0$, then $V = 1^{l(V)}$;

2. If $\binom{l(V) - 1}{|V|} \leq NUM(V)$, then $w = 1$ and we set:

$$NUM(V^*) = NUM(V) - \binom{l(V) - 1}{|V|}, |V^*| = |V| - 1, \text{ and } l(V^*) = l(V) - 1;$$

3. If $\binom{l(V) - 1}{|V|} > NUM(V)$, then $w = 0$ and we set: $NUM(V^*) = NUM(V)$, $|V^*| = |V|$, and $l(V^*) = l(V) - 1$.

Clearly, the procedure will halt when $l(V^*) = 0$. It is not hard to see that the number of steps of the above procedure is a linear function in the length of the input string V. This binary representation is very useful when the input string contains a large number of 0's. We will show this fact in the case of the input Table T considered in the previous section. We modify the binary representation of T defined in Eq. 3 in the following way:

$$(s, \alpha_1, \alpha_2, Q, [T^\#]^1, \ldots, [T^\#]^{s-1}, [T^\#]^{s+1}, \ldots, [T^\#]^n, l(U), |U|, NUM(U)) , \quad (4)$$

where s, α_1, and α_2 have been defined at the beginning of Section 2.4; the function Q is represented as a string of length 4. In order to efficiently code and decode the above sequence, we make use of the following coding operators

$$\lambda(a_1 a_2 \ldots a_r) = 00 a_1 (\neg a_1) a_2 (\neg a_2) \ldots a_r (\neg a_r) 11,$$

and $\Lambda(V) = \lambda(l(V))V$, where $V \in \{0, 1\}^*$ (note that for $\Lambda(n)$, with $n \geq 0$, we mean the output of Λ on the standard binary representation of n). We also assume that $a_1 \neq 0$ if $n \neq 0$. Given an input Table T, we construct the functions $T^\#$, Q and U (more precisely the output of these functions). Then, using the representation (4), we define

$$comp(T) = \Lambda(l(T))\Lambda(s)\Lambda(\alpha_1)\Lambda(\alpha_2)\Lambda(Q)\Lambda([T^\#]^1) \ldots$$

$$\ldots \Lambda([T^\#]^{s-1})\Lambda([T^\#]^{s+1}) \ldots \Lambda([T^\#]^n)\Lambda(l(U))\Lambda(|U|)\Lambda(NUM(U)). \quad (5)$$

Observe that from the sequence $S = \Lambda(S_1)S_2$ it is easy to compute in polynomial time (in the length of S) its "components" S_1 and S_2.

We can repeatedly apply the above decoding procedure to $comp(T)$ and find its components. Observe that the number of components is at most the length of the sequence and we can thus reconstruct in polynomial time all the components of the string in Eq. 4. We then apply the linear-time Procedure 2 on every every input of the form $(l(U), |U|, NUM(U))$ in order to obtain U. Finally, we apply Procedure 1 to reconstruct T from the boolean sequence in Eq. 3. It follows that we can apply the coding and decoding operators in order to efficiently reconstruct the binary representation in (3) and thus the table T (we denote the global decoding operator as $comp^{-1}$). This fact will be used in proving the correctness of the approximation algorithm (see Lemma 10). Using Lemma 7, it is possible to give the degree of compression achieved by the operator $comp$.

Lemma 8. *There exists a positive constant c_1 such that*

$$l(comp(T)) \leq O(n + \log l(T)) + l(T) \left(1 - c_1 \frac{(d_2 - d_1)^2}{n} \right).$$

Compressing more input Tables. As previously mentioned, our goal is to reduce the value $D = d_2 - d_1$. The key idea is to consider a sufficiently large sequence of input Tables and apply to them the compression operator shown in the previous section. Let Y be the boolean sequence which represents the input Tables (Y corresponds to the concatenation of the input Tables). We consider the partition $Y = Y_1 Y_2..Y_r$ such that $l(Y_i) = h(n)$ (for $i = 1, .., r - 1$) and $l(Y_r) \leq h(n)$, where $h(n)$ is a suitable upper-bound function which will be defined later. We will repeatedly transform Y using two different operators; however, the obtained sequence (at every step) can always be represented as a string $W = W_1 W_2 ... W_{r(W)}$ where

$$W_i = \Lambda(N_i(W))\Lambda(l_i(W))\Lambda(q_i(W))W_i^* , \quad i = 1 \ldots r(W) ;$$

here N_i and l_i are respectively the starting point and the length of the part of Y for which W_i is the coding version, and q_i denotes the type of transformation which has generated W_i. This coding structure will be strongly used to recover the bit in a generic position in Y (see the proof of Lemma 10). We will consider a suitable starting sequence and two different transformations of sequences. Thus q_i will assume three different values.

1). We start the transformation process with the following sequence $W^1 = W_1^1 \ldots W_{r(W^1)=r}^1$ where

$$W_i^1 = \Lambda((i - 1)h(n) + 1)\Lambda(l(Y_i))\Lambda(q_i = 0)\Lambda(Y_i) , \quad i = 1, \ldots, r .$$

2). A *compression* action $V = COMP(W)$, where the output sequence $V = V_1 \ldots V_{r(V)}$ is such that

$$V_i = \Lambda(N_i(W))\Lambda(l_i(W))\Lambda(q_i = 1)\Lambda(comp(W_i)) , \quad i = 1, \ldots, r(W) .$$

Notice that in this case we have $r(V) = r(W)$.

3). A *concatenation* action $V = CONCAT(W)$, where $V = V_1 V_2 \ldots V_{r(V)}$ is such that

$$V_i = \Lambda(N_{2i-1}(W))\Lambda(l_{2i-1}(W) + l_{2i}(W))\Lambda(q_i = 2)\Lambda(W_{2i-1})\Lambda(W_{2i}),$$

where $i = 1, .., \frac{r(W)}{2}$. Notice that in this case $r(V) \leq (r(W)/2) + 1$, and if $r(W)$ is not even then $V_{r(V)} = W_{r(V)}$.

The algorithm works as follows. Consider the ϵ-HSG H given in input and assume that W^t has been already generated from the sequence of input Tables. Then the algorithm checks the following condition:

$$\exists i \; : \; d_{max}(C, W_i^t, H) - d_{min}(C, W_i^t, H) \leq \epsilon \; . \tag{6}$$

If Condition (6) is verified for some index i then the algorithm returns the value $[d_{max}(C, W_i^t, H) + d_{min}(C, W_i^t, H)]/2$ which is a good approximation of $\mathbf{Pr}\,(C = 1)$ (see Lemma 6). If Condition (6) is false, the algorithm checks Condition

$$r(W^t) \leq h(n) \; . \tag{7}$$

If Condition (7) is true the algorithm returns a "failure" answer (note that in this case the total length of W^t, i.e. $l(W^t)$, is bounded by $O(h^2(n))$). When Condition (7) is false we apply the above described transformations to generate the coding sequence W^{t+1}. The type of the transformation to be applied is defined according to the following rule.

If $\exists i : l(W_i^t) > h(n)$, then $W^{t+1} = COMP(W^t)$, otherwise
$$W^{t+1} = CONCAT(W^t).$$

2.5 Complexity and correctness of the algorithm

The following lemma is a consequence of Lemma 8 and the transformation procedure described in the previous section.

Lemma 9. *Let* $H : k(n) \to n$ *be an* $\epsilon(n)$-*HSG and* $\epsilon(n)^{-1} \geq n$ *is a positive function. Choose* $h(n)$ *as a positive function such that* $n^3 \epsilon(n)^{-2} \leq h(n) \leq 2n^3 \epsilon(n)^{-2}$ *and let* $h_1(n)$ *be a positive function such that* $n \leq h_1(n) \leq 2^n$. *Then, for any* n *and for any circuit* $C : \{0,1\}^n \to \{0,1\}$ *such that* $L(C) \leq h(n)$, *and for any boolean sequence* Y *such that* $l(Y) \leq h_1(n)$, *every transformation step can be performed in polynomial time in* $h(n) \cdot h_1(n) \cdot 2^{h(n)}$, *and the maximum number* $t(n)$ *of transformations performed by the algorithm is such that* $t(n) = O(\epsilon(n)^{-2} n \log h_1(n))$.

The following two lemmas are used to prove the correctness of the algorithm, that is, by appropriately choosing the sequence Y of input Tables, the output will never be the "failure" answer. To this aim, beside being a simple boolean string, the sequence Y will also be considered as a finite boolean function $Y : \{0,1\}^{\lceil \log l(Y) \rceil} \to \{0,1\}$. We can thus consider its circuit complexity $L(Y(i))$.

Lemma 10. *With the same hypothesis and definitions of Lemma 9, if the algorithm returns a "failure" answer then there exists a polynomial* $p(n)$ *(which can be efficiently determined) such that* $L(Y(i)) \leq p((h(n)/\epsilon(n)) \log h_1(n))$.

Lemma 11. *If an* $\frac{1}{2}$-*HSG* $H : k(n) \to n$ *exists, then it is possible to construct a function* $F = \{F_i : \{0,1\}^i \to \{0,1\}, i > 0\}$ *(in polynomial time in* 2^i), *which belongs to* EXP, *and, for almost all* $i > 0$, $k(L(F_i)) \geq i/2$.

We can now state the final Theorem.

Theorem 12. *Let $q(n) \geq n$ be a positive function. If for some constant $0 < \delta < 1$ there exists a δ-HSG $H : k(n) \to n$, then there exists a deterministic algorithm A which, for any n and for any circuit $C(x_1, x_2, ..., x_n)$ of size at most $q(n)$, computes a value $A(C)$ such that*

$$|\mathbf{Pr}\,(C(x_1, x_2, ..., x_n) = 1) - A(C)| \leq \frac{1}{q(n)}.$$

The time of the algorithm A is bounded by a polynomial in $2^{O\left(k\left(q(n)^{O(1)}\right)\right)}$.

3 Complexity results

The deterministic algorithm shown in the previous section is able to use a HSG to solve the problem of approximating the fraction of 1's in the output of a boolean circuit whose size is bounded by a fixed function $q(n)$ where $n \leq q(n) \leq 2^n$. It is not hard to see that this problem is hard for the probabilistic (two-side error) class $BPTIME(q)$ since, informally speaking, the algorithm can be easily used to estimate the acceptance probability of a generic probabilistic, two-sided error Turing machine which works in time bounded by $q(|x|)$ on any possible input x (a proof of this fact is implicitly given in the construction of PSRG's introduced in [9]). Thus Theorem 1 directly implies the following result.

Corollary 13. *If a quick HSG $H : k(n) \to n$ exists, then for any time-bound $t(n)$, we have $BPTIME(t) \subseteq DTIME(2^{O(k(t^{O(1)}))})$, where $BPTIME(t)$ is the class of languages accepted by probabilistic Turing Machines running in time t and with two-sided error.*

Probably, the most interesting case is when the price $k(n)$ of the generator is a logarithmic function.

Corollary 14. *Let $k(n) = O(\log n)$. If there exists a quick HSG $H : k(n) \to n$ then $BPP = P$.*

Acknowledgemnts. We would like to thank Michael Saks, Michael Sipser, and Alexander Razborov for their patience, helpful pointers, and very interesting discussions.

References

1. Andreev A. (1995), "The complexity of nondeterministic functions", *Information and Computation*, to appear.

2. Andreev A., Clementi A., and Rolim J. (1996), "Optimal Bounds on the Approximation of Boolean Functions, with Consequences on the Concept of Hardness", *XIII Annual Symposium on Theoretical Aspects of Computer Science (STACS'96)*, LNCS, 1046, 319-329. Also available via WWW in the electronic journal ECCC (TR95-041).

3. Andreev A., Clementi A., and Rolim J. (1996), "Hitting Sets Derandomize BPP" (full version), submitted. Also available via WWW in the electronic journal ECCC (TR95-061).

4. Blum M., and Micali S. (1984), "How to generate cryptographically strong sequences of pseudorandom bits", *SIAM J. of Computing*, 13(4), 850-864.

5. Chor B., and O. Goldreich (1989), "On the Power of Two-Point Based Sampling", *J. of Complexity*, 5, 96-106.

6. Karp R., Pippenger N., and Sipser M. (1982) "Time-Randomness, Tradeoff", presented at *AMS Conference on Probabilistic Computational Complexity*.

7. Linial N., Luby M., Saks M., and Zuckerman D. (1993). "Efficient construction of a small hitting set for combinatorial rectangles in high dimension", in *Proc. 25th ACM STOC*, 258-267.

8. Nisan N. (1990), *Using Hard Problems to Create Pseudorandom Generators*, *ACM Distinguished Dissertation*, MIT Press.

9. Nisan N., and Wigderson A. (1994), "Hardness vs Randomness", *J. Comput. System Sci.* 49, 149-167 (also presented at the *29th IEEE FOCS*, 1988).

10. Sipser M. (1986), "Expanders, Randomness or Time vs Space", in *Proc. of 1st Conference on Structures in Complexity Theory*, LNCS 223, 325-329.

11. Wegener, I. (1987), *The complexity of finite boolean functions, Wiley-Teubner Series in Computer Science*.

12. Yao A. (1982), "Theory and applications of trapdoor functions", in *Proc. 23th IEEE FOCS*, 80-91.

On Type-2 Probabilistic Quantifiers*

Ronald V. Book[1], Heribert Vollmer[2], and Klaus W. Wagner[2]

[1] Department of Mathematics, University of California at Santa Barbara,
Santa Barbara, CA 93106, U.S.A.
[2] Lehrstuhl für Theoretische Informatik, Universität Würzburg,
Am Exerzierplatz 3, D-97072 Würzburg, Germany

Abstract. We define and examine several probabilistic operators ranging over sets (i.e., operators of type 2), among others the formerly studied
ALMOST-operator. We compare their power and prove that they all coincide for a wide variety of classes. As a consequence, we characterize
the ALMOST-operator which ranges over infinite objects (sets) by a
bounded-error probabilistic operator which ranges over strings, i.e. finite
objects. This leads to a number of consequences about complexity classes
of current interest. As applications, we obtain (a) a criterion for measure 1 inclusions of complexity classes, (b) a criterion for inclusions of
complexity classes relative to a random oracle, (c) a new upper time
bound for ALMOST-PSPACE, and (d) a characterization of ALMOST-
PSPACE in terms of checking stack automata. Finally, a connection between the power of ALMOST-PSPACE and that of probabilistic NC¹
circuits is given.

1 Introduction

In a fundamental paper, John Gill introduced probabilistic Turing machines and
the complexity classes they define [12]. During the run of their computation
these machines have the possibility to toss fair coins, and then continue their
work depending on the outcome. In the polynomial time case this yields the
well-known classes PP (for probabilistic polynomial time; with unbounded error
probability) and BPP (for bounded error probabilistic polynomial time) which
are regarded as natural probabilistic counterparts of the deterministic class P;
and moreover BPP is felt to be the class of "tractable" problems (since the error
bound can be made arbitrarily small).

But how do define probabilistic analogues for other (possibly not deterministic) classes? The "traditional" way is to consider *operators* in an abstract way
as we will do in this paper. This kind of randomness can best be visualized as allowing Turing machines access to a random tape, or equivalently supplying them
together with their regular input with an input sequence of random bits. Thus,
here the random bits may be multiply accessed. (This should be contrasted with

---

* Research supported by NSF Grant CCR-93-02057, DFG Grant Wa 847/1, and a
Feodor-Lynen-Fellowship from the Alexander von Humboldt Foundation. Research
performed while the second and third author were visiting the Department of Mathematics, University of California, Santa Barbara.

the machines with built-in probabilism described above: If those machines want to re-use their random bits later, they have to store them on their worktape—which might make a difference for space-bounded computations. Therefore the aforementioned built-in probabilism is also called *one-way access* to randomness, see [21].)

Well known examples for operators as just described are Wagner's counting operator C^P [30], and the corresponding bounded error operator BP^P (see e.g. [26]). It is relatively easy to see that $C^PP = PP$ and $BP^PP = BPP$, i.e. when applied to the class P these operators yield as results the classical probabilistic classes. But the operators can be applied in a general way to arbitrary classes $\mathcal{K}$, giving $C^P\mathcal{K}$ and $BP^P\mathcal{K}$. For example, the class BP^PNP has attracted some attention and has been shown to be equal to Babai's class AM (for "Arthur-Merlin," a class defined in terms of interactive proof systems, see [2]).

Yet another approach to define probabilistic computation is to consider complexity classes of the form ALMOST-$\mathcal{K}$, see e.g. [1, 5, 7, 8, 19, 21]. ALMOST-$\mathcal{K}$ is defined to be the class of all sets which are in $\mathcal{K}^A$ for almost every oracle A. For example, $L \in$ ALMOST-P if and only if the set of all A such that $L \in P^A$ has measure 1 (in the usual product measure on sets, for details see Section 2). Thus, the machines have here in this case in a sense some kind of access to a database (oracle), and they are required to work correctly for almost all such databases.

It has been observed that in a number of classes, the ALMOST- and the BP^P-operator coincide, e.g. ALMOST-P = BPP and ALMOST-NP = AM [5, 19]. However, the general relationship between the operators is open. Especially, no characterization of ALMOST-PSPACE is known.

In this paper we introduce a type-2 probabilistic quantifier, which we will denote by BP^2, and show that for a wide variety of classes, the ALMOST- and the BP^2 operators coincide, i.e., ALMOST-$\mathcal{K} = BP^2\mathcal{K}$. "Type 2" means that the operator is based on a quantifier that does not range over words but over *sets* (databases, i.e., oracles). Thus ALMOST-classes are classes accepted with bounded-error probability by machines with access to a random database. Moreover, it is not too hard to see that the type-2 operator BP^2 can often be replaced equivalently by a "classical" operators ranging over finite (i.e., type 1) objects (words). The most important special case here is the case of so called *leaf language definable* classes $\mathcal{K}$ (a definition is given in Section 2). Here we see that the BP^2 operator coincides with a type-1 operator ranging over exponentially long strings (compared to the length of the input), in contrast to the usual quantifiers where polynomially long words are considered. We will denote this new operator by BP^{exp}. Combining this with the above, we get for such $\mathcal{K}$ that ALMOST-$\mathcal{K} = BP^2\mathcal{K} = BP^{exp}\mathcal{K}$, for example ALMOST-PSPACE = BP^{exp}PSPACE or ALMOST-PP = BP^{exp}PP. Thus, in this case it turns out that working in the "ALMOST-mode" is equivalent to working with random input sequences. We think this characterization is advantageous compared to the definition since here we only have to deal with finite objects (strings) in contrast to (infinite) oracles.

We give several applications of our characterization: Since for all classes $\mathcal{K}$, $\text{BP}^{\text{P}}\mathcal{K} \subseteq \text{BP}^2\mathcal{K}$, we see that a relativizable inclusion $\mathcal{K}_1 \subseteq \text{BP}^{\text{P}}\mathcal{K}_2$ implies the measure 1 inclusion $\mathcal{K}_1 \subseteq \mathcal{K}_2$. As a consequence, we show that e.g. the set of all oracles relative to which the polynomial time hierarchy is strictly included in $\oplus \text{P}$ has measure 1 (a result which has already been proved by Regan and Royer [24]). We improve the best known EXPSPACE upper bound for ALMOST-PSPACE to $\Sigma_2^{\text{exp}} \cap \Pi_2^{\text{exp}}$. We prove that ALMOST-PSPACE allows a machine characterization in terms of checking stack automata. These automata were introduced by Oscar Ibarra in [14], where it was also shown that when working nondeterministically these machines are strictly power powerful than when working deterministically. Our results imply that the nondeterministic mode is (under reasonable complexity theoretic assumptions) even more powerful than the bounded-error probabilistic mode. Finally, we draw a connection between the power of ALMOST-PSPACE and a problem from circuit complexity by showing that proving upper bounds for probabilistic NC^1 circuits better than the up to now known BPP-bound will result in better upper bounds for ALMOST-PSPACE.

All in all we see that our systematic comparison of several ways of introducing randomness into computation allows us to improve a number of results for complexity classes of current topical interest. Along the way, we get new insights into the relationship between statements holding for a measure 1 set of oracles and statements holding for an algorithmically random oracle in the sense of Martin-Löf (see [8]), thus improving results in [7, 15].

2 Preliminaries

We assume that the reader is familiar with basic complexity theory notions, classes and reducibilities, see e.g. [3]. Let $\{0,1\}^*$ denote the set of finite binary words, whereas $\{0,1\}^\omega$ denotes the set of infinite binary words. Following common use, we identify a language, i.e. a subset of $\{0,1\}^*$, with its characteristic sequence, which is an element of $\{0,1\}^\omega$. For $w \in \{0,1\}^*$ we denote the i-th bit of w by $w(i)$. Similarly, for $A \in \{0,1\}^\omega$ we denote the i-th bit of A by $A(i)$. Using the lexicographic ordering of $\{0,1\}^*$, there is a natural bijection between $\{0,1\}^*$ and the set $\mathbb{N}$ of natural numbers. Thus, we will also write $A(w)$ for $w \in \{0,1\}^*$ and $A \in \{0,1\}^\omega$. We then mean the bit in A at "position w," i.e. $A(w) = 1$ if $w \in A$ and $A(w) = 0$ otherwise.

Let $\{M_i\}_{i \in \mathbb{N}}$ be a recursive enumeration of all oracle Turing machines. Let $M_i^A(x)$ be the result of M_i's work on input x and oracle A if this computation stops, and let $M_i^A(x)$ be undefined otherwise. Define $L(M_i) =_{\text{def}} \{ (A, x) \mid M_i^A(x) = 1 \}$ and $L(M_i^A) =_{\text{def}} \{ x \mid M_i^A(x) = 1 \}$.

A class $\mathcal{K}^{(\cdot)} \subseteq 2^{\{0,1\}^\omega \times \{0,1\}^*}$ is a *relativized class* if and only if there exists a recursive function f (the *enumeration function*) such that $M_{f(j)}^A(x)$ halts for every j, x, A with result 0 or 1 and $\mathcal{K}^{(\cdot)} = \{ L(M_{f(j)}) \mid j \in \mathbb{N} \}$. Define $\mathcal{K}^A =_{\text{def}} \{ L(M_{f(j)}^A) \mid j \in \mathbb{N} \}$ and $\mathcal{K} =_{\text{def}} \mathcal{K}^\emptyset$. For every relativized class $\mathcal{K}^{(\cdot)}$, every set in $\mathcal{K}^A$ is recursive in A. Particularly, every set in $\mathcal{K}$ is recursive. We say that a relativized class $\mathcal{K}$ is *invariant under finite variations of the oracle*, if and only if $\mathcal{K}^A = \mathcal{K}^B$ for every $A, B \in \{0,1\}^\omega$ such that $A \triangle B$ is finite. A relativized class $\mathcal{K}$ with enumeration function f is *uniformly invariant under finite variations of the oracle*, if and only if for every $u \in \{0,1\}^*$ and every $i \in \mathbb{N}$ there exists a $j \in \mathbb{N}$ such that for all oracles A we have $L(M^{u \cdot A}_{f(i)}) = L(M^A_{f(j)})$, where $u \cdot A$ is defined as $u(1)u(2)\cdots u(|u|)A(|u|+1)A(|u|+2)\cdots$. Note that the uniform invariance under finite variations of the oracle implies the (simple) invariance under finite variations of the oracle.

A special type of relativized classes are those defined by *leaf languages* (see [9, 13]). Let $\{N_i\}_{i \in \mathbb{N}}$ be a recursive enumeration of all polynomial time nondeterministic oracle Turing machines such that for every $i \in \mathbb{N}$, every oracle A and every input x, every path of N_i on input x with oracle A is time bounded by $|x|^i + i$ and produces a symbol from some finite alphabet Σ_i. Let $\beta^A_{M_i}(x)$ be the string of the such produced symbols (based on the natural order of paths of the machine). For some $B \subseteq \Sigma^*$, the class $(B)\mathrm{P}^{(\cdot)}$ is the class of all languages L for which there is some $i \in \mathbb{N}$ such that $L = \{ (A, x) \mid \beta^A_{N_i}(x) \in B \}$. In this case, B is the so called *leaf language* defining class $(B)\mathrm{P}^{(\cdot)}$. As above, define $(B)\mathrm{P}^A$ for some oracle A and $(B)\mathrm{P} = (B)\mathrm{P}^\emptyset$. Note that every $(B)\mathrm{P}^{(\cdot)}$ is uniformly invariant under finite variations of the oracle.

Let $\mu: 2^{\{0,1\}^\omega} \to [0,1]$ be the product measure based on $\mu_0: 2^{\{0,1\}} \to [0,1]$ which is defined by $\mu_0(\{0\}) = \mu_0(\{1\}) = \frac{1}{2}$. If $\Pi(\cdot)$ is some predicate with a free set variable, then we write also $\mu A(\Pi(A))$ instead of $\mu(\{ A \mid \Pi(A) \})$.

3   Bounded-error probabilistic operators

In this section we will introduce several bounded-error probabilistic operators. We start with "classical" operators, i.e. operators based on quantifiers which range over (finite) words. Let $h : \{0,1\}^* \to \mathbb{N}$ be any recursive function. For relativized classes $\mathcal{K}^{(\cdot)}$ with enumeration function f we define the operators BP_h, BP^p, and BP^exp as follows:

$L \in \mathrm{BP}_h\mathcal{K}$ iff there exists an $i \in \mathbb{N}$ such that for every x,

$$\#\{ z \mid |z| = h(|x|) \wedge (x \in L \leftrightarrow M^\emptyset_{f(i)}(x,z) = 1) \} \geq \frac{2}{3} \cdot 2^{h(|x|)}.$$

(For functions h which are large compared with the running time of $M^\emptyset_{f(i)}$ we allow random access to the bits of z via a special index tape.) We say that $L \in \mathrm{BP}^\mathrm{p}\mathcal{K}$ iff there exists a polynomial q such that $L \in \mathrm{BP}_q\mathcal{K}$, and that $L \in \mathrm{BP}^\mathrm{exp}\mathcal{K}$ iff there exists a polynomial q such that $L \in \mathrm{BP}_{2^q}\mathcal{K}$.

Type 2 operators are operators ranging over languages (oracles). More specifically, let $\mathcal{K}^{(\cdot)}$ be a relativized class with enumeration function f. (If no confucsion can arise, we will from now on omit the superscript $(\cdot)$.) Then we define type 2 operators ALMOST-, BP^2, $\widehat{\mathrm{BP}}^2$, and $\widetilde{\mathrm{BP}}^2$ as follows:

- $L \in \mathrm{ALMOST}\text{-}\mathcal{K}$ iff $\mu A(L \in \mathcal{K}^A) = 1$.

- $L \in \mathrm{BP}^2\mathcal{K}$ iff there exists an $i \in \mathbb{N}$ such that for all x,

$$\mu A(x \in L \leftrightarrow x \in L(M_{f(i)}^A)) \geq \tfrac{2}{3}.$$

- $L \in \widehat{\mathrm{BP}}^2\mathcal{K}$ iff for every polynomial p there exists an $i \in \mathbb{N}$ such that for all x,

$$\mu A(x \in L \leftrightarrow x \in L(M_{f(i)}^A)) \geq 1 - 2^{-p(|x|)}.$$

- $L \in \widetilde{\mathrm{BP}}^2\mathcal{K}$ iff there exists an $i \in \mathbb{N}$ such that $\mu A(L = L(M_{f(i)}^A)) \geq \tfrac{2}{3}$.

Whereas the operators ALMOST- and $\widetilde{\mathrm{BP}}^2$ seem to be defined inherently by infinite objects (oracles), the operators BP^2 and $\widehat{\mathrm{BP}}^2$ can easily be replaced by classical operators defined by finite objects (words). Let $\mathcal{K}$ be a relativized class with enumeration function f. For every i and x, the tree of all computations of $M_{f(i)}$ on input x for the different oracles is obviously finite. Consequently we have

Proposition 1. *For every relativized class $\mathcal{K}$, there exists a recursive function h such that $\mathrm{BP}^2\mathcal{K} = \mathrm{BP}_h\mathcal{K}$; hence every set in the class $\mathrm{BP}^2\mathcal{K}$ is recursive.*

For special complexity classes, we can make Proposition 1 more precise:

Proposition 2. *1. $\mathrm{BP}^2(B)\mathrm{P} = \mathrm{BP}^{\exp}(B)\mathrm{P}$ for every recursive set B. Particularly, $\mathrm{BP}^2\mathcal{K} = \mathrm{BP}^{\exp}\mathcal{K}$ for $\mathcal{K} = \mathrm{NP}$, co-NP, Σ_k^p, Π_k^p, Δ_k^p, Θ_k^p (for $k \geq 2$), PH, $\oplus\mathrm{P}$, PP, PSPACE, and many others.*
2. $\mathrm{BP}^2\mathrm{L} = \mathrm{BP}^p\mathrm{L}$.

We say that class $\mathcal{K}^{(\cdot)}$ has the *amplification property*, if $\mathrm{BP}^2\mathcal{K} = \widehat{\mathrm{BP}}^2\mathcal{K}$. We say that $\mathcal{K}^{(\cdot)}$ with enumeration function f has the *uniform amplification property*, if for every polynomial p and every $i \in \mathbb{N}$ there exists a $j \in \mathbb{N}$ such that for every oracle B and every input x, if $\mu A(x \in L(M_{f(i)}^{A \oplus B})) \geq \tfrac{2}{3}$, then $\mu A(x \in L(M_{f(j)}^{A \oplus B})) \geq 1 - 2^{-p(|x|)}$; and if $\mu A(x \notin L(M_{f(i)}^{A \oplus B})) \geq \tfrac{2}{3}$, then $\mu A(x \notin L(M_{f(j)}^{A \oplus B})) \geq 1 - 2^{-p(|x|)}$.

It is straightforward to verify that the following classes have the uniform amplification property: NP, co-NP, Σ_k^p, Π_k^p, Δ_k^p, Θ_k^p (for $k \geq 2$), PH, $\oplus\mathrm{P}$, PP, PSPACE, and many more.

4 Relationships between different operators

We start with the following inclusion chain between classes defined by the different type 2 operators:

Theorem 3. *If $\mathcal{K}$ is a relativizable class which is uniformly invariant under finite variations of the oracle, then*

$$\widehat{\mathrm{BP}}^2\mathcal{K} \subseteq \mathrm{ALMOST}\text{-}\mathcal{K} = \widetilde{\mathrm{BP}}^2\mathcal{K} \subseteq \mathrm{BP}^2\mathcal{K}.$$

An immediate consequence is that we obtain the equivalence between all type 2 bounded-error probabilistic quantifiers for classes which have the amplification property.

Corollary 4. *If $\mathcal{K}$ is a relativizable class which is uniformly invariant under finite variations of the oracle and has the amplification property, then*

$$\widehat{\mathrm{BP}}^2 \mathcal{K} = \text{ALMOST-}\mathcal{K} = \widetilde{\mathrm{BP}}^2 \mathcal{K} = \mathrm{BP}^2 \mathcal{K}.$$

The first equality of the just given corollary was proved independently by Merkle and Wang [18].

In the light of Proposition 1 this result says that all our bounded-error probabilistic operators of type 2, which are defined by quantifiers over infinite sets, can also be defined by quantifiers ranging over finite words. This makes these operators easier to understand and to handle. Especially for all complexity classes which are leaf language definable, we get that the ALMOST-operator coincides with the $\mathrm{BP}^{\mathrm{exp}}$ operator:

Corollary 5. *For any B such that $(B)\mathrm{P}$ has the amplification property, we have* $\text{ALMOST-}(B)\mathrm{P} = \mathrm{BP}^{\mathrm{exp}}(B)\mathrm{P}$.

Thus, we get as simple consequences all the known results about ALMOST-classes mentioned in the introduction, that is: (1) ALMOST-P $=$ BPP, (2) ALMOST-NP $=$ BP^{p}NP $=$ AM, (3) ALMOST-PH $=$ BP^{p}PH $=$ PH; but as well characterizations of ALMOST-classes which are of current topical interest, where no coincidence with a "classical" class is known, e.g. (4) ALMOST-PP $=$ BP^{exp}PP, (5) ALMOST-PSPACE $=$ BP^{exp}PSPACE. Observe that for the equalities (2) and (3), we need the coincidence of the BP^{exp} and BP^p quantifier for classes from the polynomial time hierarchy. This result builds on the pseudorandom generator construction from Nisan and Wigderson [19]. A similar newer construction of a pseudorandom generator for space-bounded computations is presented in [20]. One might first suspect that this newer generator leads to a positive settlement of the ALMOST-PSPACE $\stackrel{?}{=}$ PSPACE question, but this is not the case since this generator can only fool a machine with *one-way access* to its random bits. These questions are discussed in the appendix of [21].

The following result shows that for almost all oracles A, the class of recursive sets in $\mathcal{K}^A$ coincides with ALMOST-$\mathcal{K}$. Let REC denote the class of all recursive sets.

Theorem 6. *If $\mathcal{K}$ is a relativizable class which is uniformly invariant under finite variations of the oracle, then*

$$\mu A(\text{ALMOST-}\mathcal{K} = \mathcal{K}^A \cap \mathrm{REC}) = 1.$$

Bennett and Gill [5] showed that the class of all oracles relative to which BPP $=$ P has measure 1. Using the operator BP^2, we can generalize this result for a large variety of relativized classes instead of P.

Theorem 7. *If $\mathcal{K} = (B)\mathrm{P}$ for some recursive set B and if $\mathcal{K}$ has the uniform amplification property, then*

$$\mu A(\mathrm{BP}^2 \mathcal{K}^A = \mathcal{K}^A) = 1.$$

5 Measure 1 inclusions between complexity classes

Inclusions between classes that hold relative to oracles with probability 1 have been an important topic in complexity theory, see e.g. [5, 10, 24] and many more. From Theorem 3, we obtain the following general result:

Theorem 8. *Let $\mathcal{K}_1, \mathcal{K}_2$ be relativizable classes, where $\mathcal{K}_2$ is uniformly invariant under finite variations of the oracle and has the amplification property. Then the following holds:*

$$\mathcal{K}_1 \subseteq \mathrm{BP}^2 \mathcal{K}_2 \quad \Longleftrightarrow \quad \mu A(\mathcal{K}_1 \subseteq \mathcal{K}_2^A) = 1.$$

For classes defined via leaf languages, we find the following "lifting" for measure 1 inclusions:

Corollary 9. *Let $\mathcal{K}_1, \mathcal{K}_2$ be relativizable classes, where $\mathcal{K}_2 = (B)\mathrm{P}$ for some recursive leaf language B, and $\mathcal{K}_2$ has the amplification property. Then the following holds:*

$$\mu A(\mathcal{K}_1^A \subseteq \mathcal{K}_2^A) = \mu A(\mathcal{K}_1^A \subseteq \mathrm{BP}^2 \mathcal{K}_2^A).$$

In particular,

$$\mu A(\mathcal{K}_1^A \subseteq \mathcal{K}_2^A) = 1 \quad \Longleftrightarrow \quad \mu A(\mathcal{K}_1^A \subseteq \mathrm{BP}^2 \mathcal{K}_2^A) = 1$$

$$\mu A(\mathcal{K}_1^A \not\subseteq \mathcal{K}_2^A) = 1 \quad \Longleftrightarrow \quad \mu A(\mathcal{K}_1^A \not\subseteq \mathrm{BP}^2 \mathcal{K}_2^A) = 1$$

From this, we conclude immediatly the following easily applicable criterion to get measure 1 inclusions:

Corollary 10. *Let $\mathcal{K}_1, \mathcal{K}_2$ be relativizable classes, where $\mathcal{K}_2 = (B)\mathrm{P}$ for some recursive leaf language B, and $\mathcal{K}_2$ has the amplification property. Then, if $\mathcal{K}_1 \subseteq \mathrm{BP}^2 \mathcal{K}_2$ is relativizable, then $\mu A(\mathcal{K}_1^A \subseteq \mathcal{K}_2^A) = 1$.*

This gives us the following application which can already be found in [24] (where PH denotes the union of all classes of the polynomial time hierarchy, $\oplus \mathrm{P}$ denotes Papadimitrou and Zachos's "modest counting class" [23], and "$\subset$" denotes *strict* inclusion.):

Corollary 11. $\mu A(\mathrm{PH}^A \subset \oplus \mathrm{P}^A) = 1.$

(Observe that here, "$\subseteq$" follows from our Corollary 9, while strictness is a consequence of a result from [10].)

Corollary 12. $\mu A(\mathrm{co\text{-}NP}^A \not\subseteq \mathrm{AM}^A) = 1$

6 Random oracles

In the preceding sections, we obtained results for a class of oracles with measure 1. In this section, we want to contrast these results with results for *one single random oracle*. We denote by RAND the class of all random oracles in the sense of Martin-Löf, see [8].

The relationship between statements holding for a measure 1 set of oracles vs. those holding for a single random oracle vs. those holding for all random oracles has been examined in several papers [7, 8, 15]. Our Theorem 13 extends these results.

We recall that some $C \subseteq \{0,1\}^\omega$ is *recursively open*, if $C = W \cdot \{0,1\}^\omega$ for some recursively enumerable set $W \subseteq \{0,1\}^*$. A set C is a *recursively G_δ set*, if $C = \bigcap_{i=1}^\infty C_i$ where the $C_1, C_2, \ldots$ are recursively open. The *σ-algebra over a class* $\mathcal{K} \subseteq 2^{\{0,1\}^\omega}$ is the smallest class containing $\mathcal{K}$ closed under complementation and countable intersection.

The following result slightly generalizes a result from Kautz [15]:

Theorem 13. *If C is in the σ-algebra over the class of all recursivly G_δ sets which are closed under finite variation, then*

$$\mu(C) > 0 \iff \mu(C) = 1 \iff \text{RAND} \cap C \neq \emptyset \iff \text{RAND} \subseteq C.$$

It was observed in [8], that if $\mathcal{K}$ is a relativized class which is invariant under finite variations of the oracle, then the sets $\{ A \mid L \in \mathcal{K}^A \}$ (for any recursive set L) and $\{ A \mid L(M_i^A) \in \mathcal{K}^A \}$ (for any $i \in \mathbb{N}$) are recursively G_δ sets which are closed under finite variation. Thus, we obtain immediatly the following improvement of a result from [8, 7], where additional assumptions on $\mathcal{K}_1, \mathcal{K}_2$ were made:

Theorem 14. *Let $\mathcal{K}_1, \mathcal{K}_2$ be relativizable classes which are closed under finite variations of the oracle. Then the following statements are equivalent:*

1. $\mu A(\mathcal{K}_1^A \subseteq \mathcal{K}_2^A) > 0$.
2. $\mu A(\mathcal{K}_1^A \subseteq \mathcal{K}_2^A) = 1$.
3. $\mathcal{K}_1^A \subseteq \mathcal{K}_2^A$ *for some random oracle A.*
4. $\mathcal{K}_1^A \subseteq \mathcal{K}_2^A$ *for all random oracles A.*

Corollary 9 from Section 5 now shows immediately, that for classes $\mathcal{K}_1, \mathcal{K}_2$ fulfilling the assumptions of Corollary 9, $\mathcal{K}_1 \subseteq \text{BP}^2\mathcal{K}_2$ if and only if $\mathcal{K}_1 \subseteq \mathcal{K}_2^A$ for all random oracles A if and only if $\mathcal{K}_1 \subseteq \mathcal{K}_2^A$ for some random oracle A; and Corollary 10 shows that under the same assumptions, if $\mathcal{K}_1 \subseteq \text{BP}^2\mathcal{K}_2$ holds relativizably, then $\mathcal{K}_1^A \subseteq \mathcal{K}_2^A$ for all random oracles A. Thus, we have e.g. that $\text{PH}^A \subseteq \oplus \text{P}^A$ and co-$\text{NP}^A \not\subseteq \text{AM}^A$ for all random oracles A.

7 Type 2 operators vs. polynomially bounded operators

In this section, we want to compare type 2 operators with the familiar operators ranging over polynomially length bounded strings [33, 26]. To this end we define type 2 existential and universal operators. Let $\mathcal{K}$ be a relativized class with k oracles with enumeration function f (a machine $M_{f(i)}$ with the k oracles $A_1, ..., A_k$ uses in fact the one oracle $\bigcup_{i=1}^{k} \{1^i 0x | x \in A_i\}$).

We say that $L \in \exists^2 \mathcal{K}$ iff there exists an $i \in \mathbb{N}$ such that for all x, we have

$$(x, A_1, \cdots, A_{k-1}) \in L \iff \exists A_k (M_{f(i)}^{A_1, \cdots, A_{k-1}, A_k}(x) = 1)$$

and $L \in \forall^2 \mathcal{K}$ iff there exists an $i \in \mathbb{N}$ such that for all x,

$$(x, A_1, \cdots, A_{k-1}) \in L \iff \forall A_k (M_{f(i)}^{A_1, \cdots, A_{k-1}, A_k}(x) = 1).$$

Clearly, if $\mathcal{K}$ is a relativized class then so are $\exists^2 \mathcal{K}$ and $\forall^2 \mathcal{K}$.

Define $\Sigma_k^{\exp}$ to be the set of all languages A accepted by Σ_k machines (i.e. alternating Turing machines with $k - 1$ alternations, starting in an existential state), which on inputs of length n run in time bounded by $2^{p(n)}$ for some polynomial p. Let $\Pi_k^{\exp}$ be the class of complements of sets from $\Sigma_k^{\exp}$. Now the classes of the $\exists^2$-$\forall^2$-hierarchy (restricted to "ordinary" languages of words) can be characterized as follows [27, 22, 31]:

1. $\exists^2 \forall^2 \exists^2 \cdots Q_k^2 \overline{Q}_k^2 \cdot P = \exists^2 \forall^2 \exists^2 \cdots Q_k^2 \overline{Q}_k^p \cdot P = \exists^2 \forall^2 \exists^2 \cdots Q_k^2 \cdot \text{PSPACE} = \Sigma_k^{\exp}$
2. $\forall^2 \exists^2 \forall^2 \cdots \overline{Q}_k^2 Q_k^2 \cdot P = \forall^2 \exists^2 \forall^2 \cdots \overline{Q}_k^2 Q_k^p \cdot P = \forall^2 \exists^2 \forall^2 \cdots \overline{Q}_k^2 \cdot \text{PSPACE} = \Pi_k^{\exp}$

where $Q_k = \exists$ and $\overline{Q}_k = \forall$ if k is odd, and $Q_k = \forall$ and $\overline{Q}_k = \exists$ if k is even. The operators $\exists^p$ and $\forall^p$ are the classical polynomially length-bounded existential and universal operators [33].

Let L be the class of logspace-decidable sets. Let NC^k be the class of sets decidable by uniform circuit families of polynomial size and $O(\log^k n)$ depth [11]. (Without going into details, we remark that we adopt the uniformity condition from [4].) Let $BPNC^k$ denote the bounded error probabilistic analogue of NC^k (see [11]), i.e., $BPNC^k$ circuits have regular input gates plus gates for probabilistic bits. The probability is then taken over all possible inputs to the latter gates, where we assume (as usual) uniform distribution. We remark that $BPNC^k = BP^p NC^k$.

To compare type 2 operators with the "usual" operators, we use *translational methods,* which have a long history in complexity theory, see e.g. [6]. In all these arguments, *padding* plays a crucial role—in the just mentioned paper, tally versions of languages were used. We here introduce the following form of padding: For a language A and some integer m, define $A_m =_{\text{def}} \{ x10^{2^{|x|^m} - |x| - 1} \mid x \in A \}$.

Then, the following lemma is easy to see:

Lemma 15. *1. $A \in \Sigma_k^{\exp}$ iff there exists some $m \in \mathbb{N}$ such that $A_m \in \Sigma_k^p$.*
  2. $A \in \text{PSPACE}$ iff there exists some $m \in \mathbb{N}$ such that $A_m \in \text{L}$, iff there exists some $m \in \mathbb{N}$ such that $A_m \in NC^1$,
  3. $A \in BP^2\text{PSPACE}$ iff there exists some $m \in \mathbb{N}$ such that $A_m \in BP^p\text{L}$, iff there exists some $m \in \mathbb{N}$ such that $A_m \in BPNC^1$.
  4. $A \in \text{BPTIME}(2^{\text{Pol}})$ iff there exists some $m \in \mathbb{N}$ such that $A_m \in \text{BPP}$.

Now applying the translational results from Lemma 15 to Sipser's and Lautemann's result that BPP is included in the polynomial time hierarchy [28, 16], the following result can be obtained:

Theorem 16. $BP^2PSPACE \subseteq BPTIME(2^{Pol}) \subseteq \Sigma_2^{exp} \cap \Pi_2^{exp}$.

The up to now best known upper bound for the class ALMOST-PSPACE is ALMOST-PSPACE $\subseteq$ EXPSPACE [17]. We obtain the following improvement:

Corollary 17. ALMOST-PSPACE $\subseteq \Sigma_2^{exp} \cap \Pi_2^{exp}$.

In the theory of efficient algorithms, if no good parallel algorithm for a given problem is within reach, one tries to design quick probabilistic parallel algorithms, i.e., to prove that the problem under consideration is in $BPNC^k$ for some k. Therefore, it is of great importance to have tight upper bounds for those classes. Unfortunately, essentially only $BPNC^k \subseteq BPP$ is known. It turns out that this problem is related to that of giving upper bounds for ALMOST-PSPACE: Any upper bound for $BPNC^1$ better than BPP will give us an upper bound for ALMOST-PSPACE better than the one given in Corollary 17; for example:

Corollary 18. If $BPNC^1 \subseteq P$, then ALMOST-PSPACE $\subseteq$ EXPTIME.

8 A characterization of the class $BP^2DSPACE(s)$

In the Section 4, we saw that ALMOST-PSPACE = $BP^2PSPACE$, and we gave new upper time bounds for that class in Section 7. However, the question of whether $BP^2PSPACE$ = PSPACE remains unresolved. In this section, we give a machine characterization of $BP^2PSPACE$ which makes this equality seem unlikely to us.

A *checking stack* [14] is a stack which can be used only in two phases, first the *writing phase* where the head of the checking stack can only write new symbols on top of the stack, and second the *checking phase*, where the head of the checking stack can only read the contents of the stack (in a two-way manner).

A CS-DTM (CS-NTM, CS-PTM, CS-BPTM) is a deterministic (nondeterministic, probabilistic, bounded-error probabilistic) Turing machine with a two-way input tape, a constant number of working tapes, and a checking stack. For $s: \mathbb{N} \to \mathbb{N}$, we define $L \in CS\text{-}\mathcal{X}SPACE(s)$ if there exists a CS-$\mathcal{X}$TM (for $\mathcal{X}$ either D, N, P, or BP) such that every computation path of M on input x halts and is space-bounded by $s(|x|)$, where the workspace used in the checking stack is not taken into account. For these definitions as well as general background and results, see [32, pp. 252ff].

Now we see that ALMOST-PSPACE is exactly the class of all languages accepted by probabilistic checking stack automata working in polynomial space; more generally:

Theorem 19. *For every fully space-constructible function* $s: \mathbb{N} \to \mathbb{N}$ *such that* $s(n) \geq \log n$ *for all* n,

$$BP^2DSPACE(s) = CS\text{-}BPSPACE(s).$$

Corollary 20. *1.* $BP^{exp}PSPACE = CS\text{-}BPSPACE(Pol)$.

2. $BP^{P}L = CS\text{-}BPSPACE(log)$.

So far it is not known whether $CS\text{-}BPSPACE(s)$ coincides with one of the well-studied complexity classes. However, in this context the following results should be mentioned which can be found in [14]: For arbitrary $s \geq log$, the equations $CS\text{-}DSPACE(s) = DSPACE(s)$ and $CS\text{-}NSPACE(s) = NSPACE(2^{O(s)})$ hold. That is: Checking stack automata working nondeterministically are more powerful than those working deterministically. Our Theorem 19 now gives the following extension: For polynomial space plus checking stack, the nondeterministic computation mode is strictly more powerful than the bounded-error probabilistic mode (unless $\Sigma_2^{exp} = EXPSPACE$).

Acknowledgments. Some of the ideas presented here evolved during a seminar on Randomness and Computation, held in the Spring 95 Quarter at the University of California at Santa Barbara. Thanks are due to Zhe Dang, Todd Ebert, and Sarah Hough for helpful discussions. We thank Noam Nisan for pointing out some of the subtleties of [19, 20, 21]. We also acknowledge helpful comments from Eric Allender, Jack Lutz, Elvira Mayordomo, and Ken Regan.

References

1. K. AMBOS-SPIES, Randomness, relativizations, and polynomial reducibilities, *Proceedings of the 1st Structure in Complexity Theory Conference* (1986), Springer Lecture Notes in Computer Science Vol. 223, pp. 200–207.

2. L. BABAI, Trading group theory for randomness; *Proceedings of the 17th Symposium on Foundations of Computer Science* (1975), pp. 421–429.

3. J. L. BALCÁZAR, J. DÍAZ, J. GABARRÓ, *Structural Complexity I* (Springer Verlag, Berlin – Heidelberg – New York, ^{2}1995).

4. D. BARRINGTON, N. IMMERMAN, H. STRAUBING, On uniformity within NC^1; *Journal of Computer and System Sciences* **41** (1990), pp. 274–306.

5. C. BENNETT, J. GILL, Relative to a random oracle $P^A \neq NP^A \neq co\text{-}NP^A$ with probability 1; *SIAM J. Comput.* **10** (1981), pp. 96–113.

6. R. V. BOOK, Tally languages and complexity classes; *Information and Control* **26** (1974), pp. 186–193.

7. R. V. BOOK, On languages reducible to algorithmically random languages; *SIAM J. Comput.* **23** (1994), pp. 1275–1282.

8. R. V. BOOK, J. H. LUTZ, K. W. WAGNER, An observation on probability versus randomness with applications to complexity classes; *Mathematical Systems Theory* **27** (1994), pp. 201–209.

9. D. P. BOVET, P. CRESCENZI, R. SILVESTRI, A uniform approach to define complexity classes; *Theoretical Computer Science* **104** (1992), pp. 263–283.

10. J. Y. CAI, With probability one, a random oracle separates PSPACE from the polynomial-time hierarchy; *Journal of Computer and System Sciences* **38** (1989), pp. 68–85.

11. S. A. COOK, A taxonomy of problems with fast parallel algorithms; *Information and Control* **64** (1985), pp. 2–22.

12. J. GILL, Computational complexity of probabilistic complexity classes; *SIAM Journal on Computing* **6** (1977), pp. 675–695.

13. U. HERTRAMPF, C. LAUTEMANN, T. SCHWENTICK, H. VOLLMER, K.W. WAGNER, On the power of polynomial time bit-reductions; *Proceedings of the 8th Structure in Complexity Theory Conference* (1993), pp. 200–207.

14. O. H. IBARRA, Characterizations of some tape and time complexity classes of Turing machines in terms of multihead and auxiliary stack automata; *Journal of Computer and System Sciences* **5** (1971), pp. 88–117.

15. S. KAUTZ, Degrees of random sets; Ph. D. dissertation, Cornell University, 1991.

16. C. LAUTEMANN, BPP and the polynomial hierarchy; *Information Processing Letters* **117** (1983), pp. 215–217.

17. J. LUTZ, personal communication, 1995.

18. W. MERKLE, Y. WANG, Separations by random oracles and "Almost" classes for generalized reducibilities; *Proceedings of the 20th International Symposium on Mathematical Foundations of Computer Science* (1995), Springer Lecture Notes in Computer Science Vol. 969, pp. 179–190.

19. N. NISAN, A. WIGDERSON, Hardness vs. Randomness; *Journal of Computer and System Sciences* **49** (1994), pp. 149–167.

20. N. NISAN, Pseudorandom generators for space-bounded computation; *Journal of Combinatorica* **12** (1992), pp. 449–461.

21. N. NISAN, On read-once vs. multiple access to randomness in logspace; *Theoretical Computer Sience* **107** (1993), pp. 135–144.

22. P. ORPONEN, Complexity classes of alternating machines with oracles; *Proceedings of the 10th International Colloquium on Automata, Languages and Programming* (1983), Springer Lecture Notes in Computer Science Vol. 154, pp. 573–584.

23. C. H. PAPADIMITRIOU, S. K. ZACHOS, Two remarks on the power of counting; *Proceedings of the 6th GI-Conference on Theoretical Computer Science* (1983), Springer Lecture Notes in Computer Science Vol. 145, pp. 269–275.

24. K. W. REGAN, J. S. ROYER, On closure properties of bounded two-sided error complexity classes; *Mathematical Systems Theory* **28** (1995), pp. 229–243.

25. H. ROGERS, *Theory of Recursive Functions and Effective Computability* (McGraw-Hill, New York, NY, 1967).

26. U. SCHÖNING, Probabilistic complexity classes and lowness; *Journal of Computer and System Sciences* **39** (1989), pp. 84–100.

27. J. SIMON, On Some Central Problems in Computational Complexity; Dissertation, Cornell University (1975).

28. M. SIPSER, A complexity theoretic approach to randomness; *Proceedings of the 15th Symposium on Theory of Computing* (1983), pp. 330–335.

29. S. TODA, PP is as hard as the polynomial time hierarchy; *SIAM Journal on Computing* **20** (1991) 865–877.

30. K. W. WAGNER, Some observations on the connection between counting and recursion; *Theoretical Computer Science* **47** (1986), pp. 131–147.

31. K. W. WAGNER, High-order operators in complexity theory; manuscript.

32. K. W. WAGNER, G. WECHSUNG, *Computational Complexity* (Deutscher Verlag der Wissenschaften, Berlin, 1986).

33. C. WRATHALL, Complete sets and the polynomial-time hierarchy; *Theoretical Computer Science* **3** (1977), pp. 23–33.

Speeding-Up Single-Tape Nondeterministic Computations by Single Alternation, with Separation Results

Jiří Wiedermann*

Institute of Computer Science
Academy of Sciences of the Czech Republic
Pod vodárenskou věží 2, 182 07 Prague
Czech Republic
e-mail wieder@uivt.cas.cz

Abstract. It is shown that for any well behaved function $T(n)$, any single-tape nondeterministic Turing Machine of time complexity $T(n)$ can be simulated by a single-tape Σ_2-machine in time $T(n)/\log T(n)$. A similar result holds also for complementary, single-tape co-nondeterministic and Π_2 machines. Consequently, $\mathsf{NTIME}_1(T(n))$ is strictly contained in $\Sigma_2 - \mathsf{TIME}_1(T(n))$, and analogously $co - \mathsf{TIME}_1(T(n))$ in $\Pi_2 - \mathsf{TIME}_1(T(n))$, i.e., for single tape nondeterministic or co-nondeterministic machines adding of one more alternation leads to provably more powerful machines.

1 Introduction

Single-tape Turing machines (TMs) are generally considered as the simplest, albeit from computational point of view somewhat cumbersome, universal computing model. It is known that single-tape complexity classes differ in some details, especially in cases when polynomial factors cannot be neglected, from their counterparts for multitape TMs. However, it appears that it is just the relative simplicity of single-tape TMs that enables to obtain for them results that are so far, or due to the idiosyncracy of the machines involved, unattainable for multitape TMs. Irrespectively of the reasons of this difference, the complexity results for single-tape machines are of immense value for theoretical studies in complexity theory since they can serve at least as inspiration, or as a guide for our intuition, for similar studies in the field of more realistic models of computation.

One of the most attractive results in this respect are speed-up results that quantitatively — in terms of time complexity — measure the contribution of addition of some new computational resource to a certain kind of machines. In theory, the primary importance of speed-up results of previous type lies in the

---
* This research was supported by GA ČR Grant No. 201/95/0976 " HypercompleX", and partly by Cooperative Action IC 1000 (Project "ALTEC") of the European Union

fact that usually they enable to prove separation results w.r.t. complexity classes pertinent to machines involved in the simulation at hand.

In this context, among the central open problems that attract a lot of theoretical attention even in case of single–tape computations, are problems dealing with the "non–trivial" relationship among complexity classes related to fundamental computational resources — viz. determinism, nondeterminism, and co-nondeterminism, respectively. Unfortunately, so far the situation here has been much similar to that in the realm of multitape TMs: mostly only partial answers to only a few of the related questions have been known.

To illustrate the recent situation in the respective field, for single–tape TMs, the following results are to be reported.

In [4] or in [5] it is shown that for single–tape machine nondeterministic space is a substantially more powerful computational resource that nondeterministic time, since any nondeterministic single–tape TM of time complexity $O(T(n))$ can be simulated in linear time by another nondeterministic single–tape machine in space $O(\sqrt{T(n)})$. In [5], moreover, a tight time hierarchy for nondeterministic single tape TMs has been shown.

For the deterministic case, even speed–ups have been achieved: in [9], a deterministic single–tape $T(n)$–time bounded machine has been speeded-up by an alternating machine (i.e., using an unbounded number of alternations) down to $O(\sqrt{T(n)})$ time. Furthermore, in [3] it has been shown that the same machine, equipped with a read–only input tape, can be simulated by a multitape Σ_4–machine, i.e., by a bounded number of alternations, in time $O(T(n)^{2/3} \log T(n))$. The best result so–far, for single–tape machines, seems to be the result from [6], where the same result as before has been accomplished by using a substantially less powerful simulating machine than in [3] — namely a single–tape Σ_2–machine.

Nevertheless, even though the last mentioned result has led to a separation results between *single–tape deterministic* linear time, and *multitape nondeterministic time*, the separation of some of the related time–bounded complexity classes, for arbitrary time–bounds and for the same type of machines, has not followed.

Along these lines, the known separation result for multitape Turing machines is that of Paul et al. [8], where the separation of deterministic linear time from the nondeterministic linear time is shown. This has been achieved via a fast simulation of a multitape deterministic Turing machine by a Σ_4 machine; the speedup is very small indeed — viz. by the factor that equals to the iterated logarithm of the original running time.

Using the technique of [8] the previous result was improved by Gupta [1] who has shown that only two alternations are sufficient to achieve a $\log^* T(n)$ speedup of deterministic multitape machines. This has finally lead to a separation of $DTIME(T(n))$ from $\Sigma_2 - TIME(T(n))$.

Reviewing the previous results it is important to note that all of them have been obtained for *deterministic* machines only, and the closer look at the respective proofs reveals that they cannot be simply carried over to nondeterministic machines.

For single tape machines, it is because the idea of the respective speed–up simulations is, roughly speaking, to split nondeterministically the original deterministic computation into certain small segments whose correctness can be verified fast in parallel, using co–nondeterminism (one more alternation). It is clear that this idea works only when the machine to be simulated was of a deterministic type, since in this case the parallel verification of deterministic segments is possible. However, to simulate in this way a *nondeterministic* machine the parallel verification of non–deterministic segments would require one more alternation.

Thus, the speed–ups achieved so far have been achieved for quite a high price, namely too many powerful computational resources had to be added to get the speed–up effect. E.g., in order to achieve speed-ups in the above mentioned case of deterministic single–tape machines from [6] two alternations (i.e., first nondeterminism and then co-nondeterminism) were necessary to apply.

This paper will show a much stronger result — namely that the addition of one extra alternation is enough to speedup a single–tape nondeterministic or co-nondeterministic computation. More specifically, we will show that under certain reasonable technical conditions concerning the function $T(n)$, any $T(n)$–time bounded single–tape nondeterministic TM can be simulated by a single–tape Σ_2–machine in time $O(T(n)/\log T(n))$. The same speedup can be achieved also for the complementary machines (i.e., a single–tape Π_1 (co–nondeterministic) machine can be speeded up by a single–tape Π_2 machine). As a consequence, the separation of the related complexity classes will follow[2]: $\text{NTIME}_1(T(n)) \subset \Sigma_2 - \text{TIME}_1(T(n))$, and analogously $co - \text{TIME}_1(T(n)) \subset \Pi_2 - \text{TIME}_1(T(n))$.

These results show that the first two levels of Σ_1 and Π_1 hierarchy for (single–tape) machines with identical time bounds do not collapse. Similar results do not follow from the above mentioned results by Paul et al. [8] or Gupta [1] and in fact are not known to hold for multitape machines.

The previous results seem to be the first case where a nondeterministic computation of a universal machine has been speeded–up by a single alternation. Similarly, the corresponding separation results derived with the help of this speed-up are of a much stronger type than previous results. This is because, on one hand, they relate to machines of the same type (single–tape machines) that differ merely in their ability to use additional computational resource, and on the other hand, because the results hold for a reasonably large class of time–bounded computations.

2 Speedup of Nondeterministic Computations by Additional Alternation

The following theorem states that under certain technical assumptions, any computation of a nondeterministic, single–tape Turing machine can be speeded up

---

[2] The inclusion symbol '$\subset$' denotes the proper containment.

by a single–tape TM performing two alternations. The proof of this theorem is based on a space efficient so–called *rectangular representation* of single–tape computations that goes back to Paterson [7] (see also [2] or [4] for a similar approach). Our proof, however, differs substantially from proofs based on the similar representation in at least two important aspects:

- first, the computation of the original machine that is to be speeded up, is first transformed into an equivalent computation that enables its efficient simulation later on;
- second, by making use of a different distribution of related computational tasks into nondeterministic, and co–nondeterministic, phase of the simulating machine, the speedup of a *nondeterministic machine*, by adding a single universal alternation, is enabled.

Theorem 2.1 *Let $\sqrt{T(n)}$ be fully time and space constructible, with $T(n) \geq n^2$. Than any $T(n)$–time–bounded nondeterministic single–tape TM M can be simulated by a single-tape Σ_2–TM in time $O(T(n)/\log T(n))$.*

Proof Outline. Following [4], for $T(n)$ as above, w.l.o.g. we can assume that M is of space complexity $O(\sqrt{T(n)})$. Moreover, from [5] it follows that w.l.o.g. we can also assume that M is a *strongly accepting TM* — i.e. such that it always halts after performing $T(n)$ moves, and either accepts or rejects its input.

Next, for specific reasons that will become clearer in a sequel, we shall split the computation of M into $\Theta(T^{1/3}(n))$ *time segments*, each of length $\Theta(T^{2/3}(n))$. At the end of each time interval, we let the machine perform a complete sweep over its tape. A sweep starts at the (recorded) position of the machine head at the end of the time interval. It then proceeds by moving the head to the right end of the tape, then in the opposite direction to the left end, and finally returns to the original (recorded) head position. During a sweep the contents of the tape are not altered. The execution of a sweep will be included into the respective time segment. It is clear that the time complexity of performing a single sweep is proportional to the tape length, and thus the complexity of the resulting machine remains bounded by $O(T(n))$.

The reason for introducing sweeps over the entire tape at the end of each time segment lies in the fact, that it is possible to verify the "history" of cell rewritings during a computation by returning back in the time to the distance of only $\Theta(T^{2/3}(n))$. (This is the first idea where our proof deviates from the schema of similar proofs — say from [6].)

Now, for the machine M prepared as above we construct the rectangular representation of its computation.

To obtain this representation for any fixed computation, it is helpful to see the computation at hand as a sort of a diagram. The diagram consists of individual instantaneous descriptions (ID's) of M, for subsequent time steps $t = 0, 1, \ldots,$ written one above other. Besides the current contents of M's tape in each ID, the current head position is recorded by writing down the current state of

M's finite control over the position scanned by M's head. Also in this way the trajectory of M's head, on a given input, is recorded in our diagram.

Starting from some position j, with $0 \leq j < b(n)$, the tape of M can be partitioned into blocks of length $b(n)$ (the value of $b(n)$ to be determined later). By proceeding in this manner, the sum of length of crossing sequences over the block boundaries will be at most $T(n)/b(n)$. A position of j must exist, since in the opposite case, the respective sum would be greater than $T(n)/b(n)$ for each j. The total length of the crossing sequence, taken over all tape cells, would then be greater than $T(n)$.

The block boundaries will split the computational diagram at most into $\sqrt{T(n)}/b(n)$ vertical slots. These slots will be changed into so-called *first order rectangles* by drawing horizontal lines in between the ID's that separate the individual time segments (i.e., the line is drawn immediately after the end of each sweep as explained in the beginning of the proof).

Clearly, in this way we obtain at most $O(T^{5/6}(n)/b(n))$ first order rectangles.

First order rectangles will be split further by drawing a horizontal line at suitable points, into the *second order rectangles* whose size is maximized, subject to the satisfaction of either of the following two conditions:

- none of the two respective vertical sides are crossed by the TM head more often than $b(n)$ times;
- the total time spent by the TM head in a given second order rectangle must not exceed $b^2(n)$.

Clearly, in every vertical slot, second order rectangles can be created in each first order rectangle. There is a possible exception of "too short" slots, or also in remainders of first order rectangles that are "artificially" cut by the line separating time segments. We shall call the respective second order rectangles, that could not be created in the full size, as required by the previous two conditions, *small rectangles*.

As a result we obtain at most $O(T(n)/b^2(n))$ second order full size rectangles, since the computation within each rectangle "consumes" either $b(n)$ crossing sequence elements, or time $b^2(n)$.

We also see at most $O(T^{5/6}(n)/b(n))$ small ones. Thus the total number of second order rectangles is safely bounded by $O(T(n)/b^2(n))$.

Each second order rectangle will thus be represented by its two horizontal sides of length $b(n)$, and by the two vertical sides of length ℓ_1 and ℓ_2, respectively, with $\ell_1, \ell_2 \leq b(n)$. The horizontal sides are given by the contents of the corresponding block at the respective time steps, while the vertical sides are given by the respective crossing sequences in chronological order, for each side separately.

Hence, the size of each second order rectangle representation is at most $4b(n)$. This in a total gives $O(T(n)/b(n))$ for all rectangles.

Now, the idea of simulation is first to guess the above rectangular representation, and then to verify whether the guess was correct — i.e., whether all rectangles 'fit' together (so-called *global correctness*) and, whether each rectangle represents a valid piece of M's computation — i.e., such that starts in "partial" configuration as described by the upper horizontal side of the rectangle at hand, ends in a configuration as described by the lower side of the rectangle, and where the M's head leaves and re–enters the rectangle in accordance with the crossing sequence that corresponds to the vertical rectangle boundaries (so–called *local correctness*).

This seems to lead straightforwardly to the design of the simulation scheme in which M is simulated by a single–tape Σ_2-machine in two main phases: in the the first, nondeterministic phase, all guesses will be performed, whereas in the second, universal phase, the verification of all previous guesses will be done.

Nevertheless the implementation of the above idea is complicated by the fact that verification of local correctness requires nondeterminism. Therefore it must be performed during the first phase, i.e., sequentially. Doing this straightforwardly for each rectangle will in turn require a time of order $\Omega(T(n))$. This would prevent any speed up. The way out of this problem in our rectangular representation is to define the size of each rectangle so that there will be many equal second order rectangles. In this case it will be enough to identify, and verify only the different rectangles.

This is the second main point where our simulation departs significantly from the "standard" schema as used e.g. in [2], or in [6]. As will be shown, this departure will require also related complex preparatory actions in the first phase of simulation, and involved verification actions in the second phase of simulation.

We have already noticed that the size of a rectangle representation is at most $4b(n)$. Thus, for a given TM M there are at most $R(n) = c^{4b(n)}$ different rectangles, for some constant $c > 0$ that depends on the size of M's alphabet and on the number of its states. In order to have many equal rectangles in our rectangular representation of M's computations, $R(n)$ must be asymptotically less than their total number — i.e., less that $O(T(n)/b^2(n))$. Choose $b(n) = 1/24\log_c T(n)$. Then $R(n) = T^{1/6}(n)$, which, as we shall see later on, is quite appropriate for our purposes.

Let us describe now the simulation itself. Let S be the simulating single–tape Σ_2–TM.

The nondeterministic phase of its computation consists of the following four subphases.

Subphase 1.1 — generating a rectangular representation. For a given computation of M the rectangular representation, with $b(n)$ as above, is guessed and written down on the S's tape in the following order: time segment by time segment, and within each time segment, first order rectangle by first order rectangle, from left to right, and within each first order rectangle, second order rectangle by second order rectangle, in chronological order. Boundaries between individual (first and second order) rectangles, and time segments, respectively, are marked by special symbols on a special track.

The length of the above data, pertinent to one time segment, is at least $\Omega(\sqrt{T(n)})$ (since there must be at least $\Omega(\sqrt{T(n)}/b(n))$ second order rectangles, each of size $O(b(n))$), and at most $O(T^{2/3}(n))$ (since within the time segment of duration $T^{2/3}(n)$, at most $O(T^{2/3}(n)/b(n))$ different rectangles can be visited by M's head).

However, it is clear that the entire rectangular representation can be generated in linear time w.r.t. its length — i.e., in time $O(T(n)/\log T(n))$.

Subphase 1.2 — preparing for duplicate rectangle identification. On a special track, guess and write down all the *different rectangles* from M's rectangular representation, in the sorted order. Call the resulting string a *set-of-rectangles*. Then "copy", onto the same special track, nondeterministically, the previous string of length at most $4R(n)b(n)$, above the beginning of each time segment, $T^{1/3}(n)$ times in a row. Due to the nondeterminism involved, the entire "nondeterministic copy" operation, for all segments, can be performed in a single traversal over S's tape. Hence, the time complexity of this operation is $O(T(n)/b(n))$.

The *set-of-rectangles* strings will be subsequently, in the universal phase of simulation (see subphase 2.3), used for testing whether all the rectangles from the original rectangular representation are duplicates of the rectangles that are included in the above set.

Subphase 1.3 — local verification. Verify, whether all the rectangles from *set-of-rectangles* present a valid "piece" of M's computation. This is done by performing a nondeterministic computation as dictated by rectangle sides, for each rectangle from the first occurrence of *set-of-rectangles*. Since there are $R(n)$ rectangles, and the verification of any of them is bounded by $b^2(n)$, the verification of all rectangles requires time $O(R(n)b^2(n))$.

Subphase 1.4 — acceptance verification. Check, whether among all rectangles in the rectangular representation (i.e., not in the *set-of-rectangles* (!)) there is a rectangle in which the acceptance of M occurs. For this purpose guess the right rectangle by scanning the appropriate track of S's tape, and verify, by replaying corresponding nondeterministic computation of M, that the accepting state of M is achieved. This takes time proportional to the length of S's tape — i.e., $O(T(n)/b(n))$.

Then the universal phase follows. It consists of three subphases that are run in parallel.

Subphase 2.1 — verifying the global correctness. This phase consists in fact of two independent verification processes that can be run also in parallel:

- *verification of horizontal boundaries in rectangular representation:* For the rectangles that are the first ones in each slot we have to verify that their upper horizontal boundaries correspond to the initial content of M's tape. For the remaining rectangles we have to verify that the lower horizontal boundary of any rectangle, except the last one in the given slot, is the same as the upper horizontal boundary of a rectangle that follows the previous rectangle in a given slot.

  The correctness of rectangles from the "first row", so to speak, is easy to be verified. The verification is possible by shifting the rectangles that carry parts of the input, towards the original input, that is being kept on the special track at the beginning of the S's tape. The remaining rectangles that do not carry the input, should contain blanks in the corresponding parts. All this can be done in time $O(T(n)^{2/3}b(n))$ for each rectangle in parallel.

As far as the remaining rectangles are concerned note that due to the chosen representation of rectangles on M's tape (see subphase 1.1), the distance between the lower and the upper horizontal side of two neighbouring rectangles that find themselves within the same slot and within the same time interval is at most $O(b(n))$. The equivalence of the respective sides can be easily verified in time $O(b^2(n))$ by invoking a special parallel process for each rectangle.

When there are two neighbouring rectangles within the same slot, but in different time segments, the distance of the corresponding horizontal boundaries, that have to be compared, is at most $O(T^{2/3}(n))$. Thus, again, the necessary comparisons for all time segments can be done in parallel, in time $O(T^{2/3}(n)b(n))$.

The last verification is not completely trivial, since except the horizontal boundary of a rectangle at hand, also a counter of size $O(\log T(n)) = O(b(n))$ must be carried along each time segment. This counter counts rectangles and enables thus to identify the corresponding rectangles that are the neighbours within the same slot.

- *verification of vertical boundaries in rectangular representation:* For the kind of verification at hand it is important to realize that due to the sweeps involved at the end of each time segment, only the crossing sequences between horizontally neighbouring rectangles *within the same time segment* must be compared. Thus, it is enough for each crossing sequence element from the right side of a rectangle to find its "companion" in the left side of a horizontally neighbouring rectangle. The companion will be located at the distance of at most $O(T^{2/3}(n))$. This can be done in parallel, extra for each element, and extra for each time segment: we must only keep track of element's relative position on the vertical boundary between the respective slots within the given time segment. This amounts to shifting a counter of size $O(\log T(n))$ along the tape, to the distance of at most $O(T^{2/3}(n))$.

Subphase 2.2 — verifying the correctness of "nondeterministic copy" operation from subphase 1.2: universally split into $O(T^{1/3}(n))$ copies of S, (exactly: as many copies as there are time segments, minus one) and for each pair of neighbouring time segments verify, whether the string *set-of-rectangles* is the same in both time segments. This can be done by moving the respective string along the tape from the beginning of the first time segment at hand above the beginning of the second time segment. Since the length of two time segments on S's tape is at most $O(T^{2/3}(n))$, moving the string *set-of-rectangles* of length $R(n)$ along this distance requires time $O(T^{2/3}(n).R(n).b(n)) = O(T^{5/6}(n)b(n))$.

Subphase 2.3. — verifying duplicates of rectangles: we have to verify, whether every rectangle in a given time segment finds itself also among the rectangles within *set-of-rectangles*, that is located at the beginning of each time segment on a special track (see subphase 1.2). This can be done easily in parallel, extra for each time segment, by subsequently moving along the tape each rectangle from the rectangular representation towards the beginning of the respective segment, and by verifying, whether the segment at hand finds itself also in *set-of-rectangles*. Similarly as before the cost of this operation is of order $O(T^{5/6}(n)b(n))$.

Note that there can be some "additional" rectangles in the *set-of-rectangles* that have passed the test in subphase 1.3, nevertheless they do not find themselves within the rectangular representation of M's computations, but this is clearly harmless.

This is the end of the simulation: if all the previous verifications in Phase 1 and 2, respectively, end successfully, S accepts its input.

From the previous complexity estimates of individual parts of S's computation it is seen that the most time consuming computation has been performed in phase 1, and hence the time complexity of the whole simulation is $O(T(n)/b(n))$. □

Note that the selection of a few "parameters" in the proof of the previous theorem (like the size of a rectangle, or the length of a time segment) was somewhat arbitrary. Although there are also other possibilities, within a certain range, as it is seen from the proof and its analysis, none of them would lead to asymptotically better performance of the simulation. Our proof "works" only in the case when there are less than $O(T(n)/\log^2 T(n))$ different rectangles in the rectangular representation at hand, what leads straightforwardly to the logarithmic size of block representation. Thus our simulation cannot be improved merely by tuning some of its parameters.

Therefore, the general question, whether any asymptotically faster simulation of the above type exists remains open.

We shall end this section by restating the statement of the previous theorem in terms of relation between the respective complexity classes. To this purpose, let $\mathsf{NTIME}_1(T(n))$, and $\Sigma_2 - \mathsf{TIME}(T(n))$, respectively, denote nondeterministic and Σ_2, respectively, $T(n)$-time bounded single–tape TM complexity classes. Then the following corollary of the theorem 2.1 holds true:

Corollary 2.1 *Let $\sqrt{T(n)}$ be fully time and space constructible, with $T(n) \geq n^2$. Then*

$$\mathsf{NTIME}_1(T(n)) \subseteq \Sigma_2 - \mathsf{TIME}_1(T(n)/\log T(n))$$

Proof Outline. The above inclusion follows directly from theorem 2.1; we only have to show that we can get rid of any constant hidden in the "big O" notation in the complexity estimate of simulating machine from the theorem 2.1. A short reflection reveals that this is indeed possible, since we can speed–up our simulating machine by an arbitrary constant factor, by using the standard technique of compressing groups of its tape symbols into symbols from a larger alphabet. There are slight problems with the compression of input. The first idea to compress it in a sequential manner prior to start of the simulation would require time of order $\Omega(n^2)$, what is unacceptable since some simulating machines can run in time $O(n^2/\log n)$, say. Therefore, a more complicated process must be chosen: first, nondeterministically guess and write down, the compressed input. Later on, verify the correctness of the initial guess in the parallel computing phase. Also note that in order to achieve the above constant speed up the branching factor of nondeterministic or universal instructions of the original Σ_2 might be increased.

$\square$

3 Speed up of Co–Nondeterministic Computations by Additional Alternation

It appears that the previous results can be "reworked" to be also used for the case of co–nondeterministic computations — i.e., for computations performed by single–tape time bounded Π_1 machine. In this case, the simulating machine would be a single–tape Π_2 machine (a machine that starts its computation in a co-nondeterministic mode, and during its computation is allowed to perform one more alternation):

Theorem 3.1 *Let $\sqrt{T(n)}$ be fully time and space constructible, with $T(n) \geq n^2$. Than any $T(n)$–time–bounded co–nondeterministic single–tape TM can be simulated by a single-tape Π_2-TM in time $O(T(n)/\log T(n))$.*

We shall not present a proof here, since it is similar to the proof of theorem 2.1. In fact, the simulating machine is obtained from that from the proof of theorem 2.1 by exchanging its existential states with universal ones, and by mutually exchanging its accepting and rejecting states (recall that in the proof of theorem 2.1 the simulated machine has been w.l.o.g. considered as a strongly $T(n)$–time bounded machine — i.e., such that halts in an accepting or rejecting state on any computational path, after performing at most $T(n)$ steps).

Further, when introducing the notion of two new complexity classes, namely that of $co - \mathsf{NTIME}_1(T(n))$ and of $\Pi_2 - \mathsf{TIME}_1(T(n))$, respectively, with the obvious meaning, similarly as before, one can prove the following corollary:

Corollary 3.1 *Let $\sqrt{T(n)}$ be fully time and space constructible, with $T(n) \geq n^2$. Then*
$$co - \mathsf{NTIME}_1(T(n)) \subseteq \Pi_2 - \mathsf{TIME}_1(T(n)/\log T(n))$$

4 Separation Results

In the sequel we shall prove that the amount of speed up achieved by the previous simulation is enough to prove separation results concerning the basic complexity classes that are related to the computations at hand. To prove these results, we shall need two hierarchy theorems. The first one of them, which we shall reproduce here without a proof, is the theorem by Lorys and Liskiewicz [5]. It proves a tight hierarchy for single–tape nondeterministic time complexity classes:

Theorem 4.1 *Let $T_2(n)$ be fully time constructible, with $n^2 \in o(T_2(n))$, and let $T_1(n+1) \in o(T_2(n))$. Then*

$$\mathsf{NTIME}_1(T_1(n)) \subset \mathsf{NTIME}_1(T_2(n))$$

The next theorem shows that a similar hierarchy holds also for single–tape co–nondeterministic time:

Theorem 4.2 *Let $T_2(n)$ be fully time constructible, with $n^2 \in o(T_2(n))$, and let $T_1(n+1) \in o(T_2(n))$. Then*

$$co - \mathsf{NTIME}_1(T_1(n)) \subset co - \mathsf{NTIME}_1(T_2(n))$$

Proof Outline: Consider any language $L \in \mathsf{NTIME}_1(T_2(n)) - \mathsf{NTIME}_1(T_1(n))$. According to the previous theorem, such a language must exist, and for its complement holds $co - L \in co - \mathsf{NTIME}_1(T_2(n))$, and $co - L \notin co - \mathsf{NTIME}_1(T_1(n))$. This proves the theorem.

$\Box$

Now we can return to our separation results. First we show that single–tape nondeterministic time is strictly contained in single–tape Σ_2–time:

Theorem 4.3 *Let $T(n)$ be such that $\sqrt{T(n)}$ is a fully time and space constructible function, with $n^2 \in o(T(n))$, and $T(n+1)/\log T(n+1) \in o(T(n))$. Then*

$$\mathsf{NTIME}_1(T(n)) \subset \Sigma_2 - \mathsf{TIME}_1(T(n))$$

Proof Outline. From theorem 4.1 know the proper inclusion $\mathsf{NTIME}_1(T(n)) \subset \mathsf{NTIME}_1(T(n) \log n)$. According to corollary 2.1 the latter class is contained in $\Sigma_2 - \mathsf{TIME}_1\left(\frac{T(n)\log n}{\log(T(n)\log n)}\right) \subseteq \Sigma_2 - \mathsf{TIME}_1(T(n))$.

$\Box$

In much the same way, one can prove a similar result for the case co–nondeterministic computations. Therefore we shall state only the respective theorem, without giving its proof.

Theorem 4.4 *Let $T(n)$ be such that $\sqrt{T(n)}$ is a fully time and space constructible function, with $n^2 \in o(T(n))$, and $T(n+1)/\log T(n+1) \in o(T(n))$. Then*

$$co - \mathsf{NTIME}_1(T(n)) \subset \Pi_2 - \mathsf{TIME}_1(T(n))$$

Both previous theorems deal with the separation of the first two classes, with identical time bounds, either in Σ, or in the Π single-tape time hierarchy. Our final theorem describes the result concerning the separation of any two classes, each of them being from a different class from the two previously mentioned hierarchies.

Corollary 4.1 *Let $T(n)$ be such that $\sqrt{T(n)}$ is a fully time and space constructible function, with $n^2 \in o(T(n))$, and $T(n+1)/\log T(n+1) \in o(T(n))$. Then at least one of the following two statements holds true:*

- $co - \mathsf{NTIME}_1(T(n)) \neq \Sigma_2 - \mathsf{TIME}_1(T(n))$ and $\mathsf{NTIME}_1(T(n)) \neq \Pi_2 - \mathsf{TIME}_1(T(n))$
- $\mathsf{NTIME}_1(T(n)) \neq co - \mathsf{NTIME}_1(T(n))$ and $\Sigma_2 - \mathsf{TIME}_1(T(n)) \neq \Pi_2 - \mathsf{TIME}_1(T(n))$

Proof Outline. Assume the opposite, i.e., that $[co - \mathsf{NTIME}_1(T(n)) = \Sigma_2 - \mathsf{TIME}_1(T(n))$ or $\mathsf{NTIME}_1(T(n)) = \Pi_2 - \mathsf{TIME}_1(T(n))]$ and $[\mathsf{NTIME}_1(T(n)) = co - \mathsf{NTIME}_1(T(n))$ or $\Sigma_2 - \mathsf{TIME}_1(T(n)) = \Pi_2 - \mathsf{TIME}_1(T(n))]$. It follows that $\mathsf{NTIME}_1(T(n)) = \Sigma_2 - \mathsf{TIME}_1(T(n))$ or $co - \mathsf{NTIME}_1(T(n)) = \Pi_2 - \mathsf{TIME}_1(T(n))$ what is in a contradiction with the previous two theorems.

□

5 Conclusions

It has been shown that for a large class of time bounded classes, adding of one more alternation to a single tape nondeterministic or co–nondeterministic single tape machine leads to strictly larger complexity classes. This seems to be the first occasion where similar general results for a universal model of computation have been achieved.

A lot of work remains to be done. Among the first questions that could be amenable for the further research is the extension of the previous results to single–tape off–line TMs. This may bring us some steps further on our way towards the solution of similar problems for multitape TMs as asked e.g. in [1].

Acknowledgement. The author thanks to his colleagues Viliam Geffert and Stanislav Žák for their helpful comments on the first version of the present paper, and to an anonymous ICALP'96 referee for suggesting a correction of a little flaw in the proof of Theorem 4.3.

References

1. Gupta, S.: Alternating Time Versus Deterministic Time: A Separation. *Proc. of Structure in Complexity*, San Diego, 1993
2. Hopcroft, J. — Paul, W. — Valiant L.: On time versus space and related problems. *Proc. IEEE FOCS* **16**, 1975, pp. 57–64
3. Kannan, R.: Alternation and the power of nondeterminism (Extended abstract). *Proc. 15-th STOC*, 1983, pp. 344–346
4. Li, M. — Neuféglise, H. — Torenvliet, L. — van Emde Boas, P.: On Space Efficient Simulations. *ITLI Prepublication Series*, CS–89–03, University of Amsterdam, 1989
5. Lorys, K. — Liskiewicz, M.: Two Applications of Führers Counter to One–Tape Nondeterministic TMs. *Proceedings of the MFCS'88*, LNCS Vol. 324, Springer Verlag, 1988, pp. 445–453
6. Maass, W. — Schorr, A.: Speed-up of Turing Machines with One Work Tape and Two–way Input Tape. *SIAM J. Comput.*, Vol. **16**, no. 1, 1987, pp. 195–202
7. Paterson, M.: Tape Bounds for Tape-Bounded Turing Machines, *JCSS*, Vol. 6, 1972, pp. 115–124
8. Paul, W. — Pippenger, E. — Szemeredi, E. — Trotter, W.: On Determinism versus Nondeterminism and Related Problems. *Proc. 24-th FOCS*, 1983, pp. 429–438
9. Paul, W. — Prauss, E. J. — Reischuk, R.: On Alternation. *Acta Informatica*, 14, 1980, pp. 243–255

On ω-Generators and Codes

Sandrine JULIA *

Laboratoire I3S-U.R.A. 1376-C.N.R.S.,
Université de Nice Sophia Antipolis,
650, route des Colles, B.P. 145
06903 SOPHIA ANTIPOLIS Cedex
France

Abstract. Let C be a class of codes. Given a rational language L, deciding whether L^ω has an ω-generator in C is still an open problem except if C denotes the set of prefix codes [9]. Here, we restrict our investigations to ω-languages L^ω whose greatest ω-generator is a free submonoid. For such ω-languages, we prove that there exists an ω-generator in the sets of pure codes, circular codes, suffix codes or finite ω-codes if and only if the root of the greatest ω-generator is itself one. Furthemore, in the very precise case where the root of the greatest ω-generator is a three-element code, using a characteristic property of them [5, 6], we characterize the case where $W^\omega = L^\omega$ some ω-code W -finite or infinite-.

Introduction

Given a finite alphabet Σ, Σ^* denotes the set of words over Σ and Σ^ω the set of ω-words (i.e. infinite words). Given a language $L \subseteq \Sigma^*$, L^* is the set of words decomposable over L and L^ω is the ω-power of L: $L^\omega = \{u_1...u_n..., \forall i > 0\ u_i \in L \setminus \{\varepsilon\}\}$. L is called a generator of L^* and an ω-generator of L^ω.

A language C is a code if every word in C^* has a unique decomposition over C. By analogy, a language W is an ω-code [11] if every ω-word in W^ω has a unique decomposition over W. Clearly, an ω-code is a code. We are also interested in more traditional classes of codes whose definitions do not concern ω-words e.g., prefix, suffix or circular codes.

Let C be a class of codes. Given a rational language L, deciding whether L^* is generated by a code in C is easy: with respect to inclusion L^* has a least generator [1] as its root $(L^* \setminus \{\varepsilon\}) \setminus (L^* \setminus \{\varepsilon\})^2$ and L^* is generated by a code in C if and only if its root is in C. An ω-power does not have a least ω-generator [10] and, given a rational language L, deciding whether $K^\omega = L^\omega$ for some code K in C is still an open problem. However, this problem is solved when C is the set of prefix codes [9], a subset of the ω-codes set.

We focus on the ability of an ω-language L^ω to be ω-generated by such codes in the case where the stabilizer is the greatest ω-generator of L^ω: $stab(L^\omega) = \{u \in \Sigma^*,\ uL^\omega \subseteq L^\omega\}$ and we also require this submonoid to be free. In short we will

---

* This work was supported by the GDR-PRC *Mathématiques et Informatique*.

write: $L^\omega = C^\omega$ and $stab(L^\omega) = C^*$ for some code C.

Other results about such ω-languages can be found in [8], particularly the fact that all their minimal ω-generators are codes.

In addition, we know [10] that when L^ω is an adherence, this stabilizer is the greatest ω-generator.

For a rational language L, when the stabilizer is not the greatest ω-generator, there is a finite number of maximal ω-generators [10]. It may occur that there is a greatest ω-generator different from the stabilizer: when this submonoid is free, the following results still hold. In fact, in the general case, they provide properties for every ω-generator included in a free maximal ω-generator.

We characterize the case where an ω-language C^ω with $stab(C^\omega) = C^*$ for some code C is ω-generated by a pure code, a circular code, a prefix code, a suffix code, a biprefix or a finite ω-code as well. There exists an ω-generator in such a class $\mathcal{C}$ if and only if the root C of the stabilizer is itself in $\mathcal{C}$. Moreover if C is not an ω-code, there is no finite ω-code ω-generating C^ω

Furthermore, in the case where C is a three-element code, we prove that $W^\omega = C^\omega$ for some ω-code W -finite or infinite- if and only if C is an ω-code.

Technically, the main idea of this paper consists in linking some results obtained on factorizations of ω-words over codes to the study of ω-generators. On one hand, we use the results of [2, 3, 7] concerning the factorizations of ω-words over codes or suffix codes and on the other hand, we get the result about three-element codes thanks to a characteristic property of three-element codes pointed out by J. Karhumäki [5, 6].

1 Preliminaries

1.1 Definitions and notation

Let Σ be a finite alphabet. A word (resp. ω-word) is a finite (resp. infinite) concatenation of letters in Σ. We note ε the empty word. Σ^* is the set of words over Σ, $\Sigma^+ = \Sigma^* \setminus \{\varepsilon\}$. Σ^ω is the set of ω-words.

The subsets of Σ^* are called languages, those of Σ^ω ω-languages. Let L be a language, the language L^* is the set of words built with words in L: $L^* = \{a_1...a_n, \forall i\ 1 \le i \le n, a_i \in L\}$. In the same way, L^ω is the set of ω-words: $L^\omega = \{a_1...a_n.../\forall i > 0, a_i \in L \setminus \{\varepsilon\}\}$. L^* (resp. L^ω) is generated (resp. ω-generated) by L, and so L is called a generator (resp. ω-generator). Let X be a language and let Y be a language or an ω-language, we can define $XY = \{uv \in \Sigma^*, u \in X, v \in Y\}$ and $X^{-1}Y = \{v \in \Sigma^*, xv \in Y, x \in X\}$. If Y is a language, we can define symetrically XY^{-1}. Let Y and Y' be two ω-languages, we can also define $YY'^{-1} = \{v \in \Sigma^*, vy' \in Y, y' \in Y'\}$.

We note $|u|$ the length of the word u in Σ^*. A word u is a prefix of v and we note $u < v$ if $v \in u\Sigma^* \cup u\Sigma^\omega$.

An ω-word $a_1...a_n...$ is periodic (resp. ultimately periodic) if $\exists p \ge 1$ such that $\forall i > 0\ a_i = a_{i+p}$ (resp. $\exists p \ge 1$ such that $\exists m \ge 1\ \forall i \ge m\ a_i = a_{i+p}$).

Let L be a language, the stabilizer of L^ω is the following language: $stab(L^\omega) = \{u \in \Sigma^*, uL^\omega \subseteq L^\omega\}$. $stab(L^\omega)$ is a submonoid which can be understood as the set of all prefixes of ω-words in L^ω that could be fixed to the ω-words of L^ω to obtain yet another element in L^ω. Moreover every ω-generator of L^ω is included in $stab(L^\omega)$ and so, when $stab(L^\omega)$ is ω-generator of L^ω, it is the greatest [10]. We recall that a submonoid L^* is free if and only if its root $(L^* \setminus \{\varepsilon\}) \setminus (L^* \setminus \{\varepsilon\})^2$ is a code [1].

Definition 1. Let $L \subseteq \Sigma^*$. An *L-factorization* of a word u in L^+ is a finite sequence of words in $L \setminus \{\varepsilon\}$: $(u_1, u_2, ..., u_n)$ such that $u = u_1 u_2 ... u_n$. An *L-factorization* of an ω-word w in L^ω is an infinite sequence of words in $L \setminus \{\varepsilon\}$: $(w_1, w_2, ..., w_n, ...)$ such that $w = w_1 w_2 ... w_n ...$.

We will say indifferently L-factorization or factorization over L.

Definition 2. Let $L \subseteq \Sigma^*$. A *sub-factorization* from an L-factorization $(u_1, u_2, ..., u_n)$ of a word u in L^+ is an L^+-factorization of u: $(v_1, v_2, ..., v_p)$, with $1 \leq p < n$ and such that there exists a strictly increasing sequence of integers $(i_j)_{1 \leq j \leq p}$ with $1 \leq i_j \leq n$ verifying:
$$\begin{cases} v_1 = u_1 ... u_{i_1} \\ \forall j, \ 1 < j \leq p, \ v_j = u_{i_{j-1}+1} ... u_{i_j} \\ u_{i_p} = u_n \end{cases}$$
A *sub-factorization* from an L-factorization $(w_1, w_2, ..., w_n, ...)$ of an ω-word w in L^ω is an L^+-factorization of w: $(v_1, v_2, ..., v_m, ...)$ such that there exists a strictly increasing sequence of integers $(i_j)_{j \geq 1}$ verifying:
$$\begin{cases} v_1 = w_1 ... w_{i_1} \\ \forall j, \ j > 1, \ v_j = w_{i_{j-1}+1} ... w_{i_j} \end{cases}$$

Definition 3. Let $L \subseteq \Sigma^*$. A prefix p is said to be *relative* to an L-factorization $(u_1, u_2, ..., u_n)$ of u if there exists an i, $1 \leq i \leq n$ such that $p = u_1 ... u_i$. Idem in the case of an ω-word.

Definition 4. Let $L \subseteq \Sigma^*$. An L-factorization $(u_1, u_2, ...)$ of an ω-word is said to be *periodic* (resp. *ultimately periodic*) if there exists $p > 0$ such that $\forall i > 0$ $u_i = u_{i+p}$ (resp. $\exists j > 0 \ \forall i \geq j \ u_i = u_{i+p}$).

Finally, we recall that the rational ω-languages are the ω-languages of the form: $L = \cup_{i=1}^{n} A_i B_i^\omega$ with $n \geq 1$ and such that for every i, A_i and B_i are rational languages. This class is also the class of ω-languages recognized by Büchi automata.

1.2 Several classes of codes

Let L be a language in Σ^+. L is a *prefix code* if $L^{-1}L = \{\varepsilon\}$. L is a *code* if $L^{-1}L \cap L^* L^{*^{-1}} = \{\varepsilon\}$. L is an *ω-code* if $L^{-1}L \cap L^\omega (L^\omega)^{-1} = \{\varepsilon\}$. The set of prefix codes is strictly included in the set of ω-codes itself strictly included in the set of codes. Moreover, L is a *suffix code* if $LL^{-1} = \{\varepsilon\}$.

Definition 5. [4] Let $C \subseteq \Sigma^+$. C is a *pure code* if it is a code verifying:
$$\forall u \in \Sigma^+ \ (\forall n \geq 1 \ u^n \in C^* \Rightarrow u \in C^*)$$

Definition 6. [1] Let $C \subseteq \Sigma^+$. C is a *circular code* if and only if

$$\forall n, p \geq 1 \quad \forall u_0, ..., u_{n-1}, v_0, ...v_{p-1} \in C$$

$$\forall t \in \Sigma^* \quad \forall s \in \Sigma^+ \text{ such as } v_0 = ts$$

$$(u_0...u_{n-1} = sv_1...v_{p-1}t \Rightarrow n = p, t = \varepsilon \text{ and } \forall i \ u_i = v_i)$$

More intuitively, C is a circular code if we cannot have two words in C^* with their factorizations as shown in figure 1.

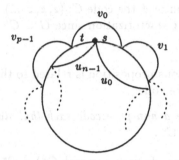

Fig. 1. Two circular factorizations

A circular code is a pure code. In the case of rational languages, a circular code is an ω-code, whereas the rational pure codes and rational ω-codes classes are not comparable [2].

1.3 Useful results

We now present some characterizations of codes, followed by a characterization of suffix codes, based on the factorizations of ultimately periodic ω-words over them.

Proposition 7. *[3, 7] Let C be a language included in Σ^+. The following assertions are equivalent:*

(i) *C is a code*
(ii) *for every $u \in C^+$, u^ω has an unique C-factorization*
(iii) *every C-factorization of each periodic ω-word is ultimately periodic*
(iv) *every C-factorization of each ultimately periodic ω-word is ultimately periodic*

Proposition 8. *[2, 3] Let C be a language included in Σ^+. The following assertions are equivalent:*

(i) *C is a suffix code*
(ii) *every C-factorization of each periodic ω-word is periodic*

2 Codes and ω-generators

From now on, we consider an ω-language C^ω with C a code verifying $stab(C^\omega) = C^*$. First of all, we give a necessary condition for a language to be an ω-generator of such a C^ω.

Lemma 9. *Let C be a code such that $stab(C^\omega) = C^*$. Any ω-generator G of C^ω contains a power of each element belonging to C.*

Proof Let C be a code such that $stab(C^\omega) = C^*$. Assume G is an ω-generator of C^ω and consider $z \in C$. $z^\omega \in G^\omega = C^\omega$ and so, according to proposition 7, it has a unique factorization over the code C: $(z, z, z, ...)$. Every G-factorization is a sub-factorization of a C-factorization since $G \subseteq C^*$. In fact, G contains a power of z. ◻

A consequence of the previous proposition is related to the ability for such a C^ω to be ω-generated by a pure code.

Proposition 10. *Let C be a non-pure code such that $stab(C^\omega) = C^*$. There is no pure code ω-generating C^ω.*

Proof Let C be a non-pure code verifying $stab(C^\omega) = C^*$. Assume K is another code such as $K^\omega = C^\omega$, we have $C \not\subseteq K$ since a code is a minimal ω-generator with respect to inclusion [10]. According to lemma 9, for each element z in $C \setminus K$, there exists an $i > 1$ such that $z^i \in K$. Hence, K is not pure because $z^i \in K^+$ although $z \notin K^+$. ◻

The following result deals with circular codes. Note that a rational circular code is an ω-code [4].

Proposition 11. *Let C be a non-circular code such that $stab(C^\omega) = C^*$. There is no circular code ω-generating C^ω.*

Proof If C is a non-circular code, then there exists a periodic ω-word α in C^ω such that $\alpha = p\alpha'$ with α' a periodic element in C^ω and $p \in \Sigma^+$. Let K be another code such that $K^\omega = C^\omega$. According to proposition 7, both α and α' have a unique ultimately periodic C-factorization and a unique K-factorization. K-factorizations are sub-factorizations of C-factorizations since $stab(C^\omega) = C^*$. Moreover, each K-factorization is ultimately periodic so there exist two respective K^+-factorization of α and α' as $(u, v, v, ..., v)$ and $(u', v', v', ..., v')$ with $\mid v \mid < \mid v' \mid$ for instance. Written in a circle, there exists a power of v whose K-factorization overlaps the K-factorization of v'. That prevents K from being a circular code. ◻

According to [9], for an ω-language L^ω whose stabilizer is ω-generator of L^ω, we have the equivalence: L^ω has a prefix code as ω-generator if and only if the stabilizer root is itself a prefix code. In the case where the stabilizer is a free submonoid, this result has a simple proof:

Proposition 12. *Let C be a code but not a prefix code such that $stab(C^\omega) = C^*$. There is no prefix code ω-generating C^ω .*

Proof Let C be a non prefix-free code such that $stab(C^\omega) = C^*$ and let P be a code ω-generator of C^ω. We consider two words u and v in C such that $u < v$. From lemma 9, each ω-generator of C^ω contains a power of each element in C. Thus, there exists an integer $i > 0$ such that $u^i \in P$.

Consider the following ω-word $(u^{i-1}v)^\omega \in C^\omega = P^\omega$. Let p be the first element of its unique P-factorization, p verifies $u^{i-1}v \leq p$. Indeed, the unique factorization of $(u^{i-1}v)^\omega$ over the code C is $(u, ...u, v, u, ..., u, v, ...)$ and its unique P-factorization is a sub-factorization of this C-factorization. Moreover, for every $j > 0$, $p \neq u^j$ and $u^{i-1}v \leq p$. In consequence u^i is a prefix of $p \in P$ and P is not a prefix code.   □

Proposition 13. *Let C be a non-suffix code such that $stab(C^\omega) = C^*$. There is no suffix code ω-generating C^ω .*

Proof From proposition 8, the fact that C is not a suffix code implies there exists a periodic ω-word w without periodic C-factorization.

Assume S is a suffix code such that $S^\omega = C^\omega$. From the same proposition, the unique S-factorization of w is periodic so there exists $u \in S^+$ such that $(u, u, ..., u, ...)$ is an S^+-factorization of w. According to proposition 7, it is a sub-factorization from its unique C-factorization, so $u \in C^+$ contradicting the fact that w does not have a periodic C-factorization.   □

The case of biprefix codes, both prefix and suffix, now appears obvious:

Corollary 14. *Let C be a non-biprefix code such that $stab(C^\omega) = C^*$. There is no biprefix code ω-generating C^ω.*

In order to handle the case when such a C^ω is ω-generated by a finite ω-code, we need the next two technical lemmas:

Lemma 15. *Let $L \subseteq \Sigma^+$ be a language and u, v two different elements in $stab(L^\omega) \setminus \{\varepsilon\}$ such that*

$$uL^\omega \cap vL^\omega \neq \emptyset$$

Each ω-code W ω-generating L^ω verifies: $\forall m \in stab(L^\omega)$ $\{mu, mv\} \not\subseteq W$.

Proof Let W be an ω-code such that $W^\omega = L^\omega$ and let u, v be two different elements in $stab(L^\omega)$ verifying $uL^\omega \cap vL^\omega \neq \emptyset$. If there is an $m \in stab(L^\omega)$ such that $\{mu, mv\} \subseteq W$, there exist $\alpha', \alpha'' \in L^\omega = W^\omega$ such that $u\alpha' = v\alpha''$, then $mu\alpha' = mv\alpha''$. We get two different W-factorizations of the same ω-word, so W is not an ω-code: a contradiction. See figure 2.   □

More precisely, we shall use:

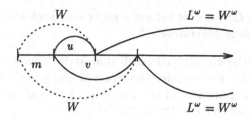

Fig. 2. Illustration for the proof of lemma 15

Lemma 16. *Let L be a language. Let u, v be two different elements in $stab(L^\omega) \setminus \{\varepsilon\}$ such that $uL^\omega \cap vL^\omega \neq \emptyset$. Each ω-code W ω-generator of L^ω verifies:*

$$\forall m \in stab(L^\omega) \ (mu \in W) \Rightarrow$$

$$(\forall s, s' \in stab(L^\omega) \setminus \{\varepsilon\} \ with \ ss' = v, \ s \neq u, \ ms \notin W)$$

Proof Let $u, v \in stab(L^\omega)$ be two different words such that $uL^\omega \cap vL^\omega \neq \emptyset$. Let $\alpha', \alpha'' \in L^\omega$ with $u\alpha' = v\alpha''$, then $mu\alpha' = mv\alpha''$. Assume there exist s and s' in $stab(L^\omega)$ such that $ss' = v$ and $s \neq u$, then $mu\alpha' = mss'\alpha''$. We know that $s'\alpha''$ is in L^ω and so we can apply the previous lemma: $uL^\omega \cap sL^\omega \neq \emptyset$. Hence, for every ω-code W ω-generator of L^ω, one has $\{mu, ms\} \not\subseteq W$. So, $mu \in W \Rightarrow ms \notin W$. □

Now we can give our first result concerning ω-codes.

Theorem 17. *Let C be a code but not an ω-code such that $stab(C^\omega) = C^*$. There is no finite ω-code ω-generating C^ω.*

Proof Assume W is an ω-code ω-generator of C^ω. Then there is an ω-word α in C^ω with two C-factorizations with different first steps u and v in C.
We build a sequence $(y_i)_{i \geq 0}$ with elements in W different from each other. We first define y_0: from lemma 9, there is an integer $n \geq 0$ such that $u^n u \in W$. We write $y_0 = u^n u$, $x_0 = u^n$, $z_0 = u$ and $z_0' = v$.
According to lemma 16, there is no prefix of $x_0 z_0'$ relative to its factorization over C which is an element of W. We then build y_{i+1} inductively from y_i:
$y_i = x_i z_i$, $x_i \in \{u, v\}^*$, $z_i, z_i' \in \{u, v\}$ and $z_i \neq z_i'$. $(x_i z_i')^\omega \in W^\omega$ has a unique C-factorization according to proposition 7: $(x_{i_1}, ..., x_{i_p}, z_i', x_{i_1}, ..., x_{i_p}, z_i', ...)$ with $(x_{i_1}, ..., x_{i_p})$ the C-factorization of x_i. Hence, there exists a prefix y_{i+1} relative to this C-factorization which is the first element of the W-factorization of this ω-word. However, from lemma 16, there is no prefix of $x_i z_i'$ relative to its C-factorization and element of W as well. Hence, we obtain: $y_{i+1} = x_{i+1} z_{i+1} \in W$, with $x_{i+1} \in \{u, v\}^+$, $z_{i+1} \in \{u, v\}$ and such that: $x_i z_i' < y_{i+1} < (x_i z_i')^\omega$. In fact, the C-factorization of $y_{i+1} = x_i z_i' c$, $c \in C^+$, contains more words than the C-factorization of y_i: they are different.
The sequence $(y_i)_{i \geq 0}$ is an infinite subset of the ω-code W. □

We do not know whether such a C^ω is ω-generated by an infinite ω-code or not. However, in the case of three-element codes, we solve this question thanks to a characteristic property of three-element codes.

3 Three-element codes and ω-generators

In [5, 6], J. Karhumäki gives a characteristic property of three-element codes. When a three-element code is not an ω-code, this characterization is linked with the notion of ω-words.

Theorem 18. [5, 6] If C is a three-element code but not an ω-code, then there exists exactly one ω-word α in C^ω which has several C-factorizations with different first steps. Moreover α is ultimately periodic and has exactly two C-factorizations with different first steps.

Before we give our main result, we need another lemma:

Lemma 19. Let C be a code and let α be an element in C^ω with at least two C-factorizations with different first steps. For each $m \in C^+$ there exists a prefix p of α such that mp is not a prefix of α.

Proof If $m \not< \alpha$, trivial; if $m < \alpha$ with $\forall p < \alpha$, $mp < \alpha$, then we obtain $m\alpha = \alpha$ which implies $m^\omega = \alpha$, but α cannot be the ω-power of a word in C^+, otherwise it would have a unique factorization over a code. □

Theorem 20. Let C be a three-element code but not an ω-code such that C^* is the greatest ω-generator of C^ω. There is no ω-code ω-generating C^ω.

Proof According to theorem 18, there exists a unique ω-word α in C^ω with two different C-factorizations with different first steps. Moreover, α is ultimately periodic.

From the proposition 7, α is ultimately periodic and its factorizations over C are also ultimately periodic. However, α has two factorizations over the code C so it does not have a periodic factorization.

Let $(u, \alpha_1, \alpha_2, ...)$ and $(v, \alpha'_1, \alpha'_2, ...)$ be the two factorizations of α with $u \neq v$ and $\forall i > 0$ $\alpha_i, \alpha'_i, u, v \in C$.

Assume that W is an ω-code ω-generator of C^ω.

We intend to build an ω-word β in C^ω which has no W-factorization.

In order to do this, we define inductively a sequence of increasing length prefixes of β. Let $\beta_1 = u$ or v such that $\beta_1 \notin W$. Assume β_n is built and verifies:

$$\begin{cases} \beta_n = u_1...u_{n-1}u_n...u_{f(n)}, \ u_i \in C \\ \forall i, \ 1 \le i \le f(n), \ u_1...u_i \notin W \\ \forall j, \ 1 \le j \le n-1, \ u_j...u_{f(n)} \not< \alpha \end{cases}$$

where $f(n)$ denotes the number of words in the C-factorization of β_n.

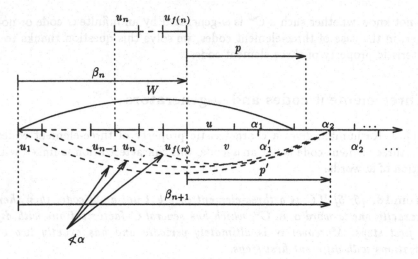

Fig. 3. Construction of β

We obtain β_{n+1} from β_n (see figure 3). According to lemma 19, there exists a prefix p of α such that $u_n...u_{f(n)}p \not< \alpha$. Consider the two C-factorizations of $\beta_n\alpha$: $(u_1, ..., u_{f(n)}, u, \alpha_1, \alpha_2, ...)$ and $(u_1, ..., u_{f(n)}, v, \alpha'_1, \alpha'_2, ...)$. Now, assume that the first element of the W-factorization of $\beta_n\alpha$ is relative to the first. Hence, there exists a prefix p' relative to $(v, \alpha'_1, \alpha'_2, ...)$ and $p \leq p'$. According to lemma 19, for each $s, s' \in C^*$ with $p' = ss'$, $\beta_n s \notin W$. At least, let $\beta_{n+1} = \beta_n p' = u_1...u_{f(n+1)}$.

β_{n+1} verifies: $\begin{cases} \forall i,\ 1 \leq i \leq f(n+1),\ u_1...u_i \notin W \\ \forall j,\ 1 \leq j \leq n,\ u_j...u_{f(n+1)} \not< \alpha \end{cases}$

Hence, we have built an ω-word β in C^ω such that $\beta \notin C^*\alpha$ which implies it has a unique C-factorization: $(u_1, ..., u_n, ...)$. Moreover, β admits no prefix relative to this factorization. A fortiori, there is no W-factorization of β. That yields a contradiction to the fact that W is an ω-generator of C^ω. $\quad\square$

Acknowledgements

We would like to thank I. Litovsky for helpful discussions.

References

1. J. Berstel and D. Perrin. *Theory of codes*. Academic Press, New-York, 1985.
2. J. Devolder. Comportement des codes vis-à-vis des mots infinis et bi-infinis - Comparaison. In D. Krob, editor, *Théories des automates et applications*, pages 75–89. Deuxièmes Journées franco-belges, Publications de l'Université de Rouen, 1992.
3. J. Devolder, M. Latteux, I. Litovsky, and L. Staiger. Codes and infinite words. *Acta Cybernetica*, 11(4):241–256, 1994.

4. J. Devolder-Muchembled. *Codes, mots infinis et bi-infinis.* PhD thesis, Université de Lille, 1993.

5. J. Karhumäki. On three-element codes. *Theoretical Computer Science,* 40:3–11, 1985.

6. J. Karhumäki. A property of three-element codes. *Theoretical Computer Science,* 41:215–222, 1985.

7. I. Litovsky. *Générateurs des langages rationnels de mots infinis.* PhD thesis, Université de Lille, 1988.

8. I. Litovsky. Free submonoids and minimal ω-generators of R^ω. *Acta Cybernetica,* 10(1-2):35–43, 1991.

9. I. Litovsky. Prefix-free languages as ω-generators. *Information Processing Letters,* 37:61–65, 1991.

10. I. Litovsky and E. Timmerman. On generators of rational ω-power languages. *Theoretical Computer Science,* 53:187–200, 1987.

11. L. Staiger. On infinitary length codes. *Theoretical Informatics and Applications,* 20(4):483–494, 1986.

On Standard Sturmian Morphisms

Aldo de Luca

Dipartimento di Matematica Università di Roma "La Sapienza"
Piazzale Aldo Moro 2, 00185, Roma, Italy

Abstract. We give a characterization of morphisms which preserve finite and infinite standard Sturmian words. The class of such morphisms coincides with the monoid $\{D, E\}^*$ of the endomorphisms of $\mathcal{A}^*$, where $\mathcal{A} = \{a, b\}$, generated by the two elementary morphisms, E which interchanges the letter a with b and D which is the Fibonacci morphism defined as: $D(a) = ab, D(b) = a$. Some new properties of these morphisms are shown. In particular, we derive a new characterization of the set PER of all words w having two periods p and q which are coprimes and such that $|w| = p + q - 2$.

1 Introduction

As is well known, Sturmian words have many applications in various different fields like Algebra and Theory of numbers, Physics (Ergodic theory, Crystallography) and Computer Science (Computer Graphics and Pattern matching); the study of the structure and combinatorics of these words became a subject of enormous interest, with a large literature on it (see, for instance, [5], [4], [9], [13] and references therein).

Sturmian words can be defined in several different but equivalent ways. Some definitions are 'combinatorial' and others of 'geometrical' nature. With regard to the first type of definition a Sturmian word is *a binary infinite word which is not ultimately periodic and is of minimal subword complexity*. A 'geometrical' definition is the following: A Sturmian word can be defined by considering the sequence of the intersections with a squared-lattice of a semi-line having a slope which is an irrational number. A horizontal intersection is denoted by the letter b, a vertical intersection by a and an intersection with a corner by ab or ba. We denote by **Sturm** the set of all Sturmian words. The Sturmian words represented by a semi-line starting from the origin are usually called *standard* or *characteristic*. They are of great interest from the language point of view since one can prove that the set of all finite subwords of a Sturmian word depends only on the slope of the corresponding semi-line [11].

The family **Stand** of infinite standard Sturmian words can be constructed by different procedures. We shall refer here to a method due to Rauzy [14]. We begin by introducing the set of *standard pairs* and then by taking the components of these pairs, the set *Stand* of all finite standard Sturmian words.

Let $\mathcal{A} = \{a, b\}$. We consider the smallest subset $\mathcal{R}$ of $\mathcal{A}^* \times \mathcal{A}^*$ which contains the pair (a, b) and is closed under the property:

$$(u, v) \in \mathcal{R} \Rightarrow (u, uv), (vu, v) \in \mathcal{R}.$$

Let us set $\mathcal{R}_0 = \{(a, b)\}$ and define for $n \geq 0$

$$\mathcal{R}_{n+1} = \{(u, v) \mid \exists (x, y) \in \mathcal{R}_n : u = x,\ v = xy \text{ or } u = yx,\ v = y\}.$$

Thus

$$\mathcal{R} = \bigcup_{n \geq 0} \mathcal{R}_n.$$

We call $\mathcal{R}$ also the set of *standard pairs*. Let us denote by $Trace(\mathcal{R})$ the set:

$$Trace(\mathcal{R}) = \{u \in \mathcal{A}^* \mid \exists v \in \mathcal{A}^* \text{ such that } (u, v) \in \mathcal{R} \text{ or } (v, u) \in \mathcal{R}\}.$$

The set *Stand* of all finite standard Sturmian words is then defined as:

$$Stand = Trace(\mathcal{R}).$$

Let us observe that by the above definition, if $(x, y) \in \mathcal{R}$ and $|x| < |y|$ (resp. $|x| > |y|$) then x (resp. y) is a prefix of y (resp. x).

It is also convenient to introduce the set S of the *unordered* standard pairs. Define for all $n \geq 0$

$$S_n = \{\{u, v\} \mid (u, v) \in \mathcal{R}_n \text{ or } (v, u) \in \mathcal{R}_n\},$$

and set:

$$S = \bigcup_{n \geq 0} S_n.$$

It follows trivially that

$$Stand = \{w \in \mathcal{A}^* \mid \exists v \in \mathcal{A}^* \text{ such that } \{v, w\} \in S\}.$$

We can label the nodes of a complete binary tree by the words on a binary alphabet $\mathcal{A} = \{a, b\}$: the root is labeled by the empty word ϵ; if w is the label of a node then the label of its left (resp. right) 'son' is wa (resp. wb). To each node we can associate a standard pair $(u, v) \in \mathcal{R}$ as follows. The root is labeled by the pair $(a, b) \in \mathcal{R}$ and if $(u, v) \in \mathcal{R}$ is the label of a node, then the label of the 'left son' is $(u, uv) \in \mathcal{R}$ and the label of the 'right son' is $(vu, v) \in \mathcal{R}$. We call this tree the *standard tree*. Let us introduce in $\mathcal{R}$ two elementary maps, or operators, a and b, defined as:

$$a(u, v) = (u, uv), \quad b(u, v) = (vu, v).$$

We can then consider a map $\sigma : \{a, b\}^* \to \mathcal{R}$ inductively defined as: for all $w \in \{a, b\}^*$

$$\sigma(\epsilon) = (a, b), \quad \sigma(wa) = a\sigma(w), \quad \sigma(wb) = b\sigma(w).$$

In this way one has for all $w \in \{a, b\}^*$

$$\sigma(w) = \tilde{w}(a, b),$$

where $\tilde{w}$ denotes the *mirror image* of w. From the above construction one has that $\sigma(w)$ is the standard pair which labels the node w in the standard tree; w is also called the *generating word* of the standard pair $\sigma(w)$. The map σ is obviously surjective. Moreover, it is also injective and then bijective [1].

We can represent any word $w \in \mathcal{A}^*$ uniquely by a finite sequence $(h_1, h_2, ..., h_n)$ of integers, where $h_1 \geq 0$, $h_i > 0$ for $1 < i \leq n$ and

$$w = a^{h_1} b^{h_2} a^{h_3} ....$$

One has that $|w| = \sum_{i=1}^n h_i$, where $|w|$ denotes the length of w. We call such a representation of the words of $\mathcal{A}^*$ the *integral representation*.

Let $\mathcal{A} = \{a, b\}$ and $\mathcal{A}^\omega$ be the set of all infinite words on $\mathcal{A}$. We consider the subset $\mathcal{A}_0^\omega$ defined as:

$$\mathcal{A}_0^\omega = \{y \in \mathcal{A}^\omega \,|\, y \notin \mathcal{A}^* x^\omega, \; x \in \mathcal{A}\}.$$

In other words $y \notin \mathcal{A}_0^\omega$ if and only if there exists a word $u \in \mathcal{A}^*$ and a letter $x \in \mathcal{A}$ such that $y = u x^\omega = u x x ....x....$ Hence any infinite word $x \in \mathcal{A}_0^\omega$ can be uniquely expressed as:

$$\mathbf{x} = a^{h_1} b^{h_2} a^{h_3} ....,$$

with $h_1 \geq 0$ and $h_i > 0$ for $i > 0$. We call the infinite sequence $(h_1, h_2, ..., h_n, ...)$ the *integral representation* of $\mathbf{x}$. To each $\mathbf{x} \in \mathcal{A}_0^\omega$ corresponds a path in the standard tree and the sequence $\{\sigma_n\}_{n \geq 0}$ of standard pairs defined as:

$$\sigma_0 = (a, b), \; \sigma_{n+1} = x_n \sigma_n, \; n \geq 0,$$

where $x_n = \mathbf{x}(n)$, $n \geq 0$. Let us denote $\sigma_n = (\alpha_n, \beta_n)$. One has then $\sigma_{n+1} = (\alpha_{n+1}, \beta_{n+1}) = (\alpha_n, \alpha_n \beta_n)$ if $x_n = a$ and $\sigma_{n+1} = (\beta_n \alpha_n, \beta_n)$ if $x_n = b$. One easily derives that the sequences $\{\alpha_n\}_{n \geq 0}$ and $\{\beta_n\}_{n \geq 0}$ of standard words converge, according to the usual topology in the set of infinite words, to the same infinite word

$$s = lim_n \alpha_n = lim_n \beta_n.$$

The sequence $(h_1, h_2, ..., h_n ...)$ is also called the *directive sequence* of s and (α_n, β_n) the *approximating bisequence* of s. All the infinite words that one can construct by this procedure constitute the set **Stand** of infinite standard Sturmian words. In previous papers we gave some different characterizations of the set *Stand*. A basic characterization can be expressed in terms of the *periodicities* of the words.

We define the set PER of all words w having two periods p and q which are coprimes and such that $|w| = p + q - 2$. Thus a word w belongs to PER if it is a power of a single letter or is a word of maximal length for which the theorem of Fine and Wilf [10] does not apply. In the sequel we assume that $\epsilon \in PER$. This is, formally, coherent with the above definition if one takes $p = q = 1$. In [9] we proved the following remarkable result:

$$Stand = \mathcal{A} \cup PER\{ab, ba\}.$$

Let us denote by PAL the set of all the words of $\mathcal{A}^*$ which are palindromes. We introduce the map $(-) : \mathcal{A}^* \to PAL$ which associates with any word $w \in \mathcal{A}^*$ the palindrome word $w^{(-)}$ defined as the shortest palindrome word having the suffix w. We call $w^{(-)}$ the *palindrome left-closure of w*. If X is a subset of $\mathcal{A}^*$ we denote by $X^{(-)}$ the set

$$X^{(-)} = \{w^{(-)} \in \mathcal{A}^* \mid w \in X\}.$$

Let us define inductively the sequence $\{X_n\}_{n \geq 0}$ of finite subsets of $\mathcal{A}^*$ as:

$$X_0 = \{\epsilon\}, \ X_{n+1} = (\mathcal{A}X_n)^{(-)}, n \geq 0.$$

Thus $s \in X_{n+1}$ if and only if there exist $x \in \mathcal{A}$ and $t \in X_n$ such that $s = (xt)^{(-)}$. We proved [7] that

$$PER = \bigcup_{n \geq 0} X_n.$$

For a word $w \in \mathcal{A}^*$ we denote by $alph(w)$ the set of the letters of $\mathcal{A}$ occurring in w. We recall the following result concerning the set PER which will be used later in this paper and whose proof is in [7].

Lemma 1. *Let $w \in PER$ be such that $Card(alph(w)) > 1$. Then w can be uniquely represented as:*

$$w = PxyQ = QyxP,$$

with x, y fixed letters in $\{a, b\}$, $x \neq y$ and $P, Q \in PAL$. Moreover, $gcd(p, q) = 1$, where $p = |P| + 2$ and $q = |Q| + 2$.

Let St be the set of all finite subwords of all infinite Sturmian words. An element $w \in St$ is called *right special* (resp. *left special*) if

$$wa, wb \in St \ (resp. \ aw, bw \in St).$$

An element $w \in St$ is called *strictly bispecial* if

$$awa, awb, bwa, bwb \in St.$$

In [9] and [7] it has been proved that PER coincides with the set of all strictly bispecial elements of St. Moreover, PER is also equal to the set of right (left) special elements of St which are palindromes.

2 Standard morphisms

Definition 2. A morphism $\phi : \mathcal{A}^* \to \mathcal{A}^*$ is called Sturmian if

$$\mathbf{x} \in \mathbf{Sturm} \implies \phi(\mathbf{x}) \in \mathbf{Sturm}.$$

The following theorem due to Mignosi and Séébold [12] gives a remarkable characterization of Sturmian morphisms:

Theorem 3. *A morphism ϕ is Sturmian if and only if $\phi \in \{E, D, G\}^*$, where E, D and G are the morphisms (elementary morphisms) defined as:*

$$D : \begin{matrix} a \to ab \\ b \to a \end{matrix}, \quad E : \begin{matrix} a \to b \\ b \to a \end{matrix}, \quad G : \begin{matrix} a \to ba \\ b \to a. \end{matrix}$$

The monoid $\{E, D, G\}^*$ of endomorphisms of $\mathcal{A}^*$ generated by the elementary morphisms E, D and G has been called also the *monoid of Sturm* [4]. The morphisms belonging to the submonoid $\{E, D\}^*$ have been called also *positive morphisms*. The following theorem due to Berstel and Séébold [2] shows that in order to have a Sturmian morphism it is sufficient to test it only on a single Sturmian word:

Theorem 4. *A morphism ϕ is Sturmian if and only if there exists a word $\mathbf{x} \in$ Sturm such that $\phi(\mathbf{x}) \in$ Sturm.*

We shall consider now morphisms which preserve the property of being a standard Sturmian word. One can introduce two notions of standard morphisms, the first for the infinite case and the second for the finite case. We shall prove in the next that these two notions are in fact equivalent.

Definition 5. A morphism ϕ is a morphism preserving infinite standard words if

$$\mathbf{x} \in \mathbf{Stand} \implies \phi(\mathbf{x}) \in \mathbf{Stand}.$$

We call such a morphism also an ω-standard morphism.

Definition 6. A morphism ϕ is a morphism preserving finite standard words if

$$w \in Stand \implies \phi(w) \in Stand.$$

We call such a morphism also a standard morphism.

The following has been proved by Crisp *et al* [6]:

Theorem 7. *Let ϕ be a morphism. If there exist $\mathbf{s}, \mathbf{t} \in$ Stand such that $\phi(\mathbf{s}) = \mathbf{t}$, then ϕ is a positive morphism.*

From this it follows, trivially, that any ω-standard morphism is positive. The following characterization of the positive morphisms has been proved by Berstel and Séébold [3]:

Theorem 8. *A morphism ϕ is positive if and only if $\{\phi(a), \phi(b)\}$ is an unordered standard pair.*

Theorem 9. *A positive morphism is standard.*

Proof. Since positive morphisms are obtained by compositions of elementary ones, it is sufficient to prove that for all $s \in Stand$, one has $D(s) \in Stand$. The fact that $E(s) \in Stand$ is trivial. If $|s| = 1$ the result is obvious. Let us then suppose $|s| > 1$. We can always write s as:

$$s = \pi xy,$$

with $\pi \in PER$ and $\{x, y\} = \{a, b\}$. By the definition of the elementary morphism D one has: $D(xy) = ayx$, so that

$$D(s) = D(\pi)ayx.$$

If we consider the standard word $t = \pi yx$, then one derives

$$D(t) = D(\pi)axy.$$

Hence, since $D(s), D(t) \in St$ it follows that $D(\pi)a$ is a right special element of St. We shall prove now that $D(\pi)a$ is also palindrome, so that one derives $D(\pi)a \in PER$ and then $D(s), D(t) \in Stand$. Indeed, one obtains by induction that if $f \in PAL$ then $D(f)a \in PAL$. The result is trivial if $|f| \leq 1$. Let us then suppose $|f| > 1$. We can write $f = zgz$ with $z \in \{a, b\}$. One has then if $z = a$:

$$D(f)a = abD(g)aba.$$

By induction $D(g)a \in PAL$ so that $D(f)a \in PAL$. If $z = b$ one has

$$D(f)a = aD(g)aa,$$

so that, by induction $D(g)a \in PAL$ and also in this case one derives $D(f)a \in PAL$.

Proposition 10. *The elementary morphism G is not standard. Indeed, one has that if $s \in Stand$, then $G(s) \in Stand$ if and only if*

$$s = b \quad or \quad s = b^r a, \quad r \geq 0.$$

Proof. If $s = b$ then $G(b) = a \in Stand$. If $s = b^r a$, $r \geq 0$, then $G(s) = a^r ba \in Stand$. Suppose now $|s| \geq 2$. We can factorize s as $s = \pi xy$ with $\pi \in PER$ and $\{x, y\} = \{a, b\}$. If $xy = ab$, then

$$G(s) = G(\pi)baa \notin Stand.$$

In the case $xy = ba$ one has: $G(s) = G(\pi)aba$. Let us suppose that $G(s) \in Stand$. This implies that $G(\pi)a \in PAL$. Hence, $\pi = bpb$, with $p \in PAL$. Thus $G(s) = aG(p)aaba$. Since $aG(p)aa \in PAL$ it follows $p = bqb$ and then by a simple iteration one derives $p = b^{|p|}$.

In order to prove the converse of Theorem 9 we need the following technical lemma concerning finite standard words and whose proof, rather technical, we omit [8].

Lemma 11. *Let $u, v, uv, vu \in Stand$. Then $\{u, v\}$ is an unordered standard pair.*

Theorem 12. *Let ϕ a standard morphism. Then ϕ is positive.*

Proof. Let ϕ a standard morphism. Since $a, b, ab, ba \in Stand$ one has:

$$\phi(a), \phi(b), \phi(ab) = \phi(a)\phi(b), \phi(ba) = \phi(b)\phi(a) \in Stand.$$

From Lemma 11 one derives that $\{\phi(a), \phi(b)\}$ is an unordered standard pair, so that from Theorem 8 one has that the morphism ϕ is positive.

Theorem 13. *Let ϕ be a positive morphism. Then ϕ is an ω-standard morphism.*

Proof. Let $s \in$ **Stand**. We can describe s by the directive sequence $(h_1, ..., h_n, ...)$, and the approximating bisequence (α_n, β_n), where $(\alpha_0, \beta_0) = (a, b)$ and at each step one applies one or the other rule

$$R1 : \begin{matrix} \alpha_{n+1} = \alpha_n \\ \beta_{n+1} = \alpha_n \beta_n \end{matrix}, \quad R2 : \begin{matrix} \alpha_{n+1} = \beta_n \alpha_n \\ \beta_{n+1} = \beta_n. \end{matrix}$$

The fact that the directive sequence is $(h_1, h_2, ..., h_n, ...)$ means that one applies rule $R1$, h_1-times, the rule $R2$, h_2-times and so on. We shall suppose, without loss of generality, that $h_1 > 0$. Let ϕ be a positive morphism. Let us set for all $n \geq 0$

$$\phi(\alpha_n) = A_n, \quad \phi(\beta_n) = B_n.$$

One has that for all $n \geq 0$ either

$$\begin{matrix} A_{n+1} = A_n \\ B_{n+1} = A_n B_n \end{matrix}, \quad \text{or} \quad \begin{matrix} A_{n+1} = B_n A_n \\ B_{n+1} = B_n. \end{matrix}$$

One has from Theorem 8 that $\{\phi(a), \phi(b)\} = \{A_0, B_0\}$ is an unordered standard pair. Without loss of generality suppose that (A_0, B_0) is a standard pair. Thus (A_0, B_0) has a finite directive sequence $(c_1, c_2, ..., c_k)$ and there exists a finite approximating bisequence (γ_i, δ_i) such that $(\gamma_0, \delta_0) = (a, b)$, $(A_0, B_0) = (\gamma_{k_0}, \delta_{k_0})$ and

$$c_1 + c_2 + ... + c_k = k_0.$$

Let $t = \phi(s)$. One has that: $t = \lim_n A_n = \lim_n B_n$. If k is even then t is an infinite standard word whose directive sequence is:

$$(c_1, c_2, ..., c_k, h_1, h_2, ..., h_n, ...).$$

If k is odd then t is an infinite standard word whose directive sequence is:

$$(c_1, c_2, ..., c_{k-1}, c_k + h_1, h_2, ..., h_n, ...).$$

We can summarize the previous results on standard morphisms in the following:

Theorem 14. *Let $\phi : A^* \to A^*$ be a morphism. The following conditions are equivalent:*

1. *There exist $s, t \in$ **Stand** such that $t = \phi(s)$,*
2. *ϕ is an ω-standard morphism,*
3. *ϕ is positive, i.e. $\phi \in \{D, E\}^*$,*
4. *$\{\phi(a), \phi(b)\}$ is an unordered standard pair,*
5. *ϕ is a standard morphism.*

3 Some properties of standard morphisms

In this section we shall consider some properties of standard morphisms which will allow us to give a new way of constructing *Stand* by means of elementary standard morphisms.

Proposition 15. *Let ϕ be a standard Sturmian morphism. Then ϕ is injective and the restriction $\phi_|$ of ϕ to Stand is an injective map $\phi_| : Stand \to Stand$.*

Proof. We know from Theorem 14 that a standard Sturmian morphism is positive, i.e. $\phi \in \{D, E\}^*$. Since E is an automorphism of A^* and D is injective, it follows that ϕ is injective in view of the fact that the injectivity is preserved under composition. Moreover, since $w \in Stand \implies \phi(w) \in Stand$ it follows that the restriction $\phi_|$ of ϕ to $Stand$ is an injective map $\phi_| : Stand \to Stand$.

In the following we shall denote by λ and μ the two standard morphisms defined as:

$$\mu = D, \quad \lambda = E \circ D,$$

where $\circ$ denotes the map composition (made from left to right). It holds the following [8]:

Proposition 16. *Any standard morphism ϕ can be decomposed as: $\phi = \psi \circ E^\alpha$, where $\psi \in \{\lambda, \mu\}^*$ and $\alpha \in \{0, 1\}$. Moreover, for any $w \in A^+$ if $\psi \neq \iota$ ($\iota = $ identity) then $\phi(w) \in aA^*$ if and only if $\alpha = 0$.*

Proposition 17. *Let s be the standard word $s = \pi xy$ with $\pi \in PER$ and $\{x, y\} = \{a, b\}$. One has: $\lambda(\pi)a$, $\mu(\pi)a \in PER$, and*

$$\lambda(s) = \lambda(\pi)axy, \quad \mu(s) = \mu(\pi)ayx.$$

Proof. By the definitions of $\mu = D$ and $\lambda = E \circ D$ one has for $\{x, y\} = \{a, b\}$, $\lambda(xy) = axy$, $\mu(xy) = ayx$, so that since λ, μ are standard morphisms,

$$\lambda(s) = \lambda(\pi)axy, \quad \mu(s) = \mu(\pi)ayx \in Stand.$$

This implies that $\lambda(\pi)a$, $\mu(\pi)a \in PER$.

Let Q be the set of rational numbers. We consider the map $\rho : \mathcal{A}^* \to Q \cup \{\infty\}$, defined as:

$$\rho(\epsilon) = 1, \quad \rho(w) = |w|_b/|w|_a, \quad \text{for } w \neq \epsilon,$$

where for $x \in \mathcal{A}$, $|w|_x$ denotes the number of occurrences of x in w. We assume that $1/0 = \infty$. For any $w \in \mathcal{A}^*$ we call $\rho(w)$ the 'slope' of w (cf. [1]).

Proposition 18. *Let $w \in \mathcal{A}^*$ and be the slope $\rho(w) = |w|_b/|w|_a$. One has then:*

$$\rho(\lambda(w)) = |w|_b/|w|, \quad \rho(\mu(w)) = |w|_a/|w|.$$

Proof. Since $\lambda(a) = a$ and $\lambda(b) = ab$ one has that $|\lambda(w)|_a = |w|_a + |w|_b = |w|$ and $|\lambda(w)|_b = |w|_b$, so that $\rho(\lambda(w)) = |w|_b/|w|$.

Since $\mu(a) = ab$ and $\mu(b) = a$ one has $|\mu(w)|_b = |w|_a$ and $|\mu(w)|_a = |w|_a + |w|_b = |w|$, so that $\rho(\mu(w)) = |w|_a/|w|$.

Let us now introduce the family $\{S_i\}_{i \geq 0}$ of sets of standard words defined, inductively, as:

$$S_0 = \{ab\}, \quad S_{i+1} = \lambda(S_i) \cup \mu(S_i), \quad i \geq 0.$$

We set

$$S = \bigcup_{i \geq 0} S_i.$$

One can prove [8] that for all $i, j, i \neq j$, $S_i \cap S_j = \emptyset$. Moreover, one easily verifies that: $S = \{\phi(ab) \mid \phi \in \{\lambda, \mu\}^*\}$, so that from the definition one has, since $ab \in Stand$, that $S \subseteq Stand$.

We shall prove (cf. Theorem 22) that $S \cup \{a\} = Stand \cap a\mathcal{A}^*$. To this end we need to prove some technical lemmas.

Lemma 19. *For $i > 0$ if $\pi xy \in S_i$ with $\{x, y\} = \{a, b\}$, then $\pi yx \in S_i$.*

Proof. The proof is by induction on the integer i. For $i = 1$ one has $S_1 = \{aab, aba\}$, so that the assertion is trivially true. Suppose now that the statement is true up to i and prove it for $i + 1$. Let $\pi xy \in S_{i+1}$. By definition there exists $w \in S_i$ such that $\pi xy = \lambda(w)$ or $\pi xy = \mu(w)$. Let us consider the first case. Since $w \in Stand$ and $\lambda(xy) = axy$ we can write $w = uxy$ with $u \in PER$. Hence,

$$\pi xy = \lambda(u) axy,$$

with $\pi = \lambda(u)a \in PER$. By induction $uyx \in S_i$, so that

$$\lambda(u)ayx = \pi yx \in S_{i+1}.$$

In the second case since $\mu(xy) = ayx$ one can write $w = uyx$, with $u \in PER$, and $\pi xy = \mu(u)axy$. By using the inductive hypothesis one has $uxy \in S_i$ and then $\mu(u)axy = \pi yx \in S_{i+1}$.

It holds the following lemma whose proof we omit [8]:

Lemma 20. *Let p/q be an irreducible fraction with p, q positive integers. Then there exist only two standard words $f = \pi x y$, $g = \pi y x$, with $\pi \in PER$, $\{x, y\} = \{a, b\}$, such that $\rho(f) = \rho(g) = p/q$. Moreover, if $p \leq q$ then $\pi \in a\mathcal{A}^*$.*

Let us recall the following lemma whose proof is in [7]:

Lemma 21. *Let $\mathcal{F}$ be the set of all irreducible fractions p/q such that $0 < p \leq q$. One has that $\mathcal{F}$ is the smallest subset of the set Q of rational numbers which contains the fraction $1/1$ and such that*

$$p/q \in \mathcal{F} \implies p/(p+q), \ q/(p+q) \in \mathcal{F}.$$

Theorem 22. $Stand \cap a\mathcal{A}^* = \{a\} \cup S.$

Proof. Since the inclusion $\supseteq$ is trivial we have to prove that $Stand \cap a\mathcal{A}^* \subseteq \{a\} \cup S$. Let $u \in S$ such that $|u| > 1$. If u has a slope $\rho(u) = |u|_b/|u|_a = p/q$ then from Proposition 18, one has

$$\rho(\lambda(u)) = p/(p+q), \ \rho(\mu(u)) = q/(p+q).$$

Since $\rho(ab) = 1/1$ then from the preceding lemma one derives that $\{\rho(u)| \ u \in S\} = \mathcal{F}$. Let now $w \in Stand \cap a\mathcal{A}^*$. If $|w| = 1$ then the result is trivial since $w = a$. Let us then suppose $|w| > 1$ and be $\rho(w) = |w|_b/|w|_a = p/q$ the slope of w. If $w = \pi x y$ with $\pi \in PER$, then there exists a morphism $\phi \in \{\lambda, \mu\}^*$ such that either $w = \phi(ab)$ or $w' = \pi y x = \phi(ab)$. In the first case one has $w \in S$. In the second case $w' \in S$. However, from Lemma 19, $w' \in S \Rightarrow w \in S$ which concludes the proof.

Corollary 23. *Let ϕ be a standard morphism. For all $w \in \mathcal{A}^*$*

$$\phi(w) \in Stand \Rightarrow w \in Stand.$$

Proof. Let ϕ be a standard morphism. From Theorem 14 one has that ϕ is positive, i.e. $\phi \in \{D, E\}^*$, so that there exists an integer k such that $\phi \in \{D, E\}^k$. The proof is by induction on k. If $\phi = \iota$ or $\phi = E$ then the result is trivial. Let $\phi = D = \mu$ and suppose that $\mu(w) = s \in Stand$. By the definition of μ it follows that $s \in Stand \cap a\mathcal{A}^*$. If $s = a$ then one has $w = b$ and the result is true. Let us then suppose that $|s| > 1$. From the preceding theorem it follows that $s \in S$ so that there exists $i \geq 0$ such that $s \in S_i$. If $i = 0$ then $s = ab$ so that since $\mu(a) = ab$ one has $w = a \in Stand$. Let us then suppose $i > 0$. There exists $u \in S_{i-1}$ for which $s = \mu(u)$, or $s = \lambda(u) = \mu(\hat{u})$, having set $\hat{u} = E(u)$. Since μ is injective it follows $w = u$ or $w = \hat{u}$. Since u, as well as $\hat{u}$, are standard words the assertion follows in this case and the base of the induction is proved. Let us now assume that the result is true up to k and prove it for $k + 1$. Let $\phi \in \{D, E\}^{k+1}$. We can write:

$$\phi = \psi \circ D \quad \text{or} \quad \phi = \psi \circ E,$$

with $\psi \in \{D, E\}^k$. For $w \in \mathcal{A}^*$ one has $\phi(w) = \mu(\psi(w))$ or $\phi(w) = E(\psi(w))$. Since $\phi(w) \in Stand$ then it follows $\psi(w) \in Stand$ and by the inductive hypothesis $w \in Stand$.

4   A characterization of the set PER

Let us define the sequence $\{Y_n\}_{n\geq 0}$ of sets, inductively, as $Y_0 = \{\epsilon\}$ and for all $n \geq 0$

$$Y_{n+1} = (\lambda(Y_n) \cup \mu(Y_n))a.$$

Theorem 24. *For all $n > 0$, $S_n = Y_n\{ab, ba\}$.*

Proof. The proof is by induction on the integer n. For $n = 1$, since $\lambda(\epsilon) = \mu(\epsilon) = \epsilon$, one has $Y_1 = \{a\}$ and $S_1 = \{aab, aba\}$. Let us then suppose that $S_i = Y_i\{ab, ba\}$ for $1 \leq i \leq n$ and prove it for $i = n+1$. Indeed one has by using the inductive hypothesis:

$$S_{n+1} = \lambda(S_n) \cup \mu(S_n) = \lambda(Y_n)\{aab, aba\} \cup \mu(Y_n)\{aba, aab\} =$$

$$(\lambda(Y_n) \cup \mu(Y_n))\{aba, aab\} = (\lambda(Y_n) \cup \mu(Y_n))a\{ab, ba\} = Y_{n+1}\{ab, ba\}.$$

Corollary 25. $PER \cap a\mathcal{A}^* = \bigcup_{n>0} Y_n = \{\phi(a) \mid \phi \in \{\lambda, \mu\}^*\}.$

Proof. From the definition of the sets Y_n one has $\bigcup_{n>0} Y_n \subseteq PER \cap a\mathcal{A}^*$. In order to prove the inverse inclusion let $w \in PER \cap a\mathcal{A}^*$. One has

$$wab, wba \in Stand \cap a\mathcal{A}^* = S \cup \{a\}.$$

Hence, there exists $i \geq 0$ such that $wab, wba \in S_i = Y_i\{ab, ba\}$. Thus $w \in Y_i \subseteq \bigcup_n Y_n$. The remaining part of the proof is trivial.

Let us define the sequence of sets $\{X_n\}_{n\geq 0}$, inductively, as:

$$X_0 = \{\epsilon\}, \ X_1 = \{a\}, \ X_{n+1} = (\mathcal{A}X_n)^{(-)}, \text{ for } n > 0,$$

where $(-)$ is the operator of left-palindrome closure. One easily derives from (cf. [7]) that

$$PER \cap a\mathcal{A}^* = \bigcup_{n>0} X_n.$$

We want to prove that for all $n \geq 0$, $X_n = Y_n$, so that the following noteworthy relation will hold:

$$Y_{n+1} = (\mathcal{A}Y_n)^{(-)} = (\lambda(Y_n) \cup \mu(Y_n))a.$$

We need to recall some definitions and prove some lemmas. Let $\psi : \mathcal{A}^* \to PER$ be the map defined as (cf. [7]): $\psi(\epsilon) = \epsilon$, $\psi(a) = a$, $\psi(b) = b$, and for all $w \in \mathcal{A}^*$ and $x \in \mathcal{A}$

$$\psi(wx) = (x\psi(w))^{(-)}.$$

Lemma 26. *For all* $n > 0$, $\psi(a\mathcal{A}^{n-1}) = X_n$ *and* X_n *is a biprefix code having* 2^{n-1} *elements.*

Proof. From the definition of ψ it follows immediately that for all $n > 0$ one has $\psi(a\mathcal{A}^{n-1}) = X_n$. Since ψ is injective (cf.[7]) one has $Card(X_n) = 2^{n-1}$. The proof that X_n is a biprefix code is in [7].

Let $a_0, a_1, ..., a_n$ be a finite sequence of integers such that $a_i > 0$, $1 \leq i < n$ and $a_0, a_n \geq 0$: we denote by $\langle a_0, a_1, ..., a_n \rangle$ the continued fraction $[a_0, a_1, ..., a_{n-1}, a_n + 1] = [a_0, a_1, ..., a_{n-1}, a_n, 1]$. The proof of the following lemma is in [1]:

Lemma 27. *Let* $u \in \mathcal{A}^*$ *have the integral representation* $(h_1, ..., h_n)$, $h_1 > 0$, *and be* $\pi = \psi(u)$. *Then the slope* $\rho(\pi xy)$ *of the standard word* πxy, *where* $\{x, y\} = \{a, b\}$, *has the development in continued fractions:*

$$\rho(\pi xy) = \langle 0, h_1, ..., h_n \rangle.$$

The proof of the following lemma is trivial:

Lemma 28. *If the irreducible fraction* p/q, $0 < p \leq q$ *has the development in continued fractions* $p/q = \langle 0, h_1, ..., h_n \rangle$, *then the irreducible fractions* $p/(p+q)$ *and* $q/(p+q)$ *have the developments in continued fractions given by:*

$$p/(p+q) = \langle 0, 1 + h_1, h_2, ..., h_n \rangle, \quad q/(p+q) = \langle 0, 1, h_1, ..., h_n \rangle.$$

Theorem 29. *For all* $n \geq 0$, $Y_n = X_n$.

Proof. The proof is by induction on n. For $n = 0$, $X_0 = Y_0 = \{\epsilon\}$. For $n = 1$, one has $X_1 = Y_1 = \{a\}$. Let us suppose the statement true up to n and then prove it for $n + 1$. We have then to show that:

$$X_{n+1} = (\lambda(X_n) \cup \mu(X_n))a.$$

Let us first prove the inclusion $\supseteq$. Let $w \in X_n$. By Lemma 26 one has $w \in \psi(a\mathcal{A}^{n-1})$. Let $u = a^{h_1} b^{h_2} ... \in a\mathcal{A}^{n-1}$ have the integral representation $(h_1, ..., h_m)$ with

$$n = \sum_{i=1}^{m} h_i,$$

and be $w = \psi(u)$. From Lemma 27 the word wxy has the slope $p/q = \langle 0, h_1, ..., h_m \rangle$. From Proposition 18 and Lemma 28 it follows that the slopes of the standard words $\lambda(wxy) = \lambda(w)axy$ and $\mu(wxy) = \mu(w)ayx$, where $\lambda(w)a, \mu(w)a \in PER$ (cf. Proposition 17), are:

$$\rho(\lambda(w)axy) = p/(p+q) = \langle 0, 1 + h_1, h_2, ..., h_m \rangle,$$

$$\rho(\mu(w)ayx) = q/(p+q) = \langle 0, 1, h_1, ..., h_m \rangle.$$

Constructions and Bounds for
Visual Cryptography*

Giuseppe Ateniese[1], Carlo Blundo[1], Alfredo De Santis[1], and Douglas R. Stinson[2]

[1] Dipartimento di Informatica ed Applicazioni,
Università di Salerno, 84081 Baronissi (SA), Italy

[2] Department of Computer Science and Engineering
and Center for Communication and Information Science
University of Nebraska-Lincoln, Lincoln NE 68588, USA

Abstract. A visual cryptography scheme for a set $\mathcal{P}$ of n participants is a method to encode a secret image SI into n images in such a way that any participant in $\mathcal{P}$ receives one image and only qualified subsets of participants can "visually" recover the secret image, but non-qualified sets of participants have no information, in an information theoretical sense, on SI. A "visual" recover for a set $X \subseteq \mathcal{P}$ consists of stacking together the images associated to participants in X. The participants in a qualified set X will be able to see the secret image without any knowledge of cryptography and without performing any cryptographic computation.

In this paper we propose two techniques to construct visual cryptography schemes for any access structure. We analyze the structure of visual cryptography schemes and we prove bounds on the size of the image distributed to the participants in the scheme. We provide a novel technique to realize k out of n visual cryptography schemes. Finally, we consider graph-based access structures, that is access structures in which any qualified set of participants contains at least an edge of a given graph whose vertices represent the participants of the scheme. Our constructions for 2 out of n visual cryptography schemes are the best possible with respect to pixel expansion and relative difference.

1   Introduction

A visual cryptography scheme for a set $\mathcal{P}$ of n participants is a method to encode a secret image SI into n images in such a way that any participant in $\mathcal{P}$ receives one image and only qualified subsets of participants can "visually" recover the secret image, but non-qualified sets of participants have no information, in an information theoretical sense, on SI. A "visual" recover for a set $X \subseteq \mathcal{P}$ consists of stacking together the images associated to participants in X. The participants

* Research of C. Blundo and A. De Santis is partially supported by Italian Ministry of University and Research (M.U.R.S.T.) and by National Council for Research (C.N.R.). Research of D. R. Stinson is supported by NSF grant CCR-9402141.

Hence, $\lambda(w)a = \psi(u_1)$, $\mu(w)a = \psi(u_2)$, where u_1 and u_2 have, respectively, the integral representations $(h_1 + 1, h_2, ..., h_m)$ and $(1, h_1, ..., h_m)$. Since

$$|u_1| = \sum_{i=1}^{m} h_i = n + 1 = |u_2|,$$

it follows that $u_1, u_2 \in a\mathcal{A}^n$, so that $\lambda(w)a, \mu(w)a \in X_{n+1}$.

Let us now prove the inverse inclusion i.e. $X_{n+1} \subseteq (\lambda(X_n) \cup \mu(X_n))a$. Since by Lemma 26, $Card(X_n) = 2^{n-1}$ and $Card(X_{n+1}) = 2^n$ it is sufficient to prove, as we shall do, that $Card((\lambda(X_n) \cup \mu(X_n))a) = 2^n$. To this end we prove that $\lambda(X_n)a \cap \mu(X_n)a = \emptyset$, so that since λ and μ are injective one easily derives the result. Let us then suppose that there exists $v \in \lambda(X_n)a \cap \mu(X_n)a$. One has then $v = \mu(w)a = \lambda(u)a$, with $w, u \in X_n$. This implies $\mu(w) = \lambda(u)$. Since w, u terminate with the letter a then the last letter of $\mu(w)$ is b, whereas the last letter of $\lambda(w)$ is a which is a contradiction.

References

1. J. Berstel and A. de Luca, Sturmian words, Lyndon words and trees, *Preprint L.I.T.P. 95/24, University of Paris 7, June 95, Theoretical Computer Science*, to appear

2. J. Berstel and P. Séébold, A characterization of Sturmian morphisms, *Lect. Notes Comp. Sci.* 1993, vol. 711, pp.281-290.

3. J. Berstel and P. Séébold, A remark on Morphic Sturmian words, *R.A.I.R.O., I.T.*, 28(1994) 255-263.

4. J. Berstel and P. Séébold, Morphismes de Sturm, *Bull. Belg. Math. Soc.* 1 (1994) 175-189.

5. T.C. Brown, Descriptions of the characteristic sequence of an irrational, *Canad. Math. Bull.*, 36 (1993) 15-21.

6. D. Crisp, W. Moran, A. Pollington and P. Shiue, Substitution invariant cutting sequences, *J. théorie des nombres de Bordeaux*, 5(1993) 123-138.

7. A. de Luca, Sturmian words: Structure, Combinatorics, and their Arithmetics, *Theoretical Computer Science, special issue on Formal Languages*, to appear.

8. A. de Luca, On standard Sturmian morphisms, *Preprint 95/18 Dipartimento di Matematica Università di Roma "La Sapienza"*.

9. A. de Luca and F. Mignosi, Some Combinatorial properties of Sturmian words, *Theoretical Computer Science*, 136(1994) 361-385.

10. M. Lothaire, *Combinatorics on words*, (Addison-Wesley, Reading, MA, 1983).

11. F. Mignosi, Infinite words with linear subword complexity, *Theoretical Computer Science*, 65(1989) 221-242.

12. F. Mignosi and P. Séébold. Morphismes sturmiens et règles de Rauzy, *J. théorie des nombres de Bordeaux*, 5(1993) 221-233.

13. M. Morse and G.A. Hedlund, Symbolic dynamics II: Sturmian trajectories, *Amer. J. Math.*, 62(1940), 1-42.

14. G. Rauzy, Mots infinis en arithmétique, in M. Nivat and D.Perrin, eds., *Automata in Infinite words*, Lecture Notes in Computer Science, vol.192 (Springer, Berlin, 1984) pp.164-171.

in a qualified set X will be able to see the secret image without any knowledge of cryptography and without performing any cryptographic computation.

The best way to understand visual cryptography is by resorting to an example. Suppose that there are 4 participants, that is $\mathcal{P} = \{1, 2, 3, 4\}$ and that the qualified sets are all subsets of $\mathcal{P}$ containing at least one of the following three sets $\{1, 2\}$, $\{2, 3\}$, and $\{3, 4\}$, that is the family of qualified sets is $\mathcal{A} = \{\{1, 2\}, \{2, 3\}, \{3, 4\}, \{1, 2, 3\}, \{1, 2, 4\}, \{1, 3, 4\}, \{2, 3, 4\}, \{1, 2, 3, 4\}\}$. We want to encode the secret image "ICALP 96". The four shares generated by a visual cryptography scheme for $\mathcal{A}$ are given in appendix. They look like random patterns and indeed no one of them gives any information, even to a very powerful machine, on the original image. To decrypt the secret image the reader should xerox each pattern on a separate transparency, stack together the transparencies associated to participants in any qualified set, and project the result with an overhead projector. If the transparencies are aligned carefully, then the reader will get the images showed in the remaining part of appendix.

This new cryptographic paradigm has been recently introduced by Naor and Shamir [8]. They analyzed the case of k out of n visual cryptography schemes in which the secret image is visible if and only if any k transparencies are stacked together. But the secret image is totally invisible if fewer than k transparencies are stacked together. A possible application of them is the following [8]. The 2 out of 2 visual cryptography scheme can be thought of as a private key cryptosystem. We encode the secret printed message into two randomly looking transparencies, one of the two image will be a printed page of ciphertext which can be sent by mail or fax, whereas the other printed transparency serves as a secret key. The original image is revealed by stacking together the two transparencies. This system is similar to the one time pad as each page of ciphertext is decoded by using a different transparency, but it is not required any cryptographic computation, the decoding is done by the human visual system.

Visual cryptography schemes with extended capabilities have been analyzed in [2]. The authors present a general technique to implement extended visual cryptography schemes which uses hypergraph colourings.

In this paper we extend the Naor and Shamir's model to *any* general access structure, where an access structure is a specification of all qualified subsets of participants. We propose two different techniques to construct visual cryptography schemes for any access structure. We analyze the structure of visual cryptography schemes and we prove bounds on the size of the image distributed to the participants in the scheme. We provide a novel technique to realize k out of n visual cryptography schemes. In particular we consider graph-based access structures, that is access structures in which any qualified set of participants contains at least an edge of a given graph whose vertices represent the participants of the scheme. Our constructions for 2 out of n visual cryptography schemes are the best possible with respect to pixel expansion and relative difference.

Due to the space limit all proofs have been omitted. The interested reader can find them in [1] along with other results on VCS and several examples.

2 The Model

Let $\mathcal{P}$ be a set of participants, a *monotone* access structure Γ on $\mathcal{P}$ is a subset $\Gamma \subseteq 2^{\mathcal{P}} \setminus \{\emptyset\}$, such that if $A \in \Gamma$ and $A \subseteq A' \subseteq \mathcal{P}$ then $A' \in \Gamma$.

The *closure* of Γ, denoted by $cl(\Gamma)$, is the set $\{C \mid B \in \Gamma \text{ and } B \subseteq C \subseteq \mathcal{P}\}$.

For a monotone access structure Γ we have $\Gamma = cl(\Gamma)$. All access structures considered in this paper are monotone.

Let Γ be an access structure, a set $C \in \Gamma$ is a *minimal* set of Γ if it does not contain any set in $\Gamma \setminus \{C\}$. A *basis* Γ_0 of Γ is the family of all minimal sets of Γ. In this paper we assume that $\mathcal{P} = \{1, \ldots, n\}$.

We will refer to a participant $P \in \mathcal{P}$ as an *essential* participant if there exists a set $X \subseteq \mathcal{P}$ such that $X \cup \{P\} \in \Gamma_0$. If a participant P is not essential then we can construct a visual cryptography scheme giving him nothing as share. In fact, a non-essential participant does not participate "actively" in the reconstruction of the image, that is the information he has is not needed by any set in $\mathcal{P}$ in order to recover the shared image. Therefore, any VCS handling non-essential participants can give to these participants nothing as share. In this paper we assume that the set of participants $\mathcal{P}$ consists only of essential participants.

For sets X and Y and for elements x and y, to avoid overburdening the notation, we often will write x for $\{x\}$, xy for $\{x, y\}$, xY for $\{x\} \cup Y$, and XY for $X \cup Y$.

We assume that the message consists of a collection of black and white pixels. Each pixel appears in n versions called *shares*, one for each transparency. Each share is a collection of m black and white subpixels. The resulting structure can be described by an $n \times m$ Boolean matrix $S = [s_{ij}]$ where $s_{ij} = 1$ iff the j-th subpixel in the i-th transparency is black. Therefore the grey level of the combined share, obtained by stacking the transparencies $i_1, \ldots, i_s$, is proportional to the Hamming weight $w(V)$ of the m-vector $V = OR(r_{i_1}, \ldots, r_{i_s})$ where $r_{i_1}, \ldots, r_{i_s}$ are the rows of S associated with the transparencies we stack. This grey level is interpreted by the visual system of the users as black or as white in according with some rule of contrast.

Definition 1. Let Γ be an access structure on a set of n participants. Two collections (multisets) of $n \times m$ boolean matrices $\mathcal{C}_0$ and $\mathcal{C}_1$ constitute a *visual cryptography scheme* (Γ, m)-*VCS* if there exist values $\alpha(m)$ and $\{t_X\}_{X \in \Gamma_0}$ satisfying:

1. *Any qualified set* $X = \{i_1, i_2, \ldots, i_p\} \in \Gamma_0$ *can recover the shared image.*
   Formally, for any $S \in \mathcal{C}_0$, the "or" V of rows $i_1, i_2, \ldots, i_p$ satisfies $w(V) \leq t_X - \alpha(m) \cdot m$; whereas, for any $S \in \mathcal{C}_1$ it results that $w(V) \geq t_X$.
2. *Any non-qualified set* $X = \{i_1, i_2, \ldots, i_p\} \notin \Gamma$ *has no information on the shared image.*
   Formally, the two collections of $p \times m$ matrices $\mathcal{D}_t$, with $t \in \{0, 1\}$, obtained by restricting each $n \times m$ matrix in $\mathcal{C}_t$ to rows $i_1, i_2, \ldots, i_p$ are indistinguishable in the sense that they contain the same matrices with the same frequencies.

Each pixel of the original image will be encoded into n pixels each one consisting of m subpixels. To share a white/black pixel the dealer randomly chooses one of the matrices in $\mathcal{C}_0/\mathcal{C}_1$ and distributes row i to participant i. The chosen matrix defines the colour of the m subpixels in each one of the n transparencies. Observe that the size of the collections $\mathcal{C}_0$ and $\mathcal{C}_1$ does not need to be the same.

The first condition is related to the contrast of the image. It states that a qualified set of users, belonging to the basis of the access structure, stacking their transparencies can correctly recover the image shared by the dealer. The value $\alpha(m)$ is called *relative difference*, the number $\alpha(m) \cdot m$ is referred to as the *contrast* of the image, and the set $\{t_X\}_{X \in \Gamma_0}$ is called the *set of thresholds*. We want that $\alpha(m) \cdot m$ be as large as possible and at least 1 subpixel over the m subpixels, that is, $\alpha(m) \geq 1/m$. The second condition is called *security*, it implies that by inspecting the shares of a non-qualified subset of participants one cannot gain any advantage in deciding whether the shared pixel was white or black.

There are few differences between the model of visual cryptography we propose and the one presented by Naor and Shamir [8]. Our model is a generalization of the one proposed in [8] as to each set $X \in \Gamma_0$ it is associated a different threshold t_X and only sets in the basis can recover the shared image. If a set of participants $X \in \Gamma$ wants to recover the shared image, then they can consider only the shares of a set $X' \subseteq X$ such that $X' \in \Gamma_0$. Notice that with our definition it is not excluded that a set of participants $X \in \Gamma \setminus \Gamma_0$ stacking their transparencies does not get the original image. Visual cryptography schemes in which any set $X \in \Gamma$ satisfies Property 1 of the Definition 1 are called *strong*.

In this paper we consider only VCS in which the collections $\mathcal{C}_0$ and $\mathcal{C}_1$ have the same size, i.e., $|\mathcal{C}_0| = |\mathcal{C}_1| = r$. Actually, this is not a restriction at all. Indeed, given an access structure Γ, we can obtain, from an arbitrary VCS for Γ, a VCS, having the same parameters m and $\alpha(m)$, with equally sized $\mathcal{C}_0$ and $\mathcal{C}_1$. Moreover, we do not consider access structures containing "isolated" participants, namely we suppose that for any $X \in \Gamma_0$ it holds that $|X| \geq 2$. If the access structure Γ contains some isolated participants, say $P_{i_1}, \ldots, P_{i_t}$, then we can always realize a VCS for Γ by "augmenting" a VCS Σ' for the access structure $\Gamma_0' = \Gamma_0 \setminus \left\{ \{P_{i_1}\}, \ldots, \{P_{i_t}\} \right\}$. It is enough to add to any $M \in \mathcal{C}_0$ and any $M' \in \mathcal{C}_1$ new rows indexed by $i_1, \ldots, i_t$ whose entries are all equal to zero and to one, respectively.

2.1 Basis Matrices

Most of the constructions in this paper are realized using two $n \times m$ matrices, S^0 and S^1 called *basis matrices* satisfying the following definition.

Definition 2. Let Γ be an access structure on a set of n participants. A visual cryptography scheme (Γ, m)-VCS with relative difference $\alpha(m)$ and set of thresholds $\{t_X\}_{X \in \Gamma_0}$ is realized using the $n \times m$ basis matrices S^0 and S^1 if the following two conditions hold.

1. If $X = \{i_1, i_2, \ldots, i_p\}$ is a qualified set (i.e., $X \in \Gamma_0$), then: The "or" V of rows $i_1, i_2, \ldots, i_p$ of S^0 satisfies $w(V) \leq t_X - \alpha(m) \cdot m$; whereas, for S^1 it results that $w(V) \geq t_X$.

2. If $X = \{i_1, i_2, \ldots, i_p\}$ is not a qualified set (i.e., $X \notin \Gamma$) then: The two $p \times m$ matrices obtained by restricting S^0 and S^1 to rows $i_1, i_2, \ldots, i_p$ are equal up to a columns permutation.

The collections $\mathcal{C}_0$ and $\mathcal{C}_1$ are obtained by permuting the columns of the corresponding matrix (S^0 for $\mathcal{C}_0$ and S^1 for $\mathcal{C}_1$) in all possible ways. Note that, in this case, the size of the collections $\mathcal{C}_0$ and $\mathcal{C}_1$ is the same and it is denoted by r. This technique has been introduced in [8]. The algorithm for the VCS based on the previous construction of the collections $\mathcal{C}_0$ and $\mathcal{C}_1$ has small memory requirements (it keeps only the basis matrices S^0 and S^1) and it is efficient (to choose a matrix in $\mathcal{C}_0$ ($\mathcal{C}_1$) it only generates a permutation of the columns of S^0 (S^1)).

3 An n Out of n Scheme

In this section we recall some of the results presented in [8] for n out of n visual cryptography schemes realizing the access structure $\Gamma = \{\mathcal{P}\}$, that is, the original message is visible if and only if all n transparencies are stacked together, but totally invisible if fewer than n transparencies are stacked together or analysed by any other method.

The construction of a general n out of n scheme is obtained by means of the construction of the basis matrices S^0 and S^1 defined as follows:
S^0 is the matrix whose columns are all the boolean n-vectors having an even number of 1; whereas, S^1 is the matrix whose columns are all the boolean n-vectors having an odd number of 1.

Lemma 3. ([8]) *The above scheme is a n out of n VCS with parameters $m = 2^{n-1}$, $\alpha(m) = 1/2^{n-1}$ and $r = 2^{n-1}!$.*

The scheme realized using the previous construction is optimal since in any n out of n visual cryptography scheme m has to be at least 2^{n-1} and $\alpha(m)$ can be at most $1/2^{n-1}$ (see [8]).

Let Γ be an access structure on a set $\mathcal{P}$ of participants. Given a subset of participants $\mathcal{P}' \subseteq \mathcal{P}$, we define the access structure *induced by $\mathcal{P}'$* as the family of sets $\Gamma[\mathcal{P}'] = \{X \in \Gamma : X \subseteq \mathcal{P}'\}$. The following lemma is immediate.

Lemma 4. *Let Γ be an access structure on a set $\mathcal{P}$ of participants and let Γ' be an induced structure of Γ. Let m' the minimum value for which there exists a (Γ', m')-VCS. For any (Γ, m)-VCS it has to be $m \geq m'$.*

4 General Constructions

In this section we will present two construction techniques to realize visual cryptography schemes for any access structure.

4.1 A Construction Using Cumulative Arrays

The first construction we consider is based on the *cumulative array* introduced in [10]. Let Γ be a monotone access structure on the set of participants $\mathcal{P} = \{1, 2, \ldots, n\}$. With Z_M we denote the collection of the maximal non-qualified sets of Γ. Hence,

$$Z_M = \{B \subseteq \mathcal{P} \mid B \notin \Gamma \text{ and } B \cup \{i\} \in \Gamma \text{ for all } \{i\} \in \mathcal{P} \setminus B\}.$$

A *cumulative map* (β, T) for the access structure Γ is a finite set T along with a mapping $\beta : \mathcal{P} \longrightarrow 2^T$ such that for $Q \subseteq \mathcal{P}$ we have that

$$\bigcup_{a \in Q} \beta(a) = T \iff Q \in \Gamma.$$

We can realize a cumulative map (β, T) for any access structure Γ based on the collection of the maximal non-qualified sets $Z_M = \{S_1, \ldots, S_t\}$ as follows. Let $T = \{T_1, \ldots, T_t\}$ and for any $i \in \mathcal{P}$ let

$$\beta(i) = \{T_j \mid i \notin S_j, 1 \leq j \leq t\}. \tag{1}$$

It is easy to see that for any $X \in \Gamma$ we have $\bigcup_{i \in X} \beta(i) = T$; whereas any set $X \notin \Gamma$ will be missing a $T_j \in T$. From the previous cumulative mapping we can obtain a *cumulative array* CA as follows. A cumulative array CA is a $|\mathcal{P}| \times |T|$ boolean matrix such that $CA(i, j) = 1$ if and only if $i \notin S_j$.

At this point we can illustrate a technique to realize visual cryptography schemes for any access structure Γ. Our technique is based on the n out of n visual secret sharing scheme of Section 3. Let Z_M be set of the maximal non-qualified sets of Γ and let $t = |Z_M|$. Let CA be the cumulative array for Γ obtained using the cumulative map (1). Let $\hat{S}^0$ and $\hat{S}^1$ be the basis matrices for a t out of t visual cryptography scheme. The basis matrices S^0 and S^1 for a visual cryptography scheme for Γ can be constructed as follows. For any fixed i let $j_{i,1}, \ldots, j_{i,g_i}$ be the integers j such that $CA(i, j) = 1$. The i-th row of S^0 (S^1) consists of the *or* of the rows $j_{i,1}, \ldots, j_{i,g_i}$ of $\hat{S}^0$ ($\hat{S}^1$).
Next theorem holds.

Theorem 5. *Let Γ be an access structure and let Z_M be the family of the maximal non-qualified sets of Γ. There exists a strong (Γ, m)-VCS with $m = 2^{|Z_M|-1}$ and $t_X = m$ for any $X \in \Gamma$.*

4.2 Constructing VCS from Smaller Schemes

In this section we present a construction for visual cryptography schemes using small schemes as building blocks in the construction of larger schemes.

Let Γ' and Γ'' be two access structures on a set of n participants $\mathcal{P}$. Suppose there exist a (Γ', m')-VCS and a (Γ'', m'')-VCS with basis matrices $S^{0,\Gamma'}$, $S^{1,\Gamma'}$ and $S^{0,\Gamma''}$, $S^{1,\Gamma''}$, respectively. We will show how to construct a VCS for the access structure $\Gamma = \Gamma' \cup \Gamma''$. From the matrices $S^{0,\Gamma'}$, $S^{1,\Gamma'}$, $S^{0,\Gamma''}$, and $S^{1,\Gamma''}$ we construct two pairs of matrices, $(\hat{S}^{0,\Gamma'}, \hat{S}^{1,\Gamma'})$ and $(\hat{S}^{0,\Gamma''}, \hat{S}^{1,\Gamma''})$, consisting of n rows as follows. Let us show how to construct $\hat{S}^{0,\Gamma'}$. For $i = 1, \ldots, n$, the i-th row of $\hat{S}^{0,\Gamma'}$ has all zeroes as entries if the participant i is not an essential participant of Γ'; otherwise, it is the row of $S^{0,\Gamma'}$ corresponding to participant i. The matrices $\hat{S}^{1,\Gamma'}$, $\hat{S}^{0,\Gamma'}$, and $\hat{S}^{1,\Gamma'}$ are constructed similarly. Finally, the basis matrices S^0 and S^1 for Γ will be realized concatenating the matrix $\hat{S}^{0,\Gamma'}$ with $\hat{S}^{0,\Gamma''}$ and the matrix $\hat{S}^{1,\Gamma'}$ with $\hat{S}^{1,\Gamma''}$, respectively (i.e., $S^0 = \hat{S}^{0,\Gamma'} \circ \hat{S}^{0,\Gamma''}$ and $S^1 = \hat{S}^{1,\Gamma'} \circ \hat{S}^{1,\Gamma''}$, where with $\circ$ we denote the operator "concatenation" of two matrices). Theorem 6 states that the scheme obtained using previous technique does realize a VCS.

Theorem 6. *Let Γ' and Γ'' be two access structures. For any (Γ', m')-VCS and (Γ'', m'')-VCS, both constructed using basis matrices, the previous construction gives a $(\Gamma' \cup \Gamma'', m' + m'')$-VCS. If the original VCS are strong so it is the resulting VCS.*

Next corollary is an immediate consequence of Theorem 6.

Corollary 7. *Let Γ be an access structure. If $\Gamma = \cup_{i=1}^{w} \Gamma_i$ and, for $i = 1, \ldots, w$, there exists a (Γ_i, m_i)-VCS constructed using basis matrices, then there exists a (Γ, m)-VCS, where $m = \sum_{i=1}^{w} m_i$. If the original VCS are strong so it is the resulting VCS.*

From Lemma 3 and Corollary 7 the following theorem holds.

Theorem 8. *Let Γ be an access structure. There exists a strong (Γ, m)-VCS where $m = \sum_{X \in \Gamma_0} 2^{|X|-1}$.*

Previous theorem states a general result on the existence of VCS for any access structure Γ. For special classes of access structures it is possible to achieve a smaller value of m as we will show in Section 6 for k out of n VCS and in Section 7 for graph-based access structures.

5   On the Structure of VCS

In this section we provide some useful properties of VCS. First, we show how to construct VCS for any non-connected access structure using VCS for its connected parts. Then, we prove that any matrix M in the collection $\mathcal{C}_0 \cup \mathcal{C}_1$ has to contain some predefined patterns (sub-matrices).

Non-Connected Access Structures An access structure Γ on a set of participants $\mathcal{P}$ is *connected* if there is no partition of $\mathcal{P}$ in two sets $\mathcal{P}'$ and $\mathcal{P}''$ such that $\Gamma_0 \subseteq 2^{\mathcal{P}'} \cup 2^{\mathcal{P}''}$. If an access structure Γ is not connected, then we can realize a VCS for Γ simply by constructing VCS for its connected parts and then by putting together the schemes in a suitable way as stated in the next theorem.

Theorem 9. *Let Γ' and Γ'' be two access structures on disjoint sets of participants $\mathcal{P}'$ and $\mathcal{P}''$, respectively. If there exist a (Γ', m')- VCS and a (Γ'', m'')- VCS, then there is a $(\Gamma' \cup \Gamma'', m)$- VCS, where $m = \max\{m', m''\}$.*

Unavoidable Patterns Let M be a matrix in the collection $\mathcal{C}_0 \cup \mathcal{C}_1$ of a (Γ, m)-VCS for the access structure Γ on a set of participants $\mathcal{P}$. For $X \subseteq \mathcal{P}$, M_X denotes the m-vector obtained considering the *or* of the vectors corresponding to participants in X; whereas $M[X]$ denotes the $|X| \times m$ matrix obtained from M by considering only the rows corresponding to participants in X.

Lemma 10. *Let X and Y be two non-empty subsets of participants and let Γ be an access structure. If $XY \in \Gamma_0$, then in any (Γ, m)- VCS, for any matrix $M \in \mathcal{C}_1$ it holds that*

$$w(M_{XY}) - w(M_X) \geq \alpha(m) \cdot m.$$

The matrices in $\mathcal{C}_0 \cup \mathcal{C}_1$ have to contain some predefined patterns referred to as *unavoidable patterns*. For instance, for any $X \in \Gamma_0$ and any matrix $M \in \mathcal{C}_1$, the matrix $M[X]$, for $i = 1, \ldots, |X|$, contains at least $\alpha(m) \cdot m$ columns with '1' in the i-th position and all zeroes in the other entries. This is an immediate consequence of Lemma 10. Indeed, by considering $X = Y \cup \{i\}$ we get $w(M_{Y \cup \{i\}}) - w(M_Y) \geq \alpha(m) \cdot m$. Therefore, there must be at least $\alpha(m) \cdot m$ columns in $M[X]$ with '1' in the row i and all zeroes in the other entries. Another unavoidable pattern contained in any matrix $M \in \mathcal{C}_0$ is the following: For any $X \in \Gamma_0$ the matrix $M[X]$ contains at least $\alpha(m) \cdot m$ columns with entries all equal to '0'. In fact from Property 1. of Definition 1 we have $w(M_X) \leq t_X - \alpha(m) \cdot m \leq m - \alpha(m) \cdot m$. Next corollaries are immediate consequences of the unavoidable patterns.

Corollary 11. *Let Γ be an access structure on a set of participants $\mathcal{P}$. For any essential participant $i \in \mathcal{P}$, in any (Γ, m)- VCS, for any matrix $M \in \mathcal{C}_0 \cup \mathcal{C}_1$ it holds that $w(M_i) \geq \alpha(m) \cdot m$.*

Corollary 12. *In any (Γ, m)- VCS where $\Gamma \neq \mathcal{P}$ it holds that $m \geq 2$.*

Corollary 13. *For any $X \in \Gamma_0$ we have $t_X \geq |X| \cdot \alpha(m) \cdot m$.*

Another consequence of the unavoidable patterns is that for the access structures based on complete graphs (i.e., access structures such that $\Gamma_0 = \{\{i, j\} : i, j \in \mathcal{P}\}$) the rows of any matrix $M \in \mathcal{C}_1$ of a VCS represent a Sperner family (see for example [6]). In fact, let $M \in \mathcal{C}_1$ be an $n \times m$ boolean matrix and let $G = \{g_1, \ldots, g_m\}$ be a ground set. For $i = 1, \ldots, n$, row i of M represents the subset $A_i = \{g_w : \text{the entry } (i, w) \text{ of } M \text{ is equal to } 1\}$ of G. Since any two rows of M contain the patterns $\begin{bmatrix} 1 \\ 0 \end{bmatrix}$ and $\begin{bmatrix} 0 \\ 1 \end{bmatrix}$, then the sets $A_1, \ldots, A_n$ constitute a Sperner family in the ground set G. Therefore, the rows of the matrix M represent a Sperner family. This will be exploited in Theorem 19 and in Section 7. The following basis matrices represent a VCS for the access structure based on the complete graph with 6 vertices. This scheme is constructed from a Sperner family in a ground set containing four elements.

$$S^0 = \begin{bmatrix} 1100 \\ 1100 \\ 1100 \\ 1100 \\ 1100 \\ 1100 \end{bmatrix} \qquad S^1 = \begin{bmatrix} 1010 \\ 1001 \\ 1100 \\ 0110 \\ 0101 \\ 0011 \end{bmatrix}$$

Next lemma states the existence of other unavoidable patterns in any matrix in the collections C_0 and C_1. Basically, it says that for any $Y \notin \Gamma$ and for any $M \in C_0 \cup C_1$, the matrix $M[Y]$ contains at least $\alpha(m) \cdot m$ columns whose entries are all equal to zero.

Lemma 14. *Let Y and Z two nonempty subsets of participants such that $ZY \in \Gamma_0$. In any (Γ, m)-VCS, for any matrix $M \in C_0 \cup C_1$ it holds that*

$$w(M_Y) \leq \min\{t_X : Y \subset X, \ X \in \Gamma_0\} - \alpha(m) \cdot m.$$

Next lemma shows the existence of unavoidable patterns in any matrix $M \in C_0$ of any strong VCS.

Lemma 15. *Let Γ be an access structure on a set $\mathcal{P}$ of participants. In any strong (Γ, m)-VCS any matrix $M \in C_0$ has at least $\alpha(m) \cdot m$ columns whose entries are all equal to zero.*

Next theorem, based on the existence of the unavoidable patterns, provides a characterization of VCS having $m = 2$.

Theorem 16. *Let Γ be a connected access structure. If there exists a $(\Gamma, 2)$-VCS, then Γ_0 is the edge-set of a complete bipartite graph.*

6 A k Out of n Scheme

A k out of n visual cryptography scheme (also called (k, n)-threshold scheme) realizes the access structure:

$$\{B \subseteq \mathcal{P} : |B| \geq k\}.$$

For k out of n schemes the original message is visible if any k of n participants stack their transparencies, but totally invisible if fewer than k transparencies are stacked together or analysed by any other method.

We can construct k out of n visual cryptography schemes by using the two techniques described in Sections 4.1 and 4.2. By using the technique based on cumulative arrays we obtain a strong k out of n VCS in which $m = 2^{\binom{n}{k-1}-1}$ and $t_X = m$ for any set X of cardinality k; whereas by using the technique of Section 4.2 we obtain a strong k out of n VCS in which $m = \binom{n}{k} \cdot 2^{k-1}$ and t_X has the same value for any set X of cardinality k.

In the following we describe a method to construct k out of n VCS achieving better results, in terms of the size of the shares, than the techniques described in Sections 4.1 and 4.2. The method we introduce is based on *perfect hashing* [5, 7, 3].

Definition 17. A *starting matrix* $SM(n, l, k)$ is a $n \times l$ matrix whose entries are elements of a set $\{a_1, \ldots, a_k\}$, with the property that, for any subset of k rows, there exists at least one column such that the entries in the k given rows of that column are all distinct.

Given a matrix $SM(n, l, k)$ we can construct a k out of n VCS as follows: The $n \times (l \cdot 2^{k-1})$ basis matrices S^0 and S^1 are constructed by replacing the symbols $a_1, \ldots, a_k$, respectively, with the 1-st,..., k-th rows of the corresponding basis matrices of the k out of k VCS described in Section 3. The scheme obtained is a k out of n VCS as the following theorem states.

Theorem 18. If there exists a $SM(n, l, k)$ then there exists a strong k out of n VCS with $m = l \cdot 2^{k-1}$.

The SM matrix is a representation of a *Perfect Hash Family* (PHF). Fredman and Komlós [5] proved that for any PHF it holds that $l = \Omega(k^{k-1}/k!) \log n$. They also proved the weaker but simpler bound $l = \Omega(1/\log k) \log n$. Melhorn [7] proved that there exist PHFs with $l = O(ke^k) \log n$. In [3] it has been provided a recursive construction for PHFs with $l = O\left((\log n)^{\log(\binom{k}{2}+1)}\right)$.

Naor and Shamir [8] showed that there exist k out of n visual cryptography schemes with $m = 2^{O(k \log k)} \cdot \log n$. Our construction produce a smaller value of m than their construction, but this has been achieved by relaxing the condition that t_X are equal as required in [8].

Next theorem provides a lower bound on m for any k out of n visual cryptography scheme.

Theorem 19. In any k out of n visual cryptography scheme it results that

$$\binom{n}{k-1} \leq \binom{m}{\lfloor m/2 \rfloor}.$$

Since $\binom{m}{\lfloor m/2 \rfloor} \leq 2^m$ and $\binom{n}{k-1} \geq (\frac{n}{k-1})^{k-1}$ we have that in any k out of n visual cryptography scheme $m = \Omega(k \log(n/k))$.

7 VCS for Graph Access Structures

In this section we present some bounds on the size of the shares for graph-based access structures. A graph-based access structure is an access structure which is the closure of the edge set of a given graph, that is, an access structure for which the set of participants can be identified with the vertex set $V(G)$ of a graph $G = (V(G), E(G))$, and the sets of participants qualified to reconstruct the secret image are precisely those containing an edge of G.

We first recall some terminology from graph theory. Given a graph $G = (V(G), E(G))$ a *vertex cover* of G is a subset of vertices $A \subseteq V(G)$ such that every edge in $E(G)$ is incident with at least one vertex in A. The *complete graph* K_n is the graph on n vertices in which any two vertices are joined by an edge. The *complete multipartite graph* $K_{a_1, a_2, \ldots, a_n}$ is a graph on $\sum_{i=1}^{n} a_i$ vertices, in which the vertex set is partitioned into subsets of size a_i $(1 \leq i \leq n)$ called *parts*, such that vw is an edge if and only if v and w are in different parts. An alternative way to characterize a complete multipartite graph is to say that the complementary graph is a vertex-disjoint union of cliques. Note that the complete graph K_n can be thought of as a complete multipartite graph with n parts of size 1.

Exploiting the construction used in Theorem 6 we can prove the following theorem.

Theorem 20. *Suppose G is a graph with a vertex cover of size v and let $\Gamma_0 = E(G)$. Then, there exists a $(\Gamma, 2v)$-VCS.*

The following result can be obtained applying Theorem 19.

Corollary 21. *Let $\Gamma_0 = E(K_n)$. There exists a (Γ, m)-VCS if and only if $n \leq \binom{m}{\lfloor \frac{m}{2} \rfloor}$.*

A modification of Corollary 21, using the well-known "splitting technique" from secret sharing schemes, together with Lemma 4, can be used to prove the following result for complete multipartite graphs.

Theorem 22. *Let $\Gamma_0 = E(K_{a_1, \ldots, a_n})$. There exists a (Γ, m)-VCS if and only if $n \leq \binom{m}{\lfloor \frac{m}{2} \rfloor}$.*

Let $\Gamma = cl(E(K_n))$. Corollary 21 proves a lower bound on the value of m for a (Γ, m)-VCS which is met with equality when the VCS for Γ is constructed from a Sperner family in a ground set of m elements. In such a scheme we have $\alpha(m) = 1/m$. If we are interested in realizing schemes achieving a greater relative difference, then we can use the following construction. The basis matrix S^1 is realized by considering all the columns of weight $\lfloor n/2 \rfloor$. Hence, $m = \binom{n}{\lfloor n/2 \rfloor}$. It is easy to see that any row in S^1 has weight equal to $\binom{n-1}{\lfloor n/2 \rfloor - 1}$. The basis matrix S^0 is realized by considering n equal rows of weight $\binom{n-1}{\lfloor n/2 \rfloor - 1}$. In such a scheme we have $\alpha(m) = (\lfloor n/2 \rfloor \lceil n/2 \rceil)/(n^2 - n)$. This is the best possible value for the relative difference as stated by the following theorem.

Theorem 23. *Let $\Gamma_0 = E(K_n)$. In any (Γ, m)-VCS it holds that*

$$\alpha(m) \leq \frac{\lfloor \frac{n}{2} \rfloor \lceil \frac{n}{2} \rceil}{n(n-1)}.$$

It is possible to construct schemes with a smallest pixel expansion achieving the bound provided by previous theorem. These schemes are based on Hadamard matrices and designs. In such schemes we have that $m = n$ and we will prove in the final version of this paper that the construction from the Hadamard matrix achieves the smallest possible m (for the given maximum possible $\alpha(m)$).

Using the splitting technique together with Lemma 4 we obtain that the bound provided by Theorem 23 holds also when $\Gamma_0 = E(K_{a_1,\ldots,a_n})$.

Acknowledgements We would like to express our gratitude to Ugo Vaccaro for illuminating discussions. Many thanks go to Carmine Di Marino who implemented the techniques presented in this paper and provided us with the images depicted in the appendix.

References

1. G. Ateniese, C. Blundo, A. De Santis, and D. R. Stinson, *Visual Cryptography for General Access Structures*. Available from *ECCC*, Electronic Colloquium on Computational Complexity (TR96–012), via WWW using http://www.eccc.uni-trier.de/eccc/.
2. G. Ateniese, C. Blundo, A. De Santis, and D. R. Stinson, *Extended Schemes for Visual Cryptography*, preprint, 1995.
3. M. Atici, S. S. Magliveras, D. R. Stinson, and W.-D. Wei, *Some Recursive Constructions for Perfect Hash Families*, Technical Report UNL, Univ. of Nebraska-Lincoln, June 1995.
4. C. Blundo, A. De Santis, D. R. Stinson, and U. Vaccaro, *Graph Decomposition and Secret Sharing Schemes*, Journal of Cryptology, Vol. 8, (1995), pp. 39-64.
5. M. L. Fredman and J. Komlós, *On the Size of Separating System and Families of Perfect Hash Functions*, SIAM J. Alg. Disc. Meth., Vol 5, No 1, March 1984.
6. J. H. van Lint and R. M. Wilson, A Course in Combinatorics, Cambridge University Press, (1992).
7. K. Melhorn, *On the Program Size of Perfect and Universal Hash Functions*, in Proc. of 23rd Annual IEEE Symposium on Foundation of Computer Science, pp. 170-175, 1982.
8. M. Naor and A. Shamir, *Visual Cryptography*, in "Advances in Cryptology – Eurocrypt '94", A. De Santis Ed., Vol. 950 of Lecture Notes in Computer Science, Springer-Verlag, Berlin, pp. 1–12, 1995.
9. P. Elias, *Zero Error Capacity Under List Decoding*, IEEE Trans. Inform. Theory, Vol. 34, 1988.
10. G. J. Simmons, W. Jackson, and K. Martin, *The Geometry of Shared Secret Schemes*, Bulletin of the ICA, 1:71–88, 1991.
11. D. R. Stinson, *Decomposition Constructions for Secret Sharing Schemes*, IEEE Trans. Inform. Theory, Vol. 40, pp. 118–125, 1994.

Visual Cryptography for "ICALP 96"

In this appendix an example of the secret image, the shares corresponding to single participants, and few groups of participants are depicted. The access structure is $\{\{1,2\}, \{2,3\}, \{3,4\}, \{1,2,3\}, \{1,2,4\}, \{1,3,4\}, \{2,3,4\}, \{1,2,3,4\}\}$.

Secret Image

Share of participant 1   Share of participant 2

Share of participant 3   Share of participant 4

Image of participants 1 and 2   Image of participants 2 and 3

Image of participants 3 and 4   Image of participants 1 and 3

On Capital Investment

Yossi Azar[1], Yair Bartal[2], Esteban Feuerstein[3], Amos Fiat[1], Stefano Leonardi[4] and
Adi Rosén[5]

[1] Department of Computer Science, Tel Aviv University. §
[2] International Computer Science Institute, Berkeley. ¶
[3] Depto. de Computacion, Facultad de Ciencias Exactas y Naturales, Universidad de
Buenos Aires & Instituto de Ciencias, Universidad de General Sarmiento. ‖
[4] International Computer Science Institute, Berkeley, & Dipartimento di Informatica
Sistemistica, Università di Roma "La Sapienza". **
[5] Department of Computer Science, University of Toronto. ††

Abstract. We deal with the problem of making capital investments in machines for manufacturing a product. Opportunities for investment occur over time, every such option consists of a capital cost for a new machine and a resulting productivity gain, i.e., a lower production cost for one unit of product. The goal is that of minimizing the total production and capital costs when future demand for the product being produced and investment opportunities are unknown. This can be viewed as a generalization of the ski-rental problem and related to the mortgage problem [3].

If all possible capital investments obey the rule that lower production costs require higher capital investments, then we present an algorithm with constant competitive ratio.

If new opportunities may be strictly superior to previous ones (in terms of both capital cost and production cost), then we give an algorithm which is $O(\min\{\log C, \log \log P, \log M\})$ competitive, where C is the ratio between the highest and the lowest capital costs, P is the ratio between the highest and the lowest production costs, and M is the number of investment opportunities. We also present a lower bound on the competitive ratio of any on-line algorithm for this case which is $\Omega(\max\{\log C, \frac{\log \log P}{\log \log \log P}, \frac{\log M}{\log \log M}\})$. This shows that the competitive ratio of our algorithm is tight (up to constant factors) as a function of C, and not far from the best achievable as a function of P and M.

§ Research supported in part by a grant from the Israel Academy of Sciences. e-mail:
{azar,fiat}@math.tau.ac.il.

¶ Research supported in part by the Rothchild Postdoctoral fellowship. email:
yairb@icsi.berkeley.edu.

‖ Research supported in part by EEC project KIT-DYNDATA. email: efeuerst@dc.uba.ar.

** Research supported in part by EU Esprit Long Term Research project ALCOM IT under
contract 20244, and by Italian Ministry of Scientific Research Project 40% "Algoritmi,
Modelli di Calcolo e Strutture Informative". email: leon@dis.uniroma1.it.

†† email: adiro@cs.toronto.edu.

1 Introduction

We consider the problem of manufacturing costs versus capital investment. A factory uses machines for producing units of some product. The production of each unit requires some fixed cost for using the machine (electricity, raw material, etc.). Over time opportunities for investment in new machines, that would replace the old ones, become available. Such opportunities could be the result of technological improvement, relocation to a cheaper market, or any other investment that would replace the facilities of the factory and would lead to lower production costs. We model all these opportunities as machines that can be bought, and then used to produce the units of the product. The factory must decide if to invest in buying new machines to reduce production costs while neither future demand for the product nor future investment opportunities are known.

Many financial problems require to take decisions without having knowledge, or while having only partial knowledge, of future opportunities. Competitive analysis of financial problems has received an increasing attention during the last years, for instance for currency exchange problems [2] or asset allocation [5].

The problem considered in this paper is a generalization of one of the basic on-line problems, the *ski-rental* problem due to L. Rudolph (see [4]), a model for the well known practical problem "rent or buy?". The ski-rental problem can be stated as follows: you don't know in advance how many times you will go skiing; renting a pair of skis costs $r; to purchase your own pair costs $p. When do you buy? It is not hard to see that the best competitive ratio is obtained if you buy when the total rental cost (thus far) is equal to the cost of buying your own pair. Another problem considered in this model in the past is the so-called mortgage problem [3], where a fluctuating mortgage rate and associated re-financing charges lead to the question, re-finance or not?

While for the ski-rental problem the only possible capital expenditure is to purchase a pair of skis, and then the "production" costs drop to zero, in the capital investment problem there may be many future capital expenditure options and the resulting productivity gains are unknown. Unlike the mortgage problem where the future demand is the servicing the entire debt (which is a known fixed value), and a capital investment has a fixed cost, in the capital investment problem future demand is unknown and capital investments may have arbitrary costs.

We consider two models for our problem, and call the first one the *convex case*. Here, we assume that to get a lower production cost, one must spend more as capital expenditures. In this case we get a constant competitive ratio. This scenario is usually true in manufacturing: purchasing a better machine costs more. However, sometimes technological breakthroughs are achieved, after which both machine costs and production costs are reduced. This matches our second model, the *non-convex case*, which allows both capital and production costs to drop.

In contrast to the convex case, for the non-convex case we present a lower bound on the competitive ratio of any on-line algorithm for the problem which is $\Omega(\max\{\log C, \log\log P/\log\log\log P, \log M/\log\log M\})$, where C is the ratio between the highest and the lowest capital costs, P is the ratio between the highest and the lowest production costs, and M is the number of investment opportunities. We complement this lower bound with an algorithm for general capital investment scenarios which is $O(\min\{\log C, \log \log P, \log M\})$ competitive.

2 The On-line Capital Investment Problem

Imagine a factory whose goal is to produce units of some commodity at low cost. From time to time, orders for units of the commodity arrive, and at times new machines become available. Every such machine is characterized by its *production cost*, and by its *capital cost*. The production cost is the cost of producing one unit of commodity using this machine. The capital cost is the capital investment necessary to buy the machine. We assume that once a machine becomes available, then it is available forever. We also assume that one can produce an unlimited number of units with any machine. An algorithm for this problem has to decide what machines to buy and when to do so, as to minimize the total cost (capital costs plus production costs).

More formally, an instance of the problem consists of a sequence of machines, and a sequence of orders of demand. Machine m_i is defined by the triplet (t_i, c_i, p_i), where t_i is the time at which the machine becomes available, c_i is its capital cost, and p_i is its production cost. Every order is defined by its arrival time. Without loss of generality we may assume that the j'th order appears at time j. Moreover, since any reasonable algorithm will not buy a new machine when there is no order pending, we can assume that for any machine i such that $j < t_i \leq j+1, t_i = j+1$. At any time t, the algorithm can buy any of the available machines (those with $t_i \leq t$), and then produce one unit of the commodity.

We say that machine m_i *dominates* machine m_j if both the production cost and the capital cost of m_i are lower than those of m_j. We call an instance of the problem *convex* if no machine presented dominates another. I.e., an instance is convex if for any two machines i, j such that $p_i < p_j$ it holds that $c_i \geq c_j$. To help distinguish between the two versions of the problem, we call the general case *non-convex*.

We note that if all machines are available at the very beginning, then all machines that are dominated by others can be removed. Thus, whenever all machines are available in advance, we are left with the convex setting. The non-convex setting only makes sense if machines appear over time and it is possible that a better machine (in terms of both capital cost and production cost criteria) will appear later.

2.1 Performance Measures

We measure the performance of an on-line algorithm for this problem by its competitive ratio [6]. Let σ be a sequence of offers of machines and orders of demand for units of the commodity to be produced.

We denote by $\mathrm{ON}(\sigma)$ the cost of the on-line algorithm ON for the problem over the sequence σ, and with $\mathrm{OPT}(\sigma)$ the cost of an *optimal* off-line algorithm that knows the entire sequence σ in advance. We parameterize the sequences by the ratio between the cost of the most expensive and cheapest machines (denoted by C), by the ratio between the highest and the lowest production cost (denoted by P), and by the total number of machines presented during the sequence (denoted by M). Denote by $\Sigma(C, P, M)$ the set of sequences that obey the above restrictions.

The competitive ratio of an algorithm may be a function of the above parameters. An on-line algorithm ON is $\rho(C, P, M)$-competitive for a set $\Sigma(C, P, M)$ of sequences if

$$\sup_{\sigma \in \Sigma(C,P,M)} \frac{\text{ON}(\sigma)}{\text{OPT}(\sigma)} \leq \rho(C, P, M).$$

3   Upper Bound for the Convex Case

In this section we study the convex case in which a machine with a lower production cost cannot be cheaper than a machine with a higher production cost. We present an on-line algorithm for the convex case with competitive ratio 7.

3.1   The Algorithm

The algorithm is defined as follows: before producing the first unit the algorithm buys the machine m_i that minimizes $p_i + c_i$ amongst all machines available at the beginning of the sequence. It then produces the first unit of commodity. The initial cost $p_i + c_i$ is considered a *production* cost.

Let α and β be positive constants satisfying $2/\alpha \leq 1$ and $1/\alpha + 2\beta \leq 1$. In particular we choose $\alpha = 2$ and $\beta = 1/4$.

Before producing any subsequent unit of commodity the algorithm considers buying a new machine. However, it is not always allowed to buy a new machine. When an amount of c is spent as capital cost to buy a machine, it is not allowed to buy another machine until the algorithm spends at least $\beta \cdot c$ on production.

When it is allowed to buy a machine, the algorithm buys the machine m_i that minimizes production cost p_i amongst all machines of capital cost at most α times the total production cost incurred since the beginning of the sequence. If no such machine is available, the algorithm does not buy a new machine.

3.2   Analysis

We prove that the competitive ratio of the above algorithm is $1 + \alpha + 1/\beta = 7$.

We use the following notation. Fix the sequence σ. Denote by $\text{ON} = \text{ON}^c + \text{ON}^p$ the total cost of the algorithm that is equal to the sum of the total capital cost ON^c and the total production cost ON^p. Let p^t be the production cost incurred by the on-line algorithm to produce unit number t. Let ON_t^p be the production cost incurred by the algorithm to produce the first t units, i.e., $\text{ON}_t^p = \sum_{i=1}^{t} p^i$. Let OPT_t be the optimal total (capital and production) cost to produce the first t units. We start by proving a bound on the total cost spent on purchasing machines, in terms of the total production cost incurred.

Lemma 1. *The total capital cost ON^c incurred by the on-line algorithm is at most $(\alpha + 1/\beta)$ its total production cost ON^p.*

Proof. The capital cost of the last machine bought is at most α times the total production cost. For every other machine, the production costs in the interval between the time this machine has been bought, and the time the next machine is bought, is at least β times the capital cost of the machine. These intervals do not overlap, and thus the total capital cost of all the machines except the last one sums to at most $1/\beta$ times the total production cost. ∎

We now relate the production cost of the on-line algorithm to the total cost of the off-line algorithm.

Lemma 2. *At any time t the production cost ON_t^p of the on-line algorithm is at most the total cost OPT_t of the off-line algorithm.*

Proof. We prove the claim by induction on the number of units produced.

For $t = 1$ the claim holds since the on-line production cost of the first unit (defined as the sum of the capital and the production costs of the first machine bought) is the minimum possible expense to produce the first unit. Therefore $ON_1^p \leq OPT_1$.

Consider unit t for $t > 1$, and assume the claim holds for any unit $t' < t$. Let m be the machine used by the on-line algorithm to produce unit t. Let m' be the machine used by the optimal off-line solution to produce unit t, p' its production cost, and c' its capital cost.

If $p' \geq p^t$ then we have $ON_t^p = ON_{t-1}^p + p^t \leq OPT_{t-1} + p' \leq OPT_t$.

If $p' < p^t$ then the on-line algorithm did not buy machine m' just before producing unit t. Let the capital cost of the last machine bought by the on-line algorithm (i.e. m) be $\bar{c}$, and assume it was bought just before unit $\bar{t}$ was produced. Since we consider the convex case we have that $p' < p^t = p^{\bar{t}}$ implies $c' \geq \bar{c}$.

As we assume that the on-line algorithm did not buy m' just before producing unit t, one of the following holds:

1. The capital cost of machine m' was too high, i.e., less than $\frac{1}{\alpha}c'$ was spent on production since the start of the sequence.
2. It was not allowed to buy any machine at this time: less than $\beta \cdot \bar{c}$ was spent on production since machine m was bought, and until unit number $t - 1$ is produced.

We consider each of these cases.

For the first case we have that $ON_t^p = ON_{t-1}^p + p^t \leq 2 \cdot ON_{t-1}^p \leq \frac{2}{\alpha}c' \leq OPT_t$. For the second case we have

$$ON_t^p = ON_{\bar{t}-1}^p + \sum_{i=\bar{t}}^{t-1} p^i + p^t \leq ON_{\bar{t}-1}^p + 2\sum_{i=\bar{t}}^{t-1} p^i < ON_{\bar{t}-1}^p + 2\beta \cdot \bar{c} .$$

We now distinguish between two cases, depending on whether machine m' is available before unit $\bar{t}$ is produced. The first case is that machine m' becomes available only after unit $\bar{t}$ is produced. In this case we have

$$ON_t^p < ON_{\bar{t}-1}^p + 2\beta \cdot \bar{c} \leq OPT_{\bar{t}-1} + 2\beta \cdot \bar{c} \leq OPT_{\bar{t}-1} + \bar{c} \leq OPT_{\bar{t}-1} + c' \leq OPT_t .$$

The second case is when machine m' is available before unit number $\bar{t}$ is produced. We have that its capital cost, c', is higher than $\alpha \cdot ON_{\bar{t}-1}^p$, otherwise the on-line algorithm would have bought this (or a better) machine at time $\bar{t}$, which contradicts $p^{\bar{t}} > p'$. Therefore we have

$$\text{ON}_{\bar{t}}^p < \text{ON}_{\bar{t}-1}^p + 2\beta \cdot \bar{c} \leq \text{ON}_{\bar{t}-1}^p + 2\beta \cdot c' \leq (1/\alpha)c' + 2\beta \cdot c' = (1/\alpha + 2\beta)c' \leq \text{OPT}_t \ .$$

∎

Combining Lemma 1 and Lemma 2 we get the following theorem.

Theorem 3. *The algorithm presented above for the convex case of the on-line capital investment problem is $(1 + \alpha + 1/\beta)$-competitive.*

4   Lower Bound for the Non-Convex Case

In contrast to the constant upper bound proved in the previous section, in this section we prove an $\Omega(\max\{\log C, \log\log P/\log\log\log P, \log M/\log\log M\})$ lower bound on the competitive ratio of any on-line algorithm, where C is the ratio between the highest and the lowest capital costs, P is the ratio between the highest and the lowest production costs, and M is the number of presented machines.

We now describe the instance of the problem on which the lower bound is achieved. We let C be some large power of 2. The capital costs of all the machines in the instance are powers of 2 between 1 and C, and their production costs will be of the form $1/\log^k C$, for some integer k.

We assign a *level* between 0 and $\log C$ to each machine; machines of level i have capital cost $c_i = 2^i$.

We say that a *phase of level i* starts when a machine of level i is presented. A phase of level i ends when one of the following occurs:

1. The on-line algorithm buys a machine of level i.
2. The on-line algorithm has reached a global cost (production and capital) in the phase greater or equal to
   - 1 for $i = 0$;
   - 2 for $i = 1$;
   - $\frac{i}{2}c_i$ for $i \geq 2$.
3. A phase of level higher than i ends.

Immediately after the end of the phase a new machine of the same level is presented and a new phase of the same level starts.

Let $n_k(i) = \frac{i!}{k!}$ for $i = 1, \ldots, \log C$, $k = 1, \ldots, \log C - 1$, and let $n_0(i) = 2i!$. When a phase of level i with an associated machine of production cost p ends in Case 1 or Case 2, a set of $i + 1$ machines are presented, one for each level $j = 0, \ldots, i$. The production cost of the appropriate machine of level j is defined to be

$$p_j = \frac{p}{(\log C)^{1 + \sum_{k=0}^{j-1} n_k(j)}}.$$

At the beginning we assume that a phase of level $i = \log C$ and $p = 1$ ends, so that a first set of machines is presented, with capital and production costs as defined above.

The sequence will be over with the end of the phase of level $\log C$ associated with the machine of capital cost C presented at the beginning. The sequence is built so that there is only one machine of capital cost C presented in the whole sequence, and that machine's production cost is at most $1/\log C$ the production cost of any other machine presented in the sequence.

We define a relation of inclusion between phases. A phase of level i contains all the phases of level $j < i$ that start simultaneously or during the level i phase. Note that no phase of level $j > i$ starts during a phase of level i.

We call a phase *active* if it is not yet ended. At every point in time one phase is active at every level.

We call a phase that ends in Case 1 or Case 2 a *complete phase* and a phase that ends in Case 3 an *incomplete phase*. If a phase of level i is complete then the i phases at lower levels that have ended as a consequence of the end of this level i phase are incomplete.

Lemma 4. *At most i machines of level $i - 1$ are presented during a phase of level i for $i \geq 2$, and at most 2 machines of level 0 are presented during a phase of level 1.*

Proof. For $i = 1$, the cost incurred by the on-line algorithm in every complete phase of level 0 is at least 1, thus after at most 2 level 0 phases the on-line cost will reach 2, and the level 1 phase will end in Case 2. For $i \geq 2$, a new machine of level $i - 1$ is presented when the on-line algorithm buys the previous one of that level or when its cost reaches $\frac{i-1}{2}c_{i-1}$. In any case, the on-line algorithm's cost for the phase of level $i-1$ is at least c_{i-1}. Hence, the maximum number x of phases of level $i-1$ is restricted to be $x c_{i-1} \leq \frac{i}{2}c_i$, which implies $x \leq i$. ∎

The production costs defined above were chosen so as to obey the property stated in the following lemma. The proof is omitted.

Lemma 5. *A machine of level i has production cost less or equal to $1/\log C$ times the production cost of any machine of level $k \leq i$ presented before the starting of the phase, and of any machine of level $k < i$ presented during the phase.*

Proof. A new machine of level i is presented when a phase of level $j \geq i$ ends. Let p be the production cost of the machine associated with the phase of level j that has just ended. The production cost of the machine associated with the new phase of level i is $p_i = \dfrac{p}{(\log C)^{1+\sum_{k=0}^{i-1} n_k(i)}}$. We prove the claim by induction. If a new phase of level i starts then a previous phase of level i has just ended. Say $\tilde{p}_i$ is the production cost of the associated machine. We know that $\tilde{p}_i \geq p \geq p_i \log C$. Since, by induction, the claim is true for the previous phase of level i, with production cost $\tilde{p}_i$, then the production cost p_i of the machine presented in the new phase of level i is less or equal than $1/\log C$ times the production cost of any machine of level $k \leq i$ presented before the start of the phase.

Let us prove the second part of the claim, i.e. that every machine presented in the phase has production cost at least $p_i \log C$. First we prove it for a phase of level $i = 1$. It contains at most 2 machines of level 0, with production cost $\frac{p}{\log C}$ and $\frac{p}{\log^2 C}$. Since $p_i = \frac{p}{\log^3 C}$, the claim is proved.

Finally, we prove the claim for $i > 1$. We prove it for the machine associated to the last phase of level $i-1$ contained in the phase of level i, by induction the machine with lowest production cost presented in the phase. In a phase of level i at most i machine of level $i-1$ are presented . A new machine is presented when the previous phase of level $i-1$ is stopped because of Case 1 or Case 2. (Recall that no phase of level higher than i ends during a phase of level i.) Hence, the production cost p'_{i-1} of the last machine of level $i-1$ is $p'_{i-1} = \dfrac{p}{(\log C)^{i(1+\sum_{k=0}^{i-2} n_k(i-1))}} \geq \dfrac{p}{(\log C)^{\sum_{k=0}^{i-1} n_k(i)}} \geq p_i \, \log C$.

■

Consider a phase of level i. Let O_i and z_i be respectively the global cost and the production cost of the on-line algorithm during that phase. The global on-line cost in a phase is given by the production cost during the phase plus the capital cost charged to the on-line algorithm for buying machines of level not higher than i (possibly including the machine of level i if the phase ends in Case 1).

We will denote by A_i the global cost of the adversary in the case in which it is committed to buy either machines presented at the beginning of the phase of level not higher than i or machines presented during the phase that, by definition of the sequence, have level lower than i. In fact this is not a restriction since the cost paid by the adversary during the unique phase of level $\log C$ is equal to the global cost of the adversary over all the sequence.

First, we state two upper bounds on the global cost A_i of the adversary during a phase. The first upper bound considers the case in which the adversary only buys the machine of level i. Observe that the machines of lower level presented at the beginning or during the phase have production cost higher than the machine of level i, and hence, in this case, can be ignored by the adversary.

Lemma 6. *If the adversary buys the machine of level i, then $A_i \leq \frac{3}{2} c_i$.*

Proof. Since the on-line algorithm has not bought the machine of level i, then the adversary produces with a production cost that is at most $\frac{1}{\log C}$ times the on-line production cost during the phase. Therefore the adversary's production cost during the phase is at most $\frac{z_i}{\log C}$. We can assume $\log C \geq 2$. For $i = 0$, we have that $A_0 \leq c_0 + \frac{z_0}{\log C} \leq c_0 + \frac{1}{\log C} \leq \frac{3}{2} c_0$. For $i = 1$, $A_1 \leq c_1 + \frac{z_1}{\log C} \leq c_1 + \frac{2}{\log C} \leq 3 = \frac{3}{2} c_1$. Finally, for $i \geq 2$, the phase ends as soon as the global cost of the algorithm during the phases reaches the value $\frac{i}{2} c_i$. Then we have that $z_i \leq \frac{i}{2} c_i \leq \frac{\log C}{2} c_i$. Therefore $A_i \leq c_i + \frac{z_i}{\log C} \leq c_i + \frac{c_i}{2} = \frac{3}{2} c_i$. ■

The second upper bound on A_i considers the case in which the adversary does not buy the machine of level i, and its global cost is composed by the sum of the costs of the phases of level $i-1$ contained in the phase of level i. A phase of level i (complete or incomplete) is partitioned into a sequence of phases of level $i-1$, whose number we indicate with s_i. The last one of those phases is possibly incomplete, while the first $s_i - 1$ are complete. Thus, we get the following lemma.

Lemma 7. *Let A_{i-1}^j, $j = 1, \ldots, s_i$ be the global cost of the adversary during the j-th phase of level $i-1$. Then $A_i \leq \sum_{j=1}^{s_i} A_{i-1}^j$.*

Theorem 8. *If an algorithm for the non-convex on-line capital investment problem is ρ-competitive then $\rho = \Omega(\log C)$.*

Proof. We first show that for any algorithm, the (unique) phase of level $\log C$ arrives to an end. For this it is enough to show that the global cost incurred by the on-line algorithm will eventually reach the value $\frac{\log C}{2} C$. This follows immediately from the fact that the production costs of all the machines that are presented in this instance are lower-bounded by the production cost of the machine of level $\log C$, that is strictly positive.

We will now show that any on-line algorithm pays a global cost (over the sequence) of at least $\frac{1}{6} \log C$ times the cost of the adversary.

We focus our attention on a phase of level i. The phase starts when a machine of level i is presented. By definition, one machine for each lower level is simultaneously presented. Observe that during this phase the on-line algorithm does not buy any machine of level higher that i. Otherwise the phase immediately ends, a new machine of level i is presented and a new phase of level i starts.

We prove the following inductive claim:

- $O_i \geq \frac{i}{6} A_i$ for a complete phase;
- $O_i \geq \frac{i}{6} A_i - \frac{c_i}{2}$ for an incomplete phase.

We prove the claim for each of the three cases in which a phase ends. Recall that in Case 1 and Case 2 the phase is complete and the first part of the claim must be proved, while in Case 3 the phase is incomplete and the second part of the claim must be proved. For Case 1 and Case 3 the proof is by induction on i. We assume that the claim holds for phases of level $i - 1$. The claim is obviously true for $i = 0$.

1. In Case 1, the on-line algorithm buys the machine of level i before the global production cost has reached the value $\frac{1}{2} c_i$. Then, the global cost of the on-line algorithm in the phase is given by the sum of the costs for each of the s_i phases of level $i - 1$ contained in the phase of level i, plus the capital cost c_i for buying the machine of level i that ends the phase. Let O_{i-1}^j be the global cost of the on-line algorithm during the j-th phase of level $i - 1$. Without loss of generality we consider that the last phase of level $i - 1$ is an incomplete phase (the inductive hypothesis is otherwise stronger). Then

$$O_i = \sum_{j=1}^{s_i} O_{i-1}^j + c_i \geq \sum_{j=1}^{s_i} \frac{i-1}{6} A_{i-1}^j - \frac{c_{i-1}}{2} + c_i$$

$$\geq \frac{i-1}{6} A_i + \frac{3}{4} c_i \geq \frac{i}{6} A_i .$$

The first inequality stems by applying the inductive hypothesis. The second inequality is obtained from Lemma 7 and the relation $c_i = 2c_{i-1}$. Finally, the last inequality follows from Lemma 6.

2. In Case 2 the global cost of the on-line algorithm has reached the value $\frac{1}{2} c_i$. Then, applying Lemma 6, it follows that

$$O_i = \frac{i}{2} c_i \geq \frac{i}{6} A_i .$$

3. In Case 3 the phase ends because a new machine of the same level is presented, i.e., a phase of a higher level ends in Case 1 or Case 2. The global cost of the on-line algorithm for an incomplete phase is obtained by summing up the global cost for every phase of level $i - 1$ contained in the incomplete phase of level i. Clearly, the capital cost of the machine of level i is not paid by the on-line algorithm. Note that in this case the last phase of level $i-1$ is also incomplete. The claim is proved as follows:

$$O_i = \sum_{j=1}^{s_i} O_{i-1}^j \geq \sum_{j=1}^{s_i} \frac{i-1}{6} A_{i-1}^j - \frac{c_{i-1}}{2}$$
$$\geq \frac{i-1}{6} A_i - \frac{c_i}{2} + \frac{c_i}{4} \geq \frac{i}{6} A_i - \frac{c_i}{2}$$

The first equality indicates the on-line global cost in the phase, while the first inequality is derived by applying the inductive hypothesis. The second inequality is obtained from Lemma 7 and the relation between the capital costs of machines of level i and $i - 1$, while the final inequality is derived from the upper bound on the adversary's global cost of Lemma 6.

Since the unique phase of level $\log C$ is a complete phase and its completion ends the sequence, then the theorem follows from the claim on complete phases. ■

The following corollary states the lower bound as a function of the ratio P between the highest and the lowest production costs, and of the maximum number of presented machines M.

Theorem 9. *If an algorithm for the non-convex on-line capital investment problem is ρ-competitive then $\rho = \Omega(\frac{\log \log P}{\log \log \log P})$ and $\rho = \Omega(\frac{\log M}{\log \log M})$.*

Proof. The claim follows by observing that in the sequence for the $\Omega(\log C)$ lower bound, the ratio between the maximum and the minimum production cost is $P = (\log C)^{(1+\sum_{k=0}^{\log C-1} n_k(\log C))} = ((\log C)^{O(\log C)!})$ and the number of machines presented is $M \leq 1 + \sum_{k=0}^{\log C-1} n_k(\log C) = O((\log C)!)$. ■

5 Upper Bound for the Non-Convex Case

In this section we present an algorithm for the general (non-convex) case of the problem. This algorithm is $O(\min\{\log C, \log \log P, \log M\})$-competitive.

5.1 The Algorithm

Given any new machine with production cost p_i, and capital cost c_i, our algorithm first rounds these costs up to the nearest power of two, i.e., if $2^{j-1} < c_i \leq 2^j$ then set $c_i = 2^j$, and if $2^{k-1} < p_i \leq 2^k$ then set $p_i = 2^k$.

The algorithm is defined as follows. Before producing the first unit buy the machine m_i that minimizes $p_i + c_i$ amongst all machines available at the beginning of the interval. It then produces the first unit of commodity. The initial cost $p_i + c_i$ is considered a *production* cost.

Before producing any subsequent unit, order all available machines by increasing production cost and (internally) increasing capital cost. Number the machines by index i, and let p_i, c_i be the production costs and capital costs, respectively. For all i, $p_i \leq p_{i+1}$, and if $p_i = p_{i+1}$, then $c_i \leq c_{i+1}$. Buy the machine with least i that satisfies the two following conditions:

- Its production cost p_i is smaller than the production cost of the current machine.
- A production cost of at least c_i has been spent since the last time a machine with capital cost c_i has been bought (or since the beginning of the run, if no such machine has been previously bought).

5.2 Analysis

We prove that the above algorithm has competitive ratio of $O(\min\{\log C, \log\log P, \log M\})$. In the following analysis we assume that all capital and production costs are indeed powers of 2, as rounded by the on-line algorithm. Clearly, an adversary that uses this modified sequence incurs a cost of at most twice the cost incurred by the real adversary that uses the real sequence.

Denote by $\text{ON} = \text{ON}^c + \text{ON}^p$ the total cost of the algorithm which is equal to the sum of the total capital cost ON^c and of the total production cost ON^p.

Lemma 10. *The total capital cost ON^c is at most $O(\log C)$ times the total production cost ON^p.*

Proof. For a given j, consider all the machines of cost 2^j that are bought. A machine of cost 2^j can be bought only after an amount of 2^j has been spent on production since the last time a machine of the same cost has been bought (or since the beginning of the sequence, if not such machine was previously bought). It follows that for any j, the total cost of the algorithm for buying machines of cost 2^j is at most ON^p. Since there are at most $\lceil \log C \rceil$ different costs for the machines, $\text{ON}^c = O(\text{ON}^p \cdot \log C)$. ∎

Lemma 11. *The total capital cost ON^c is at most $O(\log M')$ times the total production cost ON^p, where M' is the total number of machines bought.*

Proof. Let 2^l be the cost of the cheapest machine, and let 2^k be the cost of the most expensive machine such that $2^k \leq \text{ON}^p$. All machines bought by the algorithm have costs between 2^l and 2^k. For any j, $l \leq j \leq k$, let b_j be the number of machines of cost 2^j bought by the algorithm. An upper bound on the capital cost spent by the algorithm is the maximum of $Z = \sum_{j=l}^{k} b_j 2^j$ as a function of the variables $b_j, j = l, \ldots, k$ subject to constraints $b_j 2^j \leq \text{ON}^p$, and $\sum_{j=l}^{k} b_j = M'$.

We relax the problem by allowing the variables b_j to assume non-integer values. Clearly the solution to this relaxed problem is also an upper bound on ON^c. Denote by $b_j^r, j = l, \ldots, k$, the variables of the relaxed problem. For the optimal solution of the relaxed problem, there are no h and h' such that $l \leq h < h' \leq k$, $b_h > 0$ and

$b_{h'} < \frac{ON^p}{2^{h'}}$. Otherwise, there would have been a solution with higher value of the objective function Z of the relaxed problem, achieved by reducing b_h and increasing $b_{h'}$ by the same amount, until either $b_h = 0$, or $b_{h'} = \frac{ON^p}{2^{h'}}$.

From the above observation we derive an upper bound on the maximum of the objective function (and thus an upper bound on ON^c). If $\frac{ON^p}{2^k} \geq M'$ then the maximum is achieved by setting $b_k^r = M'$ and $b_j^r = 0$ for $l \leq j \leq k - 1$. In this case $\sum_{j=l}^{k} b_j 2^j \leq ON^p$, and the lemma clearly holds.

If $\frac{ON^p}{2^k} < M'$, let h^* be the maximum integer such that $\sum_{j=h^*}^{k} \frac{ON^p}{2^j} \geq M'$. An upper bound on the maximum of the objective function is obtained by assigning $b_j^r = \frac{ON^p}{2^j}$, $j = h^* + 1, \ldots, k$, $b_{h^*}^r = M' - \sum_{j=h^*+1}^{k} b_j^r \leq \frac{ON^p}{2^{h^*}}$, and $b_j^r = 0$, for $j = l, \ldots, h^* - 1$. The upper bound on the value of the objective function is

$$\sum_{j=l}^{k} b_j^r 2^j \leq \sum_{j=h^*}^{k} b_k^r 2^{k-j} 2^j = \sum_{j=h^*}^{k} b_k^r 2^k \leq ON^p \cdot (k - h^* + 1).$$

It remains to show that $k - h^* + 1 = O(\log M')$. By the definition of h^*, $\sum_{j=h^*+1}^{k} 2^{k-j} b_k^r = \sum_{j=h^*+1}^{k} b_j^r = \sum_{j=h^*+1}^{k} \frac{ON^p}{2^j} < M'$. Therefore, we get $\sum_{j=h^*}^{k} 2^{k-j} b_k^r \leq 3M'$ and thus $(2^{k-h^*+1} - 1) \leq \frac{3M'}{b_k^r}$. Since $ON^p \geq 2^k$, it follows that $b_k^r \geq 1$, and we obtain $2^{k-h^*+1} - 1 \leq 3M'$. Since $M' \geq 1$, we obtain $k - h^* + 1 = O(\log M')$. ∎

Corollary 12. *The total capital cost ON^c is at most $O(\log M)$ times the total production cost ON^p.*

Corollary 13. *The total capital cost ON^c is at most $O(\log \log P)$ times the total production cost ON^p.*

Proof. The algorithm buys a machine only if the production cost decreases. Since all production costs are powers of 2, the algorithm buys at most $O(\log P)$ machines. ∎

Lemma 14. *At any time the total production cost ON^p of the on-line algorithm is at most twice the total cost of the off-line algorithm.*

Proof. Let p^t be the production cost incurred by the on-line algorithm to produce unit number t. Let ON_t^p be the production cost incurred by the algorithm to produce the first t units, i.e., $ON_t^p = \sum_{i=1}^{t} p^i$. Let OPT_t be the lowest (optimal) cost to produce the first t units.

We prove by induction on t that $ON_t^p \leq 2 \cdot OPT_t$.

To produce the first unit the on-line algorithm buys the machine that minimizes the sum of production and capital costs. This is the minimum possible cost to produce the first unit. Thus, $ON_1^p \leq OPT_1$.

Consider unit t for $t > 1$, and assume that the claim holds for every unit number t', $t' < t$. Let m be the machine used by ON to produce unit t. Let m' be the machine used by OPT to produce unit t, p' its production cost and c' its capital cost.

If $p^t \leq p'$ then we have

$$\mathrm{ON}_t^p = \mathrm{ON}_{t-1}^p + p^t \leq 2 \cdot \mathrm{OPT}_{t-1} + p^t \leq 2 \cdot \mathrm{OPT}_{t-1} + p' \leq 2 \cdot \mathrm{OPT}_t .$$

We now consider the case in which $p' < p^t$. It follows that the on-line algorithm did not buy machine m' although it was available before unit t is produced. If this happens one of the following holds:

- The production cost incurred by the on-line algorithm by time $t - 1$ is less than c'. On the other hand, the optimal off-line algorithm buys machine m', incurring a cost of c'. It follows that

$$\mathrm{ON}_t^p = \mathrm{ON}_{t-1}^p + p^t \leq 2 \cdot \mathrm{ON}_{t-1}^p \leq 2c' \leq 2 \cdot \mathrm{OPT}_t .$$

- Some machine of cost c' was previously bought by the on-line algorithm, but the production cost incurred by the algorithm since then is less than c'. Assume that such machine was bought just before unit $\bar{t}$ was produced. As unit t is produced with production cost higher than p', we can conclude that m' was not available before unit $\bar{t}$ was produced. Thus, m' was bought by the off-line algorithm after unit $\bar{t}$ is produced. On the other hand, the on-line production cost since the production of unit $\bar{t}$ is less than c'. Therefore, we have

$$\mathrm{ON}_t^p = \mathrm{ON}_{\bar{t}-1}^p + \sum_{i=\bar{t}}^{t-1} p^i + p^t \leq \mathrm{ON}_{\bar{t}-1}^p + 2\sum_{i=\bar{t}}^{t-1} p^i \leq 2 \cdot \mathrm{OPT}_{\bar{t}-1} + 2c' \leq 2 \cdot \mathrm{OPT}_t .$$

∎

We conclude with the following theorem, whose proof is straightforward from the previous lemmata.

Theorem 15. *The competitive ratio of the on-line capital investment algorithm described above is $O(\min\{\log C, \log\log P, \log M\})$.*

References

1. A. Chou, J. Cooperstock, R. El-Yaniv, M. Klugerman and F. Leighton. The Statistical Adversary Allows Optimal Money-making Trading Strategies. In *Proceedings of the 6th Annual ACM/SIAM Symposium on Discrete algorithms*, 1995.
2. R. El-Yaniv, A. Fiat, R. Karp and G. Turpin. Competitive Analysis of Financial Games. In *Proc. of the 33rd IEEE Annual Symposium on Foundations of Computer Science*, 1992.
3. R. El-Yaniv and R.M. Karp. The Mortgage Problem. In *Proceedings of the 2nd Israeli Symposium on Theory of Computing and Systems*, pp. 304–312, June 1993.
4. R.M. Karp, On-line Algorithms Versus Off-line Algorithms: How Much is it Worth to Know the Future?. In *Proc. World Computer Congress*, 1992.
5. P. Raghavan. A Statistical Adversary for On-line Algorithms. *DIMACS Series in Discrete Mathematics and Theoretical Computer Science*, Vol 7:79-83, 1992.
6. D.D. Sleator and R.E. Tarjan. Amortized Efficiency of List Update and Paging Rules. *Communications of the ACM*. 28:202–208, February 1985.

Lower Bounds for Static Dictionaries on RAMs with Bit Operations but No Multiplication

Peter Bro Miltersen

Department of Computer Science, University of Toronto, King's College Road,
Toronto Ontario M5S 1A4, Canada. Email: *pbmilter@cs.toronto.edu.*

Abstract. We consider solving the static dictionary problem with n
keys from the universe $\{0, \ldots, m-1\}$ on a RAM with direct and indirect
addressing, conditional jump, addition, bitwise Boolean operations, and
arbitrary shifts (a *Practical* RAM). For any $\epsilon > 0$, tries yield constant
query time using space m^ϵ, provided that $n = m^{o(1)}$. We show that this is
essentially optimal: Any scheme with constant query time requires space
$\geq m^\epsilon$ for *some* $\epsilon > 0$, even if $n \leq (\log m)^2$.

1 Introduction

The static dictionary problem is the following: Given a subset S of size n of the
universe $U = \{0, \ldots, m-1\}$, store it as a data structure ϕ_S in the memory of a
unit cost random access machine, using few memory registers, each containing
$O(\log m)$ bits, so that membership queries "Is $x \in S$?" can be answered efficiently
for any value of x. The set S can be stored as a sorted table using n memory
registers. Then queries can be answered using binary search in $O(\log n)$ time.
Yao [17] first considered the possibility of improving this solution and provided
an improvement for certain values of m and n.

Fredman, Komlós and Szemerédi [13] showed that for *all* values of m and
n, there is a storage scheme using $n + o(n)$ memory cells, so that queries can
be answered in constant time. Their technique is based on the family of hash
functions $h_k(x) = (kx \bmod p) \bmod s$, i.e. multiplication and integer division is
used. Since these instructions are usually considered expensive, it is interesting
to know whether their use can be avoided.

Fich and Miltersen [12] considered this problem and showed that for a RAM
with the standard, "classical" instruction set, consisting of direct and indirect
addressing, addition, subtraction, multiplication, and conditional jump, the logarithmic query time of the sorted table/binary search solution cannot be improved
without using space at least $m/n^{o(1)}$.

Unfortunately, even though the standard RAM instruction set is well studied and elegant, results about it have not got much to do with the real world!
For instance, bitwise manipulations, such as bitwise *AND* with a *mask* of the
information in a word are not allowed. Real instruction sets certainly allow such
things and real programs take advantage of them. In fact, if multiplication,
bitwise Boolean operations and *shifts* are allowed, a solution to the static dictionary problem with linear storage space and constant query time that avoids

integer division is possible: In [9] it is shown that the family of hash functions $h_k(x) = (kx \bmod 2^b)$ div 2^a, where a and b are suitably chosen integer constants, can replace the family used by Fredman, Komlós, and Szemerédi. The function $x \to x \bmod 2^b$ is easily computed as the bitwise *AND* of the input and a mask, and the function $x \to x$ div 2^a can be computed by shifting the input a bits to the right. On the other hand, multiplication is still used and, while this is in general regarded as a cheaper instruction than integer division, it is still much more expensive to implement than, say, addition. For instance, multiplication is not in AC^0, and no linear size circuit for it is known. Thus, it is still natural to ask if it can be avoided.

In this paper, we consider lower bounds for implementing static dictionaries on a RAM with word size $w = \Theta(\log m)$, with *no* multiplication instruction, but with direct and indirect addressing, conditional jumps, bitwise Boolean operations, addition, and a shift operation, capable of shifting its first input arbitrarily many bit positions to the left or right, as indicated by its second argument. We will call such a RAM a *Practical RAM*. Note that subtraction can be computed by one bitwise negation and one addition, so we don't have to add a subtraction operation to the instruction set. As far as we know, this paper is the first one proving instruction set dependent lower bounds for this instruction set (previous papers considering shift operations [5, 4] have been restricted to one-bit shifts).

The Practical RAM is quite powerful. It is certainly sufficiently powerful to implement the classical comparison based data structures for representing sets, such as trees and heaps of all kinds. It is also sufficiently powerful to implement several fundamental algorithms and data structures where indirect addressing and/or bit manipulation is essential, such as for example tries, van Emde Boas trees [11], Gabow and Tarjan's special case Union-Find algorithm [15], Chazelle's M-structure [7], Dietz' list indexing structure [8], the $O(n \log \log n)$ sorting algorithm of Andersson *et al* [2], Andersson's $O(\sqrt{\log n})$ search structure [1], and Thorup's $O(\log \log n)$ priority queue [16]. For some of these latter applications it is essential that the shift operation is capable of shifting the word an arbitrary number of places.

The hashing schemes above are examples of data structures using a stronger instruction set. An additional example is the fusion tree [14] which also uses multiplication in a non-trivial way. Andersson's structure [1] gives a multiplication-free alternative to fusion trees. In this paper we show that no such alternative exists for static dictionaries without sacrificing either constant query time or linear (or even polynomial) storage space. Our main theorem is the following.

Theorem *There is a set $S \subseteq U$ of size $n = $ polylog m, so that for any structure ϕ_S using s memory registers, each containing $w = \Theta(\log m)$ bits, and any Practical RAM program, running in time t, accepting on input ϕ_S and $x \in S$ and rejecting on input ϕ_S and $x \in U - S$, we have*

$$t = O(1) \Rightarrow s = m^{\Omega(1)} \tag{1}$$

$$s = 2^{n^{o(1)}} \Rightarrow t = \Omega(\log \log n) \tag{2}$$

We can throw in several other low level instructions beside addition, shifts, and bitwise Boolean operations, without destroying the lower bounds. It is sufficient that the operations included obey a certain *locality*-constraint, defined in Section 3. The lower bounds are proved in Section 4.

The lower bounds compare with existing upper bounds as follows: Tries yield $O(k)$ query time using space $O(nm^{1/k})$. This means that (1) is tight (for $n = $ polylog m). The second bound may seem quite weak. Indeed, no known data structure using reasonable (say $n^{O(1)}$) space beats the $O(\log n)$ query time of the sorted table/binary search solution. However, by modifying the model slightly, we can get a matching upper bound to (2). Specifically, if we allow word size $w = \Theta((\log n)(\log m)) = O(\log^2 m)$ rather than $w = \Theta(\log m)$ and an operation fold$\oplus$ which takes as input a word $x = x[w-1]\ldots x[1]x[0]$ and outputs the word y whose i'th bit $y[i]$ is $x[i] \oplus x[i-1] \oplus \cdots \oplus x[0]$, we can get a query time of $O(\log \log n)$ using a data structure with only $O(n)$ registers. This upper bound is described in Section 2. The matching lower bound still holds for this extended machine model. To get a better lower bound, we thus have to apply different techniques. Subsequent to the research reported here, this was done by Andersson *et al* [3]. They show a lower bound of $\Omega(\sqrt{\log n / \log \log n})$ on the query time for linear space data structures on AC^0 RAMs, i.e. RAMs where all compuational instructions can be implemented in AC^0. The Practical RAM is a special case.

As we state it above, the lower bound applies to deterministic computation and worst case time. This is not as restrictive as it may seem, since we do not put any bound on the time required to construct ϕ_S from S. Moreover, the lower bound in fact also holds for Las Vegas and Monte Carlo type randomized algorithms. This simple extension of the result is omitted from this version of the paper.

Notation

Throughout the paper, U denotes the set $\{0, 1, \ldots, m-1\}$. We assume m is a power of two $m = 2^b$, i.e. b is the number of bits required to describe members of U. We shall identify members x of U with their binary notation $x[b-1]x[b-2]\ldots x[0]$, so we shall also write $U = \{0, 1\}^b$.

We shall consider random access machines with word size $w \geq b+1$, i.e. each register contains a value in the set $W = \{-2^{w-1}, \ldots, 0, 1, \ldots, 2^{w-1} - 1\}$. We shall identify members x of W with their (two-complement) binary notation $x[w-1]x[w-2]\ldots x[0]$, so we shall also write $W = \{0, 1\}^w$. By convention, for $x \in W$ and $i \notin \{0, \ldots, w-1\}$, $x[i] = 0$.

For $x, y \in W$, $x \wedge y$, $x \vee y$, $x \bar{\wedge} y$, and $x \oplus y$ denotes bitwise *AND*, *OR*, *NAND*, and *XOR*, respectively, i.e. $(x \circ y)[i] = x[i] \circ y[i]$ for $\circ \in \{\wedge, \vee, \bar{\wedge}, \oplus\}$.

For $x, y \in W$, shift(x, y) is defined by $(\text{shift}(x, y))[i] = x[i-y]$, i.e. x is shifted y bits to the left if $y > 0$ and $-y$ bits to the right if $y < 0$. Bits outside the range $\{0, \ldots, w-1\}$ are chopped off. Note that we interpret the first argument of shift as a bit vector and the second argument as the binary notation of an integer.

For $x, y \in W$, $x + y \in W$ denotes the sum of x and y modulo 2^w.

2 Upper bound

We show how a RAM with word size $w = \Theta((\log n)(\log m))$ and a fold$\oplus$ instruction can solve the static dictionary problem with a data structure using a linear number of registers and with $O(\log \log n)$ query time. We make use of the following result by Carter and Wegman [6]: The class of functions $f_A : \{0,1\}^{n_1} \to \{0,1\}^{n_2}$ given by $f_A(x) = Ax$, where A is an $n_2 \times n_1$ 0-1 matrix and the matrix product is over the field $\mathbf{Z}_2$, is a universal class of hash functions. Furthermore we use the observation by [10], that if we have black boxes computing classes of universal hash functions $f_A : U \to \{0, \ldots, 2^k - 1\}$ for various $k \leq O(\log n)$, we can use them as building blocks in Fredman, Komlós and Szemerédi's two level hashing data structure. Thus, we only have to show that for any $k \times b$ matrix A with $k \leq O(\log n)$, we can compute a word containing Ax from a word containing x in time $O(\log \log n)$. We note that any such matrix A can be stored inside a single word, by concatenating the rows of A. Let A' be such a word, where bits $ib, \ldots, (i+1)b - 1$ contain the i'th row of A, $i \in \{0, \ldots, k-1\}$.

1. Create k copies of x inside a single word by repeated shift and bitwise OR in time $O(\log k) = O(\log \log n)$ (see e.g. Andersson [1]). Call the result x'. Bits $ib, \ldots, (i+1)b - 1$ of x' contain a copy of x for $i \in \{0, \ldots k - 1\}$.
2. Compute $a = x' \wedge A'$.
3. Compute $p = \text{fold}\oplus(a)$. Note that the i'th bit of Ax is the XOR of bit $(i+1)b - 1$ and bit $ib - 1$ of p for $i \in \{0, \ldots, k-1\}$.
4. Compute $p = p \oplus \text{shift}(p, b)$. Now $(Ax)[i] = p[(i+1)b - 1]$.
5. Compute $p = p \wedge \alpha$, where α is a mask where all bits are 0, except bits $b - 1, 2b - 1, \ldots, kb - 1$ which are 1. Now all irrelevant bits in p are 0.
6. We now need to move the bits in positions $b - 1, 2b - 1, \ldots, kb - 1$ of p into positions $0, 1, 2, \ldots, k - 1$. This is done by repeated shift and bitwise OR in time $O(\log k) = O(\log \log n)$, followed by bitwise AND with a mask. We have now computed Ax in time $O(\log \log n)$ and are done.

3 Model of Computation

In order to show our lower bounds, we must define the model of computation a bit more carefully.

The random access machine has an infinite sequence of registers, indexed by the integers. Each register can contain a member of W. We denote the contents of the j'th register by $M[j]$.

Furthermore, there are a finite number of *internal* (to the CPU) registers $r_0, \ldots, r_k$, also holding a member of W, used for arithmetic and indirect addressing.

The model is parameterized by a set F of functions and a set P of predicates. Each $f \in F$ is of the form $f : W^l \to W$, $l \leq k$, possibly different l's for different f's. (To be completely precise, each f is really a *family* of functions, one for each possible word size w. This makes our asymptotic results meaningful). Each $p \in P$ is of the form $p : W^l \to \{\text{TRUE}, \text{FALSE}\}$, $l \leq k$.

The instruction set of the RAM consists of the following groups of instructions:

- *Loads and moves.* The internal registers can be assigned constants, and values can be moved between them, i.e. for each $a \in W$ and $i, j \in \{0, \ldots, k\}$, we have instructions $r_i := a$ and $r_i := r_j$.
- *Shifts.* For any $h, i, j \in \{0, \ldots, k\}$ there is an instruction $r_h := \text{shift}(r_i, r_j)$ For convenience, we assume that whenever the instruction is executed, the second argument is between $-w + 1$ and $w - 1$. This assumption is easily seen to be without loss of generality.
- *Computations.* For each $f \in F$ with $f : W^l \to W$ we have an instruction $r_0 := f(r_1, \ldots, r_l)$
- *Reads and Writes.* For any $i, j \in \{0, \ldots, k\}$, we have the instructions $M[r_i] := r_j$ and $r_i := M[r_j]$
- *Conditional jumps.* If $p \in P$ with $p : W^l \to \{\text{TRUE}, \text{FALSE}\}$, and g is a positive integer denoting a line number, we have an instruction IF $p(r_1, r_2, \ldots, r_l)$ THEN GOTO g.
- *Halts.* We have an instruction ACCEPT which halts and accepts the input and an instruction REJECT which halts and rejects the input.

A RAM *program* is a finite sequence of instructions. When running the program on input x and ϕ_S, we place x in register 0 and ϕ_S in registers $1, \ldots, s$ and start execution at the first instruction. The *program counter* indicates which instruction to execute. In most cases, the program counter is incremented at the completion of each instruction. However, if the conditional jump is performed and the predicate evaluates to true, the program counter is assigned the value g. The program counter is not updated when an accept or reject instruction is performed. The running time of a computation is the number of instructions executed until the program counter points to an accept or reject instruction.

By varying F and P we get different models. The Practical RAM described in the introduction has $F = \{+, \bar{\wedge}\}$ (the other bitwise Boolean operations can be simulated using $\bar{\wedge}$ and does not have to be included) and $P = \{\text{leq}\}$ where $\text{leq}(x, y) = \text{TRUE}$ if and only if $x \leq y$.

The only requirement we need to put on P for our lower bounds to hold is that it should be finite, i.e. its size should be independent of the word size. The only requirement we put on F is that F is finite and that there are constants $\alpha, \beta > 1$ (independent of the word size), so that each $f \in F$ is (α, β)-*local*. This concept is defined next.

Note that the shift instruction is considered separately from the computation instructions. This is because the shift-function does not obey the (α, β)-locality requirement, and must therefore be dealt with separately in the proof.

Locality

In this section we define (α, β)-local functions and derive our key lemma about them.

Notation: for $I = \{i_k > i_{k-1} > \cdots > i_1\} \subseteq \{0, 1, \ldots, w-1\}$ and $x \in \{0,1\}^w$, we shall write $x[I]$ for $x[i_k]x[i_{k-1}]\ldots x[i_1]$. For instance, $(1111011)[\{4,3,2\}] = 110$.

By an *interval* I, we mean a subset of $\{0, \ldots, w-1\}$ of the form $I = \{i, i+1, \ldots, j-1, j\}$.

Let $\alpha, \beta \geq 1$. We say that a function $f : W^l \to W$ is (α, β)-*local* if for any interval I, there are intervals $I_1, I_2, \ldots, I_l$, so that

- $\#I_i \leq \alpha(\#I)$
- If $(x_i)[I_i]$ is fixed for all i, then $f(x_1, x_2, \ldots x_l)[I]$ takes on at most β different values. More precisely, for any constants $c_1, \ldots, c_l$,

$$\#\{f(x_1, x_2, \ldots, x_l)[I] \mid \forall i : (x_i)[I_i] = c_i\} \leq \beta.$$

There are several natural examples of local functions. First, $f(x) = x \bar{\wedge} y$ is $(1,1)$-local: If the values of $x[I]$ and $y[I]$ are known, the value of $(x \bar{\wedge} y)[I]$ is uniquely determined. Similarly, the other bitwise Boolean operations are $(1,1)$-local. Also, $f(x) = x + y$ is $(1,2)$-local: If the values of $x[I]$ and $y[I]$ are known, there are at most two possible values of $(x + y)[I]$, corresponding to whether or not a carry bit was propagated to the interval I during the addition. The function $\text{fold}\oplus$ from the introduction is $(1,2)$-local: If the value of $x[I]$ is known, there are at most two possible values of $\text{fold}\oplus(x)[I]$, corresponding to the two possible parities of the bits in x preceding I. Generalizing all the previous examples, any function computed by a finite state transducer is $(1,\beta)$-local for some constant β. For any fixed k, the function $\text{shift}_k(x) = \text{shift}(x, k)$ is $(1,1)$-local: Given I, let $J = \{i - k \mid i \in I, 0 \leq i - k \leq w - 1\}$. If $x[J]$ is known, we know $\text{shift}_k(x)[I]$. Note however than the function $\text{shift}(x, y)$ is *not* (α, β)-local for any constants α, β. Therefore, the shift instruction is considered separately in the definition of the RAM.

We shall consider *circuits* C with a single input $x \in W$ and a single output $C(x) \in W$, containing gates computing (α, β)- local functions of maximum arity $k \geq 2$. Arbitrary constants $c \in W$ may also be fed into the circuit. The circuits compute functions in the natural way. The size of a circuit is the number of gates it contains.

Lemma 1. *Let C be a circuit of size t. Let an interval $I \subseteq \{0, \ldots, w-1\}$ be given. We can find a subset $M \subseteq \{0, \ldots, w-1\}$, so that*

- $\#M \leq (\alpha k)^t(\#I)$
- *For any constant c, the number of elements in the set $\{C(x)[I] \mid x[M] = c\}$ is at most β^{k^t-1}.*

Proof. We do an induction in t. For $t = 0$, we either have $C(x) = a$ for some constant a or $C(x) = x$ for all x. In the first case, the lemma holds, since the size of the set $\{C(x)[I] \mid x[M] = c\}$ will be $1 = \beta^{k^t-1}$, no matter what M is. In the second case, we put $M = I$. Then, the number of elements in the set $\{C(x)[I] \mid x[M] = c\} = \{x[I] \mid x[I] = c\}$ is $1 = \beta^{k^t-1}$.

Now assume the lemma holds for all circuits of size $< t$. Let a circuit of size t be given. The gate in the circuit computing the output has the form $f(C_1, C_2, \ldots, C_l)$ with $l \leq k$, where C_i are smaller circuits. Given I, since f is (α, β)-local, we can find $I_1, I_2, \ldots, I_l$ so that $\#I_i \leq \alpha(\#I)$ and for any constants $c_1, \ldots, c_l$, $\#\{f(x_1, x_2, \ldots, x_l)[I] \mid (x_i)[I_i] = c_i \text{ for all } i\} \leq \beta$. By induction, for each i, we can find M_i so that $\#M_i \leq (\alpha k)^{t-1}(\#I_i)$ and so that for any fixed c, the number of elements in the set $\{C_i(x)[I_i] \mid x[M_i] = c\}$ is at most $\beta^{k^{t-1}-1}$. Let $M = \cup_{i=1}^l M_i$. Note that $\#M \leq l(\alpha k)^{t-1}(\alpha(\#I)) \leq (\alpha k)^t(\#I)$. If x is fixed to a constant c on the bits marked by M, let c_i be the value of the bits marked by M_i. We then have

$$\#\{C(x)[I] \mid x[M] = c\} = \#\{f(C_1(x), C_2(x), \ldots, C_l(x))[I] \mid x[M] = c\}$$

$$\leq \beta \prod_{i=1}^l \#\{C_i(x)[I_i] \mid x[M] = c\}$$

$$\leq \beta \prod_{i=1}^l \#\{C_i(x)[I_i] \mid x[M_i] = c_i\}$$

$$\leq \beta(\beta^{k^{t-1}-1})^l$$

$$\leq \beta^{k^t-1}$$

4 The lower bound

We prove the following theorem which immediately implies the theorem stated in the introduction. Fix a RAM model with function set F and predicate set P, where for some constants α, β, every $f \in F$ is (α, β)-local.

Theorem 2. *Let any constant $c > 0$ be given. There is a constant $d > 0$ so that the following holds for sufficiently large b: There is a set $S \subseteq U$ of size $n \leq b^{c+1}$, so that for any structure ϕ_S using s memory registers, each containing $w \leq b^c$ bits, and any RAM program, running in time t, accepting on input ϕ_S and $x \in S$ and rejecting on input ϕ_S and $x \in U - S$, we have $2^{2^{dt}} \log s \geq b$.*

Let $H_{c+1} \subset U$ be the set of vectors with Hamming weight exactly $c + 1$. The set S will be some subset of H_{c+1}. The strategy of the proof is the following. Suppose we can find a data structure and a query program for any such S. We derive from the program an algorithm which accepts on $x \in S$ and rejects on $x \in H_{c+1} - S$ and does not use the data structure ϕ_S. The algorithm is described in the *computation tree* formalism, outlined below. We can make such a transformation for every possible $S \subseteq H_{c+1}$. Since the trees are self-contained, different sets must have different trees, so there must be many different trees. However, if the original RAM programs are all fast and use small data structures, each tree is going to be small, which means that there are not as many trees as there need to be. This implies the lower bound.

Let $W' = W \cup \{x\} \cup \{v_1, v_2, \ldots\}$, where $x, v_1, v_2, \ldots$ are atomic symbols. A computation tree is a rooted tree with three kinds of nodes.

- Unary *computation nodes* of the form $v_i \leftarrow f(a_1, a_2, \ldots, a_l)$, where $f \in F \cup \{\text{shift}_j\}_{j=-w+1}^{w-1}$, and the arguments a_j are taken from W'. When proceeding from the root of the tree to one of the leaves, the left hand side of the computation nodes forms the sequence $v_1, v_2, v_3, \ldots$. Furthermore, the right hand side of a computation node with left hand side v_i only contains variables v_j with $j < i$.
- Binary *if nodes* of the form "if $p(a_1, a_2, \ldots, a_l)$", where $p \in P \cup \{q\}$, where $q(j, s)$ is the predicate which is true iff $0 < j \leq s$, and the arguments a_i are taken from W'. Furthermore, arguments v_j must have appeared as the left hand side of an ancestral node.
- l-ary *case nodes* of the form "case c_0 of $c_1, c_2, \ldots, c_{l-1}$", where the arguments c_i are taken from W'. Furthermore, arguments v_j must have appeared as the left hand side of an ancestral node. The *degree* of a computation tree is defined to be the maximum arity of its case nodes.

A computation tree computes a function of a single argument in W as follows. We assign the input to the symbol "x". Then we start computing in the root of the tree and proceed towards one of the leaves. When we reach a computation node with left hand side v_i, we perform the computation in the node and assign the result of the computation to v_i (note that the operations allowed are all (α, β)-local). When we reach an if node, we proceed to the left son of the node if the predicate evaluates to true, and we proceed to the right son if the predicate evaluates to false. When we reach a case node, we find the *largest $i \geq 1$*, for which $c_0 = c_i$ is true for the current values of the arguments and proceed to the i'th son of the node. If no such i exists, we proceed to the l'th son. The value of the computation is found in the leaf finally encountered.

Given a RAM-solution to the static dictionary problem, given by a program and a data structure ϕ_S, we shall consider a computation tree which, given x, computes the configuration of the RAM after running the program for t steps on input x and ϕ_S. The configuration is not computed explicitly, instead we find in the leaf of the computation tree three objects:

- The value p of the program counter.
- A map $\tau : \{r_0, r_1, \ldots, r_k\} \to W'$, giving the contents of each of the internal registers in the following sense: When the leaf is reached, the symbols $x, v_1, \ldots$ have been assigned certain values in W. These are substituted for the symbolic values in the image of τ.
- A finite list l of length at most $t + 1$ containing items (a_i, d_i), $i = 1, 2, \ldots$ with $a_i, d_i \in W'$, which are interpreted as follows: The actual values are substituted for the symbolic ones in l. Then l determines the contents of the

random access memory as follows: For the memory location with index j, if $a_i = j$ for some i, consider the *largest* such i. Then the location contains the value d_i. If $a_i \neq j$ for all i, the location contains the value it contained before the computation started, and before the input x was given, i.e. $(\phi_S)_j$ for $1 \leq j \leq s$ and 0 for $j \leq 0$ or $j > s$.

The main obstacles encountered when converting the algorithm and data structure into a computation tree are indirect addressing and shifts with non-constant second argument. Both are dealt with using the following lemma, derived from Lemma 1.

Lemma 3. *Let $I = \{0, \ldots, i-1\}$. Let C be a circuit of size t. We can find a subset $B \subseteq H_{c+1}$, so that $\#B/\#H_{c+1} \leq 2^{O(t)}i/b$ and so that*

$$\#\{C(x)[I] \mid x \in H_{c+1} - B\} \leq 2^{2^{O(t)}}$$

Proof. According to Lemma 1, there is a set $M \subseteq \{0, \ldots, w-1\}$, with $\#M \leq (\alpha k)^t i$, so that if $x[M]$ contains only 0-bits, then $C(x)[I]$ takes on at most $\beta^{k^t-1} = 2^{2^{O(t)}}$ values. Let $B = \{x \in H_{c+1} \mid x[M]$ contains at least one 1-bit$\}$. It is easy to see that the probability that a random element in H_{c+1} is in B is $O((\#M)(c+1)/b) = 2^{O(t)}i/b$. $\quad\blacksquare$

We are now ready for the conversion lemma. It deviates slightly from what was promised in the proof sketch above, since the tree may not have an answer on a small subset of H_{c+1}.

Lemma 4. *Given a data structure ϕ_S of size s and a RAM program, there is a tree of degree $2^{2^{O(t)}}$ and depth $3t$ which on input x either correctly computes the configuration of the program after t steps on input ϕ_S and $x \in H_{c+1}$ or outputs* DON'T KNOW. *Furthermore, it only outputs* DON'T KNOW *on a fraction $v2^{O(t)}(\log w + \log s)/b$ of H_{c+1}, where v is the number of nodes in the tree.*

Proof. The proof is by induction in t. For $t = 0$, no computation takes place, so a single leaf (i.e. a tree of depth 0), containing the initial value of the program counter $p = 1$, the map $\tau(r_i) = 0$ for all i, and the list $l = [(0, x)]$, indicating that register 0 contains the input, is the tree we need.

Now assume $t > 0$ and assume that the lemma holds for $t - 1$, i.e. that we have a tree computing the configuration of the machine after $t - 1$ steps have been performed. We will extend this tree by replacing, one by one, all leaves of the tree which are not marked DON'T KNOW with subtrees representing the execution of one more instruction.

Fix a leaf, and assume that it is marked (p, τ, l). Since we know p we know which instruction is to be performed. There is a case for each type of instruction.

— Accept or reject. We leave the leaf as it is.
— Load or move. We change the map τ in the leaf to reflect the change performed. For instance, if the instruction $r_1 := r_2$ is to be performed, we let $\tau'(r_1) = \tau(r_2)$ and $\tau'(r_i) = \tau(r_i)$ for $i \neq 1$. We also update the program counter p to $p' = p + 1$. The list l is left untouched.

- Computation. If the instruction $r_0 := f(r_1, \ldots, r_j)$ is to be performed, we lookup $a_i = \tau(r_i)$ for $i = 1, \ldots, j$. We then replace the leaf with the computation node $v_h = f(a_1, a_2, \ldots, a_j)$ (where $h-1$ is the index of the computation node preceding the new node in the tree). The son of the computation node is a new leaf (p', τ', l) with $p' = p + 1$, $\tau'(r_0) = v_h$ and $\tau'(r_i) = \tau(r_i)$ for $i \neq 0$.

- Shift. Assume without loss of generality that the instruction is $r_0 := \text{shift}(r_1, r_2)$.

  Since the value of $\tau(r_2) \in \{-(w-1), w-1\}$, the $\lceil \log w \rceil + 1$ least significant bits of the value of $\tau(r_2)$ determines its value. Note that the value of $\tau(r_2)$ can be computed by a straight line program of length at most $t - 1$ given by computation nodes on the path leading to the current node, i.e. by a circuit of size at most $t - 1$. By Lemma 3, for all $x \in H_{c+1}$ except a fraction $2^{O(t)}(\log w)/b$, the $\lceil \log w \rceil + 1$ least significant bits take on at most $2^{2^{O(t)}}$ different values. Let $c_1, c_2, \ldots, c_j$ be the corresponding possible values of $\tau(r_2)$. We replace the leaf by a a "case $\tau(r_2)$ of $c_1, c_2, \ldots c_j$" node. The i'th son of the node for $i \leq j$ will be a computation node $v_h := \text{shift}_{c_i}(\tau(r_1))$. The son of this node will be a leaf $(p+1, \tau', l)$ with $\tau'(r_0) = v_h$ and $\tau'(r_i) = \tau(r_i)$ for $i \neq 0$. As the $j + 1$'st son of the case node we insert a DON'T KNOW-leaf.

- Conditional jump. If the instruction is "If $p(r_1, \ldots, r_j)$ then goto k", we replace the leaf by an if node "If $p(\tau(r_1), \ldots, \tau(r_j))$". The left son of the node is a new leaf (k, τ, l) and the right son is a new leaf $(p + 1, \tau, l)$.

- Write. Assume without loss of generality that the instruction is $M[r_0] := r_1$. We change the value in the leaf to $(p + 1, \tau, l')$, where l' is the list l with the item $(\tau(r_0), \tau(r_1))$ added to the end.

- Read. Assume without loss of generality that the instruction is $r_0 := M[r_1]$ First we check if we are reading a location where we already wrote something. Let the contents of the list l be $[(a_i, d_i)]_{i=1}^{|l|}$. We replace the leaf by a "case $\tau(r_1)$ of $a_1, a_2, \ldots a_{|l|}$". The i'th son of the case node will be a leaf $(p+1, \tau_i, l)$ with $\tau_i(r_0) = d_i$, and $\tau_i(r_j) = \tau(r_j)$ for $j \neq 0$.

  The $|l| + 1$'st son of the case node is the place we go if the location hasn't been written to. There are now two cases: Either the location is inside the data structure or it isn't. We check which is the case by placing an "if $0 < \tau(r_1) \leq s$" node at this point. The right son of the node corresponds to reading outside the data structure. We put a $(p + 1, \tau', l)$ leaf there, where $\tau'(r_0) = 0$ and $\tau'(r_i) = \tau(r_i)$ for $i \neq 0$.

  The left son of the if-node corresponds to reading inside the data structure. Note that the $\lceil \log(s + 1) \rceil$ least significant bits of the value of $\tau(r_1)$ determines where we read. Also note that $\tau(r_1)$ is the value of a circuit of size at most $t - 1$. By Lemma 3, for all $x \in H_{c+1}$ except a fraction $2^{O(t)}(\log s)/b$, the $\lceil \log(s + 1) \rceil$ least significant bits take on at most $2^{2^{O(t)}}$ different values. Let $c_1, c_2, \ldots, c_j$ be these values. We put in a "case $\tau(r_1)$ of $c_1, c_2, \ldots c_j$" node at this point. The i'th son of the node for $i \leq j$ will be a leaf $(p + 1, \tau', l)$ with $\tau'(r_0) = (\phi_S)_{c_i}$ and $\tau'(r_i) = \tau(r_i)$ for $i \neq 0$. As the $j + 1$'st son of the case node, we put a DON'T KNOW leaf.

In each case it is easy to check that the new tree correctly computes the configuration after t steps, except when an input is classified as DON'T KNOW. The number of inputs classified as DON'T KNOW which were not already classified as DON'T KNOW by the old tree is at most $v'(2^{O(t)}\max(\log s, \log w)/b)\binom{b}{c+1}$, where v' is the number of new case nodes introduced. This completes the proof of the lemma.

We are now ready to prove Theorem 2. Assume, to the contrary, that for a certain constant d, to be fixed in a little while, and all $S \subseteq H_{c+1}$, there are s, t with $2^{2^{dt}}\log s \leq b$, a data structure ϕ_S of size s, and a RAM program running in worst case time t which accepts on input $x \in S$ and ϕ_S and rejects on inputs $x \in H_{c+1} - S$ and ϕ_S.

Fix such an S. By Lemma 4, we have a tree of size $(2^{2^{O(t)}})^{O(t)} = 2^{2^{O(t)}}$ which on input $x \in H_{c+1}$ either correctly computes the configuration of the program after t steps on input x and ϕ_S or outputs DON'T KNOW. It outputs DON'T KNOW on a fraction at most $\gamma = 2^{2^{O(t)}}(2^{O(t)}(\log s + \log w)/b) = 2^{2^{O(t)}}(\log s + \log w)/b$ of H_{c+1}. By removing all information in the leaves except whether the current instruction is accept or reject, the tree either correctly determines if $x \in S$ for inputs $x \in H_{c+1}$ or outputs DON'T KNOW.

Each node in the tree can be described using $2^{2^{O(t)}}(w + O(\log t)) = 2^{2^{O(t)}}w$ bits, so the entire tree can be described using $\delta = 2^{2^{O(t)}}2^{2^{O(t)}}w = 2^{2^{O(t)}}b^c$ bits. We fix d, so that $2^{2^{dt}}\log s \leq b$ implies $\gamma \leq 1/2$ and $\delta \leq b^{c+\frac{1}{2}}$. If we present a description of the tree together with the correct answer to all the DON'T KNOW-inputs, we have described S. The total number of bits in the description is $\delta + \frac{1}{2}\binom{b}{c+1} \leq b^{c+\frac{1}{2}} + \frac{1}{2}\binom{b}{c+1}$ which is strictly less than $\binom{b}{c+1}$ for sufficiently large b.

Since this could be done for any S, we have a way of describing any subset of H_{c+1} uniquely using less than $\binom{b}{c+1}$ bits. This is impossible by the pigeon hole principle, and the proof of Theorem 2 is complete.

5 Open problems

In our view, the most interesting open problem related to this research is to get tight upper and lower bounds on the query time for linear space solutions to the static dictionary problem on Practical RAMs. Subsequent to this research and using different techniques, Andersson et al [3] improved the lower bound of $\Omega(\log\log n)$ of this paper to $\Omega(\sqrt{\log n/\log\log n})$. A polynomial gap to the $O(\log n)$ upper bound remains. We conjecture that $O(\log n)$ is in fact optimal, i.e. that tables should be sorted on Practical RAMs.

Acknowledgment

I would like to thank Faith Fich for many discussions about the material in this paper, and Amir Ben-Amram for helpful comments on terminology, in particular for suggesting the term *Practical RAM*.

References

1. A. Andersson. Sublogarithmic searching without multiplications. In *Proc. FOCS*, 1995.
2. A. Andersson, T. Hagerup, S. Nilsson, and R. Raman. Sorting in linear time? In *Proc. 27th ACM Symposium on Theory of Computing (STOC)*, pages 427–436, 1995.
3. A. Andersson, P.B. Miltersen, S. Riis, and M. Thorup. Static Dictionaries on AC^0 RAMs: Query time $\sqrt{\log n / \log \log n}$ is sufficient and necessary. Manuscript, 1996.
4. A.M. Ben-Amram and Z. Galil. When can we sort in $o(n \log n)$ time? In *Proc. 34th IEEE Symposium on Foundations of Computer Science (FOCS)*, pages 538–546, 1993.
5. N.H. Bshouty. Lower bounds for the complexity of functions in a realistic RAM model. In *Proc. Israel Symposium on the Theory of Computing and Systems*, pages 12–23, 1992.
6. J.L. Carter and M.N. Wegman. Universal classes of hash functions. *J. Comput. Syst. Sci.*, 18:143–154, 1979.
7. B. Chazelle. A functional approach to data structures and its use in multidimensional searching. *SIAM J. Comput.*, 17:427–462, 1988.
8. P.F. Dietz. Optimal algorithms for list indexing and subset rank. In *Proc. First Workshop on Algorithms and Data Structures (WADS)*, pages 39–46, 1989.
9. M. Dietzfelbinger, T. Hagerup, J. Katajainen, and M. Penttonen. A reliable randomized algorithm for the closest-pair problem. Technical Report 513, Fachbereich Informatik, Universität Dortmund, 1993.
10. M. Dietzfelbinger, A. Karlin, K. Mehlhorn, F. Meyer Auf Der Heide, H. Rohnert, and R.E. Tarjan. Dynamic perfect hashing: Upper and lower bounds. *SIAM J. Comput.*, 23:738–761, 1994.
11. P. van Emde Boas, R. Kaas, and E. Zijlstra. Design and implementation of an efficient priority queue. *Mathematical Systems Theory*, 10:99–127, 1977.
12. F. Fich and P.B. Miltersen. Tables should be sorted (on random access machines). In *Proc. 4th International Workshop on Algorithms and Data Structures (WADS)*, pages 482–493, 1995.
13. M.L. Fredman, J. Komlós, and E. Szemerédi. Storing a sparse table with O(1) worst case access time. *J. Ass. Comp. Mach.*, 31:538–544, 1984.
14. M.L. Fredman and D.E. Willard. Surpassing the information theoretic bound with fusion trees. *Journal of Computer and System Sciences*, 47:424–436, 1993.
15. H. N. Gabow and R. E. Tarjan. A linear-time algorithm for a special case of disjoint set union. *Journal of Computer and Systems Sciences*, 30:209–221, 1985.
16. M. Thorup. On RAM priority queues. In *Proceedings of the 7th ACM-SIAM Symposium on Discrete Algorithms (SODA)*, pages 59–67, 1996.
17. A.C. Yao. Should tables be sorted? *J. Ass. Comp. Mach.*, 28:615–628, 1981.

Lower Bounds for Row Minima Searching

(Extended Abstract)

Phillip G. Bradford and Knut Reinert

Max-Planck-Institut für Informatik, Im Stadtwald, D-66123 Saarbrücken, Germany.
{ bradford, kreinert }@mpi-sb.mpg.de

Abstract. This paper shows that finding the row minima (maxima) in an $n \times n$ totally monotone matrix in the worst case requires any algorithm to make $3n - 5$ comparisons or $4n - 5$ matrix accesses. Where the, so called, SMAWK algorithm of Aggarwal et al. finds the row minima in no more than $5n - 2 \lg n - 6$ comparisons.

1 Introduction

Finding the row minima (maxima) of *monotone matrices* was introduced by Aggarwal, Klawe, Moran, Shor, and Wilber [1]. They also gave an asymptotically optimal sequential algorithm for finding the row minima in *totally monotone matrices*, among other things. Row minima problems and their variants are well motivated due to their large host of applications. Two examples are the prediction of RNA secondary structure [6] and the all–farthest–neighbor problem [1].

The *row minima problem* for an $n \times n$ matrix M, whose entries belong to some totally ordered set, is to find the minimal element in each row. To be more precise, let mc(i) denote the index of the leftmost column that contains a minimal element of the i-th row. Then, the row minima problem is to find mc(i) for $1 \leq i \leq n$.

Throughout this paper, we assume that all entries of M are distinct. Clearly, the row minima problem has time complexity $\Theta(n^2)$. It turns out, however, that many problems can be reduced to the row minima problem for a matrix M that has a special form:

Definition 1. Let M be an $n \times n$ matrix. The matrix M is *monotone* if for all $1 \leq i < j \leq n$, mc(i) $\leq$ mc(j).

Given an $n \times n$ monotone matrix M of n^2 distinct elements, the *monotone row minima* problem is to get the minimal value in each row of M. Aggarwal et al. [1] gave an $O(n \log n)$ sequential algorithm for the row minima problem on a monotone matrix and showed that it is asymptotically optimal.

Definition 2. An $n \times n$ matrix M is *totally monotone* if every 2×2 minor is monotone. That is, for all $1 \leq i < k \leq n$ and $1 \leq j < l \leq n$, if $M[i,j] > M[i,l]$, then $M[k,j] > M[k,l]$.

Given a $n \times n$ *totally* monotone matrix M the *totally monotone row minima* problem, or just the *row minima* problem, is to get the minimal value in each row of M. (The row minima and row maxima problems on totally monotone matrices are symmetric, so our exposition only deals with the row minima problem.)

Aggarwal *et al.* [1] gave a sequential $O(n)$-time algorithm for the row minima problem on a totally monotone matrix. This is the SMAWK algorithm.

The SMAWK algorithm illustrates that we do not need to generate the entire $n \times n$ totally monotone matrix M, but rather we only have to compute selected entries of M when they are needed. In this paper, we attack the natural extension to this: Exactly how many matrix elements must we generate to solve the row minima problem on a totally monotone matrix?

An exact analysis by Larmore and Schieber [6] shows that Aggarwal *et al.*'s algorithm takes at most $5n$ comparisons and Larmore [5] very recently strengthened this to $5n - 2\log n - 6$ comparisons. We show that any algorithm needs at least $3n - 5$ comparisons among totally monotone matrix elements. We also give a $4n - 5$ matrix access lower bound. So any algorithm that solves the row minima problem on a totally monotone matrix must 'look at' $4n - 5$ matrix elements.

Previous lower bounds for a variation of row minima problems were given by Aggarwal *et al.* [1], who gave an $\Omega(n \log n)$ lower bound for the more general row minima problem on *monotone matrices* and Klawe [3], who gave a $\Omega(n\alpha(n))$ lower bound for the row minima problem on *partial* totally monotone matrices, where $\alpha(n)$ is the inverse Ackermann function. Larmore [4] gives a provably optimal algorithm with unknown time complexity for the *convex strictly* monotone triangular matrix searching problem. Lower bounds for related problems have also been considered. Alon and Azar [2] show that the decision tree complexity of sorting rows in totally monotone matrices is $\Omega(n^2)$. However, to our knowledge there have been no exact lower bounds for the row minima problem on a *totally* monotone matrix.

2 The Basics of Totally Monotone Matrices

For the rest of this paper assume all matrices are totally monotone. A *dead* column is a column that has been shown not to contain any row minima through only comparisons and inferences [1]. Similarly, a dead *cell* is a single matrix element in a totally monotone matrix that has been shown to contain no row minima. Showing a matrix element is dead is *killing* a matrix cell. The left side of Figure 1 shows a dead column in a totally monotone matrix.

In the left side of Figure 1, we have $A < A'$; therefore, due to the total monotonicity of M we do not have to consider any elements directly above A' as a candidate row minima. Likewise, in the left side of Figure 1, we have $B' > B$; therefore, due to the total monotonicity of M we don't have to consider any elements directly below B' as candidate row minima. That is:

Lemma 3 (Aggarwal *et al.* [1]). If M is a totally monotone matrix with the situation depicted in the left of Figure 1, that is $A < A'$ and $B' > B$, then the shaded column is dead.

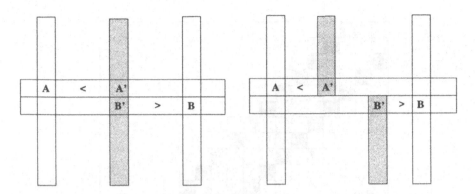

Fig. 1. Left: The Middle Column is Dead Since it Can't Contain any Row Minima in a Totally Monotone Matrix. Right: Two half-dead columns.

This lemma lies at the foundation of the SMAWK algorithm.

The right side of Figure 1 shows two *half-dead* columns. More formally, given a matrix M a column c is half-dead column from row r up to row 1 iff there is a column $c' < c$ such that $M[i, c'] < M[i, c]$, for $i : r \geq i \geq 1$. Symmetrically, a column c is half-dead column from row r down to row n iff there is a column c' where $c < c'$ is such that $M[i, c] > M[i, c']$, for $i : n \leq i \leq r$.

3 The Lower Bound

In this section we give our main result: a $3n - 5$ lower bound on the number of comparisons needed for finding the row minima of a totally monotone matrix. We begin with a trivial $2n - 2$ lower bound to find a certificate to verify a solution to the row minima problem, then this is extended giving our $3n - 5$ lower bound.

We use an adversary based argument. The adversary fixes the input for any algorithm. Of course, the input must be consistent, in that it must represent an actual $n \times n$ totally monotone matrix. The adversary starts by choosing the input such that every value in row i is smaller than the smallest value in row $i+1$. This is so comparisons of elements between different rows cannot help any algorithm. This is done without loss of generality since choosing the values in different rows of M as just stated is consonant with Definition 2.

Given an $n \times n$ totally monotone matrix M, a *certificate* for the row minima problem is a set of comparisons that indicate that all but n cells in M are dead. Certainly, finding the actual solution of the row minima problem is at least as hard as finding a certificate for a solution. Hence we will show that our adversary can make it costly for any algorithm to find a certificate.

3.1 Foundations for the Adversary

First we introduce some terminology which we illustrate in Figure 2. Along the main diagonal we have a band of width seven. The cells in the matrix which are

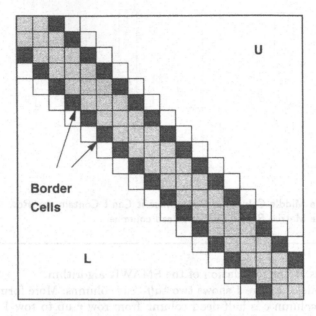

Fig. 2. A Banded Matrix with (2×2)–Boxes

to the upper right of the band are *U-cells*. Similarly the cells which are to the lower left of the band are *L-cells*. Cells within the band are *B-cells*. Killing a border cell for the first time is a *border kill*. Along the main diagonal there are intersecting (2×2)*-Boxes* which play a crucial role in our argument. Cells in (2×2)–Boxes are (2×2)*-Box-cells*. The intersecting (2×2)–Boxes in Figure 2 are in *canonical* form, since their diagonal is the same as the diagonal of M and they share one (2×2)–Box-cell. If a comparison kills a cell in a (2×2)–Box, then this comparison is a (2×2)*-Box-kill*. We imagine that each row minimum is always in some (2×2)–Box, though the (2×2)–Boxes can be moved up or down one row by the adversary to make more comparisons necessary.

The cells lying in the band but bordering all of the intersecting (2×2)–Boxes are *border cells*. The border cells are the darkest cells in the band. If a (2×2)–Box is moved up or down one row, then 'its' border cells move with it. The (2×2)–Boxes will always remain in this band of width seven.

Figure 3 shows our adversary's association of border cells with (2×2)–Box-cells. This association will turn out to be central for our adversary.

Two (2×2)–Boxes are *intersecting* if they both share one or two matrix elements. Two (2×2)–Boxes are *neighboring* if one is above the other and they share at least one matrix cell border. A *connected monotone falling chain* or a *chain* is a list of intersecting or neighboring (2×2)–Boxes that include both cells $M[1,1]$ and $M[n,n]$ and in this list of (2×2)–Boxes there is a path of (2×2)–Box-cells from $M[1,1]$ down to $M[n,n]$.

Note that the chain may contain dead cells.

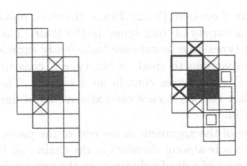

Fig. 3. Some (2×2)–Boxes and their associated border cells. Two intersecting (2×2)–Boxes never have coinciding border cells since the adversary only moves (2×2)–Boxes up or down.

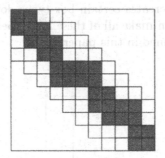

Fig. 4. The grey cells form a *chain* of (2×2)–Boxes. Such a chain would only come to light after an algorithm makes certain comparisons.

Lemma 4. For a $n \times n$ totally monotone matrix, there are exactly $3n - 2$ (2×2)–Box-cells in any chain of (2×2)–Boxes.

The basic intuition behind our adversary is that each (2×2)–Box alone requires in the worst case two comparisons to certify its row minima. A $(2 \times b)$–Box is a sub-matrix with 2 rows and b columns. A $(2 \times b)$–Box requires in the worst case $2(b - 1)$ comparisons to certify its row minima. That is, in the worst case monotonicity does not help to find the minima in a $(2 \times b)$–Box.

In canonical form (see Figure 2) even though the (2×2)–Boxes intersect, their intersection does not diminish the number of comparisons necessary for certifying minima in them. Since there are initially $n - 1$ of these (2×2)–Boxes so at the start when they run right down the diagonal we get a $2n - 2$ cost for the entire certificate. However, we will see that we can charge to almost every (2×2)–Box a third comparison via moving some of the (2×2)–Boxes up or down. In particular, if comparisons are only done among matrix elements in the chain in canonical form, then a simple adversary can keep many border cells (hence half-columns) alive that may potentially contain row minima. Furthermore, if we can move (2×2)–Boxes in the canonical chain up or down one row depending on the comparisons of an algorithm, then some of those comparisons in the original canonical chain will have been for naught while at the same time we still have to check for row minima in the moved (2×2)–Boxes. However, moving (2×2)–Boxes up or down can form a variety of 'box-types' in the chain. The adversary must ensure that these box-types have worst-case 'sufficiently expensive' certificates. Suppose it takes t comparisons to create a box-type T consisting of u (2×2)–Boxes. Assume these (2×2)–Boxes contain no dead cells. Then the adversary wants to make sure that the worst case comparison cost of finding a certificate in T plus t is at least $3u$.

The basic structure of the argument in the rest of the paper is the following. If no comparisons are made among elements in the chain and if elements of the chain are not killed as part of a dead-column, then the row minima are unknown. Suppose some comparisons are made between elements in the chain, then usually the adversary can move (2×2)–Boxes to avoid the effect of these comparisons. The moved (2×2)–Boxes creates certain box-types down the chain. Finally we show that the adversary can make all of these box-types costly enough to certify to complete the result claimed in this paper.

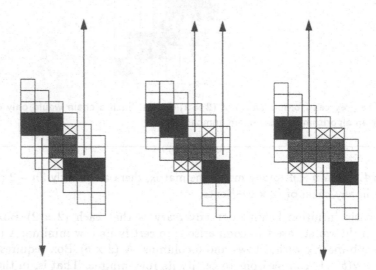

Fig. 5. The three different *isolated* box-types ((2×2)–Box, the (2×3)–Box and the (2×4)–Box) shaded in grey, with the arrows denoting the necessary half-dead columns to create these box-types and the X's denote the border cells associated with each of these $(2 \times b)$–Boxes. The black boxes are neighboring boxes in the chain.

During the run of any algorithm, there are six basic box-types that can emerge, the first three of these are depicted in Figure 5. These are *isolated* $(2 \times b)$–Boxes for $b \in \{2, 3, 4\}$. An isolated $(2 \times b)$–Box is a $(2 \times b)$–Box with no intersecting (2×2)–Box-cells from any other $(2 \times c)$–Boxes above or below. Recall that to verify a certificate in an isolated $(2 \times b)$–Box requires $2(b - 1)$ comparisons in the worst case. However, to verify a certificate in a $(3 \times b)$–Box requires less than $3(b - 1)$ comparisons in the worst case.

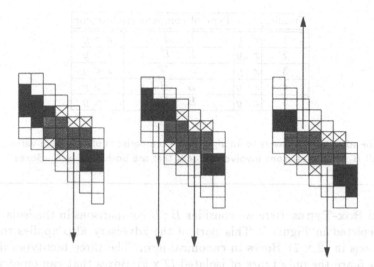

Fig. 6. The three different *non*-isolated box-types shaded in grey, with the arrows denoting the necessary half-dead columns to create these box-types. (We omit the symmetric cases.)

The three *non*-isolated box-types (without their symmetric counterparts) are in Figure 6. Non-isolated box-types consist of intersecting $(2 \times b)$–Boxes and $(2 \times c)$–Boxes. Whenever we write about *non*-isolated box-types we also mean to include their symmetric counterparts. Since non-isolated box-types 'fit inside' $(3 \times b)$–Boxes for some $b \in \{4, 5\}$ the adversary must take special care of these since finding the row minima of the middle row makes finding the row minima of the top and bottom row cheaper, even in the worst case.

Lemma 5. The isolated, non-isolated, and canonical box-types are the only box-types that can form by maintaining a chain via moving (2×2)–Boxes up or down one row.

The proof of Lemma 5 follows from the fact that adversary only moves a (2×2)–Boxes up or down one row at most.

3.2 Details of the Adversary

The adversary gives answers to an algorithm when it makes a comparison of two elements $x = M[i, j]$ and $y = M[i, k]$ from cells in row i, where $j < k$ and x and y are in either in the matrix regions U, L, or B. The adversary starts with the configuration of Figure 2 when all of the (2×2)–Boxes are in canonical form.

In Figure 7 we listed all but the $B : B$ comparisons since they are more complicated. Without loss of generality, take $B : B$ comparisons to be comparisons of two elements in the chain. Given our adversary, any comparison numbered 1 through 5 in Figure 7 can only increase the worst cost of any algorithm.

Now we focus on $B : B$ comparisons.

Number		Type of comparison	Outcome
1	$x : y$	$L \ : \ L$	$x > y$
2	$x : y$	$L \ : \ U$	$x > y$
3	$x : y$	$U \ : \ U$	$x < y$
4	$x : y$	$L \ : \ B$	$x > y$
5	$x : y$	$B \ : \ U$	$x < y$

Fig. 7. The adversaries answers to an algorithm's comparisons of cells from various regions. Assume all $B : B$ comparisons involve two cells that are both in $(2 \times b)$–Boxes.

Isolated Box-Types Here we consider $B : B$ comparisons in the isolated box-types, depicted in Figure 5. This part of the adversary also applies to $B : B$ comparisons in (2×2)–Boxes in canonical form. The three box-types depicted in Figure 5 are the only types of isolated $(2 \times b)$–Boxes that can emerge in the chain while an algorithm is run.

Assume some algorithm makes a $B : B$ comparison, say $x : y$. Let Box$_1$ be the (2×2)–Box x is in and let Box$_2$ be the (2×2)–Box y is in. These boxes can be uniquely defined: Box$_1$ and Box$_2$ must always be different boxes where one contains x and the other contains y and killing x or y will kill a border cell of Box$_1$ or Box$_2$, respectively. If we answer $x < y$, then a is the number of (2×2)–Box-cells killed. If we answer $x > y$, then b is the number of (2×2)–Box-cells killed. The numbers a and b are computed without letting a (2×2)–Box move. Then the adversary answers by the following three *Rules*:

1) If $a < b$, then answer $x < y$ and if Box$_2$ has not moved before, and if moving Box$_2$ down will not break the chain, then move Box$_2$ down one row.
2) If $a > b$, then answer $x > y$ and if Box$_1$ has not moved before, and if moving Box$_1$ up will not break the chain, then move Box$_1$ up one row.
3) If $a = b$, then answer so no border cell is killed–if possible. If both answers kill a border cell, try to move either of Box$_1$ or Box$_2$ (up or down one row) without breaking the chain. If no movement is possible answer arbitrarily.

Rules (1) and (2) are easily justified, for example consider these rules as applied to an algorithm that starts to make comparisons among elements in the canonical chain. Rule (3) is for box cells that are separated by at least one cell. In Rule (3) if both answers kill a border cell, then either answer will save one border cell. In Rule (3) if no box movement is possible, then Box$_1$ and Box$_2$ already have their 'third' comparisons accounted for.

In Figure 8, we let '2-Top,' denote the cell labeled '2' in the top of the non-isolated boxes. Further, the 4-right cell is the rightmost cell labeled '4,' *etc.*

In the last three parts of Figure 8 a comparison can be made in the neighboring (non-intersecting) dark (2×2)–Boxes invoking Rules (1),(2) and (3) making the dark (2×2)–Boxes intersect the (former) isolated box-type. These non-isolated box-types are handled with Rules (4) through (7).

Whenever a (2×2)-Box is moved, then no (2×2)-Box-cells are killed by comparison that caused the movement. So, this first comparison that cause a (2×2)-Box to move, is the "third comparison" associated with verifying the certificate in the moved (2×2)-Box. The moved (2×2)-Box has one alive border-cell remaining.

With the exception of the upper leftmost and lower rightmost (2×2) or (2×3)-Boxes (those containing cells $M[1,1]$ and $M[n,n]$ respectively), every $(2 \times b)$-Box, $b \in \{2,3,4\}$ has b border cells assigned to it.

The isolated (2×3)-Boxes and (2×4)-Boxes already have at least one or at least two dead border cells, respectively (see Figure 5). Therefore, the adversary must keep one more border cell alive per (2×3)-Box or (2×4)-Box while a certificate is verified for these boxes. This is easy to do by keeping alive a (2×2)-Box-cell directly under or directly above a border cell. Since a (2×4)-Box is made of three (2×2)-Boxes this accounts for three comparisons per (2×2)-Box in the (2×4)-Box.

Theorem 6. Given an isolated $(2 \times b)$-Box, for $b \in \{2,3,4\}$, except the leftmost and the rightmost such $(2 \times b)$-Boxes. Using Rules (1) through (3), our adversary can keep alive at least one border cell of the $(2 \times b)$-Box while an algorithm makes $2(b-1)$ comparisons among cells of the $(2 \times b)$-Box.

These border cells can always be chosen to be consistent with a totally monotone matrix. A border cell (and the associated alive half-column up or down) must eventually account for an additional comparison, otherwise the row minima are not known. Considering the border cells (of the isolated $(2 \times b)$-Box) that were killed to generate each of the isolated box-types along with the comparisons for finding a certificate an isolated $(2 \times b)$-Box and the extra comparison implied by Theorem 6, we have a total of $3(b-1)$ comparisons. That is for $b-1$ (2×2)-Boxes, three comparisons each.

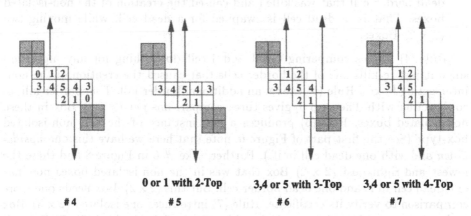

0 or 1 with 2-Top   3,4 or 5 with 3-Top   3,4 or 5 with 4-Top

#4   #5   #6   #7

Fig. 8. Details of *non*-isolated $(2 \times b)$-Boxes and $(2 \times c)$-Boxes. The leftmost box is type of non-isolated Boxes we will consider.

Non-Isolated Box-Types Here we consider non-isolated boxes as in Figure 6. These boxes consist of intersecting $(2 \times b)$–Boxes and $(2 \times c)$–Boxes. Rules (1) through (3) work for the first non-isolated box-type in Figure 6. The second non-isolated box-type can be handled very similarly to the third non-isolated box-type, so we focus on the third non-isolated box-type. The basic idea is that given a $(3 \times b)$–Box, getting a certificate for the middle row makes finding certificates for the top and bottom rows cheaper. To foil this, after some comparisons are made the adversary 'switches back' from a $(3 \times b)$–Box to a $(2 \times b')$–Box and another $(2 \times b'')$-Box that do not intersect with each other.

If an algorithm makes a $B : B$ comparison between elements in the third case of Figure 6, then we augment the adversary with the following additional rules. (See Figure 8.)

Without loss of generality, suppose that we have not made any $B : B$ comparisons between any of the (2×3)–Box cells. Take the cell numbering given in Figure 8 and the following additional rules.

4) Comparing a 1-cell and a 0-cell. Always kill the 1-cell and without loss of generality assume that no (2×3)–Box cells are dead. Apply these rules again if another $B : B$ comparison is done in such non-isolated boxes.
5) Comparing a 0-cell or a 1-cell with the 2-cell. Kill the 2-cell while shifting down the (2×2)–Box containing the 2-cell. (See Figure 8 # 5).
6) Comparing cells 3-left, 4-left or 5-cell with cell 3-right. Here kill the 3-cell and the half column above it while moving down the one (2×2)–Box. Note that this (2×2)–Box avoids the now-dead 3-cell, but it now contains the dead border cell that was killed and caused the creation of the non-isolated boxes. That is, a dead cell is swapped for a dead cell, while moving a (2×2)–Box.
7) Comparing a 3-left, 4-left or 5-cell with cell 4-right. Kill the 4-cell while moving down *two* (2×2)–Boxes that are under the now-dead 4-cell. Note that these (2×2)–Boxes avoid the now-dead 4-cell, but they now contain the dead border cell that was killed and caused the creation of the non-isolated boxes. That is, a dead cell is swapped for a dead cell, while moving two (2×2)–Boxes.

Rule (4) makes comparing 0-cell and 1-cell do nothing for any algorithm since it just re-kills one of the border cells that caused the creation of the non-intersecting boxes. Rule (5) causes an additional border kill. This border kill, in conjunction with Theorem 6, gives three comparisons per (2×2)–Box in these non-isolated boxes. Rule (6) produces a new instance of the first non-isolated box-type (See the first part of Figure 6, note that here we have this one *upside-down* and with one dead cell in it.). Further, take # 6 in Figure 8 and there the lowest and rightmost (2×2)–Box that was in the non-isolated boxes now has one dead cell in it and one dead border cell. But this (2×2)–Box needs one more comparison to verify its certificate. Rule (7) introduces one isolated (2×3)–Box and an instance of the first non-isolated box-type with one dead (2×2)-Box-cell in it and another dead border cell.

The most important idea these rules introduce is that sometimes we swap one dead-cell for another by moving one or two of the (2×2)–Boxes. That is, in the worst case we can essentially go from a $(3 \times b)$–Box (or a non-isolated box type) back to a $(2 \times b')$–Box and a $(2 \times b'')$–Box that don't intersect each other. Where one of these $(2 \times b)$–Boxes is isolated and the other is a smaller non-isolated box-type. In particular, by this swapping of (2×2)–Boxes in Rules (5),(6) and (7) we never kill more (2×2)–Box-cells, but rather we trade one dead (2×2)–Box-cell for another in order to get another sub-case we know how to deal with. This "trading" of dead cells allows us to keep enough alive border cells to continue.

All of this leads to the following lemma.

Lemma 7. Given two non-isolated (2×3)–Boxes. Rules (4) through (7) make finding a certificate in these two non-isolated boxes take at least 12 comparisons.

Lemma 7 is based on keeping a border cell (hence half-column) alive that may contain the row minima. That is, the adversary can keep a border cell alive that is consistent with a totally monotone matrix.

Lemma 7 holds with 9 comparisons in the case of a non-isolated (2×2)–Box and a (2×3)–Box (in the middle of Figure 6). That is, three comparisons per (2×2)–Box in this case.

As Figure 8 shows, if the adversary is in a situation where there is a non-isolated box, then it adjusts its strategy to charge three comparisons to each (2×2)–Box. Furthermore, consider the cells immediately surrounding the two non-isolated (2×3)–Boxes. If any of these bordering cells are killed before the first $B : B$ comparison is made between the (2×2)–Box-cells, then the adversary focuses on keeping one border cell alive. The adversary can do this by Theorem 6.

In all four parts of Figure 8 a comparison can be made in the neighboring (non-intersecting) black (2×2)–Boxes forcing the black (2×2)–Boxes to overlap the grey non-isolated boxes. This is simply another case of non-isolated boxes which the adversary deals with as the non-isolated case.

Consider if two of the situations in #7 of Figure 8 'collide.' Take the case where the lower (2×3)-Box in #7 of Figure 8 forms another instance of the non-isolated box of #4 in this figure. If the 4-right cell was already dead, then we would not have formed such a non-isolated box. On the other hand, say we compare the 4-right with the 5-cell, then the adversary will just execute Rule (7) as usual.

Completing the Lower Bound We are now ready to put everything together and prove our main result.

Theorem 8. Any algorithm solving the row minima problem on a totally monotone $n \times n$ matrix must make at least $3n - 5$ comparisons between matrix elements.

Proof. Since there are $n - 1$ (2×2)–Boxes we know by Theorem 6 and Lemma 7 that we can assign to $n - 3$ of the (2×2)–Boxes a border kill that did not kill a (2×2)–Box-cell. However all minima are within the (2×2)–Box-cells. So to find them we must kill all but n of the (2×2)–Box-cells. By Rules (1) through (7) we know that we can kill at most one (2×2)–Box-cell with a single comparison. So by Lemma 4 we have to do $2n - 2$ comparisons to kill the dead (2×2)–Box-cells. Together with the $n - 3$ comparisons for border cells as just mentioned gives a total of $3n - 5$ comparisons.

Certainly any algorithm must access all of the elements in a chain giving at least $3n - 5$ different matrix accesses. By Theorem 6 and Lemma 4 a total of $4n - 5$ matrix accesses are necessary to solve the row minima problem.

4   Conclusions

This work leaves the open problem of closing the gap between the lower and upper bounds of the row minima problem in a totally monotone matrix.

Acknowledgments

We are grateful to Shiva Chaudhuri, Volker Priebe, and Jesper Träff for their comments. Many thanks to Lawrence L. Larmore for pointing out an exact upper bound of the SMAWK algorithm. Thanks to Rudolf Fleischer, and Michiel Smid for stimulating discussions.

References

1. A. Aggarwal, M. M. Klawe, S. Moran, P. Shor, and R. Wilber: "Geometric Applications of a Matrix Search Algorithm," *Algorithmica*, Vol. 2, 195-208, 1987.
2. N. Alon and Y. Azar: "Comparison-Sorting and Selecting in Totally Monotone Matrices," *Proceedings of the Third Symposium on Discrete Algorithms (SODA '92)*, ACM Press, 403-408, 1992.
3. M. M. Klawe: "Super Linear Bounds for Matrix Searching Problems," *J. of Algorithms*, **13**, 55-78, 1992.
4. L. L. Larmore: "An Optimal Algorithm with Unknown Time Complexity for Convex Matrix Searching," *Information Processing Letters*, Vol. 36, 147-151, 1990.
5. L. L. Larmore: *Personal communication*, 1995.
6. L. L. Larmore and B. Schieber: "On-Line Dynamic Programming with applications to the Prediction of RNA Secondary Structure," *Journal of Algorithms*, **12**, 490-515, 1991.

On the Complexity of Relational Problems for Finite State Processes *

(Extended Abstract)

Sandeep K. Shukla[1]   Harry B. Hunt III[1]
Daniel J. Rosenkrantz[1]   Richard E. Stearns[1]

Department of Computer Science
University at Albany – State University of New York
Albany, NY 12222
Email: {sandeep,hunt,djr,res}@cs.albany.edu

Abstract. We study the complexity of the following three relational problems: Let $\sim$ be a binary relation on finite state processes; let p_0 be a *fixed* finite state process; and let π be a nontrivial predicate on finite state processes such that x and y are *weakly bisimilar* implies $\pi(x) = \pi(y)$. (i) P_1: Determine for processes p and q, if $p \sim q$; (ii) P_2: Determine for process p, if $p \sim p_0$; and (iii)P_3: Determine for process p, if $\pi(p) = true$. We study the complexities of these problems, when processes are represented by *sequential* transition systems and by *parallel composition* of transition systems (*with* and *without* hiding). A number of results are obtained for both representations.

1 Introduction

Motivation : Finite state processes are often represented as finite transition systems or their compositions (e.g., parallel composition) [8, 13]. We call a transition system represented by an explicit enumeration of its states and transition relation a **sequential** transition system. Transition systems represented as parallel compositions of sequential transition systems are called **parallel** transition systems. We consider parallel transition systems both *with* and *without* hiding abstraction.

Let $\sim$ be a binary relation on finite processes, and let p_0 be a fixed finite state process. Consider the following two relational problems. (i) P_1: Determine for processes p and q, if $p \sim q$, and (ii) P_2: Determine for a process p, if $p \sim p_0$. We study these problems for various relations $\sim$ defined in the literature [8, 5, 9, 7] for the following reasons:
1. Many verification problems are modelled as problem P_1 for some appropriate relation $\sim$, where the specification of a system is represented as a process q and implementation as a process p [7, 6].
2. In many industrial automated verification tools, libraries of transition systems corresponding to desired properties or operations of systems have been

---

* This research was supported by NSF Grants CCR-90-06396 and CCR-94-06611.

implemented [6]. To prove that some implementation P conforms to a given specification, one establishes relations with a fixed transition system from the library [6, 14].

For modelling concurrent/parallel systems, *parallel composition* and *hiding abstraction* have been widely used [4, 8]. For parallel transition systems with or without hiding, the state space of the system can be exponentially larger than size of the system description. As a result, problems P_1, P_2, and P_3 may be harder for processes with such succinct descriptions. We also consider problem P_1 for sequential transition systems for several relations defined in the context of I/O automata in [7]. The complexity of these relations has not been studied previously.

Results and Contributions : This paper presents two kinds of results. Tables 1 and 2 summarizes the new complexity results obtained for problems P_1 and P_2 for parallel transition systems. Table 3 summarizes our results for problem P_1 for sequential transition systems and I/O automata [7].

Type of Systems		Decision Problem	lower bound	upper bound
Unrestricted	Nondeterministic	Reachability		PSPACE
		Equivalences between bisimulation and trace		
		Preorders between simulation and trace	PSPACE-hard	
		Trace, Failure, Readiness Equivalence		PSPACE
	Deterministic	Equivalences between bisimulation and trace		PSPACE
Acyclic	Nondeterministic	Reachability	NP-hard	NP
		Equivalences between bisimulation and trace		
		Preorders between simulation and trace	co-NP-hard	
		Trace, Failure, Readiness Equivalence		co-NP
	Deterministic	Equivalence between bisimulation and Trace	co-NP-hard	co-NP

Table 1. Table showing our complexity results for the problem P_1 for processes represented by Parallel Composition **without** Hiding

Our first set of results are for processes represented by parallel transition system *with* or *without* hiding. The results obtained are as follows:

1. We show that for all these relations problem P_1 is PSPACE-hard *even* **without** hiding. In [9] it was shown that the problem P_1 is PSPACE-hard for all standard equivalences between bisimulation and trace equivalence when processes are represented by parallel composition **with** hiding. Our proof is

Types of Systems	Decision problem	lower bound	upper bound
Unrestricted	Relations between weak bisimulation and trace preorder	NSPACE(n)-hard	
	Trace preorder		NSPACE(n)
Acyclic	Relations between weak bisimulation and trace preorder	Co-NP-hard	
	Trace preorder		Co-NP

Table 2. Table showing our results for problem P_2 for processes represented by Parallel Composition **with** hiding.

Relation	Lower bound	Upper bound
Forward Simulation		P
Backward Simulation		P
Refinement Mapping	NP-hard	NP
History Relation	Graph-isomorphism-hard	NP
Prophecy Relation	Graph-Isomorphism-hard	NP

Table 3. Table showing the complexity of relations for sequential systems (e.g., I/O automata)

based on the following observation which also leads to easier proofs of most of the other results in [9]. We show that all relations between bisimulation and trace preorder are as hard as the **nonreachability problem** for finite state processes for the corresponding representation. [2]

2. We also show that the *trace, failure* and *readiness* equivalences for systems represented as in 1 are PSPACE-Complete. Thus, our *uniform* lower bound of 1 is tight. [3]

3. For problem P_2 we show that, for all relations $\sim$ between *weak* bisimulation and trace preorder, and for *any* fixed process p_0, deciding if $P \sim p_0$, is PSPACE-hard, for a process P represented by parallel composition *with* hiding. This result follows from an observation similar to the nonreachability observation. We also show that deciding $P \sim p_0$, when $\sim$ is the trace preorder relation, is in PSPACE there by showing that this uniform lower bound is also tight. We also observe that there are fixed processes for which bisimulation is

---

[2] We also show that these lower bound results hold even when all the individual transition systems in the parallel composition (without hiding) have the same two-symbol *external* action alphabet.

[3] In [12], Stockmeyer proved that bisimulation equivalence is EXPTIME-Complete for processes represented using parallel composition with hiding. It is still an open question, whether there is a **uniform** EXPTIME lower bound for all these relations when hiding is used. However, we show, without hiding, we obtain the tightest **uniform** lower bound for all these relations.

easy to decide, even for parallel composition with hiding. Hence, our uniform lower bound can not be extended to bisimulation.

4. For a process P represented by a parallel composition of acyclic sequential processes with hiding, we show that, for any fixed acyclic process p_0 and any relation $\sim$ between *weak* bisimulation and trace preorder, deciding if $P \sim p_0$ is Co-NP-hard. This proof yields a simplified Co-NP-hard lower bound proof of a stronger version of the result in [9] that, for all standard equivalences between bisimulation and trace equivalence for parallel composition of acyclic processes without hiding, problem P_1 is Co-NP-hard. We also show that for trace equivalence, problem P_1 is in Co-NP, there by proveing that this lower bound is also the best possible. We also show that the lower bound holds even for cases when the individual acyclic transition systems are deterministic with as few as six states.

We also consider problem P_3 defined as follows:
P_3: Let π be a nontrivial predicate on finite state processes such that $\pi(x) = \pi(y)$ whenever x and y are weakly bisimilar. Determine, for a process p, if $\pi(p)$. We show that for any nontrivial predicate on processes, problem P_3 for processes represented by parallel composition *with* hiding is PSPACE-hard.

In another set of results we consider *forward* and *backward* simulation relations, *refinement* mapping, *history* and *prophecy* relations as defined in [7]. We consider problem P_1 for these relations for sequential systems. The complexity of these problems has not been considered in the literature previously. We obtain polynomial time algorithms for *forward simulation* and *backward simulation* relations. We obtain an NC algorithm for deciding bisimulation equivalence between two deterministic sequential transition systems. [4] Finally, we show that finding existence of *refinement mapping* is NP-Complete, and finding existence of *history relation* and finding existence of *prophecy relation* are graph-isomorphism-hard.

2 Transition Systems, Simulations and Equivalences

This section presents relevant definitions and results required for the exposition of our results.

Definition 1. *Act* is a set of actions containing a special action τ called the **internal action** or *unobservable action*. A **transition system** $\mathcal{T}$ over *Act* is a triple $\langle S, D, s_1 \rangle$ where S is a set of states, $D \subseteq S \times Act \times S$ is a set of transitions and $s_1 \in S$ is the starting state. $\mathcal{T}$ is **finite** if both S and *Act* are finite. $ext(\mathcal{T}) = Act - \{\tau\}$ is the set of **external** or **visible** actions. If σ is a sequence over *Act*, then $\hat{\sigma}$ is the sequence over $ext(\mathcal{T})$ obtained by deleting all the τ actions from σ. If (p_1, a, p_2) is in D then we write $p_1 \xrightarrow{a} p_2$. Also if σ is a sequence of actions such that there is a transition from state p_1 to state p_2 through some intermediate steps such that the sequence of actions is σ, then we write $p_1 \xRightarrow{\sigma} p_2$ and call this an **extended step**. Given $\langle \mathcal{T} =$

---

[4] This is stated as an open problem in [3]. However, one can show the membership in NC from the results in [5] where the problem was shown to be in NL. But our technique in [11] obtains a *direct* NC algorithm for the problem.

S, D, s_1), let $\overline{D} = \{(p, a, p') \mid p \in S \wedge a \in Act \wedge p' \in S \wedge \exists \sigma \in \tau^* a \tau^*, p \overset{\sigma}{\Longrightarrow} p'\}$. We call $\overline{D}$ the **extended transition** relation of T.

Let $T_1 = \langle S, D_1, s_1 \rangle$ and $T_2 = \langle T, D_2, t_1 \rangle$ be two transition systems.

Definition 2. Let $R \subseteq S \times T$ be a binary relation between S and T. R is a **simulation** if $\forall (s, t) \in R$
$$(\forall a \in Act, \forall s' \in S, ((s, a, s') \in D_1 \Rightarrow (\exists t' \in T ((t, a, t') \in D_2 \wedge (s', t') \in R))))).$$
R is a **bisimulation** if R and R^{-1} are both simulations.

Definition 3. We define T_1 to be **bisimulation equivalent** to T_2, denoted by $T_1 \sim_{bsim} T_2$, iff there is a bisimulation R such that $(s_1, t_1) \in R$. T_1 is said to be **simulated by** T_2, denoted by $T_1 \preceq_{sim} T_2$, iff there is a simulation R such that $(s_1, t_1) \in R$.

Definition 4. $B \subseteq S \times T$ is an **weak bisimulation** relation from T_1 to T_2 if the following conditions are satisfied.

1. $(s_1, t_1) \in B$
2. $\forall (r, s) \in B, a \in Act :$ if $\exists \gamma \in \tau^* a \tau^* : r \overset{\gamma}{\Rightarrow} r'$ then $\exists s' \exists \beta \in \tau^* a \tau^* : s \overset{\beta}{\Rightarrow} s' \wedge (r', s') \in B$, and if $\exists \beta \in \tau^* a \tau^* : s \overset{\beta}{\Rightarrow} s'$ then $\exists r' \exists \gamma \in \tau^* a \tau^* : r \overset{\gamma}{\Rightarrow} r' \wedge (r', s') \in B$.

If there exists a weak bisimulation from T_1 to T_2, then we say that they are **weak bisimulation equivalent**, denoted by $T_1 \sim_{wbsim} T_2$.

Definition 5. We say γ is a **finite trace** of a transition system $T = \langle S, D, s \rangle$ if there is a finite sequence $\sigma \in Act^*$ for which there is a state $q \in S$ such that $s \overset{\sigma}{\Longrightarrow} q$ and $\gamma = \hat{\sigma}$. Let **traces**(T) denote the set of all finite traces of a transition system T. We define **trace preorder** [5] and **trace equivalence** as follows. If traces$(T_1) \subseteq$ traces(T_2) then we say that (T_1, T_2) is in the **trace preorder** and denote this by $(T_1 \preceq_{trace} T_2)$. If traces$(T_1) =$ traces(T_2) then we say that (T_1, T_2) are **trace equivalent** and denote this by $(T_1 \sim_{trace} T_2)$.

The concepts of **forward simulation, backward simulation, refinement mapping, history relation,** and **prophecy relation** were defined in [7]. Next, we adapt these definitions to transition systems.

Definition 6. Let F be a binary relation between S and T. F is a **forward simulation** (or **possibilities mapping**) from T_1 to T_2 if the following conditions are satisfied.

1. $(s_1, t_1) \in F$.
2. $\forall (s, t) \in F$
$$(\forall a \in Act, \forall s' \in S, ((s, a, s') \in D_1 \Rightarrow (\exists t' \in T (\exists \sigma \in \tau^* a \tau^* (t \overset{\sigma}{\Rightarrow} t' \wedge (s', t') \in F))))).$$

We write $T_1 \preceq_F T_2$ if there is a forward simulation from T_1 and T_2. T_1 and T_2 are **forward-simulation equivalent** denoted by $T_1 \sim_F T_2$ iff both $T_1 \preceq_F T_2$ and $T_2 \preceq_F T_1$

---

[5] A binary relation is called a **preorder** if it is reflexive and transitive.

Definition 7. Let B be a **total**[6] binary relation between S and T. B is a **backward simulation** from T_1 to T_2 if the following conditions are satisfied.

1. $(s_1, t_1) \in B$ and there is no other $t \in T$ such that $(s_1, t) \in B$.
2. $\forall (s', t') \in B$

   $(\forall a \in Act, \forall s \in S, ((s, a, s') \in D_1 \Rightarrow (\exists t \in T(\exists \sigma \in \tau^* a \tau^* (t \overset{\sigma}{\Rightarrow} t' \wedge (s, t) \in B)))))$.

We write $T_1 \preceq_B T_2$ if there is a backward simulation from T_1 and T_2. T_1 and T_2 are **backward simulation equivalent** denoted by $T_1 \sim_B T_2$ iff both $T_1 \preceq_B T_2$ and $T_2 \preceq_B T_1$

Definition 8. A **refinement** from T_1 to T_2 is a function $r : S \rightarrow T$ such that the following conditions hold.

1. $r(s_1) = t_1$.
2. If $s \overset{a}{\rightarrow} s'$ in T_1 then $\exists \sigma \in \tau^* a \tau^*$ such that $r(s) \overset{\sigma}{\Rightarrow} r(s')$ in T_2.

We write $T_1 \preceq_R T_2$ if there exists a refinement from T_1 to T_2.

A relation h over S and T is a **history** relation from T_1 to T_2 if h is a forward simulation from T_1 to T_2 and h^{-1} is a refinement from T_2 to T_1. We write $T_1 \preceq_H T_2$ if there exists a history relation from T_1 to T_2.

A relation p over S and T is a **prophecy** relation from T_1 to T_2 if p is a backward simulation from T_1 to T_2 and p^{-1} is a refinement from T_2 to T_1. We write $T_1 \preceq_P T_2$ if there exists a prophecy relation from T_1 to T_2.

Before we define parallel composition, we note that in the context of parallel composition a transition system is represented as a 4-tuple, rather than 3-tuple as in Definition 1. Here, a transition system $\langle S, D, s \rangle$ over an action alphabet Act is represented as $\langle S, s, A, \rightarrow \rangle$, where $A = Act - \{\tau\}$ and $\rightarrow = D$.[7] Although the composition we define here is in the style of CSP [4], the complexity bounds obtained in this paper also hold for variants of this style of parallel composition. For example, composition of I/O automata, composition in CCS [8] etc.

The **parallel composition** of two transition systems T_1 and T_2 denoted by $T_1 \| T_2$ is defined as follows. Here we are defining synchronous parallel composition following Hoare [4].

Definition 9. Let $T_1 = \langle Q^1, q_0^1, A^1, \rightarrow_1 \rangle$ and $T_2 = \langle Q^2, q_0^2, A^2, \rightarrow_2 \rangle$. Let $T = T_1 \| T_2 = \langle Q, q_0, A, \rightarrow \rangle$. Then $Q = Q_1 \times Q_2$, $q_0 = (q_0^1, q_0^2)$, $A = A^1 \cup A^2$. The transition relation $\rightarrow$ for T is given by the following inference rules.

1. If $a \notin A^1 \cap A^2$ then $\dfrac{q_1 \overset{a}{\rightarrow}_1 q_2}{(q_1, q) \overset{a}{\rightarrow} (q_2, q)}$ and $\dfrac{q_1 \overset{a}{\rightarrow}_2 q_2}{(q, q_1) \overset{a}{\rightarrow} (q, q_2)}$.

2. τ-transitions : $\dfrac{q_1 \overset{\tau}{\rightarrow}_1 q_2}{(q_1, q) \overset{\tau}{\rightarrow} (q_2, q)}$ and $\dfrac{q_1 \overset{\tau}{\rightarrow}_2 q_2}{(q, q_1) \overset{\tau}{\rightarrow} (q, q_2)}$.

3. If $a \in A^1 \cap A^2$ then $\dfrac{q_1 \overset{a}{\rightarrow}_1 q_1' \quad q_2 \overset{a}{\rightarrow}_1 q_2'}{(q_1, q_2) \overset{a}{\rightarrow} (q_1', q_2')}$

---

[6] A binary relation $R \subseteq S \times T$ is **total** if for all $s \in S$ there exists $t \in T$ such that $(s, t) \in R$.

[7] This notational change is needed because in parallel composition, the action alphabet differs from one transition system to another. However, it is assumed that all transition systems may have τ transitions, unless they are deterministic.

Now we define the **Hiding** operation on transition systems.

Definition 10. Let $T_1 = \langle Q^1, q_0^1, A^1, \rightarrow_1 \rangle$ be a transition system. Then $T = hide\ a\ in\ T_1$ is the transition system $\langle Q, q_0, A, \rightarrow \rangle$ where $Q = Q^1$, $A = A^1 - \{a\}$, $q_0 = q_0^1$, and transition relation $\rightarrow$ of T is defined by the following inference rules.

1. If $a' \neq a$ then $\dfrac{q_1 \xrightarrow{a'} q_2}{(q_1) \xrightarrow{a'} (q_2)}$; and 2. τ-transitions : $\dfrac{q_1 \xrightarrow{\tau} q_2}{(q_1) \xrightarrow{\tau} (q_2)}$ and $\dfrac{q_1 \xrightarrow{a} q_2}{(q_1) \xrightarrow{\tau} (q_2)}$.

Let $A \subseteq Act$ be a set of actions. Then $hide\ A\ in\ T$ means $hide\ a_1\ in(hide\ a_2\ in(...in(hide\ a_n\ in\ T)...))$.

3 Problems P_1, P_2 and P_3 for Parallel Transition Systems

Reachability Problem and Uniform Lower Bounds:

Definition 11. Let $T = \langle Q, q_0, A, \rightarrow \rangle$ be a transition system. Let $s \in Q$ be a state of T. The **reachability** problem is to decide if there is a finite sequence $\sigma \in (A \cup \{\tau\})^*$, such that $q_0 \xRightarrow{\sigma} s$. We call the complement of this problem, the **nonreachability** problem.

Theorem 13 shows that the **nonreachability** problem can be used to obtain a uniform lower bound for a number equivalences and preorders.

Definition 12. Let T_1 and T_2 be two transition systems. Let ρ and θ be two binary relations defined on transition systems. We say $\rho \Rightarrow \theta$ (ρ implies θ) if and only if whenever $(T_1, T_2) \in \rho$, also $(T_1, T_2) \in \theta$. For any three binary relations ρ, θ, ϑ, we say that the relation ϑ is **between** ρ and θ, if $\rho \Rightarrow \vartheta \Rightarrow \theta$.

Theorem 13. *Consider a particular representation (e.g., sequential or parallel transition systems) for finite transition systems. The* **nonreachability prob**-**lem** *for transition systems with that representation is $n\,polylog\,n$ time reducible to the problem of deciding any binary relation $\sim$ which is* **between** *$\sim_{bsim}$ and $\preceq_{trace}$ for transition systems of the same representation. (Note that this includes all equivalences between $\sim_{bsim}$ and $\sim_{trace}$, as well as all preoreders between $\preceq_{sim}$ and $\preceq_{trace}$).*

Now we show using Theorem 13 that for transition systems represented succinctly using parallel composition (without hiding), any equivalence $\sim$ between bisimulation equivalence and trace equivalence as well as any relation $\preceq$ between simulation preorder and trace preorder is PSPACE-hard. Moreover, we also show that trace equivalence, failure equivalence and readiness equivalence are PSPACE-Complete for parallel transition systems, which shows that the **uniform lower bound** obtained here is the strongest that holds *uniformly* for all relations in these ranges.

Reachability in Parallel Transition Systems By reducing the deterministic LBA acceptance problem [2], to the reachability problem for parallel transition systems (without hiding), we show the following:

Theorem 14. *For a transition system specified as a parallel composition of deterministic sequential transition systems, the reachability problem is PSPACE-hard. Moreover, this is true even when all the individual transition systems in the composition have the same two-symbol external action alphabet.*

Now refering to Theorem 13, we can see that for deterministic parallel transition systems all relations between bisimulation and trace equivalences are PSPACE-complete. The PSPACE-hardness follows from Theorem 14 and Theorem 13. Since for deterministic systems all these equivalences coincide [13] and since trace inequivalence is decidable in PSPACE by guessing a trace one symbol at a time, we obtain that all relations between bisimulation equivalence and preorder are in PSPACE.

For parallel composition of nondeterministic sequential transition systems without hiding, we obtain the following results (using Theorem 13 and 14.)

Theorem 15. *Given two finite state systems represented as parallel transition systems without hiding, deciding any relation between bisimulation and trace preorder is PSPACE-hard. Moreover, this uniform lower bound holds even when all the individual transition systems in the composition have the same two-symbol external alphabet.*

We show that this uniform lower bound is tight by showing that there are some relations for which the problem is in PSPACE even for parallel composition of nondeterministic transition systems (without hiding).

Definition 16. The $\tau-$closure of a state s of a transition system $\langle Q, q_0, A, \rightarrow \rangle$, is the set of states that the system could reach from s, either by executing no transition, or by a sequence of τ transitions . In other words, $\tau-\textbf{closure}(s) = \{ s' \mid \exists \sigma \in \tau^* : s \overset{\sigma}{\Longrightarrow} s' \}$. Given a set of states S, $\tau-\textbf{closure}(S) = \bigcup_{s \in S} \tau-\textbf{closure}(s)$.

Lemma 17. *Given a parallel transition system $T_1 \| T_2 \| ... T_n$, the following are true: (i) Given a state of the parallel system $s = \langle s_1, s_2, .., s_n \rangle$, $\tau-\textbf{closure}(s) = S_1 \times S_2 \times ... \times S_n$, where $S_i = \tau-\textbf{closure}(s_i)$.*
(ii) Given a set of states S of the parallel system described as the Cartesian product of sets of states of the components (i.e., $S = S_1 \times S_2 \times ... \times S_n$), $\tau-\textbf{closure}(S) = S_1' \times S_2' \times ... \times S_n'$, where $S_i' = \tau-\textbf{closure}(S_i)$.

Now we state a theorem about parallel transition systems that is the basis for the PSPACE upper bounds here. (The proof of this theorem is based on induction of the length of action sequence σ and Lemma 17). This theorem applies only when **no** hiding abstraction is allowed. Intuitively, the theorem states the following. Let the set of states that a parallel transition system could reach via a given trace σ be $S = \{ s \mid \exists \gamma : \langle q_0^1, q_0^2, ..., q_0^n \rangle \overset{\gamma}{\Rightarrow} s \wedge \hat{\gamma} = \sigma \}$. Then S is the Cartesian product of sets S_i for $i = 1, .., n$. S_i is the set obtainable by keeping track of the set of states that T_i could reach via the projection of the trace σ on the action alphabet of T_i.

Definition 18. The **projection** of a trace γ over an action alphabet A, denoted as $\gamma \uparrow A$, is the result of deleting all the symbols in γ which are not in A. (Note that $\gamma \uparrow A$ might be ϵ where ϵ is the empty string.)

Theorem 19. *Given a parallel transition system* $T_1 \| T_2 \| ... T_n$, *and* $\sigma \in (\cup_i A_i)^*$, *the set* $\{ \langle s_1, s_2, ..., s_n \rangle \mid \exists \gamma : \langle q_0^1, q_0^2, ..., q_0^n \rangle \overset{\gamma}{\Rightarrow} \langle s_1, s_2, ..., s_n \rangle \wedge \hat{\gamma} = \sigma \}$ *equals* $S_1 \times S_2 \times ... \times S_n$ *where* $S_i = \{ s \mid \exists \beta : q_0^i \overset{\beta}{\Rightarrow}_i s \wedge \hat{\beta} = \sigma \uparrow A_i \}$.

Theorem 19 enables us to obtain a PSPACE algorithm for the reachability problem as follows. Let $T_1 \| T_2 \| .... \| T_n$ be a parallel transition system. Let $(s_1, s_2, ..., s_n)$ be the state to be reached as specified in the input of the reachability problem. We can guess a string $\sigma \in (A^1 \cup A^2 \cup .... \cup A^n)^*$, one symbol at a time, and guess corresponding transitions to confirm that $(q_0^1, q_0^2, ..., q_0^n) \overset{\sigma}{\Rightarrow} (s_1, s_2, ..., s_n)$.

Applications of Theorem 19 yield PSPACE upper bounds for trace equivalence, failure equivalence and readiness equivalence for this class of systems.

Theorem 20. *For parallel transition systems without hiding, the trace equivalence, failure equivalence and readiness equivalence* [8] *problems are PSPACE-Complete.*

Complexity of Deciding Relations to a Fixed Process We use a reduction from the LBA acceptance problem to prove the following:

Theorem 21. *For any fixed process p_0, and for any binary relation on processes between **weak** bisimulation and trace preorder, problem P_2 is PSPACE-hard.*

It is easy to see that the uniform lower bound is tight since there are relations between weak bisimulation and trace preorder for which this lower bound matches the upper bound. This shows that, we have obtained the best possible lower bound one could get *uniformly* for problem P_2 for any fixed process and for all the relations between weak bisimulation and trace preorder. On the other hand, Theorem 21 does not apply to bisimulation equivalence because there are fixed processes, (for example, a two state process that has a single action from the start state to end state,) for which problem P_2 for bisimulation equivalence can be easily solved in polynomial time. Hence, Theorem 21 cannot be strengthened in that direction.

Theorem 22. *For any fixed process p_0, for trace preorder, problem P_2 is PSPACE-complete.*

---

[8] definitions of failure and readiness equivalences can be found in [13]

Complexity of P_2 for composition of acyclic processes with hiding First we present a much easier proof of a stronger version of the Co-NP-hard lower bound result in [9] for the problem P_1, for all relations $\sim$ between bisimulation and trace equivalences, for parallel composition of acyclic sequential transition systems without hiding. We then use our proof to obtain the uniform lower bound result for problem P_2 for composition of this class of systems *with* hiding. We also significantly strengthen the result in [9] by showing that for trace equivalence, the lower bound obtained is tight. To obtain the lower bound result we first prove Theorem 24 via a polynomial time reduction from the following NP-Complete problem [10].

Definition 23. Ex-1-ex-3 Monotone 3SAT: *Instance:* $C = \bigwedge_{i=1}^{m} C_i$ is a conjunction of m clauses each containing exactly 3 **positive** literals.
Question: Is there is a satisfying assignment to this instance such that each clause is satisfied by setting **exactly one** literal to *true*.

Theorem 24. *The Reachability problem for parallel composition of acyclic transition systems is NP-Complete. The NP-hard lower bound applies even when each transition system in the composition is deterministic with only six states.*

Corollary 25. *The non-reachability problem for parallel composition of acyclic transition systems is Co-NP-Complete.*

Corollary 26. *All equivalences between bisimulation and trace equivalence and all preorders between the simulation and trace preorder for parallel composition of acyclic and deterministic transition systems are Co-NP-hard.*

Since the length of the longest trace for these systems is polynomially bounded by the size of the system description, trace nonequivalence is in NP. Thus we obtain the following :

Theorem 27. *Trace equivalence for parallel composition for acyclic transition systems is Co-NP-Complete. For parallel composition of deterministic acyclic transition systems all equivalences between bisimulation and trace equivalence and all preorders between simulation and trace preorder are Co-NP-Complete.*

Now consider problem P_2 for parallel transition system with hiding, where each component is acyclic. Consider a fixed acyclic process p_0. We can use the reduction from the Ex-1-ex-3 Monotone 3SAT as in the proof of Theorem 24, in the same way we used the reduction used in theorem 14 in proving theorem 21.

Theorem 28. *For any fixed acyclic process p_0, and for any binary relation on processes between **weak** bisimulation and trace preorder, problem P_2 is Co-NP-hard for processes which are represented as parallel composition of acyclic processes with hiding abstraction.*

We can also show that this uniform lower bound is also tight for this set of relations by showing that the trace preorder is in fact Co-NP-Complete.

Theorem 29. *For any fixed process p_0, for trace preorder, problem P_2 is co-NP-complete for processes which are parallel composition of acyclic processes with hiding abstraction.*

We also observe, that this uniform lower bound can not extend to bisimulation equivalence because of the same counter example discussed above.

Complexity of Predicates that Respect Weak Bisimilarity : Suppose π is a *nontrivial* predicate on finite state processes which respects weak bisimulation relation. In other words, for any two finite processes x and y, if they are *weak bisimulation* equivalent then $\pi(x) = \pi(y)$. Another reduction from the LBA acceptance problem yields the following:

Theorem 30. *Given a nontrivial predicate π on finite processes that respects weak bisimulation, the problem of determining, for a process P represented by a parallel transition system with hiding, if $\pi(P) = true$ is PSPACE-hard.*

4 Results on Sequential Transition Systems

Due to space limitations, we discuss the results obtained for sequential transition systems, without detailing any algorithms and proofs.

For *forward* and *backward* simulation relations [7], we get polynomial time algorithms. Our algorithms involve reducing these problems to a weakly negative CNF formula [10]. The satisfiability problem for this class of CNF formulas can be easily shown to be *linear time* solvable [1]. Our reduction produces CNF formula whose size is bounded by the product of the sizes of the transition relations. Hence, we get the following theorem.

Theorem 31. Forward simulation *and* backward simulation *relations between two transition systems* $T_1 = \langle S_1, D_1, s_1 \rangle$ *and* $T_2 = \langle S_2, D_2, s_2 \rangle$ *can be decided in* $O(|D_1| \cdot |D_2|)$ *time.*

Since our reduction is a $\preceq_{logspace}$ reduction, it can also be carried out in NC. Moreover, when the given transition systems are deterministic, a similar reduction for bisimulation equivalence yields CNF formula which has only two literals per clause (i.e., an instance of the 2-SAT problem). Since 2-SAT has an NC algorithm [3], our reduction methodology yields an NC algorithm for the bisimulation of deterministic transition systems (see [11] for details).

Theorem 32. Bisimulation equivalence *of* deterministic *transition systems is in* **NC**.

Complexity of Refinement Mapping, History Relation and Prophecy Relation Now we consider the complexity of deciding the existence of refinement mapping, history relation and prophecy relation between two transition systems T_1 and T_2. For each of these decision problems, membership in NP can be established easily. Due to lack of space, proof of the following theorem is omitted. In essence, Theorem 33 is proved by reducing the **K-clique** [2] problem to the refinement mapping problem and by reducing the **graph isomorphism** [2] problem to the problems of deciding history relation and prophecy relation between transition systems.

Theorem 33. *Given two transition systems T_1 and T_2. The problem of determining whether there is a refinement mapping from T_1 to T_2 is NP-Complete. The problem of of determining whether $T_1 \preceq_H T_2$ is and the problem of determining whether $T_1 \preceq_P T_2$ are graph-isomorphism-hard.*

Acknowledgements: We thank S. S. Ravi, Frits Vaandrager, and Pierre Wolper for helpful discussions. We also thank Alexander Rabinovich and Larry Stockmeyer for making their drafts available.

References

1. W.F. Dowling and J.H. Gallier. Linear time algorithm for testing the satisfiability of propositional horn formulae. *Journal of Logic Programming*, 3:267–284, 1984.
2. M. Garey and D. Johnson. *Computers and Intractability: A Guide to the Theory of NP-Completeness*. Freeman, SanFrancisco, 1979.
3. R. Greenlaw, H. J. Hoover, and W. L. Ruzzo. *Limits to Parallel Computation: P-completeness Theory*. Oxford University Press, 1995.
4. C. A. R. Hoare. *Communicating Sequential Processes*. Prentice Hall International, 1984.
5. Dung T. Huynh and Lu Tian. On deciding some equivalences for concurrent processes. *Theoretical Informatics and Applications*, 28(1):51–71, 1994.
6. R. Kurshan. *Computer Aided Verification of Coordinating processes : An Automata Theoretic Approach*. Princeton University Press, 1994.
7. Nancy Lynch and Frits Vaandrager. Forward and backward simulations-part i: Untimes systems. *Information and Computation*, 1995.
8. R. Milner. *Communication and Concurrency*. International Series in Computer Science. Prentice Hall, 1989. SU Fisher Research 511/24.
9. A. Rabinovich. Checking equivalences between concurrent systems of finite state agents. In *ICALP, Lecture Notes in Computer Science 623*, pages 696–707, 1992. A more recent draft is available from the author (May 1995).
10. Thomas J. Schaefer. The complexity of satisfiability problems. In *Tenth Annual Symposium on Theory of Computing*, 1978.
11. S. K. Shukla, D. J. Rosenkrantz, H. B. Hunt III, S. S. Ravi, and R. E. Stearns. A uniform approach for proving polynomial time decidability simulation relations for finite state processes. Research Report TR-95-6, Department of Computer Science, SUNY Albany, 1995.
12. L. J. Stockmeyer. Dexp-time hardness of bisimulation equivalence of concurrent system of finite state processes with hiding. (Unpublished Notes), 1992.
13. R.J. van Glabbeek. The linear time - branching time spectrum. Technical Report CS-R9029, Computer Science Department, CWI, Centre for Mathematics and Computer Science, Netherlands, 1990.
14. P. Wolper. Private communications. 1995.

Deciding Finiteness of Petri Nets up to Bisimulation

Petr Jančar[1]
Dept. of Computer Science
University of Ostrava
Dvořákova 7, 701 03 Ostrava
Czech Republic

Javier Esparza[2]
Institut für Informatik
Technische Universität München
Arcisstr. 21, D-80290 München
Germany

Abstract. We study the following problems for strong and weak bisimulation equivalence: given a labelled Petri net and a finite transition system, are they equivalent?; given a labelled Petri net, is it equivalent to some (unspecified) finite transition system? We show that both problems are decidable for strong bisimulation and undecidable for weak bisimulation.

1 Introduction

The decidability of equivalence notions for infinite-state systems has been extensively studied in the last years. Among other results, it has been shown that trace equivalence is undecidable for Basic Process Algebra (BPA) and Basic Parallel Processes (BPP), while bisimulation equivalence is decidable in both cases [1, 2, 4].

For arbitrary labelled Petri nets (called just Petri nets in the rest of this introduction), all the equivalence notions commonly used in the literature are undecidable [8, 6]. Therefore, in order to obtain positive results some constraints have to be imposed on the nets accepted as problem instances.

The case in which one of the two Petri nets to be compared is bounded has been considered in [9, 11]. For interleaving equivalence notions like language, trace or bisimulation equivalence, a bounded Petri net is equivalent to a finite transition system. So, loosely speaking, a Petri net is compared against a regular behaviour. The authors of [9, 11] study the following two problems:

- The *equivalence* problem (EP): given a Petri net and a finite transition system, are they equivalent?
- The *finiteness* problem (FP): given a Petri net, is it equivalent to some (unspecified) finite transition system?

Notice that the versions of EP and FP for deterministic or general context-free grammars and language equivalence are classical results of language theory.

---

[1] This author acknowledges the support of DAAD; he is also partly supported by the Grant Agency of the Czech Republic, Grant No. 201/96/0195.
[2] Supported by the Project SAM of the Sonderforschungsbereich 342.

In [11], Valk and Vidal-Naquet showed that EP and FP are undecidable for language equivalence. In the years following the publication of their paper, other equivalences based on the notion of transition system were found to be more appropriate for the verification of concurrent systems; two of the most commonly used today are trace and bisimulation equivalence.[3] In [9], EP and FP were studied again for these two equivalences (as well as for the simulation preorders). EP was shown to be decidable in both cases, while FP was shown to be undecidable for trace equivalence. FP for bisimulation was left open.

We show in this paper that FP is decidable for bisimulation, and that both EP and FP are undecidable for weak bisimulation.

The paper is structured as follows. Sections 2 and 3 contain some preliminaries. Sections 4 and 5 are devoted to EP and FP for bisimulation, respectively. Sections 6 and 7 study the same problems for weak bisimulation. (Although the decidability of the equivalence problem was already proved in [9], we prove it again in Section 4; the new proof is arguably simpler, and all the lemmas are needed in Section 5 anyway.)

2   Preliminary definitions

Let Act be a finite set of *actions*, containing a distinguished element τ called the *invisible* or *silent* action. A (*rooted, labelled*) *transition system* over Act consists of a set of states S with a distinguished initial state and a relation $\xrightarrow{a} \subseteq S \times S$ for every action $a \in Act$.

A (*labelled, place/transition Petri*) *net* over Act is a tuple $N = (P, T, W, M_0, \ell)$ where

- P and T are finite and disjoint sets of *places* and *transitions*, respectively;
- $W: (P \times T) \cup (T \times P) \rightarrow I\!N$ is a *weight function*;
- $M_0: P \rightarrow I\!N$ is the *initial marking* of N; and
- $\ell: T \rightarrow Act$ is a *labelling*, which associates an action to each transition.

If $W(x, y) \in \{0, 1\}$ for every x, y then N is called *ordinary*.

We fix some total ordering $p_1, p_2, \ldots, p_n$ of the places of P. A marking of N, i.e., a mapping $M: P \rightarrow I\!N$ attaching $M(p)$ *tokens* to a place p, is also denoted by the vector $(M(p_1), \ldots, M(p_n))$.

A transition t is *enabled* at a marking M if $M(p) \geq W(p, t)$ for every place p. A transition t enabled at M may *fire* or *occur* yielding the marking M' given by $M'(p) = M(p) + W(t, p) - W(p, t)$ for every $p \in P$. This is denoted by $M \xrightarrow{t} M'$. For any $a \in Act$, $M \xrightarrow{a} M'$ denotes that $M \xrightarrow{t} M'$ for some transition t such that $\ell(t) = a$. $M \longrightarrow M'$ denotes that there exists some sequence $t_1 \ldots t_k$ of transitions such that $M \xrightarrow{t_1} M_1 \xrightarrow{t_2} M_2 \xrightarrow{t_3} \cdots \xrightarrow{t_k} M'$.

---

[3] The language of a Petri net is defined as the set of words corresponding to occurrence sequences which lead to some final marking, while its trace set contains the words corresponding to all occurrence sequences.

The *reachability set* of a marking M is $\mathcal{R}(M) = \{M' \mid M \longrightarrow M'\}$. The reachability set of N is $\mathcal{R}(N) = \mathcal{R}(M_0)$. The transition system of N has $\mathcal{R}(N)$ as set of states, the initial marking M_0 as initial state, and the relations $\xrightarrow{a}$ between markings as relations.

Let N_1 and N_2 be two nets. A *strong bisimulation* is a relation $\mathcal{B}$ between the markings of N_1 and N_2 such that for all $(M_1, M_2) \in \mathcal{B}$ and for all $a \in Act$:

- if $M_1 \xrightarrow{a} M_1'$ then $M_2 \xrightarrow{a} M_2'$ for some M_2' such that $(M_1', M_2') \in \mathcal{B}$; and
- if $M_2 \xrightarrow{a} M_2'$ then $M_1 \xrightarrow{a} M_1'$ for some M_1' such that $(M_1', M_2') \in \mathcal{B}$.

For every action $a \in Act$, define $\stackrel{a}{\Longrightarrow} = (\xrightarrow{\tau})^* \xrightarrow{a} (\xrightarrow{\tau})^*$. A relation $\mathcal{B}$ between the markings of N_1 and N_2 is a *weak bisimulation* if whenever $(M_1, M_2) \in \mathcal{B}$ then for all $a \in Act \setminus \{\tau\}$:

- if $M_1 \xrightarrow{a} M_1'$ then $M_2 \stackrel{a}{\Longrightarrow} M_2'$ for some M_2' such that $(M_1', M_2') \in \mathcal{B}$; and
- if $M_2 \xrightarrow{a} M_2'$ then $M_1 \stackrel{a}{\Longrightarrow} M_1'$ for some M_1' such that $(M_1', M_2') \in \mathcal{B}$.

and, moreover

- if $M_1 \xrightarrow{\tau} M_1'$ then $(M_1', M_2) \in \mathcal{B}$ or $M_2 \stackrel{\tau}{\Longrightarrow} M_2'$ for some M_2' such that $(M_1', M_2') \in \mathcal{B}$; and
- if $M_2 \xrightarrow{a} M_2'$ then $(M_1, M_2') \in \mathcal{B}$ or $M_1 \stackrel{\tau}{\Longrightarrow} M_1'$ for some M_1' such that $(M_1', M_2') \in \mathcal{B}$.

M_1 is *strongly bisimilar* to M_2, denoted by $M_1 \sim M_2$, if $(M_1, M_2) \in \mathcal{B}$ for some strong bisimulation $\mathcal{B}$. Similarly, M_1 is *weakly bisimilar* to M_2, denoted by $M_1 \approx M_2$, if $(M_1, M_2) \in \mathcal{B}$ for some weak bisimulation $\mathcal{B}$. Two nets are strongly (weakly) bisimilar if their initial markings are strongly (weakly) bisimilar.

Notice that in the case of strong bisimulation τ plays the same rôle as any other action.

We use a stratified characterisation of strong bisimulation. Given two nets N_1 and N_2, the relations $\sim_i$ between their sets of markings are inductively defined as follows. First, we stipulate that $M_1 \sim_0 M_2$ for all pairs (M_1, M_2). Then for every $n \in \mathbb{N}$ we let $M_1 \sim_{n+1} M_2$ whenever for every $a \in Act$:

- if $M_1 \xrightarrow{a} M_1'$ then $M_2 \xrightarrow{a} M_2'$ for some M_2' such that $M_1' \sim_n M_2'$; and
- if $M_2 \xrightarrow{a} M_2'$ then $M_1 \xrightarrow{a} M_1'$ for some M_1' such that $M_1' \sim_n M_2'$.

Clearly, $\sim_i$ is an equivalence relation, and $\sim_{i+1} \subseteq \sim_i$ for every $i \geq 0$. Again we apply these relations to nets by considering their initial marking.

For every action a and every marking M of a net N, the set of markings M' such that $M \xrightarrow{a} M'$ is finite, because N has finitely many transitions. It is well known that in this case we have $M_1 \sim M_2$ iff $M_1 \sim_n M_2$ for every $n \in \mathbb{N}$; moreover, the problem 'Is $N_1 \sim_n N_2$?' is decidable for every $n \in \mathbb{N}$.

We also make use of the following property:

Proposition 1. *Let $N = (P, T, W, M_0, \ell)$ be a net, and let $k = max\{W(p, t) \mid p \in P, t \in T\}$. Let M_1, M_2 be two markings of N satisfying the following property for some number n: for every place p, $M_1(p) \neq M_2(p)$ implies $M_1(p) \geq kn$ and $M_2(p) \geq kn$. Then $M_1 \sim_n M_2$.*

Proof: The occurrence of a transition removes at most k tokens from a place. Therefore, the transition systems of N with M_1 and M_2 as initial markings are isomorphic up to depth n, which implies $M_1 \sim_n M_2$. ∎ 1

Notice that every finite transition system is the transition system of some net up to renaming of the states. Using this fact we can immediately extend the definitions above to the case in which N_1 and/or N_2 is replaced by a finite transition system. We use R to denote a finite transition system, r, r' to range over its states, and r_0 to denote its initial state.

To finish the section we formally define the problems we are going to study

Strong equivalence problem: given a Petri net N and a finite transition system R, is $N \sim R$?

Strong finiteness problem: given a Petri net N, is N finite with respect to strong bisimilarity (*b-finite*, for short), i.e. is there a finite transition system R such that $N \sim R$?

We also study the *weak equivalence problem* and the *weak finiteness problem*, which are defined analogously, substituting weak bisimulation ($\approx$) for strong bisimulation ($\sim$), and *weakly b-finite* for b-finite.

3 Auxiliary results

We recall a well-known lemma, which can be easily proved by induction on $|P|$:

Lemma 2. *Given an infinite sequence of mappings $M_1, M_2, M_3, \ldots$ of the type $P \to \mathbb{N}$, for a finite set P, there are indices $1 \leq i_1 < i_2 < i_3 < \ldots$ such that $M_{i_1} \leq M_{i_2} \leq M_{i_3} \ldots$ ($\leq$ defined componentwise).*

We also need a semidecidability result. Its proof is very similar to the proof of Theorem 6.5 in [5], and is omitted here. Let P be a finite set and let $\mathcal{M} \subseteq \mathbb{N}^P$. An element $p \in P$ is *unbounded in* $\mathcal{M}$ if for every $k \geq 0$ there exists $M_k \in \mathcal{M}$ such that $M_k(p) \geq k$. The set $\mathcal{M}$ is *linear* if there exist M_b (basis), and $M_1, \ldots, M_n$ (periods) in $\mathbb{N}^P$ such that

$$\mathcal{M} = \{M_b + \sum_{i=1}^{n} a_i M_i \mid a_1, \ldots, a_n \in \mathbb{N}\}.$$

Lemma 3. *Let N be a Petri net with initial marking M_0, and let $\mathcal{M}_1, \mathcal{M}_2$ be linear sets of markings of N. It is semidecidable if the set*

$$\mathcal{M}_{12} = \{M_1 \in \mathcal{M}_1 \mid \exists M_2 \in \mathcal{M}_2 \colon M_0 \longrightarrow M_1 \longrightarrow M_2\}$$

has the same unbounded places as $\mathcal{M}_1$.[4]   ■ 3

4 Decidability of strong equivalence

The strong equivalence problem has already been shown to be decidable in [9]. The proof given there relies on some general results about semilinear sets and Presburger arithmetic. Here we look in greater detail at the $\sim_n$ equivalence classes, which enables us to give a more self contained proof. All the lemmas of this section are also used in the next one.

We consider ordinary nets, because Proposition 1 has a particularly simple form for them. The results can be easily generalised to arbitrary nets by carrying the constant k of Proposition 1 through the proofs.

We fix an ordinary Petri net $N = (P, T, W, M_0, \ell)$, and a finite transition system R having n states and initial state r_0.

We start with a simple observation. Given two states r and r' of R, we have $r \sim_n r'$ whenever $r \sim_{n-1} r'$. This is easily seen to be true by noting that the sequence of approximation equivalences $\sim_0$, $\sim_1$, ... on the states of R must stabilise within n steps (every equivalence relation on this set has at most n equivalence classes).

We examine the structure of the classes $\sim_n$ on the markings of N.

Definition 4. A marking L of N is *n-bounded* if $L \leq (n, n, \ldots, n)$. For every n-bounded marking L, we define $L^{\geq n}$ as the set of markings M such that if $L(p) < n$, then $M(p) = L(p)$, and if $L(p) = n$, then $M(p) \geq L(p)$. A marking M of N is *incompatible with R* if $M \not\sim_n r$ for every state r of R. The set of markings incompatible with R is denoted by $Inc(R)$.   ■ 4

In the sequel we use the symbols L, L_1, L_2 for n-bounded markings. Note that $M_1 \sim_n M_2$ for any $M_1, M_2 \in L^{\geq n}$ (cf. Proposition 1), and that there are only finitely many n-bounded markings. Obviously, $Inc(R) = L_1^{\geq n} \cup \ldots \cup L_k^{\geq n}$ for some $L_1, \ldots, L_k$, where each L_i is n-bounded for $1 \leq i \leq k$.

Lemma 5. *$N \sim R$ iff the following two conditions hold:*

(1) *$N \sim_n R$; and*
(2) *$\mathcal{R}(N) \cap L^{\geq n} = \emptyset$ for every n-bounded marking $L \in Inc(R)$.*

Proof: ($\Rightarrow$): (1) follows immediately from the definitions. For (2), let L be an n-bounded marking such that $L \in Inc(R)$. Then $L^{\geq n} \subseteq Inc(R)$. Since $N \sim R$,

---

[4] Ernst Mayr has observed that this problem is probably even decidable.

every marking of $\mathcal{R}(N)$ is bisimilar to some state of R, and so $\mathcal{R}(N) \cap Inc(R) = \emptyset$. Therefore $\mathcal{R}(N) \cap L^{\geq n} = \emptyset$.

($\Leftarrow$): Define $\mathcal{B} = \{(M, r) \mid M \in \mathcal{R}(N) \text{ and } M \sim_n r\}$. We show that $\mathcal{B}$ is a strong bisimulation containing (M_0, r_0). $(M_0, r_0) \in \mathcal{B}$ follows from (1). Let $(M, r) \in \mathcal{B}$ and assume $M \xrightarrow{a} M'$ for some action a. Then there exists a state r' such that $r \xrightarrow{a} r'$ and $M' \sim_{n-1} r'$. We have $M' \in \mathcal{R}(N)$ and $M' \in L^{\geq n}$ for some n-bounded marking L. It follows from (2) that $L \notin Inc(R)$, and hence R contains a state r'' such that $M' \sim_n r''$. So we get $r' \sim_{n-1} r''$. By our observation at the beginning of the section, $r' \sim_n r''$, and hence $M' \sim_n r'$. This implies $(M', r') \in \mathcal{B}$, and so $\mathcal{B}$ is a strong bisimulation. ∎ 5

Theorem 6. *It is decidable whether $N \sim R$ or not.*

Proof: It suffices to prove that (1) and (2) in Lemma 5 are decidable. (1) is clearly decidable. Since there exist finitely many n-bounded markings, the set of n-bounded markings $L \in Inc(R)$ is effectively constructible. The decidability of the reachability problem for submarkings [3] easily implies the decidability of the problem $\mathcal{R}(N) \cap L^{\geq n} = \emptyset$ for a given n-bounded marking L. So (2) is decidable as well. ∎ 6

5 Decidability of strong finiteness

In this section we prove that the strong finiteness problem is decidable. Again, the proof is carried out for ordinary nets, but it can be easily generalised.

Let N be a Petri net, and let M be a marking of N. M is infinite with respect to strong bisimilarity (*b-infinite* for short) if there exist infinitely many markings $M_1, M_2, M_3, \ldots$ reachable from M such that $M_i \not\sim M_j$ for $i \neq j$. The net N is *b-infinite* if its initial marking is b-infinite. Since the strong equivalence problem is decidable, the strong finiteness problem is semidecidable. Therefore, it suffices to show that b-infiniteness is semidecidable.

We fix an ordinary Petri net $N = (P, T, W, M_0, \ell)$ for the rest of the section, and introduce a little notation. Let $M_1 \colon P_1 \to \mathbb{N}$ and $M_2 \colon P_2 \to \mathbb{N}$ be two mappings, where P_1, P_2 is a partition of P such that $P_2 \neq \emptyset$. (M_1, M_2) denotes the marking of N whose projection on P_1 is M_1, and whose projection on P_2 is M_2. We say 'a marking (M_1, M_2) of N' instead of 'a partition $P_1, P_2 \neq \emptyset$ of P and mappings $M_1 \colon P_1 \to \mathbb{N}$, $M_2 \colon P_2 \to \mathbb{N}$'. $(M_1, -)$ denotes the marking M_1 of the net obtained from N by removing all places that do not belong to P_1, together with their incident arcs. We make abuse of language, and speak of 'the marking $(M_1, -)$ of N'. We also say 'a marking $(M_1, -)$ of N' instead of 'a partition $P_1, P_2 \neq \emptyset$ of P and a mapping $M_1 \colon P_1 \to \mathbb{N}$'.

An argument similar to that of Proposition 1 proves that $M' \geq (i, \ldots, i)$ for some number i implies $(M, M') \sim_i (M, -)$.

Lemma 7. *If $(M, M_1) \sim (M, M_2) \sim (M, M_3) \sim \ldots$ and $M_1 < M_2 < M_3 \ldots$ ($<$ defined componentwise), then $(M, M_1) \sim (M, -)$.*

Proof: For every $i \geq 0$ there surely exists an index j such that $M_j \geq (i, i, \ldots, i)$. Then $(M, -) \sim_i (M, M_j)$ holds and, since $(M, M_1) \sim (M, M_j)$, we also have $(M, -) \sim_i (M, M_1)$. Therefore, $(M, -) \sim_i (M, M_1)$ for every $i \geq 0$, and so $(M, -) \sim (M, M_1)$. ■ 7

Lemma 8. *N is b-infinite iff there exists a marking $(M, -)$ of N that satisfies either*

(1) *$(M, -)$ is b-infinite and there exists a chain $M_1 < M_2 < M_3 \ldots$ such that $(M, M_i) \in \mathcal{R}(N)$ for every $i \geq 1$, or*

(2) *$(M, -)$ is b-finite and there exists a chain $M_1 < M_2 < M_3 \ldots$ such that $(M, M_i) \in \mathcal{R}(N)$ and $(M, M_i) \not\sim (M, -)$ for every $i \geq 1$.*

Proof: ($\Rightarrow$): If N is b-infinite, then there exists an infinite set of pairwise non-bisimilar reachable markings. By Dickson's Lemma, we can extract from this set an infinite subsequence $(M, M_1), (M, M_2), (M, M_3), \ldots$ such that $M_1 < M_2 < M_3 < \ldots$. So either (1) or (2) holds, according to whether $(M, -)$ is b-finite or b-infinite.

($\Leftarrow$): Define $\mathcal{M} = \{(M, M_i) \mid i \geq 1\}$. Assume (1) holds. If $\mathcal{M}$ contains infinitely many pairwise non-bisimilar markings, then N is b-infinite, and we are done. Otherwise, $\mathcal{M}$ contains infinitely many pairwise bisimilar markings. By Lemma 7, all these markings are bisimilar to $(M, -)$ and hence b-infinite. So N is b-infinite. Assume now that (2) holds. If $\mathcal{M}$ contains infinitely many pairwise bisimilar markings, then by Lemma 7 all of them are bisimilar to $(M, -)$, contradicting the condition $(M, M_i) \not\sim (M, -)$ for every $i \geq 0$. So $\mathcal{M}$ contains infinitely many pairwise non-bisimilar markings, which implies that N is b-infinite. ■ 8

Theorem 9. *It is decidable whether N is b-finite or not.*

Proof: We proceed by induction on the number of places of N. If N has no places, then it is clearly b-finite. Assume now that N has some places. By Theorem 6, the b-finiteness of N is semidecidable. So it suffices to prove that the b-infiniteness of N is also semidecidable, or, equivalently, that conditions (1) and (2) of Lemma 8 are semidecidable.

For that, we enumerate all markings $(M, -)$ of N for all partitions P_1, P_2 such that $P_2 \neq \emptyset$. Given a marking $(M, -)$, we can decide by induction hypothesis if it is b-finite or b-infinite; moreover:

(1) The existence of a chain $M_1 < M_2 < M_3 \ldots$ such that $(M, M_i) \in \mathcal{R}(N)$ for every $i \geq 1$ is semidecidable.

Let $\mathcal{M}_1$ be the set of all markings of the form (M, M'); it is obviously a linear set where exactly the places of P_2 are unbounded. Now the desired semidecidability follows from Lemma 3 (putting e.g. $\mathcal{M}_2 = \mathcal{M}_1$).

(2) If $(M, -)$ is b-finite, then the existence of a chain $M_1 < M_2 < M_3 \ldots$ such that $(M, M_i) \in \mathcal{R}(N)$ and $(M, M_i) \not\sim (M, -)$ for every $i \geq 1$ is semidecidable.

Assume that $(M, -)$ is b-finite. Then by exhaustive search and Theorem 6 a finite transition system R can be found such that $(M, -) \sim R$. Let n denote the number of states of R.

Say a chain is *adequate* if it satisfies the conditions of (2). We prove that there exists an adequate chain iff there exists an n-bounded marking L of N satisfying the following two conditions

(a) $L \in Inc(R)$; and

(b) there exists a chain $M_1 < M_2 < M_3 \ldots$ and markings $M_1', M_2', M_3', \ldots \in L^{\geq n}$ such that $M_0 \longrightarrow (M, M_i) \longrightarrow M_i'$ for every $i \geq 1$.

($\Rightarrow$): Let $M_1 < M_2 < M_3 \ldots$ be an adequate chain. There exists an index i_0 such that $M_i \geq (n, n, \ldots, n)$ for every $i \geq i_0$. For $i \geq i_0$ we have $(M, M_i) \not\sim (M, -)$ by assumption (and so $(M, M_i) \not\sim R$), but $(M, M_i) \sim_n (M, -)$ (and so $(M, M_i) \sim_n R$). By Lemma 5 there exists an n-bounded marking $L_i \in Inc(R)$ such that $\mathcal{R}((M, M_i)) \cap L_i^{\geq n} \neq \emptyset$.

By the pigeonhole principle there exists an n-bounded marking L and infinitely many indices $i_1 < i_2 < i_3 \ldots$ such that $L = L_{i_1} = L_{i_2} \ldots$. Clearly, L satisfies (a) and the subchain $M_{i_1} < M_{i_2} < M_{i_3} \ldots$ satisfies (b).

($\Leftarrow$): Let M_i be an arbitrary marking of the chain given by (b). We prove $(M, M_i) \not\sim (M, -)$, which shows that the chain is adequate. Since $M_0 \longrightarrow (M, M_i) \longrightarrow M_i'$ for some marking $M_i' \in L^{\geq n}$, we have $\mathcal{R}((M, M_i)) \cap L^{\geq n} \neq \emptyset$. By (a) and Lemma 5 we have $(M, M_i) \not\sim R$, which together with $(M, -) \sim R$ implies $(M, M_i) \not\sim (M, -)$.

It remains to prove the semidecidability of conditions (a) and (b) for a given n-bounded marking L. Condition (a) is clearly decidable. For condition (b), apply Lemma 3 with $\mathcal{M}_1$ as the set of markings of N of the form (M, M'), and $\mathcal{M}_2 = L^{\geq n}$. Note that $\mathcal{M}_1$ and $\mathcal{M}_2$ are linear sets. ∎ 9

6 Undecidability of weak equivalence

We show that the weak equivalence problem is undecidable. In fact, we prove that neither of the problems $N \approx R$ and $N \not\approx R$ is semidecidable; as a corollary of this result, we also find a fixed transition system R_{fix} with 7 states such that $N \approx R_{fix}$ is undecidable. The proof is by reduction from the

> **Containment problem:** given two Petri nets N_1, N_2 having the same number of places, and a bijection f from the places of N_1 onto the places of N_2, is $f(\mathcal{R}(N_1)) \subseteq \mathcal{R}(N_2)$? (where f is extended to markings and sets of markings in the obvious way.)

The undecidability of this problem was proved by Rabin by means of a reduction from Hilbert's 10th problem. A reduction from the halting problem for counter machines can be found in [8].

Let $N_1 = (P_1, T_1, W_1, M_{10}, \ell_1)$, $N_2 = (P_2, T_2, W_2, M_{20}, \ell_2)$ be two Petri nets, and let $f: P_1 \to P_2$ be a bijection. We construct another net N which is weakly bisimilar to the state r_1 of the finite transition system R shown in Figure 1 if $f(\mathcal{R}(N_1)) \not\subseteq \mathcal{R}(N_2)$ and weakly bisimilar to the state r_5 if $f(\mathcal{R}(N_1)) \subseteq \mathcal{R}(N_2)$.

(The state r_0 of R is used in the next section.) Without loss of generality, we assume that (a) the sets of places and transitions of N_1 and N_2 are disjoint, (b) $f(M_{10}) = M_{20}$, (c) $|\mathcal{R}(N_2)| \geq 2$, and (d) $(0, \ldots, 0) \notin \mathcal{R}(N_1) \cup \mathcal{R}(N_2)$.

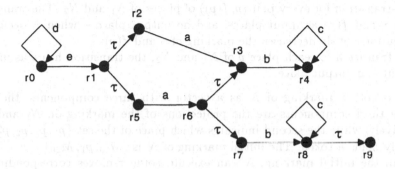

Fig. 1. The transition system R

Instead of giving a formal definition of N, which would be tedious, we describe it informally at a level of detail which suffices to follow our arguments. We need the following notion: a place p is a *run place* of a set T of transitions if $W(p, t) = 1 = W(t, p)$ for every $t \in T$. The transitions of T can only occur when p holds at least one token.

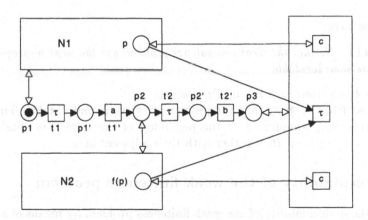

Fig. 2. Scheme net N

Figure 2 shows a schema of the net N. To construct N, we first take the union of N_1 and N_2 (the sets of places and transitions of N_1 and N_2 are disjoint by assumption (a)), labelling all transitions by τ. Then we add some further places and transitions: the place p_1 is a run-place of T_1 (graphically represented by a double pointed white arrow), and contains initially one token. This token can be moved by a new τ-transition t_1 to a place p'_1, and then by an a-transition t'_1 to p_2, which is a run-place of T_2. From p_2, the token can be moved by another τ-transition t_2 to p'_2 and by a b-transition t'_2 to p_3, which is a run-place of an additional set of transitions. This set contains:

- a τ-transition for every pair $(p, f(p))$ of places of N_1 and N_2. The transition has p and $f(p)$ as input places, and no output place – when it occurs, it simultaneaously decreases the marking of p and $f(p)$;
- a c-transition for each place p of N_1 and N_2; the transition has p as unique input and output place.

We denote a marking of N as a vector with three components: the first and the third components are the projections of the marking on N_1 and N_2, respectively, while the second indicates which place of the set $\{p_1, p'_1, p_2, p'_2, p_3\}$ currently holds a token. The initial marking of N is (M_{10}, p_1, M_{20}).

From the initial marking, N can execute some τ-moves corresponding to transitions of N_1. If at some moment t_1 occurs (this is also a τ-move), then a marking (M_1, p'_1, M_{20}) is reached, and the marking M_1 becomes 'frozen'. Then, after an a-move, N can execute some τ-moves corresponding to transitions of N_2. Again, if at some moment t_2 occurs, then a marking (M_1, p'_2, M_2) is reached, and the marking M_2 becomes 'frozen'.

The following Proposition is easy to prove making use of the asumptions (a) to (d):

Proposition 10. *If $f(\mathcal{R}(N_1)) \subseteq \mathcal{R}(N_2)$, then $N \approx r_5$. If $f(\mathcal{R}(N_1)) \nsubseteq \mathcal{R}(N_2)$, then $N \approx r_1$.* ■ 10

We now have:

Theorem 11. *Neither the weak equivalence problem nor the weak non-equivalence problem are semidecidable.*

Proof: Follows from $r_1 \napprox r_5$ and Proposition 10. ■ 11

It follows from $r_1 \napprox r_5$ that the problem $N \approx r_5$ is undecidable. The transition system R_{fix} announced at the beginning of the section is obtained by removing r_0, r_1, r_2 from R, together with their adjacent arcs.

7   Undecidability of the weak finiteness problem

We show the undecidability of the weak finiteness problem by means of a reduction from the

Halting problem for 2-counter machines and 0-input: does a given 2-counter machine C halt when both counters are initialised to 0?

whose undecidability is well known. Let C be an arbitrary 2-counter machine. We construct another 2-counter machine C' which for input $(x, 0)$ computes as follows: first, it checks if $x = 2^k$ for some $k \geq 0$; if this is the case, then it sets the counters to 0 and simulates C, otherwise it halts. We have:

- if C halts for $(0,0)$, then C' halts for every input $(x,0)$, $x \geq 0$;
- if C does not halt for $(0,0)$ then C' halts for $(x,0)$ iff x is not a power of 2.

We now use a procedure described in [8]: it accepts the 2-counter machine C' and outputs two nets N_1, N_2 with distinguished places c_{11}, c_{12} and c_{21}, c_{22} (initially unmarked), together with a bijection f between the places of N_1 and N_2 satisfying $f(c_{11}) = c_{21}$ and $f(c_{12}) = c_{22}$. N_1 and N_2 satisfy the following property for every $x \geq 0$:

$$C' \text{ halts for input } (x,0) \quad \text{iff} \quad f(\mathcal{R}(N_1^{x,0})) \not\subseteq \mathcal{R}(N_2^{x,0})$$

where $N_i^{x,0}$ denotes the result of changing the initial marking of N_i by putting exactly x tokens on c_{i1}.

We apply the construction of the Section 6 to the nets N_1 and N_2, and obtain this way a net N. We modify this net in the following way. First, we remove the token from p_1. Second, we add some new places and transitions:

- a place p_0, initially marked with one token;
- a d-transition with p_0 as only input place, and p_0, c_{11} and c_{21} as output places (i.e., p_0 is a run-place for this transition);
- an e-transition, with p_0 as input place and p_1 as output place.

Let N' be the result of this modification. From its initial marking, N' can repeatedly execute the d-transition, by which it puts an arbitrary number of tokens x on the places c_{11} and c_{21}. Then, it may execute the e-transition. After that, the place p_1 carries a token, and N' behaves like the net we would obtain by applying the construction of the last section to $N_1^{x,0}$ and $N_2^{x,0}$.

Proposition 12. N' is weakly b-finite iff the counter machine C halts for $(0,0)$.

Proof: ($\Rightarrow$): If C does not halt for $(0,0)$ then C' halts for $(x,0)$ iff x is not a power of 2. Therefore $f(\mathcal{R}(N_1^{x,0})) \subseteq \mathcal{R}(N_2^{x,0})$ if $x = 2^k$ for some $k \geq 0$, and $f(\mathcal{R}(N_1^{x,0})) \not\subseteq \mathcal{R}(N_2^{x,0})$ otherwise. Let R be the transition system of Figure 1 with states $r_0, \dots, r_9$. We have for any x and the unique marking M reached after executing the sequence d^x in N':

(1) if x is not a power of 2 then there is M' s.t. $M \overset{e}{\Longrightarrow} M'$ and $M' \approx r_1$;
(2) if x is a power of 2 then there is no such M'.

We prove that N' is weakly b-infinite by contradiction. Assume that N' is weakly bisimilar to some transition system R' with n states. Let r_0' be the initial state of R', and let r' be a state such that $r_0' \overset{u}{\longrightarrow} r'$, where u is a sequence of actions whose projection on the set of visible actions is d^{2^n}. By the pumping lemma, $r_0' \overset{vw^i x}{\longrightarrow} r'$ for sequences v, w, x and for every $i \geq 0$, where $u = vwx$ and the projection of w on the set of visible actions is a nonempty sequence of d's. By (1) and (2) we have that there is r'' in R' s.t. $r' \overset{e}{\Longrightarrow} r''$ and $r'' \approx r_1$, and at the same time there is no such $r'' - $ a contradiction. ($\Leftarrow$): If C halts for $(0,0)$, then C' halts for every input $(x,0)$, $x \geq 0$. Therefore after the occurrence of the e-transition we always have $f(\mathcal{R}(N_1^{x,0})) \not\subseteq \mathcal{R}(N_2^{x,0})$, independently of the value of x. It is then clear that $N' \approx r_0$. So N' is weakly b-finite. ∎ 12

Theorem 13. *Neither the weak finiteness problem nor the weak infiniteness problems are semidecidable.*

Proof: By Proposition 12, C does not halt for $(0,0)$ iff N' is weakly b-infinite. So the weak infiniteness problem is not semidecidable. We can also change C' in the following way: if x is not a power of 2, then C' enters an infinite loop. In this case, C does not halt for $(0,0)$ iff the net N' is weakly b-finite. So the weak finiteness problem is not semidecidable either. ∎ 13

8 Conclusions

We have shown that the finiteness problem is decidable for Petri nets, while the weak equivalence and weak finiteness problems are undecidable. The finiteness problem for Basic Process Algebra has been recently studied in [10]. The results are similar to ours: undecidable for trace equivalence, but decidable for bisimulation equivalence. Finally, since BPPs are bisimilar to a particular class of Petri nets (called communication-free nets in [4]), the decidability of the finiteness problem for BPP and bisimulation follows as a corollary of our results.

References

1. S. Christensen, Y. Hirshfeld, and F. Moller. Bisimulation Equivalence is Decidable for all Basic Parallel Processes. In *Proceedings of CONCUR '93*, LNCS 715, 143–157, 1993.
2. S. Christensen, H. Hüttel, and C. Stirling. Bisimulation Equivalence is Decidable for all Context-free Processes. In *Proceedings of CONCUR '92*, LNCS 630, 138–147, 1992.
3. J. Esparza and M. Nielsen. Decidability Issues for Petri Nets – a Survey. *Bulletin of the EATCS*, 52:245–262, 1994, and Journal of Information Processing and Cybernetics 30(3):143–160, 1995.
4. Y. Hirshfeld. Petri Nets and the Equivalence Problem. In *Proceedings of CSL '93*, LNCS 832, 165–174, 1994.
5. P. Jančar. Decidability of a Temporal Logic Problem for Petri Nets. *Theoretical Computer Science*, 74:71–93, 1990.
6. P. Jančar. All Action-based Behavioural Equivalences are Undecidable for Labelled Petri Nets. *Bulleting of the EATCS*, 56:86–88, 1995.
7. P. Jančar. Decidability Questions for Equivalences in Petri Nets. Technical report, University of Ostrava, habilitation thesis, 1995.
8. P. Jančar. Undecidability of Bisimilarity for Petri Nets and Some Related Problems. *Theoretical Computer Science*, 148:281–301, 1995.
9. P. Jančar and F. Moller. Checking Regular Properties of Petri Nets. In *Proceedings of CONCUR '95*, LNCS 962, 348–362, 1995.
10. S. Mauw and H. Mulder. Regularity of BPA-systems is Decidable. In *Proceedings of CONCUR '94*, LNCS 836, 34–47, 1994.
11. R. Valk and G. Vidal-Naquet. Petri Nets and Regular Languages. *Journal of Computer and system Sciences*, 23(3):299–325, 1981.

Mobile Processes with a Distributed Environment *

Chiara Bodei, Pierpaolo Degano and Corrado Priami

Dipartimento di Informatica, Università di Pisa
Corso Italia 40, I-56125 Pisa, Italy - {chiara,degano,priami}@di.unipi.it

EXTENDED ABSTRACT

Abstract. We introduce *local* environments for mobile processes, expressed in π-calculus. Each local name is equipped with its *relative address*, i.e., with the information needed to point back to the process that generated it. Relative addresses are built upon the labels of the proved transition system of π-calculus. A router is specified that guarantees sound exportation of names.

1 Introduction

The specification of systems of mobile processes is receiving more and more attention [9, 1, 12, 14, 10]. Efficiency considerations suggest implementations that provide each process composing the system with its own *local* environment. In the π-calculus [9] view, this amounts to saying that each process has its own space of private names. Possibly, some of these names are communicated to another process and so they become shared by different local environments.

Here, we give a new structural operational semantics to π-calculus that considers names *localized* to their owners. In other words, each sequential mobile process has its local space of names and a local name manager that generates a fresh name, whenever necessary. When a name is exported, it is equipped with the information needed to point back to the local environment where it has been installed as fresh. More precisely, while deducing a communication (or an extrusion), the name exported records the path from the receiving process to the one that generated the name (*not* to the sender). We call this path *relative address*. In this way, names generated by different environments will be certainly kept distinct. There is no need of a global, thus inefficient, check that a name involved in a transition captures names already in use. Also, α-conversions are no more necessary to enforce disjointness of local environments (cfr. the semantic definitions of the calculi for mobile processes).

Our transition system is necessarily more concrete and detailed than the original one, because it is closer to an implementation. However, the two are strongly related. A transition is present in our transition system if and only if a variant of its is present in the original one. (In passing: variants are banned from

* Work partially supported by ESPRIT Basic Research Action n.8130 - LOMAPS

our transition system, which is therefore *inherently* finitely-branching, rather than finitely-branching up to α-conversions.)

The relative addresses we use to handle names locally are defined according to the abstract syntax of systems. We foresee no difficulty in adopting a different addressing mechanism, when a more basic description of the topology of the network is available. Actually, we only need to re-define the router.

Needless to say, our results can be easily transferred to real programming languages, like *Facile* [6, 13], that already has a proved operational semantics [2], or *CML* [8, 11]. We hope that our proposal could help the design of language implementations, giving hints on how these can be made truly distributed, both with respect to control and to data. As they are, our extended names are quite unreadable. However, they are to be used as *internal* names for specifications closer to efficient implementations, rather than in high-level specifications, where a global space of names and global checks are acceptable.

2 π-Calculus

In this section we briefly recall π-calculus [9, 7], a model of concurrent communicating processes based on the notion of *naming*.

Definition 1. Let $\mathcal{N}$ be a countable infinite set of *names* ranged over by $a, b, \ldots$, $x, y, \ldots$ with $\mathcal{N} \cap \{\tau\} = \emptyset$. *Processes* (denoted by $P, Q, R, \ldots \in \mathcal{P}$) are built from names according to the syntax

$$P ::= 0 \mid \pi.P \mid P + P \mid P|P \mid (\nu x)P \mid [x = y]P \mid P(y)$$

where π may be either $x(y)$ for *input*, or $\overline{x}y$ for *output* (where x is the *subject* and y the *object*) or τ for *silent* moves. Hereafter, the trailing 0 will be omitted.

The late operational semantics of π-calculus is defined in *SOS* style, and the labels of transitions are τ for silent actions, $x(y)$ for bound input, $\overline{x}y$ for free output, and $\overline{x}(y)$ for bound output. We will use μ as a metavariable for the labels of transitions (it is distinct from π, the metavariable for prefixes, though it coincides in two cases). We recall the notion of free names $fn(\mu)$, bound names $bn(\mu)$, and names $n(\mu) = fn(\mu) \cup bn(\mu)$ of a label μ. Also two functions are defined giving th esubject and the object of I/O actions.

μ	Kind	$fn(\mu)$	$bn(\mu)$	$sbj(\mu)$	$obj(\mu)$
τ	Silent	$\emptyset$	$\emptyset$	$\emptyset$	$\emptyset$
$\overline{x}y$	Free Output	$\{x, y\}$	$\emptyset$	$\{x\}$	$\{y\}$
$x(y), \overline{x}(y)$	Bound Input and Output	$\{x\}$	$\{y\}$	$\{x\}$	$\{y\}$

Hereafter, we denote singleton without brackets. Functions fn, bn and n are extended in the obvious way to processes. Below we assume that the *structural congruence* $\equiv_P$ on processes defined as the least congruence satisfying $(\mathcal{P}/_{\equiv_P}, +, 0)$

is a commutative monoid, and $[x = x]P \equiv_P P$. Note that the $|$ is *neither* associative *nor* commutative.

A *variant* of $P \xrightarrow{\mu}_L Q$, is a transition which only differs in that P and Q have been replaced by structurally congruent processes, and μ has been α-converted, where a name bound in μ includes Q in its scope.

We report the late transition system of π-calculus in Tab. 1. The transition in the conclusion of each rule, as well as in the axiom, stands for all its variants. The table only reports one rule for the binary operator $|$; it also has a symmetric rule.

$$Act : \mu.P \xrightarrow{\mu}_L P \qquad\qquad Ide : \frac{P\{y/x\} \xrightarrow{\mu}_L P'}{Q(y) \xrightarrow{\mu}_L P'}, Q(x) = P$$

$$Par : \frac{P \xrightarrow{\mu}_L P'}{P|Q \xrightarrow{\mu}_L P'|Q}, bn(\mu) \cap fn(Q) = \emptyset \qquad Sum : \frac{P \xrightarrow{\mu}_L P'}{P+Q \xrightarrow{\mu}_L P'}$$

$$Res : \frac{P \xrightarrow{\mu}_L P'}{(\nu x)P \xrightarrow{\mu}_L (\nu x)P'}, x \notin n(\mu) \qquad Open : \frac{P \xrightarrow{\bar{x}y}_L P'}{(\nu y)P \xrightarrow{\bar{x}(y)}_L P'}, y \neq x$$

$$Close : \frac{P \xrightarrow{\bar{x}(y)}_L P', Q \xrightarrow{x(y)}_L Q'}{P|Q \xrightarrow{\tau}_L (\nu y)(P'|Q')} \qquad Com : \frac{P \xrightarrow{\bar{x}y}_L P', Q \xrightarrow{x(z)}_L Q'}{P|Q \xrightarrow{\tau}_L P'|Q'\{y/z\}}$$

Table 1. Late transition system for π-calculus.

3 Handling names

Consider for a while the parallel composition as the primary operator of π-calculus, insisting that it is not commutative. Then, build abstract syntax trees of processes as binary trees whose nodes are $|$ operators and whose leaves are the sequential components (notion made precise later) of the whole process. Call them *trees of (sequential) processes* (see Fig. 1). Assume that their left (right) branches denote the left (right) component of parallel composition, and label their arcs as $||_0$ ($||_1$). Therefore, any sequential component of a process is uniquely identified by a string over $\{||_0, ||_1\}^*$. The string corresponds to a path from the root, the top-level $|$ of the whole process, to a leaf. Intuitively, the string is the address of the sequential component relative to the root. We will make use of this information to specify a distributed name manager that handles names locally to sequential processes. Of course, a distributed environment rules out the equations that handle restrictions globally, e.g. $(\nu x)(P|Q) \equiv_P (\nu x)P|Q$ if $x \notin fn(Q)$.

Although the distinction is not so sharp, in π-calculus names can be divided in free or bound. Bound names may become free through either input actions or extrusions. When a bound name becomes free, an expensive α-conversion may be needed to avoid captures of free names (see [9] for a detailed discussion).

To avoid a gobal management of fresh names, we have to solve two problems. Names have to be generated *locally* and to be brand new in that local environment. Furthermore, when a name is exported to other local environments via communications or extrusions, we must guarantee that it *captures no other free name* around.

First, we introduce a new indexed set of *localized names* (for simplicity, natural numbers), and we associate a counter with every sequential process. When needed, the first name not in use is taken and the counter is increased. If firing a prefix originates new sequential processes, the counter is distributed to them all. Clearly, this mechanism guarantees that a newly generated name is unique in its scope and does not capture other names therein.

The second problem arises when two different sequential processes, say G and R, have generated two names syntactically equal, say n, that are although semantically distinct. Suppose now that G sends to R its n. To distinguish between the two different instances of n in the local environment of R, the name generated by G will be enriched with the *address* of G *relative* to R. The relative address can be decomposed into two parts according to the minimal common predecessor P of G and R. Hence, a relative address is a string $\vartheta \bullet \vartheta' \in \{||_0, ||_1\}^* \bullet \{||_0, ||_1\}^*$, where ϑ represents the path from P to R, and ϑ' the path from P to G. Consider Fig. 1, and let G be P_3 and R be P_1. The address of P_3 relative to P_1 is $||_0||_1 \bullet ||_1||_1||_0$. We will inductively build relative addresses while deducing transitions according to the inference rules of the proved transition system of π-calculus [5]. It suffices to record the application of inference rules involving the | in the label of a deduced transition.

Slightly more complex is when a process receives a name and sends it to another process. The name must arrive to the new receiver with the address of the generator (*not* of the sender) relative to the new receiver. This is done by composing relative addresses. Consider again Fig. 1 where P_1 sends to P_2 a name that was generated by P_3 (i.e. with relative address $||_0||_1 \bullet ||_1||_1||_0$). The rules for communication provide us with the address of P_2 relative to P_1, i.e. $||_1||_0 \bullet ||_0||_1$. The composition of the two relative addresses, written $||_1||_0 \bullet ||_0||_1 \star ||_0||_1 \bullet ||_1||_1||_0$, will result in $||_0 \bullet ||_1||_0$, and $\star$ is the router defined in the next section.

Actually, the relative positions of the minimal common predecessors of the possible pairs of G, R and S in an abstract syntax tree are only three, up to symmetries. This three cases are depicted in Fig. 2. Note that the other cases are obtained when some nodes coincide, and symmetries do not alter relative addresses.

Fact. *Given a tree of processes T and three of its processes, there are exactly three possible placements for them in T, up to symmetries and coincidence of processes.*

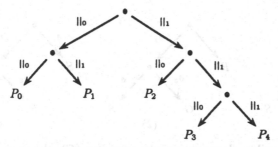

Fig. 1. The tree of (sequential) processes of $(P_0|P_1)|(P_2|(P_3|P_4))$

4 A router

Here we show how relative addresses are updated when names are exported. Some notation could help.

Definition 2. Let $\vartheta, \vartheta', \vartheta_i, \ldots \in \{||_0, ||_1\}^*$ and let ϵ be the empty string. Then, $\mathcal{A} = \{||_0, ||_1\}^* \cdot \{||_0, ||_1\}^*$ is the set of *relative addresses*, provided that $\vartheta_0 \cdot \vartheta_1$, with $\vartheta_0 \vartheta_1 \neq \epsilon$, implies $\vartheta_0 = ||_i \vartheta_0'$ and $\vartheta_1 = ||_{i \oplus 1} \vartheta_1'$, where $\oplus$ is sum modulo 2.

Note that $||_i \vartheta_0' \cdot ||_{i \oplus 1} \vartheta_1'$ makes it explicit that the two components of the relative address describe the two distinct paths outgoing from the same node in a (binary) tree of processes. This node is the minimal common predecessor of the generator of a name and its user.

Hereafter, we say that two addresses $\vartheta_0 \cdot \vartheta_1, \vartheta_2 \cdot \vartheta_3 \in \mathcal{A}$ can be composed through the router $\star$, only if ϑ_1 is a suffix of ϑ_2, or vice versa. This corresponds to the three situations depicted in Fig. 2, that are the only possible ones in our setting. Furthermore, we assume that whenever we write $\vartheta_0 \cdot \vartheta_1 \star \vartheta_2 \cdot \vartheta_3$ the two addresses can be composed. We will make sure later that $\star$ is defined whenever used (see Corollary 17).

Definition 3. Router $\star : (\mathcal{A} \times \mathcal{A}) \longrightarrow \mathcal{A}$ is defined by the following three exaustive cases:

1. $\vartheta_0 \cdot \vartheta \star \vartheta_2 \vartheta \cdot \vartheta_3 = \vartheta_2 \vartheta_0 \cdot \vartheta_3$ with $\vartheta_2 \neq \epsilon$
2. $\vartheta_0 \cdot \vartheta_1 \vartheta \star \vartheta \cdot \vartheta_3 = \vartheta_0 \cdot \vartheta_1 \vartheta_3$ with $\vartheta_1 \neq \epsilon$
3. $\vartheta' \vartheta_0 \cdot \vartheta \star \vartheta \cdot \vartheta' \vartheta_3 = \vartheta_0 \cdot \vartheta_3$

The following theorem states that router $\star$ is always defined when needed, although it is partial.

Theorem 4. *Router $\star$ is total on the possible relative addresses arising from the relative positions of a generator, a sender and a receiver of a name.*

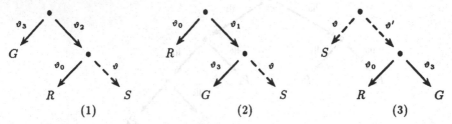

Fig. 2. The three possible placements of the generator (G), the sender (S) and the receiver (R) of a name.

The above theorem also says that router $\star$ correctly computes the address of R relative to G. Pictorially, $\star$ discards the dashed paths in Fig. 2.

We state below some useful properties of $\star$. Actually, $\langle \mathcal{A} \times \mathcal{A}, \star \rangle$ would be a group, if $\star$ were total. This is quite natural a property of routers: given a space of addresses and an interconnection topology, a router must always connect two sites in both ways, provided that there is a path between them. The first property says that $\star$ has a neutral element and an inverse on $\mathcal{A} \times \mathcal{A}$. Its proof is immediate from Def. 3. The second property states that $\star$ is also associative.

Proposition 5. $\forall \vartheta_i \cdot \vartheta_j \in \mathcal{A}$ we have that

1. $\epsilon \cdot \epsilon \star \vartheta_i \cdot \vartheta_j = \vartheta_i \cdot \vartheta_j \star \epsilon \cdot \epsilon = \vartheta_i \cdot \vartheta_j$, i.e. $\epsilon \cdot \epsilon$ is the neutral element of $\star$;
2. $\vartheta_j \cdot \vartheta_i \star \vartheta_i \cdot \vartheta_j = \epsilon \cdot \epsilon$, i.e. the inverse of $\vartheta_i \cdot \vartheta_j$ is equal to $\vartheta_j \cdot \vartheta_i$

Proposition 6. Whenever defined, $\star$ is associative, i.e.

$$(\vartheta_0 \cdot \vartheta_1 \star \vartheta_2 \cdot \vartheta_3) \star \vartheta_4 \cdot \vartheta_5 = \vartheta_0 \cdot \vartheta_1 \star (\vartheta_2 \cdot \vartheta_3 \star \vartheta_4 \cdot \vartheta_5).$$

Identity of names is not lost during exportations, because a name always encodes a pointer to its generator. More formally, sending a name from a process S to R, and then sending the same name from R to S is an involution. Again, this is mandatory for a router.

Proposition 7. $\vartheta_1 \cdot \vartheta_0 \star (\vartheta_0 \cdot \vartheta_1 \star \vartheta_2 \cdot \vartheta_3) = \vartheta_2 \cdot \vartheta_3$.

5 Operational semantics

We define a late operational semantics of π-calculus that originates and handles names locally according to the discussion in the previous sections. We start with the set of names.

Definition 8. Let $\mathcal{N}' = \mathcal{A} \cdot \{\mathcal{N} \cup I\!\!N\}$ be a countable set of *names*, ranged over by $r, s, u, \ldots$, with $\mathcal{N}' \cap \{\tau\} = \emptyset$. We assume that $\forall a \in \mathcal{N} \ \vartheta \cdot \vartheta a \equiv_{\mathcal{N}'} a$.
The new syntax of π-calculus is the one in Def. 1, with r, s and s_i in place of x, y and y_i. In particular, π can be either $r(x)$ or $\bar{r}s$ or τ.

The standard notions on actions (still ranged over by μ) and on names (subject, object, free, bound), as well as the structural congruence on processes, are extended to the new syntax in the obvious way.

Following the ideas of [4, 5], we encode in the labels of transitions the parallel structure of processes to identify the sequential component that acts. Actually, these labels encode a portion of the proof of the transitions, so we call them *proof terms* as in [5].

Definition 9. *Labels* of transitions (with metavariable θ) are defined as $\mu \bullet \vartheta$. The set of labels, or *proof terms*, is denoted by Θ.

As discussed above, to handle names locally, we enrich processes with a counter implemented as a family of operators $n \Rightarrow$ in the style of the causal transition system of [3]. The intuitive meaning of $n \Rightarrow P$ is that P has generated $n - 1$ new names and the next one will be n. This *extended processes* are ranged over by t and we assume on them the least structural congruence $\equiv$ that satisfies the following clauses

- $n \Rightarrow 0 \equiv 1 \Rightarrow 0$
- $n \Rightarrow (\nu r)P \equiv (\nu r)n \Rightarrow P$
- $n \Rightarrow (P|Q) \equiv (n \Rightarrow P)|(n \Rightarrow Q)$

If the axioms are oriented from left to right, we obtain a terminating and confluent rewriting system. Hereafter, we feel free to consider processes in normal form whenever convenient.

We now start considering the problem of sending names. This requires the composition via $\star$ of the address of the name itself with the address of the sender relative to the receiver, as intuitively discussed in Sect. 3. We now lift composition of addresses to exportation of names. Recall that $a \equiv_{\mathcal{N}'} \epsilon \bullet \epsilon a$.

Definition 10. Let $r = (\vartheta_r \bullet \vartheta'_r)n$ (resp. a) be a name. Name r exported at the relative address $\vartheta \bullet \vartheta'$ is $\vartheta \bullet \vartheta' \star r = (\vartheta \bullet \vartheta' \star \vartheta_r \bullet \vartheta'_r)n$ (resp. a).

Note that $\vartheta \bullet \vartheta'$ is the address of the sender relative to the receiver. Recall that names also encode a pointer to their generator (above, $\vartheta_r \bullet \vartheta'_r$).

The following three auxiliary definitions are used in the operational semantics. We start with a selector of the sequential components of a process t at address ϑ, denoted by $t@\vartheta$. Then, when considering process t, we will write $r@\vartheta$ to say that the local environment where r is defined is the one of $t@\vartheta$.

Definition 11. Let $t@\vartheta$ be the sub-process at depth ϑ defined inducing on the syntax as

- $t@\vartheta = t$ if $\vartheta = \epsilon$ or t is either of $0, \pi.t, t + t', (\nu x)t, [r = s]t, t(y)$;
- $(t_0|t_1)@||_i\vartheta = t_i@\vartheta$, where i is either 0 or 1.

Also, if $r \in n(t)$ then $r@\vartheta$ implies $r \in n(t@\vartheta)$.

Now, we introduce an equivalence relation on names. Intuitively, two names of two sequential processes are equivalent if they coincide when both are sent to the same process. For simplicity, we check equivalence of names by sending them to a common predecessor of the sequential processes.

Definition 12. Given $r \in n(t@\vartheta)$ and $s \in n(t@\vartheta')$, let

$$\bar{r}@\vartheta \simeq r@\vartheta \simeq s@\vartheta' \simeq \bar{s}@\vartheta' \quad \Leftrightarrow \quad \epsilon.\vartheta \star r = \epsilon.\vartheta' \star s.$$

Note that $\simeq$ is an equivalence relation, where reflexivity, symmetry and transitivity come out from the corresponding properties of $=$. Furthermore, two equivalent names must have the same action part because our router does not affect actions, but only addresses. The following theorem shows that if two names are equivalent in a node of a tree of processes, then they are equivalent everywhere.

Theorem 13. Let $r \in n(t@\vartheta_r)$ and $s \in n(t@\vartheta_s)$.
Then, $r@\vartheta \simeq s@\vartheta'$ iff $\forall \vartheta.\vartheta_r, \vartheta.\vartheta_s$ relative addresses, $\vartheta.\vartheta_r \star r \in n(t@\vartheta)$ and $\vartheta.\vartheta_s \star s \in n(t@\vartheta)$ and $\vartheta.\vartheta_r \star r = \vartheta.\vartheta_s \star s$.

The following theorem suggests how names may be computed when exported in the operational semantics definitions. It obviously relies on router $\star$.

Theorem 14. Given r, $r@\vartheta \simeq s@\vartheta'$ iff $s = \vartheta'.\vartheta \star r$.

A few auxiliary definitions follow. We extend standard substitution of one name for another so that the new substitution updates the involved names while descending a tree of processes. In this way the names receive the right address at every node of the tree.

Definition 15. Let $\{-/-\}$ be the standard substitution. Then the *routed substitution* $\{\!|-/-|\!\}$ is defined inducing on the syntax as follows:

- $0\{\!|u'/u|\!\} = 0$
- $(n \Rightarrow P)\{\!|u'/u|\!\} = n\{u'/u\} \Rightarrow P\{\!|u'/u|\!\}$
- $(\bar{r}s.P)\{\!|u'/u|\!\} = (\bar{r}s)\{u'/u\}.P\{\!|u'/u|\!\}$
- $(r(a).P)\{\!|u'/u|\!\} = \begin{cases} r\{u'/u\}(a).P & \text{if } a \in \{u, u'\} \\ r\{u'/u\}(a).(P\{\!|u'/u|\!\}) & \text{otherwise} \end{cases}$
- $(P + Q)\{\!|u'/u|\!\} = P\{\!|u'/u|\!\} + Q\{\!|u'/u|\!\}$
- $((\nu\, r)P)\{\!|u'/u|\!\} = \begin{cases} (\nu\, r)P & \text{if } r \in \{u, u'\} \\ (\nu\, r)(P\{\!|u'/u|\!\}) & \text{otherwise} \end{cases}$
- $[r = s]P\{\!|u'/u|\!\} = [r = s]\{u'/u\}P\{\!|u'/u|\!\}$
- $P(r)\{\!|u'/u|\!\} = P\{\!|u'/u|\!\}(r\{u'/u\})$
- $(t_0|t_1)\{\!|u'/u|\!\} = t_0\{\!|u'_0/u_0|\!\}|t_1\{\!|u'_1/u_1|\!\}$ where $u'_i@\|_i \simeq u'$ and $u_i@\|_i \simeq u$

We define also $t\{\!|/|\!\}_{@\vartheta}$, that applies the substitution to $t@\vartheta$, as

- $t\{\!|r/s|\!\}_{@\vartheta} = t\{\!|r/s|\!\}$
  if $\vartheta = \epsilon$ or t has either form $0, \pi.t, t + t', (\nu x)t, [r = s]t, t(y)$
- $(t_0|t_1)\{\!|r/s|\!\}_{@\vartheta} = \begin{cases} t_0\{\!|r/s|\!\}_{@\vartheta'}|t_1 & \text{if } \vartheta = \|_0\vartheta' \\ t_0|t_1\{\!|r/s|\!\}_{@\vartheta'} & \text{if } \vartheta = \|_1\vartheta' \end{cases}$

Hereafter we will omit $@\epsilon$ in routed substitutions and in selectors $t@\vartheta$ and $r@\vartheta$.

Consider for instance $(2 \Rightarrow \overline{x}y.w(z).z \mid 2 \Rightarrow x(z).(\nu\,x)\overline{w}x.(x(y) \mid \overline{x}z))\{\!|\bullet 1/x|\!\}$ that will be used in the example before Theorem 18. It results in

$$(2 \Rightarrow \overline{x}y.w(z).z\{\!|\!\mid_0 \bullet 1/x|\!\}) \mid (2 \Rightarrow x(z).(\nu\,x)\overline{w}x.(x(y) \mid \overline{x}z)\{\!|\!\mid_1 \bullet 1/x|\!\}) =$$
$$(2 \Rightarrow \overline{\mid\mid_0 \bullet 1}y.w(z).z) \mid (2 \Rightarrow \mid\mid_1 1(z).[(\nu\,x)\overline{w}x.(x(y) \mid \overline{x}z)]\{\!|\!\mid_1 \bullet 1/x|\!\}) =$$
$$(2 \Rightarrow \overline{\mid\mid_0 \bullet 1}y.w(z).z) \mid (2 \Rightarrow \mid\mid_1 \bullet 1(z).(\nu\,x)\overline{w}x.(x(y) \mid \overline{x}z))$$

Note that name x to be substituted is not enriched with a relative address by $\{\!| - / - |\!\}$ in the case of parallel composition because $\vartheta \bullet \vartheta' x \equiv_{\mathcal{N}'} x$. Furthermore the substitution takes the binders of names into account correctly. In fact, the distribution of $\{\!| \bullet 1/x|\!\}$ stops when the new binder $(\nu\,x)$ is encountered.

Our version of the late proved transition system for π-calculus is reported in Tab. 2. Some comments are in order. We omit $@\epsilon$ from labels of transitions. Rules *In* and *Open* generate a new name $\bullet n$ and increment the counter of the sequential components that move. In the case of *In*, the new name is distributed to the residual through substitution $\{\!| \bullet n/x|\!\}$ that enriches $\bullet n$ with the correct relative addresses. As for *Open*, the new name must be distributed to t, the whole process under restriction. Thus, we use the name $\bullet \vartheta n$ as it is known at t. The rules for communication locally check if the channel is the same, through the equivalence of names $r_0@\mid\mid_0\vartheta_0 \simeq r_1@\mid\mid_1\vartheta_1$. Then, the receiver distributes the value read to the sequential component $t_i@\vartheta_i$ (which fired the input) by using the information encoded in the proof term of the transition. Note that *Close* introduces the correct restriction by relying on equivalence of names. Finally, the index x in the transition relation is needed in the case of rule *Res*. It ensures that the placeholder x replaced in rule *In* by $\bullet n$ was not restricted.

We remark that all side conditions of our rules are simply needed to compute the names through our router $\star$ and that they involve *no global* condition on names. In particular, we drop the side conditions of π-calculus on free and bound names. The only rule which applies substitutions to a context larger than a single process is *Open*. However, only the operand of ν is affected. This larger information is the minimum needed to advice the owners of a name that it is no more private to them.

We now report an example of derivation of a transition. Consider the process

$$2 \Rightarrow w(z).z \mid 3 \Rightarrow (\nu\,x)\overline{w}x.(x(y) \mid \overline{x}y).$$

We deduce a communication between $w(z)$ and $\overline{w}x$, by using rule *Close*. The derived transition will be used in the example before Theorem 18.

$$2 \Rightarrow w(z).z \xrightarrow{w(\bullet 2)}_{z} 3 \Rightarrow \bullet 2, \quad \dfrac{3 \Rightarrow \overline{w}x.(x(y) \mid \overline{x}y) \xrightarrow{\overline{w}x}_{\tau} (3 \Rightarrow x(y) \mid 3 \Rightarrow \overline{x}y)}{(\nu x)(3 \Rightarrow \overline{w}x.(x(y) \mid \overline{x}y) \xrightarrow{\overline{w}(\bullet 3)}_{\tau} (\nu x)(4 \Rightarrow \mid\mid_0 \bullet 3(y) \mid 4 \Rightarrow \overline{\mid\mid_1 \bullet 3}y)}$$

$$2 \Rightarrow w(z).z \mid 3 \Rightarrow (\nu\,x)\overline{w}x.(x(y) \mid \overline{x}y) \xrightarrow{\tau}_{\tau} (\nu \bullet \mid\mid_1 3)(3 \Rightarrow \mid\mid_1 \bullet \mid\mid_0 3 \mid (4 \Rightarrow \mid\mid_0 \bullet 3(y) \mid 4 \Rightarrow \overline{\mid\mid_1 \bullet 3}y)$$

Consider the target of the transition above, in which $||_0 \cdot 3@||_1||_0 \simeq \overline{||_1 \cdot 3@||_1}||_1 \simeq \cdot||_1 3$. Thus $(\nu \cdot ||_1 3)$ enforces communication between the rightmost processes. The following theorem ensures that no name in Tab. 2 is left unknown.

Theorem 16. *All names in the conclusions and conditions in Tab. 2 are defined.*

The above theorem guarantees also that every call to router $\star$ is well defined.

Corollary 17. *The arguments of every call made in Tab. 2 to router $\star$ can be composed.*
Also, all names exported are prefixed with relative addresses.

Finally we define the actual transition relation as

$$\boxed{\text{if } t \xrightarrow[x]{\mu \circ \vartheta} t' \text{ then } t \xrightarrow{\mu} t'}$$

Given a process P, its computations will start from the extended process $1 \Rightarrow P$. Any other extended process $n \Rightarrow P$ works as well, due to Theorem 18 below, that proves our transition system equivalent to the classical late one of π-calculus. For instance, consider the computation depicted in Fig. 3.

$$1 \Rightarrow a(x).(\overline{x}y.w(z).z \mid x(z).(\nu x)\overline{w}x.(x(y) \mid \overline{x}z))$$

$$\downarrow a(\cdot 1)$$

$$(2 \Rightarrow \overline{||_0 \cdot 1}y.w(z).z \mid 2 \Rightarrow ||_1 \cdot 1(z).(\nu x)\overline{w}x.(x(y) \mid \overline{x}z))$$

$$\downarrow \tau$$

$$(2 \Rightarrow w(z).z \mid 3 \Rightarrow (\nu x)\overline{w}x.(x(y) \mid \overline{x}y))$$

$$\downarrow \tau$$

$$(\nu \cdot ||_1 3)(3 \Rightarrow ||_1 \cdot ||_0 3 \mid (4 \Rightarrow ||_0 \cdot 3(y) \mid 4 \Rightarrow \overline{||_1 \cdot 3}y))$$

Fig. 3. A computation of $1 \Rightarrow a(x).(\overline{x}y \mid x(z).(\nu x)\overline{x}v.(x(y) \mid \overline{x}z))$.

The first transition in Fig. 3 shows the generation of the new name $\cdot 1$. The application of the routed substitution $\{\!|\cdot 1/x|\!\}$ introduced by rule In is reported in the example after Def. 15. The counter $2 \Rightarrow$ is distributed to the components of the parallel composition. The second transition is a communication along a channel called $\overline{||_0 \cdot 1}$ by the sender, and called $||_1 \cdot 1$ by the receiver. Indeed, $||_0 \cdot 1@||_0 \simeq ||_1 \cdot 1@||_1$. Also, the counter of the residual of the receiver is increased because rule In generates a new name $\cdot 2$ that will be replaced with y by substitution $\{\!|y/\cdot 2|\!\}$ in the conclusion of rule Com_0 (see the example after Def. 15).

The derivation of the last transition is exemplified before Theorem 16, and uses a *Close* rule.

The following theorem says that our transition system is a more concrete version of the original one. In its statement, we use an auxiliary function *FC* that erases counters from extended processes, and has the obvious definition by structural induction.

Theorem 18. $P_0 \xrightarrow{\mu}_L P_1$ *iff* $t \xrightarrow{\mu'} t'$ *and* $FC(t) = P \xrightarrow{\mu'}_L P' = FC(t')$ *is a variant of* $P_0 \xrightarrow{\mu}_L P_1$, *where* $\longrightarrow_L$ *is the late transition relation of π-calculus.*

References

1. M. Boreale and D. Sangiorgi. A fully abstract semantics of causality in the π-calculus. In *Proceedings of STACS'95, LNCS*. Springer Verlag, 1995.
2. Roberta Borgia, Pierpaolo Degano, Corrado Priami, Lone Leth, and Bent Thomsen. Understanding mobile agents via a non-interleaving semantics for Facile. Technical Report ECRC-96-4, European Computer-Industry Research Centre, 1996.
3. Ph. Darondeau and P. Degano. Causal trees. In *Proceedings of ICALP'89, LNCS 372*, pages 234–248. Springer-Verlag, 1989.
4. P. Degano and C. Priami. Proved trees. In *Proceedings of ICALP'92, LNCS 623*, pages 629–640. Springer-Verlag, 1992.
5. P. Degano and C. Priami. Causality for mobile processes. In *Proceedings of ICALP'95, LNCS 944*, pages 660–671. Springer-Verlag, 1995.
6. A. Giacalone, P. Mishra, and S. Prasad. Operational and algebraic semantics for Facile: A symmetric integration of concurrent and functional programming. In *Proceedings ICALP'90, LNCS 443*, pages 765–780. Springer-Verlag, 1990.
7. R. Milner. The polyadic π-calculus: a tutorial. Technical Report ECS-LFCS-91-180, University of Edinburgh, 1991.
8. R. Milner, D. Berry, and D. Turner. A semantics for ML concurrency primitives. In *Proceedings of POPL'92*, 1992.
9. R. Milner, J. Parrow, and D. Walker. A calculus of mobile processes (I and II). *Information and Computation*, 100(1):1–77, 1992.
10. F. Orava and J. Parrow. An algebraic verification of a mobile network. *Formal Aspects of Computing*, pages 497–543, 1992.
11. J. Reppy. *Higher order concurrency*. PhD thesis, Cornell University, TR 92-1285, 1992.
12. D. Sangiorgi. *Expressing Mobility in Process Algebras: First-Order and Higher-Order Paradigms*. PhD thesis, University of Edinburgh, 1992.
13. B. Thomsen, L. Leth, S. Prasad, T.-M. Kuo, A. Kramer, F. Knabe, and A. Giacalone. Facile Antigua Release Programming Guide. Technical Report ECRC-93-20, European Computer-Industry Research Centre, 1993.
14. D. Walker. Objects in the π-calculus. *Information and Computation*, 1994. To appear.

$$Out : n \Rightarrow \overline{r}s.P \xrightarrow{\frac{\overline{r}s}{\tau}} n \Rightarrow P \qquad In : n \Rightarrow r(x).P \xrightarrow{\frac{r(\bullet n)}{x}} (n+1) \Rightarrow P\{\!|\bullet n/x|\!\}$$

$$Tau : n \Rightarrow \tau.P \xrightarrow{\frac{\tau}{\tau}} n \Rightarrow P \qquad Ide : \frac{n \Rightarrow P\{\!|r/y|\!\} \xrightarrow{\frac{\mu\odot\vartheta}{x}} t'}{n \Rightarrow Q(r) \xrightarrow{\frac{\mu\odot\vartheta}{x}} P'}, \; Q(y) = P$$

$$Par_0 : \frac{t_0 \xrightarrow{\frac{\mu\odot\vartheta}{x}} t'_0}{t_0|t_1 \xrightarrow{\frac{\mu\odot\|_0\vartheta}{x}} t'_0|t_1} \qquad Par_1 : \frac{t_1 \xrightarrow{\frac{\mu\odot\vartheta}{x}} t'_1}{t_0|t_1 \xrightarrow{\frac{\mu\odot\|_1\vartheta}{x}} t_0|t'_1}$$

$$Sum : \frac{t \xrightarrow{\frac{\mu\odot\vartheta}{x}} t'}{t + t'' \xrightarrow{\frac{\mu\odot\vartheta}{x}} t'}$$

$$Res : \frac{t \xrightarrow{\frac{\mu\odot\vartheta}{x}} t'}{(\nu u)t \xrightarrow{\frac{\mu\odot\vartheta}{x}} (\nu u)t'}, \; \mu \neq \tau \Rightarrow \begin{cases} u \notin \{r,s,x\} \\ sbj(\mu)@\vartheta \simeq r \\ obj(\mu)@\vartheta \simeq s \end{cases}$$

$$Com_0 : \frac{t_0 \xrightarrow{\frac{\overline{r}_0 s_0\odot\vartheta_0}{x}} t'_0, \, t_1 \xrightarrow{\frac{r_1(\bullet m)\odot\vartheta_1}{y}} t'_1}{t_0|t_1 \xrightarrow{\tau} t'_0|t'_1\{\!|s_1/\bullet m|\!\}_{\odot\vartheta_1}}, \; \begin{cases} s_0@\|_0\vartheta_0 \simeq s_1@\|_1\vartheta_1 \\ r_0@\|_0\vartheta_0 \simeq r_1@\|_1\vartheta_1 \end{cases}$$

$$Com_1 : \frac{t_1 \xrightarrow{\frac{r_1(\bullet m)\odot\vartheta_1}{x}} t'_1, \, t_0 \xrightarrow{\frac{\overline{r}_0 s_0\odot\vartheta_0}{y}} t'_0}{t_1|t_0 \xrightarrow{\tau} t'_1|t'_0\{\!|s_1/\bullet m|\!\}_{\odot\vartheta_1}}, \; \begin{cases} s_0@\|_0\vartheta_0 \simeq s_1@\|_1\vartheta_1 \\ r_0@\|_0\vartheta_0 \simeq r_1@\|_1\vartheta_1 \end{cases}$$

$$Open : \frac{t \xrightarrow{\frac{\overline{r}s@\vartheta}{x}} t'}{(\nu u)t \xrightarrow{\frac{\overline{r}(\bullet n)@\vartheta}{x}} (t'\{n+1/n\}_{\odot\vartheta})\{\!|\bullet\vartheta n/u|\!\}}, \; \begin{cases} s@\vartheta \simeq u \\ r \neq s \\ t@\vartheta = n \Rightarrow P \end{cases}$$

$$Close_0 : \frac{t_0 \xrightarrow{\frac{\overline{r}_0(s_0)@\vartheta_0}{x}} t'_0, \, t_1 \xrightarrow{\frac{r_1(\bullet m)@\vartheta_1}{x}} t'_1}{t_0|t_1 \xrightarrow{\tau} (\nu u)(t'_0|t'_1\{\!|s_1/\bullet m|\!\}_{\odot\vartheta_1})}, \; \begin{cases} s_0@\|_0\vartheta_0 \simeq s_1@\|_1\vartheta_1 \\ r_0@\|_0\vartheta_0 \simeq r_1@\|_1\vartheta_1 \\ u \simeq s_0@\|_0\vartheta_0 \end{cases}$$

$$Close_1 : \frac{t_1 \xrightarrow{\frac{r_1(\bullet m)@\vartheta_1}{x}} t'_1, \, t_0 \xrightarrow{\frac{\overline{r}_0(s_0)@\vartheta_0}{x}} t'_0}{t_1|t_0 \xrightarrow{\tau} (\nu u)(t'_1|t'_0\{\!|s_1/\bullet m|\!\}_{\odot\vartheta_1})}, \; \begin{cases} s_0@\|_0\vartheta_0 \simeq s_1@\|_1\vartheta_1 \\ r_0@\|_0\vartheta_0 \simeq r_1@\|_1\vartheta_1 \\ u \simeq s_0@\|_1\vartheta_1 \end{cases}$$

Table 2. Late proved transition system of π-calculus.

The Meaning of Negative Premises in Transition System Specifications II

(extended abstract)[†]

R.J. van Glabbeek[*]

Computer Science Department, Stanford University
Stanford, CA 94305, USA.
rvg@cs.stanford.edu

This paper reviews several methods to associate transition relations to transition system specifications with negative premises in Plotkin's structural operational style. Besides a formal comparison on generality and relative consistency, the methods are also evaluated on their taste in determining which specifications are meaningful and which are not.

1 Transition system specifications

In this paper V and A are two sets of *variables* and *actions*. Many concepts that will appear are parameterised by the choice of V and A, but as in this paper this choice is fixed, a corresponding index is suppressed.

Definition 1 (*Signatures*). A *function declaration* is a pair (f, n) of a *function symbol* $f \notin V$ and an *arity* $n \in \mathbb{N}$. A function declaration $(c, 0)$ is also called a *constant declaration*. A *signature* is a set of function declarations. The set $\mathbb{T}(\Sigma)$ of *terms* over a signature Σ is defined recursively by:

- $V \subseteq \mathbb{T}(\Sigma)$,
- if $(f, n) \in \Sigma$ and $t_1, \ldots, t_n \in \mathbb{T}(\Sigma)$ then $f(t_1, \ldots, t_n) \in \mathbb{T}(\Sigma)$.

A term $c()$ is often abbreviated as c. A Σ-*substitution* σ is a partial function from V to $\mathbb{T}(\Sigma)$. If σ is a substitution and S any syntactic object (built from terms), then $S[\sigma]$ denotes the object obtained from S by replacing, for x in the domain of σ, every occurrence of x in S by $\sigma(x)$. In that case $S[\sigma]$ is called a *substitution instance* of S. S is said to be *closed* if it contains no variables. The set of closed terms is denoted $T(\Sigma)$.

Definition 2 (*Transition system specifications*). Let Σ be a signature. A *positive* Σ-*literal* is an expression $t \xrightarrow{a} t'$ and a *negative* Σ-*literal* an expression $t \xrightarrow{a}\!\!\!\!\!/\ $ or $t \xrightarrow{a}\!\!\!\!\!/\ t'$ with $t, t' \in \mathbb{T}(\Sigma)$ and $a \in A$. For $t, t' \in \mathbb{T}(\Sigma)$ the literals $t \xrightarrow{a} t'$ and $t \xrightarrow{a}\!\!\!\!\!/\ $, as well as $t \xrightarrow{a} t'$ and $t \xrightarrow{a}\!\!\!\!\!/\ t'$, are said to *deny* each other. A *transition rule* over Σ is an expression $\frac{H}{\alpha}$ with H a set of Σ-literals (the *premises* or *antecedents* of the the rule) and α a Σ-literal (the *conclusion*). A rule $\frac{H}{\alpha}$ with $H = \emptyset$ is also written α. An *action rule* is a transition rule with a positive conclusion. A *transition system specification (TSS)* is a pair (Σ, R) with Σ a signature and R a set of action rules over Σ. A TSS is *standard* if its rules have no antecedents of the form $t \xrightarrow{a}\!\!\!\!\!/\ t'$, and *positive* if all antecedents of its rules are positive.

---

[*]This work was supported by ONR under grant number N00014-92-J-1974.

[†]Full version available as Report STAN-CS-TN-95-16, and by ftp from boole.stanford.edu.

The first systematic study of transition system specifications with negative premises appears in BLOOM, ISTRAIL & MEYER [2]. The concept of a (positive) TSS presented above was introduced in GROOTE & VAANDRAGER [9]; the negative premises $t \not\xrightarrow{a}$ were added in GROOTE [8]. The notion generalises the *GSOS rule systems* of [2] and constitutes the first formalisation of PLOTKIN's *Structural Operational Semantics (SOS)* [10] that is sufficiently general to cover most, if not all, of its applications. The premises $t \not\xrightarrow{a} t'$ are added here, mainly for technical reasons.

The following definition tells when a transition is provable from a TSS. It generalises the standard definition (see e.g. [9]) by (also) allowing the derivation of transition rules. The derivation of a transition $t \xrightarrow{a} t'$ corresponds to the derivation of the transition rule $\frac{H}{t \xrightarrow{a} t'}$ with $H = \emptyset$. The case $H \neq \emptyset$ corresponds to the derivation of $t \xrightarrow{a} t'$ under the assumptions H.

Definition 3 (*Proof*). Let $P = (\Sigma, R)$ be a TSS. A *proof* of a transition rule $\frac{H}{\alpha}$ from P is a well-founded, upwardly branching tree of which the nodes are labelled by Σ-literals, such that the root is labelled by α, and if β is the label of a node q and K is the set of labels of the nodes directly above q, then

1. either $K = \emptyset$ and $\beta \in H$,

2. or $\frac{K}{\beta}$ is a substitution instance of a rule from R.

If a proof of $\frac{H}{\alpha}$ from P exists, then $\frac{H}{\alpha}$ is *provable* from P, notation $P \vdash \frac{H}{\alpha}$. A closed negative literal α is *refutable* if $P \vdash \beta$ for a literal β denying α.

Definition 4 (*Transition relation*). Let Σ be a signature. A *transition relation* over Σ is a relation $T \subseteq T(\Sigma) \times A \times T(\Sigma)$. Elements (t, a, t') of a transition relation are written as $t \xrightarrow{a} t'$. Thus a transition relation over Σ can be regarded as a set of closed positive Σ-literals (*transitions*).

A closed literal α *holds* in a transition relation T, notation $T \models \alpha$, if α is positive and $\alpha \in T$ or $\alpha = (t \not\xrightarrow{a} t')$ and $(t \xrightarrow{a} t') \notin T$ or $\alpha = (t \not\xrightarrow{a})$ and $(t \xrightarrow{a} t') \in T$ for no $t' \in T(\Sigma)$. Write $T \models H$, for H a set of closed literals, if $T \models \alpha$ for all $\alpha \in H$. Write $T \models p$, for p a closed proof, if $T \models \alpha$ for all literals α that appear as node-labels in p.

A positive TSS specifies a transition relation in a straightforward way as the set of all provable transitions. But as pointed out in GROOTE [8], it is much less trivial to associate a transition relation to a TSS with negative premises. Several solutions are proposed in [8] and BOL & GROOTE [3]. Here I will present these solutions from a somewhat different point of view, and also review a few others.

$$P_1 \qquad \boxed{\begin{array}{cc} \dfrac{c \not\xrightarrow{a}}{c \xrightarrow{b} c} & \dfrac{c \not\xrightarrow{b}}{c \xrightarrow{a} c} \end{array}}$$

The TSS P_1 can be regarded as an example of a TSS that does not specify a well-defined transition relation (under any plausible definition of 'specify').[1] So unless a systematic way can be found to associate a meaning to TSSs like P_1, one has to accept that some TSSs are meaningless. Hence there are two questions to answer:

$$\textit{Which TSSs are meaningful,} \tag{1}$$
$$\textit{and which transition relations do they specify?} \tag{2}$$

---

[1] All my examples P_i consider TSSs (Σ, R) in which Σ consists of the single constant c only.

In this paper I present 8 possible answers to these questions, each consisting of a class of TSSs and a mapping from this class to transition relations. Two such solutions are *consistent* if they agree which transition relation to attach to a TSS in the intersection of their domains. Solution S' *extends* S if the class of meaningful TSSs according to S' extends that of S and the two are consistent, i.e. seen as partial functions $S \subseteq S'$.

Logic programming

The problems analysed in [8] in associating transition relations to TSSs with negative premises had been encountered long before in logic programming, and most of the solutions reviewed in the present paper stem from logic programming as well. However, the proof theoretic approach to Solution 7, as well as Solutions 6 and 8 and some comparative observations, are, as far as I know, new here.

The connection with logic programming may be best understood by introducing *proposition system specifications (PSSs)*. These are obtained by replacing the set A of actions by a set of *predicate declarations* (p, n) with $p \notin V$ a *predicate symbol* (different from any function symbol) and $n \in \mathbb{N}$. A literal is then an expression $p(t_1, \ldots, t_n)$ or $\neg p(t_1, \ldots, t_n)$ with $t_i \in \mathbb{T}(\Sigma)$. A PSS is now defined in terms of literals in a same way as a TSS. A *proposition* is a closed positive literal, and a *proposition relation* or *closed theory* a set of propositions. The problem of associating a proposition relation to a PSS is of a similar nature as associating a transition relation to a TSS, and in fact all concepts and results mentioned in this paper apply equally well to both situations.

If I would not consider TSS involving literals of the form $t \xrightarrow{a}$, a TSS would be a special case of a PSS, and it would make sense to present the paper in terms of PSSs. The main reason for not doing so is to do justice to the rôle of literals $t \xrightarrow{a}$ in denying literals of the form $t \xrightarrow{a} t'$. However, as elaborated in the full paper, every TSS can be encoded as a PSS and vice versa, in such a way that all concepts of this paper are preserved under the translations.

A logic program is just a PSS obeying some finiteness conditions. Hence everything I say about TSSs applies to logic programming too. Consequently, this paper can in part be regarded as an overview of a topic within logic programming, but avoiding the logic programming jargon. However, I do not touch issues that are relevant in logic programming, but not manifestly so for transition system specifications. For these, and many more references, see APT & BOL [1].

2 Model theoretic solutions

Solution 1 (*Positive*). A first and rather conservative answer to (1) and (2) is to take the class of positive TSSs as the meaningful ones, and associate with each positive TSS the transition relation consisting of the provable transitions.

Before proposing more general solutions, I will first recall two criteria from BLOOM, ISTRAIL & MEYER [2] and BOL & GROOTE [3] that can be imposed on solutions.

Definition 5 (*Supported model*). A transition relation T *agrees* with a TSS P if:

$$T \models t \xrightarrow{a} t' \Leftrightarrow \text{there is a closed substitution instance } \frac{H}{t \xrightarrow{a} t'} \text{ of a rule of } P \text{ with } T \models H.$$

T is a *model* of P if "$\Leftarrow$" holds; T is *supported* by P if "$\Rightarrow$" holds.

The first and most indisputable criterion imposed on a transition system T specified by a TSS P is that it is a model of P. This is called being *sound* for P in [2]. This criterion says that the rules of P, interpreted as implications in first-order or conditional logic, should evaluate to true statements about T. The second criterion, of being supported, says that T does not contain any transitions for which it has no plausible justification to contain them. In [2] being supported is called *witnessing*. Note that the universal transition relation on $\mathsf{T}(\Sigma)$ is a model of any TSS. It is however rarely the intended one, and the criterion of being supported is a good tool to rule it out. Next I check that Solution 1 satisfies both criteria.

Proposition 1 Let P be a positive TSS and T the set of transitions provable from P. Then T is a supported model of P. Moreover T is the least model of P.

Starting from Proposition 1 there are at least three ways to generalise Solution 1 to TSSs with negative premises. One can generalise either the concept of a proof, or the least model property, or the least supported model property of positive TSSs. Starting with the last two possibilities, observe that in general no least model and no least supported model exists. A counterexample is given by the TSS P_1 (given earlier), which has two minimal models, $\{c \xrightarrow{a} c\}$ and $\{c \xrightarrow{b} c\}$, both of which are supported.

Solution 2 (*Least*). A TSS is meaningful iff it has a least model (this being its specified transition relation).

Solution 3 (*Least supported*). A TSS is meaningful iff it has a least supported model.

These two solutions turn out to have incomparable domains. The TSS P_2 below has $\{c \xrightarrow{a} c\}$ as its least model, but has no supported models. On the other hand P_3 has two minimal models, namely $\{c \xrightarrow{b} c\}$ and $\{c \xrightarrow{a} c\}$, of which only the latter one is supported. This is its least supported model.

$$P_2 \quad \boxed{\begin{array}{c} c \xrightarrow{a}\!\!\!/ \\ \hline c \xrightarrow{a} c \end{array}} \qquad\qquad P_3 \quad \boxed{\begin{array}{c} c \xrightarrow{b}\!\!\!/ \\ \hline c \xrightarrow{a} c \end{array}}$$

Obviously Solution 1 is extended by both solutions above. However, Solutions 2 and 3 turn out to be inconsistent with each other. P_4 has both a least model and a least supported model, but they are not the same.

$$P_4 \quad \boxed{\begin{array}{ccc} c \xrightarrow{a}\!\!\!/ & c \xrightarrow{b} c & c \xrightarrow{b} c \\ \hline c \xrightarrow{a} c & c \xrightarrow{a} c & c \xrightarrow{b} c \end{array}} \qquad P_5 \quad \boxed{\begin{array}{c} c \xrightarrow{a} c \\ \hline c \xrightarrow{a} c \end{array}}$$

Solution 2 is not very productive, because if fails to assign a meaning to the perfectly reasonable TSS P_3. Moreover, it can be criticised for yielding unsupported transition systems, as in the case of P_2. However, in P_4 the least model $\{c \xrightarrow{a} c\}$ appears to be a better choice than the least supported model $\{c \xrightarrow{a} c, c \xrightarrow{b} c\}$, as the 'support' for transition $c \xrightarrow{b} c$ is not overwhelming. Thus, to my taste, Solution 3 is somewhat unnatural.

In BLOOM, ISTRAIL & MEYER [2] the following solution is applied.

Solution 4 (*Unique supported*). A TSS is meaningful iff it has a unique supported model.

The positive TSS P_5 above has two supported models, $\emptyset$ and $\{c \xrightarrow{a} c\}$, and hence shows that Solution 4 does not extend Solution 1.

Although for the kind of TSSs considered in [2] (the *GSOS rule systems*) this solution coincides with all acceptable solutions mentioned in this paper, in general it suffers from the same drawback as Solution 3. The least supported model of P_4 is even the unique supported model of this TSS. My conclusion is that the criterion of being supported is too weak to be of any use in this context.

This conclusion was also reached by FAGES [5] in the setting of logic programming, who proposes to strengthen this criterion. Being supported can be rephrased as saying that a transition may only be present if there is a nonempty proof of its presence, starting from transitions that are also present. However, these premises in the proof may include the transition under derivation, thereby allowing for loops, as in the case of P_4. Now the idea behind a *well-supported model* is that the *absence* of a transition may be assumed a priory, as long as this assumption is consistent, but the *presence* of a transition needs to be proven without assuming the presence of (other) transitions. Thus a transition may only be present if it admits a valid proof, starting from negative literals only.

Definition 6 (*Well-supported*).[2] A transition relation T is *well-supported* by a TSS P if:

$$T \models t \xrightarrow{a} t' \quad \Leftrightarrow \quad \begin{array}{l} \text{there is a closed proof } p, \text{ with } T \models p, \text{ of a} \\ \text{transition rule } \dfrac{N}{t \xrightarrow{a} t'} \text{ without positive antecedents.} \end{array}$$

Note that "$\Leftarrow$" is trivial, and a well-supported transition relation is surely supported.

My concept of well-supportedness can easily be seen to coincide with the one of FAGES [5]. It is closely related to the earlier concept of *stability*, developed by GELFOND & LIFSCHITZ [7] in logic programming, and adapted for TSSs by BOL & GROOTE [3].

Definition 7 (*Stable transition relation*). A transition relation T is *stable* for a TSS P if:

$$T \models t \xrightarrow{a} t' \quad \Leftrightarrow \quad \begin{array}{l} \text{there is a closed transition rule } \dfrac{N}{t \xrightarrow{a} t'} \text{ without positive antecedents} \\ \text{with } P \vdash \dfrac{N}{t \xrightarrow{a} t'} \text{ and } T \models N. \end{array}$$

Proposition 2 The concept of stability of Definition 7 coincides with that from [3]. Moreover, T is stable for P iff it is a well-supported model of P.

The following two solutions are adaptations of Solutions 3 and 4, were the requirement of being supported has been replaced by that of being well-supported. The second is taken from [3].

Solution 5 (*Stable*). A TSS is meaningful iff it has a least stable transition relation.

Solution 5 (*Stable*). A TSS is meaningful iff it has a unique stable transition relation.

The particular numbering of these two solutions is justified by the following.

Proposition 3 A TSS has a least stable transition relation iff it has a unique stable transition relation. Moreover, Solution 5 (*stable*) extends Solution 1 (*positive*) and is consistent with Solution 2 (*least*) and 3 (*least supported*).

---

[2] The full version of this paper, which appeared as Stanford report STAN-CS-TN-95-16, contained an incorrect definition of well-supportedness (but leading to the same notion of a well-supported model). As observed by Jan Rutten, Proposition 3 in that version, stating that well-supported transition relations are supported, was false. The mistake had no other bad consequences.

Solution 5 improves Solutions 3 and 4 by rejecting the TSS P_4 as meaningless. It also improves Solution 2 by rejecting the TSS P_2 (whose least model was not supported). Surprisingly however, Solution 5 not only differs from the earlier solutions by being more fastidious; it also provides meaning to perfectly acceptable TSSs that were left meaningless by Solutions 2, 3 and 4.

$$P_6 \quad \boxed{\dfrac{c \not\xrightarrow{a}}{c \xrightarrow{b} c} \qquad \dfrac{c \xrightarrow{a} c}{c \xrightarrow{a} c}}$$

An example is the TSS P_6. There is clearly no satisfying way to obtain $c \xrightarrow{a} c$. Hence $c \not\xrightarrow{a}$ and consequently $c \xrightarrow{b} c$. $\{c \xrightarrow{b} c\}$ is indeed the unique stable transition relation of this TSS. However, P_6 has two minimal models, both of which are supported, namely $\{c \xrightarrow{b} c\}$ and $\{c \xrightarrow{a} c\}$.

It is interesting to see how the various solutions deal with *circular* rules, such as $\dfrac{c \xrightarrow{a} c}{c \xrightarrow{a} c}$, and rules like $\dfrac{c \not\xrightarrow{a}}{c \xrightarrow{a} c}$. The support-based solutions (3 and 4) may use a circular rule to obtain a transition that would be unsupported otherwise (Example P_4). This is my main argument to reject these solutions. In addition they may (or may not) reject TSSs as meaningless because of the presence of such a rule (Example P_6). On the other hand, Solutions 2 and 5 politely ignore these rules. To my taste, there are two acceptable attitudes towards circular rules: to ignore them completely (as done by Solutions 1, 2 and 5), or to reject any TSS with such a rule for being ambiguous, unless there is independent evidence for a transition $c \xrightarrow{a} c$. A strong argument in favor of the first approach is the existence of useful rules of which only certain substitution instances are circular (cf. [3]). A solution that caters to the second option will be proposed in the next section.

Solution 2 can treat a rule $\dfrac{c \not\xrightarrow{a}}{c \xrightarrow{a} c}$ as equivalent to $c \xrightarrow{a} c$ (namely if there are no other closed terms than c, cf. P_2), which gives rise to unsupported transition relations. Solutions 3, 4 and 5 do not go so far, but use such a rule to choose between two otherwise equally attractive transition relations. This is illustrated by the TSS P_7, which determines the transition system $\{c \xrightarrow{a} c\}$ according to each of the solutions 2–5.

$$P_7 \quad \boxed{\dfrac{c \not\xrightarrow{a}}{c \xrightarrow{b} c} \qquad \dfrac{c \not\xrightarrow{b}}{c \xrightarrow{a} c} \qquad \dfrac{c \not\xrightarrow{a}}{c \xrightarrow{a} c}} \qquad P_8 \quad \boxed{c \xrightarrow{a} c \qquad \dfrac{c \not\xrightarrow{a}}{c \xrightarrow{a} c}}$$

Ignoring rules like $\dfrac{c \not\xrightarrow{a}}{c \xrightarrow{a} c}$ is unacceptable, as this would yield unsound transition relations (non-models). But it could be argued that any TSS with such a rule should be rejected as meaningless, unless there is independent evidence for a transition $c \xrightarrow{a} t$, as in P_8. This would rule out P_7. Solutions that cater to this taste will be proposed next.

3 Proof theoretic solutions

In this section I will propose solutions based on a generalisation of the concept of a proof. Note that in a proof two kinds of steps are allowed, numbered 1 and 2 in Definition 3. Step 1 just allows hypotheses to enter, in case one wants to prove a transition rule. This step can not be used when merely proving transitions. The essence of the notion is step 2. This step reflects the postulate that the desired transition relation must be a model of the given TSS. As a consequence those and only those transitions are provable that appear in any model. When generalising the notion of a proof to derive negative literals it makes sense to import more postulates about the desired transition relation. Note that a model T of a TSS P is supported iff

$$T \not\models t \xrightarrow{a} t' \iff \text{for each closed substitution instance } \tfrac{H}{t \xrightarrow{a} t'} \text{ of a rule of } P \text{ one has } T \not\models H.$$

and well-supported iff

$$T \not\models t \xrightarrow{a} t' \iff \text{for each set of negative closed literals } N \text{ with } P \vdash \tfrac{N}{t \xrightarrow{a} t'}, \text{ one has } T \not\models N.$$

Therefore I propose the following two concepts of provability.

Definition 8 (*Supported proof*). A *supported proof* of a closed literal α from a TSS P is like a closed (positive) proof (see Definition 3), but admitting steps of the form

3. β is negative and for each closed substitution instance of a rule of P whose conclusion denies β, a literal in K denies one of its antecedents.

α is *s-provable*, notation $P \vdash_s \alpha$, if a supported proof of α from P exists.
A literal is *s-refutable* if a denying literal is *s*-provable.

Definition 9 (*Well-supported proof*). A *well-supported proof* of a closed literal α from a TSS P is like a closed (positive) proof (Definition 3), but admitting steps of the form

3. β is negative and for every set N of negative closed literals such that $P \vdash \tfrac{N}{\gamma}$ for γ a closed literal denying β, a literal in K denies one in N.

α is *ws-provable*, notation $P \vdash_{ws} \alpha$, if a well-supported proof of α from P exists.
A literal is *ws-refutable* if a denying literal is *ws*-provable.

Note that these proof-steps establish the validity of β when K is the set of literals established earlier. In case K and N are sets of closed literals and a literal in K denies one in N, one has $T \not\models N$ for any transition relation T with $T \models K$. Thus step 3 from Definition 9 allows one to infer $t \xrightarrow{a\;/} t'$ whenever it is manifestly impossible to infer $t \xrightarrow{a} t'$, or $t \xrightarrow{a\;/}$ whenever for any term t' it is manifestly impossible to infer $t \xrightarrow{a} t'$. This practice is sometimes referred to as *negation as failure* [4]. Definition 8 allows such an inference only if the impossibility to derive $t \xrightarrow{a} t'$ can be detected by examining all possible proofs that consist of one step only. This corresponds with the notion of *negation as finite failure* of CLARK [4]. The extension of these notions (especially $\vdash_{ws}$) from closed to open literals α, or to transition rules $\tfrac{H}{\alpha}$, is somewhat problematic, and not needed in this paper. The following may shed more light on $\vdash_s$ and $\vdash_{ws}$. From here onwards, statements hold with or without the text enclosed in square brackets.

Proposition 4 Let P be a TSS. Then $P \vdash_s t \xrightarrow{a\;/} [t']$ iff every closed substitution instance $\tfrac{H}{t \xrightarrow{a} t'}$ of a rule of P has an *s*-refutable antecedent. Moreover $P \vdash_{ws} t \xrightarrow{a\;/} [t']$ iff every set N of closed negative literals with $P \vdash \tfrac{N}{t \xrightarrow{a} t'}$ contains an *ws*-refutable literal.

Proposition 5 For P a TSS and α a closed literal $P \vdash \alpha \;\Rightarrow\; P \vdash_s \alpha \;\Rightarrow\; P \vdash_{ws} \alpha$.

Definition 10 For P a TSS and α a closed literal, write $P \models_s \alpha$ if $T \models \alpha$ for any supported model T of P and $P \models_{ws} \alpha$ if $T \models \alpha$ for any well-supported model T of P. A notion $\vdash_x$ is called

- *consistent* if there is no TSS deriving two literals that deny each other.
- *sound* w.r.t. $\models_x$ if for any TSS P and closed literal α, $P \vdash_x \alpha \Rightarrow P \models_x \alpha$.

- *complete* w.r.t. $\models_x$ if for any TSS P and closed literal α, $P \vdash_x \alpha \Leftarrow P \models_x \alpha$.

Proposition 6 $\vdash_{[w]s}$ is consistent, $\vdash_{ws}$ is sound w.r.t. $\models_{ws}$ and $\vdash_s$ is sound w.r.t. $\models_s$.

However, $\vdash_s$ and $\vdash_{ws}$ are not complete w.r.t. $\models_{[w]s}$. A trivial counterexample concerns TSSs like P_2 that have no [well-]supported models. $P_2 \models_{[w]s} \alpha$ for any α, which by Proposition 6 (consistency) is not the case for $\vdash_{[w]s}$. A more interesting counterexample concerns the TSS P_7, which has only one [well-]supported model, namely $\{c \xrightarrow{a} c\}$. In spite of this, $P_7 \not\vdash_{[w]s} c \xrightarrow{a} c$ and $P_7 \not\vdash_{[w]s} c \xrightarrow{b} \!\!\!\!\!\not\;\;$.

As argued in the previous section, there is a point in excluding P_7 from the meaningful TSSs, since there is insufficient evidence for the transition $c \xrightarrow{a} c$. Here the incompleteness of $\vdash_{[w]s}$ w.r.t. $\models_{[w]s}$ comes as a blessing rather than a shortcoming.

3.1 Solutions based on completeness

I will now introduce the concept of a *complete* TSS: one in which any transition is either provable or refutable. Just as in the theory of logic there is a distinction between the completeness of a logic (e.g. first-order) and the completeness of a particular theory (e.g. arithmetic), here the completeness of a TSS is something different from the completeness of a proof-method $\vdash_x$. Let x be s or ws.

Definition 11 (*Completeness of a TSS*). A TSS P is *x-complete* if for any transition $t \xrightarrow{a} t'$ either $P \vdash_x t \xrightarrow{a} t'$ or $P \vdash_x t \xrightarrow{a}\!\!\!\!\not\;\; t'$. By 'complete' I will mean 'ws-complete'.

Solution 6 (*Complete with support*). A TSS is meaningful iff it is s-complete. The associated transition relation consists of the s-provable transitions.

Solution 7 (*Complete*). A TSS is meaningful iff it is (ws-)complete. The associated transition relation consists of the ws-provable transitions.

The TSS P_6 is complete, but not complete with support. P_3 is even complete with support. In BOL & GROOTE [3] a method called *reduction* for associating a transition relation with a TSS was proposed, inspired by the *well-founded models* of VAN GELDER, ROSS & SCHLIPF [6] in logic programming. In the full version of this paper I show that this solution coincides with Solution 7. Solution 7 can therefore be regarded as a proof theoretical characterisation of the ideas from [6, 3]. Solution 6 may be new.

Proposition 7 The set of $[w]s$-provable transitions of any TSS is well-supported. Moreover, the set of $[w]s$-provable transitions of a $[w]s$-complete TSS P is a model of P.

Proposition 8 Solution 6 [7] is strictly extended by Solution 4 [5].

Proof: Suppose P is $[w]s$-complete. By Proposition 7 the $[w]s$-provable transitions constitute a [well-]supported model of P, and by Proposition 6 (soundness) this is the only such model. Strictness follows from the TSS P_7, which has an unique [well-]supported model, but is left meaningless by Solutions 6 and 7. □

At the end of Section 2 I recommended two acceptable attitudes towards rules like $\frac{c \xrightarrow{a} c}{c \xrightarrow{a} c}$. In the full paper I show that Solution 7 ignores such rules completely (which is one option), whereas Solution 6 rejects a TSS with such a rule, unless there is independent evidence for a transition $c \xrightarrow{a} c$ (the other option). Moreover, Solutions 6 and 7 reject any TSS containing rules like $\frac{c \xrightarrow{a}\!\!\!\not\;\;}{c \xrightarrow{a} c}$, unless there is independent evidence for a transition $c \xrightarrow{a} t$. As shown by counterexample P_7 all model theoretic solutions fail this test.

3.2 Attaching meaning to *all* transition system specifications

In this section I will associate a transition relation to arbitrary TSSs. As illustrated by P_1 and P_2, such a transition relation can not always be a supported model. I will insist on soundness (being a model), and thus have to give up support.

Let me first decide what to do with P_1. Since the associated transition relation should be a model, it must contain either $c \xrightarrow{a} c$ or $c \xrightarrow{b} c$. For reasons of symmetry I cannot choose between these transitions, so the only way out is to include both. There is no reason to include any more transitions. Hence the transition system associated to P_1 should be $\{c \xrightarrow{a} c, \, c \xrightarrow{b} c\}$.

In the full paper I reject a model theoretic solution that gives this result. Among the proof theoretic solutions the best I could find was

Solution 8 (*Irrefutable*). Any TSS is meaningful. The associated transition relation consists of the *ws*-irrefutable transitions.

This solution is inspired by the following proposition.

Proposition 9 The set of x-irrefutable transitions of any TSS consititutes a model.

In the case of P_1 Solution 8 yields the desired result $\{c \xrightarrow{a} c, \, c \xrightarrow{b} c\}$ and likewise P_2, P_3 and P_4 yield $\{c \xrightarrow{a} c\}$. The transition relation of P_7 is the same as the one of P_1. This indicates that Solution 8 is inconsistent with Solutions 2–5. I don't consider this to be a problem, as the model theoretic allocation of a transition relation to P_7 was not very convincing.

4 Compositionality

In concurrency theory it is common practice to group together representations of concurrent systems in equivalence classes. As system representations often closed terms over some signature are considered. The equivalence relation employed is then formulated in terms of the transition relation between closed terms obtained from a given TSS over that signature. All equivalence relations employed in concurrency have the properties that systems for which the reachable parts of the transition relation are isomorphic are equivalent, and that a system without outgoing transitions (a *deadlock*) cannot be equivalent to a system with an outgoing a-transition.

In order to allow modular reasoning it is important to use an equivalence relation that is a *congruence*. This means that the meaning (the associated equivalence class) of a closed term $f(t_1, ..., t_n)$ is completely determined by the meaning of the subterms $t_1, ..., t_n$. The most popular equivalence relation is *bisimulation equivalence*. In BOL & GROOTE [3] it was established that for complete TSSs whose rules satisfy a syntactic criterion (the *well-founded ntyft/ntyxt format*, developed earlier in [9, 8]), bisimulation equivalence is guaranteed to be a congruence, and so are many other equivalence relations. Moreover, a counterexample was given against the extension of this result to TSSs that are meaningful according to Solution 5 (stable). Of course the example concerned an incomplete TSS in well-founded ntyft/ntyxt format with a unique stable transition relation for which bisimulation is not a congruence. This TSS also has a unique supported model, and thus shows that the congruence theorem does not generalise to Solution 4 either. Here I show that also Solution 8—or any other proof theoretic solution giving meaning to all TSSs for that matter—does not lend itself to such a generalisation, indicating that Solution 7 (complete) is the most general one for which this nice result holds. My counterexample concerns the following TSS S over a signature with constants c, d and e and a unary function f.

$$S \quad \boxed{\; c \xrightarrow{a} f(c) \qquad \dfrac{x \xrightarrow{a} y \;\; \xcancel{\xrightarrow{a}}}{f(x) \xrightarrow{a} c} \qquad d \xrightarrow{a} e \;}$$

This TSS is surely in the well-founded ntyft/ntyxt format. The transitions $c \xrightarrow{a} f(c)$, $d \xrightarrow{a} e$ and $f(d) \xrightarrow{a} c$ are $[w]s$-provable, and with the exception of $f(c) \xrightarrow{a} c$, all other transitions are $[w]s$-refutable. As the validity of $f(c) \xrightarrow{a} c$ is left undetermined, the TSS is incomplete (has no meaning according to Solution 7). It also has no meaning under Solution 5 (stable). The proof theoretic approach offers only one choice, namely whether or not to include the transition $f(c) \xrightarrow{a} c$. Each of these possibilities yields a transition relation for which no equivalence relation used in concurrency theory is a congruence. Solution 8 (irrefutable) includes the transition $f(c) \xrightarrow{a} c$. Now c and $f(c)$ are equivalent (the reachable part of the transition relation from each of them is an a-loop), but $f(c)$ and $f(f(c))$ are inequivalent ($f(f(c))$ deadlocks). Taking only the provable transitions (instead of the irrefutable ones) would exclude the transition $f(c) \xrightarrow{a} c$. In that case c and d are equivalent, but $f(c)$ and $f(d)$ are not.

5 Conclusion

This paper dealt with the problem of associating a transition relation to a given TSS. The related problem of finding a good TSS to specify a given transition relation is left for future research. I presented 8 answers to the question of which transition system specifications are meaningful and which transition relations they specify. The relations between these 8 solutions, as well as the two solutions (9 and 10) proposed in [8], are indicated below. There $S_1 \longrightarrow S_2$ indicates that solution S_2 extends S_1, as defined in Section 1,

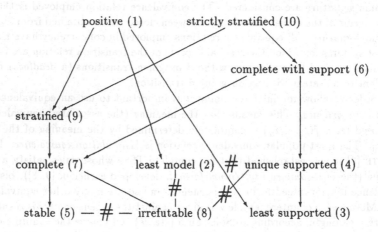

and $S_1 \# S_2$ indicates that S_1 and S_2 are inconsistent. By the definition of extension and consistency, $S_1 \longrightarrow S_2 \longrightarrow S_3$ implies $S_1 \longrightarrow S_3$ (*transitivity*) and $S_1 \# S_2 \longrightarrow S_3$ implies $S_1 \# S_3$ (*conflict heredity*). All extensions are strict and there are no more extensions or inconsistencies than indicated in the figure (or derivable by transitivity and conflict heredity). Strictness, the absence of further extensions and the inconsistencies follow from the information collected in the table below, which indicates which of the TSSs P_1–P_8 given in this paper are meaningful according to each of the solutions. A '−' indicates that the TSS is meaningless, a '+' that it has the same meaning as given by Solution 8, and a '*' that is has a meaning different from the one given by Solution 8.

Solution		P_1	P_2	P_3	P_4	P_5	P_6	P_7	P_8
1	positive	−	−	−	−	+	−	−	−
2	least	−	+	−	+	+	−	*	+
3	least supported	−	−	+	*	+	−	*	+
4	unique supported	−	−	+	*	−	−	*	+
5	stable	−	−	+	−	+	+	*	+
6	complete with support	−	−	+	−	−	−	−	+
7	complete	−	−	+	−	+	+	−	+
8	irrefutable	+	+	+	+	+	+	+	+
9	stratified	−	−	+	−	+	+	−	−
10	strictly stratified	−	−	+	−	−	−	−	−

Evaluation of the solutions

Solution 9 (stratified) stems from PRZYMUSINSKI [11] and is the perhaps the best known solution in logic programming. A variant that only allows TSSs with a unique supported model is Solution 10 (strictly stratified), proposed by GROOTE [8].

Solution 1 is the classical interpretation of TSSs without negative premises, and Solutions 2 (least model) and 3 (least supported model) are two straightforward generalisations. Solution 4 (unique supported model) stems from BLOOM, ISTRAIL & MEYER [2], where it was used to ascertain that TSSs in their so-called GSOS format are meaningful (such TSSs have unique supported models). My counterexample P_4 shows that Solution 4 yields contraintuitive results and is therefore not suited to base such a conclusion on. Fortunately, TSSs in the GSOS format are even strictly stratified, which is one of the most restrictive criteria for meaningful TSSs considered. Solution 3 can be rejected on the same grounds as Solution 4 and Solution 2 is not very useful because it leaves most TSSs with negative premises meaningless (cf. P_3).

Solution 5 (unique stable transition relation) stems from GELFOND & LIFSCHITZ [7] and is generally considered to be the most general acceptable solution available. Counterexample P_7 however suggests that this solution may yield debatable results, although to a lesser extent than Solutions 3 and 4.

Solution 7 (well-founded, positive after reduction, complete) is essentially due to VAN GELDER, ROSS & SCHLIPF [6]. It is the most general solution without undesirable properties. In BOL & GROOTE [3], where this solution has been adapted to TSSs, an example in the area of concurrency is given (the modelling of a priority operator in basic process algebra with abstraction) that can be handled with Solution 7, but not with Solution 9. This example can neither be handled by Solution 6, showing that the full generality of Solution 7 can be useful in applications.

My presentation of Solution 7 differs so much from the original one [6, 3] that I gave it a new name. It is based on a concept of provability incorporating the notion of *negation as failure* of CLARK [4]. In the full paper I establish the correspondence between my version and the one from [6, 3]. There I also illustrate how my proof-theoretic characterisation can be useful in applications.

Solutions 6 (complete with support) and 8 (irrefutable) may be new. The first is based on a notion of provability that is somewhat simpler to apply, and only incorporates the notion of *negation as finite failure* [4]. Moreover, it only yields unique supported models, like Solution 10 (and 4). Solution 8 appears to be the best way to associate a transition relation to arbitrary TSSs. However, it has the disadvantage that it sometimes yields unstable transition relations, and even unsupported models. A good example from concurrency theory of an incomplete TSS is Basic Process Algebra with a priority operator, unguarded recursion and renaming, as defined in GROOTE [8]. This TSS has no supported models. Solution 8 does give a meaning to this TSS, but it appears rather arbitrary and not very useful. In particularly, recursively defined processes do no longer satisfy their defining equation, which makes algebraic reasoning virtually impossible. Also the absence of a congruence theorem as demonstrated in Section 4 is a bad property of this Solution. Hence, Solution 7 (complete) remains the most general completely acceptable answer to (1) and (2).

Acknowledgments This paper benefited greatly from the insightful comments of Roland Bol. Also my thanks to the audience of the PAM seminar for useful feedback. Finally, Jan Rutten is gratefully acknowledged for spotting a mistake in a previous version of this paper (see Footnote 2).

References

[1] K.R. APT & R. BOL (1994): *Logic programming and negation: A survey.* Journal of Logic Programming 19–20, pp. 9–71.

[2] B. BLOOM, S. ISTRAIL & A.R. MEYER (1995): *Bisimulation can't be traced.* JACM 42(1), pp. 232–268.

[3] R.N. BOL & J.F. GROOTE (1991): *The meaning of negative premises in transition system specifications (extended abstract).* In J. Leach Albert, B. Monien & M. Rodríguez, editors: *Proceedings 18th ICALP*, Madrid, LNCS 510, Springer-Verlag, pp. 481–494. Full version to appear in JACM.

[4] K.L. CLARK (1978): *Negation as failure.* In H. Gallaire & J. Minker, editors: *Logic and Databases*, Plenum Press, New York, pp. 293–322.

[5] F. FAGES (1991): *A new fixpoint semantics for general logic programs compared with the well-founded and the stable model semantics.* New Generation Computing 9(4), pp. 425–443.

[6] A. VAN GELDER, K. ROSS & J.S. SCHLIPF (1991): *The well-founded semantics for general logic programs*, JACM 38(3), pp. 620–650.

[7] M. GELFOND & V. LIFSCHITZ (1988): *The stable model semantics for logic programming.* In R. Kowalski & K. Bowen, editors: *Proceedings 5th International Conference on Logic Programming*, MIT Press, Cambridge, USA, pp. 1070–1080.

[8] J.F. GROOTE (1993): *Transition system specifications with negative premises.* Theoretical Computer Science 118(2), pp. 263–299.

[9] J.F. GROOTE & F.W. VAANDRAGER (1992): *Structured operational semantics and bisimulation as a congruence.* Information and Computation 100(2), pp. 202–260.

[10] G.D. PLOTKIN (1981): *A structural approach to operational semantics.* Report DAIMI FN-19, Computer Science Department, Aarhus University.

[11] T.C. PRZYMUSINSKI (1988): *On the declarative semantics of deductive databases and logic programs.* In Jack Minker, editor: *Foundations of Deductive Databases and Logic Programming*, Morgan Kaufmann Publishers, Inc., pp. 193–216.

Average Case Analyses of List Update Algorithms, with Applications to Data Compression

Susanne Albers* and Michael Mitzenmacher**

Abstract. We study the performance of the Timestamp(0) (TS(0)) algorithm for self-organizing sequential search on discrete memoryless sources. We demonstrate that TS(0) is better than Move-to-front on such sources, and determine performance ratios for TS(0) against the optimal offline and static adversaries in this situation. Previous work on such sources compared online algorithms only to static adversaries. One practical motivation for our work is the use of the Move-to-front heuristic in various compression algorithms. Our theoretical results suggest that in many cases using TS(0) in place of Move-to-front in schemes that use the latter should improve compression. Tests on a standard corpus of documents demonstrate that TS(0) leads in fact to improved compression.

1 Introduction

We study deterministic online algorithms for self-organizing sequential search. Consider a set of n items $x_1, x_2, \ldots, x_n$ that are stored in an unsorted linear linked list. At any instant of time, an algorithm for maintaining this list is presented with a *request* that specifies one of the n items. The algorithm must serve this request by *accessing* the requested item. That is, the algorithm has to start at the front of the list and search linearly through the items until the desired item is found. Serving a request to the i-th item in the list incurs a cost of i. Immediately after a request, the requested item may be moved at no extra cost to any position closer to the front of the list; this can lower the cost of subsequent requests. At any time two adjacent items in the list may be exchanged at a cost of 1; these moves are called *paid exchanges*. The goal is to serve a *sequence of requests* so that the total cost incurred on that sequence is as small as possible. A list update algorithm typically works *online*, i.e., when serving a present request, the algorithm has no knowledge of future requests.

Early work on the list update problem assumes that a request sequence is generated by a probability distribution $\mathbf{p} = (p_1, p_2, \ldots, p_n)$. A request to item x_i occurs with probability p_i; the requests are generated independently. More recent research on the list update problem was inspired by Sleator and Tarjan [14]

* Max-Planck-Institut für Informatik, Im Stadtwald, 66123 Saarbrücken, Germany. Part of this work was done while the author was at the International Computer Science Institute, Berkeley; E-mail: albers@mpi-sb.mpg.de

** Computer Science Division, UC Berkeley, Berkeley, CA 94720. This work was supported in part by the Office of Naval Research and in part by NSF Grant CCR-9505448; E-mail: mitzen@cs.berkeley.edu

who suggested to compare the performance of an online algorithm to that of an *optimal offline* algorithm. An optimal offline algorithm knows the entire request sequence in advance and can serve it with minimum cost. An online algorithm A is called c-competitive if, for all request sequences, the cost incurred by A is at most c times the cost incurred by the optimal offline algorithm.

The following on-line algorithms have been investigated extensively:

Move-to-front (MTF): Move the requested item to the front of the list.

Transpose (T): Exchange the requested item with the immediately preceding item in the list.

Frequency count (FC): Maintain a frequency count for each item in the list. Whenever an item is requested, increase its count by 1. Maintain the list so that the items always occur in nonincreasing order by frequency count.

In this paper we again investigate the list update problem under the assumption that a request sequence is generated by a probability distribution $\mathbf{p} = (p_1, p_2, \ldots, p_n)$; that is, it is generated by a discrete memoryless source. We note that this assumption is suitable in many applications, and many of our techniques generalize to other models as well. Our work is motivated by the goal to present a universal algorithm that is competitive on any request sequence (in the Sleator Tarjan model) but also performs especially well on distributions. Previous results have shown that MTF is such an algorithm, whereas algorithms T and FC are not. Our main contribution is to show that there is an algorithm that has an even better overall performance than MTF. The algorithm we analyze belongs to the Timestamp(p) family of randomized online algorithms [1]. For $p = 0$, the algorithm is deterministic and can be formulated as follows:

Algorithm TS(0): Insert the requested item, say x, in front of the first item in the list that has been requested at most once since the last request to x. If x has not been requested so far, leave the position of x unchanged.

In [1] it was shown that TS(0) is 2-competitive, as is Move-to-front [14]. Here we demonstrate that TS(0) performs better on distributions, both by developing a formula for the expected cost per request, and by comparing TS(0) with the optimal static and dynamic offline algorithms.

Since our results show that TS(0) performs better than MTF on distributions, we consider applying the algorithm in the context of data compression, where MTF has been used to develop a locally adaptive data compression scheme [3]. Here we prove that for all distributions $\mathbf{p} = (p_1, p_2, \ldots, p_n)$, the expected number of bits needed by a TS(0)-based encoding scheme to encode one symbol is linear in the entropy of the source. Our implementations also demonstrate that in practice TS(0)-based schemes can achieve better compression than MTF schemes.

1.1 Comparison with previous work

We briefly review the main results in the model that a request sequence is generated by a probability distribution. The performances of MTF, T, and FC have generally been compared to that of the *optimal static ordering*, which we call

STAT. The optimal static ordering arranges the items x_i in nonincreasing order by probabilities p_i and serves a request sequence without changing the relative position of items. For any algorithm A, let $E_A(\mathbf{p})$ denote the asymptotic expected cost incurred by A in serving one request in a request sequence generated by the distribution $\mathbf{p}$. Rivest [12] showed that for all $\mathbf{p}$, $E_{FC}(\mathbf{p})/E_{STAT}(\mathbf{p}) = 1$. However, the algorithm FC has the drawback that it adapts very slowly to changing probability distributions. Chung et al. [5] analyzed the MTF rule and proved $E_{MTF}(\mathbf{p})/E_{STAT}(\mathbf{p}) \leq \frac{\pi}{2} \approx 1.5708$ for all $\mathbf{p}$. This bound is tight because Gonnet et al. [7] showed that one can find $\mathbf{p}_0$ with $E_{MTF}(\mathbf{p}_0)/E_{STAT}(\mathbf{p}_0) \geq \alpha$ for any α arbitrarily close to $\pi/2$.

Sleator and Tarjan proved that the MTF algorithm is 2-competitive. They also showed that the algorithms T and FC are not c-competitive for any constant c. The competitive ratio of 2 is the best ratio that a deterministic online algorithm for the list update problem can achieve [11].

In classical data compression, it is often assumed that a discrete memoryless source generates a string S to be compressed. The string S consists of *symbols*, where each symbol is an element in the alphabet $\Sigma = \{x_1, x_2, \ldots, x_n\}$. Each symbol is equal to x_i with probability p_i. Bentley et al. [3] showed how a list update algorithm can be used to develop a data compression scheme. The idea is to convert the string S of symbols into a string I of integers. Whenever the symbol x_i has to be compressed, an encoder looks up the current position of x_i in a linear list of symbols it maintains, outputs this positions and updates the list. A decoder that receives the string I can recover the original message by looking up in its own linear list, for each integer j it reads, the symbol that is currently stored at position j. The decoder also updates its list. The string I of integers is actually transmitted using a variable length prefix code. Bentley et al. showed that, for all $\mathbf{p} = (p_1, p_2, \ldots, p_n)$, the expected number of bits needed to encode one symbol in a string S using the MTF rule is linear in the entropy of the source. By Shannon's source coding theorem, this is optimal, up to a constant factor.

Recently, Grinberg et al. [8] proposed a modification of the MTF encoding, which they call *MTF encoding with secondary lists*. They implemented the new compression scheme but their simulations do not show an explicit comparison between MTF and MTF with secondary lists. Also recently, a fast and efficient compression scheme that uses MTF encoding as a subroutine has been developed [4]. This algorithm appears competitive with those used in standard compression tools, and thus the examination of alternatives to MTF may lead to better practical compression algorithms.

1.2 Our results

We compare the expected cost incurred by TS(0) to that of the optimal offline algorithm, which we shall denote by OPT. We recall that OPT may rearrange the list after each request and is not forced to serve a request sequence using the optimal static ordering.

First we develop a formula for TS(0)'s expected cost on a distribution $\mathbf{p} = (p_1, p_2, \ldots, p_n)$. This formula implies that if we have a distribution $\mathbf{p}$ with $p_i = \frac{1}{n}$, for all i, then MTF and TS(0) have the same expected cost. On all other distributions, TS(0) has a smaller expected cost. Then we compare TS(0) to the optimal offline algorithm OPT and show $E_{TS}(\mathbf{p})/E_{OPT}(\mathbf{p}) \leq 1.5$ for all distributions $\mathbf{p}$. This is a performance MTF cannot match because $E_{MTF}(\mathbf{p}_0)/E_{STAT}(\mathbf{p}_0) > 1.57$ for some $\mathbf{p}_0$, and when MTF is compared to OPT the ratio might even be worse. We also show that, for any $\mathbf{p}$ and any $\epsilon > 0$, the cost of TS(0) is at most $1.5 + \epsilon$ times the cost of OPT with high probability on sufficiently long sequences. It is worthwhile to note that 1.5 is the best lower bound currently known on the competitiveness that can be achieved by randomized list update algorithms against the oblivious adversary [15]. Thus, the performance ratio of TS(0) on distributions is at least as good as the performance ratio of randomized algorithms on any input. Finally we evaluate TS(0) against the optimal static ordering and show, for all $\mathbf{p}$, $E_{TS}(\mathbf{p})/E_{STAT}(\mathbf{p}) \leq 1.34$.

Given these results, we examine the potential for TS(0) in compression algorithms. As previously mentioned, we prove that for all distributions $\mathbf{p} = (p_1, p_2, \ldots, p_n)$, the expected number of bits needed by a TS(0)-based encoding scheme to encode one symbol is linear in the entropy of the source. Our upper bounds are better than similar upper bounds for MTF-encoding in this case. For *any* input sequence (that might not necessarily be generated by a distribution) we can show that the average number of bits needed by a TS(0)-based encoding to encode one symbol is linear in the "empirical entropy" of the sequence. We also provide evidence that TS(0) can be useful in practice by implementing TS(0) compression algorithms and testing them on the standard Calgary Compression Corpus files [16]. In almost all of our tests, TS(0) encoding achieves a better compression ratio than MTF encoding.

2 Analyses for the list update problem

2.1 The expected cost of TS(0)

Theorem 1. *For any probability distribution* $\mathbf{p} = (p_1, p_2, \ldots, p_n)$,

a) the asymptotic expected cost incurred by TS(0) in serving a request to item
$$x_i, \ 1 \leq i \leq n, \ \text{is } e_{TS}(x_i) = \frac{1}{2} + \sum_{j=1}^{n} \frac{p_j^3 + 3p_j^2 p_i}{(p_i + p_j)^3}.$$

b) $E_{TS}(\mathbf{p}) = \sum_{1 \leq i \leq j \leq n} \frac{p_i p_j}{p_i + p_j} \left(2 - \frac{(p_i - p_j)^2}{(p_i + p_j)^2} \right).$

Proof. Part a): The cost $e_{TS}(x_i)$ is 1 plus the the expected number of items x_j, $x_j \neq x_i$, that precede x_i in the list. Let A_{ji} be the event that x_j precedes x_i in the list when TS(0) serves a request to x_i. We compute the asymptotic probability $Prob(A_{ji})$ using the following lemma. The proof of the lemma is omitted here.

Lemma 2. *Consider any point in the request sequence where there have been at least three requests for x_i and x_j. Then x_j precedes x_i in the list if and only if a majority of the last three requests for x_i and x_j have been for x_j.*

Lemma 2 implies that the event A_{ji} occurs if and only if the last three requests for x_i and x_j are (B_1) $x_jx_jx_j$; (B_2) $x_jx_jx_i$; (B_3) $x_jx_ix_j$; or (B_4) $x_ix_jx_j$. It is not hard to verify that $Prob(B_1) = p_j^3/(p_i + p_j)^3$ and $Prob(B_k) = p_ip_j^2/(p_i + p_j)^3$, for $k = 2, 3, 4$. Therefore, $Prob(A_{ji}) = (p_j^3 + 3p_j^2p_i)/(p_i + p_j)^3$ and

$$e_{TS}(x_i) = 1 + \sum_{\substack{j=1 \\ j \neq i}}^{n} Prob(A_{ji}) = 1 + \sum_{\substack{j=1 \\ j \neq i}}^{n} \frac{p_j^3 + 3p_j^2p_i}{(p_i + p_j)^3} = \frac{1}{2} + \sum_{\substack{j=1 \\ j \neq i}}^{n} \frac{p_j^3 + 3p_j^2p_i}{(p_i + p_j)^3}.$$

Part b): The asymptotic expected cost incurred by TS(0) on one request is

$$E_{TS}(\mathbf{p}) = \sum_{i=1}^{n} p_i e_{TS}(x_i) = \frac{1}{2} + \sum_{i=1}^{n} \sum_{j=1}^{n} p_i \left(\frac{p_j^3 + 3p_j^2p_i}{(p_i + p_j)^3} \right)$$

$$= \sum_{1 \leq i \leq j \leq n} \frac{p_ip_j}{p_i + p_j} \left(\frac{p_i^2 + 6p_ip_j + p_j^2}{(p_i + p_j)^2} \right) = \sum_{1 \leq i \leq j \leq n} \frac{p_ip_j}{p_i + p_j} \left(2 - \frac{(p_i - p_j)^2}{(p_i + p_j)^2} \right). \ \square$$

Corollary 3. *For any probability distribution* $\mathbf{p} = (p_1, p_2, \ldots, p_n)$,

$$E_{MTF}(\mathbf{p}) - E_{TS}(\mathbf{p}) = \sum_{1 \leq i \leq j \leq n} p_ip_j \frac{(p_i - p_j)^2}{(p_i + p_j)^3}.$$

Proof. Rivest [12] showed $E_{MTF}(\mathbf{p}) = \sum_{1 \leq i \leq j \leq n} 2\frac{p_ip_j}{p_i+p_j}$. Using part b) of Theorem 1, the result follows immediately. $\square$

2.2 Performance against dynamic offline algorithms

Theorem 4. *For any probability distribution* $\mathbf{p} = (p_1, p_2, \ldots, p_n)$,

$$E_{TS}(\mathbf{p}) \leq \frac{3}{2} E_{OPT}(\mathbf{p}).$$

Proof. The analysis consists of two main parts. In the first part we show that, given a fixed request sequence σ, the cost incurred by TS(0) and OPT on σ can be divided into costs that are caused by each unordered pair $\{x, y\}$ of items x and y, $x \neq y$. This technique of evaluating cost by considering pairs of items was also used in [2, 10, 1]. In the second part of the analysis we show that, for each pair $\{x, y\}$, the asymptotic expected cost paid by TS(0) is at most $\frac{3}{2}$ times the asymptotic expected cost incurred by OPT.

In the following we will always assume that serving a request to the i-th item in the list incurs a cost of $i - 1$ rather than i. If $E_{TS}(\mathbf{p}) \leq \frac{3}{2} E_{OPT}(\mathbf{p})$ holds in this $(i - 1)$-*cost model*, then the inequality obviously also holds in the i-cost model. Now consider a fixed request sequence $\sigma = \sigma(1)\sigma(2), \ldots, \sigma(m)$ of length

m. For an algorithm $A \in \{TS(0), OPT\}$, let $C_A(t, x)$ denote the cost caused by item x when A serves request $\sigma(t)$. That is, $C_A(t, x) = 1$ if x precedes the item requested by $\sigma(t)$ in A's list at time t; otherwise $C_A(t, x) = 0$. For any pair $\{x, y\}$ of items $x \neq y$, let $p(x, y)$ be the total number of paid exchanges incurred by A in moving x in front of y or y in front of x. The cost incurred by A on σ can be written as

$$C_A(\sigma) = \sum_{\substack{\{x,y\} \\ x \neq y}} \left(\sum_{\substack{t \in [1,m] \\ \sigma(t)=x}} C_A(t, y) + \sum_{\substack{t \in [1,m] \\ \sigma(t)=y}} C_A(t, x) + p(x, y) \right)$$

$$= \sum_{\substack{\{x,y\} \\ x \neq y}} \left(\sum_{\substack{t \in [1,m] \\ \sigma(t) \in \{x,y\}}} (C_A(t, y) + C_A(t, x)) + p(x, y) \right).$$

Now, for any unordered pair $\{x, y\}$ of items x and y, with $x \neq y$, let σ_{xy} be the request sequence that is obtained from σ if we delete all requests that are neither to x nor to y. Let $C_{TS}(\sigma_{xy})$ be the cost incurred by TS(0) if it serves σ_{xy} on a two item list that consists of only x and y. In [1] it was shown that if TS(0) serves σ on the long list, then the relative position of x and y changes in the same way as if TS(0) serves σ_{xy} on the two item list. Therefore,

$$C_{TS}(\sigma_{xy}) = \sum_{\substack{t \in [1,m] \\ \sigma(t) \in \{x,y\}}} (C_{TS}(t, y) + C_{TS}(t, x))$$

$$C_{TS}(\sigma) = \sum_{\substack{\{x,y\} \\ x \neq y}} C_{TS}(\sigma_{xy}). \tag{1}$$

Note that TS(0) does not incur paid exchanges and hence $p(x, y) = 0$ for all pairs $\{x, y\}$. The optimal cost $C_{OPT}(\sigma)$ can be written in a similar way:

$$C_{OPT}(\sigma_{xy}) \leq \sum_{\substack{t \in [1,m] \\ \sigma(t) \in \{x,y\}}} (C_{OPT}(t, y) + C_{OPT}(t, x)) + p(x, y)$$

$$C_{OPT}(\sigma) \geq \sum_{\substack{\{x,y\} \\ x \neq y}} C_{OPT}(\sigma_{xy}). \tag{2}$$

Here, only inequality signs hold because if OPT serves σ_{xy} on the two items list, then it can always arrange x and y optimally in the list, which might not be possible if OPT serves σ on the entire list. Equation (1) and equation (2) allow us to compare $C_{TS}(\sigma)$ and $C_{OPT}(\sigma)$ by simply comparing $C_{TS}(\sigma_{xy})$ and $C_{OPT}(\sigma_{xy})$ for each pair $\{x, y\}$ of items. In the following we concentrate on one particular pair $\{x, y\}$ of items $x \neq y$. Let $p = \frac{p_x}{p_x + p_y}$ and $q = \frac{p_y}{p_x + p_y}$.

For an algorithm $A \in \{TS(0), OPT\}$, let $E_A(\sigma_{xy}^t)$ be the asymptotic expected cost incurred by items x and y if A serves a single request in σ_{xy}, given that the request sequence σ is generated by p. For example, $E_{TS}(\sigma_{xy}^t)$ is the asymptotic value of $E[C_{TS}(t, y) + C_{TS}(t, x)]$, where $\sigma(t)$ is a request in σ_{xy}. We will show that

$$E_{TS}(\sigma_{xy}^t) \leq \frac{3}{2} E_{OPT}(\sigma_{xy}^t).$$

This proves the theorem.

We first evaluate $E_{TS}(\sigma_{xy}^t)$. TS(0) incurs a cost of 1 on a request in σ_{xy} if x is requested and y precedes x in TS(0)'s list or if y is requested and x precedes y in TS(0)'s list. Otherwise, TS(0) incurs a cost of 0. By Lemma 2, y precedes x in TS(0)'s list if and only if the majority of the last three requests for x and y have been for y, i.e., if the last three requests have been (B_1) yyy; (B_2) yyx; (B_3) yxy; or (B_4) xyy. Thus, the asymptotic probability that y precedes x in TS(0)'s list is $q^3 + 3q^2p$. Similarly, the asymptotic probability that x precedes y in TS(0)'s list is $p^3 + 3p^2q$. Thus

$$E_{TS}(\sigma_{xy}^t) = p(q^3 + 3q^2p) + q(p^3 + 3p^2q) = pq(p^2 + 6pq + q^2).$$

Next we determine $E_{OPT}(\sigma_{xy}^t)$. Consider OPT's movements when it serves σ_{xy} on the two item list. We may assume without loss of generality that whenever there are two consecutive requests to the same item, OPT moves that item to the front of the list after the first request if it is not already there. Thus, OPT incurs a cost of 1 on a request in σ_{xy} if x is requested and the last requests in σ_{xy} were of the form $yy(xy)^i$ for some $i \geq 0$, or if y is requested and the last requests in σ_{xy} were of the form $xx(yx)^i$ for some $i \geq 0$. Therefore,

$$E_{OPT}(\sigma_{xy}^t) = p(q^2/(1 - pq)) + q(p^2/(1 - pq)) = pq/(1 - pq).$$

We conclude that $E_{TS}(\sigma_{xy}^t) \leq (1-pq)(p^2+6pq+q^2)E_{OPT}(\sigma_{xy}^t)$. The expression $(1 - pq)(p^2 + 6pq + q^2)$ is maximal for $p = q = \frac{1}{2}$ and hence $E_{TS}(\sigma_{xy}^t) \leq \frac{3}{2}E_{OPT}(\sigma_{xy}^t)$. $\square$

We also note that for long enough sequences on discrete memoryless sources, TS(0) will be at worst $1.5 + \epsilon$ competitive with high probability. We omit the proof of this theorem here.

Theorem 5. *For every distribution* $\mathbf{p} = (p_1, p_2, \ldots, p_n)$ *and* $\epsilon > 0$ *there exist constants* c_1, c_2, *and* m_0 *dependent on* $\mathbf{p}$, n, *and* ϵ *such that for a request sequence* σ *of length* $m \geq m_0$:

$$Prob\{C_{TS}(\sigma) > (1.5 + \epsilon)C_{OPT}(\sigma)\} \leq c_1 e^{-c_2 m}$$

2.3 Performance against static offline algorithms

Recall that the expected cost incurred by TS(0) in serving one request in a request sequence generated by $\mathbf{p} = (p_1, p_2, \ldots, p_n)$ is

$$1 + \sum_i \sum_{j \neq i} \frac{p_i p_j^3 + 3p_j^2 p_i^2}{(p_i + p_j)^3} = \sum_{i,j} \frac{p_i p_j (p_i^2 + 6p_i p_j + p_j^2)}{2(p_i + p_j)^3} + \frac{1}{2}.$$

We can now apply the techniques presented in [5] to bound the ratio between $E_{TS}(\mathbf{p})$ and $E_{STAT}(\mathbf{p})$. We assume $p_1 \geq p_2 \geq \ldots \geq p_n$. As $E_{STAT}(\mathbf{p}) = \sum_i i p_i = \frac{1}{2} \sum_{i,j} \min(p_i, p_j) + \frac{1}{2}$, we have

$$\frac{E_{TS}(\mathbf{p})}{E_{STAT}(\mathbf{p})} = \frac{\sum_{i,j} \frac{p_i p_j (p_i^2 + 6p_i p_j + p_j^2)}{(p_i + p_j)^3} + 1}{\sum_{i,j} \min(p_i, p_j) + 1} < \frac{\sum_{i,j} \frac{p_i p_j (p_i^2 + 6p_i p_j + p_j^2)}{(p_i + p_j)^3}}{\sum_{i,j} \min(p_i, p_j)}.$$

Our proof will rely on the following theorem:

Theorem 6. *If $p_i > 0 \, (1 \leq i \leq n)$, then*

$$\frac{\sum_{i,j} \frac{p_i p_j (p_i^2 + 6 p_i p_j + p_j^2)}{(p_i + p_j)^3}}{\sum_{i,j} \min(p_i, p_j)} \leq 1.34.$$

This theorem is in turn based on a continuous version in which the summations above are approximated by integrals whose ratio equals the ratio of the summations in the limiting case. The proof depends on the following version of Hilbert's inequality:

Theorem 7 Hilbert's inequality. *For $p, q > 1$ satisfying $\frac{1}{p} + \frac{1}{q} = 1$, suppose that $K(x, y)$ is non-negative and homogeneous of degree -1, and that*

$$\int_0^\infty K(x, 1) x^{-1/p} dx = \int_0^\infty K(1, y) y^{-1/q} dy = C.$$

Then

$$\int_0^\infty dx \left(\int_0^\infty K(x, y) g(y) dy \right)^q \leq C^q \int_0^\infty g^q(y) dy.$$

Lemma 8. *Suppose f is an integrable function on $(0, \infty)$ with $\int_0^\infty f dx = 0$. Let $H(x, y) = \frac{\partial^2 G}{\partial x \partial y}$ be homogeneous of degree -1, and $H^+(x, y) = max\{H(x, y), 0\}$. Then*

$$\frac{\int_0^\infty \int_0^\infty G(x, y) f(x) f(y) dx dy}{\int_0^\infty \int_0^\infty \min(x, y) f(x) f(y) dx dy} \leq \int_0^\infty H^+(x, 1) x^{-1/2} dx.$$

Theorem 6 will follow from Lemma 8 by choosing f to be 1 in arbitrarily small intervals around the p_i and 0 elsewhere, and by setting $G(x, y) = \frac{xy(x^2 + 6xy + y^2)}{(x+y)^3}$.

Proof. We set $F(x) = \int_{-\infty}^x f(x) dx$. Then $\int_0^\infty \int_0^\infty \min(x, y) f(x) f(y) dx dy = \int_0^\infty F^2(x) dx$. [Lemma 2 of [5]] Similarly,

$$\int_0^\infty \int_0^\infty G(x, y) f(x) f(y) dx dy = \int_0^\infty f(x) \left[G(x, y) F(y) |_0^\infty - \int_0^\infty \frac{\partial G}{\partial y} F(y) dy \right] dx$$

$$= -\int_0^\infty \int_0^\infty \frac{\partial G}{\partial y} f(x) F(y) dx dy = \int_0^\infty \int_0^\infty \frac{\partial^2 G}{\partial x \partial y} F(x) F(y) dx dy$$

$$= \int_0^\infty \int_0^\infty H(x, y) F(x) F(y) dx dy \leq \int_0^\infty \int_0^\infty H^+(x, y) F(x) F(y) dx dy$$

$$\leq \left[\int_0^\infty F^2(x) dx \right]^{1/2} \left[\int_0^\infty dx \left[\int_0^\infty H^+(x, y) F(y) dy \right]^2 \right]^{1/2}$$

$$\leq \int_0^\infty F^2(x) dx \int_0^\infty H^+(x, 1) x^{-1/2} dx.$$

The second to last step follows from the Cauchy-Schwarz inequality, and the last step utilizes Hilbert's inequality. The lemma follows immediately. $\square$

We set $G(x, y) = \frac{xy(x^2 + 6xy + y^2)}{(x+y)^3}$. Note that $\frac{\partial^2 G(x,y)}{\partial x \partial y} = \frac{-6xy(x^2 - 6xy + y^2)}{(x+y)^5}$; since this can be negative, and applying the Cauchy-Schwarz inequality requires the functions inside the integral to be positive, we must use $H^+(x, y)$ in place of $\frac{\partial^2 G(x,y)}{\partial x \partial y}$. Calculating the required integral is a simple exercise, and we find $\int_0^\infty H^+(x, 1) x^{-1/2} dx \approx 1.338765....$

3 Analyses and simulations for data compression

3.1 Theoretical results

Let $B_{TS}(\mathbf{p})$ be the expected number of bits that TS(0) needs to encode one symbol in an input sequence that is generated by $\mathbf{p} = (p_1, p_2, \ldots, p_n)$. We assume $p_i > 0$ for all i. In order to analyze $B_{TS}(\mathbf{p})$, we have to specify how an integer i should be encoded. We use a variable length prefix code by Elias [6] which encodes the integer i using $1 + \lfloor \log i \rfloor + 2\lfloor \log(1 + \log i) \rfloor$ bits (all logarithms are base 2). Bentley et al. [3] showed that, using this prefix code, the expected number of bits needed by the MTF algorithm is $B_{MTF}(\mathbf{p}) \leq 1 + H(\mathbf{p}) + 2\log(1 + H(\mathbf{p}))$, for all $\mathbf{p}$. Here $H(\mathbf{p}) = \sum_{i=1}^{n} p_i \log(\frac{1}{p_i})$ is the entropy of the source. We prove similar bounds for TS(0).

Theorem 9. *For any* $\mathbf{p} = (p_1, p_2, \ldots, p_n)$, $B_{TS}(\mathbf{p}) \leq 1 + \overline{H}(\mathbf{p}) + 2\log(1 + \overline{H}(\mathbf{p}))$, *where* $\overline{H}(\mathbf{p}) = \sum_{i=1}^{n} p_i \log(\frac{1}{p_i}) + \log(1 - \sum_{1 \leq i \leq j \leq n} \frac{p_i p_j (p_i - p_j)^2}{(p_i + p_j)^2})$.

As $0 \leq \sum_{1 \leq i \leq j \leq n} \frac{p_i p_j (p_i - p_j)^2}{(p_i + p_j)^2} < 1$, we have the following corollary:

Corollary 10. *For any* $\mathbf{p} = (p_1, p_2, \ldots, p_n)$, $B_{TS}(\mathbf{p}) \leq 1 + H(\mathbf{p}) + 2\log(1 + H(\mathbf{p}))$, *where* $H(\mathbf{p}) = \sum_{i=1}^{n} p_i \log(\frac{1}{p_i})$ *is the entropy of the source.*

Proof of Theorem 9: Let $f(j) = 1 + \log j + 2\log(1 + \log j)$. Consider a fixed symbol x_i, $1 \leq i \leq n$. For $j = 1, \ldots, n$, let q_{ij} be the asymptotic probability that x_i is at position j in TS(0)'s list. The expected number of bits to encode the symbol x_i is $\sum_{j=1}^{n} q_{ij} f(j)$, which, by Jensen's [9] inequality, is at most $f(\sum_{j=1}^{n} q_{ij} j)$. Jensen's inequality states that for any concave function f and any set $\{w_1, w_2, \ldots, w_n\}$ of positive reals, $\sum_{i=1}^{n} w_i f(y_i) \leq f(\sum_{i=1}^{n} w_i y_i)$. Note that $q_{ij} j$ is the asymptotic expected position $e_{TS}(x_i)$ of symbol x_i in TS(0)'s list. Therefore, $B_{TS}(p) \leq \sum_{i=1}^{n} p_i f(e_{TS}(x_i))$. In the following we show that

$$\sum_{i=1}^{n} p_i \log(e_{TS}(x_i)) \leq \overline{H}(p). \qquad (3)$$

Using this inequality, we can easily derive Theorem 9 because $B_{TS}(\mathbf{p}) \leq \sum_{i=1}^{n} p_i f(e_{TS}(x_i)) \leq 1 + \sum_{i=1}^{n} p_i \log(e_{TS}(x_i)) + 2\sum_{i=1}^{n} \log(1 + p_i \log(e_{TS}(x_i)))$ $\leq 1 + \overline{H}(p) + 2\log(1 + \overline{H}(p))$.

We now show inequality (3). By Theorem 1a), we have $e_{TS}(x_i) = \frac{1}{2} + \sum_{j=1}^{n} \frac{p_j^3 + 3p_j^2 p_i}{(p_i + p_j)^3} = \frac{1}{2} + \sum_{j=1}^{n} \frac{p_j}{p_i + p_j} + \sum_{j=1}^{n} \frac{p_i p_j^2 - p_i^2 p_j}{(p_i + p_j)^3}$.

Since $\frac{1}{2} + \sum_{j=1}^{n} \frac{p_j}{p_i + p_j} = \frac{1}{p_i}(\frac{1}{2}p_i + \sum_{j=1}^{n} \frac{p_i p_j}{p_i + p_j}) \leq \frac{1}{p_i}(p_i + \sum_{\substack{j=1 \\ j \neq i}}^{n} p_j) = \frac{1}{p_i}$,

$$\sum_{i=1}^{n} p_i \log(e_{TS}(x_i)) \leq \sum_{i=1}^{n} p_i \log\left(\frac{1}{p_i} + \sum_{j=1}^{n} \frac{p_i p_j^2 - p_i^2 p_j}{(p_i + p_j)^3}\right)$$

$$= \sum_{i=1}^{n} p_i \log\left(\frac{1}{p_i}\right) + \sum_{i=1}^{n} p_i \log\left(1 + \sum_{j=1}^{n} \frac{p_i^2 p_j^2 - p_i^3 p_j}{(p_i + p_j)^3}\right)$$

$$\leq \sum_{i=1}^{n} p_i \log\left(\frac{1}{p_i}\right) + \log\left(1 + \sum_{i=1}^{n} p_i \sum_{j=1}^{n} \frac{p_i^2 p_j^2 - p_i^3 p_j}{(p_i + p_j)^3}\right).$$

The last step follows again from Jensens's inequality. We conclude $\sum_{i=1}^{n} p_i \log(e_{TS}(x_i)) \leq \sum_{i=1}^{n} p_i \log(\frac{1}{p_i}) + \log(1 - \sum_{1 \leq i \leq j \leq n} \frac{p_i p_j (p_i - p_j)^2}{(p_i + p_j)^2})$. □

So far we have assumed that an input sequence S to be compressed is generated by a probability distribution $\mathbf{p} = (p_1, p_2, \ldots, p_n)$. Now consider *any* input sequence S. Let m be the length of S, and let m_i, $1 \leq i \leq n$, be the number of occurrences of the symbol x_i in the string S. Let $A_{TS}(S)$ and $A_{MTF}(S)$ be the average number of bits needed to encode one symbol in the string S using the TS(0) algorithm and the MTF algorithm, respectively. Again, we assume that an integer j is encoded by means of the Elias encoding that requires $1 + \lfloor \log j \rfloor + 2\lfloor \log(1 + \log j) \rfloor$ bits. Bentley *et al.* [3] show that for any input sequence S, $A_{MTF}(S) \leq 1 + H(S) + 2\log(1 + H(S))$ where $H(S) = \sum_{i=1}^{n} \frac{m_i}{m} \log(\frac{m}{m_i})$ is the "empirical entropy" of S. The empirical entropy is interesting because it corresponds to the average number of bits per symbol used by the optimal static Huffman encoding of a sequence; this result implies that MTF encoding is, at worst, almost as good as static Huffman encoding. We can show a similar bound for a variation of TS(0), where after the first occurrence of a symbol it is moved to the front of the list. The proof of the following theorem is omitted.

Theorem 11. *For any input sequence S, $A_{TS}(S) \leq 1 + H(S) + 2\log(1 + H(S))$, where $H(S) = \sum_{i=1}^{n} \frac{m_i}{m} \log(\frac{m}{m_i})$.*

3.2 Simulation results

Our theoretical work suggests that a compression scheme similar to Move-to-front using the TS(0) scheme may provide better performance. In effect, TS(0) is a conservative version of MTF encoding; like MTF-encoding, it responds well to locality of references by moving recently requested items to the front, but it responds more slowly. Understanding this intuition is important to understand where TS(0)-encoding can improve on MTF-encoding: when the locality is very strong, then MTF encoding will perform better, since it responds more aggressively. On the other hand, TS(0) encoding is more effective when the input to be compressed resembles a string generated by a distribution, possibly with a large number of rare items each with a small probability of appearing.

We have tested our theoretical results by implementing simple versions of TS(0)-encoders and decoders for text compression. Our tests use standard documents from the Calgary Compression Corpus [16]. The current goal of these tests is not to develop an all-purpose functional compression system, but merely to demonstrate the potential gains from using TS(0) in place of MTF. The compression is performed by turning the document into a token stream. The tokens are then encoded by their position in the list using standard variable-length prefix encodings given by Elias [6]; each integer i requires $1 + 2\lfloor \log i \rfloor$ bits. We can compare the compression of MTF and TS(0) compression by varying the adaptive discipline of the list.

File	TS(0)		MTF		Original	
	Bytes	% Orig.	Bytes	% Orig.	Bytes	% Orig.
bib	99121	89.09	106478	95.70	111261	100.00
book1	581758	75.67	644423	83.83	768771	100.00
book2	473734	77.55	515257	84.35	610856	100.00
geo	92770	90.60	107437	104.92	102400	100.00
news	310003	82.21	333737	88.50	377109	100.00
obj1	18210	84.68	19366	90.06	21504	100.00
obj2	229284	92.90	250994	101.69	246814	100.00
paper1	42719	80.36	46143	86.80	53161	100.00
paper2	63654	77.44	69441	84.48	82199	100.00
pic	113001	22.02	119168	23.22	513216	100.00
progc	33123	83.62	35156	88.75	39611	100.00
progl	52490	73.26	55183	77.02	71646	100.00
progp	37266	75.47	40044	81.10	49379	100.00
trans	79258	84.59	82058	87.58	93695	100.00

Table 1. MTF vs. TS(0) : Byte-based compression

File	TS(0)		MTF		Original	
	Bytes	% Orig.	Bytes	% Orig.	Bytes	% Orig.
bib	34117	30.66	35407	31.82	111261	100.00
book1	286691	37.29	296172	38.53	768771	100.00
book2	260602	42.66	267257	43.75	610856	100.00
news	116782	30.97	117876	31.26	377109	100.00
paper1	15195	28.58	15429	29.02	53161	100.00
paper2	24862	30.25	25577	31.12	82199	100.00
progc	10160	25.65	10338	26.10	39611	100.00
progl	14931	20.84	14754	20.59	71646	100.00
progp	7395	14.98	7409	15.00	49379	100.00

Table 2. MTF vs. TS(0) : Word-based compression

In the first test ASCII characters (that is, single bytes) constitute the tokens, and the list is initialized in order of character frequency in standard text. The results of Table 1 demonstrate that TS(0)-encoding outperforms MTF-encoding significantly on the sample documents; the improvement is typically 6 – 8 %. Moreover, in all cases TS(0)-encoding beats MTF-encoding. However, this character-based compression scheme performs worse than standard UNIX utilities, such as pack and compress.

In order to make TS(0) and MTF encoding comparable to the standard UNIX utilities, we have to use words as the tokens, which we do in our second test. A word is taken to be a sequence of non-white space characters between white space. This technique assumes that the decompressor has a dictionary consisting of a list of all words in the document; in practice, this dictionary (in compressed or uncompressed form) can be included as part of the compressed document. For convenience, we placed no memory limitation on the compressor or decompressor; that is, the length of the list was allowed to grow as large as necessary. In practice one might wish to devise a more memory-efficient scheme, using the list as a cache as in [3]. The results of Table 2 reflect the compression achieved, including only the token stream and not the dictionary. As one might expect, the gains from TS(0) in this situation are less dramatic but still noticeable.

References

1. S. Albers. Improved randomized on-line algorithms for the list update problem. In *Proc. of the 6th Annual ACM-SIAM Symposium on Discrete Algorithms*, pages 412–419, 1995.
2. J.L. Bentley and C.C. McGeoch. Amortized analyses of self-organizing sequential search heuristics. *Communication of the ACM*, 28:404–411, 1985.
3. J.L. Bentley, D.S. Sleator, R.E. Tarjan and V.K. Wei. A locally adaptive data compression scheme. *Communication of the ACM*, 29:320–330, 1986.
4. M. Burrows and D.J. Wheeler. A block-sorting lossless data compression algorithm. DEC SRC Research Report 124, 1994.
5. F.R.K. Chung, D.J. Hajela and P.D. Seymour. Self-organizing sequential search and Hilbert's inequality. *Proc. 17th Annual Symposium on the Theory of Computing*, pages 217–223, 1985.
6. P. Elias. Universal codeword sets and the representation of the integers. *IEEE Transactions on Information Theory*, 21:194–203, 1975.
7. G.H. Gonnet, J.I. Munro and H. Suwanda. Towards self-organizing linear search. In *Proc. 19th Annual IEEE Symposium on Foundations of Computer Science*, pages 169–174, 1979.
8. D. Grinberg, S. Rajagopalan, R. Venkatesan and V.K. Wei. Splay trees for data compression. In *Proc. of the 6th Annual ACM-SIAM Symposium on Discrete Algorithms*, pages 522–530, 1995.
9. G.H. Hardy, J.E. Littlewood and G. Polya. *Inequalities*. Cambridge University Press, Cambridge, England, 1967.
10. S. Irani. Two results on the list update problem. *Information Processing Letters*, 38:301–306, 1991.
11. R. Karp and P. Raghavan. From a personal communication cited in [13].
12. R. Rivest. On self-organizing sequential search heuristics. *Communication of the ACM*, 19:63–67, 1976.
13. N. Reingold, J. Westbrook and D.D. Sleator. Randomized competitive algorithms for the list update problem. *Algorithmica*, 11(1):15–32, 1994.
14. D.D. Sleator and R.E. Tarjan. Amortized efficiency of list update and paging rules. *Communication of the ACM*, 28:202–208, 1985.
15. B. Teia. A lower bound for randomized list update algorithms. *Information Processing Letters*, 47:5–9, 1993.
16. I.H. Witten and T. Bell. The Calgary/Canterbury text compression corpus. Anonymous ftp from ftp.cpsc.ucalgary.ca : /pub/text.compression/corpus/text.compression.corpus.tar.Z.

Self-Organizing Data Structures with Dependent Accesses

Frank Schulz and Elmar Schömer

Universität des Saarlandes, FB 14, Informatik, Lehrstuhl Prof. Dr. G. Hotz,
Postfach 151150, 66041 Saarbrücken, Germany.
Email: {schulz,schoemer}@cs.uni-sb.de

Abstract. We consider self-organizing data structures in the case where
the sequence of accesses can be modeled by a first order Markov chain.
For the simple-k- and batched-k–move-to-front schemes, explicit formulae for the expected search costs are derived and compared. We use a
new approach that employs the technique of expanding a Markov chain.
This approach generalizes the results of Gonnet/Munro/Suwanda.
In order to analyze arbitrary memory-free move-forward heuristics for
linear lists, we restrict our attention to a special access sequence, thereby
reducing the state space of the chain governing the behaviour of the data
structure.
In the case of accesses with locality (inert transition behaviour), we find
that the hierarchies of self-organizing data structures with respect to the
expected search time are reversed, compared with independent accesses.
Finally we look at self-organizing binary trees with the move-to-root rule
and compare the expected search cost with the entropy of the Markov
chain of accesses.

1 Introduction and summary

We consider the dictionary problem when the query source can be modelled by an
ergodic Markov chain. In (Hotz [7]) it has been shown that a search graph of size
$\Theta(n^2)$ can be constructed such that the expected search time is asymptotically
equal to the entropy of the Markov chain. When the transition behaviour of the
chain exhibits the phenomenon of locality, that expected search time can also
be achieved with $O(n)$ memory.

Here we want to continue that line of research. We examine approximation
schemes that are restricted to use only linear amounts of memory. In literature,
these techniques are known as self-organizing data structures.

Consider a linear list containing the elements $1, 2, \ldots, n$. At each access the
list has to be searched sequentially, and the search cost is assumed to be proportional to the number of comparisons needed to find the requested element.

Heuristics for self-organization perform a reordering of the list and thereby
hope to reduce the search cost for the next accesses. An example is the move-to-
front rule (MTF) which places the requested element at the head of the list and
slides the other elements back one position. Another memory-free heuristic is
the transposition rule (TR) which moves the accessed element one place towards
the head of the list.

1.1 Known results for independent accesses

In the case of independent accesses, the requests are assumed to follow a fixed but unknown probability distribution $(p_1, \ldots, p_n)$ such that element i is accessed with probability p_i. Since no ordering is used for sequential search, we may assume $p_1 \geq \cdots \geq p_n$. When the list is ordered by decreasing access probabilities, the expected search cost is minimized, yielding $M = \sum_{i=1}^n i p_i$. Let $b(j, i)$ denote the asymptotic probability that element j stands in front of element i in the list when the move-to-front heuristic is used. It is easy to see that for $i \neq j$, we have $b(j, i) = p_j / (p_i + p_j)$. It has been shown that the expected search cost under the move-to-front rule is

$$\mu = \sum_i p_i \cdot \left(1 + \sum_{j \neq i} b(j, i)\right) = 1 + 2 \sum_{i < j} \frac{p_i p_j}{p_i + p_j} \leq \frac{\pi}{2} \cdot M ,$$

(see Burville/Kingman [2], Chung/Hajela/Seymour [4]).

A class of heuristics that are allowed to use some memory consists of the k-in-a-row strategies, in combination with a memory-free rule R (Gonnet/Munro/Suwanda [5]). With the simple-k-heuristic, the action of R is performed only if the same element has been requested k times in a row. The batched-k-heuristic groups the accesses into blocks of length k and calls R only if all requests within the block are the same. The k-in-a-row strategies use $O(\log k + \log n)$ memory in order to store the counter value between 1 and k and the element requested last. Let $b_k(j, i)$ and $b_k'(j, i)$ denote the asymptotic probability that j stands in front of i in the list when the simple-k-move-to-front rule or the batched-k-move-to-front rule respectively are applied. It has been shown that

$$b_k(j, i) = \frac{p_j^k \sum_{l=0}^{k-1} p_i^l}{p_j^k \sum_{l=0}^{k-1} p_i^l + p_i^k \sum_{l=0}^{k-1} p_j^l} \tag{1}$$

$$b_k'(j, i) = \frac{p_j^k}{p_i^k + p_j^k} , \tag{2}$$

which allows for the calculation of the expected search time μ. It has been demonstrated that considerable improvements of the search time are achieved even for small values of k, and that $\mu \to M$ as $k \to \infty$. Furthermore, for fixed k batched-k-MTF performs better than simple-k-MTF (see Gonnet/Munro/Suwanda [5]).

The move-to-front heuristic is an example from the class of memory-free move-forward rules. Such a rule is specified by a sequence $1 = m_1 \leq m_2 \leq \cdots \leq m_n$ with $m_i < i$ for all $i = 2, \ldots, n$. The interpretation is that an element at position i of the list is moved to position m_i upon request, leaving the relative ordering of the other elements unchanged. Examples of these rules are:

MOVE-UP(k) the requested element is placed at the head of the list if its position is $i \leq k$, and moved k places up if its position is $i > k$

POS(k) the requested element is moved one place up if its position is $i \leq k$, and moved to position k if its position is $i > k$

SWITCH(k) the accessed element is moved to the head of the list if its position is $i \leq k$, and moved one place up if its position is $i > k$.

A partial ordering of the move-forward rules can be defined as follows: given two rules R and R' specified by $m_1, \ldots, m_n$ and $m'_1, \ldots, m'_n$ respectively, $R \leq R'$ holds if $m'_i \leq m_i$ for all $i = 1, \ldots, n$. The following spectra of rules can be compared:

TR $\leq R \leq$ MTF   for all move-forward rules R

MOVE-UP(k) $\leq$ MOVE-UP($k + 1$)

POS($k + 1$) $\leq$ POS(k)

SWITCH(k) $\leq$ SWITCH($k + 1$)

For the special access distribution $p_1 = \alpha$, $p_2 = \cdots = p_n = \beta$, it has been shown by Lam [12] that the expected search times $\mu(R)$ and $\mu(R')$ of two move-forward rules R and R' obey

$$R \leq R' \implies \mu(R) \leq \mu(R') \ , \tag{3}$$

which entails the corresponding relations of expected search time for the spectra described above (see Kan/Ross [8], Lam [12], Phelps/Thomas [14], Tenenbaum/Nemes [16]).

1.2   Known results for dependent accesses

In the case of dependent accesses we assume that the sequence of requests can be modeled by an ergodic first-order Markov chain with transition matrix $P = (p_{ij})$ and stationary distribution $(q_1, \ldots, q_n)$. Let $M = (m_{ij})$ denote the matrix of mean first-passage times. It has been shown that the asymptotic probability $b(j, i|i)$ that j stands in front of i in the list when i is requested and the move-to-front rule is given by $b(j, i|i) = 1/(q_i(m_{ij} + m_{ji}))$, and hence the expected search time is

$$\mu = \sum_{i=1}^{n} q_i \cdot \left(1 + \sum_{j \neq i} b(j, i|i)\right) = 1 + 2 \sum_{i<j} \frac{1}{m_{ij} + m_{ji}}$$

(see Lam/Leung/Siu [13]). Since the heuristic does not have a memory, a query source with memory can always outperform it. There are situations, however, in which the heuristic performs very well. These situations can be described by the phenomenon of *locality*: a small subset of elements is requested for a long time before the access sequence switches to another small subset.

Recently it has been demonstrated that in some of these cases, the move-to-front heuristic is optimal in that its expected search time is not greater than the expected search time of any other sequential search strategy, even if that search strategy may use arbitrary amounts of memory (Chassaing [3]).

1.3 New results

We consider the simple-k- and batched-k-move-to-front schemes and derive formulae for the asymptotic probabilities $b_k(j, i|i)$ and $b'_k(j, i|i)$ that j stands in front of i when i is requested. This is achieved by expanding the chain of accesses to k-tuples and looking for the occurrence of tuples which consist solely of i's or j's. In the special case of independent accesses the results cited above are obtained.

As an example, we calculate the expected search cost for the access sequence defined by

$$p_{ij} = \left\{ \begin{array}{ll} \alpha & : \quad i = j \\ \beta & : \quad i \neq j \end{array} \right\}$$

with $0 \leq \alpha < 1$ and $\beta = (1 - \alpha)/(n - 1)$. The phenomenon of locality appears for $\alpha > \beta$. In this case, we observe that the expected search time gets worse for increasing k and approaches $(n + 1)/2$ as $k \to \infty$. Furthermore, simple-k-MTF performs better than batched-k-MTF. Hence the ordering of strategies with respect to the expected search time is reversed, compared with independent accesses.

As for the analysis of the class of memory-free move-forward rules, we restrict our attention to the same access sequence. This enables us to reduce the state space of the Markov chain governing the behaviour of the data structure. Given two move-forward rules R, R' and $\alpha \geq \beta$, the expected search times satisfy

$$R \leq R' \implies \mu(R') \leq \mu(R) .$$

As a consequence, we find again that the hierarchy of heuristics is exactly reversed, compared with independent accesses (see (3)). For the first time, memory-free heuristics other than move-to-front are analyzed in the case of Markovian access sequences.

Finally we look at self-organizing binary search trees, and compare the expected search time with the entropy of the chain of accesses.

2 k-in-a-row heuristics

In order to analyze the simple-k-MTF heuristic, we observe

Lemma 1. *Using simple-k-move-to-front, the element j stands in front of i if the last sequence of k subsequent accesses to j has occurred after the last sequence of k subsequent accesses to i.*

At any instant in time, the current form of the data structure depends on its history. Therefore, we want to employ backward analysis and look at the past. Hence we reformulate the lemma:

When l is requested, j stands in front of i if we trace back the chain of accesses starting from l and encounter k subsequent accesses to j before we encounter k subsequent accesses to i.

Two techniques from Markov chain theory are needed: reversal of time and expansion of the chain (see for example Kemeny/Snell [10]).
When P is an ergodic Markov chain with state space $S = \{1, \ldots, n\}$ and stationary distribution $(q_1, \ldots, q_n)$, the time-reversed chain $\hat{P}$ is defined by its transition probabilities

$$\hat{p}_{ij} = \frac{q_j p_{ji}}{q_i} .$$

For $k > 1$, the expanded chain $\tilde{P}$ has the state space

$$\tilde{S} = \{[i_1 \ldots i_k] \mid i_j \in S, p_{i_1 i_2} \cdot \ldots \cdot p_{i_{k-1} i_k} > 0\} .$$

A transition can be viewed as sliding a window of size k over the original sequence:

$$\tilde{p}_{[i_1 \ldots i_k][j_1 \ldots j_k]} = p_{i_k j_k} \cdot \delta_{[i_2 \ldots i_k][j_1 \ldots j_{k-1}]} ,$$

where

$$\delta_{[a_1 \ldots a_j][b_1 \ldots b_j]} = \left\{ \begin{array}{ll} 1 & : \quad a_i = b_i \text{ for } i = 1, \ldots, j \\ 0 & : \quad \text{otherwise} \end{array} \right\}$$

denotes the Kronecker symbol. The fundamental matrix of an ergodic chain is defined as $Z = (z_{ij}) = (I - P + Q)^{-1}$, where Q is the limiting transition matrix of the chain [10]. From this, the first passage times can be calculated as $m_{ij} = (\delta_{ij} - z_{ij} + z_{jj})/q_j$. We want to express the first passage times of the time-reversed expanded chain with quantities of the original chain. This gives:

$$\hat{m}_{[i_1 \ldots i_k][j_1 \ldots j_k]} = z_{j_k j_1}/q_{j_1} - z_{j_k i_1}/q_{i_1}$$
$$+ \frac{1 - \sum_{l=1}^{k-2} q_{j_1} p_{j_1 j_2} \cdots p_{j_l j_{l+1}} (\delta_{j_{l+1} i_1} \cdots \delta_{j_k i_{k-l}}/q_{i_1} - \delta_{j_{l+1} j_1} \cdots \delta_{j_k j_{k-l}}/q_{j_1})}{q_{j_1} p_{j_1 j_2} \cdots p_{j_{k-1} j_k}} .$$

For an ergodic chain, the probability of starting in state l and reaching state j without passing through the taboo state i is given by

$${}_i f^*_{lj} = \frac{m_{li} + m_{ij} - m_{lj}}{m_{ij} + m_{ji}}$$

(see Kemperman [11]). Let $b([j \ldots j], [i \ldots i] \mid [l_1 \ldots l_{k-1}i])$ be the asymptotic probability that $[j \ldots j]$ appears before $[i \ldots i]$ when we start the time-reversed expanded chain in state $[l_1 \ldots l_{k-1}i]$. With the above formula, this can be calculated from

$$b([j \ldots j], [i \ldots i] | [l_1 \ldots l_{k-1}i]) = \frac{\hat{m}_{[l_1 \ldots l_{k-1}i][i \ldots i]} + \hat{m}_{[i \ldots i][j \ldots j]} - \hat{m}_{[l_1 \ldots l_{k-1}i][j \ldots j]}}{\hat{m}_{[i \ldots i][j \ldots j]} + \hat{m}_{[j \ldots j][i \ldots i]}} .$$

Given these probabilities, and taking the mean over all $l_1, \ldots, l_{k-1}$, we get the asymptotic probability $b_k(j, i|i)$ that j stands in front of i when i is accessed and the simple-k-move-to-front rule is used. The probability of occurence of $[l_1 \ldots l_{k-1}i]$ is the probability to trace back the sequence from i and encountering $l_{k-1}, \ldots, l_1$ subsequently. Hence

$$b_k(j, i|i) = \sum_{l_1 \ldots l_{k-1}} \hat{p}_{il_{k-1}} \cdot \hat{p}_{l_{k-1} l_{k-2}} \cdots \hat{p}_{l_2 l_1} \cdot b([j \ldots j], [i \ldots i] | [l_1 \ldots l_{k-1}i]) .$$

Let $p_{ij}^{(t)}$ denote the probability of going from i to j in exactly t steps. Combining these considerations, we get the following result.

Theorem 2.

$b_k(j, i|i) =$

$$\frac{q_j p_{jj}^{k-1} \left[\sum_{t=0}^{k-2} p_{ii}^t + p_{ii}^{k-1} \sum_{t=1}^{k-1} (p_{ji}^{(t)} - p_{ii}^{(t)}) + p_{ii}^{k-1} \left(q_i m_{ji} + \sum_l q_l p_{li}^{(k-1)} (m_{il} - m_{jl}) \right) \right]}{q_i p_{ii}^{k-1} \sum_{t=0}^{k-2} p_{jj}^t + q_j p_{jj}^{k-1} \sum_{t=0}^{k-2} p_{ii}^t + q_i q_j p_{ii}^{k-1} p_{jj}^{k-1} (m_{ij} + m_{ji})}$$

See [15] for the details of the derivation. The expected search time under the simple-k-move-to-front rule is $\mu = 1 + \sum_{j \neq i} q_i \cdot b_k(j, i|i)$.

For independent accesses, $p_{ij} = q_j$, hence $p_{ij}^{(t)} = q_j$ for $t \geq 1$ and $m_{ij} = 1/q_j$, thus formula (1) for $b_k(j, i)$ is obtained.

For the batched-k-move-to-front, we observe

Lemma 3. *Employing batched-k-move-to-front, the element j stands in front of i if the last block of accesses to j has occured after the last block of accesses to i.*

In order to apply backward analysis, we use the reformulation:

When l is requested, j stands in front of i if we trace back the chain of accesses starting from the current block (the block containing l) and encounter a block of j's before a block of i's has appeared.

Simplifying the calculation, we use the transition matrix that combines k steps into one: $\bar{P} = P^k$. Let $\bar{Z} = (\bar{z}_{ij}) = (I - \bar{P} + Q)^{-1}$ be the corresponding fundamental matrix, where Q is the limiting transition matrix of the ergodic chain (Kemeny/Snell [10]) .

Using a similar technique as above, we define an expanded chain with transition probabilities

$$\tilde{p}_{[i_1 \ldots i_k][j_1 \ldots j_k]} = p_{i_k j_1} \cdot p_{j_1 j_2} \cdots p_{j_{k-1} j_k} .$$

Then we calculate the asymptotic probability $b([j \ldots j], [i \ldots i]|l)$ that $[j \ldots j]$ appears before $[i \ldots i]$ appears, when the time-reversed expanded chain is started from $[l \ldots]$. As no tuples overlap, it is sufficient to know the first element l of the current block.

$$b([j \ldots j], [i \ldots i]|l) = \frac{\hat{m}_{[l \ldots][i \ldots i]} + \hat{m}_{[i \ldots i][j \ldots j]} - \hat{m}_{[l \ldots][j \ldots j]}}{\hat{m}_{[i \ldots i][j \ldots j]} + \hat{m}_{[j \ldots j][i \ldots i]}}$$

In the current block, the current access to i can appear at position $1, 2, \ldots, k$, each with probability $1/k$. When it occurs at position m within the current block, we go $m - 1$ steps back to reach the first element of the block. This element is l with probability $\hat{p}_{il}^{(m-1)}$. Hence

$$b'_k(j, i|i) = \frac{1}{k} \sum_{m=1}^{k} \sum_l \hat{p}_{il}^{(m-1)} \cdot b([j \ldots j], [i \ldots i]|l)$$

Theorem 4.

$$b'_k(j, i|i) = \frac{q_j p_{jj}^{k-1} - q_j p_{ii}^{k-1} p_{jj}^{k-1} \sum_x \left(\bar{z}_{jx} - \bar{z}_{ix}\right)\left(p_{xi} - \sum_{m=1}^{k} p_{xi}^{(m)}/k\right)}{q_i p_{ii}^{k-1} + q_j p_{jj}^{k-1} - p_{ii}^{k-1} p_{jj}^{k-1} \sum_x \left(\bar{z}_{jx} - \bar{z}_{ix}\right)\left(q_j p_{xi} - q_i p_{xj}\right)}$$

For independent accesses, formula (2) for $b'_k(j, i)$ follows as a corollary.

We want to illustrate the results with the access sequence

$$p_{ij} = \begin{Bmatrix} \alpha & : & i = j \\ \beta & : & i \neq j \end{Bmatrix} \tag{4}$$

with $0 \leq \alpha < 1$ and $\beta = (1 - \alpha)/(n - 1)$. For $\alpha > 1/n$, this sequence exhibits the phenomenon of elementwise locality, the degree of locality can be adjusted via the parameter α.

Theorem 5. *For the access sequence (4), the simple-k-move-to-front rule produces the expected search cost*

$$\mu = 1 + \frac{n(n-1)(1-\alpha^k)}{2(n-\alpha^{k-1})},$$

and the batched-k-move-to-front rule

$$\mu' = 1 + \frac{n-1}{2} \cdot \left(1 - \frac{\alpha^{k-1}(n\alpha - 1)\left((n-1)^k - (n\alpha - 1)^k\right)}{k(n - n\alpha)\left((n-1)^k - (n\alpha - 1)^k + \alpha^{k-1}(n\alpha - 1)(n - 1)^{k-1}\right)}\right).$$

One easily sees that for $\alpha > 1/n$ (locality of references), the expected search time is monotone increasing in k. For all $\alpha < 1$, it approaches $(n + 1)/2$ as $k \to \infty$. The expected search time in dependence of α looks as follows:

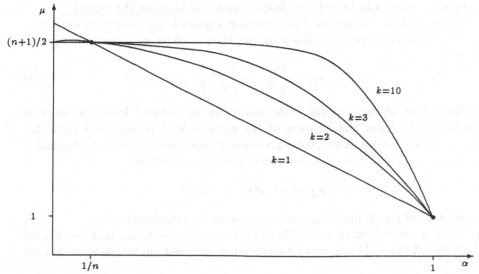

Comparing μ and μ', and using Bernoulli's inequality, we find

Theorem 6. *For the access sequence (4) and all $n \geq 2$, $k \geq 2$, the simple-k rule performs better than the batched-k rule in the situation of locality:*

$$\alpha > \beta \implies \mu < \mu' \ .$$

Again, the order is reversed, compared with independent accesses.

3 Move-forward rules

In order to analyze arbitrary move-forward rules, one has to look at the Markov chain that describes the states of the data structure and the transitions between them. Each permutation of the list is a state, and a transition from state π to state σ is possible if there is an element i such that σ is obtained when the action of the heuristic is applied to π upon request of i.

In the case of independent accesses that are identically distributed with the distribution $(p_1, \ldots, p_n)$, this transition has the probability $p_{\pi\sigma} = p_i$.
When accesses are dependent and governed by the transition matrix $P = (p_{ij})$, we have to consider the extended Markov chain with states $(\pi|i)$, where π is the current permutation of the list and i is the element that will be requested next. Then the transitions have the probability

$$P_{(\pi|i)(\sigma|j)} = \begin{cases} p_{ij} & : \quad \pi \text{ is permuted to } \sigma \text{ after the request of } i \\ 0 & : \quad \text{otherwise} \end{cases} .$$

In the situation of independent accesses, several authors have restricted the accesses to the distribution $p_1 = \alpha, p_2 = \cdots = p_n = \beta$ (see [8, 12, 14, 16]). The benefit is that it now suffices to know the position of element 1. Thus, the size of the state space can be reduced from $n!$ to n (by lumping the chain).
For dependent accesses, we take a similar approach. By restricting our attention to the access sequence defined by the Markov chain with transition probabilities

$$p_{ij} = \begin{cases} \alpha & : \quad i = j \\ \beta & : \quad i \neq j \end{cases} , \tag{5}$$

we claim that the size of the state space can be reduced from $nn!$ to n. It is sufficient to know the position of the element that is requested next. Let $q(i_1, \ldots, i_n | i_j)$ be the asymptotic probability that the list is in configuration $\pi = (i_1, \ldots, i_n)$ and i_j is requested next. Then the definition

$$q(j) = n! \cdot q(i_1, \ldots, i_n | i_j)$$

is well-defined for all move-forward rules, as we will demonstrate.
Let R be a move-forward rule defined by $1 = m_1 \leq \cdots \leq m_n$ with $m_i < i$ for $i \geq 2$. Let $W(k) = \{l \mid m_l = k\}$ be the set of all positions whose elements are

moved to position k upon request. The steady-state equations for the extended Markov chain are

$$q(i_1,\ldots,i_n|i_j) = \sum_{k=1}^{n} \sum_{l\in W(k)} p_{i_k i_j} \cdot q(i_1,\ldots,i_{k-1},i_{k+1},\ldots,i_l,i_k,i_{l+1},\ldots,i_n|i_k)$$

with the convention $q(i_1,\ldots,i_{k-1},i_{k+1},\ldots,i_l,i_k,i_{l+1},\ldots,i_n|i_k) = q(i_1,\ldots,i_n|i_1)$ in the case $k = l = 1$.

Using the above reduction and noting $\sum_{l=1}^{n} q(l) = 1$, we obtain

$$q(j) = \beta + (\alpha - \beta) \sum_{l\in W(j)} q(l) \qquad \text{for } j = 1,\ldots,n \qquad (6)$$

This system of equations is consistent and possesses a unique solution. The $n \times n$ transition matrix P of the extended but reduced chain looks as follows:

$$P = \begin{pmatrix} \alpha & \beta & \beta & \beta & \cdots & \beta \\ \vdots & \beta & \beta & \beta & \cdots & \beta \\ \alpha & \beta & \beta & \beta & \cdots & \beta \\ \beta & \alpha & \beta & \beta & \cdots & \beta \\ \beta & \vdots & \beta & \beta & \cdots & \beta \\ \beta & \alpha & \beta & \beta & \cdots & \beta \\ & & \vdots & & & \\ \beta & \cdots & \beta & \alpha & \beta & \cdots \end{pmatrix} \qquad (7)$$

In column k, exactly the elements p_{ik} with $i \in W(k)$ are α, all others are β.

Since $q(j)$ is the probability that the element on position j will be requested next, the expected search time is $\mu = \sum_{j=1}^{n} jq(j)$.

Theorem 7. *Given two move-forward rules R, R', their expected search times under the access sequence (5) and $\alpha \geq \beta$ obey*

$$R \leq R' \implies \mu(R') \leq \mu(R) .$$

For the proof we need the concepts of vector dominance and monotone matrices (see Keilson/Kester [9] for the background, Lam [12] for the application to independent accesses).

A probability vector $q = (q(1),\ldots,q(n))$ dominates another probability vector $q' = (q'(1),\ldots,q'(n))$, denoted by $q' \prec q$, if

$$\sum_{j=i}^{n} q'(j) \leq \sum_{j=i}^{n} q(j) \qquad \text{for all } i = 1,\ldots,n .$$

A Markov chain with transition matrix $P = (p_{ij})$ is stochastically monotone if for all fixed s, the partial sum of the ith row, $P_{is} = \sum_{j=1}^{s} p_{ij}$, is monotone non-increasing in i.

Lemma 8. *Let P and P' be two ergodic and stochastically monotone Markov chains with stationary probabilities q and q' respectively. Then*

$$qP' \prec q \implies q' \prec q.$$

(see [9, 12]).

Now given two move-forward rules $R \leq R'$ under the access sequence (5), we show that in the situation $\alpha \geq \beta$, their transition matrices P, P' given by (7) are stochastically monotone and their stationary probabilities obey $qP' \prec q$. With the lemma, we get $q' \prec q$: $\sum_{j=i}^{n} q'(j) \leq \sum_{j=i}^{n} q(j)$ for all i. Summing over i yields $\sum_{i=1}^{n} iq'(i) \leq \sum_{i=1}^{n} iq(i)$, which is $\mu(R') \leq \mu(R)$.

From the definition of P it is easy to see that P is stochastically monotone in the case $\alpha \geq \beta$.

Since q is the stationary distribution for P, $q = qP$, and $qP' \prec q$ is equivalent to $qP' \prec qP$. Let

$$V(i) = \bigcup_{j=i}^{n} W(j) = \{l \mid m_l \geq i\} .$$

We demonstrate $qP' \prec qP$ in the situation $\alpha \geq \beta$. For all $i = 1, \ldots, n$, the following is equivalent

$$\sum_{j=i}^{n} (qP')_j \leq \sum_{j=i}^{n} (qP)_j$$

$$\sum_{j=i}^{n} \left(\beta + (\alpha - \beta) \sum_{l \in W'(j)} q(l) \right) \leq \sum_{j=i}^{n} \left(\beta + (\alpha - \beta) \sum_{l \in W(j)} q(l) \right)$$

$$\sum_{j=i}^{n} \sum_{l \in W'(j)} q(l) \leq \sum_{j=i}^{n} \sum_{l \in W(j)} q(l)$$

$$\sum_{l \in V'(i)} q(l) \leq \sum_{l \in V(i)} q(l)$$

which is true since $m'_l \leq m_l$ for all $l = 1, \ldots, n$ entails $V'(i) \subseteq V(i)$ for all $i = 1, \ldots, n$. This completes the proof. $\quad\square$

4 Self-organizing binary search trees

In this section we briefly consider self-organizing search trees with the move-to-root heuristic. For independent accesses that are distributed with the distribution $(p_1, \ldots, p_n)$, the move-to-root heuristic achieves the expected search time

$$\mu = 1 + \sum_{i<j} \frac{p_i p_j}{p_i + \cdots + p_j} \leq 1 + 2\ln 2 \cdot H(p)$$

where $H(p) = -\sum_i p_i \log_2 p_i$ is the entropy of the access distribution (see [1] or [6]).

For dependent accesses, we want to apply backward analysis. Hence we reformulate the lemma given in (Allen/Munro [1]).

Lemma 9. *Let $i < j$. When k is requested, j is an ancestor of i in the tree, if we trace back the sequence of accesses starting from k and encounter j before we encounter any of the elements $i, \ldots, j-1$. Similarly, when k is requested, i is an ancestor of j, if we trace back the access sequence starting from k and encounter i before we encounter any of the elements $i+1, \ldots, j$.*

In the case of move-to-front, we had to deal with singleton taboo sets. Now the taboo sets $H = \{i, \ldots, j-1\}$ and $H = \{i+1, \ldots, j\}$ respectively are of size $|j - i|$.

The probability $d(j, i|i)$ that j is an ancestor of i in the tree when i is accessed is given by

$$d(j, i|i) \;=\; {}_H \hat{f}_{ij}^* \;=\; \frac{\det((\delta_{ab} - 1)\hat{m}_{ab} + \hat{m}_{ib} + \hat{m}_{aj} - \hat{m}_{ij})_{a,b \in H}}{\det((\delta_{ab} - 1)\hat{m}_{ab} + \hat{m}_{jb} + \hat{m}_{aj})_{a,b \in H}}$$

with $H = \{i, \ldots, j-1\}$ and $\hat{m}_{ij}$ being the mean first passage time in the time-reversed chain of accesses (for $d(i, j|j)$ similarly), see (Kemperman [11]). The expected search time is $\mu = 1 + \sum_{j \neq i} q_i \cdot d(j, i|i)$. Unfortunately we did not succeed in deriving a nice expression giving μ in terms of the Markov chain of accesses.

We restrict our attention to a special sequence of requests. Given a probability distribution $q = (q_1, \ldots, q_n)$, all $q_i > 0$, we define

$$P = \lambda \cdot Q + (1 - \lambda) \cdot I \tag{8}$$

where Q is the matrix of independent transitions with distribution q, I is the identity matrix, and $0 < \lambda \leq 1$. The stationary distribution of the ergodic chain P is q.

Theorem 10. *Let $\mu(Q)$ be the expected search cost for independent accesses with distribution q. The expected search cost under the move-to-root heuristic for the access sequence P defined by (8) is*

$$\mu(P) = \lambda \cdot \mu(Q) + (1 - \lambda) \ . \tag{9}$$

This is exactly the same formula that was found for linear lists and the move-to-front rule under that access sequence (Lam/Leung/Siu [13]). See [15] for the derivation.

As a corollary, for all distributions q with $q_i > 0$ and all $\varepsilon > 0$ we can define a Markov chain with stationary distribution q such that the expected search cost under the move-to-root heuristic is $\mu(P) < 1 + \varepsilon$, by choosing λ sufficiently small, e.g. $\lambda = \min(1, \varepsilon/(\mu(Q) - 1))$.

Theorem 11. *For the Markov chain P defined by (8), the expected search cost is bounded by*

$$
\begin{aligned}
\mu(P) \;&\leq\; 1 + 2\ln 2 \cdot \Big(H(P) + 1 - H(\lambda, 1 - \lambda)\Big) \\
&\leq\; 1 + 2\ln 2 \cdot \Big(H(P) + 1\Big) \\
&\leq\; 3 + 2\ln 2 \cdot \Big(\mu_{\text{opt}} + \log \mu_{\text{opt}}\Big)
\end{aligned}
$$

where $H(P)$ is the entropy of the chain, $H(.,.)$ is the entropy function and μ_{opt} is the asymptotically optimal search time which can be realised upon knowledge of P (see Hotz [7]).

The factor $2\ln 2$ is the same as for independent accesses. This is interesting since neither $H(P)$ nor μ_{opt} reflect the locality of accesses. It would be desirable to establish a bound on the expected search cost of move-to-root for a broader range of access sequences.

References

1. B. Allen and I. Munro. Self-organizing binary search trees. *Journal of the ACM*, 25(4):526–535, 1978.
2. P. J. Burville and J. F. C. Kingman. On a model for storage and search. *Journal of Applied Probability*, 10:697–701, 1973.
3. Ph. Chassaing. Optimality of move-to-front for self-organizing data structures with locality of references. *Annals of Applied Probability*, 3(4):1219–1240, 1993.
4. F. R. K. Chung, D. J. Hajela, and P. D. Seymour. Self-organizing sequential search and Hilbert's inequalities. In *Proceedings of the 17th Annual ACM Symposium on Theory of Computing*, pages 217–223, New York, 1985.
5. G. H. Gonnet, J. I. Munro, and H. Suwanda. Exegesis of self-organizing linear search. *SIAM Journal of Computing*, 10(3):613–637, 1981.
6. G. H. Gonnet, and R. Baeza-Yates. *Handbook of Algorithms and Data Structures*. Addison-Wesley, 1991.
7. G. Hotz. Search tress and search graphs for markov sources. *Journal of Information Processing and Cybernetics*, 29:283–292, 1993.
8. Y. C. Kan and S. M. Ross. Optimal list order under partial memory constraints. *Journal of Applied Probability*, 17:1004–1015, 1980.
9. J. Keilson and A. Kester. Monotone matrices and monotone markov processes. *Stochastic Processes and Applications*, 5:231–241, 1977.
10. J. G. Kemeny and J. L. Snell. *Finite Markov Chains*. Van Nostrand, Princeton, 1960.
11. J. H. B. Kemperman. *The First Passage Problem for a Stationary Markov Chain*. University Press, Chicago, 1961.
12. K. Lam. Comparison of self-organizing linear search. *Journal of Applied Probability*, 21:763–776, 1984.
13. K. Lam, M. Y. Leung, and M. K. Siu. Self-organizing files with dependent accesses. *Journal of Applied Probability*, 21:343–359, 1984.
14. R. I. Phelps and L. C. Thomas. On optimal performance in self-organizing paging algorithms. *Journal Inf. Optimization Science*, 1:80–93, 1980.
15. F. Schulz. Self-organizing data structures with dependent accesses. MSc Thesis supervised by Prof. Dr. G. Hotz, University of Saarbrücken, 1995 (in German). See http://hamster.cs.uni-sb.de/~schulz/
16. A. M. Tenenbaum and R. M. Nemes. Two spectra of self-organizing sequential search algorithms. *SIAM Journal of Computing*, 11:557–566, 1982.

Lopsided Trees: Analyses, Algorithms, and Applications
(extended abstract)

Vicky Siu-Ngan Choi[1] and Mordecai Golin[1]

Hong Kong University of Science and Technology
Clearwater Bay, Kowloon, Hong Kong
email: {vicky,golin}@cs.ust.hk

Abstract. A lopsided tree is a rooted, ordered, tree in which the length c_i of an edge from a node to its ith child depends upon the value of i. In this paper we examine three problems on such trees and show how to apply the results to the analysis of problems in data compression (Varn-codes) and distributed computing (broadcasting in the Postal model of communication).
Keywords: Varn Codes, Fibonacci Recurrences, Mellin Transforms, Postal Model.

1 Introduction

In this paper we discuss some combinatorial properties of lopsided trees. A tree is said to be *lopsided* if it is a rooted, ordered (i.e., the children of a node are ordered) tree with maximum arity r, in which the length of an edge from a parent to its ith child is c_i where $c_1 \leq c_2 \leq \cdots \leq c_r$ are r fixed positive reals. Figure 1 illustrates an example of one infinite and two finite lopsided trees. The name *lopsided trees* was only coined in 1989 by Kapoor and Reingold [10] but the trees themselves have been implicitly present in the literature at least since 1961 when Karp [11] used them to model minimum-cost prefix-free (Huffman) codes in which the length of the edge of the letters in the encoding alphabet were unequal; c_i represented the length of the ith letter in the encoding alphabet (the trees were later used in [8] to design a more efficient algorithm for the same problem).

For fixed $c_1 \leq c_2 \leq \cdots \leq c_r$ we study three problems on these trees:

- Given n, efficiently constructing the lopsided tree with n leaves that has minimal cost, where the cost of a tree is its external path length, i.e., the sum of the lengths of the paths from the root to all of the leaves.
- Expressing the cost of the minimal cost tree as a function of n and the c_i.
- Calculating exactly how many nodes of depth $\leq x$ exist in the infinite tree.

To motivate these problems we first introduce the concept of a Varn-code [18] [17]. Suppose that we wish to construct a prefix-free encoding of n symbols using an encoding alphabet of r letters, $\Sigma = \{\alpha_1, \ldots, \alpha_r\}$ in which the length of character α_i is c_i, where the c_is may all be different. As an example consider the Morse code alphabet $\Sigma = \{., -\}$ in which the length of a "dash" may be longer than that of a "dot". By a prefix-free encoding we mean a set of n strings $\{\omega_1, \ldots, \omega_n\} \subseteq \Sigma^*$ in which no ω_i is a prefix of any ω_j.

If a symbol is encoded using string $\omega = \alpha_{i_1}\alpha_{i_2}\ldots\alpha_{i_l}$ then $cost(\omega) = \sum_{j \leq l} c_{i_j}$ is the length of the string. For example if $r = 2$, $\Sigma = \{0,1\}$ and $c_1 = c_2 = 1$ then the cost of the string is just the number of bits it contains as in standard Huffman encoding.

Now suppose that the n symbols to be encoded are known to occur with equal frequency. The *cost* of the code is then defined to be $\sum_{i \leq n} cost(\omega_i)$ (which divided by n is the average cost of transmitting a symbol). Given $c_1 \leq c_2 \leq \cdots \leq c_r$ a *Varn-code* for n symbols is a minimum-cost code. Varn codes have been extensively studied in the compression and coding literature; [17] contains an up-to-date description of what is currently known about them.

Such codes can be naturally modelled by lopsided trees in which the length of the edge from a node to its ith child is c_i; we call such an edge an ith *edge*. Suppose that v is a leaf in a lopsided tree and the unique path from the tree's root to v first traverses an i_1^{st} edge then an i_2^{nd} edge and so on up to an i_l^{th} edge. We can then associate with this leaf the codeword $\omega = \alpha_{i_1}\alpha_{i_2}\ldots\alpha_{i_l}$. The cost of this codeword is exactly the same as the depth of v in the tree, i.e., $\sum_{j \leq l} c_{i_j}$. Using this correspondence, every tree with n leaves corresponds to a set of n codewords and vice-versa; the cost of the code is exactly equal to the external path length of the tree which we will henceforth call the *cost* of the tree. This correspondence is extensively used, for example, in the analysis of Huffman codes. A lopsided tree with minimal cost for n leaves will be called an *optimal tree*.

With this correspondence and notation we see that the problems of constructing a Varn code and calculating its cost are equivalent to those of constructing an optimal tree and calculating its cost. Under these two different guises these problems have been extensively studied in both the coding/compression and computer science communities. Algorithms for finding such trees appear in [18] [6] [14] [9]; the current fastest one runs in $O(n \log^2 r)$ time [9]. The bottleneck in all of these algorithms is that there are many possible trees with n leaves and all the algorithms need to construct large sets of such trees in order to find the minimal-cost one. Analyses of the costs of Varn codes or, equivalently, of the costs of optimal lopsided trees appear in [16] [13] [4] [5] [1] [10] [17]. As the authors of these papers mention, their various analyses are only tight for some special cases but in most cases only provide loose upper and lower bounds.

The major result of this paper is a new way of looking at optimal lopsided trees. We assume that $c_1 \leq c_2 \leq \cdots \leq c_r$ are fixed and examine how the structure of optimal lopsided trees evolve as n increases. We prove that the trees evolve in a very regular and understandable fashion. This will permit us to know what an optimal tree for n nodes looks like without having to search through a large collection of lopsided trees.

A first consequence of this result is a pair of new algorithms for constructing optimal trees or Varn-codes that both improve upon the algorithms in the literature. The first algorithm runs in $O(n \log r)$ time beating the old $O(n \log^2 r)$ bound. The second algorithm runs in $n + O(C \log r \log^r n)$ time where C is a constant dependent upon the c_i. It is therefore not directly comparable with the algorithms in the literature (although, for fixed $c_1, \ldots, c_r$, it beats them for large enough n.).

A second and more important consequence of this result is an exact analysis of the cost of optimal trees or Varn-codes. Because we know exactly how the structure of the optimal tree evolves as n grows we are able to calculate how the cost of the optimal tree increases with n. This provides, once and for all, a unified analysis that gives asymptotically exact bounds for Varn-code costs in all cases.

In order to perform the above calculation it is necessary to calculate exactly how many nodes in the infinite tree have depth at most x for every x. This sub-problem

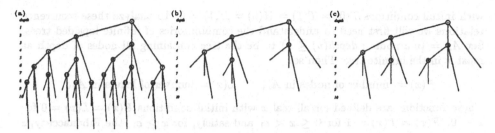

Fig. 1. $(c_1, c_2, c_3) = (2, 3, 6)$: (a) The infinite tree; (b) A tree with 11 leaves, 5 internal nodes and cost 83; (c) A tree with 11 leaves, 6 internal nodes, and cost 82.

reduces to analyzing Fibonacci-type recurrences of the form

$$L(x) = L(x - c_1) + L(x - c_2) + \cdots L(x - c_r).$$

While the solution to the case $r = 2$ is implicit in the work of Fredman and Knuth [7] and, later, Pippenger [15], the general r-ary case does not seem to have been addressed previously. We show how to use Mellin transform and singularity analysis techniques to solve these equations. Surprisingly, the solutions to these equations will have qualitatively different behaviors depending upon whether the c_i are rationally related to each other or not. This difference in behavior will be reflected in the ways in which the trees evolve.

The analyses of these equations are also of independent interest. We illustrate this by describing an immediate application in the derivation of exact bounds on the time needed for broadcasting in the Postal-Model of computation, improving the bounds given in [2].

Note: In this extended abstract we only present sketches of the various proofs. Details may be found in [3] or the full version of this paper.

1.1 The Recurrence Relations

We start by describing the infinite lopsided tree. In what follows we assume that the r parameters $0 < c_1 \leq c_2 \leq \cdots \leq c_r$ are given and fixed.

Definition 1

1. *The* infinite lopsided r-ary tree *is the infinite, rooted, r-ary tree such that the length of the edge connecting a node to its ith child is c_i. A* lopsided tree *is any subtree T of the infinite lopsided r-ary tree containing the root.*

2. *The* depth *of a node u is the sum of the lengths of the edges on the path connecting the root of the infinite tree to u. The* external path length *or* cost *of lopsided tree T is $C(T) = \sum_v$ a leaf of T $depth(v)$. The* height *of T is $H(T) = \max_{u \in T} depth(u)$.*

See Figure 1. Let $T(n)$ be the minimal cost of a lopsided tree with n leaves and let $H(n)$ be the minimum height of a lopsided tree with n leaves. It is not difficult to see that $T(n)$ and $H(n)$ satisfy the following recurrence relations:

$$T(n) = \min_{\substack{n_1 + n_2 + \cdots + n_r = n \\ n_1, n_2, \ldots, n_r \geq 0}} \sum_{1 \leq i \leq r} [T(n_i) + c_i n_i] \qquad H(n) = \min_{\substack{n_1 + n_2 + \cdots + n_r = n \\ n_1, n_2, \ldots, n_r \geq 0}} \left[\max_{n_i \neq 0} (H(n_i) + c_i) \right]$$

$$(1)$$

with initial conditions $T(0) = T(1) = H(0) = H(1) = 0$. To analyze these recurrence relations we will first need to understand the combinatorics of infinite lopsided trees. Set $A_x = \{v \text{ a node} : depth(v) \leq x\}$ to be the tree containing all nodes of depth at most x in the infinite tree. Then set

$$F(x) = \text{ number of nodes in } A_x, \qquad L(x) = \text{number of leaves in } A_x.$$

These functions are defined for all real x with initial conditions $F(x) = L(x) = 0$ for $x < 0$, $F(x) = L(x) = 1$ for $0 \leq x < c_1$ and satisfy, for $x \geq c_1$, the Fibonacci-type recurrences

$$F(x) = 1 + F(x - c_1) + F(x - c_2) + \cdots + F(x - c_r) \qquad (2)$$
$$L(x) = L(x - c_1) + L(x - c_2) + \cdots + L(x - c_r). \qquad (3)$$

In reality $L(x) = F(x) - F(x - c_1)$ can be verified by inspection so it will be enough for us to analyze $F(x)$; an analysis of $L(x)$ follows automatically. For later use we also point out that the recurrence relation for $L(x)$ also describes the *maximum number of leaves in a tree of height x*.

2 Basics

In this section we analyze $F(x)$ and therefore $L(x)$. The crucial point in this analysis is understanding at what depths nodes can appear in the infinite tree. If, for example, $(c_1, c_2) = (10, 15)$ then nodes can only have depths that are multiples of 5 and, deep enough in the tree, nodes will appear on *every* level with depth a multiple of 5. If though, $(c_1, c_2) = (3, \pi)$, then nodes appear on all depths that can be written in the form $a_1 c_1 + a_2 c_2$, $a_1, a_2 \geq 0$ integers, and general theorems about irrational numbers [12] imply that the depth difference between (neighboring) successive levels upon which nodes appear tends to zero. To formalize this distinction we introduce the following definitions:

Definition 2 Let $(c_1, \ldots, c_r)$ be a tuple of r positive reals:

1. The tuple is *rationally related* if there exists $d > 0$ and positive integers $(c_1{}', \ldots, c_r{}')$ such that
$$(c_1, \ldots, c_r) = d \cdot (c_1{}', \ldots, c_r{}') \qquad \text{and} \qquad gcd(c_1{}', \ldots, c_r{}') = 1.$$

2. If $(c_1, \ldots, c_r)$ is rationally related we define the *gcd* of the tuple by $gcd(c_1, \ldots, c_r) = d$.

3. If $(c_1, \ldots, c_r)$ is not rationally related it is said to be *irrationally related*.

We can now analyze $F(x)$ exactly. As noted in the introduction, the analysis of the particular case $r = 2$ is already implicit in the analyses of [7] and [15].

Theorem 1 *Let $F(x)$, $L(x)$ be defined as previously: Let α be the smallest real positive root of the equation $Q(z) = 1 - z^{c_1} - z^{c_2} - \ldots z^{c_r}$ and $\varphi = 1/\alpha$. Let $c = \left(\sum_{i=1}^{r} c_i \varphi^{-c_i} \right)$. Then,*

1. *If $(c_1, \ldots, c_r)$ is rationally related*

$$F(x) = D(x)\varphi^x + o(\varphi^x), \qquad L(x) = E(x)\varphi^x + o(\varphi^x)$$

where $D(x) = \frac{d}{c(1 - \varphi^{-d})} \varphi^{-d\{x/d\}}$ and $E(x) = \frac{d}{c(1 - \varphi^{-d})} \varphi^{-d\{x/d\}} \left(1 - \varphi^{-c_1} \right)$ is a periodic function with period d. $\{x\} = x - \lfloor x \rfloor$ represents the fractional part of x.

2. If $(c_1, \ldots, c_r)$ is irrationally related

$$F(x) = \frac{1}{c \ln \varphi} \varphi^x + o(\varphi^x), \qquad L(x) = \frac{1 - \varphi^{-c_1}}{c \ln \varphi} \varphi^x + o(\varphi^x).$$

Proof.(Sketch) In this extended abstract we omit the full proof of this theorem and only sketch the techniques utilized. In the rational case we let $G(z) = \sum_{n=0}^{\infty} F(n)z^n$ be the generating function of $F(n)$ and find the asymptotics of $F(n)$ by isolating the singularities of $G(z)$. For the irrational case we set $f(t) = 0$ for $t \leq 1$ and $f(t) = F(\ln t)$ for $t \geq 1$ and set $G(s) = \int_1^{\infty} f(t)dt/t^{s+1}$ ($G(-s)$ is the Mellin transform of $f(t)$). Standard Mellin-transform inversion techniques such as those described in [7, Lemma 4.3] yield the asymptotics of $f(t)$ and therefore $F(x)$. $\square$

3 The Minimum Height of a Tree

Recall that $H(n)$ was defined to be the minimum height of a tree containing n leaves. $L(x)$ is the maximum number of leaves in a tree of height x. Let $\overline{H(n)} = \min\{x : L(x) \geq n\}$. While it is not always true that $\overline{H(n)} = H(n)$ it can be shown that $\lim(\overline{H(n)} - H(n)) = 0$ as $n \to \infty$ so inverting the solution for $L(x)$ given in Theorem 1 yields the following (Proof deleted in this abstract):

Theorem 2 *Let $(c_1, \ldots, c_r)$ be an r-tuple of positive reals and $H(n)$ defined as above. Let α be the smallest real positive root of the equation $Q(z) = 1 - z^{c_1} - z^{c_2} - \ldots z^{c_r}$, $\varphi = 1/\alpha$ and $c = \left(\sum_{i=1}^{r} c_i \varphi^{-c_i}\right)$.*

1. *If $(c_1, \ldots, c_r)$ is rationally related with $d = \gcd(c_1, \ldots, c_r)$ then*

$$H(n) = d \left\lceil \frac{1}{d} \left(\log_\varphi n - \log_\varphi \left(\frac{d\left(1 - \varphi^{-c_1}\right)}{c\left(1 - \varphi^{-d}\right)} \right) \right) \right\rceil + O(1).$$

Furthermore there exists $N > 0$ such that if $n > N$ then the $O(1)$ term is one of d, 0, or $-d$.

2. *If $(c_1, \ldots, c_r)$ is irrationally related then*

$$H(n) = \log_\varphi n - \log_\varphi \frac{1 - \varphi^{-c_1}}{c \ln \varphi} + o(1).$$

3.1 Applications

In [2] Bar-Noy and Kipnis introduce the *Postal Model* of message passing for distributed systems. In this model, counting time from when a sender first starts sending a message, a sender requires one unit of time before completing the work of sending a message and being able to start sending another one, but a recipient requires λ units of time to receive and process the message before being able to broadcast himself: $\lambda \geq 1$ is a parameter representing the *latency* of the system. Bar-Noy and Kipnis demonstrated that, in t time units, the maximum number of recipients that can receive a broadcast message in a one-to-many broadcast scheme in this model satisfies:

$$F_\lambda(t) = \begin{cases} 1 & \text{if } 0 \leq t < \lambda \\ F_\lambda(t-1) + F_\lambda(t-\lambda) & \text{if } t \geq \lambda \end{cases} \tag{4}$$

Therefore the minimum time that it takes to broadcast to n recipients in the model satisfies $f_\lambda(n) = \min\{t : F_\lambda(t) \geq n\}$.

Notice, though, that $F_\lambda(t)$ is exactly the function that we have labelled $L(t)$ with parameters $c_1 = 1$, $c_2 = \lambda$. The recurrence relations are exactly the same; to check that the initial conditions match it is enough to note that for $1 \leq t < \lambda$, $L(t) = L(t-1)$ so $L(t) = 1$ for $0 \leq t < \lambda$. Therefore $f_\lambda(t)$ is precisely the function $H(x)$. Applying Theorem 2 yields the exact solution

$$f_\lambda(n) = \ln_\varphi n + O(1)$$

where $\varphi = 1/\alpha$, α the smallest positive root of $1 - z - z^\lambda$.

This improves the upper bound of $f_\lambda(n) \leq 2\lambda + \frac{2\lambda \log n}{\log(\lceil \lambda \rceil + 1)}$ given in [2].

4 Trees With Minimal External Path Length

Suppose now that $(c_1, \ldots, c_r)$ are fixed. In this section we examine trees that have minimal external path length among all trees with n leaves for those edge costs. Recall that $T(n)$ was defined to be the minimal cost (external path length) of a tree with n leaves. A tree is said to be *optimal* if $C(T) = T(n)$. In this section we examine how the structure of the optimal tree changes as n grows. As stated in the introduction, in this extended abstract we omit most of the proofs of the theorems and direct interested readers [3] and the full version of this article.

We start by labelling the nodes of the infinite tree as $1,2,3,\ldots$, in order of increasing depth, breaking ties arbitrarily. That is, if u and v are two nodes with $depth(u) < depth(v)$, then $u < v$; if $depth(u) = depth(v)$ break ties arbitrarily. See Figure 1 for an example.

Definition 3 *Let V be any set of nodes. Set*

$$LEAF(V) = \{u : parent(u) \in V, u \notin V\}.$$

If $n \leq |LEAF(V)|$, let

$$LEAF_n(V) = \text{the } n \text{ smallest labelled nodes in } LEAF(V).$$

It is obvious that an optimal tree must be *proper* (each of its internal nodes must have at least two children) otherwise some internal node can be replaced by its child, decreasing the cost. For any given n, let m be the number of internal nodes of some proper tree having n leaves. The total number of edges in the tree is $n + m - 1$ so $2m \leq n + m - 1 \leq rm$, or $\lceil \frac{n-1}{r-1} \rceil \leq m \leq n - 1$.

Definition 4 *Let $\lceil \frac{n-1}{r-1} \rceil \leq m \leq n - 1$, and $V_m = \{1, 2, \ldots, m\}$. Set $T_n^m = V_m \cup LEAF_n(V_m)$.*

By definition, if $u \in T_n^m$ is not a root, then $parent(u) \in T_n^m$ so T_n^m is a tree (it is called a *shallow tree* in [9].). See Figure 2. We now see how the T_n^m help us find optimal trees. The following facts derived from [9] provide a criterion for identifying optimal trees.

Lemma 1 *For a fixed n, set $m_{min} = \lceil \frac{n-1}{r-1} \rceil \leq m \leq n - 1$, and $m_{max} = \min\{m \geq m_{min} : T_n^{m+1}$ is not proper$\}$. Then there is some m, $m_{min} \leq m \leq m_{max}$ such that T_n^m is optimal. Furthermore, the sequence of tree costs $C(T_n^m)$, $m_{min} \leq m \leq m_{max}$ is convex, i.e. for $m_{min} < m < m_{max}$,*

$$(C(T_n^{m+1}) - C(T_n^m)) - (C(T_n^m) - C(T_n^{m-1})) \geq 0.$$

In particular, this implies the existence of m_0 with $T_n^{m_0}$ optimal such that

$$C(T_n^{m_{min}}) > \cdots > C(T_n^{m_0-1}) > C(T_n^{m_0}) \leq C(T_n^{m_0+1}) \leq \cdots \leq C(T_n^{m_{max}}).$$

This lemma provides a *local* test of the optimality of any particular proper T_n^m; simply compare it to its predecessor and successor. This will be crucial later.

4.1 The Evolution of Optimal Trees

We need the following combinatorial fact stated here without proof:

Lemma 2 *Let* $x_m = \frac{\sum_{i=1}^{m} c_i}{m-1}$ *for* $m = 2, \ldots, r$. *Then there exists* $k \geq 2$ *such that*

$$x_2 \geq x_3 \geq \cdots \geq x_{k-1} \geq x_k < x_{k+1} < \cdots < x_r.$$

For later use, set $h = x_k$.

Return now to the infinite tree and let $l_0, l_1, l_2, \ldots$, be the consecutive levels upon which nodes appear, i.e., $l_0 = 0$, $l_1 = c_1$, etc..

Now let $m_j = F(l_j)$ be the number of nodes higher than or on depth l_j as described in section 1.1. With these definitions, we can now state our main result. See Figure 2 for examples of the trees described by it.

Theorem 3 *Given* $c_1 \leq c_2 \ldots \leq c_r$, *let* k *and* h *be defined in Lemma 2 and let* l_j *and* m_j *as defined above. Set*

$$A_j = \{v \in LEAF(V_{m_j}) : depth(v) \leq l_j + h\}$$

and $a_j = |A_j|$ *the number of nodes in* A_j, *set*

$$B_j = A_j \cup \{v \in LEAF(V_{m_j}) : l_j + h < depth(v) \leq l_{j+1} + h\}$$

and $b_j = |B_j|$ *the number of nodes in* B_j.

1. *If* $n = a_j$, *then the tree* $T_{a_j}^{m_j} = V_{m_j} \cup A_j$ *is optimal;*
2. *If* $a_j < n \leq b_j$, *then the tree* $T_n^{m_j}$ *is optimal; furthermore,* $T_{b_j}^{m_j} = V_{m_j} \cup B_j$.
3. *If* $b_j < n < a_{j+1}$, *then*
   (a) $n = b_j + p(k-1)$, *then* $T_n^{m_j+p}$ *is optimal;*
   (b) $n = b_j + p(k-1) + q$ *with* $q < k-1$, *then one of* $T_n^{m_j+p}$ *and* $T_n^{m_j+p+1}$ *is optimal.*

Proof. (Idea) Recall that Lemma 1 gives a criterion for deciding whether a tree is optimal or not. The proof of the theorem is just a case-by-case application of this lemma proving that the trees described in the Theorem are all optimal. □

Intuitively, what we are doing in this theorem is reversing the problem. Instead of asking "how many internal nodes are in an optimal tree with n leaves" we instead ask, and answer, the question "suppose T is an optimal tree with m internal nodes, how many leaves n_m can T have?". Since, if T is optimal the m internal nodes in T must be the highest m nodes in the infinite tree, this provides a tool for constructing the optimal tree with n leaves. Simply sweep a horizontal line down from the root until $j + 1$ levels have been swept over with $a_j \leq n < a_{j+1}$. Then use the theorem to construct the optimal tree.

4.2 An Algorithm For Constructing Optimal Trees

In this section we describe how to use the results of the previous section to develop an algorithm that, given n, uses the sweep line idea of the last paragraph to construct an optimal tree with n leaves. This algorithm will run in $O(n \log r)$ time, improving upon the best previously known algorithms which ran in $O(nr^2)$ [14] and $O(n\log^2 r)$ [9] time respectively. There are two parts to our algorithm. The first part constructs the tree

$$T = \begin{cases} T_n^{m_j}, & \text{if } a_j \leq n \leq b_j; \\ T_{b_j+p(k-1)}^{m_j+p}, & \text{if } b_j < n < a_{j+1}, \text{ where } p = \lfloor \frac{n-b_j}{k-1} \rfloor. \end{cases}$$

where a_j and b_j are as defined in Theorem 3. According to that theorem, if $a_j \leq n \leq b_j$, then $T = T_n^{m_j}$ is an optimal tree for n leaves and the algorithm concludes; otherwise $b_j < n < a_{j+1}$ and the optimal tree for n leaves is the minimum of $\{T_n^{m_j+p}, T_n^{m_j+p+1}\}$, where $p = \lfloor \frac{n-b_j}{k-1} \rfloor$. The second part of our algorithm constructs $T_n^{m_j+p}$ and $T_n^{m_j+p+1}$ from the tree $T = T_{b_j+p(k-1)}^{m_j+p}$ which had been constructed in the first part and then returns the one with minimum cost.

Let T be the current tree, $EX(T)$ the set of leaves in T, and $UN(T)$ the set of nodes which are the children of internal nodes in T but not in T themselves. We now describe two operations on T.

$Branch(T, k)$: Take the minimum node in $EX(T)$ and make it an internal node with k children;

$Add(T)$: Add the minimum node in $UN(T)$ to T.

Clearly, each $Branch(T, k)$ adds $(k-1)$ leaves to T, while each $Add(T)$ adds 1 leaf to T.

Recall the definition of trees $T_{a_t}^{m_t}$ and $T_{b_t}^{m_t}$ in Theorem 3; $T_{a_t}^{m_t}$ is obtained from $T_{b_{t-1}}^{m_{t-1}}$ by making all leaves on level t (i.e. of depth l_t) into internal nodes with k children. Thus $a_t = b_t + \alpha_t \cdot (k-1)$, where α_t is the number of nodes on level t. See Figure 2. Let $T = T_{b_{t-1}}^{m_{t-1}}$, and set $n = n - b_{t-1}$ to be the number of leaves remaining to be created. Define procedure $BRANCHING(T, n)$ which returns tree $T_{b_{t-1}+p\cdot(k-1)}^{m_{t-1}+p}$, where $p = \min\{\lfloor \frac{n}{k-1} \rfloor, \alpha_t\}$. Thus if $n \geq \alpha_t$ $BRANCHING(T, n)$ returns $T_{a_t}^{m_t}$.

$BRANCHING(T_{b_{t-1}}^{m_{t-1}}, n)$ can be implemented by starting with $T = T_{b_{t-1}}^{m_{t-1}}$ and repeating $Branch(T, k)$ until the depth of the minimum node in $EX(T)$ is larger than l_t (i.e. equals l_{t+1}) or the remaining number of leaves n is less than $(k-1)$.

Note that $T_{b_t}^{m_t}$ is obtained from $T_{a_t}^{m_t}$ by adding all of the nodes in $UN(T_{a_t}^{m_t})$ whose depths are less than or equal to $l_{t+1} + h$. See Figure 2. Let $T = T_{a_t}^{m_t}$ and set $n = n - a_t$ to be the number of leaves remaining to be created. Define procedure $ADDING(T, n)$ which returns $T_{a_t+q}^{m_t}$, where $q = \min\{n, b_t - a_t\}$.

$ADDING(T_{a_t}^{m_t}, n)$ can be implemented by starting with $T = T_{a_t}^{m_t}$ and repeating $Add(T)$ until the depth of the minimum node in $UN(T)$ is larger than $l_{t+1} + h$ or the remaining number of leaves n is zero.

Given $(c_1, \ldots, c_r)$ and n, our algorithm first computes k and h and creates the tree $T = T_{b_{-1}}^{m_{-1}}$ containing the single, root, node. As long as the remaining number of leaves n is greater than or equal to $(k-1)$, it calls $BRANCHING(T_{b_{t-1}}^{m_{t-1}}, n)$; if the tree returned is $T_{a_t}^{m_t}$ (*this can be discovered by checking if the depth of the minimum leaf in the tree is larger than l_t, if it is not, that implies there are still some leaves on level t and so it is not $T_{a_t}^{m_t}$*) then call $ADDING(T_{a_t}^{m_t}, n)$, else exit the iteration. At the completion of this first part of the algorithm, it is not difficult to see that the tree constructed is

$$T = \begin{cases} T_n^{m_j}, & \text{if } a_j \le n \le b_j; \\ T_{b_j+p(k-1)}^{m_j+p}, & \text{if } b_j < n < a_{j+1}, \text{ where } p = \lfloor \frac{n-b_j}{k-1} \rfloor. \end{cases}$$

We now discuss how to efficiently implement *BRANCHING* and *Adding* which reduces to how to efficiently implement $Branch(T, k)$ and $Add(T)$. Procedure $Branch(T, k)$ requires identifying the minimum node in $EX(T)$; procedure $Add(T)$ requires identifying the minimum node in $UN(T)$. To reduce the number of nodes that must be considered in finding minimum nodes in $EX(T)$ and $UN(T)$, we use the following representation of nodes. Notice that any non-root node v in the tree is the ith child of some node u in the tree for some $1 \le i \le r$: v can thus be uniquely expressed by $child_i(u)$. Let u_i be the minimum node such that $child_i(u_i)$ is in $EX(T)$. The minimum node in $EX(T)$ is then the minimum among $\{child_i(u_i) : i = 1, \ldots, r\}$. Our algorithm finds the minimum node in $EX(T)$ by maintaining these r particular children in a priority queue in $O(\log r)$ time. Similarly, let v_i be the minimum node such that $child_i(v_i)$ is in $UN(T)$. We can find the minimum node in $UN(T)$ by maintaining $\{child_i(v_i) : i = 1, \ldots, r\}$ in a priority queue in $O(\log r)$ time. Thus, the time complexity for $BRANCHING(T_{b_{t-1}}^{m_{t-1}}, n) = p_t(k-1) + p_t \cdot O(\log r)$ where $p_t = \min\{\lfloor \frac{n-b_{t-1}}{k-1} \rfloor, \alpha_t\}$ and the time complexity of $ADDING(T_{a_t}^{m_t}, n) = q_t \cdot O(\log r)$ where $q_t = \min\{n - a_t, b_t - a_t\}$. Since $\sum_{t=1}^{j}(p_t + q_t) < n$, the time complexity of the first part of the algorithm is $O(n \log r)$.

The second part of our algorithm constructs $T_n^{m_j+p}$ and $T_n^{m_j+p+1}$ from $T_{b_j+p(k-1)}^{m_j+p}$, and returns the minimum one. This can be straightforwardly implemented in $O(r \log r)$ time. which is dominated by the $O(n \log r)$ of the first part. Therefore, the time complexity of the algorithm is $O(n \log r)$.

We have glossed over some of the details of the analysis (such as showing that the priority queues can be updated efficiently). The reader wanting more complete details is referred to the full version of the abstract or [3]. Those references also demonstrate that, if all of the *Branch* operations within one *BRANCHING* are batched together as are all of the *Add* operations within one *ADDING*, then the algorithm can be made to run in $n + O(r \log r \log^r n)$ time (where $O()$ hides constants dependent upon the values $(c_1, \ldots, c_r)$).

4.3 Exact Asymptotics of $T(n)$

Combining Theorems 1 and 3 we can derive the exact asymptotics of $T(n)$, the cost of the optimal lopsided tree with n leaves, as $n \to \infty$. As described previously this is equivalent to exactly analyzing the costs of Varn Codes.

Theorem 4 *Let $T(n)$ be as defined in (1). Let α be the smallest real positive root of the equation $Q(z) = 1 - z^{c_1} - z^{c_2} - \ldots z^{c_r}$ and $\varphi = 1/\alpha$. Let $c = \sum_{i=1}^{r} c_i \varphi^{-c_i}$. Let k and $h = x_k$ be as defined in Lemma 2.*

1. *If $(c_1, \ldots, c_r)$ is rationally related with $gcd(c_1, \ldots, c_r) = 1$ define*

$$K = \frac{1}{c(1 - \varphi^{-1})} \left(\varphi^{\lfloor h \rfloor} - \sum_{i=1}^{k} \varphi^{\lfloor h \rfloor - c_i} + (k-1) \right)$$

$$A = \frac{(k-1)}{c(1 - \varphi^{-1})K}, \qquad R = \log_\varphi((1 - A)\varphi + A).$$

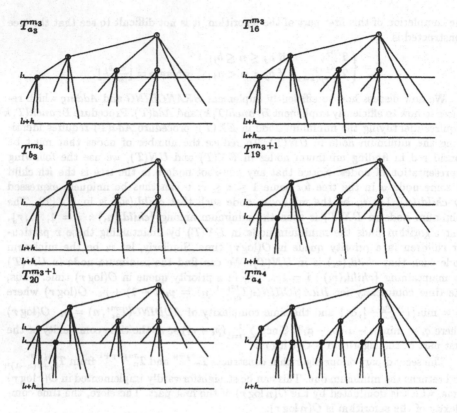

Fig. 2. Examples with $r = 5$, $(c_1, c_2, c_3, c_4, c_5) = (3, 5, 5, 8.75, 10)$, so $k = 3$, $h = 6.5$, $m_3 = 5$, $a_3 = 14$, $b_3 = 17$, $a_4 = 25$, $m_4 = 9$. For reference, $C(T_{20}^6) = 217.75$, while $C(T_{20}^7) = 218.25$.

Then

$$T(n) = n \log_\varphi n + B\left(\{\log_\varphi \frac{n}{K}\}\right) n + D\left(\{\log_\varphi \frac{n}{K}\}\right) n + o(n)$$

where

$$B(\theta) = h + 1 - \log_\varphi K - \theta - \left(\frac{1}{\varphi - 1} + \{h\}(1 - A)\right) \varphi^{1-\theta}$$

$$D(\theta) = \begin{cases} \{h\}(\varphi^{R-\theta} - 1) & \theta \le R \\ 0 & \theta > R \end{cases}$$

See Figure 3. If $(\overline{c_1}, \ldots, \overline{c_r}) = d(c_1, \ldots, c_r)$ with $gcd(c_1, \ldots, c_r) = 1$ then $\overline{T}(n) = d \cdot T(n)$ where $\overline{T}(n)$ is the cost function defined by $(\overline{c_1}, \ldots, \overline{c_r})$.

2. If $(c_1, \ldots, c_r)$ is irrationally related define $K = \frac{1}{c \ln \varphi}\left(\varphi^h - \sum_{i=1}^k \varphi^{h-c_i} + (k-1)\right)$.
   Then

$$T(n) = n \log_\varphi n + \left(h - \log_\varphi K - \frac{1}{\ln \varphi}\right) n + o(n).$$

548

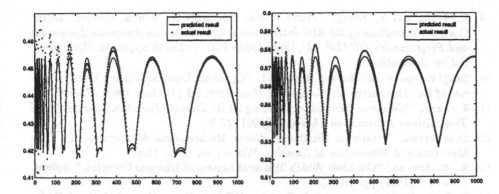

Fig. 3. The left diagram has $(c_1, c_2, c_3, c_4, c_5) = (2, 3, 4, 7, 11)$, the right $(2, 3, 5, 7, 11)$. Predicted cost plots $B(\{\log_\varphi \frac{n}{K}\}) + D(\{\log_\varphi \frac{n}{K}\})$. Actual cost plots $\frac{1}{n}(T(n) - n\log_\varphi n)$.

Acknowledgements: The authors would like to thank Dr. Jacob Ecco for introducing them to the Morse Code puzzle which sparked this investigation. They would also like to thank Assaf Schuster for pointing them towards the Postal-Model and Rudolf Fleischer for conversations concerning the construction algorithm. The work of both authors was partially supported by HK RGC Grant HKUST 181/93E. and HKUST652/95E

References

1. Doris Altenkamp and Kurt Mehlhorn, "Codes: Unequal Probabilities, Unequal Letter Costs," *Journal of the Association for Computing Machinery*, **27**(3) (July 1980) 412-427.

2. Amotz Bar-Noy and Shlomo Kipnis, "Designing Broadcasting Algorithms in the Postal Model for Message Passing Systems," *SPAA 1992 (to appear in Mathematical Systems Theory, 1994)*,.

3. V. S.-N. Choi, Lopsided Trees: Analyses and Algorithms. M.Phil. Thesis, Hong Kong University of Science and Technology. Department of Computer Science. Technical Report HKUST-CS95-36. (1995).

4. I. Csiszár, "Simple Proofs of Some Theorems on Noiseless Channels," *Inform. Contr.*, bf 14 (1969) pp. 285-298

5. I. Csiszár, G. Katona and G. Tsunády, "Information Sources with Different Cost Scales and the Principle of Conservation of Energy," *Z. Wahrscheinlichkeitstheorie verw*, **12**, (1969) pp. 185-222

6. N. Cot, "Complexity of the Variable-length Encoding Problem," *Proceedings of the 6th Southeast Conference on Combinatorics, Graph Theory and Computing*, (1975) 211-224.

7. M. L. Fredman and D. E. Knuth, "Recurrence Relations Based on Minimization," *Journal of Mathematical Analysis and Applications*, **48** (1974) 534-559.

8. M. Golin and G. Rote, "A Dynamic Programming Algorithm for Constructing Optimal Prefix-Free Codes for Unequal Letter Costs," *Proceedings of the 22nd International Colloquim on Automata Languages and Programming (ICALP '95)*, (July 1995).

9. M. Golin and N. Young, "Prefix Codes: Equiprobable Words, Unequal Letter Costs," *Proceedings of the 21st International Colloquium on Automata Languages and Programming (ICALP '94)*, (July 1994). Full version to appear in *SIAM Journal on Computing*.

10. Sanjiv Kapoor and Edward Reingold, "Optimum Lopsided Binary Trees," *Journal of the Association for Computing Machinery*, **36** (3) (July 1989) 573-590.

11. R. Karp, "Minimum-Redundancy Coding for the Discrete Noiseless Channel," *IRE Transactions Information Theory*, **7** (1961) 27-39.

12. Ivan Niven, "Irrational Numbers," *Carus Mathematical Monographs*, **11**. The Mathematical Association of America, Washington D.C., 1956.

13. R. M. Krause, "Channels Which Transmit Letters of Unequal Duration," *Inform. Contr.*, **5** (1962) pp. 13-24,

14. Y. Perl, M. R. Garey, and S. Even. "Efficient Generation of Optimal Prefix Code: Equiprobable Words Using Unequal Cost Letters," *Journal of the Association for Computing Machinery*, 22(2):202–214, April 1975.

15. Nicholas Pippenger, "An Elementary Approach to Some Analytic Asymptotics," *SIAM Journal of Mathematical Analysis*, **24**(5) (September 1993) 1361-1377.

16. C.E. Shannon "A Mathematical Theory of Communication," *Bell System Technical Journal* **27** (1948) 379-423, 623-656.

17. Serap A. Savari, "Some Notes on Varn Coding," *IEEE Transactions on Information Theory*, 40(1) (Jan. 1994) pp. 181-186.

18. B.F. Varn "Optimal Variable Length Codes (Arbitrary Symbol Costs and Equal Code Word Probabilities)," *Informat. Contr.*, **19** (1971) pp. 289-301

Optimal Logarithmic Time Randomized Suffix Tree Construction

Martin Farach*
Rutgers University

S. Muthukrishnan†
Univ. of Warwick

Abstract

The suffix tree of a string, the fundamental data structure in the area of combinatorial pattern matching, has many elegant applications. In this paper, we present a novel, simple sequential algorithm for the construction of suffix trees. We are also able to parallelize our algorithm so that we settle the main open problem in the construction of suffix trees: we give a Las Vegas CRCW PRAM algorithm that constructs the suffix tree of a *binary string* of length n in $O(\log n)$ time and $O(n)$ work with high probability. In contrast, the previously known work-optimal algorithms, while deterministic, take $\Omega(\log^2 n)$ time.

We also give a work-optimal randomized comparison-based algorithm to convert any string over an unbounded alphabet to an equivalent string over a binary alphabet. As a result, we obtain the *first work-optimal algorithm* for suffix tree construction under the unbounded alphabet assumption.

1 Introduction

Given a string s of length n, the *suffix tree* T_s of s is the compacted trie of all the suffixes of s. For many reasons, this is the fundamental data structure in combinatorial pattern matching. It has a compact $O(n)$ space representation which has several elegant uses [Apo84]. Furthermore, Weiner [Wei73], who introduced this powerful data structure, showed that T_s can be constructed in $O(n)$ time. This sequential construction and its analysis are nontrivial. Considerable effort has been put into producing simplified $O(n)$ time suffix tree constructions, however, to date, though innovative, all such algorithms are variants of one another

*Department of Computer Science, Rutgers University, Piscataway, NJ 08855, USA. (*farach@cs.rutgers.edu*, *http://www.cs.rutgers.edu/~farach*). Supported by NSF Career Development Award CCR-9501942 and an Alfred P. Sloan Research Fellowship. Work done while this author was a Visitor Member of the Courant Institute of NYU.

†*muthu@dcs.warwick.ac.uk*; Partly supported by DIMACS (Center for Discrete Mathematics and Theoretical Computer Science), a National Science Foundation Science and Technology Center under NSF contract STC-8809648. Partly supported by Alcom IT.

and continue to rely on the classical approaches of Weiner [Wei73] and Mc-Creight [McC76]. Efficient parallel construction of suffix trees has also proven to be a challenge. In fact, a classical open problem in stringology is to design a fast, work-optimal parallel algorithm for constructing this data structure. While some recent progress has been made in designing work-optimal algorithms (at the expense of the running time), in this paper, we settle the following open problem: our *main result* is a novel simplified suffix tree contruction which when parallelized yields an $O(\log n)$ time, $O(n)$ space and $O(n)$ work Las Vegas type suffix tree construction algorithm. We believe that our algorithm, which departs completely from the Weiner/McCreight approach, is interesting in both its sequential and parallel versions.

In what follows, we will explain our results in more detail before giving a technical overview of our algorithms.

1.1 Our Results.

Given the versatility of suffix trees, it is not surprising that they have been extensively studied. Since Weiner presented his classical construction, many simplified constructions have been developed [CS85, McC76]. Sequential construction of suffix trees continues to be an active area of research [DK95, Kos94].

All these algorithm are inherently sequential since they scan the string in monotonic order, updating the partial data structure as each symbol is scanned. In [AIL+88], it is shown how to build a suffix tree in $O(\log n)$ time and $O(n \log n)$ work. This algorithm is not optimal; moreover, it uses polynomial space.[1] Designing work-optimal algorithms for this problem remained a challenge. Recent results have shown that one can achieve work optimality at the expense of the running time. In [SV94], a work optimal $O(\log^2 n)$ time algorithm was presented using polynomial space. In [Har94] a different work-optimal algorithm is presented which uses linear space; this algorithm takes $O(\log^4 n)$ time.[2]

In this paper, we settle the following double challenge: our main result is a novel suffix tree algorithm whose sequential version is quite simple and whose (considerably non-trivial) parallelization yields an $O(\log n)$ time and $O(n)$ work parallel algorithm for suffix tree construction that uses $O(n)$ space. Our algorithm is randomized of the Las Vegas type.

An additional issue of string processing interest is that of *alphabet dependence*. String processing algorithms that work under the *unbounded alphabet* assumption access the string using only order comparisons. Sequential suffix tree construction algorithms that work under this model take $\Theta(n \log \sigma)$ time if the number of distinct symbols in s is σ; in contrast, the best known parallel algorithms in this model do $\Theta(n \log n)$ work and are therefore sub-optimal. In this paper, we give a simple randomized work/time optimal reduction from the case when alphabet set is unbounded to the case when it is binary. Our reduction works in the ordered-comparison model. Consequently we derive the

---

[1] Throughout, we will use "polynomial space" to mean $\Theta(n^{1+\epsilon})$ space for some $\epsilon > 0$.

[2] This algorithm works on the CREW PRAM while all other algorithms mentioned in this paper, including ours, use the stronger Arbitrary CRCW PRAM.

first known work-optimal algorithms for suffix tree construction under the unbounded alphabet assumption. Our reduction above can be used with any of the work optimal algorithms of [SV94, Har94] for binary strings, however using it with our algorithm in this paper gives the fastest work optimal suffix tree construction algorithm under the unbounded alphabet assumption.

1.2 Technical Overview.

Suffix tree construction for binary strings. Let $s \in \{0, 1\}^*$. Then, the suffix tree of s is the compacted trie[3] of all the suffixes of $s\$$ where $\$ \neq 0, 1$. At a high level, our algorithm works as follows. First we recursively determine the *odd tree*, T_o, which is the compacted trie of all the suffixes beginning at the odd positions. From the odd tree, we then derive the *even tree*, T_e, the compacted trie of all the suffixes at the even positions. Finally, we merge T_o and T_e to obtain the suffix tree of s. The technical crux of this approach in both sequential and parallel setting lies in merging the trees.

The approach of constructing the suffix tree by merging the odd and even trees, one of which is recursively constructed, is not new. It was used in a modified form in [SV94] to obtain an $O(\log^2 n)$ time work-optimal suffix tree construction. There too merging two partial suffix trees (such as T_o and T_e) formed the technical crux. There the authors developed a novel symmetry-breaking naming scheme to solve that problem; however their algorithm for merging the two trees takes $\Omega(n \log^* n)$ work[4]. That was sufficient for them to obtain an work optimal algorithm for suffix tree construction (taking $O(\log^2 n)$ time) by relying on the Four Russian's technique of constructing certain exhaustive tables for "short" strings.

In this paper, we directly solve the problem of merging the two partial suffix trees T_o and T_e by using several structural properties of these trees which we identify and prove here. Our algorithm for this takes $O(\log n)$ time and $O(n)$ work and is therefore work-optimal. Using this algorithm as such to construct suffix trees in parallel we can get an algorithm that takes $O(\log n \log \log n)$ time and $O(n)$ work. Note that this is already the fastest work-optimal algorithm known for this problem. However to get our main result, which is a $O(\log n)$ time work-optimal algorithm for this problem, we use additional ideas which are explained in detail in Section 3. We merely note that in order to achieve this result we had to reüse the structural properties of the partial trees we identified to design an algorithm for merging them in only $O(\frac{\log n}{\log \log n})$ time (rather than the $O(\log n)$ time quoted above), again in $O(n)$ work, provided the Euler tours of the two trees are given.

Finally, as a curiosity, we mention that our parallel algorithm makes use of a number of tools that have been developed in the parallel algorithms community recently – see the list of lemmas in [FM96]. However, our sequential algorithm does not need these tools and is very simple.

---

[3] A compacted trie differs from a trie in that degree 2 paths are replaced by edges labeled by the appropriate substring.

[4] $\log^* n = \min\{i | \log^{(i)} n \leq 2\}$.

Suffix tree construction for strings over unbounded alphabet. A standard approach to cope with the strings drawn from an unbounded alphabet is to map the distinct symbols of the string to a small alphabet and then use the suffix tree construction algorithm that works for strings over small alphabet. We can thus formalize the following *renaming* problem. Given an array of size n with σ distinct elements, map distinct elements in the array to range $1 \cdots \sigma$ in the (ordered) comparison model; here σ is unknown at the beginning. Renaming is a generic reduction from string problems over unbounded alphabet to those over (integer) alphabet $1 \cdots \sigma$.

Sequentially renaming can be done optimally using a balanced binary tree such as a 2–3 tree. Known parallel algorithms for 2–3 trees [PVW83], however, do not give optimal parallel bounds for renaming. The best known algorithms for renaming use parallel sorting and therefore take $O(n \log n)$ work and $O(\log n)$ time. In this paper, we present the first known simultaneously time and work-optimal algorithm for renaming. Our algorithm is randomized and of Las Vegas type; it takes $O(\log \sigma)$ expected time and $O(n \log \sigma)$ expected work. Our algorithm is obtained by random sampling reminiscent of randomized parallel sorting algorithms [Rei81]. Since σ is unknown and we seek an algorithm with time dependent on only σ, considerable technical difficulties arise in our setting which we overcome (we omit mentioning any further details here; See [FM96]).

Map. We describe the sequential version of our algorithm in Section 2. We describe our parallel suffix tree construction in Section 3. We have only sketched the details in most of the steps, and the reader is referred to [FM96] for details.

2 The Sequential Algorithm

We will use one basic tool: KR naming. With this tool, we can check the equality of any two substrings in constant time.

Lemma 2.1 ([KR87]) *Given a string s of length n, there exists a function $f(i, j)$ which can be computed in linear work and $O(\log n)$ time such that for any i, j, k, if $f(i, i + k) \neq f(j, j + k)$ then, $s[i \ldots i + k] \neq s[j \ldots j + k]$, and if $f(i, i + k) = f(j, j + k)$ then, $s[i \ldots i + k] = s[j \ldots j + k]$ with probability at least $1 - 1/n^3$.*

Now our algorithm has three main steps:

Compute T_o: To construct the odd tree T_o recursively, we make the following observation. Consider a string s' generated from s in which $s'[i] = f(s[2i - 1], s[2i])$ where f is the randomized fingerprint function of Lemma 2.1. Let T' be the suffix tree of s'. Then, T_o is closely related to T'. The difference is the following. Consider any two odd suffixes. The least common ancestor (lca) of their corresponding leaves in T_o gives their longest common prefix, while the lca of their corresponding leaves in T' is the longest common prefix of *even* length. Therefore, T_o can be derived from T' by local patching in linear time. Since $|s'| = |s|/2$, this recursive procedure does not increase the complexity of our algorithm.

Compute T_e: To construct the even tree T_e, we note that every even suffix is a single character followed by an odd suffix. From T_o, we know the longest common prefix of any two odd suffixes, so we also now know the longest common prefix of any two even suffixes. To construct a tree, we only need to know an in-order traversal of the leaves and the relative depth for the lca of adjacent leaves. The latter we know already, and we can get the in-order traversal of the leaves by a stable integer sorting of the in-order leaves of T_o using the character preceding an odd suffix as its sort key, once again, in linear time.

Merging T_o and T_e: Merging T_o and T_e will give the suffix tree of s, so this step is the heart of the algorithm. We will merge the two trees by a coupled DFS, so at each recursive step, we will be merging a subtree of T_o with one from T_e. For purposes of illustration, we will consider subtrees rooted at edges as well as those rooted at nodes. Thus, we must describe what our algorithm does when merging pairs rooted at node/node, node/edge and edge/edge. Since we start by merging at the roots of both trees, we start with a node/node pair.

NODE/NODE: Suppose we are merging v_o and v_e. The only relevant information is the first character on the edges leading to their children. For any children c_i of v_o and c_j of v_e which begin with the same character, merge their subtrees via an Edge/Edge merge. Any child whose edge begins with a unique character is not merged. Its subtree is simply included in the final tree unchanged.

NODE/EDGE: Suppose we are merging node v and edge e, and that the first character on e is c. If v does not have an edge to a child which begins in c, then root e's subtree at v. Otherwise, let d be the child of v whose edge starts with a c. Merge the edge to d with e and root the resulting tree at v. Leave all other subtrees below v unchanged.

EDGE/EDGE: In both of the above cases, we recursively merge two trees rooted at edges. Let e_o and e_e be the edges we are merging. We have two cases. First, suppose the string of one edge is a proper prefix of the string of the other edge. WLOG, we will assume e_o is a prefix of e_e. We can check this in constant time via fingerprints. Then we add a node b in e_e after $|e_o|$ characters. Now, let d be the node at the bottom of e_o. We know that we can proceed by merging b and d, as above.

The other case is much more interesting. The edges e_o and e_e share a common prefix but neither is a prefix of the other. We cannot afford to find out how long the common prefix is for these two edges, since it would take $\Omega(\log n)$ to do so by direct methods, and there could be $\Omega(n)$ such pairs of edges to merge. Instead of directly determining this longest string, we will introduce a *refinement node* into e_o and e_e which will act as a place-holder. The placement of the refinement nodes within the edge pair gives us enough information to complete the structural merging of the even and odd trees. After placing the refinement node, we have bottomed out the recursion.

REFINEMENT: However, we have yet to determine exactly where the edges need to be "broken" to introduce the refinement node; in other words, the refinement nodes can be thought of as sliding nodes whose exact placement we defer until after merging the trees.

Consider some refinement node r. By construction, it has out-degree 2, and one of its descendant trees, t_o comes from the odd tree, while the other, t_e, comes from the even tree. Let l_i be an arbitrary leaf in t_o and l_j be an arbitrary leaf in t_e, where, for any k, l_k represents the kth suffix of s. We can *restate* the problem of placing the refinement node within the edge pair as that of determining the longest common prefix of l_i and l_j. Now, we can conclude that the length of r (which is all we are after) is 1 more than the length of $\text{lca}(l_{i+1}, l_{j+1}) = v$. But we may not know the length of v if v is itself a refinement node. So we see that $|r| = 1 + |v|$, and conclude that we have a dependency tree (more generally, a forest) between the length of various nodes in the tree. We define $d(r) = v$ where r and v are as above. The length of r depends on its depth in the tree defined by $d(\cdot)$. Finally, we determine, by DFS on the d tree, the depth of all refinement nodes in their trees and thus deduce their lengths.

The above coupled-DFS merging procedure, along with the final DFS refinement procedure take $O(n)$ time, thus completing our construction. The output of this algorithm is correct with high probability. In [FM96] we show an algorithm that checks the correctness of the output in $O(\log n)$ time and $O(n)$ work on a PRAM, and hence is sufficient to check the output of this algorithm, as well as the following parallel algorithm.

3   The Parallel Algorithm

The parallel algorithm has the same three basic steps as the sequential algorithm in Section 2. However, we now have the added constraint that we want an $O(\log n)$ time and $O(n)$ work algorithm. The sequential algorithm naturally decomposes into $\log n$ recursive phases, and each phase seems to take $\Omega(\log n)$ time, since basically any tree manipulation has list ranking as bottle-neck. In addition, we will need a radically different approach to merging the two trees since a DFS based approach is not well suited to parallelization.

Given this scheme, we make the following three contributions.
1. We provide a work-optimal algorithm for merging the odd and even trees. This step is the crux of our technical contribution and it relies on several structural properties of the odd and even suffix trees that we isolate and prove (See Section 3.1). Our algorithm works in $O(\log n)$ time w.h.p. Combined with a

straightforward $O(\log n)$ time algorithm for deriving the even tree from the odd tree, this gives a work-optimal $O(\log^2 n)$ time suffix tree algorithm. Note that this already provides an alternate algorithm matching the running time of the fastest known optimal algorithm for the suffix tree construction problem [SV94], while taking only linear space.

2. Consider the following lemma:

Lemma 3.1 *There exists an algorithm to compute the suffix tree of a string of length n in $O(\log n)$ time, $O(n \log n)$ work w.h.p. and $O(n \log^2 n)$ space.*

Proof: The algorithm in [AIL+88] uses a bulletin board which can be replaced by the constant time hashing scheme of [GMV91]. ∎

We may now bottom out the recursion after $O(\log \log n)$ levels. We use this lemma to derive a $O(\log n \log \log n)$ time suffix tree construction algorithm, while still preserving the linear work and space. We note that this "bottoming out" cannot be done in the known work-optimal constructions of either [Har94] or [SV94] to speed up their algorithms.

3. Each step of our recursion takes $\Theta(\log n)$ time. This is not surprising since we perform several tree manipulation operations at each stage, each of which relies on computing Euler tours. Computing the Euler tour in turn has the well-known bottleneck of list ranking, which takes $\Theta(\log n)$ time as noted above. We overcome this bottleneck in our algorithm by *introducing the key idea of maintaining the Euler tour* (rather than the tree itself) through the levels of recursion.

This, on the one hand, implies that Euler tours need not be computed at each level and therefore the $\Theta(\log n)$ bottleneck is averted. On the other hand, maintaining Euler tours and manipulating them give rise to new problems, which can be illustrated as follows. Suppose we have a tree and its Euler tour and we wish to reörder the children at each node (as we are required to do in our algorithm). In a tree representation, such a rearrangement is a purely local computation at the nodes. In the Euler tour, however, such a change involves global reörganization. We demonstrate how to perform this reörganization as well as the other tasks at each level in $O(\log n/ \log \log n)$ time with optimal work. This in turn gives our main result, namely:

Theorem 3.2 *Given a binary string s of length n, its suffix tree can be computed in $O(\log n)$ time and $O(n)$ work w.h.p. using $O(n)$ space.*

In what follows, we describe how each of the three steps is implemented so as to derive the Euler tour of the suffix tree of s. In fact, *all trees will be assumed to be in their Euler tour representation*. Note that it is straightforward to generate the pointer version of the suffix tree given its Euler tour in $O(\log \log n)$ time and $O(n)$ work.

Compute T_o: Derive s' from s as in the sequential version, and recurse to produce T'. Deriving T_o from T' is technically quite difficult since T' is in an Euler Tour representation. We make this transformation via the procedure UNNAME. We will show the following in [FM96].

Lemma 3.3 *Generating s' can be done in $O(1)$ time and $O(n)$ work. Procedure* UNNAME *can be implemented in $O(\log n/\log\log n)$ time and $O(n)$ work w.h.p. if s is drawn from an alphabet of size $O(\text{polylog } n)$ and in $O(\log n/\log\log n)$ time and $O(n\log\log n)$ work w.h.p. if it is drawn from an alphabet of size $O(\text{poly } n)$; in both cases, the space is linear.*

By recursing for only $2\log\log n$ levels, we get:

Lemma 3.4 *All calls to procedure* UNNAME *over all recursive levels take $O(\log n)$ time and $O(n)$ work w.h.p. and use $O(n)$ space.*

Proof: Note that at every recursive level s' is at most half the length of s. Also, if the alphabet size of s is σ, then the alphabet size of s' is at most σ^2. At each stage, we hash the tuples of s' to ensure that in fact the alphabet is in the range $1\ldots\sigma^2$. For the first $\log\log\log n$ levels, the alphabet size of the strings remains polylogarithmic and therefore, by Lemma 3.3, the work performed during this step is linear. At this stage, the string under consideration is of length at most $O(n/\log\log n)$. Therefore, we can afford to use the suboptimal large alphabet version of the procedure UNNAME up to $2\log\log n$ levels while still performing only linear work. Note that our alphabet never grows bigger than n. ∎

Compute T_e: The sequential version is easy to parallelize. We need only be careful about parallel stable integer sorting.

Lemma 3.5 *Procedure* ODD-EVEN *can be implemented in $O(\log n/\log\log n)$ time and $O(n)$ work w.h.p. if s is drawn from an alphabet of size $O(\text{polylog } n)$ and in $O(\log n/\log\log n)$ time and $O(n\log\log n)$ work w.h.p. if it is drawn from an alphabet of size $O(\text{poly } n)$; in both cases, the space is linear. All calls to procedure* ODD-EVEN *over all recursive levels take $O(\log n)$ time and $O(n)$ work w.h.p. and use $O(n)$ space.*

Merging T_o and T_e: We change the focus of our algorithm from the sequential version. In the sequential algorithm, we considered merging subtrees rooted at nodes or edges. In the parallel version, we will consider the set of strings which can be obtained by tracing down from the root from a trie. We can partition this set into *node strings*, which are strings that end at nodes, and *edge strings*, which are strings that end between two nodes. Our parallel algorithm will rely on finding strings which are shared by both tries and then performing the actual structural merging.

The node string/edge string bipartition means that strings shared by the two tries come in three flavors: *node-node*, *node-edge*, and *edge-edge*. As it turns out, common strings of the node-node flavor are easy to find, and node-edge strings require only a little more work. As in the sequential case, the true challenge is to find common strings of the edge-edge flavor. We will end up once again with refinement nodes, but will need to avoid computing the depth in the d tree, since this would take $O(\log n)$ time. We will present an algorithm for refinement node computation which takes only $O(\alpha(n))$ time[5] while still performing $O(n)$ work (For details, see Section 3.1).

Summarizing the discussion above, we need three procedures: NODE-NODE, NODE-EDGE and EDGE-EDGE. In addition, we need a procedure to generate the merged tree once we have found common strings. We call this last procedure MERGE-PATHS and use these four together to implement MERGE-TREES. See Section 3.1.

Lemma 3.6 *Procedure* MERGE-TREES *can be implemented in $O(\log n/\log\log n)$ time and $O(n)$ work w.h.p. if s is drawn from an alphabet of size $O(\text{polylog } n)$ and in $O(\log n/\log\log n)$ time and $O(n\log\log n)$ work w.h.p. if it is drawn from an alphabet of size $O(\text{poly } n)$; in both cases, the space is linear. All calls to procedure* MERGE-TREES *over all recursive levels take $O(\log n)$ time and $O(n)$ work w.h.p. and use $O(n)$ space.*

From the preceding sequence of lemmas, we can conclude Theorem 3.2. In what follows, we will describe one of the important procedures, namely that for tree merging.

3.1 Tree Merging

Recall that procedure MERGE-TREE is implemented by first finding common strings represented by the odd and even trees (using procedures NODE-NODE, NODE-EDGE, and EDGE-EDGE) and finally building the merged tree around this core of common strings (using procedure MERGE-PATHS). Note that any prefix of a common string is also a common string, thus, to represent all common strings, it suffices to determine common strings which are *maximal*. In the discussion that follows, such maximal strings will be referred to as *explicit* strings and their prefixes will be referred to as *implicit*. Procedure MERGE-TREE relies on the following analysis of strings common to both trees.

The simplest of common strings is a Node-Node string. Define an *anchor pair* to be a pair (u, v), $u \in V(T_o)$, $v \in V(T_e)$ such that their strings are the same; each such node is called an *anchor*.

Property 1 *The pair $(\text{lca}(u_1, u_2), \text{lca}(v_1, v_2))$ is also an anchor pair if (u_1, v_1) and (u_2, v_2) are any two anchor pairs.*

Consider any node in one of the trees which has an anchor descendant. Clearly the string that corresponds to this node is represented in the other tree as well. Therefore, the set of all anchor pairs implicitly defines all such common strings. So in searching for Node-Edge strings, we need only consider nodes with no anchor descendants.

Let a *side tree* be a maximal component such that no node in the side tree is an anchor node or has an anchor descendant. Define a *side tree pair* to be a pair of side trees such that the parents of their roots form an anchor pair, and the first

---

[5]Here $\alpha(\cdot)$ is the inverse Ackerman function.

character on the edge for the anchor pair to the roots is the same. Node-Edge
strings which occur in side trees must occur in a side tree pair. Within a side
tree pair, define a node to be *active* if its string is a Node-Edge string, that is, if
it occurs in an edge of the paired side tree. It can be easily shown that the least
common ancestor of two active nodes is an anchor unless one is an ancestor of
the other. Since there are no anchors in a side tree, this implies the following
property.

Property 2 *Active nodes within a single side tree are linearly ordered; in fact,
they form a path from the root of the side tree.*

Fix a side tree pair and consider all its active nodes. Let u be the active node
with the longest string in either of the trees. Let $\overline{vw}$ be the edge in the other
tree such that the string of v is a prefix of the string of u which is a prefix of the
string of w. Suppose we insert a node u' into the edge $\overline{vw}$ such that the string
of u' equals the string of u. We will call such a pair (u, u') a *pseudo-anchor
pair* and call each node a *pseudo-anchor*. Observe that once we have created a
pseudo-anchor pair, it implicitly represents *all* Node-Edge strings in its side tree
pair.

It remains to consider Edge-Edge strings. In fact, we need only consider those
which are not implicitly represented by anchors or pseudo-anchors. These Edge-
Edge strings occur within edge pairs that have the following property: either
the edge pairs come off pseudo-anchor pairs, or they are the edges connecting
the roots of a side tree pair to their anchor pair parents, in the case where there
are no active nodes in that side tree pair. Recall that the side tree pairs were
limited to hang off of anchor nodes; the stated property for edge pairs gives a
similarly useful restriction to the location of relevant Edge-Edge pairs.

Finally, we must find the longest shared Edge-Edge string within an edge pair,
since such a string implicitly represents *all* Edge-Edge strings within an edge pair.
As before, we will introduce a refinement node into each edge. The placement
of the refinement nodes within the edge pair gives us enough information to
complete the structural merging of the even and odd trees, and as before, we
finish merging the trees before resolving the lengths of the strings at refinement
nodes. Just as in the sequential algorithm, we can define and construct $d(r)$
for any refinement node r, and as before, the length of r depends on its depth
in the tree defined by $d(\cdot)$. It would appear that such a depth computation
takes $O(\log n)$ time by tree contraction. We circumvent this by the following
observation. Define a refinement node r to be *deep* if $d(r)$ is a refinement node;
otherwise, it is *shallow*.

Property 3 *If deep node r has leaf descendant l_i, then the refinement node
ancestor of l_{i+1} is $d(r)$.*

Thus, finding the depth of r in the $d(\cdot)$ tree is reduced to finding the smallest
$j > i$ such that the refinement node ancestor of l_j is shallow. The linear ordering
of the suffixes allows us to find such a j for all i in $o(\log n)$ time with optimal
work. See procedure REFINE in [FM96].

Implementation Details. To implement MERGE-TREE we need only implement procedures NODE-NODE, NODE-EDGE, EDGE-EDGE, MERGE-PATHS, and REFINE.

- NODE-NODE: We hash the fingerprints of the node strings in one tree and look up the fingerprints of the node strings of the other tree. This gives the Node-Node strings. Time: $O(\log^* n)$.
- NODE-EDGE: Since the active nodes are linearly ordered, searching for the pseudo-anchor is reminiscent of finding the leftmost 1 in a 0/1 array. The tree structure adds complications which we overcome by random sampling of the leaves. Time: $O(1)$.
- EDGE-EDGE: Once we have found the pseudo-anchors, inserting the refinement nodes is trivial. Time: $O(1)$.
- MERGE-PATHS: We resort to techniques which we use to implement UNNAME and which is outlines in [FM96]. This step requires integer sorting and thus is both the time and work bottleneck.
- REFINE: We easily compute the d pointers and have each refinement node determine if it is shallow or deep. Each leaf l_i determines if it has a shallow ancestor, and if so, marks $A[i] = 1$ in some work array A. All other positions of A get marked with 0. Now an all-nearest-ones computation gives the depth of each refinement node in the d tree. Time: $O(\alpha(n))$.

Taken together we achieve the bounds stated in Lemma 3.6. Arguments for time and work bounds, proofs of the properties mentioned in this section and technical details of the implementation are in [FM96].

4 Discussion

In this paper, we settle a long standing open problem in stringology by giving a log time, linear work suffix tree construction algorithm. Just as importantly, we have given an algorithm with a very simple sequential version, one which differs considerably from the amortization based algorithm known heretofore. We leave as an open problem to give a linear work logarithmic time deterministic algorithm for this problem. Although several sequential uses of suffix trees are known, optimal parallel algorithms for these applications of suffix trees are open. Only recently optimal use of suffix trees was developed for some applications, namely in dictionary matching and compression [FM95]. But optimal parallel usability of suffix trees for other applications remains to be explored.

References

[AIL+88] A. Apostolico, C. Iliopoulos, G.M. Landau, B. Scieber, and U. Vishkin. Parallel construction of a suffix tree with applications. *Algorithmica*, 3:347–365, 1988.

561

[Apo84]   Alberto Apostolico. The myriad virtues of subword trees. In A. Apostolico and Z. Galil, editors, *Combinatorial Algorithms on Words*, pages 85–96. Springer-Verlag, Berlin, 1984.

[CS85]   M. T. Chen and J. Seiferas. Efficient and elegant subword tree construction. In A. Apostolico and Z. Galil, editors, *Combinatorial Algorithms on Words*, chapter 12, pages 97–107. NATO ASI Series F: Computer and System Sciences, 1985.

[DK95]   A. Delcher and S. Kosaraju. Large-scale assembly of dna strings and space-efficient construction of suffix trees. *Proc. of the 27th Ann. ACM Symp. on Theory of Computing*, 1995.

[FM95]   M. Farach and S. Muthukrishnan. Optimal parallel dictionary matching and compression. *7th Annual ACM Symposium on Parallel Algorithms and Architectures*, 1995.

[FM96]   M. Farach and S. Muthukrishnan. Optimal logarithmic time randomized suffix tree construction. Technical Report, DIMACS, 1996.

[GMV91]   J. Gil, Y. Matias, and U. Vishkin. Towards a theory of nearly constant time parallel algorithms. *Proc. of the 32nd IEEE Annual Symp. on Foundation of Computer Science*, pages 698–710, 1991.

[Har94]   R. Hariharan. Optimal parallel suffix tree construction. *Proc. of the 26th Ann. ACM Symp. on Theory of Computing*, 1994.

[Kos94]   S. Kosaraju. Real-time suffix tree construction. *Proc. of the 26th Ann. ACM Symp. on Theory of Computing*, 1994.

[KR87]   R.M. Karp and M.O. Rabin. Efficient randomized pattern-matching algorithms. *IBM Journal of Research and Development*, 31:249–260, 1987.

[McC76]   E. M. McCreight. A space-economical suffix tree construction algorithm. *Journal of the ACM*, 23:262–272, 1976.

[PVW83]   W. Paul, U. Vishkin, and H. Wagener. Parallel computation on 2-3 trees. Technical Report TR-70, Courant Institue, NYU, 1983.

[Rei81]   Rüdiger Reischuk. A fast probabilistic parallel sorting algorithm. In *Proc. of the 22nd IEEE Annual Symp. on Foundation of Computer Science*, pages 212–219, 1981.

[SV94]   S. C. Sahinalp and U. Vishkin. Symmetry breaking for suffix tree construction. *Proc. of the 26th Ann. ACM Symp. on Theory of Computing*, 1994.

[Wei73]   P. Weiner. Linear pattern matching algorithm. *Proc. 14 IEEE Symposium on Switching and Automata Theory*, pages 1–11, 1973.

Improved Parallel Approximation of a Class of Integer Programming Problems

Noga Alon*[1] and Aravind Srinivasan**[2]

[1] School of Mathematical Sciences, Raymond and Beverly Sackler Faculty of Exact Sciences, Tel Aviv University, Tel Aviv 69978, Israel. E-mail: noga@math.tau.ac.il.

[2] Dept. of Information Systems & Computer Science, National University of Singapore, Singapore 119260, Republic of Singapore. E-mail: aravind@iscs.nus.sg.

Abstract. We present a method to derandomize RNC algorithms, converting them to NC algorithms. Using it, we show how to approximate a class of NP-hard integer programming problems in NC, to within factors better than the current-best NC algorithms (of Berger & Rompel and Motwani, Naor & Naor); in some cases, the approximation factors are as good as the best-known sequential algorithms, due to Raghavan. This class includes problems such as global wire-routing in VLSI gate arrays. Also for a subfamily of the "packing" integer programs, we provide the first NC approximation algorithms; this includes problems such as maximum matchings in hypergraphs, and generalizations. The key to the utility of our method is that it involves sums of *superpolynomially many* terms, which can however be computed in NC; this superpolynomiality is the bottleneck for some earlier approaches.

1 Introduction

Research in the derandomization of randomized algorithms has at least three motivations. First, though randomized algorithms perform well empirically, computers do not use "real" random sources: it has been shown that if algorithms such as randomized Quicksort are not implemented carefully when used with some existing pseudorandom generators, their expected running times can be high [12]. In fact, there have been reports of Monte-Carlo simulations giving quite different results under different random-number generators [8], and direct implementations of certain RNC algorithms taking longer time than expected due to the pseudorandom nature of computer-generated "random" bits [10, 11].

---

* Work was done in part while visiting the Institute for Advanced study, School of Mathematics, Princeton, NJ 08540, USA, supported in part by the Sloan Foundation, grant No. 93-6-6 and by the Fund for Basic Research administered by the Israel Academy of Sciences.

** Work done in part at the National University of Singapore, at DIMACS (supported in part by NSF-STC91-19999 and by support from the N.J. Commission on Science and Technology), and at the Institute for Advanced Study, Princeton (supported in part by grant 93-6-6 of the Alfred P. Sloan Foundation).

Second, especially in critical applications, it is preferable to have absolute certainty if possible, rather than probabilistic guarantees. Finally, such research makes progress toward settling the complexity-theoretic question of how much computational power randomness provides.

We present a method to derandomize a class of RNC algorithms. We use it to derive good NC approximation algorithms for a class of integer programming problems, matching the approximation factors of the best-known RNC algorithms (and, in some cases, the best-known sequential algorithms also) and improving on the guarantees provided by known NC algorithms. Our work builds on some existing tools for derandomization [4, 22]. A key property of our method is that while, as in [14, 5, 16], it uses the method of conditional probabilities in parallel, its structure enables it to handle a conditional estimator that is a sum of *superpolynomially many* terms, which is a bottleneck for the techniques of [5, 16]. The bottleneck arises from the fact that the work of [14, 5, 16] essentially assigns one processor to each term of the conditional estimator, thus giving them the power of handling estimators which have only polynomially many terms. We expect our method to be useful in other contexts too.

Our first application is in approximately solving a class of integer programming (IP) problems–*minimax integer programs*–by solving their linear programming (LP) relaxations (approximately in parallel, via [15]) and then employing randomized rounding as in [21]; our task is to do the rounding in NC.

For any non-negative integer k, let $[k]$ denote the set $\{1, 2, \ldots, k\}$.

Definition 1. A minimax integer program (MIP) in our case has variables W and $\{x_{i,j} : i \in [\ell], j \in [n_i]\}$, for some integers $\{n_i\}$. Let $N = \sum_{i \in [\ell]} n_i$ and let x denote the N-dimensional vector of the variables $x_{i,j}$ (arranged in any fixed order). An MIP seeks to minimize W, subject to:

(i) Equality constraints: $\forall i \in [\ell] \sum_{j \in [n_i]} x_{i,j} = 1$;

(ii) a system of linear inequalities $Ax \leq \mathbf{W}$, where $A \in [0, 1]^{m \times N}$ and $\mathbf{W}$ is the m-dimensional vector with the variable W in each component;

(iii) Integrality constraints: $x_{i,j} \in \{0, 1\} \ \forall i, j$, and

(iv) W can be any non-negative real.

Note, from constraints (i) and (iii) of MIPs, that for all i, any feasible solution will make the set $\{x_{i,j} : j \in [n_i]\}$ have precisely one 1, with all other elements being 0; MIPs thus model many "choice" scenarios. Consider, e.g., global routing in VLSI gate arrays [21]; this can be generalized as follows (Chapter 3 of [19]). We are given an undirected graph G with m edges, a set of pairs of vertices $\{(s_i, t_i) : 1 \leq i \leq \ell\}$, and $\forall i \in [\ell]$, a set P_i of paths in G, each connecting s_i to t_i. The objective is to connect each s_i with t_i using exactly one path from P_i, so that the maximum number of paths which use any edge in G, is minimized; an MIP formulation is obvious. Similarly, vector-selection [21] and many discrepancy-type problems, are modeled by MIPs; many MIP instances, e.g., global routing, are NP-hard. This has led to the study of efficient approximation algorithms for MIPs. A useful approach for this has been to start with the *linear programming (LP) relaxation* of a given MIP, which lets $x_{i,j} \in [0, 1]$ for each i, j. Thus, such an LP relaxation is a linear program and hence is solvable in polynomial time.

When the optimum C^* of the LP relaxation of a given MIP is at most $O(\log m)$, we present an NC approximation algorithm for the MIP which has a better approximation guarantee than does previous work [5, 16], and matches that of the known sequential method [20], to within a $(1 + o(1))$ factor. Concretely, if u denotes $\max\{1, \log((\log m)/C^*)\}$, we derive NC algorithms that deliver integral feasible solutions with objective function value $O((\log m)/u)$, for families of MIP instances where C^* is $O(\log m)$. However, a better *existential* result–that the integrality gap of the LP relaxation of *sparse* MIPs is better than that proven by [20]–is known [24]; the results of [20] are the current-best *constructive* approximations. If $C^* = O(\log m)$, we always improve on the approximation factor of [5, 16] by at least a $\log^\epsilon m$ factor for some fixed $\epsilon > 0$. This improvement increases with decreasing C^*; e.g., if $C^* = O(1)$, the improvement is $\Theta(\log^\epsilon m \log\log m)$. An application to an MIP that models a problem in telephone network planning is presented in Sect. 4.

Our method also applies to the class of *packing integer programs* (PIPs), which model many problems in combinatorial optimization; most of these again are NP-hard. A PIP seeks to maximize $c^T \cdot x$ subject to $Ax \leq b$, where $A \in [0, 1]^{m \times n}$, b is an m-vector and c is an n-vector such that *the entries of b and c are non-negative*, with the integrality constraint $x_j \in \{0, 1, \dots, d_j\}$ for every entry x_j of x; some of the d_js could also be infinite. Here, if b_i is at most $O(\log(m + n))$, we derive the first NC approximation algorithms with any "reasonable" performance guarantee; our guarantees again match those of the best-known RNC algorithms (these RNC algorithms are directly got by combining [15] and [21]). However, *sequential* algorithms with better approximation guarantees are known now [25, 24]. The LP relaxation of a given PIP lets each x_j be a real lying in $[0, d_j]$. Let C^* be the optimal objective function value of such an LP relaxation; note that C^* is at least as big as the optimal value of the PIP. Let $B = \min_i b_i$. For PIPs, the work of [21, 20] presents sequential algorithms that deliver integral solutions of value $\Omega(C^*/m^{1/B})$ and $\Omega(C^*/m^{1/(B+1)})$ respectively, if $A \in [0, 1]^{m \times n}$ and $A \in \{0, 1\}^{m \times n}$. If each b_i is at most $O(\log(m + n))$, we present an NC algorithm that matches this bound. This is the first NC approximation algorithm with any reasonable performance guarantee when the b_is are all $O(\log(m + n))$ (the algorithms of [5, 16] will not necessarily satisfy the constraints, even if $b_i = O(\log(m+n))$ for a few values of i). However, the results of [5, 16] are generally better when all the b_is grow faster than $\log(m + n)$.

An important class of PIPs is matching problems on hypergraphs. Recall that a hypergraph $H = (V, E)$ is a collection of subsets E (hyperedges) of a finite set V (vertices). A *matching* in H is a collection of hyperedges from E such that no vertex occurs in more than one edge; a basic and well-known NP-hard problem is to find a matching of maximum cardinality in the given hypergraph. A generalization of this notion is that of k-matchings, for integral $k \geq 1$ [13]: here, we allow each vertex to be present in at most k edges in the subcollection. The k-matching problem is naturally written as a PIP with $b_i = k$ for all i, and thus our method applies if $k = O(\log(m + n))$. Thus even for the basic case of $k = 1$, our method yields the first NC approximation algorithms with any reasonable performance guarantee; we present integral feasible solutions of value $\Omega(C^*/\sqrt{m})$, where m is the number of vertices in the hypergraph. This matches the sequential bound of [20, 1]. Similar results hold for k-matching, when $k = O(\log(m + n))$. Our method provides certain approximations even if all the b_is are not $O(\log(m + n))$: see Theorem 8.

In brief, our approach is as follows. Given an integer program, we first solve its LP relaxation approximately in parallel, using results of [15]. We next show how to do the "randomized rounding" of the resulting fractional solutions in NC. For the IPs we consider, the work of [22] shows that it suffices if the random choices are $O(\log(m + n))$–wise independent. This requires work with certain functions S_k: these are introduced, and a useful simple property of these is spelt out, in Sect. 2. We then use the idea of an "approximate method of conditional probabilities" introduced in [4], that lets us work with polynomial-sized almost $O(\log(m+n))$–wise independent sample spaces, rather than $O(\log(m+n))$–wise independent ones. This lets us do the rounding efficiently.

Thus, the contributions of this work are to present a parallel derandomization technique and to apply it to derive improved NC approximation algorithms for a class of IP problems; some of these are the first such NC algorithms.

2 Preliminaries

We denote "random variable" by "r.v.". For real x and any positive integer r, we define, as usual, $\binom{x}{r} = x(x-1)\cdots(x-r+1)/r!$; $\binom{x}{0} = 1$. We start with a recent tool [22]: a new look at the Chernoff-Hoeffding (CH) bounds [7, 9]. Define, for $z = (z_1, z_2, \ldots, z_n) \in \Re^n$, a family of symmetric polynomials $S_j(z), j = 0, 1, \ldots, n$, where $S_0(z) \equiv 1$, and for $1 \leq j \leq n$, $S_j(z) = \sum_{1 \leq i_1 < i_2 \cdots < i_j \leq n} z_{i_1} z_{i_2} \cdots z_{i_j}$. Then, a small extension of a basic theorem of [22] that we will need is

Theorem 2. ([22]) *Given r.v.s* $X_1, \ldots, X_n \in [0, 1]$, *let* $X = \sum_{i=1}^n X_i$ *and* $\mu = E[X]$. *(a) For any* $\delta > 0$, *any nonempty event* Z *and any* $k \leq \mu(1 + \delta)$, $Pr(X \geq \mu(1 + \delta)|Z) \leq E[Y_k|Z]$, *where* $Y_k = S_k(X_1, \ldots, X_n)/\binom{\mu(1+\delta)}{k}$. *(b) If the* X_is *are independent and* $k = \lceil \mu\delta \rceil$, *then* $Pr(X \geq \mu(1 + \delta)) < E[Y_k] \leq G(\mu, \delta)$, *where* $G(\mu, \delta) = (e^\delta/(1 + \delta)^{1+\delta})^\mu$.

Proof. Suppose $r_1, r_2, \ldots r_n \in [0, 1]$ satisfy $\sum_{i=1}^n r_i \geq a$. Then, it is proven in [22] that for any non-negative integer $k \leq a$, $S_k(r_1, r_2, \ldots, r_n) \geq \binom{a}{k}$. This clearly holds even given the occurrence of any (positive probability) event Z. Hence, $Pr(X \geq \mu(1 + \delta)|Z) \leq Pr(Y_k \geq 1|Z)$, which in turn is at most $E[Y_k|Z]$ by Markov's inequality. A proof of (b) is given in [22]. $\square$

A simple property of the functions S_k helps us. Suppose r.v.s $X_1, X_2, \ldots, X_n$ take on only non-negative values. Focus on a generic term $E[X_{i_1} X_{i_2} \cdots X_{i_k}]$ in the expansion of $E[S_k(X_1, \ldots, X_n)]$. Suppose that for some parameter $\epsilon > 0$, we are able to show that this term is within a relative error of $\pm\epsilon$ from the value of such a term, had the X_i been independent (with the same marginals–individual distributions–as the variables X_i). Then, $E[S_k(X_1, \ldots, X_n)]$ would be at most $(1 + \epsilon)$ times what it would be in the "independent" case; this is since the coefficient of $E[X_{i_1} X_{i_2} \cdots X_{i_k}]$ in $E[S_k(X_1, \ldots, X_n)]$ is *non-negative* (one). This property is not enjoyed by other derandomization tools such as the k-th moment inequality.

Definition 3. For $x \in (0, 1)$, define $D(\mu, x)$ such that $G(\mu, D(\mu, x)) = x$ (D is well-defined).

Fact 1 is easily checked; we will primarily be interested in its first case.

Fact 1 *There is a constant $c > 0$ such that $D(\mu, x) = \Theta(\frac{\log(x^{-1})/\mu}{\log(\log(x^{-1})/\mu)})$ if $\mu \leq c \ln(x^{-1})$, and $\Theta(\sqrt{\log(x^{-1})/\mu})$ otherwise.*

A simple but very useful fact we will need is that though $S_k(r_1, \ldots, r_n)$ has superpolynomially many (in n) terms if $k = k(n) \to \infty$, it is efficiently computable:

Lemma 4. *For any $\mathbf{r} = (r_1, \ldots, r_n)$ and $1 \leq k \leq n$, $S_k(\mathbf{r})$ is computable in NC.*

Proof. The polynomial $f_{\mathbf{r}}(z) = \prod_{i=1}^{n}(z - r_i)$ is $\sum_{k=0}^{n}(-1)^k S_k(r_1, \ldots, r_n) z^{n-k}$. Evaluating f at $(n + 1)$ distinct points and solving a system of linear equations gives us $S_k(r_1, \ldots, r_n)$. $\square$

We also recall a key property of small-bias probability spaces ([17]; see also [3, 2, 6]): a "d-wise ρ-biased" sample space S for n-bit vectors has the property that if $\mathbf{X} = (X_1, \ldots, X_n)$ is sampled uniformly at random from S, then

$$\forall I \subseteq [n], |I| \leq d, \quad \forall b_1, b_2, \ldots b_{|I|} \in \{0, 1\} \quad |Pr_{\mathbf{X} \in S}[\wedge_{i \in I} X_i = b_i] - 2^{-|I|}| \leq \rho. \tag{1}$$

Such spaces of cardinality $O((d \log n/\rho)^2)$, for instance, are constructed explicitly in [3]; these simplify the constructions of [17].

3 LP Relaxations and Randomized Rounding

We now recapitulate the idea of randomized rounding [21]; consider, e.g., the problem of global routing in VLSI, defined in the introduction. Letting r_{ij} be the jth path in P_i and x_{ij} be the indicator variable for path r_{ij} being chosen to join s_i and t_i, we get an MIP, denoted (IP1), of minimizing C, subject to (i) $x_{ij} \in \{0, 1\}$, $\forall i, 1 \leq i \leq \ell$, $\sum_j x_{ij} = 1$, and (ii) $\forall e \in E, \sum_{e \in r_{ij}} x_{ij} \leq C$. The LP relaxation (LP1) of (IP1), relaxes each x_{ij} to lie in $[0, 1]$. Solve (LP1), let $\{x_{ij}^*\}$ be the values of the variables in the optimal solution, and let C^* be the optimal objective function value. The key *randomized rounding* idea of [21] is, for each i independently of the others, to choose the path r_{ij} with probability x_{ij}^*. (The extension of this idea to general MIPs is obvious: independently for each i, randomly round exactly one $x_{i,j}$ to 1, guided by the "probabilities" $\{x_{i,j}^*\}$.)

Now for any edge $e \in E$, $Pr(\text{More than } C^*(1 + D(C^*, 1/m)) \text{ paths use } e) < 1/m$ from the definition of D and thus,

$$Pr(\exists e \in E : \text{More than } C^*(1 + D(C^*, 1/m)) \text{ paths use } e) < m/m = 1; \quad (2)$$

hence, there exists a rounding method with objective function value at most $C^*(1 + D(C^*, 1/m))$. This is derandomized *sequentially* in [20] via the method of conditional probabilities.

In the parallel setting, suppose we are given a *positive linear program* (PLP) of input size N—an LP problem where the coefficient matrix, the r.h.s. vector, and the objective function coefficients are all non-negative, with the variables also constrained to lie in nonnegative ranges. Given any $\epsilon > 0$ with $\epsilon = \log^{-O(1)} N$, an NC algorithm of [15] finds a feasible solution to (PLP) (if one exists), at which the objective function value is within a relative error of $\pm\epsilon$ from optimal. Thus for global routing, for instance, by trying out all values $1, 2, \ldots, \ell$ (and a finer range, if C^* is very small: note that $C^* \geq \max_{i \in [\ell]} 1/|P_i|$) for C in parallel, a fractional solution which is at most $(1 + \log^{-\Theta(1)}(m + n))$ the optimal fractional solution, can be found in NC. We need a little work here, to first transform (LP1) to a packing formulation, to apply the algorithm of [15]. Given a candidate value $a \leq \ell$ for C, consider the LP

(LP1') Maximize $\sum_{i,j} x_{i,j}$ subject to: (i) $0 \leq x_{i,j} \leq 1$ for each i, j; (ii) $\forall i$, $\sum_j x_{i,j} \leq 1$, and (iii) $\sum_{e \in r_{i,j}} x_{i,j} \leq a$ for each $e \in E$.

It is easily checked that this is a formulation equivalent to (LP1), if we wish to try out the case $C = a$. In all our IP applications, finding such an approximately good fractional solution is handled similarly.

But the rounding in parallel is trickier, and that is where we apply our method. To our knowledge, the best current method is the *parallel lattice approximation* algorithm of [16]. Given a matrix $A \in [0,1]^{m \times n}$ and a vector $p \in [0,1]^n$, the lattice approximation problem is to find a *lattice point* $q \in \{0,1\}^n$ such that $\|A \cdot (p - q)\|_\infty$ is "small" [20]. Letting $c_i = (Ap)_i$, [16] shows how to find $q \in \{0,1\}^n$ in NC such that for each i, $|A \cdot (p-q)| = O(c_i^{1/2+\epsilon} \sqrt{\log m} + \log^{1/(1-2\epsilon)} m)$, for any fixed $\epsilon > 0$. This is not as good as in the sequential domain. In particular for our problem, randomized rounding and its sequential derandomization [20] guarantee a solution with value $C_1 = C^*(1 + D(C^*, 1/m))$. Note from Definition 3 and Fact 1 that if $C^* = o(\log m)$, then $C_1 = C^* + O(\frac{\log m}{\log(\log m/C^*)})$; if $C^* = a \log m$, then $C_1 \leq C^*(1 + g(a))$, where g is a positive decreasing function which goes to 0 as $a \to \infty$. Thus, C_1 is better than the value $C^* + O((C^*)^{1/2+\epsilon} \sqrt{\log m} + \log^{1/(1-2\epsilon)} m)$ guaranteed by [16]; similar remarks hold for all MIPs. We demonstrate our method by showing how to match the sequential approximation guarantee of [20] (to within a $(1 + o(1))$ factor) in parallel, if $C^* = O(\log m)$. However, lattice approximation also has an advantage over our approach as it bounds $(Aq)_i - (Ap)_i$ from below also.

4 Approximating Minimax Integer Programs

We now illustrate our method by showing how to achieve an objective function value of $C_3 = C_2(1 + o(1))$ in parallel for global routing, if $C^* = O(\log m)$ ($C_2 = C^*(1 + D(C^*, 1/(2m)))$ here). Note that $C_3 = C_1(1 + o(1))$. We discuss global routing just for concreteness--our results hold for all MIPs. We may assume that each x^*_{ij} is a rational of the form $a/2^b$, where $b = O(\log(m + n))$; as in [16], it is easily seen that such a perturbation affects C^* little. Thus, for each i, we may partition $R = \{0, 1, \ldots, 2^b - 1\}$ into subsets S_{ij}, with $|S_{ij}| = 2^b x^*_{ij}$; we imagine picking a uniformly random b-bit number y_i independently for each pair (s_i, t_i), and choose path r_{ij} for (s_i, t_i) iff $y_i \in S_{ij}$. For each edge $e \in E$, note that (2) refers to a sum of *independent* random bits, one for each (s_i, t_i) path possibly using e; each such random bit Z_i (corresponding to the (s_i, t_i) path using e) will be one iff y_i lies in some fixed subset of R. Looking at our problem slightly more generally as follows, sheds light on our other applications also.

The $O(\log n)$th moment estimator problem. We are given n *independent* r.v.s $y_1, \ldots, y_n$, each of which takes values *uniformly* in $R = \{0, 1, \ldots, 2^b - 1\}$ where $b = O(\log(m + n))$. We are also given, for each $j \in [n]$, a finite set of binary r.v.s $\{z_{jt} : t = 1, 2, \ldots\}$ where z_{jt} is 1 iff y_j lies in some fixed subset R_{jt} of R. Also given are m random variables

$$C_i = \sum_{j=1}^n a_{ij} z_{j,f(i,j)}, \ i \in [m],$$

where $a_{ij} \in [0, 1]$ and f is some arbitrary function. Given that $E[C_i] = c_i = O(\log(m+n))$ for each i, the problem is to find a setting for the y_i's in NC, such that $\forall i, C_i \le d_i = c_i(1 + D(c_i, 1/(2m)))(1 + o(1))$. Letting $y_j = y_{j,b-1} y_{j,b-2} \cdots y_{j,0}$ for each j, we will show how to set, in stages $s = 0, 1, \ldots, b - 1$, the vector $v_s = (y_{1,s}, y_{2,s}, \ldots, y_{n,s})$; since $b = O(\log(m+n))$, the sequentiality in the stages is fine. For each $i \in [m]$, let $k_i = \lceil c_i D(c_i, 1/(2m)) \rceil$; note, crucially, that

$$k_i = O(\log(m + n)), \text{ since } c_i = O(\log(m + n)). \tag{3}$$

Let B denote the "bad" event $(\exists i \in [m] : C_i \ge d_i)$. By Theorem 2(a),

$$Pr(B|Z) \le E[X|Z] \text{ for any nonempty event } Z, \tag{4}$$

where X is defined by the *pessimistic estimator*

$$X = \sum_{i=1}^m \left(S_{k_i}(a_{i1} z_{1,f(i,1)}, a_{i2} z_{2,f(i,2)}, \ldots, a_{in} z_{n,f(i,n)}) \Big/ \binom{d_i}{k_i} \right). \tag{5}$$

Given any positive $\epsilon = (m + n)^{-O(1)}$, we now show how to set the vectors $v_0 := w_0, \ldots, v_{b-1} := w_{b-1}$ in that order, so that the desired inequalities hold, up to some function of ϵ. Let $k = \max_i k_i$. Fix a k-wise $\rho = (2^{-k}\epsilon)$ biased sample space for n-bit vectors, S. Note that $|S| = \text{poly}(m, n)$, since $2^k/\epsilon$ is;

this is crucially where the $O(\log(m + n))$ bound on each c_i is needed (see (3)). For $t = 0, 1, \ldots, b - 1$, let $V_t = (v_t, \ldots, v_{b-1})$, U_t and U_S denote the uniform distributions on $\{0, 1\}^{n(b-t)}$ and S respectively, and let $BU_t = U_S \times U_{t+1}$. If D' is a probability distribution, let $H \sim D'$ denote r.v. H being distributed according to D'. We will pick $w_0, w_1, \ldots$ such that for $t = -1, 0, \ldots, b - 1$,

$$E[X|v_0 = w_0, \ldots, v_t = w_t, \text{ and } V_{t+1} \sim U_{t+1}] \leq (1/2)(1 + \epsilon)^{t+1}. \quad (6)$$

From (4), we see that establishing this for $t = b - 1$ will prove the algorithm if $\epsilon = o(1/\log(m + n))$, since $(1/2)(1 + \epsilon)^b < 1$. Inequality (6) is established by induction on t, as in [4]; the basis $t = -1$ holds, since by definition of D and Theorem 2, $E[X|V_0 \sim U_0] \leq 1/2$. Assume that (6) is true for t; we now show how to pick $w_{t+1} \in S$ such that it holds for $t+1$. Suppose we pick V_{t+1} according to BU_{t+1}. Focus on some term

$$S_{k_i}(a_{i1}z_{1, f(i,1)}, a_{i2}z_{2, f(i,2)}, \ldots, a_{in}z_{n, f(i,n)})$$

in X, which is the sum of $\binom{n}{k_i}$ sub-terms. Then, from (1), we can see that the expectation of each such sub-term is at most

$$(2^{-k_i} + \rho)/2^{-k_i} \leq (1 + \epsilon)$$

times what it would have been, had V_{t+1} been picked according to U_{t+1}. Thus by induction hypothesis, $E[X|v_0 = w_0, \ldots, v_t = w_t, V_{t+1} \sim BU_{t+1}] \leq (1/2)(1 + \epsilon)^{t+1}$; hence, $\exists w \in S : E[X|v_0 = w_0, \ldots, v_t = w_t, v_{t+1} = w, V_{t+2} \sim U_{t+2}] \leq (1/2)(1 + \epsilon)^{t+1}$. Finding w reduces to computing

$$E[X|v_0 = w_0, \ldots, v_t = w_t, v_{t+1} = w, V_{t+2} \sim U_{t+2}] \quad (7)$$

for each $w \in S$ and picking the w with the smallest conditional expectation; we can search over all $w \in S$ in parallel since $|S| = \text{poly}(m, n)$. All this is as in [4].

The key point now is that each term

$$E[S_{k_i}(a_{i1}z_{1, f(i,1)}, \ldots, a_{in}z_{n, f(i,n)})|v_0 = w_0, \ldots, v_t = w_t, v_{t+1} = w, V_{t+2} \sim U_{t+2}] \quad (8)$$

in (7) can be computed efficiently in NC as follows. For each $z_{i,j}$, $p_{ij} = Pr(z_{i,j} = 1|v_0 = w_0, \ldots, v_t = w_t, v_{t+1} = w, V_{t+2} \sim U_{t+2})$ can be computed easily in NC. Now, (8) equals

$$S_{k_i}(a_{i1}p_{1, f(i,1)}, a_{i2}p_{2, f(i,2)}, \ldots, a_{in}p_{n, f(i,n)}),$$

since $v_{t+2}, \ldots v_{b-1}$ are assumed to be picked *independently*. Thus, by Lemma 4, (8) and hence (7), can be computed in NC. Therefore, we get

Theorem 5. *The $O(\log n)$th moment estimator problem is in NC.*

Corollary 6. *Given any MIP, let C^* be its fractional optimum. If $C^* = O(\log m)$, then a feasible integral solution with objective function value at most $C^*(1 + D(C^*, 1/(2m)))(1 + o(1))$, can be found in NC.*

Another problem modeled by MIPs is a generalization of the problem of telephone network planning in bidirectional SONET rings: the "ring loading problem" [23]. Given a ring and a traffic demand between every pair of vertices, all traffic between them must be routed one way or the other around the ring. For a given set of traffic demands, the problem is to route the traffic so that the maximum traffic on any link is minimized; the generalization to arbitrary networks involves, for every pair of vertices, an allowed set of paths of which exactly one must be chosen. If f is the maximum traffic demand between any pair of vertices, and m denotes the number of edges in the network, we can formulate an IP for this problem and scale down the inequality corresponding to each edge by a factor of f, to ensure that the coefficient matrix has entries in $[0, 1]$. Thus, our method delivers better approximation factors here than does [16], if the optimum objective function value is within an $O(\log m)$ factor of f.

5 NC Approximation of a Class of Packing Integer Programs

For PIPs, as in [20], we assume without loss of generality that the entries of c are in $[0, 1]$. We also assume, for simplicity, that $x_i \in \{0, 1\}$ for each i; our results also hold for the general case where $x_i \in \{0, 1, \ldots, d_i\}$. Our method applies when each b_i is at most $O(\log(m+n))$, matching the sequential guarantee. Given such an NC algorithm, why can't we use it for arbitrary PIPs (with no required bound on the b_is), by scaling down constraint i suitably if $b_i > c\log(m + n)$ for some desired constant c? The answer is that the resulting tail bounds will not be good enough: Fact 1 shows that $D(\mu_1, x) > D(\mu_2, x)$ if $\mu_1 < \mu_2$. Thus, we assume for now that $b_i = O(\log(m+n))$ for each i, and consider the general case in Theorem 8.

The idea of [21, 20] is to solve the LP relaxation of (PIP) as usual, then *scale down* each LP optimal variable by a suitable $r \geq 1$ so that after a randomized rounding is performed, all the constraints will be satisified *and* the objective function will not decrease by "too much", with positive probability. Let C^* be the value of the LP optimum. Suppose the CH bounds say that for a sum X of independent r.v.s taking values in $[0, 1]$ with $E[X] = \mu$, $Pr(X \leq E[X](1 - F(\mu, x))) < x$ for $0 < x < 1$. For any given positive ϵ, we pick r such that $r \geq \max_{i \in [m]}(1 + D(b_i/r, \epsilon/(m + 1)))$; such an r exists, since the function $t \mapsto D(b_i/t, \epsilon/(m + 1))$ decreases monotonically to 0 for $t \in (1, \infty)$. Reasoning similarly as for MIPs, with a probability of at least $1 - \epsilon$, all constraints will be satisfied *and* the objective function value will be at least $(C^*/r)(1 - F(C^*/r, \epsilon/(m + 1)))$. In the parallel setting, we can find a feasible solution $\{x_i^* : i \in [n]\}$ to within $(1 - \log^{-O(1)}(m + n))$ of C^* via [15], as before. Let r be such that $r \geq \max_{i \in [m]}(1 + D(b_i/r, 1/(m \ln m)))$. We scale the values $\{x_i^*\}$ down by r and can assume, as before, that each x_i^*/r is rational with denominator 2^b, where $b = O(\log(m + n))$. We show how to do the rounding in NC now, by posing our problem as follows.

The $O(\log n)$th moment estimator problem for packing. We are given n *independent* r.v.s $y_1, \ldots, y_n$, each of which takes values *uniformly* in $R = \{0, 1, \ldots 2^b - 1\}$ where $b = O(\log(m + n))$. We are also given, for each $j \in [n]$, a binary r.v. z_j, where z_j is 1 iff y_j lies in some fixed subset R_j of R. Also given are $(m + 1)$ random variables

$$G_i = \sum_{j=1}^{n} a_{ij} z_j, \ i \in [m], \text{ and } H = \sum_{j=1}^{n} c_j z_j,$$

where $a_{ij}, c_j \in [0, 1]$. Now given that $E[G_i] = g_i = O(\log(m + n))$ for each i and that $E[H] = h$, the problem is to find a setting for the y_i's in NC, such that $G_i \leq h_i = g_i(1 + D(g_i, 1/(m \ln m)))(1 + o(1))$ for each i, and $H \geq h - \sqrt{h}(1 - 2/\ln m)^{-0.5}$. We first show that such a setting exists. As before, let $k_i = \lceil g_i D(g_i, 1/(m \ln m)) \rceil$ $(= O(\log(m + n))$ again), and $k \doteq \max_{i \in [m]} k_i$. We have $Pr(H \leq h - \sqrt{h}(1 - 2/\ln m)^{-0.5}) \leq E[(H - h)^2](1 - 2/\ln m)/h$ by Chebyshev's inequality. Let A and C respectively denote

$$\sum_{i=1}^{m} S_{k_i}(a_{i1} z_1, \ldots a_{in} z_n) / \binom{h_i}{k_i} \text{ and } (H - h)^2(1 - 2/\ln m)/h,$$

and B be the "bad event" $(\exists i : G_i \geq h_i, \text{ or } H \leq h - \sqrt{h}(1 - 2/\ln m)^{-0.5})$. For any positive probability event Z, $Pr(B|Z) \leq E[X|Z]$, where $X = A + C$; X is our *(pessimistic) conditional estimator* now. From Defn. 3, we have $E[A] \leq \sum_{i=1}^{m}(1/(m \ln m)) = 1/\ln m$; the fact $E[(H - h)^2] \leq h$ implies that $E[C] \leq 1 - 2/\ln m$. Hence, $Pr(B) \leq E[X] \leq 1/\ln m + 1 - 2/\ln m < 1$ and hence, a "good" setting for the y_i's exists. Let $y_{ij}, S, \rho, v_t, U_t, U_S$, and BU_t be as for the $O(\log n)$th moment estimator problem. As before, we can show by induction on t, $t = -1, 0, \ldots, b - 1$, how to pick $v_0 = w_0, v_1 = w_1 \ldots$, such that for some constant $a > 0$, $E[X|v_0 = w_0, \ldots, v_t = w_t, \text{ and } V_{t+1} \sim U_{t+1}]$ is at most $E[X] + a(t + 1)2^k \rho$. We omit the details. Thus by picking $\rho = (2ab2^k \ln m)^{-1}$, say, we can ensure that $E[X|v_0 = w_0, \ldots, v_{b-1} = w_{b-1}] < 1$, implying a "good" setting for the y_i's.

Theorem 7. *The $O(\log n)$th moment estimator problem for packing is in NC. Thus, PIPs with the r.h.s. constants bounded by $O(\log(m + n))$ can be approximated in NC to within a $(1 + o(1))$ factor of the sequential bounds of [20, 18].*

Scaling allows us to handle general PIPs. For any fixed $c > 1$ and given an arbitrary PIP, suppose we want an NC approximation algorithm which produces a feasible solution that is at least $1/c$ times the sequential guarantee. Fact 1 shows us a function h such that $D((h(c) \log m)/c, 1/(m \ln m)) \leq c - 1$; thus if we have a sum X of independent r.v.s each lying in $[0, 1]$ such that $E[X] = (h(c) \log m)/c$, then $Pr(X \geq h(c) \log m) \leq 1/(m \ln m)$. Thus if we scale down each inequality with r.h.s. $b_i > h(c) \log m$ by $b_i/(h(c) \log m)$, then our method will produce a feasible solution in NC, which is at least $1/c$ times the sequential guarantee. Thus we can handle general PIPs also, but only to within any constant factor $c > 1$ of the best-known sequential algorithms, instead of an $(1 + o(1))$ factor.

Theorem 8. *For any constant $c > 1$, PIPs can be approximated in NC to within an $1/c$ factor of the sequential guarantee of [20, 18].*

To conclude, our derandomization technique provides good parallel approximation algorithms for a family of integer programs; we expect our technique to be used further and extended in future. A basic way in which this technique is useful is in allowing superpolynomially many terms in the conditional estimator. However, a major limitation of our technique is that it works best only when all terms in the conditional estimator are positive (note that S_k involves a sum of *positive* terms, when its arguments are all positive). This is why we cannot use other useful tools for bounding tail probabilities such as the k-th moment inequality. In particular, our technique currently cannot be used for approximating covering integer programs, which seek to minimize $c^T \cdot x$ subject to $Ax \geq b$, where $A \in \Re_+^{m \times n}$, $b \in \Re_+^m$, and $c \in \Re_+^n$, with the entries x_i of x being constrained to be non-negative integers in some range. It would be nice to get good NC approximatation algorithms for covering integer programs. Another open question is to come up with NC algorithms that match the improved bounds for minimax, packing and covering integer programs due to [25, 24]. In fact, not even RNC algorithms are known for these improved bounds; also, the result of [24] on MIPs is non-constructive, and does not imply even a sequential (randomized) polynomial-time algorithm. Derandomization usually converts a fairly efficient randomized algorithm (which, however, assumes a source of "perfect randomness") into a less efficient deterministic procedure. In a given setting, the cost of this loss of efficiency has to be weighed against the benefit of absolute certainty. The work of [8, 10, 11] mentioned in Sect. 1, suggests that de-randomization techniques might prove their worth in critical applications.

Acknowledgements. We thank Prabhakar Raghavan, David Zuckerman, and the referees for their helpful comments.

References

1. R. Aharoni, P. Erdős, and N. Linial. Optima of dual integer linear programs. *Combinatorica*, 8:13–20, 1988.
2. N. Alon, J. Bruck, J. Naor, M. Naor, and R. Roth. Construction of asymptotically good, low-rate error-correcting codes through pseudo-random graphs. *IEEE Trans. Info. Theory*, 38:509–516, 1992.
3. N. Alon, O. Goldreich, J. Håstad, and R. Peralta. Simple constructions of almost k–wise independent random variables. *Random Structures and Algorithms*, 3(3):289–303, 1992.
4. N. Alon and M. Naor. Derandomization, witnesses for Boolean matrix multiplication and construction of perfect hash functions. To appear in *Algorithmica*.
5. B. Berger and J. Rompel. Simulating ($\log^c n$)-wise independence in NC. *Journal of the ACM*, 38:1026–1046, 1991.
6. S. Chari, P. Rohatgi, and A. Srinivasan. Improved algorithms via approximations of probability distributions. In *Proc. ACM Symposium on Theory of Computing*, pages 584–592, 1994.

7. H. Chernoff. A measure of asymptotic efficiency for tests of a hypothesis based on the sum of observations. *Annals of Mathematical Statistics*, 23:493–509, 1952.

8. A. M. Ferrenberg, D. P. Landau, and Y. J. Wong. Monte Carlo simulations: Hidden errors from "good" random number generators. *Physical Review Letters*, 69(23):3382–3384, 1992.

9. W. Hoeffding. Probability inequalities for sums of bounded random variables. *American Statistical Association Journal*, 58:13–30, 1963.

10. T.-s. Hsu. *Graph augmentation and related problems: theory and practice*. PhD thesis, Department of Computer Sciences, University of Texas at Austin, October 1993.

11. T.-s. Hsu, V. Ramachandran, and N. Dean. Parallel implementation of algorithms for finding connected components. In *DIMACS International Algorithm Implementation Challenge*, pages 1–14, 1994.

12. H. J. Karloff and P. Raghavan. Randomized algorithms and pseudorandom numbers. *Journal of the ACM*, 40(3):454–476, 1993.

13. L. Lovász. On the ratio of optimal integral and fractional covers. *Discrete Mathematics*, 13:383–390, 1975.

14. M. Luby. Removing randomness in parallel computation without a processor penalty. *J. Comput. Syst. Sci.*, 47(2):250–286, 1993.

15. M. Luby and N. Nisan. A parallel approximation algorithm for positive linear programming. In *Proc. ACM Symposium on Theory of Computing*, pages 448–457, 1993.

16. R. Motwani, J. Naor, and M. Naor. The probabilistic method yields deterministic parallel algorithms. *J. Comput. Syst. Sci.*, 49:478–516, 1994.

17. J. Naor and M. Naor. Small–bias probability spaces: efficient constructions and applications. *SIAM J. Comput.*, 22(4):838–856, 1993.

18. S. A. Plotkin, D. B. Shmoys, and É. Tardos. Fast approximation algorithms for fractional packing and covering problems. In *Proc. IEEE Symposium on Foundations of Computer Science*, pages 495–504, 1991.

19. P. Raghavan. *Randomized Rounding and Discrete Ham–Sandwich Theorems: Provably Good Algorithms for Routing and Packing Problems*. PhD thesis, University of California at Berkeley, July 1986. Also available as Computer Science Department Report UCB/CSD 87/312.

20. P. Raghavan. Probabilistic construction of deterministic algorithms: approximating packing integer programs. *J. Comput. Syst. Sci.*, 37:130–143, 1988.

21. P. Raghavan and C. D. Thompson. Randomized rounding: a technique for provably good algorithms and algorithmic proofs. *Combinatorica*, 7:365–374, 1987.

22. J. P. Schmidt, A. Siegel, and A. Srinivasan. Chernoff-Hoeffding bounds for applications with limited independence. *SIAM Journal on Discrete Mathematics*, 8:223–250, 1995.

23. A. Schrijver, P. Seymour, and P. Winkler. The ring loading problem. In preparation, 1994.

24. A. Srinivasan. An extension of the Lovász Local Lemma, and its applications to integer programming. In *Proc. ACM/SIAM Symposium on Discrete Algorithms*, pages 6–15, 1996.

25. A. Srinivasan. Improved approximations of packing and covering problems. In *Proc. ACM Symposium on Theory of Computing*, pages 268–276, 1995.

Efficient Collective Communication in Optical Networks*

J.-C. Bermond,[1] L. Gargano,[2] S. Perennes,[1] A. A. Rescigno,[2] and U. Vaccaro[2]

[1] I3S, CNRS, Université de Nice, 06903 Sophia Antipolis Cedex, France
[2] Dipartimento di Informatica, Università di Salerno, 84081 Baronissi (SA), Italy.

Abstract. This paper studies the problems of broadcasting and gossiping in optical networks. In such networks the vast bandwidth available is utilized through *wavelength division multiplexing*: a single physical optical link can carry several logical signals, provided that they are transmitted on different wavelengths. In this paper we consider both *single–hop* and *multihop* optical networks. In single–hop networks the information, once transmitted as light, reaches its destination without being converted to electronic form in between, thus reaching high speed communication. In multihop networks a packet may have to be routed through a few intermediate nodes before reaching its final destination. In both models we give efficient broadcasting and gossiping algorithms, in terms of time and number of wavelengths. We consider both networks with arbitrary topologies and particular networks of practical interest. Several of our algorithms exhibit optimal performances.

1 Introduction

Motivations. Optical networks offer the possibility of interconnecting hundreds to thousands of users, covering local to wide area and providing capacities exceeding those of traditional technologies by several orders of magnitude. Optical–fiber transmission systems also achieve very low bit error rate compared to their copper–wire predecessors, typically 10^{-9} compared to 10^{-5}. Optics is thus emerging as a key technology in state–of–the–art communication networks and is expecting to dominate many applications. The most popular approach to realize these high–capacity networks is to divide the optical spectrum into many different channels, each channel corresponding to a different wavelength. This approach, called *wavelength–division multiplexing* (WDM) [11] allows multiple data streams to be transferred concurrently along the same fiber–optic, with different streams assigned separate wavelengths.

The major applications for such networks are video conferencing, scientific visualisation and real-time medical imaging, high–speed super-computing and

---

* The work of J-CB and SP was partially supported by the French GDR/PRC Project PRS and by Galileo Project, the work of LG, AAR, and UV was partially supported by the Italian Ministry of the University and Scientific Research, Project: "Algoritmi, Modelli di Calcolo e Strutture Informative", and by Galileo Project.

distributed computing [18, 39, 43]. We refer to the books of Green [18] and McAulay [29] for a presentation of the physical theory and applications of this emerging technology.

In order to state the new algorithmic issues and challenges concerning data communication in optical networks, we need first to describe the most accepted models of optical networks architectures.

The Optical Model. In WDM optical networks, the bandwidth available in optical fiber is utilised by partitioning it into several channels, each at a different wavelength. Each wavelength can carry a separate stream of data. In general, such a network consists of routing nodes interconnected by point–to–point fiber optic links. Each link can support a certain number of wavelengths. The routing nodes in the network are capable of routing a wavelength coming in on an input port to one or more output ports, independently of the other wavelengths. The same wavelength on two input ports *cannot* be routed to a same output port. WDM ligthwave networks can be classified into two categories: *switchless* (also called *broadcast–and–select* or *non–reconfigurable*) and *switched* (also called *re-configurable*). Each of these in turn can be classified as either *single–hop* (also called *all-optical*) or *multihop* [39]. In switchless networks, the transmission from each station is broadcast to all stations in the network. At the receiver, the desired signal is then extracted from all the signals. These networks are practically important since the whole network can be constructed out of passive optical components, hence it is reliable and easy to operate. However, switchless networks suffer of severe limitations that make problematic their extension to wide area networks. Indeed it has been proven in [1] that switchless networks require a large number of wavelengths to support even simple traffic patterns. Other drawbacks of switchless networks are discussed in [39]. Therefore, optical switches are required to build large networks.

A switched optical network consists of nodes interconnected by point–to–point optic communication lines. Each of the fiber–optic links supports a given number of wavelengths. The nodes can be terminals, switches, or both. Terminals send and receive signals. Switches direct their input signals to one or more of the output links. Each link is bidirectional and actually consists of a pair of unidirectional links [39].

In this paper we consider switched networks with *generalised switches*, as done in [1, 3, 10, 38]. In this kind of networks, signals for different requests may travel on a same communication link into a node v (on different wavelengths) and then exit v along different links. Thus the photonic switch can differentiate between several wavelengths coming along a communication link and direct each of them to a different output of the switch. The only constraint is that no two paths in the network sharing a same optical link have the same wavelength assignment. In switched networks it is possible to "reuse wavelengths" [39], thus obtaining a drastic reduction on the number of required wavelengths with respect to switchless networks [1]. We remark that optical switches do not modulate the wavelengths of the signals passing through them; rather, they direct the incoming waves to one or more of their outputs.

Single–hop networks (or all-optical networks) are networks where the information, once transmitted as light, reaches its final destination directly without being converted to electronic form in between. Maintaining the signal in optic form allows to reach high speed in these networks since there is no overhead due to conversions to and from the electronic form. However, engineering reasons [39] suggest that in some situations the multihop approach can be preferable. In these networks, a packet from a terminal node may have to be routed through a few terminal nodes before reaching its final destination. At each terminal node, the packet is converted from light to electronic form and retransmitted on another wavelength. See [32, 33] for more on these questions. In the present paper we consider both switched single–hop and switched multihop networks.

Our Results. In this paper we initiate the study of the problem of designing efficient algorithms for collective communication in switched optical networks.

Collective communication among the processors is one of the most important issues in multi-processor systems. The need for collective communication arises in many problems of parallel and distributed computing including many scientific computations [9, 12, 15] and database management [17, 44]. Due to the considerable practical relevance in parallel and distributed computation and the related interesting theoretical issues, collective communication problems have been extensively studied in the literature (see the surveys [20, 25, 16]). In this paper we will consider the design of efficient algorithms for two widely used collective communication operations: *Broadcasting* and *Gossiping* (also called all-to–all broadcasting). Formally, the broadcasting and gossiping processes can be described as follows.

Broadcasting: One terminal node v, called the source, has a block of data $B(v)$. The goal is to disseminate this block so that each other terminal node in the network gets $B(v)$.

Gossiping: Each terminal node v in the network has a block of data $B(v)$. The goal is to disseminate these blocks so that each terminal node gets all the blocks $B(u)$, for each terminal u in the network.

Although our work seems to be the first to address the problem of collective communication in switched optical networks, there is a substantial body of literature that has considered related problems. Optical routing in arbitrary networks has been recently considered in [1, 3, 30, 38]. Above papers contain also efficient algorithms for routing in networks of practical interest. Routing in hypercube based networks has been considered by [3, 34, 38]. Lower bounds on the number of wavelengths necessary for routing permutations have been given in [34, 4, 37]. Gossiping in broadcast–and–select optical networks has been considered in [1]. Other work related to ours is contained in [13, 23, 14, 24, 25]. In these papers the problem of designing efficient broadcasting and gossiping algorithms in traditional networks has been considered under the assumption that data exchange can take place through edge–disjoint paths in the network.

In this paper we consider both single-hop and multihop networks. In case of single-hop networks we design broadcasting and gossiping algorithms that do not need buffering at intermediate nodes. The algorithms have to guarantee that there is a path between each pair of nodes requiring communication and no link will carry two different signals on the same wavelength. For our purposes, a wavelength will be an integer in the interval $[1, W]$. Generally, we wish to minimise the quantity W, since the cost of switching and amplification devices depends on the number of wavelengths they handle. For single–hop networks we obtain:

- *Optimal* broadcasting algorithms for *all* maximally edge–connected graphs;
- *optimal* gossiping algorithm for rings and hypercubes, *quasi-optimal* algorithms for toruses;
- upper and lower bounds on the number of wavelengths necessary to gossip in arbitrary graphs in terms of the edge–expansion factor.

For multihop networks we derive non–trivial tradeoffs between the number of wavelengths and the number of hops (rounds) necessary to complete the process. We obtain, among several results:

- Asymptotically tight bounds for bounded degree networks;
- Tight bounds for hypercubes, meshes, and toruses.

Some of our results generalise previously known ones; indeed the results of [13] and [23] can be seen as particular cases of our results, when only *one* wavelength is available.

Due to the space limits, all proofs are omitted. We refer to the full version [8] for all omitted proofs.

2 Notations and Definitions

We represent the network as a graph $G = (V(G), E(G))$. For physical reasons, each edge in G is to be considered bidirectional and consisting of a pair of uni-directional optical links [39, 30]. In graph–theoretic language, this is equivalent to say that the network should be represented by a *directed symmetric* graph. For sake of simplicity, we prefer to consider G as an *undirected* graph. However, we will be always careful to count the number of signals crossing an edge taking into account their directions, that is, our algorithms will always assign *different* wavelengths to signals crossing an edge in the *same* direction. We will use the term graph and network interchangeably. The number of vertices of G will be always denoted by n. Given $v \in V(G)$, we denote with $d(v)$ the *degree* of v, with d_{max} and d_{min} we denote the maximum and minimum degree of G, respectively.

Processes are accomplished by a set of calls; a call consists of the transmission of a message from some node x to some destination node y along a path from x to y in G. Each call requires one round and is assigned a fixed wavelength. A node can be involved in an arbitrary number of calls during each round, but we require that if two calls share an edge in the same direction during the same round then they must be assigned different wavelengths.

Given a network G, a node $x \in V(G)$, and an integer t, we denote by $\mathbf{wb}(G, x, t)$ the minimum possible number of *wavelengths* necessary to complete the broadcasting in G in at most t rounds, when x is the source of the broadcast;

we set $\mathtt{wb}(G,t) = \max_{x \in V(G)} \mathtt{wb}(G,x,t)$. Analogously, with $\mathtt{wg}(G,t)$ we shall denote the minimum possible number of *wavelengths* necessary to complete the gossiping process in G in at most t rounds.

Given G, a node $x \in V(G)$, and an integer w, we denote by $\mathtt{tb}(G,x,w)$ the minimum possible number of *rounds* necessary to complete the broadcasting process in G using up to w wavelengths per round, when x is the source of the broadcast; we set $\mathtt{tb}(G,w) = \max_{x \in V(G)} \mathtt{tb}(G,x,w)$. We denote by $\mathtt{tg}(G,w)$ the minimum possible number of *rounds* necessary to complete the gossiping process using up to w wavelengths per round.

The *edge-expansion* $\beta(G)$ of G [26], (also called *isoperimetric number* in [31, 42] and *conductance* in [27]) is the minimum over all subsets of nodes $S \subset V(G)$ of size $|S| \leq n/2$, of the ratio of the number of edges having exactly one endpoint in S to the size of S.

A graph G is k–edge–connected if k is the minimum number of edges to be removed in order to disconnect G, G is *maximally edge–connected* if its edge–connectivity equals its minimum degree.

A *routing* for a graph G is a set of $n(n-1)$ paths $R = \{R_{x,y} \mid x,y \in V(G),\ x \neq y\}$, where $R_{x,y}$ is a path in G from x to y. Given a routing R for the graph G, the load of an edge $e \in E(G)$, denoted by $\mathtt{load}(R,e)$, is the number of paths of R going through e in either directions. The *edge-forwarding index* of G [21], denoted by $\pi(G)$, is the minimum over all routings R for G of the maximum over all the edges of G of the load posed by the routing R on the edge, that is, $\pi(G) = \min_R \max_{e \in E(G)} \mathtt{load}(R,e)$. It is known that [42]

$$\pi(G) \geq \frac{n}{\beta(G)}. \tag{1}$$

Unless otherwise specified, all logarithms in this paper are in base 2.

3 Single–Hop Networks

In this section we consider the number of wavelengths necessary to realize the broadcasting and gossiping processes in single–hop (all–optical) networks.

In the single–hop model it is sufficient to study the number of wavelengths necessary when only *one* communication round is used. Indeed, any one–round algorithm that uses w wavelengths can also be executed in t rounds using $\lceil w/t \rceil$ wavelengths per round, that is,

$$\mathtt{wg}(G,t) \leq \left\lceil \frac{\mathtt{wg}(G,1)}{t} \right\rceil, \qquad \mathtt{wb}(G,t) \leq \left\lceil \frac{\mathtt{wb}(G,1)}{t} \right\rceil. \tag{2}$$

On the other hand, the assumption of a single–hop system implies that if we have a realization of a process in t rounds using up to w wavelengths per round, we can easily obtain a new realization using wt wavelengths and one round. Therefore, in the sequel of this section we will focus on one–round algorithms; we will write $\mathtt{wb}(G)$ and $\mathtt{wg}(G)$ to denote $\mathtt{wb}(G,1)$ and $\mathtt{wg}(G,1)$, respectively.

3.1 Broadcasting

Given a graph G and a node $v \in V(G)$, when v is the source of the broadcasting process there must exist at least $(n-1)/d(v)$ calls of the $n-1$ originated at v that share a same edge incident on v. Therefore,

Lemma 3.1 *For each graph G on n nodes* $\text{wb}(G) \geq \left\lceil \frac{n-1}{d_{\min}(G)} \right\rceil$.

We give now an upper bound that allows to determine the exact value of $\text{wb}(G)$ for all maximally edge–connected graphs and, therefore, for most of the used interconnection networks.

Theorem 3.1 *For each k–edge-connected graph G on n nodes* $\text{wb}(G) \leq \left\lceil \frac{n-1}{k} \right\rceil$.

From Lemma 3.1 and Theorem 3.1 we get

Corollary 3.1 *If G is maximally edge-connected then* $\text{wb}(G) = \left\lceil \frac{n-1}{d_{\min}(G)} \right\rceil$.

The above corollary gives the exact value of the number of wavelengths necessary to broadcast in one round in various classes of important networks. By Mader's theorem [28], Corollary 3.1 gives the exact value of $\text{wb}(G)$ for the wide class of vertex–transitive graphs. In particular, we have

- for the d-dimensional hypercube H_d $\text{wb}(H_d) = \lceil (2^d - 1)/d \rceil$;
- for the $r \times s$ mesh $M_{r,s}$ $\qquad \text{wb}(M_{r,s}) = \lceil (rs - 1)/2 \rceil$;
- for the d dimensional torus C_m^d $\qquad \text{wb}(C_m^d) = \lceil (m^d - 1)/(2d) \rceil$;
- for any Cayley graph G of degree d $\text{wb}(G) = \lceil (n-1)/d \rceil$.

For other classes of graphs G for which the edge connectivity is equal to $d_{\min}$ and, therefore, for which $\text{wb}(G) = \left\lceil \frac{n-1}{d_{\min}(G)} \right\rceil$ by Corollary 3.1, see the survey paper [7].

3.2 Gossiping

In this section we study the minimum possible number of wavelengths necessary to perform gossiping in single–hop networks in exactly one round. In the following lemma we put in relation $\text{wg}(G)$ with $\pi(G)$.

Lemma 3.2 *For each graph G it holds that* $\text{wg}(G) \geq \pi(G)/2$.

Minimising the number of wavelengths is in general not the same problem as that of realizing a routing that minimises the number of paths sharing a same edge. Indeed, our problem is made much harder due to the further requirement of wavelengths assignment on the paths. In order to get equality in Lemma 3.2 one should find a routing R achieving the bound $\pi(G)/2$ for which the associated *conflict graph*, that is, the graph with a node for each path in R and an edge between any two paths sharing an edge in the same direction, is $\pi(G)/2$–vertex colorable. We also notice that the problem of determining the edge–forwarding index of a graph is NP-complete [41].

In the rest of this section we will put in relation the minimum possible number of wavelengths necessary to perform gossiping in G in one round with the edge–expansion of G. From Lemma 3.2 and (1) we get the lower bound $\text{wg}(G) = \Omega(n/\beta(G))$. We now show that gossiping can be efficiently realized in any bounded degree graph with a number of wavelengths within a $(\log^2 n)/\beta(G)$ factor from the optimal. In order to gossip in one round one has to choose a path for each pair of nodes and use these paths concurrently, this is equivalent to the problem of embedding the nodes of the complete graph K_n in G and route the edges of K_n as paths in G. For a bounded degree graph G, Leighton and Rao [26] showed that this problem can be efficiently solved with congestion $O(\frac{n \log n}{\beta(G)})$ and dilation $O(\frac{\log n}{\beta(G)})$. Since each vertex in the conflict graph of the resulting routing has degree upper bounded by (congestion $\times$ dilation)$= O(\frac{n \log^2 n}{\beta^2(G)})$, the greedy colouring algorithm can be used to colour the vertices of the conflict graph with $O(\frac{n \log^2 n}{\beta^2(G)})$ colours, that is, it can be used to assign $O(\frac{n \log^2 n}{\beta^2(G)})$ wavelengths to the paths of the routing so that no two paths sharing an edge have the same wavelength assignment. Summarising,

Theorem 3.2 *In any bounded degree graph G on n nodes* $\text{wg}(G) = O\left(\frac{n \log^2 n}{\beta^2(G)}\right).$

Computing $\beta(G)$ seems an hard computational problem (see [31]), therefore it can be useful also to relate $\text{wg}(G)$ with easily computable parameters of G. In particular, we can obtain bounds on $\text{wg}(G)$ in terms of the spectrum of matrices associated to G. Recalling that the Laplacian of a graph with adjacency matrix A and degree function $d(\cdot)$ is the $n \times n$ matrix with entries $d(x)\delta_{x,y} - A_{x,y}$, where $\delta_{x,y}$ is the Kronecker symbol, from Lemma 2.1 of [2], Theorem 4.2 of [31], Lemma 3.2, Theorem 3.2, and formulæ (1) of the present paper we get:

Theorem 3.3 *Let λ be the second smallest eigenvalue of the Laplacian associated to G. We have* $\text{wg}(G) = \Omega\left(\frac{n}{\sqrt{\lambda(2d_{\max} - \lambda)}}\right)$ *and* $\text{wg}(G) = O\left(\frac{n \log^2 n}{\lambda^2}\right).$

We show now that for some classes of important networks the lower bound on $\text{wg}(G)$ given in Lemma 3.2 can be efficiently reached.

In case of the path P_n on n nodes it is not hard to prove that the shortest path routing gives a set of paths that can be coloured with an optimal number of colours $\pi(P_n)/2 = \frac{1}{2}\left\lfloor \frac{n^2}{2} \right\rfloor$, so that all paths sharing an edge in the same direction have different colours. In the next theorems we determine $\text{wg}(\cdot)$ for the ring, the torus, and the hypercube.

Theorem 3.4 *Let C_n be the ring on n nodes. Then*

$$\text{wg}(C_n) = \left\lceil \frac{1}{2} \left\lfloor \frac{n^2}{4} \right\rfloor \right\rceil.$$

Theorem 3.5 *Let C_k^2 be the $k \times k$ torus. If k is odd then*

$$\mathbf{wg}(C_k^2) = k\lfloor k^2/4 \rfloor/2,$$

if k is even then

$$k^3/8 \leq \mathbf{wg}(C_k^2) \leq (k+1)(k^2/8 + k/2).$$

Theorem 3.6 *Let H_d be the d–dimensional hypercube. We have*

$$\mathbf{wg}(H_d) = 2^{d-1}.$$

4 Multihop Networks

In this section we show that by exploiting the capabilities of the multihop optical model, a drastic reduction on the number of wavelengths can be obtained with respect to (2).

In the following, we will be mostly interested in investigating broadcasting algorithms. Indeed, as it is well known, the gossiping process can be accomplished by first accumulating all blocks at one node and then broadcasting the resulting message from this node. Since accumulation corresponds to the inverse process of broadcasting we get the obvious result

Lemma 4.1 *For each graph G and number of wavelengths w*

$$\mathbf{tb}(G, w) \leq \mathbf{tg}(G, w) \leq 2\,\mathbf{tb}(G, w).$$

4.1 Lower Bounds

Lemma 4.2 *For each graph G on n nodes of minimum degree $d_{\min}$ and maximum degree $d_{\max}$*

$$\mathbf{tb}(G, w) \geq \left\lceil \frac{\log(1 + (n-1)d_{\max}/d_{\min})}{\log(wd_{\max} + 1)} \right\rceil. \tag{3}$$

Lemma 4.3 *Given a graph G on n nodes of maximum degree d, let $\mathbf{t}_0 = \mathbf{tb}(G, w)$. It is possible to perform gossiping on G in t rounds using w wavelengths only if*

$$2(n-1)\frac{(wd+1)^{t-\mathbf{t}_0} - 1}{wd} + (2\mathbf{t}_0 - t)(wd+1)^{t-1} \geq \pi(G)/(2w).$$

4.2 Upper Bounds

In order to obtain our general upper bound on the number of rounds to broadcast in G with a fixed number of wavelengths, we need the following covering property.

Definition 4.1 *An s–tree cover for a graph $G = (V.E)$ is a family $\mathcal{F}$ of subtrees of G such that:*

1. *$\bigcup_{F \in \mathcal{F}} V(F) = V$;*
2. *For each $F, F' \in \mathcal{F}$ it holds $|V(F) \cap V(F')| \leq 1$;*
3. *For each $F \in \mathcal{F}$ it holds $|V(F)| \leq s$.*

The s–tree cover number of G is the minimum size of an s–tree cover for G.

The following result upper bounds the s–tree cover number of any graph; its proof also furnishes an efficient way to determine an s–tree cover which attains the bound.

Lemma 4.4 *For each graph G on n nodes and bound s, the s–tree cover number of G is upper bounded by $2n/s$.*

Before giving the upper bound on the broadcasting time in general graphs, we notice the following application of Lemma 4.4 to the function $\mathbf{wb}(\cdot)$.

Theorem 4.1 *For each k–edge connected graph G on n nodes*

$$\left\lceil \frac{\sqrt{1 + (n-1)d_{\max}/d_{\min}} - 1}{d_{\max}} \right\rceil \leq \mathbf{wb}(G, 2) \leq \left\lceil \sqrt{\frac{2n}{k}} \right\rceil.$$

By using Lemma 4.4 we can prove a general upper bound on $\mathbf{tb}(G, w)$ for any $w \geq 2$; in the case $w = 1$ the bound $\mathbf{tb}(G, 1) \leq \lceil \log n \rceil$ has been given in [13].

Theorem 4.2 *For each graph G on n nodes and number of wavelengths $w \geq 2$*

$$\mathbf{tb}(G, w) \leq \lceil \log n / (\log(w + 1) - 1) \rceil.$$

By Lemma 4.2 and Theorem 4.2 we get

Corollary 4.1 *For each bounded degree graph G on n nodes*

$$\mathbf{tb}(G, w) = \Theta(\log_{w+1} n).$$

We give now a sharper bound on the broadcasting time in the d-dimensional hypercube in terms of the maximum number of wavelengths. In the special case $w = 1$ it is proved in [23] that $\mathbf{tb}(H_d, 1) = \Theta(d/\log d)$.

Theorem 4.3 *For each d and number of wavelengths w*

$$\left\lceil \frac{d}{\log(wd + 1)} \right\rceil \leq \mathbf{tb}(H_d, w) \leq c(d, w) \frac{d}{\lfloor \log(wd + 1) \rfloor} + 2$$

with $c(d, w) \leq 4$ and $\lim_{d \to \infty} c(d, w) \leq \begin{cases} 1 & \text{if } \log w = o(2^d), \\ 1 + \frac{\log e}{e} & \text{otherwise.} \end{cases}$

For meshes and toruses we have the following result

Theorem 4.4 *Let M_{k_1,k_2} and C_{k_1,k_2} be the $k_1 \times k_2$ mesh and torus, respectively, on the $n = k_1 k_2$ nodes in the set $\{(x_1, x_2) \ : \ 0 \leq x_i < k_i, \ i = 1, 2\}$. For each w, k and $k_1, k_2 \leq k$*

$$\left\lceil \frac{\log(2n-1)}{\log(4w+1)} \right\rceil \leq \mathtt{tb}(M_{k_1,k_2}, w) \leq \left\lceil \frac{\log k}{\log\lfloor \sqrt{4w+1} \rfloor} \right\rceil + 1,$$

$$\left\lceil \frac{\log n}{\log(4w+1)} \right\rceil \leq \mathtt{tb}(C_{k_1,k_2}, w) \leq \left\lceil \frac{\log k}{\log\lfloor \sqrt{4w+1} \rfloor} \right\rceil.$$

5 Conclusions and Open Problems

In this paper we have initiated the study of efficient collective communication in switched optical networks. Although we have obtained a number of results, several open problems can be investigated for future lines of research. We list the most important of them here.

• The computation complexity of the quantities $\mathtt{wb}(G, t)$, $\mathtt{wg}(G, t)$, $\mathtt{tb}(G, w)$, $\mathtt{tg}(G, w)$ deserves to be investigated. It is likely that for some of them it is NP–hard. In this view, approximation algorithms in the sense of [40] and [19] could be interesting to design.

• Our algorithm require a centralised control. This seems not to be a severe limitation in that the major applications for optical networks require connections that last for long periods once set up; therefore, the initial overhead is acceptable as long as sustained throughput at high data rates is subsequently available [38]. Still distributed algorithms are worth investigating.

• We did not consider fault tolerant issues here. See the recent survey [35] for an account of the vast literature on fault–tolerance in traditional networks.

References

1. A. Aggarwal, A. Bar-Noy, D. Coppersmith, R. Ramaswami, B. Schieber, M. Sudan, "Efficient Routing and Scheduling Algorithms for Optical Networks", in: *SODA '94*, (1994), 412–423.
2. N. Alon and V. D. Milman, "λ_1, Isoperimetric Inequalities for Graphs, and Superconcentrators", *J. Combinatorial Theory*, Series B, vol. 38, (1985), 73–88.
3. Y. Aumann and Y. Rabani, "Improved Bounds for All Optical Routing", in: *SODA '95*, (1995), 567–576.
4. R. A. Barry and P. A. Humblet, "Bounds on the Number of Wavelengths Needed in WDM Networks", in: *LEOS '92 Summer Topical Mtg. Digest*, (1992), 21–22.
5. R. A. Barry and P. A. Humblet, "On the Number of Wavelengths and Switches in All–Optical Networks", to appear in: *IEEE Trans. on Communications*.
6. C. Berge, *Graphs*, North–Holland.
7. J.-C. Bermond, N. Homobono, and C. Peyrat, "Large Fault–Tolerant Interconnection Networks", *Graphs and Combinatorics*, vol. 5, (1989), 107–123.

584

8. J.-C. Bermond, L. Gargano, S¿ Perennes, A.A. Rescigno, and U. Vaccaro, "Efficient Collective Communication in Optical Networks", manuscript available from the authors.

9. D. P. Bertsekas, and J. N. Tsitsiklis, *Parallel and Distributed Computation: Numerical Methods*, Prentice–Hall, Englewood Cliffs, NJ, 1989.

10. K. W. Cheng, "Acousto-optic Tunable Filters in Narrowband WDM Networks", *IEEE J. Selected Areas in Comm.*, vol. 8, (1990), 1015–1025.

11. N.K. Cheung et al., *IEEE JSAC*: Special Issue on Dense WDM Networks, vol. 8 (1990).

12. J. J. Dongarra and D. W. Walker, "Software Libraries for Linear Algebra Computation on High Performances Computers", *SIAM Review*, vol. 37, (1995), 151–180.

13. A.M. Farley, "Minimum–Time Line Broadcast Networks", *NETWORKS*, vol. 10 (1980), 59-70.

14. R. Feldmann, J. Hromkovic, S. Madhavapeddy, B. Monien, P. Mysliwietz, "Optimal Algorithms for Dissemination of Information in Generalised Communication Modes", in: *PARLE '92*, Springer LNCS 605, (1992) 115–130.

15. G. Fox, M. Johnsson, G. Lyzenga, S. Otto, J. Salmon, and D. Walker, *Solving Problems on Concurrent Processors, Volume I*, Prentice Hall, Englewood Cliffs, NJ, 1988.

16. P. Fraignaud, E. Lazard, "Methods and Problems of Communication in Usual Networks", *Discrete Applied Math.*, 53 (1994), 79–134.

17. L. Gargano and A. A. Rescigno, "Communication Complexity of Fault–Tolerant Information Diffusion", *Proceeding of SPDP '93*, Dallas, TX, 564–571, 1993

18. P. E. Green, *Fiber-Optic Communication Networks*, Prentice–Hall, 1992.

19. G. Kortsarz and D. Peleg, "Approximation Algorithms for Minimum Time Broadcast", *SIAM J. Discrete Math.*, vol. 8, (1995), 401–427.

20. S. M. Hedetniemi, S. T. Hedetniemi, and A. Liestman, "A Survey of Gossiping and Broadcasting in Communication Networks", *NETWORKS*, 18 (1988), 129–134.

21. M.C. Heydemann, J. C. Meyer, and D. Sotteau, "On Forwarding Indices of Networks", *Discrete Applied Mathematics*, vol. 23, (1989), 103–123.

22. M.-C. Heydemann, J.-C. Meyer, J. Opatrny, and D. Sotteau, "Forwarding indices of consistent routings and their complexity", *NETWORKS*, vol. 24, (1994), 75–82.

23. C.-T. Ho and M.-Y. Kao, "Optimal Broadcast in All-Port Wormhole-Routed Hypercubes", *IEEE Trans. Par. and Distr. Sys.*, vol. 6, No. 2, (1995), 200–204.

24. J. Hromkovič, R. Klasing, W. Unger, H. Wagener, "Optimal Algorithms for Broadcast and Gossip in the Edge-Disjoint Path Modes", in: *Proc. of SWAT '94*, Springer LNCS 824,(1994), pp. 219-230.

25. J. Hromkovič, R. Klasing, B. Monien, and R. Peine, "Dissemination of Information in Interconnection Networks (Broadcasting and Gossiping)", to appear in: F. Hsu, D.-Z. Du (Eds.) *Combinatorial Network Theory*, Science Press & AMS.

26. F. T. Leighton and S. Rao, "An Approximate Max-Flow Min-Cut Theorem for Uniform Multicommodity Flow Problems with Applications to Approximation Algorithms", in: *Proceedings of FOCS '88*, (1988), 422–431.

27. L. Lovász, *Combinatorial Problem and Exercises*, 2nd edition, Elsevier, 1993.

28. W. Mader, "Minimale n-fach Kantenzusammenhangende Graphen", *Math. Ann.*, 191 (1971), 21-28.

29. A. D. McAulay, *Optical Computer Architectures*, John Wiley, 1991.

30. M. Mihail, K. Kaklamanis, S. Rao, "Efficient Access to Optical Bandwidth", in: *Proceedings of FOCS '95*, (1995), 548–557.

31. B. Mohar, "Isoperimetric Number of Graphs", *J. Combinatorial Theory*, Series B, vol. 47, (1989), 274–291.
32. B. Mukherjee, "WDM–Based Local Lightwave Networks, Part I: Single–Hop Systems", *IEEE Networks*, vol. 6 (1992), 12–27.
33. B. Mukherjee, "WDM–Based Local Lightwave Networks, Part II: Multihop Systems", *IEEE Networks*, vol. 6 (1992), 20–32.
34. R.K. Pankaj, *Architectures for Linear Lightwave Networks*, PhD Thesis, Dept. of Electrical Engineering and Computer Science, MIT, Cambridge, MA, 1992.
35. A. Pelc, "Fault Tolerant Broadcasting and Gossiping in Communication Networks", *Technical Report*, University of Quebec.
36. S. Personick, "Review of Fundamentals of Optical Fiber Systems", *IEEE J. Selected Areas in Comm.*, vol. 3, (1983), 373–380.
37. G. R. Pieris and G. H. Sasaki, "A Linear Lightwave Beneš Network", to appear in: *IEEE/ACM Trans. on Networking*.
38. P. Raghavan and E. Upfal, "Efficient Routing in All–Optical Networks", in: *Proceedings of STOC '94*, (1994), 133-143.
39. R. Ramaswami, "Multi-Wavelength Lightwave Networks for Computer Communication", *IEEE Communication Magazine*, vol. 31, (1993), 78–88.
40. R. Ravi, "Rapid Rumour Ramification: Approximating the Minimum Broadcasting Time", *Proc. FOCS '94*, (1994), 202–213.
41. R. Saad, "Complexity of the Forwarding Index Problem", *SIAM J. Discrete Math.*, (1995), xxx.
42. P. Solé, "Expanding and Forwarding", *Discr. Appl. Math.*, vol. 58, (1995), 67–78.
43. R.J. Vitter and D.H.C. Du, "Distributed Computing with High-Speed Optical Networks", *IEEE Computer*, vol. 26, (1993), 8–18.
44. O. Wolfson and A. Segall, "The Communication Complexity of Atomic Commitment and Gossiping", *SIAM J. on Computing*, 20 (1991), 423–450.

Shared-Memory Simulations on a Faulty-Memory DMM

Bogdan S. Chlebus[1] and Anna Gambin[12] and Piotr Indyk[3]

[1] Instytut Informatyki, Uniwersytet Warszawski, Banacha 2, Warszawa 02-097, Poland. E-mail: chlebus@mimuw.edu.pl
[2] Fachbereich Informatik LS-2, Dortmund Universität, D-44221 Dortmund 50, Germany. E-mail: aniag@poseidon.informatik.uni-dortmund.de
[3] Department of Computer Science, Stanford University, Stanford, CA 94305, USA. E-mail: indyk@cs.stanford.edu

Abstract. We study the Distributed Memory Machine (DMM) with faults in memory. The DMM consists of n synchronized processors together with n memory units (MUs). A MU can be accessed by at most one processor at a time. The total number of memory faults is assumed to be at most a fixed fraction of the total number of words.

We develop two fast randomized simulations of the PRAM on such a faulty DMM. A simulation consists of two phases: the preprocessing is followed by the simulation proper done in a step-by-step fashion. One simulation is of an $n \cdot \log n$-processor PRAM and it operates with the optimal expected slowdown $\mathcal{O}(\log n)$, the other is of a PRAM with $n/\log n$ processors and has the slowdown $\mathcal{O}(\log \log n)$.

1 Introduction

There is a variety of existing parallel architectures, what is in contrast with the sequential computing where the RAM model is dominant. The RAM can be generalized directly to a parallel computer by combining a number of processors and global memory, each processor having random access to any memory word. The obtained parallel model PRAM is theoretical but powerful and convenient to work with. The Distributed Memory Machine (DMM) is a more realistic model. It is weaker than the PRAM, the difference is that the memory words are organized in memory units (MUs), and only one processor may access a MU at a time. Any processor can contact with arbitrary MU in a single step. This could be realized by the optical technology, and the DMM is essentially equivalent to the Optical Communication Parallel Computer (OCPC).

We present several simulations of the PRAM on a DMM with faults. The DMM is assumed to have faulty memory words, otherwise everything is assumed to be operational. In particular the communication between the processors and the MUs is reliable, and a processor may always attempt to obtain an access to any MU, and, having been granted it, may access any memory word in it, even if all of them are faulty. The only restriction on the distribution of faults among memory words is that their total number is bounded from above by a fraction of the total number of memory words in all the MUs. In particular, some MUs may contain only operational cells, some only faulty cells, and some mixed cells.

Models of Computation. A *Parallel Random Access Machine* (PRAM) consists of a number of synchronized processors and the global shared memory. Each processor has a random access to any memory word for both reads and writes. We work with the EREW PRAM which does not allow concurrent access to a memory word by more than one processor.

A *Distributed Memory Machine* consists of a set of n synchronized processors, denoted P_i, each equipped with its local memory of $\mathcal{O}(1)$ words of size $\mathcal{O}(\log n)$, and n *Memory Units* (MU), denoted M_i, for $1 \leq i \leq n$. The size m of a MU, that is, the number of memory words it stores, is a parameter of the model. Throughout the paper we assume that $m = n^\gamma$, for some constant $\gamma > 0$. A memory unit can be accessed by any processor, but only one at a time. On the low level, the communication of a processor with a MU is in two stages. The first is of reserving an access: there is a special *connection register* of the MU which shows the result of an attempt to be connected. If only one processor makes such an attempt then the connection register shows a confirmation of access. Once a processor has been granted access, it uses the MU as a RAM, for one memory access. If more than one processors attempt to access a MU then a special collision symbol appears in the connection register and is read by all the involved processors; in such a case no processor is granted access. This variant of the DMM is usually referred to as the *1-collision DMM* (see [14]), we write simply Collision DMM.

We consider a DMM with possibly faulty memory words. This model has an additional feature, not present in the ideal fully operational model, that it is able to recognize memory errors. Namely, each processor of the faulty DMM is equipped with a special *fault-detection register*, and after each attempt to access a memory word, the register stores a bit whose value depends on whether the accessed memory word was operational or not (this is analogous to the faulty-memory PRAM of [2, 3, 6]). On the high level, the processors access memory as follows. If the operation is a read, then the processor specifies the MU M and an address of a word x in M; it receives back one of the three possible kinds of messages:
1) conflict for access to M;
2) access to M granted, word x faulty;
3) access to M granted, x operational, x stores value v.
If the operation is a write, then the processor specifies additionally a value v' to store in x.

The errors are *static*, what means that the status of a memory word (operational or faulty) does not change in the course of a computation. It is assumed that there is a constant $0 < q < 1$ such that the total number of faulty memory words does not exceed $q \cdot n \cdot m$. The distribution of errors across MUs can be *arbitrary*; it is possible that a particular MU does not contain any operational memory words at all. The distribution of errors is not known in advance, hence the processors begin a computation with preprocessing the MUs, to create a mechanism of access to the (possibly) faulty memory. We refer to such a preprocessing as the *memory formatting*. Each processor has a private local memory consisting of $\mathcal{O}(1)$ *registers*, which is assumed to be fully operational. Since the memory is faulty, the input data cannot be stored there. They are either in the processors' local memories or are provided to the processors after formatting.

An *Optical Communication Parallel Computer* consists of a completely connected set of processors. A processor may send a message to any other processor in one step. The message is received successfully provided it was the only one sent during the step to the receiver. The operational Collision DMM and OCPC are equivalent. If these machines may have memory faults, then the DMM is weaker because the processors cannot exchange messages directly, the only way to communicate is via the memory.

Related Research. Simulations of the PRAM on a DMM were given by Czumaj, Meyer auf der Heide and Stemann [4], Dietzfelbinger and Meyer auf der Heide [5], Karp, Luby and Meyer auf der Heide [8], Mehlhorn and Vishkin [13], Meyer auf der Heide, Scheideler and Stemann [15]; see also the survey article [14] and the references therein. All these simulations assumed a fully operational DMM.

There has been a lot of research done recently with the aim of designing either specific fault-resilient PRAM algorithms or PRAM simulations on machines prone to processor errors. The applied approaches differed in the nature of faults (static versus dynamic, deterministic versus random, fail-stop versus restartable), the properties of the underlying model (synchronous versus asynchronous) and the efficiency criteria (time versus work). Kanellakis and Shvartsman [7] introduced the fail-stop PRAM and developed many deterministic and robust (that is, work efficient) algorithms. Kedem, Palem and Spirakis [10] and Kedem *et al.* [9] designed robust general PRAM simulations for probabilistic fail-stop errors.

Randomized fast simulations of the operational PRAM on PRAM with memory faults were designed by Chlebus, Gambin and Indyk [2]; the models of faults considered were both static and dynamic, for each case two simulations were developed: one operating with a constant expected slowdown and the other in the logarithmic expected slowdown, depending on the number of processors available. Deterministic logarithmic-time-slowdown simulations of a PRAM on a synchronous PRAM with both processor and memory failures were developed by Chlebus, Gąsieniec and Pelc [3]. Indyk [6] studied computations exploiting bit operations and resilient to memory faults.

Overview of the Simulations. The simulations are randomized and Monte Carlo; this means that they always operate within the stated time bounds, and are correct with a large probability. A simulation consist of two phases: preprocessing and the proper part. The preprocessing is like disk formatting, we refer to it as "memory formatting", its goal is to provide an access mechanism to the operational memory cells. The simulation proper is done in a step-by-step way, the time of simulating one step is referred to as the *slowdown* of the simulation.

Two simulations are presented. One is of a $n \cdot \log n$-processor PRAM, it has the optimal (expected) slowdown $\mathcal{O}(\log n)$. The other simulation is of a $n/\log n$-processor PRAM, and the slowdown is $\mathcal{O}(\log \log n)$. The formatting algorithms operate in time sublinear in m, more precisely in time $\mathcal{O}(\sqrt{m} \cdot \log n)$. All the above resource bounds hold with high probability.

Probability. A property Φ depending on a natural number k is said to hold *with a high probability* (abbreviated to w.h.p.) if, for some constant $\delta > 0$, Φ holds with the probability at least $1 - k^{-\delta}$, as long as k is sufficiently large. Whenever this phrase is used, it is understood that δ can be made arbitrarily large by manipulating other constants, often assumed only tacitly.

Let $\mathcal{H}$ be a family of hash functions $h : U \rightarrow [1..n]$, where $|U| \leq n^c$, for some constant c. $\mathcal{H}$ is said to be *k-universal* if for any fixed sequence of k distinct keys $\langle x_1, \ldots, x_k \rangle$ and a function h selected from $\mathcal{H}$ at random, $\mathbf{P}[h(x_1) = y_1, \ldots, h(x_k) = y_k] = n^{-k}$, for any $y_1, \ldots, y_k$. We will have the memory of a DMM hashed by random $\mathcal{O}(\log n)$-universal hash functions $h : [1..n \cdot m] \rightarrow [1..n]$ selected from the family of functions $\mathcal{H}_S$ defined by Siegel [16]. Each such a function needs space $\mathcal{O}(n^\epsilon)$ to be stored, for $\epsilon < 1$, and can be evaluated in time $\mathcal{O}(1)$. If the memory is operational, then a representation of such function is stored in an array, otherwise the function is distributed throughout the memory.

Paper Organization. Section 2 includes some low-level techniques useful in a faulty-memory environment: testing blocks of memory, broadcasting and graph processes. The high-level techniques are presented in the next two sections, they are the major building blocks of the simulations. In Section 3 we discuss communication issues between the processors and MUs, and present an adaptation and extension of the algorithm originally developed by Anderson and Miller [1] and Valiant [17]. Section 4 includes the description of an implementation of a dictionary in a faulty-memory environment. Having developed the tools, the simulations are presented in Section 5. The proofs will be described in the final version.

2 Basic Algorithms

It is convenient to have a uniform addressing across all the MUs of the DMM. We simply refer to the jth word of the ith MU as the (global) xth word, where $x = i \cdot m + j$, and denote it by $M[x]$. Let $[a..b]$ be the set of integers i such that $a \leq i \leq b$. Suppose that there is given a sequence $\langle d_i \rangle$, for $1 \leq i \leq k$, of numbers in $[1..m \cdot n]$. We associate with it the sequence $\langle D_i \rangle$ of *address functions*, for $1 \leq i \leq k$, where $D_i(y) = (my + d_i) \bmod (m \cdot n)$, for $1 \leq y \leq n$. Throughout the paper, the number k is equal to $d \cdot \log n$, where d is a constant (usually) depending on q.

The algorithms are presented in framed boxes. Whenever they consist of a list of steps, this always means that there is a global synchronization between the steps.

Finding Useful Memory. The total number of operational memory words is at least $(1 - q) \cdot m \cdot n$, where the constant q in known by the algorithm. Let us define a MU to be *useful* if there are at least $\frac{1-q}{2} \cdot m$ operational memory words among all the m words in the MU. There are at least $\frac{1-q}{1+q} \cdot n$ such MUs. Each processor P_i can determine in time $\mathcal{O}(\log n)$ w.h.p. whether the MU M_i is useful by checking a random sample of size $\mathcal{O}(\log n)$ of memory cells if they are operational.

ALGORITHM: SAMPLE MODULE

Each processor P_i selects randomly and attempts to read $a \cdot \log n$ memory words in the MU M_i. A MU is *accepted* (as useful) if there are at least $\frac{5}{8}(1 - q) \cdot a \cdot \log n$ operational words in the sample.

Lemma 1. *For any constant $q > 0$ there is a suitable parameter $a > 0$ in algorithm* SAMPLE MODULE *such that w.h.p. $\Theta(n)$ MUs are accepted and all the accepted MUs are useful.*

To make all the processors participate in computations, a useful MU is assigned to each processor. This is done in such a way that each MU is assigned to $\mathcal{O}(1)$ processors, hence there is a $\mathcal{O}(1)$-time delay for a processor to access its MU. A description of the algorithm ASSIGN MODULES is given.

ALGORITHM: ASSIGN MODULES

S-1. Each processor runs SAMPLE MODULE. If it accepts the MU (as useful) then the processor is referred to as *good*.

S-2. Each good processor P_i writes the address b_i of one of the operational cells of M_i to $c\sqrt{m} \cdot \log n$ different operational memory cells of M_i.

S-3. The processors which are not good try to find a useful MU. They are partitioned into $1/\epsilon$ groups, each of size at most ϵn, where ϵ is chosen such that ϵn is smaller than half of the number of useful MUs. Each group performs the computation in different phases. During the kth phase, each processor P_i from the kth group repeats the following steps, until it finds some value b_j:

1. P_i attempts to connect to a random MU until granted access to some M_j.

2. P_i reads $\sqrt{m}$ random memory cells from M_j, in order to find b_j. If it is successful, P_i keeps attempting to connect to this MU till the end of the phase to block other processors.

Lemma 2. *The algorithm* ASSIGN MODULES *assigns useful modules to all the processors in time $\mathcal{O}(\sqrt{m} \cdot \log n)$ w.h.p..*

Some λ items may be stored in a useful module in a list in time $\mathcal{O}(\lambda + \log n)$ w.h.p. : each processor P_i keeps reading randomly chosen memory words from M_i until λ operational ones are found, adding every new word to the list. This algorithm operates in time $\mathcal{O}(\lambda + \log n)$ w.h.p..

We assume from now on that each MU is useful (in the applications the preprocessing will guarantee that each processor is assigned to a useful MU w.h.p., with $\mathcal{O}(1)$ processors per one MU).

Broadcasting. The operation of *broadcasting* propagates the messages initially known by a few processors to all the other processors. More specifically, we develop an algorithm to propagate $\mathcal{O}(\log n)$ messages. In the beginning each of processors P_i, for $1 \leq i \leq d \log n$ and some $d > 0$, knows one such a message.

ALGORITHM: BROADCAST

S-1. Each processor P_i, for $i = 1, \ldots, d \log n$, writes its value to $a\sqrt{m}$ different operational cells of M_i.

S-2. Each processor P_i repeats the following action, called a *round*, until a message is found, but no more than $b \log n$ times.
   1. Select and contact some memory unit M_k at random,
   2. If granted access then read $\sqrt{m}$ randomly chosen memory cells from M_k.

S-3. Each processor P_i that found a message, writes it to $a\sqrt{m}$ different operational cells of M_i.

S-4. Each processor P_i repeats a round until finding a message, but at most $c \log n$ times. If a message is found, P_i performs Step 3 and stays idle during the following rounds.

S-5. Each processor P_i writes the message it knows into $c\sqrt{m} \log n$ randomly chosen memory cells of M_i.

S-6. Each processor performs the following procedure $f \log^2 n$ times :
   1. Choose some module M_k at random,
   2. Read $\sqrt{m}/\log n$ random memory cells of M_k,
   3. If a new message has been found then add it to the list.

Lemma 3. *The algorithm* BROADCAST *propagates successfully $d \log n$ messages among all the processors in time $O(\sqrt{m} \cdot \log n)$ w.h.p..*

Graph Processes. A directed acyclic graph $G = (V, E)$ is given. A *node process* is associated with each node: it is a sequence of trials, each with the constant probability p of success (which causes termination of the process), the probability p is the same for all the nodes. The *graph process* proceeds in *steps*, at one step there is one trial for each (operating) node process. The process starts at nodes with in-degree 0 (*input nodes*) by initializing the associated node process. In general, the process of a node is initialized, if the processes of all its predecessors have terminated. The graph process terminates if all the node processes have terminated. The trials in a step need not to be independent, but any collection of trials from distinct steps must be independent. Graph processes were investigated in [12].

Let us assume that each node of G has the in-degree bounded by t and the out-degree bounded by s. G is *layered* if the set V can be partitioned into disjoint *layers* $V_1, \ldots, V_d$, such that for any edge $(v, w) \in E$ there is i such that $v \in V_i$ and $w \in V_{i+1}$. Let $n = \max_i |V_i|$. A sequence $w_1, \ldots, w_c$ of nodes is a *column* if its elements belong to distinct layers such that the layer of w_i has a greater index than that of w_j provided $i > j$. There exists a partition of V into at most n columns.

The graph represents a computation in the following sense. A node corresponds to computing a value, and an edge from v_1 to v_2 means that the value computed in v_1 is needed at v_2 to compute its value. Also, at a given step, the value at node v can be computed with at least some constant probability p provided that the nodes connected with v by incoming edges have completed their evaluations. We consider such a graph of computation of an n-processor EREW PRAM algorithm, where the processors are associated with nodes in a dynamic way, and a processor can compute the incoming and outgoing edges in its local registers, knowing the current node. The *input* and *output* nodes are distinguished. The input nodes are set to some values, and the goal is to evaluate the output nodes in parallel. Each processor is assigned a column and evaluates its consecutive nodes.

Lemma 4. *Graph of computations can be evaluated on a n-processor DMM with faulty memory by referencing only $2n$ memory addresses, in time $(s + t)^2(5c/p)(d + \log P)$ with the probability $1 - P^{-c}$, where $P = n(t + s)^d$.*

3   Realizing h-Relations in Faulty Memory

A set of access requests among processors and MUs is a *h-relation* if each processor sends at most h requests and each MU is to receive at most h requests. The task of *realizing* a relation is that of satisfying all the requests. Anderson and Miller [1] and Valiant [17] proposed an algorithm to realize $\mathcal{O}(\log n)$-relations. We refer to it as the *AMV algorithm*. The algorithm is in two phases: queue stage and the cleanup stage. We present an adaptation of the AMV algorithm to a faulty-memory environment. The algorithm will be used in a PRAM-step simulation after some preprocessing has already been done. During the preprocessing, some $\mathcal{O}(\log n)$ address functions will have been generated randomly and made known to all the processors.

There are two kinds of memory requests that we will need to be able to realize. The *simple* one is when a processor knows the MU it needs to access and the address of an operational memory cell in the MU. The *probing* request is when a processor knows some *virtual address* x, it is an integer $1 \le x \le n$, and it needs to access some operational memory cell of the form $M[D_t(x)]$. In the simulations, all the processed requests are either simple or probing. We describe and analyze AMV for the case of probing requests, this subsumes the case of simple requests. A code for the queue stage is given. The *queue* of a processor or a MU is the set of its requests not realized yet at the moment. The requests are stored in the array A with $\log n$ entries. After satisfying the request from $A[i]$, this position is kept blank, and the array is compressed at the end of each round. In the original AMV-algorithm, the rightmost request was put in the place of the just realized one, hence the array stored the still-not-realized requests in a contiguous part. Our modification simplifies the analysis and is crucial in proving Lemma 5. The requests with the virtual address x are said to be *in queue to the virtual unit* G_x. For a selected function D_t, the queue of G_i belongs to the queue to the MU with number $(i + \lfloor \frac{d_i}{m} \rfloor) \bmod n$. Notice that if two processors attempt to realize two distinct virtual addresses then there is no conflict between them for access to a MU. Anderson and Miller [1] showed that the expected number of requests still not satisfied when the QUEUE STAGE algorithm (with simple

```
┌─────────────────────────────────────────────────────────────┐
│ ALGORITHM: QUEUE STAGE                                       │
│ t = 0 ; {t is the index for address functions }             │
│ for k = log log n downto 1 do                               │
│     for j = 1 to a · 2^k do                                 │
│         t = t + 1 ;                                          │
│         set i to a random number in [1..2^k] ;              │
│         if A[i] is not blank then                           │
│             attempt to read M[D_t(A[i])] ;                  │
│             if communication with MU was successful          │
│             and the cell is operational then                │
│                     read and store the information ;         │
│                     set A[i] to blank ;                      │
│     if there are still more than 2^{k-1} requests           │
│         then become idle                                    │
│         else compress array A to the size 2^{k-1}           │
└─────────────────────────────────────────────────────────────┘
```

requests) terminates is $\mathcal{O}(n)$. The algorithm QUEUE STAGE for probing requests has the same property. We strengthen this fact and show that the QUEUE STAGE algorithm for probing requests satisfies all but $\mathcal{O}(n)$ requests w.h.p..

Lemma 5. *The number of requests still not realized after the algorithm* QUEUE STAGE *is finished is* $\mathcal{O}(n)$ *w.h.p.*.

The cleanup stage starts by redistributing the memory references still not satisfied among all the processors in such a way that each processor receives $\mathcal{O}(1)$ of them. Next the requests are sorted on their MU addresses. To this end, we simulate the randomized hypercubic sorting developed by Leighton and Plaxton [11], which sorts n elements in time $\mathcal{O}(\log n)$ w.h.p.. The algorithm can be represented as a graph with a simple hypercubic structure, hence the connections between the processors may be generated on-line and Lemma 4 applies. This shows:

Theorem 6. *A DMM with n processors can realize a $\mathcal{O}(\log n)$-relation of probing memory requests in time $\mathcal{O}(\log n)$ w.h.p.*.

4 Dictionary in Faulty Memory

A dictionary supports the operations of insertion and lookup. We develop a sequential implementation of a dictionary, to be run in each MU. There is a memory of size m available, having at most $q \cdot m$ faulty cells w.h.p.. The keys are from the universe $U = [1..n \cdot m]$. Let $MU[x]$ be the xth word of MU in which a dictionary is implemented. A description for the dictionary preprocessing is given.

Lemma 7. *The dictionary preprocessing can be accomplished in time $\mathcal{O}(m^\epsilon)$ w.h.p., for any fixed $0 < \epsilon < 1$.*

A code for the dictionary operation is given.

ALGORITHM: DICTIONARY PREPROCESSING

S-1. Build a list of length $d \log n$.

S-2. Select $d \log n$ random numbers $d_1, d_2, \ldots$ from the interval $[0..m-1]$ and store them in the list. These numbers define a family $\mathcal{D}$ of *local* address functions as follows: $D_i(x) = (x + d_i) \bmod m$.

S-3. Generate two random functions f and g from $\mathcal{H}_S$, where $f : U \to [1..m]$ and $g : [1..m] \to [1..m]$. Store the functions as *s-entries* $< i, s_i >$, each such a pair in the operational cells of the form $MU[D_k(i)]$, for all indexes of address functions.

Let $\mathcal{R}$ be a sequence of dictionary operations, where the number of keys is at most βm, for a constant $\beta > 1$. The requests from $\mathcal{R}$ are coming in packets of length $\mathcal{O}(\log n)$, all the keys in one packet are distinct. To evaluate f or g at a point, $\mathcal{O}(1)$ s-entries are needed. They have been distributed throughout the memory during preprocessing. If, at the moment t of processing a packet, we need a s-entry with index i, then the cells $MU[D_t(i)], MU[D_{t+1}(i)], \ldots$ are searched, until it is found. Each address function is used at most once while processing the packet.

ALGORITHM: DICTIONARY OPERATION $x \in U$ is the key in a dictionary operation.

$y := f(x)$;

while $g(y)$ is faulty **do** $y := y + 1$;

{ now $MU[g(y)]$ is operational and is the head of a list, maybe empty}

if the operation is a lookup **then**

    traverse the list until the key is found or the end encountered

else {insertion}

    repeat

        select a memory address v at random

    until $MU[v]$ is operational **and** not occupied ;

    store x in $MU[v]$ and add $MU[v]$ to the beginning of the list

---

Theorem 8. *The time to process a packet $\mathcal{L}$ of dictionary requests from $\mathcal{R}$ is $O(\log n)$ w.h.p..*

5 Simulations

Optimal Simulation. A simulation of a PRAM with $n \log n$ processors and $\mathcal{O}(n \cdot m)$ memory words is developed. The processors of the PRAM are partitioned into n groups of $\log n$ elements, a group is simulated by one processor of the DMM. Each step of the simulated PRAM is executed in time $\mathcal{O}(\log n)$ w.h.p. on the DMM, what is optimal. A description of formatting is given.

Theorem 9. *The algorithm OPTIMAL FORMAT, run on a DMM with n processors and the capacity of each MU equal to $m = n^\gamma$, preprocesses all the MUs in time $\mathcal{O}(\sqrt{m} \cdot \log n)$ w.h.p..*

During the simulation proper, the memory is hashed by a random function h in $\mathcal{H}_S$. Two kinds of access requests to MUs are generated: probing and simple. The

---

ALGORITHM: OPTIMAL FORMAT

S-1. Each processor P_i is assigned to a useful MU by the procedure ASSIGN MODULES, together with an operational cell b_i. The cell b_i is to be used by other processors to communicate with P_i.

S-2. The processors $P_1, \ldots, P_{d \log n}$ generate $d \log n$ random numbers $d_1, d_2, \ldots,$ determining the address functions $D_1, D_2, \ldots,$ and broadcast them to all the processors. The processors store them in lists.

S-3. Each processor P_i places records consisting of number i, the number of the assigned MU and b_i in the words $M[D_k(i)]$, for all the address functions D_k.

S-4. All the processors are organized as a full binary tree. A processor gets to know about its neighbors in the tree, say i_1 and i_2, by reading the information stored in words $M[D_k(i_1)]$ and $M[D_k(i_2)]$, for all indexes k.

S-5. The root selects a random hash function $h : [1..mn] \to [1..n]$ from $\mathcal{H}_S$ and propagates its description on the tree to all the processors. Each processor stores the s-entries in a dictionary.

---

former are to find the number of the useful MU storing address $h(x)$, the latter are sent to the useful MUs, and then back, if the request is a read operation. A description of a simulation of a PRAM step is given.

Step 1 of the algorithm OPTIMAL STEP is accomplished by probing requests, and step 2 by simple ones. The following are the key facts:

(1) The number of PRAM addresses that hash to the same MU by h is $\mathcal{O}(m)$ w.h.p..

(2) For each step of the simulated PRAM, the number of requests directed to each MU is $\mathcal{O}(\log n)$ w.h.p..

They provide the assumptions under which the efficiency and correctness of realizing h-relations and supporting dictionary requests are proved in Sections 3 and 4.

ALGORITHM: OPTIMAL STEP

S-1. Each processor finds the physical number of the MU of the $h(x)$-th processor, for all the PRAM cells x that it needs to access.

S-2. Each processor sends the access request to the MU of the $h(x)$-th processor, for all the cells x that it needs to access. Simultaneously, the processor receives requests sent to it from other processors.

S-3. Each processor performs dictionary operations on its assigned MU, corresponding to the received read and write requests.

S-4. Each processor sends back the answers to the read requests.

This shows:

Theorem 10. *A DMM with n processors and the capacity of each MU equal to $m = n^\gamma$ possibly faulty memory cells can simulate, after preprocessing, a single step of a PRAM with $n \log n$ processors and $\mathcal{O}(n \cdot m)$ memory cells with the delay $\mathcal{O}(\log n)$ w.h.p..*

Faster Simulation. A simulation of a PRAM with $\Theta(n/\log n)$ processors and $\Theta(mn/\log n)$ memory words is developed. Each step of the simulated PRAM is executed in time $\mathcal{O}(\log\log n)$ w.h.p. on the DMM. A description of formatting is given.

ALGORITHM: FAST FORMAT
S-1. The processors perform Steps 1 through 4 of OPTIMAL FORMAT.
S-2. The processors in each group form a tree.
S-3. $p\log n$ hash functions $h_i : [1..mn] \to [1..n]$ are selected randomly from $\mathcal{H}_S$. The kth function is stored by every kth processor in a group of processors in a dictionary.

Theorem 11. *The time of the FAST FORMAT run on a DMM with n processors and the capacity of each MU equal to $m = n^\gamma$ is $\mathcal{O}(\sqrt{m} \cdot \log n)$ w.h.p..*

Let us denote $\lambda = p\log n$. Let the simulated PRAM have n/λ processors and $c_m \cdot n \cdot m/\lambda$ memory cells. The processors are partitioned into n/λ groups $\mathcal{P}_i$ of λ elements. The ith group simulates the ith processor of the PRAM. We assume that all the processors in group $\mathcal{P}_i$ know a PRAM memory address x to access. To simulate the PRAM step, the processors in $\mathcal{P}_i$ reference the memory words $M[h_k(x)]$, for $1 \leq k \leq \lambda$, and then identify the one with the latest update. In order to reduce the number of collisions in each step, only a fraction of a group is activated, say $c_p\lambda$ processors, and after $1/c_p$ steps all the tasks are completed. A description of the simulation of a step is given. All the processors in a group know the next PRAM address to access after the last step of the algorithm FAST STEP.

ALGORITHM: FAST STEP
Access by the ith PRAM processor to address x.
S-1. The processors in $\mathcal{P}_i$ make read attempts: the kth processor attempts to read $M[h_k(x)]$.
S-2. All the successful processors identify the value v with the maximum time stamp among the retrieved ones. This is accomplished by running a maximum-finding algorithm on the tree, which finds the maximum of λ keys on λ processors in time $\mathcal{O}(\log\lambda)$.
S-3. The value v computed is communicated to all the processors in the group via the tree. Then all the processors perform the required local computations of the PRAM. The memory cells $M[h_k(x)]$ are attempted to be accessed, the successful processors perform the writes.

Theorem 12. *The algorithm FAST STEP simulates a single step of a PRAM in time $\mathcal{O}(\log\log n)$ w.h.p..*

Acknowledgement: Thanks are due to Artur Czumaj for his comments.

References

1. R.J. Anderson, and G.L. Miller, Optical Communication for Pointer Based Algorithms, *Tech. Rep.* CRI 88-14, Comp. Sci. Dpt., USC, Los Angeles, 1988.
2. B.S. Chlebus, A. Gambin, and P. Indyk, PRAM Computations Resilient to Memory Faults, in *Proceedings of the 2nd Annual European Symposium on Algorithms,* 1994, LNCS 855, pp. 401–412.
3. B.S. Chlebus, L. Gąsieniec, and A. Pelc, Fast Deterministic Simulation of Computations on Faulty Parallel Machines, in *Proceedings of the 3rd Annual European Symposium on Algorithms,* 1995, LNCS 979, pp. 89–101.
4. A. Czumaj, F. Meyer auf der Heide, and V. Stemann, Shared Memory Simulations with Triple-Logarithmic Delay, in *Proceedings of the 3rd Annual European Symposium on Algorithms,* 1995, LNCS 979, pp. 46–59.
5. M. Dietzfelbinger, and F. Meyer auf der Heide, Simple, Efficient Shared Memory Simulations, in *Proceedings of the 5th Annual ACM Symposium on Parallel Algorithms and Architectures,* 1993, pp. 110–119.
6. P. Indyk, On Word-Level Parallelism in Fault-Tolerant Computing, in *Proceedings of the 13th Annual Symposium on Theoretical Aspects of Computer Science,* 1996.
7. P.C. Kanellakis, and A.A. Shvartsman, Efficient Parallel Algorithms Can Be Made Robust, *Distributed Computing,* 5 (1992) 201-217.
8. R. Karp, M. Luby, and F. Meyer auf der Heide, Efficient PRAM Simulations on Distributed Memory Machine, in *Proceedings of the 24-th Annual ACM Symposium on Theory of Computing,* 1992, pp. 318–326.
9. Z. M. Kedem, K. V. Palem, A. Raghunathan, and P. Spirakis, Combining Tentative and Definite Executions for Very Fast Dependable Parallel Computing, in *Proceedings of the 23rd Annual ACM Symposium on Theory of Computing,* 1991, pp. 381-390.
10. Z. M. Kedem, K. V. Palem, and P. Spirakis, Efficient Robust Parallel Computations, in *Proceedings of the 22nd Annual ACM Symposium on Theory of Computing,* 1990, pp. 138-148.
11. T. Leighton, and G. Plaxton, A (Fairly) Simple Circuit That (Usually) Sorts, in *Proceedings of the 31st Annual Symposium on Foundations in Computer Science,* 1990, pp. 264–274.
12. C. Martel, A. Park, and R. Subramonian, Work-Optimal Asynchronous Algorithms for Shared Memory Parallel Computers, *SIAM J. Comput* (1992) 1070–1099.
13. K. Mehlhorn, and U. Vishkin, Randomized and Deterministic Simulations of PRAMs by Parallel Machines with Restricted Granularity of Parallel Memories, *Acta Informatica* 21 (1984) 339–374.
14. F. Meyer auf der Heide, Hashing Strategies for Simulating Shared Memory on Distributed Memory Machines, in *Proceedings of the 1st Heinz Nixdorf Symposium "Parallel Architectures and their Efficient Use,"* 1992, LNCS 678, pp.20–29.
15. F. Meyer auf der Heide, C. Scheideler, and V. Stemann, Exploiting Storage Redundancy to Speed Up Randomized Shared Memory Simulations, in *Proceedings, 12th Annual Symposium on Theoretical Aspects of Computer Science,* 1995, LNCS 900, pp. 267–278.
16. A. Siegel, On Universal Classes of Fast High Performance Hash Functions, Their Time-Space Tradeoff, and Their Applications, in *Proceedings of the 30th Annual Symposium on Foundations of Computer Science,* 1989, pp. 20–25.
17. L.G. Valiant, General Purpose Parallel Architectures, in *"Handbook of Theoretical Computer Science,"* J. van Leeuwen (Ed.), Elsevier, 1990, vol. A, pp. 869–941.

Fast Deterministic Backtrack Search[*]

Kieran T. Herley[1]   Andrea Pietracaprina[2]   Geppino Pucci[3]

[1] Department of Computer Science, University College Cork, Cork, Ireland
[2] Dipartimento di Matematica Pura e Applicata, Università di Padova, Padova, Italy
[3] Dipartimento di Elettronica e Informatica, Università di Padova, Padova, Italy

Abstract. The *backtrack search problem* involves visiting all the nodes of an arbitrary binary tree given a pointer to its root, subject to the constraint that the children of a node are revealed only after their parent is visited. This note describes a fast, deterministic backtrack search algorithm for a p-processor COMMON CRCW-PRAM that visits any n-node tree of height h in time $O\left((n/p + h)(\log\log\log p)^2\right)$. This upperbound compares favorably with a natural $\Omega(n/p + h)$ lower bound for this problem. Our approach embodies novel, efficient techniques for dynamically assigning tree-nodes to processors to ensure that the work is shared equitably among them.

1 Introduction

Many computations that arise from divide-and-conquer, branch-and-bound, functional expression evaluation and other algorithmic techniques involve the systematic exploration of a tree whose structure reflects that of the underlying computation. In this paper, we devise novel techniques for tree exploration on shared-memory parallel machines. Specifically, we present an efficient deterministic PRAM algorithm for the *backtrack search problem*, which involves visiting all of the nodes of an arbitrary binary tree given its root and subject to the constraint that the children of a node are revealed only after the node is visited.

Throughout the paper, we use the symbols n and h to denote the size of the tree under consideration and its height, respectively. Since $\Omega(n)$ work is needed to visit n nodes and since any tree of height h contains a path of h nodes whose visit times must form a strictly increasing sequence, it follows that any backtrack search algorithm requires $\Omega(n/p + h)$ time on a p-processor machine. We obtain the following result that is within a triply-logarithmic factor of this natural lower bound.

Theorem 1. *There is an algorithm running on a p-processor COMMON CRCW-PRAM that performs backtrack search on any n-node binary tree of height h in $O\left((n/p + h)(\log\log\log p)^2\right)$ time, in the worst case.*

---

[*] This research was supported, in part, by MURST and CNR of Italy and by the ESPRIT III Basic Research Programme of the EC under contract No. 9072 (project GEPPCOM).

Our result is the first efficient, deterministic PRAM algorithm that places no restrictions on the structure, size or height of the tree to which it is applied. The algorithm performs an optimal number of $O(n/p+h)$ "node-visiting" steps: intuitively, the $O((\log\log\log p)^2)$ multiplicative factor in the running time of Theorem 1 captures the average overhead per step required to ensure that the workload is equitably distributed among the processors.

Efficient backtrack search algorithms have been obtained previously for the completely-connected network of processors [KZ93, LAB93], the butterfly network [Ran94] and the two-dimensional mesh [KP94], under a variety of cost models. (A number of related problems such as *branch-and-bound* [Ran90, KZ93, LAB93, KP94] and *dynamic tree embedding* [LNRS92, AL91, BGLL91] have also been addressed in the literature.)

Both Ranade's butterfly algorithm [Ran94] and the Karp-Zhang algorithm [Ran90, KZ93] are randomized and explore the tree in $O(n/p+h)$ steps, with high probability. It should be noted that the butterfly algorithm focuses on the number of "node-visiting" steps and does not fully account for overhead due to manipulations of local data structures. The (synchronous) algorithm of Kaklamanis and Perisano [KP94] is deterministic and visits any tree in $O(\sqrt{ph})$ time on a $\sqrt{p} \times \sqrt{p}$ mesh, provided that the tree has size $O(p)$. Since n may well be significantly larger than p in many of the intended applications, this restriction on the tree-size may prove to be a significant limitation.

To specify the backtrack search problem more formally, we assume that the root of the tree is stored in cell 0 of the PRAM's (unbounded) shared memory, and that the other tree nodes are stored in arbitrary (unknown) memory locations. Each node stores the pointers to its children (if any). We say that a node is *visited* when its memory location is accessed and the pointers to its children are revealed. No assumption is made about the structure of the tree or about its location within memory, nor indeed does the algorithm depend on the parameters n and h. Note that while our formulation implicitly suggests that the tree is present in memory before the execution of the algorithm, this is not an essential requirement: all of our techniques apply equally to the setting where the tree is dynamically generated as the algorithm executes.

One straightforward strategy to solve the backtrack search problem is to visit the tree in a "breadth-first", level-by-level fashion. An algorithm based on such a strategy would proceed in phases, where each phase visits all the nodes at a certain level and evenly redistributes their children among the processors, in order to balance the work to be done in the next phase. The redistribution could, for example, be accomplished deterministically using parallel prefix [JáJ92]. However, a more efficient approach on the COMMON CRCW-PRAM would be to use Goldberg and Zwick's algorithm for *approximate prefix sums* [GZ94] which takes an arbitrary sequence of p integer values $a_0, a_1, \ldots, a_{p-1}$ and produces values $b_0, b_1, \ldots, b_{p-1}$ such that $\sum_{j=0}^{i} a_j \leq b_i \leq (1+\epsilon)\sum_{j=0}^{i} a_j$, with $\epsilon = o(1)$, and $b_i \geq b_{i-1} + a_i$. This algorithm is deterministic and runs in $O(\log\log p)$ time using $p/\log\log p$ processors. The use of this strategy for redistribution yields an $O(n/p + h\log\log p)$ algorithm for the backtrack search problem, for any values of n, h and p. Our algorithm significantly outperforms this simple strategy for small trees, namely those of size $n = o(ph \log \log p/(\log \log \log p)^2)$.

The general structure of our algorithm is described in Section 2, while Section 3 and Section 4 describe two routines that are the key components of the algorithm. Finally, Section 5 discusses how the algorithm can be employed to evaluate arbitrary bounded-degree DAGs.

2 A High-Level View of the Algorithm

Our algorithm proceeds in a quasi-breadth-first fashion. Let the tree nodes be partitioned into h *levels*, where the nodes of one level are all at the same distance from the root. The exploration process is split into *stages*, each of which visits a *stratum* of the tree consisting of $\ell = \Theta(\log \log p)$ consecutive levels. At the beginning of a stage, all nodes at the top level of the stratum are evenly distributed among the processors. Note that the straightforward strategy mentioned in the Introduction visits any stratum of size $m = \Omega(p\ell^2)$ optimally. Therefore, we focus on techniques to cope efficiently with smaller strata.

Consider a stage visiting a stratum with $m = O(p\ell^2)$ nodes. For convenience, we number the levels of the stratum from 0 to $\ell - 1$, from top to bottom. The stage explores all the nodes in these levels. At any point during the exploration, the set of unvisited nodes whose parents have been visited is called the *frontier*. (The initial frontier contains all the nodes in the top level of the stratum.) Let $F(j)$ denote those frontier nodes at level j, for $0 \le j < \ell$, so that $F = \cup_{j=0}^{\ell-1} F(j)$. In order to evaluate the progress that the algorithm is making, we define a *weight* function on the frontier F as

$$w(F) = \sum_{j=0}^{\ell-1} |F(j)| 3^{\ell-j} .$$

Note that the contribution of a frontier node to $w(F)$ is exponentially decreasing in its level. Therefore, visiting frontier nodes at lower numbered levels rather than nodes further down the tree results in a more substantial decrease in the weight function. The exploration is not guaranteed to proceed in a regular, breadth-first manner, due to the unpredictable structure of the tree. However, we make use of certain weight-biased, load-balancing techniques to ensure that the exploration of heavier nodes in a stratum are not unduly delayed by the exploration of those of lesser weight.

A stage consists of two parts. In the first part, a sequence of *visiting steps* is performed to explore nodes in the stratum until the frontier weight is less than or equal to p. The visiting steps are executed in batches of ℓ and a weight estimate is computed after the execution of each batch, using the approximate prefix sums algorithm, whose complexity is dominated by that of the visiting steps. The second part of the stage completes the exploration of the stratum by assigning $2^{\ell-j}$ distinct processors to each node of $F(j)$, which will visit all of its descendants in $O(\log \log p + \ell)$ time in a straightforward fashion. Note that since the frontier weight is at most p, less than p processors are needed in the second part. At the end of the stage, the children of nodes at level $\ell - 1$, are evenly distributed among the processors and the next stage can begin.

In order to determine the total running time of a stage, we need to give a bound on the number of visiting steps. Let F_t be the frontier at the beginning of the tth visiting step. The step is called *full*, if it visits $\Omega(p)$ nodes in F_t, and it is called *reducing* if it visits at least half of the nodes in $\bigcup_{j=0}^{i} F(j)$, for each i in the range $0 \le i < \ell$. Sections 3 and 4 will show how to perform a visiting step in time $O((\log \log \log p)^2)$ while ensuring that it is always either full or reducing. (A visiting step can both full and reducing.) Clearly, there are at most $O(m/p)$ full visiting steps in the stage, whereas the number of reducing steps is bounded by the following lemma.

Lemma 2. *If* $m = O(p\ell^2)$, *then* $O(\ell)$ *reducing visiting steps are sufficient to reduce the frontier weight to at most* p.

Proof. The proof is based on the following property.

Claim. Let $x_0, x_1, \cdots, x_{n-1}$ and $y_0, y_1, \cdots, y_{n-1}$ be two sequences of nonnegative integers such that $\sum_{j=0}^{i} x_j \le \sum_{j=0}^{i} y_j$, for all $0 \le i < n$. Then,

$$\sum_{i=0}^{n-1} x_i/3^i \le \sum_{i=0}^{n-1} y_i/3^i .$$

Proof of Claim. The proof is by induction on n. The case $n = 1$ is trivial. Suppose that the property holds for some $n \ge 1$ and consider sequences of $n+1$ elements. Assume that $x_n > y_n$, since otherwise the inductive step is immediate. It is easy to see that

$$\sum_{i=0}^{n} \frac{x_i}{3^i} \le \sum_{i=0}^{n-2} \frac{x_i}{3^i} + \left(\frac{x_{n-1} + x_n - y_n}{3^{n-1}} \right) + \frac{y_n}{3^n} .$$

Note that $\sum_{i=0}^{n-2} x_i + (x_{n-1} + x_n - y_n) \le \sum_{i=0}^{n-1} y_i$, therefore, by applying the induction hypothesis, we have that

$$\sum_{i=0}^{n-2} \frac{x_i}{3^i} + \left(\frac{x_{n-1} + x_n - y_n}{3^{n-1}} \right) \le \sum_{i=0}^{n-1} \frac{y_i}{3^i} ,$$

which, combined with the first inequality, proves the claim.

Consider a reducing visiting step. Let F be the frontier prior to the execution of the step and let n_j be the number of nodes in $F(j)$ visited in the step, $0 \le j < \ell$. By definition of reducing visiting step we have

$$\sum_{j=0}^{i} \frac{|F(j)|}{2} \le \sum_{j=0}^{i} n_j ,$$

for any i, $0 \leq i < \ell$, and the claim shows that

$$\frac{3^\ell}{2} \sum_{j=0}^{\ell-1} \frac{|F(j)|}{3^j} \leq 3^\ell \sum_{j=0}^{\ell-1} \frac{n_j}{3^j} .$$

Thus, the visited nodes account for at least half the total frontier weight. Since the combined weight of the children of any node is at most two thirds of the weight of their parent, it follows that the weight reduction must be at least one third of the total weight of the visited nodes, *i.e.*, at least one sixth of the frontier weight W prior to the execution of the visiting step. Thus, the frontier weight W' following the completion of the step is at most $(5/6)W$.

Let W_t denote the total frontier weight following the completion of the tth visiting step. The above discussion has shown that $W_t \leq (5/6)W_{t-1}$, for $t > 0$. Since the frontier at the beginning of the stage contains $O(p\ell^2)$ nodes, we have $W_0 = O(p\ell^2 3^\ell)$, which implies that the frontier weight is less than or equal to p after $t = O(\ell)$ reducing steps. This proves the lemma.

Consequently, the strategy outlined above visits any stratum of size $m = O(p\ell^2)$ in time $O((m/p + \ell)(\log\log\log p)^2)$. Since strata of size $m = \Omega(p\ell^2)$ can be also visited in $O(m/p)$ time using the straightforward strategy, we can interleave the two strategies and obtain an algorithm that visits any stratum in time $O((m/p + \ell)(\log\log\log p)^2)$. This result immediately yields a backtrack search algorithm with the running time stipulated in Theorem 1.

3 Implementation of a Visiting Step

In this section, we describe the implementation of a visiting step which enforces the property that the step be always full or reducing. For convenience, we regard the processors as being conceptually arranged into ℓ *rows* and $q = p/\ell$ *columns*. At the beginning of the visiting step, the processors of the ith row store all current frontier nodes at level i, $0 \leq i < \ell$. Each processor maintains a data structure, called a *tree ring*, to store the tree nodes assigned to it. A tree ring is structured as a forest of complete binary trees of different sizes[1]. The leaf vertices in a tree ring hold nodes of the tree being visited, and each internal vertex contains pointers to its children and to the leftmost and rightmost leaf vertex in its subtree. All the trees in the same tree ring have different sizes and their roots are organized in a doubly-linked list, ordered by tree size. A similar data structure was used in [CV88]. We will argue in the next section that when the stratum size is $O(p\ell^2)$, the maximum size of a tree ring is at most K, where K is a value that is polynomial in ℓ, and therefore the heights of the trees in a tree ring are always $O(\log \ell)$.

A visiting step consists of two substeps, *VISIT* and *BALANCE*, which are described in the following paragraphs.

---

[1] To avoid confusion discussing the elements of the tree being visited and the trees employed in the tree rings, we will use the term *node* exclusively in connection with the former and reserve the term *vertex* for the latter.

VISIT. Within each column do the following:

1. Select the $\min\{s, 5c\ell\}$ heaviest nodes from the tree rings in the column, where s is the total number of nodes held in the column and c is a constant to be specified later;
2. Distribute these nodes evenly among the processors in the column;
3. Let each processor visit the nodes it receives;
4. Insert the children of these just-visited nodes into the appropriate tree ring within the column.

BALANCE. This substep is executed in parallel by each row and aims to partially balance the nodes stored by the processors in the row. We define the *degree* of a processor as the number of tree nodes contained in its tree ring. Let f_i be the sum of the degrees of all processors in row i, for $0 \le i < \ell$. BALANCE redistributes the nodes among the tree rings in such a way that upon completion at most $\min\{f_i, q\}/(2K)$ processors have degree larger than $c\lceil f_i/q \rceil$ in row i, for any $0 \le i < \ell$. Moreover, BALANCE never increases the maximum processor degree of any row. The actual implementation of the BALANCE substep is rather involved and is discussed in Section 4.

Let F be the frontier at the beginning of a visiting step. The following lemma shows that a visiting step is always either full or reducing.

Lemma 3. *If $|F| \ge 4p$ then $\Omega(p)$ nodes are visited in the step. Otherwise, for each $0 \le i < \ell$, at least half of the nodes in $\bigcup_{j=0}^{i} F(j)$ are visited.*

Proof. Recall that the BALANCE substep executed at the end of the visiting step preceding the one under consideration has left at most $\min\{|F(j)|, q\}/(2K)$ processors in row j of degree larger than $c\lceil |F(j)|/q \rceil$, for each j, $0 \le j < \ell$. We call the tree nodes stored by these processors *bad nodes* and all the other ones *good nodes*. Since K is an upper bound to the degree of any processor, we have that the total number of bad nodes in the first i levels of the frontier is

$$K \sum_{j=0}^{i} \frac{\min\{|F(j)|, q\}}{2K} \le \frac{1}{2}\left|\bigcup_{j=0}^{i} F(j)\right| ,$$

for any $0 \le i < \ell$. Thus the bad nodes at level i or lower account for at most half the total number of frontier nodes in those levels.

Suppose $|F| \ge 4p$ and let $r \le q$ be the number of columns holding fewer than $5c\ell$ nodes. Since a column holds at most $\sum_{j=0}^{\ell-1} c\lceil F(j)/q \rceil \le c(|F|/q + \ell)$ good nodes, the number of good nodes is bounded as follows:

$$\frac{|F|}{2} \le |\{\text{good nodes}\}| \le 5cr\ell + (q - r)c(|F|/q + \ell) ,$$

which implies $r \le q(5c - 2)/(5c - 1)$. Since c is constant greater than one, we conclude that $\Theta(q)$ columns hold at least $5c\ell$ nodes. Thus, the visiting step will visit $\Theta(\ell q) = \Theta(p)$ nodes, hence the step is full.

Consider now the case $|F| \leq 4p$. Since the number of good nodes in each column is at most $c(|F|/q + \ell) \leq 5c\ell$, it follows that the total number of nodes to be visited in the step is at least equal to the total number of good nodes. From the observation made above, we know that if we visited only the good nodes, then for any $0 \leq i < \ell$ we would visit at least half of the frontier nodes at level i or lower and hence the step would be reducing. Since we explore up to $5c\ell$ nodes in each column and we select the heaviest nodes available, if s good nodes are not visited it can only be because that number of heavier bad nodes are visited in their place, thus ensuring that the step is reducing.

In order to implement the visiting step described above, we need efficient primitives to operate on the tree rings. Consider a stage visiting a stratum of size $m = O(p\ell^2)$. Note that at the beginning of the stage the degree of each processor is $O(\ell^2)$, and that after each VISIT substep the degree increases by at most an $O(\ell)$ additive term. Since the BALANCE substep does not increase the maximum degree and $O(m/p + \ell) = O(\ell^2)$ visiting steps are executed overall, we can conclude that the maximum degree of any processor, will always be $O(\text{poly}(\ell))$. As a consequence, throughout the stage each tree ring contains at most $O(\log \ell)$ trees of $O(\text{poly}(\ell))$ size and $O(\log \ell)$ height each. Under these conditions, it is easy to show that:

1. Given $s = O(\ell)$ nodes evenly distributed among $\Theta(s)$ processors, a tree ring whose trees contain these nodes as leaves, can be constructed by the processors in $O(\log \ell)$ time.
2. Two tree rings can be merged into one tree ring in $O(\log \ell)$ time.
3. Any number of leaves $s = O(\ell)$ can be extracted in $O(\log \ell)$ time by $O(s)$ processors from a complete binary tree when initially only a pointer to the root is known by one of the processors. Moreover, the rest of the tree can be split into subtrees forming a tree ring within the same time bound.

It can be easily argued that the VISIT substep can be implemented by using standard prefix operations within each column, and by employing the above primitives to manipulate the tree rings. The above discussion provides a proof of the following lemma.

Lemma 4. *For strata of size* $O(p\ell^2)$, *VISIT can be executed in* $O(\log \log \log p)$ *time.*

4 Implementation of BALANCE

Recall that $K = O(\text{poly}(\ell))$ is a fixed upper bound to the degree of any processor when a stratum of size $O(p\ell^2)$ is visited, hence we can assume that K is known to all processors. Since BALANCE is executed in parallel and independently by all rows, we can concentrate on the operations of an arbitrary row, say row k. Let f_k denote the total number of tree nodes stored by the processors of this row at the beginning of the BALANCE substep. The purpose of the substep is to redistribute these nodes among the processors in such a way that, after the redistribution, the number of processors of degree greater than $c\lceil f_k/q\rceil$ is at most $\min\{f_k, q\}/(2K)$. (It should be noted that the value f_k is not known to the processors.) A crucial feature of the implementation of BALANCE is that nodes are not physically exchanged between the processors, which would be too costly for our purposes, but instead they are virtually moved by manipulating the corresponding tree rings, with a cost logarithmic in the number of nodes being moved.

The algorithm for the BALANCE substep is based on a balancing strategy introduced by Broder *et al.* in [BFSU92], which makes use of a special kind of expander defined below.

Definition 5 [BFSU92]. A graph $G = (V, E)$ is called (a, b)-*extrovert*, for some a, b with $0 < a, b < 1$, if for any set $S \subseteq V$, with $|S| \leq a|V|$, at least $b|S|$ of its vertices have strictly more neighbors in $V - S$ than in S.

The existence of extrovert graphs can be proved through the probabilistic method [BFSU92].

We identify the PRAM processors with the nodes of a d-degree (a, b)-extrovert graph $G = (V, E)$, where a, b and d are constants. Let $\gamma = (4d+3)/(4d+4)$ and $\tau = \lceil \log_{1/\gamma} aK/2\rceil + 1$. The algorithm consists of τ *phases*, numbered from 0 to $\tau - 1$. In each phase, a number of nodes are marked as *dormant*, and will not participate in subsequent phases. The remaining nodes are said to be *active*. At the beginning of Phase 0 all frontier nodes are active. For $0 \leq i < \tau$, Phase i performs the following actions.

1. Each processor with more than $K\gamma^i/2$ active nodes declares itself *congested*.
2. Let $\alpha = 1 + a$ and $\sigma = \lceil \log_{1/(1-b)}(\alpha^\tau 2K/a)\rceil$. A DAG D is built as a directed version of a certain subgraph of G. The construction proceeds by performing σ steps of the procedure given in [BFSU92]. Initially, the DAG is empty. In the first step, every congested processor v with at least $\lfloor d/2\rfloor + 1$ non-congested neighbors enters D by "acquiring" the edges directed from v to its non-congested neighbors. In the subsequent $\sigma - 1$ steps, every congested processors v not yet in D checks whether $\lfloor d/2\rfloor + 1$ of its neighbors are either non-congested or already in D and, if so, enters D by acquiring edges to these neigbours. The construction guarantees that each congested processor in the DAG has outdegree greater than its indegree, while each non congested processor in the DAG has outdegree 0.
3. Each congested processor not in D with more than $K\gamma^{i+1}$ active nodes, marks all but $K\gamma^{i+1}$ of its active nodes as dormant.
4. Let j be such that $2^j \leq K\gamma^i/(2d+2) < 2^{j+1}$. Note that $2^j > K\gamma^i/(4d+4)$. Each processor in D sends a tree containing 2^j active nodes to every neighbor in D. Such a tree can be regarded as a tree ring itself.
5. Each processor creates a single tree ring by merging the tree rings it receives into its tree ring.

In what follows, we show that at the end of the τ phases the number of processors of degree more than $c\lceil f_k/q\rceil$ is at most $\min\{f_k, q\}/(2K)$.

Lemma 6. *For $0 \leq i < \tau$, at the beginning of Phase i each processor holds at most $K\gamma^i$ active nodes and at most $K(1-\gamma^i)$ dormant nodes.*

Proof. By induction on i. Immediate for $i = 0$. Suppose that the proposition is true up to index $i - 1$. By induction, each processor starts Phase $i - 1$ with at most $K\gamma^{i-1}$ active nodes and at most K nodes overall. A congested processor that does not make it into the DAG D is not involved in any movement of nodes in the phase, therefore it does not change its overall degree, and, at the end of the phase, at most $K\gamma^i$ of its nodes are active. A congested processor that does succeed in entering the DAG looses at least $2^j \geq K\gamma^{i-1}/(4d+4)$ active nodes. Therefore, the number of active nodes at the end of the phase is at most

$$K\gamma^{i-1} - \frac{K\gamma^{i-1}}{4d+4} \leq K\gamma^i \ ,$$

and its overall degree is decreased. Finally, a non congested processor begins the phase with at most $K\gamma^{i-1}/2$ active nodes and receives at most $d2^j \leq dK\gamma^{i-1}/(2d+2)$ new active nodes, which adds up to

$$\frac{K\gamma^{i-1}}{2} + \frac{dK\gamma^{i-1}}{2d+2} \leq K\gamma^i \ .$$

Moreover, the number of dormant nodes is unchanged, namely $K(1 - \gamma^{i-1}) \leq K(1 - \gamma^i)$.

The above lemma implies that the maximum degree of any processor is always bounded above by K. We refer to the processors containing dormant nodes as *renegades*. Let $R(j)$ denote the set of renegades at the beginning of Phase j and $C(j)$ the set of processors that declare themselves congested in the phase. Define $r_j = |R(j)|$ and $c_j = |C(j)|$, for $0 \leq j < \tau$. Let $\tau' = \left\lfloor \log_{1/\gamma} \left(\frac{Ka}{3\lceil f_k/q \rceil} \right) \right\rfloor$ and note that $\tau' < \tau$. We have

Lemma 7. *For $0 \leq j \leq \tau'$, we have*

$$r_j \leq a^j (1-b)^\sigma \min\{f_k, q\} \ .$$

Proof. By induction on j. Immediate for $j = 0$ since $e_0 = 0$. Suppose that the property holds up to index $j - 1$ and consider index j. Note that the renegades at the beginning of Phase j will be given by the set $R(j-1)$ plus a set $C' \subseteq C(j-1)$ containing congested processors that did not make it into the DAG during Phase $j - 1$. Let us give an upper bound to $|C'|$. Note that $c_{j-1} \leq a\min\{f_k, q\} \leq aq$, since otherwise congested processors would account for more than

$$\frac{K\gamma^{j-1}a\min\{f_k,q\}}{2} > \frac{K\gamma^{\tau'}a\min\{f_k,q\}}{2} \geq f_k$$

active nodes, which is impossible. By the extrovertness of the graph G, after the first step of DAG construction, the number of processors not in D are at most $c_{j-1}(1 - b)$. In general, it can be shown that after the tth DAG construction step, the number of congested processors not in D is at most $c_{j-1}(1-b)^t$. This implies that $|C'| \leq c_{j-1}(1-b)^\sigma$, hence the number of renegades at the beginning of Phase j will be

$$r_j \leq c_{j-1}(1-b)^\sigma + r_{j-1}$$
$$\leq a(1-b)^\sigma \min\{f_k, q\} + \alpha^{j-1}(1-b)^\sigma \min\{f_k, q\} \quad \text{(by induction)}$$
$$\leq \alpha^j(1-b)^\sigma \min\{f_k, q\} .$$

Lemma 8. *Let $c = 2/(a\gamma) = \Theta(1)$. By the end of the BALANCE procedure, the number of processors of degree more than $c\lceil f_k/q \rceil$ is at most $\min\{f_k, q\}/(2K)$. Moreover, the procedure is executed in time $O((\log\log\log p)^2)$ on the COMMON CRCW-PRAM.*

Proof. At the beginning of Phase τ', each processor has at most $K\gamma^{\tau'} \leq c\lceil f_k/q \rceil$ active nodes (by Lemma 6), and the number of renegades is

$$r_{\tau'} \leq \alpha^{\tau'}(1-b)^\sigma \min\{f_k, q\} \leq \frac{\min\{f_k, q\}}{2K}$$

(by Lemma 7 and the choice of σ). Moreover, an argument similar to the one used to prove Lemma 6 shows that the maximum degree of processors that are not renegades at the end of Phase τ' will not increase above the $c\lceil f_k/q \rceil$ threshold in the subsequent $\tau - \tau' - 1$ phases.

We now evaluate the running time. In each phase, every DAG construction step is accomplished in constant time and the extraction and merging operations on the tree rings take $O(\log K)$ time. Noting that $\tau = O(\log K)$ and $\sigma = O(\tau + \log K) = O(\log K)$, we conclude that the overall running time is

$$O(\tau \log K) = O(\log^2 K) = O(\log^2 \ell) = O((\log\log\log p)^2).$$

5 Evaluation of Bounded-Degree DAGs

This section sketches how some of the ideas involved in the backtrack search algorithm may be used to solve the DAG evaluation problem. In a *computation DAG*, nodes with zero indegree are regarded as *inputs*, while other nodes represent operators whose operands are the values computed by nodes adjacent along their incoming arcs. Nodes with zero outdegree are regarded as *outputs*. A node is deemed *ready* when its operands have been evaluated and, hence, the node itself can be evaluated. The *DAG evaluation problem* consists of evaluating all output nodes.

We define the layer of a node in the DAG in the obvious way: the inputs are at layer zero and the layer of every other node is one plus the maximum layer of its predecessors. It is straightforward to evaluate a DAG layer by layer using the approximate prefix sums algorithm of [GZ94] to evenly distribute the nodes at each layer among the processors. If the DAG has bounded degree (*i.e.* each node has indegree and outdegree bounded by some constant), such a strategy yields an algorithm with complexity $O(n/p + h \log \log p)$, where n denotes the number of nodes and h the number of layers in the DAG.

Notice the similarity between the DAG evaluation and the backtrack search problems. In fact, the only main difference is that in the former the underlying graph is not necessarily a tree and, therefore, no parent/child relationship among the nodes can be established based only on the graph structure. Nevertheless, we can still visit (*i.e.*, evaluate) the DAG by proceeding as in the backtrack search algorithm, with the exploration process implicitly defining a forest of trees that spans the DAG.

More precisely, let us first observe that our backtrack search algorithm can be employed to visit any forest of bounded-degree trees in $O((n/p+h)(\log \log \log p)^2)$ time, where n is the total number of nodes and h the maximum tree-height in the forest. Consider now the evaluation of a bounded-degree DAG and apply the backtrack search strategy. The evaluation starts at the inputs and is performed stratum by stratum, where each stratum consists of $\ell = O(\log \log p)$ layers. For a stratum of size $\Omega(p\ell^2)$, we adopt the straightforward strategy mentioned above, which yields optimal running time. For the evaluation of a smaller stratum, we execute a number of *evaluating steps* (the analogous of the visiting steps of the backtrack algorithm) until all but few nodes in the stratum remain to be evaluated. In an evaluating step, only ready nodes can be evaluated. For convenience, we assume that each node is labeled with the number of the step that evaluates it, and that the node adopts as its *parent* whichever of its predecessors has the larger label, with ties being broken arbitrarily. The node-parent relationship defines a forest of trees that spans the stratum, where the top nodes are the roots. The weight of each node is based on a priority associated with the node as follows: the priority of a node is 0 if it is at the top of the stratum, otherwise it is one plus the priority of its parent. In other words, the priority of a node is its level number within the forest of trees and these priorities play the role the level number did in the backtrack search algorithm.

Each evaluating step guarantees that either $\Omega(p)$ ready nodes are evaluated or a constant fraction of the ready nodes of priority less than or equal to any fixed threshold are evaluated. Arguing as in the analysis of the backtrack search algorithm, we can conclude that, for a stratum of size $m = O(p\ell^2)$, $O(m/p + \ell)$ evaluating steps suffice to reduce the total weight to less than p. At this point, the evaluation of the stratum can be completed in optimal time in a direct fashion. Since each evaluating step can be implemented in $O((\log \log \log p)^2)$ time, the overall complexity of the algorithm is $O((n/p + h)(\log \log \log p)^2)$.

References

[AL91] B. Aiello and T. Leighton. Coding theory, hypercube embeddings and fault-tolerance. In *Proceedings of 3rd ACM Symposium on Parallel Algorithms and Architectures*, pages 125–136, 1991.

[BFSU92] A.Z. Broder, A.M. Frieze, E. Shamir, and E. Upfal. Near-perfect token distribution. In *Proceedings of the 19th International Colloquium on Automata, Languages and Programming*, pages 308–317, July 1992.

[BGLL91] S.N. Bhatt, D. Greenberg, F.T. Leighton, and P. Liu. Tight bounds for on-line tree embeddings. In *Proceedings of the 2nd ACM-SIAM Symposium On Discrete Algorithms*, pages 344–350, January 1991.

[CV88]   R. Cole and U. Vishkin. Approximate parallel scheduling. Part I: The basic technique with applications to optimal parallel list ranking in logarithmic time. *SIAM Journal on Computing*, 17(1):128–142, 1988.

[GZ94]   T. Goldberg and U. Zwick. Optimal deterministic approximate parallel prefix sums and their applications. In *Proceedings of 3th Israel Symposium on Theory and Computing Systems*, 1994.

[JáJ92] J. JáJá. *An Introduction to Parallel Algorithms*. Addison Wesley, Reading MA, 1992.

[KP94]   C. Kaklamanis and G. Persiano. Branch-and-bound and backtrack search on mesh-connected arrays of processors. *Mathematical Systems Theory*, 27:471–489, 1994.

[KZ93]   R.M. Karp and Y. Zhang. Randomized parallel algorithms for backtrack search and branch and bound computation. *Journal of the ACM*, 40:765–789, 1993.

[LAB93] P. Liu, W. Aeillo and S. Bhatt. An atomic model for message-passing. In *Proceedings of 5th ACM Symposium on Parallel Algorithms and Architectures*, pages 154–163, 1993.

[LNRS92] T. Leighton, M. Newman, A. G. Ranade and E. Schwabe. Dynamic tree embeddings in butterflies and hypercubes. *SIAM Journal on Computing*, 21:639–654, 1992.

[Ran90] A.G. Ranade. A simpler analysis of the Karp-Zhang parallel branch-and-bound method. Technical Report No. 586, Computer Science Division, University of California at Berkeley, Berkeley, California, 1990.

[Ran94] A.G. Ranade. Optimal speed-up for backtrack search on a butterfly network. *Mathematical Systems Theory*, 27:85–101, 1994.

Agent Rendezvous:
A Dynamic Symmetry-Breaking Problem

Xiangdong Yu[1]* and Moti Yung[2]

[1] Computer Science Department, Columbia University.
[2]IBM Research Division, T.J. Watson Center, Yorktown Heights, NY 10598.

Abstract. We consider the problem of a rendezvous (coordinated meeting) of distributed units (intelligent agents in network computing or autonomous robots). The environment is modeled as a graph, the node labeling of which may not be "common knowledge" to the units, due to protocol and naming convention mismatch, machine faults, status change, or even hostility of the environment. Meeting of such units is likely to be a basic procedure in the area of distributed "intelligent agent" computing and in the domain of coordinated tasks of autonomous robots. The crux of the problem which we present here and initiate research on, is the breaking of potential symmetry while the units dynamically move. The units are more intelligent (computing power, control and memory) than simple (traditional) pebbles or tokens, and our algorithms will make use of this capability for speeding up the convergence to a common place (e.g., we will allow units to meet exchange information and depart). We consider both randomized protocols and deterministic (but non-uniform) protocols; the problem is unsolvable by a uniform deterministic algorithm. The deterministic procedure employs ideas from design theory and achieves $\tilde{O}(n)$ time, while the randomized methods are based on random walks and may achieve $\tilde{O}(n/k)$ time where k is the number of agents.

1 Introduction

We consider units with independent computational power and decision ability (such as robots, mobile processes, or intelligent agents). Our concern is the coordination of their behavior in a distributed environment. In particular, we study the problem of rendezvous (or reunion) after the units are lost from each other in a distributed network (an environment with only distributed control). This models a number of computational scenarios. A typical analogous case among humans is when a planned meeting point of travelers (say in an airport) turns out to be non-uniquely defined (e.g. meeting at a "car rental company counter" is decided upon, but in reality there are a few counters in different terminal buildings). Meeting in such a case can be tricky, and often ends up in a failure or unexpected delays.

* Work partially supported by NSF grant CCR-93-16209 and CISE Institutional Infrastructure Grant CDA-90-24735.

When the travelers are robots exploring an area, the confusion about rendezvous may further be caused by their limited capability of image recognition and lack of global identification of vicinities. In a communication network, it has been suggested recently to employ "intelligent agents technology" for certain computational tasks that require the involvement of subnetwork exploration (a mission). Examples of such computations are a "poll"" or a "census" that require many agents to collect local information (for speed-up and load-balancing of the extensive task). Here, the capability of the agent to move freely is assumed to be backed by a universal protocol ported to every network node (as part of the "intelligent agent technology"). Each node can check the header of a process, temporarily allocate resources to run it, and pass it to another node as it wants. The node does not necessarily have resources permanently dedicated to a process; in other words, the nodes are temporary hosts (like hotels) while the processes travel around between hosts (like human travelers).

However, besides this universal protocol (that supports the agents/robots), the nodes do not provide any other services to the processes. In particular, in the context of network security, the remote subnetwork may even be "hostile" and keep the agents from knowing the IDs of its nodes, or deliberately give false IDs. Obviously, such loss of identification will cause difficulty and confusion in determining a rendezvous. Other reasons resulting in identification confusion include faulty nodes, incompatible naming conventions and machine types. We remark that the need for minimization of services to external intelligent agents and giving them partial information as well as limiting their local influence on the node's resources has been realized recently as important prevention measure, against the intelligent agent technology becoming a tool for network penetration by hackers and virus writers.

In our model we, therefore, assume that each unit has managed to realize the topology of the network (e.g., through a map or by exploration) but marks the nodes using names or IDs which are local to the unit. Therefore the units do not necessarily have a common idea about which node is unique (such as the one with minimum ID) and usable as a default rendezvous location.

Our goal is to design a protocol which is to be given to the units before they are sent on their mission to the exploration area, so that they can reunite quickly when they have to. We note that this problem becomes trivial in many cases, such as:

- Communication to a central control is handy.
- Public bulletin boards are available at nodes.
- There is a unique node known to everyone, such as the one with a unique degree, the North most one, the unique articulation node, etc. In this paper we only consider networks that are symmetric to every node, such as complete graph, cycle, and hypercube.
- There is a leader in the group of units. They can meet by letting the leader traverse the network and every other one wait at its initial node, or the other way around, every one traverse the network and the leader wait at one node. Both methods works $O(n)$ time where n is the number of nodes in the network.

However, we rule out these shortcuts since if central control is not available or becomes faulty, or parts of the environment changes (change their ID) we would nevertheless want to perform the task. Thus, we try to find solutions that work on distributed symmetric environment.

Naturally, the rendezvous problem can not be solved by using a uniform deterministic algorithm. For example, if the network is a cycle of four nodes $\{a, b, c, d\}$, unit X marks them as $a = 1, b = 2, c = 3, d = 4$ and unit Y marks them as $a = 3, b = 4, c = 1, d = 2$, and initially X is at a and Y is at c, then no matter what kind of (identical) deterministic protocol X and Y use, they can never meet. Thus, we have to break the symmetry in this environment. One way to break the symmetry is via randomized protocols. As we will also see, if different units use different (predetermined) deterministic algorithms, the problem can still be solved quite efficiently, employing design theoretic ideas (e.g., [2]). We employ variants on random walks combined with other traversals. A typical random walk by unit A on graph G is a randomized procedure carried out by A which starts at node v and at any step it picks with equal probability a neighbor node of its current node then moves to that node. Random walk on graphs is by itself a fruitful research area and also finds numerous applications in computer science starting with Aleliunas et. al. [1]. A related work in a distributed system setting is by Tetali and Winkler [6] who showed that self-stabilizing token management (initiated in [3]) can be done efficiently through random walk of tokens. In our rendezvous algorithms, random walks are executed by k units simultaneously and interrupted by meetings. The rule of random movement may be slightly different from that described above, in symmetric graphs. Unlike the token case of [6], we exploit more intelligent units that can be united and redistributed in the course of the algorithm. Other graph theoretic models and problems that deal with different aspects of motion of units are in [5, 4].

1.1 Model and Definitions

Definition 1. A remote area is defined as an n-node undirected graph $G = < V, E >$. A marking of G is a 1-1 function $M : V \mapsto \{1, 2, \ldots, n\}$. A unit is defined as an object that can carry information, finish computational tasks at a node in G, and move from one node to another along an edge in G. We assume any unit takes one time step to pass any edge, and ignore the time cost of any computational task at any node.

Time measure is a little subtle here. We may associate an arbitrary passing time with each edge, but that may break the symmetry of the graph and provide a unique location as a default rendezvous. So we simply assume it takes the same amount of time (one step) for any unit to pass through any edge. This assumption also implies that the amount of information a unit carries does not affect its passing time through any edge. Under this model, the network is "synchronous" and a global clock can be assumed accessible at each node. Given that a typical unit has to spend considerable time in each node, we only require a coarse synchronization.

Definition 2. A mission center is defined as a set U of h units, each of which has a unique ID. A mission team R_k is a subset of U containing k units $r_1, r_2, \ldots, r_k$. A rendezvous problem is the dispatch of a team R_k to a remote area G for a mission and a requirement that all the members of the team meet at some node in G after the mission. Moreover, the following assumptions are made about a rendezvous problem:

- Each unit in R_k knows there are k units in the mission but does not know who they are (The subset R may be determined arbitrarily or change locally during the mission, and decision about each unit is made separately). Consequently, there is no agreed upon leader.
- The time when they should start to reunite in G is approximately specified. By "approximately" we mean that there can be some discrepancies between different units' starting times of the rendezvous procedure. For example, when unit r_1 starts to reunite at time 123, unit r_2 has not finished its mission yet and must wait until time 145 to start. In this sense, different units' operations are not synchronized. We denote by Δ the maximum discrepancy between two starting times, and assume it is known to every unit.
- G is not known before the mission. By the time a unit starts to reunite with others it knows G but marks G's nodes in its own way (not known by other units in R).
- The location where a unit starts the rendezvous procedure is arbitrary and unknown beforehand.
- During the rendezvous procedure, the entire information a unit knows is carried with it and all the decisions are made based on it and possibly some random bits.
- Units can only meet at nodes. (not on edges. Here we make assumption that they become computationally active only at nodes). A typical meeting is a computational action happening at a certain node and a certain time step, where units acquire information carried by other units and reach an agreement together. Meetings are the only information sources; no central control or public bulletin board can be used as means of communication without meeting.

Note that the unit knows its ID r_i but not its sequence number i, while we, in the analysis know and use both. r_i and i may be used interchangeably when there is no confusion.

1.2 Our Results

We first present a deterministic solution for a general graph. In particular, we show that it is possible to tell each $u \in U$ a deterministic algorithm A_u, such that for any R_k, units in R_k can reunite in time $O(\Delta + n \log \frac{|U|}{k-1})$.

Then we describe random walk based algorithms for three well-known symmetric graphs. Note that this kind of algorithm is not very interesting for general graphs, since it may take $\Omega(n^3)$ time for a unit to random walk from one node to another.the specific graphs, though, we have better time bounds as compared to the deterministic algorithm mentioned above. (When we say certain event happens with *high probability*, it means probability $1 - O(1/n^c)$ for some constant c large enough.)

For randomized algorithm, we only consider the case where k is small as compared to n (which is reasonable). More specifically, we assume $k < \sqrt{n}$. The randomized results are:

1. $G = K_n$, the n-node complete graph.There is a random walk based algorithm AK where units in R reunite in time $O((n/k)\log n)$ with high probability.
2. $G = H_n$, the $n = 2^N$-node hypercube. The time bound becomes $O((n/k)\log^2 n)$.
3. $G = C_n$, the n-node cycle. The time bound is $O((n/k)\log n) + n/2$.

2   A Deterministic Algorithm

We first study the case of $\Delta = 0$, i.e., all units in R start the rendezvous procedure at the same time. Let G be a connected graph. Then a unit can traverse G and come back to its origin along a DFS tree in time $2(n - 1)$. In this deterministic solution, every unit $u \in U$ is given an algorithm A_u, which can be represented as a binary string B_u of $\lceil \log_2(\frac{|U|}{k-1}+1) \rceil$ bits. The B_u's can be chosen in such a way that any k of them are not all identical, by partitioning U into groups of size $k - 1$ and associate one string with all members of a group. A_u is run in $|B_u|$ phases and the i-th bit of B_u is used by u as its guide for the i-th phase in the following manner:

1. If the bit is 1, then u traverses G in time $2(n - 1)$ and comes back to its start node.
   If the bit is 0, then u waits in its current node for $2(n - 1)$ steps. It records the ID and start location of the traversing unit with the minimum ID (they can exchange and coordinate this information).
2. If u was waiting and has caught a traversing unit, then it goes to the start location it recorded, in time $n - 1$.
   If u was traversing, then wait for $n - 1$ steps.
3. If the bit is 1, then u traverses again, but stops in the middle if a waiting node asks it to do so.
   If the bit is 0, then u waits again (in the new location), and stops any traversing unit.
4. If u finds another unit in its node with an opposite bit, then we claim that all units are in the node: rendezvous is done. Otherwise, go to next phase.

Theorem 3. *When $\Delta = 0$, rendezvous can be done by the above deterministic algorithm in time $O(n \log \frac{|U|}{k-1})$.*

Proof:

The total number of steps in one phase is bounded by $5n$. Given the time for one phase, the entire procedure runs in time $5n|B_u| = O(n \log \frac{|U|}{k-1})$. Since there are at least two different B_u's in any R_k, there must be one phase (corresponding to one bit position in the binary strings) in which both traversing and waiting units exist. In this case, the rendezvous is guaranteed to happen. To see this note that in step 1 all the waiting units will be "caught" by the traversing unit with minimum ID, then in step 2 they will all go to the location of this unit, then in step 3 all traversing units of step 1 will traverse again and will all stop at the same location by the waiting units. ∎

The harder case is when $\Delta > 0$. Now, one unit's phase 1 may be another unit's phase 2, 3, or even Δ/n. In other words, the binary strings are not necessarily aligned and the required different bits at the same phase may not appear. The lemma below provides a new way of choosing B_u to meet our requirements.

Lemma 4. *For any $\delta \geq 0$ and $x > \delta$, there is a set C of at least $2^x/x$ $2x$-bit binary strings such that any two strings in C always mismatch (i.e. have different bits) in at least one position when they are aligned with heads off by no more than δ positions.*

Proof: Take any x-bit string B, denote by $\tilde{B}$ the equivalence class of all the x-bit strings obtainable from B by a circular shift, i.e., cut off a prefix of B and append it as a suffix. Obviously $|\tilde{B}| \leq x$ and the total number of the equivalence classes is at least $2^x/x$. Now, pick any two strings a and b such that $\tilde{a} \neq \tilde{b}$. If $|a| > \delta$, then, as long as the starting bits of string aa and bb disalign by no more than δ positions, there will be one bit in aa which is different from its aligned bit in bb. To see this, observe that the second a (or b) is entirely covered by the other string, and the aligned substring in the other string belongs to $\tilde{b}$ (or $\tilde{a}$), is thus different. So, C can be formed by picking any one string from each of the equivalence classes. ∎

The number x must be chosen such that the number of circular equivalence classes is at least $|U|/(k-1)$. Solving $2^x/x = |U|/(k-1)$, an $x = (1+\epsilon) \log \frac{|U|}{k-1}$ suffices for any $\epsilon > 0$. Note that we also require $x > \delta$.

Theorem 5. *For any $\Delta > 0$, rendezvous can be done deterministically in time $O(\Delta + n \log \frac{|U|}{k-1})$.*

Proof: We choose $\delta = \Delta/n$ and $x = \delta + 1.5 \log \frac{|U|}{k-1}$ from Lemma 4, and pick any $|U|/(k-1)$ strings from the resulting C each of which is to be held by $k-1$ units.

Because any set of k units (robots) must hold at least two different strings, we want to design a protocol such that the rendezvous is guaranteed to happen within the phase when the first bit difference occurs. Thus, if each phase is of length $O(n)$, the total reunion time will be bounded by $2x \cdot O(n) = O(\Delta + n \log \frac{|U|}{k-1})$.

The idea extends the case of $\Delta = 0$. Each bit in the string represents an action phase, with 1 meaning active and 0 meaning passive. It is expected that if at least two robots have different bits on any of their corresponding phases, then during this phase, certain active robot will at first catch all the passive ones, then certain passive one will catch all the active ones. The difficulty is that even the corresponding phases of different robots may start at different times. Note that the word "corresponding" is a little tricky here. If it means two phases overlap on at least half of their lengths, then, it may be the case that phase a of robot A overlaps with the first half of phase b of robot B, and phase c of robot C overlaps with the second half of b, while a and c hardly overlap. Therefore, the correspondence between a and b and between b and c does not imply transitively the correspondence between a and c. Such inconsistency is solved by choosing one robot as the base, We say that: whichever phase overlaps a base phase by more than half it is considered as in the group identified by the base phase. Two phases are corresponding to each other if and only if they are in the same group.

Note that the "base" robot is defined for the purpose of analysis. The robots themselves do not know or need to know who is the "base". We choose the base to be the one with the minimum ID in the team of k robots. We want to assure that all the robots that overlap in the first half of the base reunite at the base's origin during the first half, and then do the similar thing for the second half.

The algorithm each robot executes runs in time cycles. Each phase is $36n$ steps long, divided into 6 cycles. Each cycle is divided into a section of $5n$ steps and a section of n steps. A robot has a "home" at the beginning of a cycle, during the $5n$-step section, it will decide (in a way to be described below) where its home in next cycle is. Then, it will migrate to the new home during the n-step section, thus finishing one cycle.

If the robot is assigned to be active (bit 1), its $5n$-section will be a moving one, consisting of a complete traversal of the network at the beginning of the section as well as one at the end of the section, and idle steps in between. The key property is that each traversal takes less than $2n$ steps and thus the number of idle steps is at least n. Therefore, if the $5n$ section is broken up by taking away any n-subsection at least one traversal is left intact. For a robot which is assigned to be passive (bit 0), the $5n$-section is a waiting one, i.e., it stays at home for $5n$ step.

Each node has an ID, and an effective ID variable which is initialized to its ID at the beginning of the rendezvous procedure. Whenever two or more robots meet, they choose the minimum of the effective ID's held among them as their common new effective ID. All the active ones merge into one super-unit, i.e., they always act and move together afterwards; and all the passive ones also merge into one. Then, any robot r, active or passive, during its next n-section, will move to the home of the robot from which r got its current effective ID. This is how a robot decides on its new home in a cycle.

Next we will see how rendezvous could happen in a base phase in which at least two robots hold different bits. There are two cases: (1) the base is active in the phase; or (2) it is passive in the phase. For case (1), in the first cycle, the base must catch all the waiting ones. This is because, even if the $5n$-section of the base is broken into two pieces by the n-section of a passive robot, at least one of the pieces will contain a complete traversal, thus guaranteeing to "hit" the passive ones. In the worst case, the n-section of a passive one starts at a position near the $2n$-th step in the active one's $5n$-section, ends near $3n$-th step, and the base could not hit the passive during the first nearly $2n$ steps. So the hit will happen during the last $2n$ steps of the base's $5n$-section, and the migration of the passive one will happen within $3n + 6n = 9n$ steps from the start of the base phase. Next, this passive one holding the base's ID will play the role of the base as above, but in a dual fashion: it does not traverse, it waits at home and changes the effective ID of whoever passes by. For a similar reason, its $5n$-section will cover a complete traversal of any active robot, and will thus make them migrate to its home within $9n$ steps. Therefore, all the robots that overlap at the first half ($18n$ steps) of the base phase will reunite within this half. Similar argument works for the second half, and for case (2).

∎

3 Random Walk Based Algorithms

We are interested in this section in randomized algorithms. Note that a randomized algorithm that simulates the one in the previous section where in each phase a unit decides to either draw a 0 or a 1, may take $O(n \log k)$ expected time. We are interested in specific underlying environments (e.g., parallel machine and local area network configurations) whereby using random walk (combined with meeting and splitting of unit groups), we can do better for certain n, k parameter range.

3.1 Rendezvous on A Complete Graph

We let every unit belong to a group at any moment during its rendezvous procedure. A group consists of one or more units. Different groups are disjoint. Each group has a leader which is the unit with minimum ID in it; we also use this ID to identify the group.

At the beginning of the procedure, every unit belongs to a group containing only itself. The algorithm is essentially a procedure of group merging. After complete rendezvous, all the units will be in the same group. Units in a group perform random walk most of the time, but they all agree on a meeting to be held at a certain node and a certain future time. The node and time currently associated with a group g are denoted by $msite(g)$ (represented using the leader's marking) and $mtime(g)$, respectively. Each unit r in group g maintains a vector of variables for its current group: $\mathbf{w}_r = (g_r, t_r, z_r, y_r)$, where $g_r = g$ is the group ID, t_r is a time counter, recording the number of steps since group g has started the procedure, it is incremented after each step; $z_r = msite(g)$, and $y_r = mtime(g)$. r may encounter units from other groups in the random walk and it must record information about the potential merging group. For this purpose, r maintains another vector of variables $\mathbf{p}_r = (m_r, c_r, d_r)$ where m_r is that group's ID, c_r is the time counter difference between these two groups, and $d_r = msite(m_r)$. When a unit at node v wants to start the rendezvous procedure at time t_0, it runs the following algorithm.

Rendezvous Algorithm.

1. Set $t_r = 0$, $z_r = v$, and $S = 16c(n/(.9k)^2)\log n$, where c is a constant. t_r will be incremented by one at each step.
2. Set $m_r = \infty$. Then start random walk:
   (a) Pick with uniform probability $1/n$ a node $v' \in V$, go to (or stay at) v'.
   (b) For each unit r' encountered now at v', do: if $g_{r'} \neq g_r$ and $g_{r'} < m_r$, then set $(m_r, c_r, d_r) = (g_{r'}, t_{r'} - t_r, z_{r'})$.
   (c) If t_r is divisible by S then exit random walk and go to z_r. Otherwise, go to step 2a.
   (d) Now at z_r. At this time g may already be expanded. If the number of units in g is k, rendezvous is done. If it is greater than or equal to αk, for some constant $\alpha > 0$ to be determined in the proof below, go to step 3.
   (e) If no unit in g encountered any unit out of g during the previous random walk, goto step 2; Otherwise, find $m_{x0} = min\{m_x | x \in g\}$, and set $(g_r, t_r, z_r) = (m_{x0}, t_r + c_r, d_{x0})$ to make it look just the same as other units in m_{x0}, then goto step 2.
3. Partition V into $|g|$ subsets of size about $n/|g|$, which is $O(n/k)$ since $|g| \geq \alpha k$, and assign one unit of g into each subset to do a random walk, i.e., every step go to a node chosen from the subset with uniform probability about $|g|/n$, if encounter a unit not in g, tell it to go to z_g, the rendezvous place of g.

Lemma 6. *Consider the experiment W_k^n of throwing k balls independently to n holes with equal probability $1/n$. Let $P_t(K)$ be the probability that there is not any two balls being thrown into the same hole in any of t independent W_k^n trials. Then $P_t(k) = 1/n^c$ for some $c > 0$ when $t \geq (16cn/k^2)\log n$.*

Proof: We follow standard combinatorial arguments. $P_1(k) = n!/(n^k(n-k)!)$ and $P_t(k) = P_1^t(k)$. Let $q = \lfloor k/2 \rfloor$, we can bound $P_t(k)$ as follows,

$$P_1 \leq (n-q)(n-q-1)\ldots(n-k+1)/n^q$$
$$\leq (n-q)^q/n^q$$
$$= (1 - q/n)^q$$

Take $t = (16cn/k^2)\log n$, notice $0 < q/n \leq 1/2$ and $2q/k \geq (k-1)/k \geq 1/2$, we have,

$$P_t \leq (1 - q/n)^{tq}$$
$$= (1 - q/n)^{(n/q)(2q/k)^2 \log n^{4c}}$$
$$< e^{-(1/4)\log n^{4c}}$$
$$= 1/n^c.$$

Lemma 7. *When the procedure is in stage 2a, with high probability the following event $C(t)$ happens: there is at least one encounter during time interval $I = (t_1 = t - S, t_2 = t + S)$.*

Proof: In any step, a unit is either in a meeting or in a random walk. If x units are in random walks, this step can be viewed as a trial W_x^n. This trial is independent of similar trials in other steps.

Consider the following two events:

E_1: There are at least S steps in interval I each of which containing a $W_{.9k}^n$ trial.

E_2: No encounter happens in I.

Note that if E_2 happens, each unit will join exactly two meetings in I. This is because the meetings of the same group are scheduled at least S steps far from each other, and if no encounter happens, units in each group will not join meetings other than their own group's.

Assume at time t_1 the group configuration is $\{(g_i, z_i, y_i) | 1 \leq i \leq l\}$. Denote by $U(s)$ the number of units that are not in a meeting at step s. If E_2 happens, we have $\sum_{t_1 < s < t_2} (k - U(s)) = 2k$. Hence there are at least S s'es such that $U(s) \geq 0.9k$. (otherwise $\sum_{t_1 < s < t_2} (k - U(s)) > (t2 - t_1 - 2 - S)(k - 0.9k) = 0.1(S - 2)k > 2k$). In other words, E_2 implies E_1, or $Prob[E2/\overline{E1}] = 0$.

By Lemma 6, $Prob[E_2/E_1] = 1/n^{\Omega(1)}$. We get,

$$Prob[E_2] = Prob[E_2/E_1] \cdot Prob[E_1] = 1/n^{\Omega(1)}.$$

Thus $\overline{E_2}$, i.e., the event that an encounter occurs in I, will happen with high probability, ∎

Lemma 8. *With high probability, after $O(k \log n)$ iterations of stage 2a, either rendezvous is done, or the procedure goes to step 3.*

Proof: From Lemma 7 we know that two iterations of stage 2a will contain an encounter with high probability. Assume when the encounter occurs all the groups are $g_1, g_2, \ldots, g_l$, among which $g_1, g_2, \ldots, g_{j-1}$ are in meeting. So the encounter must happen between two units in set $A = \cup_{j \leq i \leq l} g_i$. Since $|A| > 0.9k$, there are at least $0.81k^2$ pairs of units from A. If the size of every $g_i \subseteq A$ is smaller than $0.4k$, then the total number of pairs belonging to a same g_i is upper bounded by $(|A|/0.4k)0.16k^2 \leq 0.4k^2$ and there will be at least $0.81k^2 - 0.4k^2 = 0.41k^2$ pairs between units from different groups. Since each pair appears with equal probability, there is more than 50% of a chance that two different groups get in touch, which in turn will result in at least one merge in time S. To make the 50% chance into high probability, the above action must be repeated $c \log_2 n$ times and in total, $O(\log n)$ iterations of stage 2a are needed for one merge. The above discussion tells us that the α in step 2d of the algorithm should be chosen to be 0.4.

Since it takes $k - 1$ merges to achieve the rendezvous, after $O(k \log n)$ iterations of Stage 2a, with high probability, either the rendezvous happens, or a group with size greater than $0.4k$ will be formed, causing the algorithm to go to Step 3.

∎

Lemma 9. *After the algorithm gets into Step 3, with high probability, the rendezvous can be done in time $O((n/k) \log n)$.*

Proof: The number of units that have not been caught by g is bounded by $k - |g| \leq 0.6k$. In every step, each of which has a chance of $|g|/n > 0.4k/n$ of being caught by g. The probability that it has not been caught after $O((n/k) \log n)$ steps can be $1/n^{c+1}$, and the probability that ANY one of them has not been caught is bounded by $0.6k/n^{c+1} < 1/n^c$, which means, with high probability, all of them will be caught by then.

∎

Theorem 10. *The rendezvous is achieved by the above algorithm in expected time $O((n/k) \log n)$.*

Proof: Combined the time bounds in the above two lemmas, and the assumption that $k < \sqrt{n}$. We get the theorem. ∎

For the rendezvous problem, any graph with diameter D can be thought of as a complete graph whose edge has passing time D. By relaxing the above algorithm's time step from 1 to D, we get the following bound for general graphs.

Corollary 11. *For any graph with diameter D, the rendezvous can be done in time $O(nD/k) \log n$.*

3.2 Rendezvous on Hypercube

Since diameter $D = \log n$ for hypercube. From Corollary 11 we immediate we get $RT(H_n) = O((n/k) \log^2 n)$.

3.3 Rendezvous on Cycle

No matter how the units are distributed on C_n, there must be two units r_i and r_j such that $dist(r_i, r_j)$, the smaller number of edges between them, is less than or equal to $l = \lceil n/k \rceil$. Hence, if r_i waits and r_j moves to r_i along the distance path, they will meet for sure. The rendezvous on a cycle can be solved by the following simple algorithm:

1. Each unit chooses one of three actions in the following way: Move l steps to the left with probability $1/4$; Move l steps to the right with probability $1/4$; Wait at current node for l steps with probability $1/2$. Then with probability $1/4$, r_i and r_j will encounter.

2. If a unit did not encounter any other unit, goto stage 1. Otherwise, the pair of units with the minimum IDs among all units in a meeting start to concurrently traverse the cycle from the meeting node in opposite directions, when they meet again, they will have covered the entire cycle.

3. If in stage 1 a unit encounters a traversing unit, it quits its current action and follows the traversing unit. If two traversing units belonging to different pairs encounter, the one belonging to the lower ranked pair follows the one with the higher rank, where a pair is given higher rank if its smaller ID is smaller than that of the other pair.

If stage 1 is repeated $c \log n$ times, an encounter will occur with high probability. Then the traversal will be started and finished in $n/2$ steps. Since a stage 1 takes time $O(n/k)$, we have,

Theorem 12. *k units can achieve rendezvous in C_n in time $O((n/k) \log n) + n/2$.*

4 Conclusion

In this paper we introduce the problem of coordinating the rendezvous of intelligent mobile agents in a relatively unknown and potentially hostile environment (with no common agreement). We presented initial results in this model. This problem may have implications in robotics, and network-wide computing by agents acting in environments that are relatively unknown (e.g. in secure networks). A number of open questions arise: how about environments where units have only local knowledge (limited vicinity is known)? how about limited memory and limited resource agents (we allowed them to maintain the topology)? How about other special graphs (planar, Euclidean)?

References

1. R.Aleliunas, R.M.Karp, R.J.Lipton, L.Lovasz, and C.Rackoff, "Random Walks, Universal Traversal Sequences, and the Complexity of Maze Problems", 20th FOCS, 1979, pp. 218-223.
2. I. Anderson, "Combinatorial Design", John Willey, 1990.
3. A. Israeli, and M. Jalfon, "Token Management Scheme and Random Walks Yield Self-Stabilizing Mutual Exclusion", PODC 90, 119-131.
4. C.H. Papadimitriou, P. Raghavan, M. Sudan and H. Tamaki "Motion Planning on a Graph" FOCS 94.
5. C.H. Papadimitriou and M. Yanakakis, "Shortest Path without a Map" TCS 84, 1991, 127-150.
6. P. Tetali and P.Winkler, "On a Random Walk Problem Arising in Self-Stabilizing Token Management", PODC 91, 273-280 .

Efficient Asynchronous Consensus with the Value-Oblivious Adversary Scheduler

Yonatan Aumann[1]* and Michael A. Bender[2]**

[1] Department of Mathematics and Computer Science
Bar-Ilan University
Ramat-Gan, Israel
Email: aumann@bimacs.cs.biu.ac.il

[2] Aiken Computation Laboratory
Harvard University
Cambridge, MA 02138
Email: bender@das.harvard.edu

Abstract. We consider the power given to adversary scheduler of an asynchronous system and define the *value-oblivious scheduler*. At each step this scheduler determines the next processor to operate based on the full history of the *dynamics* of the execution; the scheduler is oblivious to the intermediate *values* the processors manipulate. We argue that the value-oblivious scheduler captures the possible sources of asynchrony in real systems.

Assuming the value oblivious adversary, we study the asynchronous consensus problem in the shared-memory setting with atomic reads and writes. We present a probabilistic algorithm that obtains consensus in $O(n \log^2 n)$ *total work*. Here, total work is defined as the total number of steps performed by *all* processors collectively. Thus, the amortized work per processor is $O(\log^2 n)$.

1 Introduction

The Problem. The consensus problem on n processors is defined as follows. Initially, each processor P_i has a private input value $v_{P_i} \in \{0, 1\}$. The goal is to obtain one *consensus value* v, agreed upon by all processors, such that $v \in \{v_{P_i}\}_{i=1}^{n}$. If all processors operate synchronously, then consensus is easily attainable. In reality, however, processors are unreliable and may fail or become exceedingly slow. Even in a fault-free situation processors may encounter delays as a result of page faults, exhausting scheduling quotas, interrupts and network communication delays. Thus, we seek a protocol that obtains consensus using whatever time slices it is awarded by the participating processors.

Efficiency. The first requirement from a consensus protocol is that it be *wait-free*: each processor must obtain the consensus value after a bounded number, T, of its *own* steps, regardless of the speeds of the other processors. Most previous work focuses on minimizing the value of T. This measure, however, only considers the worst case bound *per processor* and does not address the question of the total efficiency of the protocol. In particular, there is no advantage if more processors

* This work was done while the author was at Harvard University, supported in part by ONR contract ONR-N00014-91-J-1981.

** The author was supported by NSF contract CCR-9313775.

cooperate in the protocol. Thus, Attiya, Lynch, and Shavit [6] observe that many wait-free protocols are slow in "normal" executions. They suggest also measuring the number of *rounds* in executions having no or relatively few faulty processors.

In this paper, we employ a strong, unified complexity measure, previously used in the context of A-PRAM (*Asynchronous* PRAM) computations (see [16]). This measure directly gauges the efficiency with which the protocol uses whatever cycles are provided by the processors. In the We count the *total number of instruction cycles* used by the protocol. We present a protocol for n asynchronous processors to obtain consensus in $O(n \log^2 n)$ total number of instruction cycles. After this many instructions, consensus is obtained regardless of the distribution of the instructions among the processors and regardless of the processors' relative speeds and interleaving activity. In particular, if the system is near synchronous, then each processor performs only $O(\log^2 n)$ steps; if only one processor is fast, then this processor reaches a decision after $O(n \log^2 n)$ steps; if processors alternate arbitrarily between running slowly and quickly, then the protocol still makes optimal use of all instruction cycles. The algorithm is randomized and the behavior is guaranteed in both the Las Vegas and Monte Carlo senses.

The protocol runs in the shared-memory setting, and assumes a *value-oblivious adversary scheduler*. This adversary model, similar to that described by Rabin in the context of mutual exclusion [18], is somewhat weaker than the adversary usually considered for consensus algorithms. We describe the model in detail.

The Model. The consensus problem can be viewed as a game between the processors, which seek to attain consensus, and an adversary scheduler, which tries to prevent them from doing so. At each step, the scheduler decides which processor operates next. How much information should be made available to the adversary for making this decision? Most previous papers assume a dynamic adversary scheduler (e.g. [12, 10, 11, 1, 3, 14, 4]) having full knowledge of the entire state of the system at all previous steps.

In the present work, we make a distinction between two separate components of the system's history: the *control history* and the *content history*. The *control history* includes all information regarding the *dynamics* of the computation. In particular, the control history specifies the active processors in previous steps, the operations performed by these processors (read, write, add etc.), and the addresses/locations of the operands of the operations. We define the *value-oblivious scheduler*, which has full knowledge of control history, but is not provided with the actual *values* stored in the memory locations (the *content history*). The scheduler can gain information regarding the value stored in a memory location only insofar as this value affects the dynamics of the computation. In addition, in order to account for pipeline effects and delays due to page faults, we assume that the scheduler has the *operation lookahead* capability: it knows the next operation to be performed by each of the processors. We also assume that the scheduler has full knowledge of the processors' input values and their programs. In fact, we allow the adversary to have full knowledge of the values stored in all memory locations other than those used to store intermediate values of our algorithm.

We argue that the power given to the value-oblivious scheduler faithfully captures the possible sources of asynchronous behavior of real systems. Common causes of asynchrony, such as hardware failures, clock skews, operating system interrupts, cache misses, page faults, and network delays, all depend on the dynamics of the execution. However, on modern computers, most integer ALU operations run in the same number of cycles regardless of the values of the operands. (Our algorithms do not use division, where an interrupt may depend on the computed value.) Similarly, the time to load and store data from memory does not depend on the data. Thus, the value-oblivious adversary captures all the real-world sources of asynchrony.

Previous Related Work. Fischer, Lynch, and Paterson [12] prove the impossibility of deterministic asynchronous consensus in the message-passing model, even if only one processor fails. Chor, Israeli, and Li [10] and Loui and Abu-Amara [15] show that the same holds in the shared-memory model (see also [11]). Randomized solutions for the standard, full knowledge adversary scheduler, are presented in [10, 1, 3, 8, 4]. All of these solutions require $\tilde{O}(n)$ operations per-processor. Our model and results differ in several ways. On the one hand, we consider a weaker adversary scheduler. In addition, we allow multiple-writer registers, which are not used in the above. On the other hand, our work bounds are much stricter. Even for the best of the previous results, the total number of instructions required for completion may be as high as $O(n^2 \log n)$. If most processors are active, then our protocol completes in $O(\log n)$ steps, versus $O(n \log n)$ from the above. The methods used in [10, 1, 3, 8, 4] focus on the number of *rounds* and are ill-suited for our *work* bounds. Thus, we introduce new techniques that allow the protocol to make optimal usage of *instruction cycles*, regardless of the particular processor executing the cycle.

Saks, Shavit, and Woll [19] consider the problem of obtaining a solution that is both wait-free in the general case and fast in the near-synchronous case. Using the *interleaving algorithms* method ([6]) they provide a protocol that with f faulty processors terminates in $O(\log n)$ rounds if $f < \sqrt{n}$, and in $O(\frac{n^3}{n-f})$ rounds, otherwise. The length of the *round* is determined by the speed of the slowest nonfaulty processor. The [19] algorithm provides a good solution for the case that most processors are synchronous and only a small number are faulty. However, it does not provide the flexibility that our work bounds provide for the general asynchronous case. We allow all processors, faulty and nonfaulty, to behave in an arbitrary asynchronous fashion. We emphasis, however, that the [19] result assumes the stronger full-knowledge adversary model and single writer registers, while ours assumes the value-oblivious scheduler and multi writer registers.

Recently, Chandra obtained improved work bounds for consensus in an adversary model similar to ours. In [9], Chandra presents a consensus protocol with expected $O(\log^2 n)$ steps per processor. Thus, the total work is identical to ours, but his result also provides a bound on the work *per processor*. We note however, that the expected contention on the multi-writer registers in [9] can be as high as $O(n)$. In our algorithm, the expected contention is $O(1)$, and with high probability never do more than $O(\log n)$ processors try to access the same register concurrently. Thus, our results also hold in the model that processors queue for shared registers, or the model that all concurrent accesses to the same register fail.

Definitions and Notations. The system consists of n *processors* $\{P_i\}_{i=1}^n$ sharing a common memory space. Reading and writing shared memory cells are atomic operations. W.l.o.g. we assume the computation proceeds in discrete *time steps*, and that at each time step exactly one processor performs one atomic operation. A *schedule S*, is a function $S : \mathbb{N} \to \{1, \ldots, n\}$ that associates one active processor $P_{S(t)}$ with each time step t. The schedule is determined by a dynamic *scheduler S*; the scheduler must determine $S(t)$ before time step t.

We define the information available to the scheduler for making this decision. For each time step $t' < t$, let $A(t')$ be a the triplet $(S(t'), O(t'), L(t'))$, where $P_{S(t')}$ is the processor active at t', $O(t')$ the operation performed by $P_{S(t')}$ at time t', and $L(t')$ is the set of memory locations/registers accessed in this operation. The sequence $(A(1), \ldots, A(t-1))$ is the *control history* for time t. In addition, for processor P_i and time t let $N(t, i)$ be the action P_i will perform as soon as it is active. (In particular, $N(t, i) = A(t)$ if the scheduler sets $S(t) = i$.) We call the set $\{N(t, i)\}_{i=1}^{n}$ the *operation lookahead set* for time t. We assume that the scheduler is provided with the following: (i) the entire content of the memory before starting the execution of the algorithm, (ii) the control history for time t, and (iii) the operation lookahead for time t. Efficiency is measured by the total number of steps performed in the system, summed up over all processors. This total may be arbitrarily distributed among the processors. We say that an event E occurs *with high probability (w.h.p.)* if for any $c > 0$ there exists a proper choice of constants such that $\Pr[E] \geq 1 - n^{-c}$.

The Phase Clock. Many of our algorithmic constructs operate in *phases*, each requiring of $\Theta(n)$ or $\Theta(n \log n)$ work. Measuring work in the asynchronous setting is nontrivial. Aumann and Rabin [7] present a general *Phase Clock* structure which provides the necessary tool. We briefly review the main features of the [7] "clock" structure. At any point in time the Phase Clock has a *value* which is either an integer, or "undefined". The integral values are non-decreasing in time. The value of the clock is initialized to 0. The Clock-Update procedure, which takes $O(1)$ operations, allows the processors to participate in advancing the clock. For any n and α there exists a Phase Clock such that: (i) after every $O(n)$ invocations of Clock-Update, the value of the clock advances from one integral value to the next, (ii) the clock remains in each integral value for at least αn invocations. The Read-Clock procedure, which takes $O(\log n)$ operations, allows processors to get the current value of the clock, if it exists. Regardless, Read-Clock never returns a value greater than that in the clock. As we shall see later, by interleaving clock updates among the rest of the system's work, we get a measure of the total work performed in the system.

Outline. The main consensus procedure is presented in Section 3. The procedure necessitates $O(\log n)$ common *weak random coins* (defined later). In Section 2 we show how processors collectively generate these weak random coins. Combining the procedures of Sections 2 and 3 yields a randomized consensus algorithm in the Monte-Carlo sense (i.e. with high probability it is successful, but may also fail). In Section 4 we show how to convert the algorithm to one with a Las Vegas guarantee. The conversion is based on a novel construct we introduce, which we call the *Write-Once Register (WOR)*.

2 Weak Random Coins

A weak random coin has at least a constant probability of falling on heads and at least a constant probability of falling on tails, but it can also have an undefined value. In this section we show how the processors collectively flip $O(\log n)$ coins, half of which are guaranteed to be weak random coins.

Set $d = \lceil \log(n/\log n) \rceil$ and $N = 2^d$. We employ the following data structure. With each weak random coin r_i, we associate a fixed $N \times (d+1)$ array B_i. We view each array B_i as a butterfly structure of $d+1$ levels, as follows. For $k = 0, \ldots, d$, cells $B_i[k][\cdot]$ constitute the *k-th level* of the butterfly . Each cell $B_i[k][j]$ of level k, is linked to two *predecessor* cells in the previous level: $B[k-1][j]$ and $B[k-1][j^{(k)}]$. Here $j^{(k)}$ is the number whose binary representation is identical to that of j except for the k-th bit. We call the cells of level 0 the *inputs* and those of level d the *outputs*.

A pseudocode description of the algorithm is given hereunder. All cells $B_i[k][j]$ are assumed to be initialized to 0. The cells are (over)written one level at a time. A cell that is overwritten is said to be *filled*. First, the processors fill the inputs of all butterflies randomly with either 0 or 1, with $\Pr[\text{writing } 1] = 1/N$. Once most inputs are filled, subsequent levels of all butterflies are filled in sequence. To fill level k, active processors repeatedly choose at random a cell of the level and write to it the logical OR of its two predecessors. The Phase Clock of [7] is used to keep track of the level currently being filled. At the first level, p Processors read the clock after each write in the first level, and every $\log n$ writes, at subsequent levels. **ClockUpdate** is invoked following each round.

```
Weak-Random-Coin (for processor P_i)
 1     k ← Read-Clock                                      { O(log n) operations }
 2     if k > d then abort                                 { algorithm completed }
 3     if k = 0 then                                       { phase 0, fill inputs }
 4         choose i, j at random        { random butterfly B_i and cell B_i[0][j] }
 5         choose r ∈ {0,1} at random with Pr[r = 1] = 1/N
 6         B_i[0][j] ← r
 7     else                                                { phase k > 0 }
 8         for t = 1 to log n do
 9             choose i, j at random   { random butterfly B_i and cell B_i[k][j] }
10             B_i[k][j] ← B_i[k − 1][j]OR.B_i[k − 1][j^(k−1)]
11     Clock-Update                                        { O(1) operations }
12     goto line 1
```

Analysis. We denote by *phase* k, $0 \le k \le d$, the time interval when the clock has value k. A processor should fill level k in phase k.

Lemma 1. *W.h.p. the weak random coin algorithm completes in $O(n \log^2 n)$ total work.*

Proof. There are $O(\log n)$ levels and $O(n \log n)$ writes per level. Each write requires $O(1)$ work. □

Recall that there are $c \log n$ butterflies and that at the clock remains in each integral value for at least αn clock updates.

Lemma 2. *For any c there exists an α such that w.h.p. the following holds: (1) The total number of inputs not filled during phase 0 is at most $2n$. (2) For any $k > 0$ the total number of cells of level k not filled during phase k is at most $n/\log n$.*

Proof. During phase 0 the processors update the clock at least αn times. Since line 11 is reached at least αn times, line 4 is reached at least αn times. Level 0 contains cn cells, and each time a processor reaches line 4 it chooses a cell at random. Thus, by a version of the Chernoff bounds, for a sufficiently large α, w.h.p. $(c-1)n$ cells are chosen. After a processor chooses a cell, the adversary can stop the processor from advancing to line 5. However, there are only n processors. Thus, the adversary can only stop n chosen cells from being filled. Hence the total number of filled cells in the level is at least $(c-2)n$. The proof for other levels is similar, except that w.h.p. each cell is chosen $O(\log n)$ times. □

Consider an input-output pair of cells $(B_i[0][j], B_i[d][j'])$. Exactly one directed path $c_0, c_1, \ldots, c_{d-1}, c_d$ extends from input $B_i[0][j]$ to output $B_i[k][j']$.

Definition 3. Let $c_0, c_1, \ldots, c_d$ be an input-output path in B_i. We say that c_0 *successfully affects* c_d if c_0 is filled in phase 0 and each cell c_k is filled during phase k. We say an input cell c_0 is *strongly affecting* if it successfully affects at least $3/4$ of the outputs of B_i. We say that B_i is *strongly input dependent* if at least half of its inputs are strongly affecting and there are at most $N/4$ writes to its inputs after the completion of phase 0.

The following lemma follows from the structure of the butterfly using a counting argument. The proof is omitted.

Lemma 4. *There exists a constant c such that w.h.p. at least $c \log n/2$ of the B_i's are strongly input dependent.*

Lemma 5. *Consider the state the system after the procedure completes. There is a constant p such for any $r \in \{0, 1\}$ and any strongly input dependent B_i, $\Pr[3/4 \text{ of the outputs } B_i \text{ are } r] \geq p$.*

Proof. Consider the set of values stored in the inputs of B_i when phase 0 completes and the set of values of the late writes to the inputs of B_i. These are the only values that affect the outputs. Since B_i is strongly input dependent, these two sets contain at most $5n/4 \log n$ values. Each single value is 1 with probability $\log n/n$. Thus, there is a constant probability that none are 1. In this case all outputs of B_i are 0's. On the other hand, consider the set of strongly affecting inputs of B_i that are not written after phase 0 ends. There are at least $n/4 \log n$ such inputs. There is a probability p bounded from 0 by a constant, that the last value written to at least one of these cells is a 1. In this case at least $3/4$ of the outputs get the value 1. Note that the dynamics of the execution are independent of the values appearing in the cells. Thus, to thwart the coin flip, the adversary must determine which cells are not filled independent of the input values. Thus, whether a cell is strongly affecting is independent of the values of these cells. □

Reading the Coin Flip Value. Processor P_i obtains the value of the coin flip r_i by sampling $O(\log n)$ of the output cells of B_i (level d). The majority of the sample is the coin flip value. By Lemma 5, if B_i is strongly input dependent then w.h.p. all samples of B_i give the same value, and this value constitutes a weak random coin flip.

3 The Consensus Procedure

We now explain how to obtain consensus using the weak random coin procedure of the previous section. Let C be an array of size γn, where γ is a constant to be determined later. All cells of C are assumed to be initialized to Λ. The pseudocode of the algorithm appears hereunder. The algorithm runs in *phases*, each consisting of $\Theta(n)$ write cycles. The phases are determined using the clock structure of [7]. During the first phase (phase 0), each processor P_i repeatedly picks a cell of C at random and copies its private input v_{P_i} into this cell. By the end of the phase most cells of C contain values. Now, consider phase $k > 0$. First, processor P_i samples $O(\log n)$ locations of C. Let β be the fraction of 1's in the sample. Then, from β and r_k (the value of the k-th weak random coin), processor P_i determines the *value*, s, of the sample as follows:

$r_k = 0$	$r_k = 1$
$s = \begin{cases} 1, \beta > \frac{1}{3}; \\ 0, \text{otherwise}; \end{cases}$	$s = \begin{cases} 1, \beta > \frac{2}{3}; \\ 0, \text{otherwise}. \end{cases}$

Processor P_1 then writes s to a random location in C. This procedure is repeated for $c \log n = O(\log n)$ phases.

```
Consensus-Procedure (for processor P_i)
1     k ← Read-Clock
2     if k > c log n then abort                           { algorithm completed }
3     if k = 0 then                                              { first phase }
4            s ← local input v_{P_i}                        { copy local input }
5     else                                                { subsequent phases }
6            sample b log n locations of C. Let β be the fraction of 1's in the sample.
7            obtain value of weak random coin r_k
8            s ← (β > (1+r_k)/3)                             { threshold decision }
9     choose j ∈ [1,...,γn] at random
10    C[j] ← s                        { write value in random cell of current level }
11    Clock-Update
12    goto line 1
```

Analysis.

Lemma 6. *W.h.p. the consensus procedure completes in $O(n \log^2 n)$ steps.*

Proof. The are $O(\log n)$ phases. Each contains $O(n)$ clock updates. There is $O(\log n)$ work between consecutive updates. □

Recall that the array is of size γn and the clock advances every αn updates.

Lemma 7. *For any γ there exists an α such that w.h.p. the following holds: for any k at least 0.9 of the cells of C are written during phase k.*

Proof. Line 11 is reached at least αn times during the phase. There are n processors. Thus, lines 3-10 are executed at least $(\alpha - 2)n$ times. □

For $1 \leq k \leq c \log n$, let M_k be the number of cells in C that store the value 1 when phase k starts. Set $m_k = M_k / \gamma n$.

Lemma 8. *There exist constants b and γ such that for all $k > 0$ the following holds. If $m_k \geq 1/2$ (res. $m_k < 1/2$) and $r_k = 0$ (res. $r_k = 1$) then w.h.p. all but at most n of the writes to C during phase k are with the value 1 (res. 0).*

Proof. Suppose $m_k \geq 1/2$ and $r_k = 0$. Let t_k be the starting time of phase k. At most n writes from phase k have samples (line 6) before t_k. We call these *late writes*, and say that they may produce any value. We prove that all writes having their entire sample within phase k output the value 1. The proof is by induction. Consider a single round of the protocol (i.e. execution of lines 1-10) that has its write (line 10) at time t. Assume by induction that, apart from the late writes, all writes in phase k occurring prior to t output the value 1. Then, at all times $t_k \leq t' < t$, the number of 1's in C is at least $M_k - n$. Hence, with γ sufficiently large, if $m_k \geq 1/2$, then *throughout* the interval $[t_k, t)$ the fraction of 1's in C is at least 0.4. Thus, w.h.p. any $b \log n$ sample taken during this interval finds at least $1/3$ of the cells with value 1. Since $r_k = 0$, it outputs the value 1. The argument for $M_k < 1/2$ and $r_k = 1$ is analogous. $\square$

Lemma 9. *W.h.p. there exists a value $v \in \{0,1\}$ such that any time after the last phase completes, at least 0.8 of the cells in C store the value v.*

Proof. Consider phase k. Suppose $m_k \geq 1/2$. By Lemmas 5 and 4 there is a constant probability that $r_k = 0$. If $r_k = 0$ then by Lemma 8 w.h.p. all but n writes during the phase write the value 1. By Lemma 7 at least 0.9 of the cells of C are rewritten during the phase. Thus, for γ sufficiently large, the fraction of ones in C at the beginning of phase $k + 1$ is at least 0.8. Hence, by induction, in all subsequent phases, w.h.p. all samples find the fraction of 1's to be at least 0.7 and thus output a 1 (regardless of the coin flip). Thus, the values in 0.8 of the cells of C remains 1. Similarly, if $m_k < 1/2$, there is a constant probability that $r_k = 1$. Note that since the adversary does not know the value r_k, m_k is independent of r_k. Thus, after $O(\log n)$ phases, w.h.p. only one value appears in 0.8 of the filled cells. $\square$

Lemma 10. *Any value other than Λ appearing in C is one of the processors' original input value.*

Proof. In the first phase C is filled with input values. From then on, values are only copied. $\square$

To obtain the consensus value, the processors either sample C, or else use the error-free protocol of the next section.

4   The Las Vegas Protocol

The algorithm described in the previous section has a (polynomially-small) probability of failure. In this section we show how to transform the algorithm into a Las Vegas algorithm having $O(n \log^2 n)$ expected work.

The Write-Once Register

The Las Vegas algorithm relies on a basic building block we call the *Write-Once Register (WOR)*. The WOR is a deterministic data structure that can only store a single value. If overwritten, the WOR obtains a "garbage" value. We show how to construct a WOR from three elementary read/write registers, assuming only atomic reads and writes.

Let w_1, w_2, and w_3, be the three atomic read/write registers constituting the WOR w. Each w_i has two fields: $w_i.value$ and $w_i.P\text{-}id$. When a processor writes a WOR register it attaches its own processor identification number ($P\text{-}id$) to the value it writes. The registers are assumed to be initialed to Λ. To write to a WOR, a processor writes the three registers in sequence, checking before each write that no other processor has attempted writing the same WOR. The WOR is said to *hold a value* iff all three registers constituting the WOR are identical. If the registers are not identical, the value of the WOR is defined to be "garbage".

```
Write-WOR(v, P_id)
  1     for i = 1 to 3
  2         for j = 1 to i - 1
  3             if w_j ≠ (v, id) then abort
  4         for j = i to i
  5             if w_j ≠ Λ then abort
  6         w_i ← (v, id)

Read-WOR()
  7     read w_1, w_2 and w_3 in order
  8     if w_1 = w_2 = w_3 then
  9         return w_1.value
 10     else
 11         return "garbage"
```

Lemma 11. *Consider a WOR w and suppose that each processor attempts to write to the WOR at most once. Suppose that two separate reads of w return the values v_1 and v_2. If $v_1, v_2 \notin \{\Lambda, \text{"garbage"}\}$ then $v_1 = v_2$.*

Proof. Assume the contrary. W.l.o.g. suppose processor P_1 wrote v_1, and P_2 wrote v_2. Also, w.l.o.g. assume that P_1 wrote to w_1 before P_2. Then, for P_1 to write w_3, it must be that when P_1 reads w_1 in line 3, with $i = 3$, then $w_1 = (v_1, P_1)$. Therefore, P_2 must write value v_2 to w_1 after P_1 reaches line 3 with $i = 3$. Thus, when P_2 reaches line 4, with $i = 2$ P_1 has already written w_2. The test in line 5 must therefore fail. □

The Certified Agreement Procedure

We now present a procedure by which processors are either convinced that consensus is achieved or else are warned of a failure and can rerun the protocol. In Section 3 we showed that w.h.p. a consensus value appears in 80% of the γn cells of the array C. To obtain a *proof* that consensus is achieved, the processors collectively estimate the number of 1's and 0's in C. The procedure employs a butterfly structure W with WOR registers at the first and last levels. Intermediate levels are composed of regular registers. Each level of W has γn nodes and there are $\log \gamma n + 1$ levels. Processors fill the levels of W in sequence. For the first level, the input level, processors copy the values from C, substituting a (-1) for 0 and a 0 for Λ. In subsequent levels, node $W[k][j]$ is written with the sum of the values appearing in its two predecessors. A garbage value is treated as a 0. We employ the [7] clock structure to keep track of the level currently being filled, where we tune the clock so that at least $\alpha n \log n$ clock updates are required to advance the clock by one. A description of the algorithm appears hereunder. All reads and writes to the first and last levels are WOR procedures.

```
Estimate-Counting (for processor Pᵢ)
 1    k ← Read-Clock
 2    if k > log γn then
 3        abort                                          { algorithm completed }
 4    choose j ∈ [1, ..., γn] at random
 5    if k = 0 then                                            { fill inputs }
 6        if C[j] = Λ then                                  { cell not filled }
 7            W[0][j] ← 0                                { substitute 0 for Λ }
 8        else                                          { C[j] has a value }
 9            W[0][j] ← 2C[c log n][j] − 1          { substitute a −1 for 0 }
10    else                                              { subsequent levels }
11        W[k][j] ← W[k − 1][j] + W[k − 1][j^(k)]      { sum of predecessors }
12    Clock-Update
13    goto line 1
```

Once the last level, the output level, is filled, the processors obtain the consensus value by searching for an output with an absolute value greater than $\frac{1}{2}\gamma n$. We call such an output a *strongly-dominated output*. The sign of the strongly-dominated output determines the consensus value. If the processor cannot find a strongly-dominated output it decides that a failure has occurred and reruns the algorithm (line 15). A psuedocode of the process appears in the next page. Note that it is possible that one processor detects a failure, while another exists with a value, v. In this case we guarantee that all processors rerunning the protocol enter with v as their private input value. The array V, initialed to Λ, and lines 5-6, 9-14, provide this guarantee.

Analysis

Due to lack of space, most proofs in this section are omitted.

Lemma 12. *W.h.p. at most 1/6-th of the inputs and 1/6-th of the outputs get the value "garbage".*

Lemma 13. *There exist constants γ, α such that w.h.p. the following holds. For any $0 \leq k \leq \log \gamma n$, at most $n/\log n$ of the cells of level k are not filled during phase k.*

The proof is similar to the proof of Lemma 2. A *strongly-dominated output* is an output that has absolute value greater than $\gamma n/2$.

```
Obtaining-the-Consensus-Value (for processor P_i)
1    repeat                                              { scan outputs }
2         choose j ∈ [1, ..., γn] at random.
3         w ← W[log γn][j]
4         if |w| > ½γn then                      { strongly-dominated output }
5              V[i] ← ½ + w/(2|w|)    { write sign of w to candidate value array }
6              if w = W[log γn][j] then           { W[log γn][j] unchanged }
7                   exit with consensus value v = ½ + w/(2|w|)
8    until all outputs are sampled
9    for j = 1 to γn do          { this line is reached only in case of a failure }
10        overwrite W[log γn][j] with "garbage"³      { erase all outputs }
11   s ← v_{P_i}                                       { use private input }
12   for j = 1 to n do                                 { scan the array V }
13        if V[j] ≠ Λ then                             { a candidate exists }
14             s ← V[j]                                { use candidate value }
15   Start a new agreement protocol using s as the private input value
```

Lemma 14. *There exist constants γ (of the array size) and α (of advancing the clock), such that if 80% of the cells of C have the value 1 (res. 0) then w.h.p. at least half of the outputs are strongly dominated and hold a positive (res. negative) value.*

The next lemma follows from Lemma 10.

Lemma 15. *If all private inputs to the consensus procedure are 1 (res. 0) then any strongly-dominated output is positive (res. negative).*

Corollary 16. *All processes successfully terminating at line 7 exit the same consensus v. This value is one of the input values.*

Corollary 16 completes the proof that w.h.p. in $O(n \log^2 n)$ steps consensus is achieved. We now consider the low-probability event of failure and prove that consensus is always obtained, and in expected $O(n \log^2 n)$ steps. We call the algorithm executed in line 15, the *alternate consensus execution.*

Lemma 17. *If any processor exits the procedure with a consensus value v in line 7, then any processor participating in the alternate execution enters with the value v.*

Proof. Suppose P_1 exits with the value 1 and P_2 participates in the alternate execution. Let $w = W[\log \gamma n][j]$ be the strongly-dominated output found by P_1. In lines 9–10 processor P_2 erases all outputs. Thus, P_1 must have read w for the second time before P_2 reached line 10. By this time, P_1 has already written the value 1 in $V[1]$. By Lemma 15 no other value can be written to any other entry in V. Thus, at line 13, P_2 finds the value 1 in V, and uses it as its input value for the alternate execution (line 15).

The following lemma follows from Lemma 15 and line 14.

Lemma 18. *Any value used as an input value to the alternate execution is one of the original input value.*

---

³ Here, the WOR write procedure *is not followed*. The WOR is overwritten regardless of what is already there.

We thus obtain:

Theorem 19. *Using the above algorithm w.h.p. consensus is obtained in $O(n \log^2 n)$ total number of steps. The algorithm never fails, and $O(n \log^2 n)$ steps is also the expected number of steps.*

References

1. K. Abrahamson. On achieving consensus using shared memory. In *Proceedings of the 7th Annual ACM Symposium on the Principles of Distributed Computing*, pages 291–302, 1988.
2. J. Aspnes. Time- and space-efficient randomized consensus. In *Proceedings of the 9th ACM Symposium on Principles of Distributed Computing*, pages 325–331, 1990.
3. J. Aspnes and M. Herlihy. Fast randomized consensus using shared memory. *Journal of Algorithms*, 11(3):441–461, September 1990.
4. J. Aspnes and O. Waarts. Randomized consensus in expected $O(n \log^2 n)$ operations per processor. In *Proceedings of the 33rd Annual Symposium on the Foundations of Computer Science*, pages 137–146, 1992.
5. H. Attiya, D. Dolev, and N. Shavit. Bounded polynomial randomized consensus. In *Proceedings of the 8th ACM Symposium on Principles of Distributed Computing*, pages 281–294, 1989.
6. H. Attiya, N. Lynch, and N. Shavit. Are Wait-Free Algorithms Fast? In *31st FOCS*, pages 55–64, 1990.
7. Y. Aumann and M.O. Rabin. Clock construction in fully asynchronous parallel systems and pram simulation. *Theoretical Computer Science*, 128:3–30, 1994.
8. G. Bracha and O. Rachman. Randomized consensus in expected $O(n^2 \log n)$ operations. In *Proceedings of the 5th International Workshop on Distributed algorithms*. Springer-Verlag, 1991.
9. T.D. Chandra. Polylog Randomized Wait-Free Consensus. To appear in *Proceedings of the 15th PODC*, 1996.
10. B. Chor, A. Israeli, and L. Ming. On processor coordination using asynchronous hardware. In *Proceedings of the 6th ACM Symposium on Principles of Distributed Computing*, pages 86–97, 1987.
11. D. Dolev, S. Dwork, and L. Stockmeyer. On the minimal synchronism needed for distributed consensus. *Journal of the ACM*, 34(1):77–97, January 1987.
12. M.J. Fischer, N.A. Lynch, and M.S. Paterson. Impossibility of distributed commit with one faulty process. *Journal of ACM*, 32(2):374–382, April 1985.
13. M. Herlihy. Impossibility results for asynchronous PRAM. In *Proceedings of the 3rd ACM Symposium on the Parallel Architectures and Algorithms*, pages 327–336, 1991.
14. M. Herlihy. Wait-free synchronization. *ACM Transactions on Programming Languages and Systems*, 13(1):124–149, January 1991.
15. M. G. Loui and H. Abu-Amara. Memory Requirements for Agreement Among Unreliable Asynchronous Processors. *Advances in Computing Research*, vol. 4, pages 163–183, 1987.
16. C. Martel, R. Subramonian, and A. Park. Asynchronous PRAMs are (almost) as Good as Synchronous PRAMs. In *Proceeding of 31st FOCS*, pages 590–599, 1990.
17. S. Plotkin. Sticky bits and universality of consensus. In *Proceedings of the 8th ACM Symposium on Principles of Distributed Computing*, pages 159–176, 1989.
18. M.O. Rabin. N-Process mutual exclusion with bounded waiting by $4 \log_2 N$-valued shared variable. *Jour. Comp. Sys. Sc.*, 25:66–75, 1982.
19. M. Saks, N. Shavit, and H. Woll. Optimal Time Randomized Consensus - Making Resilient Algorithms Fast in Practice. In *Proceedings of the 2nd Annual ACM-SIAM Symposium on Discrete Algorithms*, pages 351–362, 1991.

A Formal Framework for Evaluating Heuristic Programs [*]

Lenore Cowen,[1] [**] Joan Feigenbaum,[2] Sampath Kannan[3]

[1] Dept. of Math. Sciences and Dept. of CS,
Johns Hopkins University, Baltimore, MD 21218.
cowen@cs.jhu.edu
[2] AT&T Research, Room 2C-473,
600 Mountain Avenue, Murray Hill, NJ 07974.
jf@research.att.com
[3] Dept. of Computer and Information Science,
University of Pennsylvania, Philadelphia, PA 19104.
kannan@central.cis.upenn.edu

Abstract. We address the question of how one evaluates the usefulness of a heuristic program on a particular input. If theoretical tools do not allow us to decide for every instance whether a particular heuristic is fast enough, might we at least write a simple, fast companion program that makes this decision on some inputs of interest? We call such a companion program a *timer* for the heuristic. Timers are related to program checkers, as defined by Blum [3], in the following sense: Checkers are companion programs that check the *correctness* of the output produced by (unproven but bounded-time) programs on particular instances; timers, on the other hand, are companion programs that attempt to *bound the running time* on particular instances of correct programs whose running times have not been fully analyzed. This paper provides a family of definitions that formalize the notion of a timer and some preliminary results that demonstrate the utility of these definitions.

1 Introduction

We address the question of how one evaluates the usefulness of a heuristic program on a particular input of interest. Our intuitive notion of a "heuristic program" is one that is known to produce correct answers but whose running time is not analyzable or has not been analyzed. For example, a heuristic that computes an NP-hard function might, for the problem size at hand, finish in under an hour on some instances, take several hours on some other instances, and run for an entire week on the rest. If our theoretical tools do not allow us to characterize

---

[*] Most of this work first appeared in an AT&T Bell Laboratories Technical Memorandum on December 1, 1994.

[**] Supported in part by an NSF Mathematical Sciences Postdoctoral Fellowship and a consulting agreement with AT&T Bell Laboratories.

the three classes of instances precisely, might we at least write a companion program that, on some relevant instance, takes five or ten minutes to tell us that we should give up on using this heuristic on this instance, unless we are willing to wait all week? In a related example, we may have code for two different heuristics for the same NP-hard function; can we write a fast program that, on some inputs of interest, tells us that one heuristic will finish significantly sooner than the other? We call such a companion program a *timer*.

This paper proposes a formal framework for the evaluation of heuristic programs and provides initial evidence of the effectiveness of the framework. Let f be a function defined on a domain $D = \cup_{n \geq 0} D_n$ and H be a heuristic program that computes f. The timer E is also defined on domain D. Let d be a "deadline function" defined on the natural numbers. $T_H(x)$ is the running time of H on input x. For a particular $x \in D_n$, we are interested in whether or not $T_H(x) \leq d(n)$. If the timer decides that $T_H(x) > d(n)$, it outputs STOP; if it decides that $T_H(x) \leq d(n)$ or if it cannot decide one way or the other, it outputs GO.

Ideally, a timer would output STOP if and only if $T_H(x) > d(n)$. However, this ideal is not attainable in many realistic situations, and we want the scope of our study to include timers that are useful even though they do not achieve the ideal. We take as our point of departure the following basic principle: A timer E should not render the heuristic program H less useful than H is on its own; therefore, E should not tell us to STOP on instances on which H meets our deadline. On the other hand, E should add some value to H; therefore, on at least some of the instances on which H does not meet the deadline, E should tell us to STOP. Thus there might be "bad" instances for this heuristic that the timer "misses," but it cannot miss them all. At the same time, it never calls a "good" instance bad. A family of definitions that capture this notion formally is presented in Section 2.

Our definition of timers addresses a question raised by Valiant in 1979. In an early paper on the complexity of counting [20], Valiant suggests the use of "counting as runtime prediction." He claims that significant gains in the understanding the performance of a heuristic algorithm could be made by an algorithm that, for any input, predicts the heuristic's runtime in much less time than would be required to run the heuristic. This is exactly the notion that we capture in our definitions of timers, but we do not always require that the prediction hold "for any input," because such a requirement might be unrealistic.

Timers are related to the concept of program checking introduced by Blum [3]. A checker is a companion program that checks the *correctness* of the output produced by an (unproven but bounded-time) program on a particular instance. Timers, on the other hand, are companion programs that attempt to *bound the running time* on a particular instance of a correct program whose running time has not been fully analyzed.

Our work on timers is in part a continuation of the research program on checking: Recall that Blum says of his definition of a program checker that

in the above [definition], it is assumed that any program . . . for a problem

π halts on all instances of π. This is done in order to help focus on the problem at hand. In general, however, programs do not always halt, and the definition of 'bug' must be extended to cover programming errors that slow a program down or cause it to diverge altogether [3, pp. 2–3].

It is exactly when a timer says STOP that it has detected a "bug" of this form.

Program checking was introduced with a practical motivation, but it has had a profound impact on complexity theory [15, 18, 2]. We hope that the study of timers, also motivated by practical concerns, will lead to interesting theoretical results.

The next section contains our family of definitions. Section 3 gives examples of timers drawn from diverse problem areas in computer science. Finally, in Section 4, we propose directions for future work, including some alternative ways to formalize the intuitive notion of timer.

2 Definitions

Let f, H, E, d, $D = \cup_n D_n$, and $T_H(x)$ be as in Section 1. The heuristic program H is assumed to be correct, but nothing is assumed about its time complexity.

Definition 1. Let E and H be deterministic programs. E is a **timer for (H, d)** if

1. For all n and all $x \in D_n$, if $E(x) = $ STOP, then $T_H(x) > d(n)$.
2. For all n, if there is at least one $x \in D_n$ for which $T_H(x) > d(n)$, then there is at least one $x \in D_n$ for which $E(x) = $ STOP.

We note that there are several situations in which timer design is trivial, including the following three.

1. For a deadline function $d(n)$, there is a trivial timer that runs in time $d(n)$: It simply simulates H for $d(n)$ steps. To disallow this, we will insist that the timer run in time $o(d(n))$. If the running time of H is superpolynomial, we may insist on satisfying the stricter requirement that the timer be polynomial-time.
2. If H works by partitioning the input space into "easy cases" and "hard cases," testing in time $o(d(n))$ whether an input is an easy case, and finishing in time less than or equal to $d(n)$ exactly on these cases, then a trivial timer E would simply perform the same test as H and output GO exactly when the input falls into the easy case.
3. If H always (resp. never) finishes in time $d(n)$, then a timer E that always outputs GO (resp. STOP) is a third type of trivial timer.

As discussed in Section 1, timers are in some way analogous to program checkers as defined in [3]. With Definition 1 in hand, we can point out two respects in which timers and checkers are fundamentally different. A checker is an oracle machine that calls the program H whose output is being checked, whereas a nontrivial timer cannot call H as a subroutine. Secondly, a checker, by definition, must work for any program H that purports to compute the function f, whereas a timer is, by definition, a companion of a specific heuristic program H.

We regard Definition 1 as a version of the weakest possible requirements that a timer must satisfy to be worthy of the name. Such a definition could be useful in proving interesting negative results. In dealing with real heuristic programs, however, we would like to have timers that recognize a substantial fraction of the bad instances in each subdomain, rather than just a single bad instance. This more pragmatic requirement is formalized in Definition 2.

Definition 2. As in Definition 1, E and H are deterministic. Let $g(n)$ be a polynomial. E is a **g-strong timer** for $(\boldsymbol{H}, \boldsymbol{d})$ if

1. (Same as item 1 of Definition 1.)
2. There is a constant $c > 0$ such that, for all n, E has the following property: If the set X of all $x \in D_n$ such that $T_H(x) > g(n)d(n)$ is nonempty, then E says STOP on at least $\max(1, c|X|)$ of the instances in X.

We may interpret Definition 2 to mean that there are two thresholds, separated by a (usually small) polynomial multiplicative factor. If $T_H(x)$ is under the first threshold, the timer never says STOP; if it is between the thresholds, the timer can say STOP or GO; finally, among the instances on which it is over the second threshold, the timer says STOP on at least a constant fraction.

Definition 3. Suppose that at least one of E and H is probabilistic. The probabilities in items 1 and 2 below are computed over the coin-toss sequences of the relevant programs. E is a **probabilistic timer** for $(\boldsymbol{H}, \boldsymbol{d})$ if there are polynomials $p(n)$ and $q(n)$ such that

1. For all n and all $x \in D_n$, if $\text{Prob}(E(x) = \text{STOP}) \geq 1/p(n)$, then $\text{Prob}(T_H(x) > d(n)) \geq 1 - 1/q(n)$.
2. For all n, if there is at least one $x \in D_n$ for which $\text{Prob}(T_H(x) > d(n)) \geq 1 - 1/q(n)$, then there is at least one x for which $\text{Prob}(E(x) = \text{STOP}) \geq 1/p(n)$.

We also define timers that actually do satisfy the ideal discussed at the beginning of this section.

Definition 4. Suppose that E and H are deterministic. E is a **complete timer** for $(\boldsymbol{H}, \boldsymbol{d})$ if, for all n and all $x \in D_n$, $E(x) = \text{STOP}$ if and only if $T_H(x) > d(n)$.

Similarly, we also define **g-complete**, **g-strong probabilistic**, **strong**, and **complete probabilistic** timers, and we give examples of some of these in Section 3. Other variations on the notion are possible. Analogously there are different definitions of "one-way function," some useful in complexity theory [6, 14] and some in cryptography [9, 5, 8].

One straightforward class of timers arises as follows. If there are easy-to-compute implicit parameters, such the number of edges of a graph, the diameter of a graph, etc., on which the running time of a heuristic H depends, then a simple strategy for a timer is to evaluate these implicit parameters and decide whether to run H or not. In Sections 3.4 below, we exhibit timers that *approximate* implicit parameters that govern the running time of the heuristic but may be hard to compute.

We conclude this section with some basic negative results about timers. First we exhibit a heuristic for which there is no nontrivial complete timer.

Proposition 5. Let d be a fully time-constructible deadline function. Then there is a heuristic H such that any complete timer for (H, d) must be trivial.

Proof: Consider the "universal heuristic" H that takes as input a (program, input) pair (P, x) and simulates P on x. Simple diagonalization shows that there cannot be a nontrivial complete timer E for (H, d). Suppose there were such an E. Because it is nontrivial, its running time is $o(d(n))$. Consider the program P_E that behaves as follows on input x. P_E first computes $E((P_E, x))$. If E outputs GO (i.e., indicates that $T_H((P_E, x)) \leq d(n)$), then P_E runs for an additional $d(n)$ steps; if E outputs STOP (i.e., indicates that $T_H((P_E, x)) > d(n)$), then P_E halts. E cannot be a complete timer, because it is incorrect on (P_E, x): It outputs STOP exactly when $T_H((P_E, x)) \leq d(n)$. ∎

The reason that the construction in Proposition 5 does not provide a counterexample to the weaker Definition 1 is that there is no particular input length on which the timer is always wrong. An encoding trick is used in Proposition 6 to overcome this.

Proposition 6. Let d be a fully time-constructible deadline function. Then there is a heuristic H such that any timer for (H, d) must be trivial.

Proof: Let $\{P_i\}_{i \geq 1}$ be an enumeration of all programs and $f : Z^+ \times Z^+ \to Z^+$ be a one-to-one function. Let H, E, and P_E be as in Proposition 5, except that an input (P_i, x), where x is of length n, must be encoded as a string of length $f(i, n)$ before it is presented to H. If j is the index of the diagonalizing program P_E, then E will be wrong on all inputs of length $f(j, \cdot)$ and hence will not satisfy Definition 1. ∎

3 Examples of Timers

In this section, we describe several examples of timers. Our examples are chosen to satisfy a variety of the definitions given in the previous section.

3.1 Bubble Sort

Let H be a standard implementation of Bubble Sort, such as the one given in Knuth [13]. Let $d(n) = \Omega(n^2)$. (This is only interesting, of course, if $d(n)$ is less than the worst-case running time of bubble sort; if it's not, then there's a trivial timer for H that just says GO on all inputs.) For input sequence $(x_1, x_2, \ldots, x_n)$, we denote by $b(i)$ the *inversion number* of x_i, i.e., the number of indices j such that $j < i$ and $x_j > x_i$. Let $M(i) \equiv \sum_{k=0}^{b(i)-1} i - k$ and $M \equiv \max_{1 \leq i \leq n} M(i)$.

Then it is clear from the description of H given in [13] that each pass of bubble sort reduces each non-zero inversion number by one. To get the inversion number of x_i to 0, we need $b(i)$ passes, and, if all the passes stopped at x_i, the total cost would be $\sum_{j=0}^{b(i)-1}(i-j)$.

Thus M, the largest of the $M(i)$'s, is a lower bound on the running time of H on input $(x_1, x_2, \ldots, x_n)$. We show that, when bubble-sort takes $\Omega(n^2)$ time, M is also $\Omega(n^2)$. If bubble-sort takes time $c_1 n^2$, there exist constants c_2 and c_3 such that at least $c_2 n$ of the passes process at least $c_3 n$ of the elements each. Then it is clear that, if i is the index at which the last of these $c_2 n$ passes stops, and x_k is the element that is in position i at that point, then $M(k)$ is $\Omega(n^2)$. We use these facts to define a linear-time, deterministic, $O(1)$-complete timer E for (H, d).

Let $c \geq 2$ be a constant. E considers c segments of input elements, namely $(x_1, \ldots, x_{\frac{n}{c}}), (x_{\frac{n}{c}+1}, \ldots, x_{\frac{2n}{c}}), \ldots, (x_{n-\frac{n}{c}+1}, \ldots, x_n)$. (If n is not a multiple of c, the last segment can be shorter than the rest.) For $1 \leq l \leq c$, E first finds the minimum element x_{i_l} in the lth segment; it then computes $M(i_l)$. Let M' be the maximum, over l, of $M(i_l)$. E says STOP if and only if $M' \geq d(n)$.

It is clear that E runs in linear time. To prove that E is an $O(1)$-complete timer for (H, d), note first that, if E says STOP, then the real running time of H on input $(x_1, x_2, \ldots, x_n)$ is at least $M \geq M' \geq d(n)$. Next, we must show that there is a constant c' such that E says STOP whenever the real running time is at least $c'd(n)$. It suffices to show that M' is at least a constant fraction of M since we have already shown that M is $\Omega(n^2)$ whenever the real running time of H is $\Omega(n^2)$. Suppose that the i for which $M = M(i)$ is in the lth segment. Then $M(i_l) \leq M'$. The input element x_{i_l} is less than or equal to x_i, and thus any element that comes earlier in the input than x_{i_l} and contributes to $b(i)$ also contributes to $b(i_l)$. More precisely, $i_l \geq i - n/c$ and $b(i_l) \geq b(i) - n/c$. Thus $M' \geq M(i_l) \geq M - (n/c)^2$, which is what we wanted to show.

The constant implied in the statement that E is an $O(1)$-complete timer depends on the constant c; similarly, the meaningful range of values for c depends on the implied constant in $d(n) = \Omega(n^2)$.

3.2 Euclid's Algorithm for GCD

A timer for Euclid's GCD algorithm is omitted from this extended abstract because of space limitations. Currently we do not know how to construct a strong timer for Euclid's algorithm.

3.3 Proving primality

Let D_n be the set of n-bit integers and H be a probabilistic program that, when given an integer x, searches for a proof that x is prime. Suppose that

H proceeds by running the sophisticated algorithm of Adleman and Huang [1] for n^c steps[4] and then, if no proof of primality is found, switching to a simple-minded trial-division algorithm that takes exponential time but always decides correctly whether a number is prime or composite. Let $d(n) = n^c$. A program E that runs the Miller-Rabin compositeness test on x (which takes time $O(n^4)$) and outputs STOP if and only if the test finds a proof of compositeness is a complete probabilistic timer for (H, d).

The crucial fact about this example is that the best-known algorithms for proving primality are considerably slower than the best-known algorithms for proving compositeness. The idea can be generalized to any language $L \in$ RP $\cap$ coRP with RP algorithm A and coRP algorithm B such that one of A or B is significantly faster than the other.

3.4 Timing Enumeration Algorithms

In this section, we present a general method of building timers for "enumeration" (or "listing") programs. A (deterministic) *listing program* for a parameterized family S of combinatorial structures is a program that takes as input a parameter value p and gives as output a list $S(p)$. For example, the program could take as input a graph G and output the list $S(G)$ of all perfect matchings in G; in this example, the family S is the set of all perfect matchings, and the parameter values are graphs G. Similarly, a listing program could take as input a graph G and output the list of all spanning trees of G. For an excellent introduction to the theory of listing, see Goldberg [4].

We restrict attention to listing programs that run in *polynomial total time*, i.e., in time polynomial in n (the length of the input) and C (the length of the output). This restriction is imposed in order to rule out certain simple-minded listing programs that have trivial complete timers.[5]

Listing programs conform to our intuitive notion of "heuristics," because the running time of such a program on input p is in general very hard to calculate; the length of the list $S(p)$ is obviously a lower bound on this running time, but this length is often hard to compute. Interestingly, Valiant [20] mentioned listing algorithms for enumeration, but since his work predated the results of Jerrum and Sinclair that we use in this section by nearly a decade, he was not able to resolve the question of whether approximate counting ever helps in runtime prediction. The number of spanning trees of a graph G can be computed exactly in deterministic polynomial time [12], but the number of perfect matchings is a $\#P$-complete function [19]. A general method of building timers for listing

---

[4] The expression n^c is used here as a symbolic representation of the running time of the Adleman-Huang algorithm. c has not been calculated precisely [1] but $c > 50.$; for our purposes it suffices that $c > 4$.

[5] The two listing problems that we examine in detail happen to have algorithms with the more stringent *polynomial delay* property, first defined by [11]. In a polynomial-delay algorithm, the time it takes to generate the first output configuration and the time between any two consecutive output configurations are both bounded by a polynomial in the size of the input

programs is to compute (either exactly or approximately) the length l of the list $S(p)$ and then to output STOP if and only if the estimate is significantly greater than $d(|p|)$. Both the type of timer that the method yields and the meaning of "significantly" depend on the particular listing problem.

Let H be any *polynomial total time* listing program[6] for spanning trees (e.g., the one of Read and Tarjan [17]); this means that the running time of H is $poly(n, l)$, where n is the size of the input graph G, and l is the number of spanning trees. Let A be the algorithm of [12] that computes l in time $M(n)$, where $M(n)$ is the time to compute the determinant of an $n \times n$ matrix. For any deadline function $d(n)$ such that $M(n) = o(d(n))$, the algorithm that runs A and outputs GO if and only if $A(G) \leq d(n)$ is a nontrivial timer for (H, d).

The timer E that we give for programs that list perfect matchings uses the same basic idea as the one for spanning trees, but it differs in some details. Jerrum and Sinclair [10] give a probabilistic method for approximating the number of perfect matchings in a $2n$ vertex graph that runs in time $O(q^3 n^5 \log^2 n)$, where q is a known upper bound on M_{n-1}/M_n, the ratio of near-perfect matchings to perfect matchings. Even if a good upper bound on q is not known a priori, [10] shows how, given a candidate upper bound c_1, the algorithm can be modified to halt within a small constant factor of the time bounds reported above, with q replaced by c_1; with high probability, the modified algorithm either produces a good estimate for the number of perfect matchings or reports that M_{n-1}/M_n is greater than c_1 and halts. The graphs that pass the test for a bound on the ratio M_{n-1}/M_n (or contain 0 perfect matchings) are called *q-amenable*.

Procedure E

```
{
    if (G has no perfect matching) Output GO
    else
        Choose a polynomial q(n) such that the running time of the
            Jerrum-Sinclair algorithm is less than d(n).
        if (no such q exists) Output GO
        if (G is NOT q-amenable) Output GO.
        else
            Run the Jerrum-Sinclair algorithm for O(q³n⁵ log² n) steps
            if (Estimated number of perfect matchings > 2d(n)) Output STOP
            else Output GO.
}
```

Theorem 7. Let H be any listing program for perfect matchings that runs in

---

[6] We require H to be polynomial total time in order to avoid listing programs that have trivial timers. For example, many combinatorial listing problems can be solved by simple-minded programs that *always* take exponential time, even on instances in which the length of the list is subexponential; if H is such a program, the algorithm that always says STOP is a trivial timer for (H, d), where d is any subexponential deadline function.

total time $Cg(n)$, where C is the number of matchings in the input graph, and let $d(n) = \Omega(n^{35} \log^2 n)$. Then E is a g-strong probabilistic timer for (H, d).

Proof. We first argue that E is a nontrivial probabilistic timer for (H, d). E says STOP only if the estimate for the number of perfect matchings is greater than $2d(n)$. With high probability this estimate is within a factor of 2. Since the number of matchings is a lower bound on the time it takes to list them, H will run for more than $d(n)$ steps with high probability. Also, for n large enough, we know that there exists a graph G such that H will not finish by deadline $d(n)$ on input G, because there exist graphs with an exponential (in n) number of matchings. So it remains to show that there exists a G for which E answers STOP with high probability, i.e. there exists a family of graphs with more than a polynomial number of matchings, for which the ratio M_{n-1}/M_n is not too large. This is satisfied by the simple observation in [10] that all bipartite graphs on n vertices with minimal degree $n/4$ are q-amenable, for $q(n) = n^2$. It is easy to construct such graphs with a superpolynomial number of perfect matchings. Thus choosing $d(n)$ to be $\Omega(n^{11} \log^2 n)$ makes E a nontrivial timer.

To show that E is a strong timer, we recall the following result of Jerrum and Sinclair about the fraction of graphs that have a bounded ratio M_{n-1}/M_n: If $p \geq (1 + \epsilon)n^{-1} \log n$, then with probability $1 - O(n^{-k})$ (where k is a constant depending on ϵ) the random graph $G_{n,p}$ is $q(n)$-amenable, where $q(n) = n^{10}$. Thus E with deadline $\Omega(n^{35} \log^2 n)$ will output STOP on almost all graphs that have more than $d(n)$ matchings. Because of the bound on the total running time of H, any instance on which H takes time $d(n)g(n)$ or greater must have at least $d(n)$ perfect matchings. Because E outputs STOP on almost all such graphs, E is a g-strong probabilistic timer. ∎

A listing program for perfect matchings that runs in time $Cg(n)$, for some polynomial g, can be obtained using the "recursive listing" technique described in [4, §2.1.1]. Finally, we remark that Jerrum and Sinclair's algorithm continues to be sped up, but the constants and the degrees of the polynomials even in the best current incarnations still render it impractical for use as a timer; it is of theoretical interest, however, because it provides a strong probabilistic timer for a class of heuristics that do not seem to have timers that are complete or deterministic.

3.5 Timing Iterative Numerical Algorithms

We now present an example of a timer from the realm of numerical analysis. Our example involves one of the simplest iterative methods; however, it is easy to see how to generalize to other iterative methods. The timer will come from a lower bound on the rate of convergence.

Let $F : R \rightarrow R$ be a continuously differentiable contraction function, with $1 > c' \geq |F'(x)| \geq c > 0$, for all $x \in R$. (See [16] for definitions.) Further assume that $F(0)$ is not 0.

Define H to be the following algorithm that takes F, ϵ, and an initial point $x^{(0)}$ as input and computes an ϵ-approximation to the (unique) fixed point of F. Define the iterative sequence

$$x^{(k+1)} = F(x^{(k)}) \quad (k = 0, 1, \ldots)$$

H computes this sequence, beginning with $x^{(0)}$, until $|x^{(k+1)} - x^{(k)}| < \epsilon$.

That this process will converge to the fixed point of F and that this point is unique, is a well-known fixed-point theorem [16].

Now consider the following procedure E. We will show, for a certain class of functions F, that E is in fact a timer. Let d be the given deadline.

Procedure E

```
{  l_F = (known) lower bound on the amount of time
         it takes to compute F(x), for any x in the domain
   k = d/l_F

   Find c and c', with 1 > c' ≥ |F'(x)| ≥ c > 0 for all x in R
   α = 1/(1-c') |F(0)|
   if (x^(0) > α and c^k(|x^(0)| - α) > ε) Output STOP
   else Output GO
}
```

Theorem 8. Let s be the unique fixed point of F, satisfying the above conditions. If $x^{(0)} > \alpha$ and $c^k(|x^{(0)}| - \alpha) > \epsilon$, then $|x^{(k)} - s| > \epsilon$.

Proof. By the triangle inequality, $|s| - |F(0)| = |F(s)| - |F(0)| \le |F(s) - F(0)|$. The assumption that $c' \ge |F'|$ implies that $|F(0) - F(s)| \le c'|s|$. Solving for $|s|$, $|s| \le \frac{1}{1-c'}|F(0)|$. By definition, the righthand side is just α; so $|s| \le \alpha$. Now let $\epsilon_0 = |x^{(0)} - s|$, and similarly, $\epsilon_i = |x^{(i)} - s|$. Take $x^{(0)} > \alpha$. Then $|\epsilon_{i+1}| = |F(x^{(i)}) - s| = |F(x^{(i)}) - F(s)|$. By the fundamental theorem of calculus, this is equal to $|\int_s^{x_i} F'(x)dx| \ge c|x_i - s| = c\epsilon_i$ by the assumption that $|F'| \ge c$. Iterating, we have $|\epsilon_k| \ge c^k|\epsilon_0| = c^k|x^{(0)} - s|$. Because we have shown that $|s| \le \alpha$,

$$|x^{(k)} - s| \ge c^k|x^{(0)} - s| \ge c^k(|x^{(0)}| - \alpha) > \epsilon, \text{ by assumption.} \quad \blacksquare$$

To complete the formal proof that E is a timer, we need to show that, if there are some bad instances, there are some that E finds. However, E is not yet completely specified. Step (2) assumed that E finds some c and c', with $1 > c' \ge \max_{x \in R} |F'(x)| \ge c > 0$. How successful E can be in accomplishing step (2), depends on the class of functions F considered.

As an example, for $x > 0$, let $f(x) = e^{-c_1 x} + c_2 x$, for $1 > c_2 > c_1 > 0$. The input to the fixed point algorithm we wish to time are the values of c_2, c_1, an initial point $x^{(0)}$ and ϵ. Since $0 < c_2 - c_1 \le f'(x) \le c_2 < 1$, the timer will run procedure E with $c = c_2 - c_1$, and $c' = c_2$. The timer E will be nontrivial for most reasonable values of the deadline d, since each step involves just a few multiplications and divisions, whereas the fixed-point algorithm computes F once in each iteration, with high precision.

4 Discussion and Future Directions

In Section 2, we presented a family of definitions that capture many of the properties that one naturally wants in timers. cAs additional variations, one could instead choose different thresholds for parameters such as the fraction (currently required to be a constant) of bad instances on which a strong timer must say STOP or the probability (currently required to be inverse-polynomial) with which, if a probabilistic timer says STOP, it must really be timing a bad instance. The choices presented in Section 2 should be reevaluated as more timers are exhibited.

More fundamentally, one could view timers from an overall perspective that is dual to the one we've presented. The guiding principle for the family of definitions presented in Section 2 is that a timer may not err when it says STOP but may err when it says GO. This viewpoint makes sense in scenarios in which the heuristic is run "offline," and its being slow on a particular instance is undesirable but not fatal. If the heuristic were part of a real-time system, one may only want to run on an instance if one could guarantee that it would finish before its deadline. In that case, timers should be defined so that they may not err when they say GO but may err when they say STOP. It is clear how one to alter the definitions presented in Section 2 so that they capture this dual notion.

We have focussed so far on timing a heuristic on a given input. Our framework could be extended to focus on whether the heuristic uses an acceptable amount of space, communication bandwidth, number of processors, or any other crucial resource.

More concretely, we believe that numerical analysis is a natural application domain in which to use timers. For example, there are many iterative methods for the solution of systems of linear equations. Standard numerical analysis texts show how to *upper* bound the number of iterations that various methods will require to solve $Ax = b$ in terms of, e.g., the spectral radius of A. For timer design, however, we need to compute *lower* bounds on the number of iterations, and these bounds may depend crucially on the initial iterate $x^{(0)}$. The construction of such timers is an important goal, both practically (because the timers could be deployed in numerical linear algebra packages and help guide users' choices of iterative methods for particular problem instances) and theoretically (because the lower bounds may require new analytical results).

5 Acknowledgments

We thank Robert Cowen, Nick Reingold, and Bart Selman for fruitful discussions during the formative stages of this work. We thank Jong-Shi Pang for assistance with Theorem 8 and Leslie Goldberg for references on listing algorithms.

References

1. L. Adleman and M. Huang, Recognizing primes in random polynomial time, in *Proc. 19th Symposium on Theory of Computing*, ACM, New York, 1987, pp. 462–469.
2. L. Babai, L. Fortnow, and C. Lund, Nondeterministic exponential time has two-prover interactive protocols, Computational Complexity, 1 (1991) 3–40.
3. M. Blum, Program result checking: a new approach to making programs more reliable, in *Proc. 20th International Colloquium on Automata, Languages, and Programming*, Lecture Notes in Computer Science, vol. 700, Springer, Berlin, 1993, pp. 2–14. First appeared in preliminary form in report 88-009, International Computer Science Institute Technical, Berkeley CA, 1988.
4. L. Goldberg, *Efficient Algorithms for Listing Combinatorial Structures*, Cambridge University Press, Cambridge UK, 1993.
5. S. Goldwasser and S. Micali, Probabilistic encryption, J. Comput. and Sys. Scis., 28 (1984) 270–299.
6. J. Grollman and A. Selman, Complexity measures for public-key cryptosystems, SIAM J. Comput., 17 (1988) 309–335.
7. J. Gustedt and A. Steger, Testing hereditary properties efficiently on average, in *Proc. of ORDAL '94*.
8. J. Hastad, Pseudo-random generators under uniform assumptions, in *Proc. 22nd Symposium on the Theory of Computing*, ACM, New York, 1990, pp. 395–404.
9. R. Impagliazzo, L. Levin, and M. Luby, Pseudo-random generation from one-way functions, in *Proc. 21st Symposium on the Theory of Computing*, ACM, New York, 1989, pp. 12–24.
10. M. Jerrum and A. Sinclair, Approximating the permanent, SIAM J. Comput., 18 (1989) 1149–1178.
11. D. Johnson, C. Papadimitriou, and M. Yannakakis, On generating all maximal independent sets, Inf. Proc. Letters, 27 (1988) 119–123.
12. G. Kirchoff, Uber die Auflosung der Gleichungen, auf welche man bei der Untersuchung der linearen Verteilung galvanische Strome gefuhrt wird, Ann. Phys. Chem., 72 (1847) 497–508.
13. D. Knuth, *Sorting and Searching*, The Art of Computer Programming, vol. 3, Addison-Wesley, Reading, 1973.
14. K. Ko, T. Long, and D. Du, On one-way functions and polynomial-time isomorphisms, Theor. Comp. Sci., 47 (1986) 263–276.
15. C. Lund, L. Fortnow, H. Karloff, and N. Nisan, Algebraic methods for interactive proof systems, J. ACM, 39 (1992) 859–868.
16. J.R. Munkres, *Topology: A First Course*, Prentice-Hall, Englewood Cliffs, 1975.
17. R. Read and R. Tarjan, Bounds on backtrack algorithms for listing cycles, paths, and spanning trees, Networks, 5 (1975) 237–252.
18. A. Shamir, *IP = PSPACE*, J. ACM, 39 (1992) 869–877.
19. L. Valiant, The complexity of computing the permanent, Theor. Comp. Sci., 8 (1979) 189–201.
20. L. Valiant, Negative results on counting, in *Proc. 4th Theoretical Computer Science GI Conference*, Lecture Notes in Computer Science, vol. 67, 1979, Springer, Berlin, pp. 40-45.

Improved Scheduling Algorithms for Minsum Criteria

(Extended Abstract)

Soumen Chakrabarti*   Cynthia A. Phillips**   Andreas S. Schulz***

David B. Shmoys[†]   Cliff Stein[‡]   Joel Wein[§]

Abstract. We consider the problem of finding near-optimal solutions for a variety of $\mathcal{NP}$-hard scheduling problems for which the objective is to minimize the total weighted completion time. Recent work has led to the development of several techniques that yield constant worst-case bounds in a number of settings. We continue this line of research by providing improved performance guarantees for several of the most basic scheduling models, and by giving the first constant performance guarantee for a number of more realistically constrained scheduling problems. For example, we give an improved performance guarantee for minimizing the total weighted completion time subject to release dates on a single machine, and subject to release dates and/or precedence constraints on identical parallel machines. We also give improved bounds on the power of preemption in scheduling jobs with release dates on parallel machines.

We give improved on-line algorithms for many more realistic scheduling models, including environments with parallelizable jobs, jobs contending for shared resources, tree precedence-constrained jobs, as well as shop scheduling models. In several of these cases, we give the first constant performance guarantee achieved on-line. Finally, one of the consequences of our work is the surprising structural property that there are schedules that simultaneously approximate the optimal makespan and the optimal weighted completion time to within small constants. Not only do such schedules exist, but we can find approximations to them with an on-line algorithm.

1 Introduction

Recently there has been significant progress in giving approximation algorithms to minimize average weighted completion time for a variety of $\mathcal{NP}$-hard scheduling problems [16, 11, 18]. Constructing a schedule to minimize average completion time in a one-machine or parallel machine scheduling environment has long

* soumen@cs.berkeley.edu. Computer Science Division, U. C. Berkeley, CA 94720. Supported partly by ARPA/DOD (DABT63-92-C-0026), DOE (DE-FG03-94ER25206), and NSF (CCR-9210260, CDA-8722788 and CDA-9401156). Part of the work was done while visiting IBM T. J. Watson Research Center.

** caphill@cs.sandia.gov. Sandia National Labs, Albuquerque, NM. This work was performed under U.S. Department of Energy contract number DE-AC04-76AL85000.

*** schulz@math.tu-berlin.de. Department of Mathematics, Technical University of Berlin, 10623 Berlin, Germany. Supported by the graduate school Algorithmische Diskrete Mathematik (DFG), grant We 1265/2-1.

† shmoys@cs.cornell.edu. School of Operations Research and Industrial Engineering, Cornell University, Ithaca, NY 14853. Research partially supported by NSF grant CCR-9307391.

‡ cliff@cs.dartmouth.edu. Department of Computer Science, Sudikoff Laboratory, Dartmouth College, Hanover, NH. Research partially supported by NSF Award CCR-9308701, a Walter Burke Research Initiation Award and a Dartmouth College Research Initiation Award.

§ wein@mem.poly.edu. Department of Computer Science, Polytechnic University, Brooklyn, NY, 11201. Research partially supported by NSF Research Initiation Award CCR-9211494 and a grant from the New York State Science and Technology Foundation, through its Center for Advanced Technology in Telecommunications.

been known to be polynomial-time solvable [21, 4, 12]; when one adds release dates, precedence constraints, or weights, essentially all versions of the problem become $\mathcal{NP}$-hard, and until [16, 11, 18] very little was known about approximation algorithms with good performance guarantees.

Recent progress on these problems follows from two basic approaches. In the first approach a linear programming relaxation of the scheduling problem is solved, and then a schedule is constructed simply by list scheduling in a natural order dictated by an optimum solution to the linear program. The second approach is a general on-line framework, in which one attempts to pack the most profitable jobs into successive intervals of geometrically increasing size.

In this paper we improve and extend these techniques. We develop new analytical tools that lead to improved off-line and on-line approximation algorithms for many of the problems considered in [16, 11, 18]. By *on-line*, we mean that the algorithm constructs the schedule in time, and that the existence of a job is known only at its release date. We also extend the on-line techniques to a number of different scheduling models. Finally, we extend the on-line technique to yield *bicriteria* scheduling algorithms: we give very general algorithmic and structural results for schedules that simultaneously come within a small factor of both optimal schedule length and average weighted completion time.

Notation and models. We will consider a number of scheduling models; in the most basic we have a set J of n jobs and m identical parallel machines. Each job j has a positive integral processing requirement (size) p_j and must be processed for that amount of time on one of the machines. In *preemptive schedules*, a job may be interrupted and continued later on another machine. In *nonpreemptive schedules*, a job must be processed in an uninterrupted fashion.

We will be interested in constrained scheduling problems, in which each job j may have a release date r_j before which it cannot be processed, and/or there may be a partial order $\prec$ on the jobs, where $j \prec k$ means that job k cannot start before job j completes. For most of our models a job is processed on only one machine at a time, and a machine can process only one job at a time. We will at times consider *parallelizable jobs*, where job j can run on a number of processors simultaneously, with some specified speedup.

We denote the completion time of a job j in a schedule S as C_j^S; the S will be dropped when clear from the context. Most often we seek to minimize the total completion time $\sum_{j \in J} C_j$, or the total *weighted* completion time: a weight w_j is associated with each job and the goal is to minimize $\sum_{j \in J} w_j C_j$. If we divide either objective function by n, we obtain the average completion time and average weighted completion time objectives. In contrast, the makespan is $\max_j C_j$. All of the problems we consider are $\mathcal{NP}$-hard, and thus we seek approximation algorithms. We define a ρ-approximation algorithm to be a polynomial-time algorithm that delivers a solution of quality at most ρ times optimal.

Discussion of results. The first part of the paper has its roots in the observation that one may construct a nonpreemptive parallel machine schedule by list scheduling in order of the completion times of a preemptive schedule while at most losing a factor of $(3 - \frac{1}{m})$ in average weighted completion time [16]. This proved to be an important observation, for a generalization of this idea proved to be a powerful tool in rounding solutions to linear programming formulations of a number of scheduling problems [11, 18].

In this paper we give improved rounding techniques for these problems; specifically, we give an algorithm that takes a preemptive parallel machine schedule of average completion time C and converts it to a nonpreemptive schedule of average completion time $\frac{7}{3}C$. As a corollary, this gives the same bound on the power of preemption in this scheduling environment; allowing preemption improves the average completion time by at most a factor of $\frac{7}{3}$. When applied to the solution of a linear programming formulation considered by Schulz [18] and independently by Queyranne and by Hall, Shmoys & Wein, this technique yields a 3.5-approximation algorithm for the nonpreemptive scheduling of parallel machines subject to release dates to minimize average completion time; this improves on the previous best bound of $(4 - \frac{1}{m})$.

In the second part of the paper we give a framework for designing on-line algorithms that minimize the average weighted completion time, by improving and extending a result of Hall, Shmoys, & Wein [11]. By incorporating an idea of Goemans & Kleinberg [8] that exploits randomization in an elegant way, we can improve the performance guarantee of the resulting algorithms; the resulting bounds are quite strong, and in certain cases even improve upon off-line bounds achieved via the linear programming formulations previously discussed. Furthermore, we show that this framework actually produces a schedule that is simultaneously near-optimal with respect to both the average weighted completion time objective, and with respect to the makespan objective.

The on-line framework requires a dual ρ-approximation algorithm for the problem of scheduling a maximum weight subset of jobs to complete by a given deadline D. We show that, given a dual ρ-approximation algorithm for this problem, there is an algorithm that yields a schedule that is simultaneously within a factor of 4ρ and 2.89ρ of the minimum makespan and total weighted completion time, respectively. We also give a structural theorem, that shows that, for a very general class of scheduling models, there exist schedules that are simultaneously within a factor of 2 of the minimum total weighted completion time, and within a factor of 2 of the minimum makespan. Such simultaneous bounds were only known in very restrictive scheduling environments.

In the third part of the paper we design the above-mentioned dual approximation routine for a number of scheduling problems. As a result, we give the first (off-line or on-line) constant-approximation algorithms for open shop scheduling and job shop scheduling with a fixed number of machines. We also give algorithms for a number of models that capture realistic elements of parallel computing, including resource constraints, forest precedence constraints, and parallelizable jobs. Many of our on-line results are the first constant approximation algorithms for the corresponding problems, on-line or off-line, while others are on-line algorithms whose performance is close to the best known off-line results.

2 Tighter bounds for parallel machines with release dates

In this section, we consider preemptive and nonpreemptive schedules for minimizing average completion time on parallel machines with release dates. We will show that the ratio between the preemptive and nonpreemptive average completion times is at most $\frac{7}{3}$ by giving an algorithm to convert any preemptive schedule to a nonpreemptive schedule with average completion time at most $\frac{7}{3}$ times greater. This improves on the bound of $(3-\frac{1}{m})$, given by Phillips, Stein and Wein [16]. We then use this technique to obtain a 3.5 approximation algorithm for $P|r_j|\sum C_j$, improving upon the best previous bound of $(4-\frac{1}{m})$ [18].

We use the algorithm CONVERT, introduced by [16], which takes a preemptive schedule P and list schedules the jobs nonpreemptively in the order of their completion times in P. Each job in turn is scheduled as early as possible without violating its release date, and without disturbing the jobs that have already been scheduled. We will show that this produces a schedule N that has average completion time at most $\frac{7}{3}$ that of P. For any schedule S, let $C^S = \sum_{j \in J} C_j^S$.

Theorem 1. *Given a preemptive schedule P for scheduling jobs with release dates on parallel machines, algorithm CONVERT produces a nonpreemptive schedule whose average completion time is at most $\frac{7}{3}C^P$.*

Proof. Let $\mathcal{L}$ be the last m jobs to finish in P and let $\mathcal{K}$ be the remaining jobs. In P, multiple jobs in $\mathcal{L}$ can finish on the same machine. Simple movement of job pieces, however, yields a new schedule of no greater total completion time where the last work done on each machine is the last piece of some job in $\mathcal{L}$. We will assume, for this proof, that P is of this form. The jobs in $\mathcal{L}$ are the last m jobs to start in N. Still, some machines may not process any jobs in $\mathcal{L}$. We define a modified schedule, N', that guarantees that each job in $\mathcal{L}$ runs last on some machine in N'. To obtain N' from N, for each machine that has more than one job from $\mathcal{L}$ in N, all but the first to run are removed and each one is appended to the schedule of some machine that has no such job. Since N is a list schedule, job completion times cannot decrease, i.e., we have $C_j^{N'} \geq C_j^N$, for all jobs j. We further divide $\mathcal{L}$ into two sets $\mathcal{L}_1$ and $\mathcal{L}_2$. The set $\mathcal{L}_2$ contains those jobs $j \in \mathcal{L}$ for which $C_j^{N'} = r_j + p_j$ and the set $\mathcal{L}_1 = \mathcal{L} - \mathcal{L}_2$. We let $C^{L_1} = \sum_{j \in \mathcal{L}_1} C_j^P$, $C^K = \sum_{j \in \mathcal{K}} C_j^P$, and $C^{L_2} = \sum_{j \in \mathcal{L}_2} C_j^P$.

Next we show that given a preemptive schedule P for parallel machine scheduling with release dates, algorithm CONVERT produces a nonpreemptive schedule N in which $C^N \leq 2C^K + 3C^{L_1} + 2C^{L_2}$. Phillips, Stein, and Wein[16] show that $C_j^N \leq 2C_j^P + p_j$. If we apply this bound to the jobs in $\mathcal{L}_1 \cup \mathcal{K}$, apply the bound $C_j^N \leq C_j^{N'} = r_j + p_j$ to the jobs in $\mathcal{L}_2$, and sum over all jobs, we get that

$$C^N \leq \sum_{j \in \mathcal{K} \cup \mathcal{L}_1} (2C_j^P + p_j) + \sum_{j \in \mathcal{L}_2} (r_j + p_j) \leq 2(C^K + C^{L_1}) + \sum_{j \in J} p_j + \sum_{j \in \mathcal{L}_2} r_j$$

$$\leq 2(C^K + C^{L_1}) + \sum_{j \in J} p_j + C^{L_2} . \tag{1}$$

But $\sum_{j \in J} p_j \leq C^{L_1} + C^{L_2}$, because each unit of processing contributes to the completion time of at most one of the jobs in $\mathcal{L}_1 \cup \mathcal{L}_2$ and so plugging into (1), we obtain the desired bound.

Now we show that $C^{N'} \leq 3C^K + C^{L_1} + 2C^{L_2}$. Since $C_j^{N'} = C_j^N$, for jobs in $\mathcal{K}$, we can, as above, apply the bound $C_j^{N'} \leq 2C_j^P + p_j$ for jobs $j \in \mathcal{K}$. We can also, by definition, apply the bound $C_j^{N'} = r_j + p_j$ for jobs $j \in \mathcal{L}_2$.

Now we bound the completion times of jobs in $\mathcal{L}_1$, which are each on a different machine. Consider a particular job $j \in \mathcal{L}_1$ on some machine M. Let t be the last time that machine M was idle before running j. Let k be the job that ran immediately after that idle time. Note that $r_k = t$, or else we could have run k earlier. Let $j_1, \ldots, j_\ell$ be the jobs that run on M between k and j. Then

$$C_j^{N'} = r_k + p_k + p_{j_1} + \ldots + p_{j_\ell} + p_j . \tag{2}$$

Since j is in $\mathcal{L}_1$ and not in $\mathcal{L}_2$, we know that there is a job running immediately before j and hence $k \neq j$, and $k \in \mathcal{K}$. If we sum (2) over all jobs in $\mathcal{L}_1$, each job in $\mathcal{K} \cup \mathcal{L}_1$ contributes to the right-hand side at most once, since each job in $\mathcal{L}_1$ is on a different machine. The jobs in $\mathcal{L}_2$ don't contribute at all, since they are run on different machines than the ones which run jobs in $\mathcal{L}_1$. Thus, summing over $\mathcal{L}_1$, we get $\sum_{j \in \mathcal{L}_1} C_j^{N'} \leq \sum_{k \in \mathcal{K}} r_k + \sum_{j \in \mathcal{L}_1 \cup \mathcal{K}} p_j$. Combining this with the bounds above for the remaining jobs we get $C^N \leq C^{N'} \leq 3C^K + 2C^{L_2} + C^{L_1}$. Balancing the two cases proves the theorem. $\qquad\square$

This theorem improves upon the previous best bound of $(3 - \frac{1}{m})$ for the ratio of average completion time in preemptive vs. nonpreemptive schedules; however, the best known preemptive algorithm is a 2-approximation algorithm [16]; therefore a direct application of this theorem does not give an approximation algorithm with an improved performance guarantee. We can, however, apply it to a different relaxation of the nonpreemptive schedule and obtain an improved approximation bound; we also believe that the ideas will prove useful in sharpening the analysis of other linear programming formulations for scheduling problems.

Schulz [18], and independently Queyranne and Hall, Shmoys and Wein, have given a $(4 - \frac{1}{m})$-approximation algorithm for nonpreemptively scheduling parallel machines with release dates to minimize average weighted completion time, which is based on the LP-relaxation of a formulation in *completion time variables:* with each job j is associated a variable C_j. We can show that by letting the solution to the LP-relaxation play the role of the preemptive completion time, we obtain an improved approximation algorithm.

Theorem 2. *There is a 3.5-approximation algorithm to minimize the average completion time of jobs with release dates on parallel machines.*

In Section 3 we give different techniques that yield better randomized performance guarantees.

3 An on-line framework for bicriteria scheduling

In this section, we will improve and extend a result of Hall, Shmoys, & Wein [11], that gives a framework for designing on-line algorithms that minimize the total weighted completion time. By incorporating an idea of Goemans & Kleinberg [8] that exploits randomization in an elegant way, we can improve the performance guarantee of the resulting algorithms. Furthermore, we show that this framework actually produces a schedule that is simultaneously near-optimal with respect to both the total weighted completion time objective, and the maximum completion time objective. The result of Goemans & Kleinberg improves upon a result of Blum et al., who present a similar bicriteria result for traveling-salesman type problems [3]; we show that their approach applies to a very general class of scheduling problems.

Our framework Greedy-Interval is based on algorithms for the *maximum scheduled weight problem:* given a deadline D, a set of jobs available at time 0, and a weight for each job, construct a schedule that maximizes the total weight of jobs completed by time D. We require a *dual ρ-approximation algorithm*, DualPack, which produces a schedule of length $\leq \rho D$ and whose total weight is at least the optimal weight for the deadline D.

Greedy-Interval uses DualPack in the following way. Let $\tau_\ell = \alpha 2^\ell$, where we shall set the constant $\alpha \in [1/2, 1)$ later. In iteration $\ell = 1, 2, \ldots$ we construct the schedule for the time interval $(\rho\tau_\ell, \rho\tau_{\ell+1}]$. Let J_ℓ denote the set of jobs that have been released by τ_ℓ, but not scheduled in the previous iterations. We call DualPack for the set of jobs J_ℓ (modified so that each is available at time 0) and the deadline $D = \tau_\ell$. Since $\rho\tau_{\ell+1} - \rho\tau_\ell = \rho\tau_\ell$, we can translate the schedule produced by the dual approximation algorithm into the specified interval; furthermore, each job is scheduled no earlier than its true release date.

Fix an optimal schedule with respect to $\sum w_j C_j$, in which each job j completes at time C_j^*; let B_j denote the start of the interval $(\tau_{\ell-1}, \tau_\ell]$ in which job j completes. Hall, Shmoys, & Wein [11] show that Greedy-Interval finds a schedule of total weighted completion time at most $4\rho \sum_{j=1}^n w_j B_j \leq 4\rho \sum_{j=1}^n w_j C_j^*$.

The value B_j, for a particular $j \in \{1, \ldots, n\}$, is determined by the choice of α and the value of C_j^*. If, for our choice of α, $\sum_j w_j B_j = \beta \sum_j w_j C_j^*$, where $\beta < 1$, we get an improved performance guarantee of $4\beta\rho$. In fact, we shall simply choose α at random: choose X uniformly in the interval $(0,1]$ and set $\alpha = 2^{-X}$. Consequently, the value β is a random variable. A routine calculation gives that

$$E[B_j] = C_j^* \int_0^1 2^{-x} dx = \frac{1}{2\ln 2} C_j^* . \tag{3}$$

By linearity of expectation, this implies that, for any instance, the expectation of the ratio between the total weighted completion time for the schedule found and for the optimum is at most $\frac{2\rho}{\ln 2} < 2.89\rho$.

Next consider the makespan of the Greedy-Interval schedule. Let $C_{\max}^*$ denote the optimal makespan, and let $B_{\max}$ denote the start of the interval that contains $C_{\max}^*$; i.e., $B_{\max} = \alpha 2^K$. We know that the algorithm must terminate by iteration

$K + 1$, since the dual approximation algorithm must return all of the jobs when $D = 2B_{\max} > C^*_{\max}$. Hence, the makespan of the schedule is at most $\rho\alpha 2^{K+2} = 4\rho B_{\max} \leq 4\rho C^*_{\max}$. That is, we have shown a performance guarantee of 4ρ. If we choose α as above, then the expected performance ratio is also similarly improved, since $E[B_{\max}] = \frac{1}{2\ln 2} C^*_{\max}$.

Theorem 3. *Given a dual ρ-approximation algorithm* DualPack, *the randomized framework* Greedy-Interval *is, simultaneously, an on-line 4ρ-approximation algorithm to minimize the total weighted completion time, and an on-line 4ρ-approximation algorithm to minimize the makespan. Furthermore, for every instance, each objective function is expected to be within a factor of $\frac{2\rho}{\ln 2}$ of its respective optimum.*

The use of randomization is a relatively mild one. If we are interested in off-line results, then we can run the algorithm with many choices of α, select the best output, and thereby achieve the following deterministic results.

Corollary 4. *Given a dual ρ-approximation algorithm* DualPack, *the framework* Greedy-Interval *yields a 2.89ρ-approximation algorithm to minimize the total weighted completion time, and to minimize the makespan.*

This result yields the best known performance guarantee for minimizing the total weighted completion time subject to release date constraints in either a single-machine or a parallel-machine environment, improving results of [11, 18, 10]. Observe that in Corollary 4, we did not state that the two bounds could be achieved by the same schedule. Indeed, we do not know how to achieve these simultaneously, since a choice of α that is good for the first criterion might be bad for the latter. Nonetheless, if we weaken either one of the two bounds to 4ρ, then they can be achieved simultaneously by a deterministic polynomial-time algorithm.

We now compare our algorithmic results to existence theorems about schedules that are simultaneously near-optimal for these two objective functions. We can also show a general result that holds for any of the scheduling environments we consider in this paper, and a wide variety of other models: let S and T denote optimal schedules with respect to the makespan and total weighted completion time objectives, respectively; there exists a schedule with makespan at most $2C^S_{\max}$ and total weighted completion time at most $2\sum_j w_j C^T_j$.

We shall show how to construct the desired schedule from S and T. Let T' be the schedule induced from T by considering only those jobs j for which $C^T_j \leq C^S_{\max}$. Let S' be the schedule induced from S by considering only jobs for which $C^S_j > C^S_{\max}$. Construct the schedule N by first scheduling according to T' and then according to S'. Clearly, the makespan of N is at most $2C^S_{\max}$. Furthermore, each job j scheduled in T' completes at the same time in N as in T. For each job j scheduled in S', $C^T_j > C^S_{\max}$, and yet $C^N_j \leq 2C^S_{\max}$. Hence, $C^N_j < 2C^T_j$ for each job j.

This proof can be refined to yield somewhat better constants; the details will be given in the complete version of the paper. In contrast to this result, Hurkens

& Coster [13] have given a family of unrelated parallel machine instances for which all minimum total completion time schedules have makespan that is an $\Omega(\log n)$ factor greater than the optimal makespan.

3.1 Applying randomization to other interval-based algorithms

We can also use the randomized technique of Goemans and Kleinberg [8] to improve LP-based results of Hall, Schulz, Shmoys, & Wein [10]. In particular, we improve upon a 7-approximation algorithm for the problem of minimizing the total weighted completion time on scheduling parallel machines subject to precedence constraints and release dates. The algorithm of [10] first solves an LP relaxation of this problem to obtain an optimal solution $\widetilde{C}_j$. Assume that the jobs are indexed by nondecreasing LP value; the LP solution satisfies the following properties: (1) for each $j = 1, \ldots, n$, $\sum_{k=1}^{j} p_k/m \leq 2\widetilde{C}_j$; (2) for each $j = 1, \ldots, n$, $r_j \leq \widetilde{C}_j$; (3) $\widetilde{C}_j + p_k \leq \widetilde{C}_k$ whenever $j \prec k$; (4) $\sum_j w_j \widetilde{C}_j \leq \sum_j w_j C_j^*$.

Given this solution, the algorithm partitions the time horizon into intervals. Set $\tau_\ell = \alpha 2^\ell$, $\ell = 1, \ldots, L$, where we will judiciously choose $\alpha \in [0.5, 1)$. (In [10], this value was, in essence, 0.5.) We partition the time horizon into the intervals $(\tau_{\ell-1}, \tau_\ell]$, $\ell = 1, 2, \ldots, L$, where L is chosen so that each $\widetilde{C}_j$ value is contained in some interval. Partition the jobs into sets such that J_ℓ is defined to be the set of jobs j for which $\widetilde{C}_j$ lies within the ℓth interval, $\ell = 1, \ldots, L$. We construct disjoint schedules for each set J_ℓ by performing ordinary list scheduling on each set, using any list ordering that is consistent with the precedence constraints and ignoring any release dates. Set $\overline{\tau}_\ell = \tau_{\ell+1} + \sum_{k=1}^{\ell} t_k$, where $t_\ell = \sum_{j \in J_\ell} p_j/m$, $\ell = 1, \ldots, L$. We schedule the fragment for J_ℓ between $\overline{\tau}_{\ell-1}$ and $\overline{\tau}_\ell$, $\ell = 1, \ldots, L$.

It is relatively straightforward to show that this yields a feasible schedule. For each $j \in J_\ell$, we can bound the completion time of j in this schedule by $\beta_j + 6\tau_{\ell-1}$, where β_j is the length of some chain that ends with job j. This yields the performance guarantee of 7 of [10]. Let $B_j = \tau_{\ell-1}$ as in the on-line case. Thus, if α is set equal to 2^{-X} where X is selected uniformly from $(0, 1]$, then the expected value of $6B_j$ is equal to $\frac{3}{\ln 2}\widetilde{C}_j$. This implies that the expected ratio between the total weighted completion of the schedule found and the optimum is at most $1 + \frac{3}{\ln 2} < 5.328$. Of course, we can derandomize the algorithm as well.

Theorem 5. *There is a 5.33-approximation algorithm for nonpreemptively scheduling on parallel machines with release dates and precedence constraints.*

Applying techniques used in [10], it is also quite straightforward to generalize this theorem to the setting in which machine i runs at speed s_i. In that model, the resulting algorithm has a performance guarantee that is at most $\min\{2.89 + 2.45\frac{\max_i s_i}{\min_i s_i}, 2.89 + 5.31\sqrt{m-1}\}$.

4 Dual packing algorithms and applications

In order to apply our on-line framework of Section 3, we need to construct the corresponding subroutine $\mathsf{DualPack}(J, D)$. In this section, we shall present this subroutine for a number of scheduling environments.

Several of our results will involve parallelizable jobs. A parallelizable job is described by a processing time p_j and a number of machines m_j. A *non-malleable* job j must be run on exactly m_j processors, and has a specified running time p_j. A *perfectly malleable* job j may be run on μ processors, where $\mu = 1, \ldots, m_j$, and then has running time $p_j m_j / \mu$. A *malleable* job j is a common generalization where, for each possible number of processors $\mu = 1, \ldots, m$, there is a specified running time $p_j(\mu)$.

Our implementation of DualPack(J, D) has the same outline in each of our applications. First, we prune from J the jobs that are impossible to complete within D time units, either because of precedence constraints or because their processing time is too large. We shall restrict our attention to precedence constraints $\prec$ that are out-trees; in-trees can be handled analogously. To prune based on precedence constraints, define PathToRoot(j) $= p_j$ if j is a root, and PathToRoot(j) = PathToRoot(Parent(j)) $+ p_j$ otherwise. These quantities can be computed in linear time, and the appropriate jobs can then be eliminated. Second, we use a routine Knapsack to solve a modified knapsack problem in order to find a set of jobs of sufficiently large weight to schedule. Third, we schedule those jobs selected using a known makespan algorithm.

The input to the routine Knapsack(J, S) consists of a set J of n items (jobs), where item j has weight w_j and size s_j, and a knapsack of size S; in addition, we might be given precedence constraints on the items; if $j_1 \prec j_2$ we forbid the packing of j_2 unless j_1 is also packed. The knapsack problem is to find the maximum weight set that can be packed into the knapsack; let the optimal value be denoted W^*. The routine Knapsack(J, S) finds a set of total weight at least W^* that has total size at most $(1 + \epsilon)S$, where $\epsilon > 0$ is an arbitrarily small constant. To achieve this, we round down each s_j by units of $\epsilon S / n$ and then use dynamic programming as in [14].

In each of the following subsections, we will give an implementation of DualPack for a specific problem; throughout, we denote the deadline by D and the set of jobs from which we choose by J. All of these are bicriteria results; we report only the min-sum result and omit proofs for brevity.

4.1 Malleable jobs

In this subsection we give an algorithm for malleable parallelizable jobs without precedence constraints. The best off-line performance guarantee known for the non-malleable special case, without release dates, is 8.53, due to Turek *et al* [22]; by applying an idea of Ludwig & Tiwari [15], this can be extended to the malleable case. Our *on-line* min-sum algorithm with release dates has a performance guarantee of $12 + \epsilon$, and if we allow randomization, a nearly identical guarantee of 8.67.

We implement DualPack as follows: for each job j, we find the value of μ such that $p_j(\mu) \le D$ for which $\mu p_j(\mu)$ is minimized. Jobs for which there is no such μ are removed from J; otherwise, let m_j be the value for which this minimum is attained. If job j is scheduled by DualPack, it will be run on m_j machines. We set the weight and size of job $j \in J$ to be w_j and $m_j p_j$, respectively, and then call

Knapsack, setting $J' = \mathsf{Knapsack}(J, mD)$. Finally, we adapt the list scheduling algorithm of Garey & Graham [7] to schedule J'.

Theorem 6. *The above* DualPack *routine is a dual* $(3 + \epsilon)$*-approximation algorithm for the maximum scheduled weight problem. This gives a deterministic on-line* $(12+\epsilon)$*-approximation algorithm for scheduling malleable jobs on parallel machines, and a randomized on-line algorithm with expected performance within* 8.67 *of optimal.*

Using the multidimensional knapsack routine in §4.4, together with adaptations of graph labeling results of [7], we can also generalize non-malleable jobs to resource constrained jobs. In this model, each job j holds $r_{ij} \leq 1$ units of resource type $i, i = 1, \ldots, m$, while it runs for duration p_j. At most 1 unit of each resource is available at any time. We get an on-line $O(m)$-approximation algorithm for the min-sum problem; only makespan results were known previously.

4.2 Sequential jobs with tree precedence

Whereas Hall, Shmoys and Wein gave versions of DualPack for m parallel machines, they were not able to handle precedence-constrained jobs. We present here a routine DualPack that can handle forest precedence constraints. In contrast to the previous subsection, we now consider sequential tasks; each job must be scheduled to run on exactly one machine. Our DualPack routine is as follows. We remove from J all jobs j with $\mathsf{PathToRoot}(j) > D$, and let $J' = \mathsf{Knapsack}(J, mD)$. We then list schedule J'.

Theorem 7. *The above* DualPack *routine is a dual* $(2 + \epsilon)$*-approximation algorithm for the maximum scheduled weight problem. This gives a deterministic on-line* $(8 + \epsilon)$*-approximation algorithm to minimize the average weighted completion time of sequential jobs with release dates and forest precedence on parallel machines, and a randomized on-line algorithm with expected performance within* 5.78 *of optimal.*

4.3 Perfectly malleable jobs with tree precedence

We shall consider perfectly malleable jobs, as in Feldmann *et al* [6], and out-tree precedence constraints. (A study of precedence and non-malleability is initiated in [5].) Our DualPack routine is as follows. We remove from J any j with $\mathsf{PathToRoot}(j) > D$, set $J' = \mathsf{Knapsack}(J_\ell, mD)$, and list schedule J' as in [6]: let $\phi = (\sqrt{5} - 1)/2$ be the golden ratio; whenever there is a job j with all of its predecessors completed and the number of busy processors is less than ϕm, schedule the job on the minimum of m_j and the number of free processors.

Theorem 8. *The above* DualPack *routine is a dual* $(2 + \phi + \epsilon)$*-approximation algorithm for the maximum scheduled weight problem. This gives a deterministic on-line* 10.48*-approximation minsum algorithm for perfectly malleable jobs with forest precedence constraints, and a randomized algorithm with expected performance within* 7.58 *of optimal.*

4.4 Minsum shop scheduling

In shop scheduling each job consists of *operations*, each of which must run on a specified machine; for each job, no two of its operations may run concurrently. In the *job shop* problem the operations must proceed in a specified sequence, while in the *open shop* problem they can be scheduled in any order. Minimizing the makespan of job shops is perhaps the most notorious of difficult $\mathcal{NP}$-hard scheduling problems; even small instances are difficult to solve to optimality [1] and the best approximation algorithms give polylogarithmic performance guarantees [19].

We give the first constant-factor approximations for min-sum shop scheduling with a fixed number of machines m. No approximation algorithms were known for minimizing average completion time in shop scheduling, except for the recent results by Schulz that give an m-approximation algorithm for *flow shop scheduling*, the special case of job shop in which the order is the same for each job [18]. We assume for ease of presentation that each job has at most one operation on a machine; let p_{ij} be the size of the operation of job j on machine i.

We again give a version of DualPack for this problem. Let the maximum weight of jobs that can be scheduled by D be W^*, and the optimal set of jobs be J^*. We note that two lower bounds on the makespan of a shop schedule (job or open) are the maximum job size $P_{\max} = \max_{j \in J^*}\{\sum_i p_{ij}\}$, and the maximum machine load, $\Pi_{\max} = \max_i\{\sum_{j \in J^*} p_{ij}\}$. Let W' be the optimum of the following multidimensional knapsack problem: maximize $\sum_j w_j x_j$ such that $\sum_j x_j p_{ij} \leq D$ for $i = 1, \ldots, m$, and $\sum_i x_j p_{ij} \leq D$ for $j = 1, \ldots, n$, where $x_j \in \{0, 1\}$; note that $W' \geq W^*$.

Lemma 9. *For fixed m there is an algorithm for the multidimensional knapsack problem that produces a solution of weight $W'' \geq W'$ for which $\sum_j x_j p_{ij} \leq (1 + \epsilon)D$ and $\sum_i x_j p_{ij} \leq D$.*

We now show that having found a set of jobs of total weight $W'' \geq W^*$, we can schedule this set within makespan $(2 + \epsilon)D$. For an open shop instance a nonpreemptive schedule of length $P_{\max} + \Pi_{\max} \leq (2 + \epsilon)D$, and a preemptive schedule of length $\max(P_{\max}, \Pi_{\max}) \leq (1 + \epsilon)D$ can be constructed [9, 2]. For job shop scheduling we can adapt the known $(2 + \epsilon)$-approximation makespan algorithm for fixed m [19].

Theorem 10. *There is a randomized $(2.89+\epsilon)$-approximation algorithm for preemptive open shop scheduling with a fixed number of machines to minimize average weighted completion time, and there are randomized $(5.78+\epsilon)$-approximation algorithms for nonpreemptive open shop and job shop scheduling on a fixed number of machines to minimize average weighted completion time.*

Acknowledgments We thank Jon Kleinberg for pointing out that the randomized interval selection technique could be applied to Greedy-Interval, and Leslie Hall for helpful discussions.

References

1. D. Applegate and W. Cook. A computational study of the job-shop scheduling problem. *ORSA Journal of Computing*, 3:149–156, 1991.
2. I. Bárány and T. Fiala. Többgépes ütemezési problémák közel optimális megoldása. *Szigma-Mat.-Közgazdasági Folyóirat*, 15:177–191, 1982.
3. A. Blum, P. Chalasani, D. Coppersmith, B. Pulleyblank, P. Raghavan, and M. Sudan. The minimum latency problem. In STOC 26, pages 163–172, 1994.
4. J.L. Bruno, E.G. Coffman, and R. Sethi. Scheduling independent tasks to reduce mean finishing time. *Communications of the ACM*, 17:382–387, 1974.
5. S. Chakrabarti and S. Muthukrishnan. Resource scheduling for parallel database and scientific applications. To appear in SPAA 96, June 1996.
6. A. Feldmann, M. Kao, J. Sgall, and S. Teng. Optimal online scheduline of parallel jobs with dependencies. In STOC 25, pages 642–653, 1993.
7. M.R. Garey and R.L. Graham. Bounds for multiprocessor scheduling with resource constraints. *SIAM Journal on Computing*, 4:187–200, 1975.
8. M. Goemans and J. Kleinberg. An improved approximation ratio for the minimum latency problem. In SODA 7, pages 152–157, 1996.
9. T. Gonzalez and S. Sahni. Open shop scheduling to minimize finish time. *Journal of the ACM*, 23:665–679, 1976.
10. L.A. Hall, A.S. Schulz, D.B. Shmoys, and J. Wein. Scheduling to minimize average completion time: Off-line and on-line algorithms. Joint journal version of [11] and [18]; in preparation.
11. L.A. Hall, D.B. Shmoys, and J. Wein. Scheduling to minimize average completion time: Off-line and on-line algorithms. In SODA 7, pages 142–151, 1996.
12. W. Horn. Minimizing average flow time with parallel machines. *Operations Research*, 21:846–847, 1973.
13. C.A.J. Hurkens and M.J. Coster. On the makespan of a schedule minimizing total completion time for unrelated parallel machines. Unpublished manuscript.
14. D.S. Johnson and K.A. Niemi. On knapsacks, partitions, and a new dynamic programming technique for trees. *Mathematics of Operations Research*, 8:1–14, 1983.
15. W. Ludwig and P. Tiwari. Scheduling malleable and nonmalleable parallel tasks. in SODA 5, pages 167–176, 1994.
16. C. Phillips, C. Stein, and J. Wein. Scheduling jobs that arrive over time. In WADS 4, pages 86–97, 1995.
17. M. Queyranne and A.S. Schulz. Polyhedral approaches to machine scheduling. Preprint 408/1994, Dept. of Mathematics, Technical University of Berlin, 1994.
18. A.S. Schulz. Scheduling to minimize total weighted completion time: Performance guarantees of LP-based heuristics and lower bounds. Preprint 474/1995, Dept. of Mathematics, Technical University of Berlin. To appear in IPCO V, June 1996.
19. D.B. Shmoys, C. Stein, and J. Wein. Improved approximation algorithms for shop scheduling problems. *SIAM Journal on Computing*, 23:617–632, 1994.
20. D.B. Shmoys and É. Tardos. Scheduling parallel machines with costs. In SODA 4, pages 448–455, 1993.
21. W.E. Smith. Various optimizers for single-stage production. *Naval Research Logistics Quarterly*, 3:59–66, 1956.
22. J. Turek, U. Schwiegelshohn, J. Wolf, and P. Yu. Scheduling parallel tasks to minimize average response time. In SODA 5, pages 112–121, 1994.

On the Complexity of String Folding

Mike Paterson[1] and Teresa Przytycka[2]

[1] Department of Computer Science, University of Warwick, Coventry CV4 7AL, UK
[2] Department of Mathematics and Computer Science, Odense University, DK 5230
Odense M, Denmark

Abstract. A *fold* of a finite string S over a given alphabet is an embedding of S in some fixed infinite grid, such as the square or cubic mesh. The *score* of a fold is the number of pairs of matching string symbols which are embedded at adjacent grid vertices. Folds of strings in two- and three-dimensional meshes are considered, and the corresponding problems of optimizing the score or achieving a given target score are shown to be NP-hard.

1 Introduction

The motivation for the string-folding problems considered here lies in computational biology. Prediction of the three-dimensional structure of a protein from its known linear sequence of amino acids is an important practical open problem, which seems to be extremely challenging. The way in which a protein folds determines many of its biological and chemical properties. A natural approach is to look for a spatial configuration achieving a minimum free energy level. The energy is determined by such factors as the number of chemical bonds established between amino acid residues in the sequence and the number of hydrophobic interactions.

While most people would expect that finding a minimum energy configuration would be computationally intractable, previous results to this effect are very limited. Ngo and Marks [6] consider the problem of embedding a string of atoms of length exponential in the input size. The string is described by giving its length (in binary) and the locations and descriptions of the small number of special atoms along the chain which do *not* have the default geometric characteristics. The energy which is to be minimised is based on the dihedral angles of the (non-default) bonds. The proof uses a transformation from the PARTITION problem (see [3]). Unger and Moult [7] use a model like ours in that they embed a string over an arbitrary alphabet into the three-dimensional mesh, but their "distance function" is very artificial. In effect it forces the active subsequence of a string to lie in one straight line. It is then easy to design a transformation from the OPTIMAL LINEAR ARRANGEMENT problem (see [3]). Fraenkel's construction [2] is much more elaborate. He uses a model with charged atoms (his alphabet is $\{-1, 0, 1\}$), where the interactions are taken between all pairs of atoms embedded at adjacent vertices of the (two- or three-dimensional) mesh. He uses a reduction from 3-DIMENSIONAL MATCHING (see [3]). The most

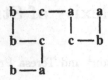

Fig. 1. Fold of **bacbbcacba** in $\mathbb{Z}^2$ with a score of 4

significant limitation of this result, in comparison with our problem, is that the object to be embedded is a (rather exotic) graph rather than just a string.

We too prove an NP-completeness result for a much-simplified model, in which we attempt to capture rather more of the essential character of the protein-folding problem. The protein molecule is represented by a string of symbols, a bond can be made only between a pair of identical symbols, and we seek an embedding of the given string in a grid so as to maximise the number of pairs of matching symbols at adjacent grid points. This version of the folding problem involves a mixture of combinatorial, geometric and topological considerations. Hart and Istrail [4] have given an approximation algorithm for the "hydrophobic-hydrophilic" model over the same grids. This corresponds to our model, but with a binary alphabet in which the only matches counted are between adjacent 1's.

2   Preliminaries

A fixed infinite grid $\mathbb{G}$ is given. In our paper this graph will be either the two-dimensional square mesh $\mathbb{Z}^2$ or the three-dimensional cubic mesh $\mathbb{Z}^3$, though other grids such as the triangular mesh or tetrahedral meshes would also be of interest. A given finite string S, of length n say, is to be embedded in $\mathbb{G}$. A *fold of S in* $\mathbb{G}$ is an injective mapping from $[1, \ldots, n]$ to $\mathbb{G}$ such that adjacent integers map to adjacent nodes of $\mathbb{G}$. (Each node of $\mathbb{Z}^2$ has four neighbours.) The *score* of a fold of S in $\mathbb{G}$ is the number of bonds in the fold, where a *bond* is a pair of identical symbols mapped to adjacent nodes of $\mathbb{G}$. For convenience we do not count a pair of successive identical symbols in S as forming a bond. In Figure 1 we show a fold of $S =$ **bacbbcacba** in $\mathbb{Z}^2$ which has a score of 4. Readers may like to verify that this score is maximal and that the fold achieving it is unique up to the obvious symmetries. The maximum score for a fold of S in $\mathbb{Z}^3$ is 5 however.

We define the following recognition version of the problem of finding an optimal fold.

STRING-FOLD

Instance: A finite string S, an integer k, and a grid $\mathbb{G}$.
Question: Is there is a fold of S in $\mathbb{G}$ with a score of at least k?

Note that the alphabet of symbols is not fixed but is implicitly part of the instance. We have been unable to extend our results to deal with a fixed alphabet.

A variety of different models for bonding are possible. We may wish to represent "neutral" elements which can form no bonds. However, any symbol which occurs exactly once in S obviously has this property and can be regarded as a neutral or *blank* symbol. For notational clarity and convenience we will use a single new symbol, $*$, for such blanks. Also we may want bonds to be formed between pairs of "complementary" symbols, a_+ and a_- say. This feature comes automatically, though somewhat artificially, for string folding in bipartite grids such as Z^d, since alternate symbols in the string must map into grid nodes of opposite parity. Since adjacent grid nodes have opposite parity, we can regard a symbol a as being an a_+ or an a_- according to whether it occurs in an even or odd position in S.

Our main results are that STRING-FOLD is NP-complete when G is either Z^2 or Z^3. These will be proved in Sections 3 and 4. The proofs involve transformations from two known NP-complete problems, 3SAT [1, 3] and "planar" 3-satisfiability, P3SAT [5, 3]. These useful technical problems are defined as follows.

3SAT

Instance: A set $X = \{x_1, \ldots, x_m\}$ of variables and a collection $B = \{C_1, \ldots, C_k\}$ of clauses over X such that each clause has three literals.

Question: Is there a truth assignment for X such that each clause in B has at least one true literal?

The planarity condition for P3SAT is given in terms of the following associated graph. Given clauses B and variables X as above, the graph $G(B) = (V, E)$ is given by $V = B \cup X$, and $E = E_1 \cup E_2$ where:

$E_1 = \{(C_i, x_j) \mid x_j \in C_i \text{ or } \neg x_j \in C_i\}$ are the so-called *variable-clause* edges and

$E_2 = \{(x_j, x_{j+1}) \mid 1 \le j \le m\} \cup \{(x_m, x_1)\}$ are the so-called *variable-variable* edges.

P3SAT

Instance: A set X of variables and a collection B of clauses as in 3SAT, such that $G(B)$ is planar.

Question: Is B satisfiable?

3   STRING-FOLD in Z^2

In this section, we show that the STRING-FOLD problem for the grid Z^2 is NP-complete, by using a transformation from P3SAT.

Given a planar formula B with k clauses, we construct a string $S = s_1 s_2 \ldots s_n$ such that S can be embedded with score $f + k$ if and only if B is satisfiable, where f is a value that follows from the construction of the string. The string S is composed from several substrings designed separately. In combining substrings, a symbol $*$ from one string will sometimes be replaced by a symbol from another string so that the corresponding substrings fit together. A substring composed entirely of $*$'s is called a *flexible substring*.

Conflict graph. Given a set S of substrings of S, a *conflict graph* for S is a graph G, whose vertex set is the set of pairs (i, j), $1 \le i < j \le n$, such that

$s_i = s_j$ and i, j have opposite parity. Thus the vertices of G represent potential bonds. The edge set of G has the property that if $((i, j), (i', j'))$ is an edge of G then, for every fold of S, the bonds (i, j) and (i', j') are mutually exclusive.

Note that we do not require that a conflict graph be maximal, i.e., that every pair of conflicting bonds is represented by an edge of G. Any conflict graph can be used to give an upper bound for the maximum score of a fold of S, since the following property follows immediately from the definition.

Lemma 1. *If G is a conflict graph for a set S of substrings of S then in any fold of S the bonds of S form an independent set in G.*

Our basic tool for constructing conflict graphs is given by the following easy lemma.

Lemma 2. *Let $S = Ua_1Vb_1Wa_2Xb_2Y$, where U, V, W, X, Y are substrings, and $a_1 = a_2$, $b_1 = b_2$ are pairs of symbols of opposite parity. If $|W| > (|V| + |X|)^2$ and $\min\{|U|, |Y|\} > (|V| + |W| + |X|)^2$ then, in any fold of S, the bonds (a_1, a_2) and (b_1, b_2) exclude each other.*

We will apply Lemma 2 in cases where V and X have lengths bounded by a small constant, and W can easily be made large enough to satisfy the first inequality. Since we can extend U and Y suitably just by adding extra blank symbols at the beginning and end of the string S, the inequalities in the statement of the lemma can be easily satisfied.

Shift-lines. A *shift-line of order* m is a pair of strings,

$$S_1 = a_1a_1 * A_1a_2a_2A_2A_2 \ldots a_ia_iA_iA_i \ldots a_ma_mA_mA_m * *A_{m+1},$$

$$S_2 = A_1 * *A_2A_2a_1a_1 \ldots A_iA_ia_{i-1}a_{i-1} \ldots A_mA_ma_{m-1}a_{m-1}A_{m+1} * a_ma_m,$$

such that S_1 and $\overline{S_2}$ are substrings of S with S_1 preceding $\overline{S_2}$, where $\overline{S}$ denotes the reversal of S and the occurrences of the first symbols of S_1 and S_2 have the same parity.

Lemma 3. *The maximum score of a shift-line of order m is $2m$. Furthermore the maximum score is achieved only by a fold whose bonds are formed by the pairs of upper case letters or by a fold using the pairs of lower case letters.*

Proof. It is easy to confirm that the score $2m$ is obtained by a fold that has one of the two types of bonds specified in the statement of the lemma (see Figure 2(a) and (b)).

To show that $2m$ is the maximum possible score and that it is attained by no other set of bonds, we use the conflict graph argument. Denote by s^i the ith occurrence of a symbol s in the string S_1 or in the string S_2, and construct a conflict graph G_m for $\{S_1, S_2\}$ as follows (see Figure 2(c)). G_m has $4m$ vertices $\{A_1, A_2^1, \ldots, A_m^1, A_m^2, A_{m+1}, a_1^1, a_1^2, \ldots a_m^1, a_m^2\}$ where vertex A_i^t (resp. a_i^t), $t = 1, 2$, corresponds to the bond $\{A_i^t, A_i^t\}$ (resp. $\{a_i^t, a_i^t\}$). For any i, $1 < i \leq m$, the graph induced by $A_i^1, A_i^2, a_{i-1}^1, a_{i-1}^2, a_i^1, a_i^2$ is a complete bipartite graph with the

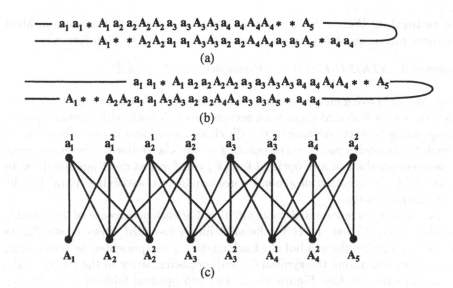

Fig. 2. (a),(b) Two optimal folds of a shift-line, and (c) the conflict graph (for $m = 4$)

bipartition corresponding to upper case and lower case letters. Furthermore A_1 is adjacent to a_1^1 and a_1^2 and A_{m+1} to a_m^1 and a_m^2. By Lemma 2, the resulting graph G_m is a conflict graph.

We claim that G_m has precisely two maximum size independent sets: $I_1 = \{a_1^1, a_1^2, \ldots a_m^1, a_m^2\}$ and $I_2 = \{A_1, A_2^1, A_2^2, \ldots A_m^1, A_m^2, A_{m+1}\}$. To show this, let I be an arbitrary independent set. Regarding Figure 2(c) as a $2 \times 2m$ array, no two elements of I can be in the same column, and so $|I| \le 2m$. Furthermore, if $|I| = 2m$ then every column of the array contains exactly one element of I. If I has some element from each row of the array, then it must have a pair of such elements lying in adjacent columns. However this is impossible since any such pair is adjacent in G_m. This contradiction concludes the proof. □

The fact that there are only two possible sets of bonds for an optimal fold of a shift-line does not imply that there are only two embeddings of a shift-line. Each optimal embedding of a shift-line can be visualised as a double chain of matched intervals of length two interleaved with double flexible strings of length two. This flexibility would allow a shift-line which starts horizontally to move up or down in a series of steps, while maintaining its horizontal orientation. We also observe that it was not essential for the shift-line to be constructed from two continuous substrings. Between any pair of adjacent matching symbols in a shift-line we can insert a short subsequence, built over a set of symbols disjoint from the symbols in the shift-line. Provided the inequalities for Lemma 2 are satisfied, Lemma 2 still holds. Similarly, a longer subsequence could be inserted provided that there are several bonds due to be formed between matching pairs of symbols in the initial and final parts of the subsequence. Then, if the ends of the subsequence are not embedded closely enough together to apply Lemma 2, enough of these bonds would be lost to negate any possible advantage gained by a non-standard embedding of the shift-line.

The reduction. We are ready to prove the main theorem of this section. Most structures will be presented by example and picture rather than formally.

Theorem 4. *STRING-FOLD is NP-complete for the grid* $\mathbb{Z}^2$.

Proof. We will have a unique symbol C_i corresponding to each clause C_i, and we design the string S so that there is an occurrence of C_i, each with the same parity, corresponding to each variable in the ith clause, and there is one occurrence of C_i with the opposite parity corresponding to the clause itself. The occurrences are designed so that, in any optimal fold, C_i can form at most one bond. In an optimal fold, the creation of a bond corresponds to a satisfying literal for the corresponding clause.

Each variable is represented by a shift-line. For each occurrence of the variable in a clause, say C_i, we insert in the shift-line a so-called *exposer* (see Figure 3(a),(b)) containing the symbol C_i. Each clause C_i is represented by a substring called a *trap* containing the symbol C_i with opposite parity to the parity of the C_i's in the exposers (see Figure 3(c)). The two optimal folds of the shift-line allow or prevent an extra bond between the C_i from the exposer and the C_i from the trap.

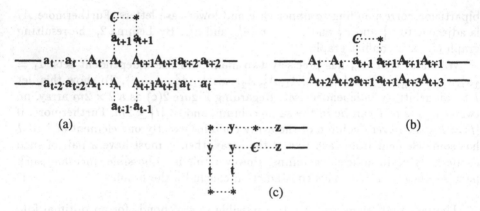

Fig. 3. (a),(b) Two optimal folds of a shift-line containing an exposer that correspond to assignments to the variable which satisfy and do not satisfy clause C_i respectively. For a *negated* variable the pair $C_i *$ would be inserted between symbols denoted by *upper case* letters. (c) Representation of a variable trap for clause C_i.

It remains to show how the substrings corresponding to variables and clauses are composed to form the string S. First we construct the string $T = [S_1^i]_{i=1}^n *$ $[\overline{S_2}^{n+1-i}]_{i=1}^n$ where S_1^i, S_2^i is a shift-line representing x_i. We think of $[S_1^i]_{i=1}^n$ as the *top side* of T and of $[\overline{S_2}^{n+1-i}]_{i=1}^n$ as the *bottom side* of T. The shift-lines are constructed using disjoint sets of symbols.

Next, we add exposers and traps to T. Consider a fixed planar embedding of the graph $G(B)$. See Figure 4. The traps and exposers that correspond to the

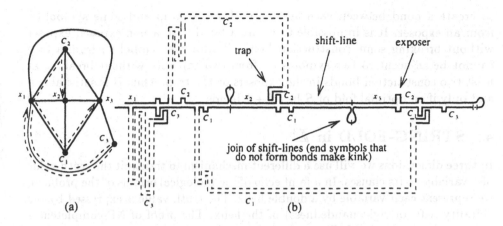

Fig. 4. (a) The graph $G(B)$ for formula $B = C_1 \wedge C_2 \wedge C_3$, where $C_1 = x_1 \vee \neg x_2 \vee \neg x_3$, $C_2 = x_1 \vee x_2 \vee \neg x_3$ and $C_3 = \neg x_1 \vee x_3$, and (b) its representation on $\mathbf{Z}^2$. The embedding of flexible strings realizing the connection between the base of a trap and the exposer corresponds to non-intersecting paths in $G(B)$ denoted in diagram (a) with dashed lines.

clauses in the interior of the cycle $(x_1, \ldots, x_n)$ in the embedding of $G(B)$ will be added to one side (say the top side of T), and traps and exposers corresponding to the clauses embedded in the exterior of this cycle on the opposite side. The order of placing the exposers on each side of the shift-line for x_i is defined by the cyclic ordering of edges adjacent to x_i. On the top side of the shift-line the exposers follow the cyclic order of the variable–clause edges about x_i inside the variable cycle, from the edge $(x_i, x_{(i-1) \bmod n})$ to $(x_i, x_{(i+1) \bmod n})$, and similarly for the bottom side and the edges outside the variable cycle.

A clause C_i is represented by a trap containing symbol C_i. The trap is attached using flexible strings to any of the shift-lines corresponding to a variable that occurs in C_i. We call the place of attachment of a trap the *base* of the trap. The trap is attached to the corresponding shift-line next to the exposer with the symbol C_i (i.e., the substring separating the base of the trap and the exposer cannot contain other traps or exposers — it can contain only shift-line symbols).

We assume that all shift-lines will be embedded horizontally along the first coordinate axis. The lengths of the flexible substrings from bases to traps are sufficient for each trap to reach any exposer corresponding to occurrences of variables in its clause, independently of other traps reaching their exposers.

By the planarity of $G(B)$, there are no topological obstructions to such simultaneous connection between pairs of trap-bases and corresponding exposers (see Figure 4(a)). However we have to take into account that the flexible strings use some area and have fixed length. This presents no significant problem and details are omitted from this abstract.

Let f be the number of bonds in an optimal fold of the string S' obtained from S by replacing all symbols C_i $(1 \le i \le k)$ with $*$. The bonds of S' are called *construction bonds*. A fold of S can have $f + k$ bonds only if it is possible to create a bond between each symbol C_i from its trap and some symbol C_i from an exposer. It is impossible to create a bond with a non-exposed exposer without breaking some construction bond. Similarly, a symbol C_i from a trap cannot be adjacent to two symbols C_i from two exposers without breaking at least two construction bonds in the exposers or the trap. Thus B is satisfiable if and only if an optimal fold of S has $f + k$ bonds. □

4 STRING-FOLD in $\mathbb{Z}^3$

In three dimensions we will use a different mechanism to transmit the truth value of a variable to its clauses. In a faint echo of the biological origins of the problem, we represent each variable by a double helix. The truth value is expressed by the chirality (left- or right-handedness) of the helix. The proof of NP-completeness uses a reduction from 3SAT.

The intended layout of the string S is as a doubled string, following the overall shape of a "comb", in which each "tooth" consists of a helix corresponding to a variable. The teeth are attached to the spine of the comb in such a way that, although they are constrained to lie parallel to each other in a regular planar array, there is an independent choice of chirality for each tooth. There are docking points on each tooth corresponding to each clause in which a literal of that variable occurs. Depending on the chirality of the embedding chosen for that tooth, a docking site is either exposed on the top or bottom surface of the comb and available for docking, or hidden in the crevice formed with an adjacent tooth.

For each clause, there is, attached to the comb at any suitable place, a long flexible loop at the end of which is a "ligand" corresponding to that clause. The loop is long enough for the ligand to dock with any one exposed docking site corresponding to that clause on a tooth. The target score is such that it can be attained if and only if there is a choice of chirality for each tooth (i.e., truth value for each variable) such that at least one docking site for each clause is exposed (i.e., each clause is satisfied by at least one literal). An impression of the overall structure is shown in Figure 5.

We proceed to describe the components in more detail. A string $S = S_1 \overline{S_2}$ is called a (rooted) helix of order m if $S_1 = (s_0^1, s_1^1, s_2^1, \ldots s_{6m+2}^1) = (1, [4i, 4i - 1, *, 4i+1, 4i+2, 4i+1]_{i=1}^m, *, 4m+3)$, and $S_2 = (s_0^2, s_1^2, \ldots s_{6m+1}^2) = (2, 3, [4i, 4i+1, 4i, 4i+3, 4i+2, *]_{i=1}^m)$. (See the bold lines in Figure 6(a).) Note that the parity of s_0^1 is the same as the parity of s_0^2.

Lemma 5. *The maximum score of an embedding of a helix of order m is $6m$. Furthermore, given a fixed embedding of the first and last edges of the helix such that (s_0^1, s_1^1) and (s_0^2, s_1^2) lie parallel along opposite edges of a unit cube, there are exactly two folds that achieve the maximum score. These consist of a right-handed and a left-handed double helix.*

Proof. The score of an embedding of S is maximized if all pairs of identical symbols of different parity are adjacent. Such an embedding is possible (see Figure 6(b)) and we will argue that there are only two different optimal embeddings.

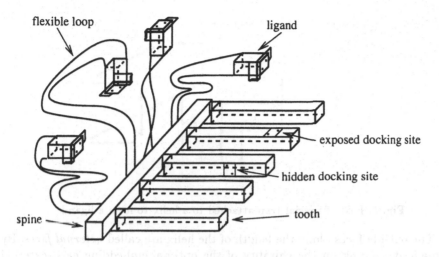

Fig. 5. Overall construction

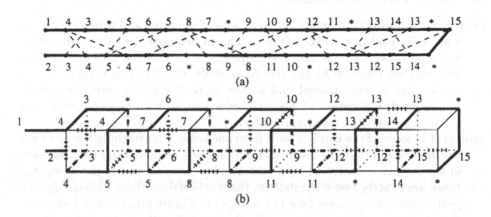

(a)

(b)

Fig. 6. The double helix of order $m = 3$

The second optimal embedding is similar to the one presented in Figure 6(b) but the two strings twist around each other in the opposite sense.

Let the *bond graph* $B(S) = (V, E)$ be the graph with set of vertices V equal to the set of elements of S, and such that $(x, y) \in E$ if and only if x, y are either two consecutive elements of S, or x and y are equal symbols with different parity, which implies that one is from S_1 and the other from S_2. (See Figure 6(a).) Then any optimal fold of S corresponds to a grid embedding of $B(S)$ such that each edge is embedded on an edge of the grid. In the full paper we complete the proof of Lemma 5 with an inductive argument. □

Now we are ready to describe the details of the overall construction. All elements of the "comb" are built up using helices. We think of each helix as a sequence of adjacent cubes. The face defined by the last two elements of S_1 and the last element of S_2 (in Figure 6(b) the rightmost face) is called the *extremal*

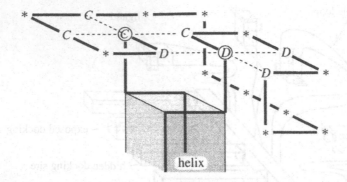

Fig. 7. Fold of special trap attached to a helix to form a ligand

face. The outside faces along the length of the helix are called *external faces.* By Lemma 5, depending on the chirality of the optimal embedding each external face is in one of two pairwise perpendicular positions. Using this observation, the "comb" is designed as follows:

teeth: Each variable x corresponds to a tooth of the comb and is represented by a helix. For each occurrence of the variable in a clause C_i, a pair of "diagonal" $*$ symbols (i.e., two $*$'s that are endpoints of an external face diagonal) in the helix are replaced by the pair of symbols (C_i, D_i). This is done in such a way that in any optimal fold all the pairs of symbols corresponding to non-negated occurrences of x are embedded on external faces perpendicular to those for negated occurrences of x.

spine: The spine of the comb is built from one helix. For attaching teeth, we use the observation that any optimal embedding of a helix contains external faces of the form $A = (a, *, b, \hat{*})$ such that $a, *, b$ are consecutive in the string and there are exactly two a symbols in the whole string, these having opposite parity (for example, see face $(11, *, 13, *)$ in Figure 6(b)). Since faces with the above properties occur periodically on a helix, the teeth can be placed regularly along the base parallel to one another. Let the new tooth be formed from the string $\hat{1}T\hat{2}$, a variable helix with initial and final symbols $\hat{1}$ and $\hat{2}$. We replace the substring $a * b$ by $a\,\hat{2}\,\hat{1}\,T\,\hat{2}\,a\,b$. The new bond (a, a) forces the new symbol a to be embedded in the same position as the $*$ symbol replaced. To make sure that the tooth is not loose, we replace the $\hat{*}$ in A by $\hat{1}$.

ligand: Referring to Figure 6(b), we see that the extremal face has a diagonal pair of $*$ symbols. We let each ligand corresponding to a clause $C = C_i$ be a helix at least as long as the distance between the teeth, to which is attached the special trap illustrated in Figure 7. (We use (C, D) for (C_i, D_i).) The subsequence $*(4m + 3)*$ which occurs on the extremal face of the helix is replaced by $\tau = C\,D*C*C********D*D*C\,D$. The sequence τ has four occurrences of C and four of D and, as Figure 7 shows, τ can be embedded so that all three potential (C, C) bonds and all three potential (D, D) bonds are simultaneously achieved. The effect of such an embedding is to leave the circled occurrences of C and D with only one free edge in the grid with which to make a further bond with an exposed symbol on a tooth. The target score will be set so as to require four (C, C) bonds and four (D, D) bonds. The ligand is attached to the comb using flexible strings long enough to reach any exposed pair (C, D). In this way a ligand can dock without a penalty with any (but with at most one) such pair exposed by a variable tooth.

This completes the description of the structure used to establish our second main result.

Theorem 6. *STRING-FOLD is NP-complete for the grid* Z^3.

Proof. One aspect of the proof is much simpler than in the corresponding theorem for Z^2. The target score which is set requires the simultaneous formation of the *maximum possible* number of bonds for every alphabet symbol. We have shown that for most parts of the structure this requirement imposes an embedding which is unique up to mirror symmetry.

Without loss of generality we can take one fixed embedding for the spine of the comb. Each tooth is attached to fixed points on the spine by a pair of edges, which are the first and last edges in these helices. By Lemma 5 there are just two embeddings of each tooth relative to the spine. The choice between these two embeddings determines which pair of opposite long faces of the tooth are exposed in the plane of the comb and which are hidden in the gaps between successive teeth.

Each ligand has several distinct embeddings, but the essential active part of each is the diagonal pair (C, D), circled in Figure 7, which is constrained to occur at the extremal face of the helix of the ligand. This helix is designed to be too long for the active pair to be able to reach any docking site which is on a hidden surface of a tooth (see Figure 5). The flexible loops which attach the ligands to the spine of the comb offer no obstruction to achieving any docking.

The target score is reached if and only if every ligand achieves its bonds, and this is possible if and only if there is an orientation of each tooth so that every ligand has a corresponding exposed docking site with which to dock. This last condition is equivalent to there being a choice of truth value for each variable in the instance of 3SAT such that each clause is satisfied by at least one of its literals. □

5 Open Problems and Conclusion

To obtain our results, we needed to allow an alphabet of unbounded size. The principal open problem that remains is to resolve the complexity of STRING-FOLD in Z^2 and Z^3 for the "hydrophobic-hydrophilic" model considered by Hart and Istrail [4]. This corresponds to a binary alphabet in which only one symbol forms bonds. An intermediate problem, which still seems challenging, is to extend our NP-hardness results to some fixed finite alphabet.

The grids Z^2 and Z^3 that we have used are bipartite, and parity arguments were helpful in maintaining control over the possible embeddings. This feature is not in keeping with the biological motivation and more realistic models. It would be a significant advance to extend our results to triangular and tetrahedral grids, which do not have the convenience of bipartiteness.

We expect our results to be of interest more to computer scientists than biologists since our model is very restricted and omits so many of the important characteristics of the protein-folding problem. The grid we impose does not capture the subtlety of molecular geometry: the model of bonds is much too simple.

It is not clear whether the biological motifs (e.g., the docking of ligands and the double helix) arose naturally in our solution or suggested themselves subconsciously because of the biological background to the problem. To prove our results we have needed to construct several two- or three-dimensional structures by suitably "programming" in a one-dimensional string. This aspect of our work demonstrates, we believe for the first time, the design of sequences with precisely controlled optimal folded forms of some complexity. We hope that our examples and open questions will stimulate others to tackle string-folding problems in a more biologically realistic model.

Acknowledgements

We are grateful to Aviezri Fraenkel for introducing this type of problem to us, and to William Hart and Sorin Istrail for sending us an early copy of their paper.

This research was supported in part by the EU under contracts No. 7141 (project ALCOM II) and No. 20244 (project ALCOM-IT). Most of this work was done while the second author was visiting the University of Warwick.

References

1. S.A. Cook. The complexity of theorem-proving procedures. *Proc. 3rd ACM Symp. on Theory of Comp.* (1971), 151–158.
2. A.S. Fraenkel. Complexity of protein folding. *Bull. Math. Biology* 55, (1993), 1199–1210.
3. M. Garey and D. Johnson. Computers and Intractability. (W.H. Freeman and Co. 1979).
4. W.E. Hart and S. Istrail. Fast protein folding in the hydrophobic-hydrophilic model within three-eighths of optimal. *Proc. 27th ACM Symp. on Theory of Comp.* (1995), 157–168.
5. D. Lichtenstein. Planar formulae and their uses. *SIAM J. Computing* 11, (1982), 329–343.
6. J.T. Ngo and J. Marks. Computational complexity of a problem in molecular structure prediction. *Protein Engineering* 5, (1992), 313–321.
7. R. Unger and J. Moult. Finding the lowest free energy conformation of a protein is an NP-hard problem: proof and implications. *Bull. Math. Biology* 55, (1993), 1183–1198.

A Polynomial-Time Algorithm for Near-Perfect Phylogeny

David Fernández-Baca[1]*, Jens Lagergren[1]**

[1] Department of Computer Science, Iowa State University, Ames, IA 50011.
E-mail: fernande@cs.iastate.edu
[2] Department of Numerical Analysis and Computing Science, Royal Institute of
Technology, S-100 44 Stockholm, Sweden. Email: jensl@nada.kth.se

Abstract. We define a parameterized version of the Steiner tree problem in phylogeny where the parameter measures the amount by which a phylogeny differs from "perfection." This problem is shown to be solvable in polynomial time for any fixed value of the parameter.

1 Introduction

A fundamental problem in biology is that of inferring the evolutionary history of a set of species, each of which is specified by the set of *traits* or *characters* that it exhibits [10, 11]. In mathematical terms, the problem can be expressed as follows. Let C be a set of *characters*, and for every $c \in C$, let A_c be the set of allowable *states* for character c. Let $m = |C|$ and $r_c = |A_c|$. A *species* s is an element of $A_1 \times \cdots \times A_m$; $s(c)$ is referred to as the *state of character c for s*. A *phylogeny* for a set of n distinct species S is tree T with the following properties:

(C1) $S \subseteq V(T) \subseteq A_1 \times \cdots \times A_m$,
(C2) Every leaf in T is in S.

The *length* of an edge (u, v) in a phylogeny T is the Hamming distance between u and v; i.e., the number of character states in which u and v differ. Let Length(T) be the sum of the lengths of the edges of T. The *Steiner tree problem in phylogeny* (STP) is to find a phylogeny T of minimum length for a given set of species S.

STP and many of its variants are known to be NP-hard. However, it is not hard to obtain an approximation algorithm for STP that has relative error at most 2; also, algorithms with better approximation ratios are known, see [4]. STP is related to another well-known problem. Let us say that a phylogeny is *perfect* if, in addition to (C1) and (C2), it satisfies:

---
* Supported in part by the National Science Foundation under grants CCR-9211262 and CCR-9520946.
** Supported by the grants from NFR and TFR.

(C3) For every $c \in C$ and every $j \in \mathcal{A}_c$, the set U of all $\boldsymbol{u} \in V(T)$ such that $\boldsymbol{u}(c) = j$ induce a subtree of T.

The *perfect phylogeny problem* is to determine if a given set of species S admits a perfect phylogeny. It is not hard to verify that condition (C3) implies that the length of a perfect phylogeny for S is exactly $\sum_{c \in C} (r_c - 1)$. One can in fact, obtain the following result [2] (see also Gusfield [12]).

Theorem 1 *Let T^* be a phylogeny for S. Then $Length(T^*) \geq \sum_{c \in C} (r_c - 1)$ and T^* is a perfect phylogeny if and only if $Length(T^*) = \sum_{c \in C} (r_c - 1)$.*

The perfect phylogeny problem was shown to be NP-complete by Bodlaender et al. [6] and, independently, by Steel [19]. This has motivated the study of the fixed-parameter versions of the problem where either m or r is fixed. Both versions have been shown to be polynomially solvable; the first by McMorris, Warnow, and Wimer [17]; the second by Agarwala and Fernández-Baca [2] — the time bound of the algorithm for the latter has since been improved by Kannan and Warnow [16].

If a set of species admits a perfect phylogeny, the underlying set of characters C is said to be *compatible*; for this reason, the perfect phylogeny problem is often called the *character compatibility problem*. It has been observed that most sets of characters are incompatible, and thus a natural problem is to find a maximum-cardinality subset of C that is compatible. This problem is, unfortunately, equivalent to CLIQUE, and hence not only NP-hard, but also extremely hard to solve approximately, see [3].

The difference between m and the maximum-cardinality compatible subset of C is one measure of the "degree of compatibility" of a set of species; here we study a new measure of incompatibility that we believe is equally natural. Theorem 1 states that the length of a perfect phylogeny (assuming one exists) gives a lower bound on the length of any phylogeny for S [2]. Motivated by this observation, let us define the *penalty* of a phylogeny T, denoted Penalty(T), as $Length(T) - \sum_{c \in C} (r_c - 1)$. Obviously, STP can be rephrased as the problem of finding a phylogeny T such that Penalty(T) is minimum. We will be interested in the fixed-parameter version of the problem, namely, given a set of species S and an integer q, does S have a phylogeny with penalty at most q? We will show that for each fixed q and r, the resulting "near-perfect" phylogeny problem can be solved in polynomial time. The running time of our algorithm is a polynomial whose degree depends on the parameters, making the algorithm practical only for small values of the parameters. On the other hand, the flexibility of allowing one or more characters to violate condition (C3) by some fixed amount may significantly extend the range of applicability of character-based methods.

Before we continue, we should point out that the use of the word "species" in this paper should not lead the reader to conclude that phylogenies are only useful in biology. Indeed, the concept of an evolutionary tree is relevant to the study of the development of natural languages. Notable progress has recently been achieved in this area through techniques originally developed for the perfect phylogeny problem [20].

2   Definitions

To simplify our terminology, we will assume that no state appears on two characters. While this assumption is violated by data given by, say, DNA sequences, it can easily be enforced by replacing each state σ on every character c by the state (σ, c).

Let T denote a phylogeny for the set of species S. Without loss of generality, we shall only consider trees where every two adjacent nodes are distinct. Clearly, a tree that does not satisfy this condition can easily be transformed into another one that does and which has the same penalty. Observe that the contribution of a state to Penalty(T) equals the number of connected components associated with that state minus one. (The connected components associated with a state σ, on a character c, are the connected components of the subtree of T induced by $\{s | s(c) = \sigma\}$.) Any state that is associated with more than one connected component in T will be called a *penalty state*; a character that has penalty states will be called a *penalty character*. Observe that if Penalty(T) $= q$, then there can be at most q penalty states and, hence, at most q penalty characters. The number of ways in which penalty states can be distributed among the various characters is at most

$$\sum_{i=0}^{q} \binom{m}{i} \leq qm^q.$$

This number is polynomial for each fixed q.

A *character partition* with respect to a character c is a partition (S_1, S_2) of S such that no species in S_1 shares a state of c with any species of S_2; the subsets S_1 and S_2 shall be referred to as *character subfamilies*. A *penalty character partition* is a partition (S_1, S_2) of S such that no species in S_1 shares a state of *any* penalty character with any species of S_2; the subsets S_1 and S_2 shall be referred to as *penalty character subfamilies*.

The removal of any edge e from T defines an *edge partition* of S (with respect to T); this is a partition (S_1, S_2), where S_1 and S_2 are the subsets of S contained in each of the connected components of $T - e$. The subsets S_1 and S_2 shall be referred to as *edge subfamilies*. An edge of T will be said to be *good* if its edge partition is also a character partition; otherwise, the edge is *bad*. It will turn out that the only edge partitions that will be of interest to us are those where T is a minimum-penalty tree.

Definition 1 Suppose T is a phylogeny for S and let p be some node in T. We shall say that the state of p on a non-penalty character c is *forced* if p lies on the path between nodes a and b in S such that $a(c) = b(c)$.

Note that if the state of a character of a node is unforced, several assignments may be possible. We shall call an assignment of states to the penalty characters a *penalty state assignment*.

Definition 2 Two vectors a and b are said to be *compatible* on a set of characters A if and only if $a(c) = b(c)$ for every $c \in A$.

3 Phylogenies and subphylogenies

A *subphylogeny* for a subfamily $P \subseteq S$ is a phylogeny for P containing a special node x, called a *connection*; the latter will be viewed as the root of the subphylogeny. Suppose we know which characters are penalty characters and which are not. The state of x on a non-penalty character c is forced if a state on c is shared between P and $S \setminus P$; in this case $x(c)$ must equal that shared state. If T is a subphylogeny, then T_v denotes the maximum subtree of T containing only the descendants of v (v included).

Our algorithm enumerates all possible choices for penalty characters; for each such choice it enumerates all character subfamilies and tries to identify which of them are edge subfamilies with respect to some optimal tree T^* having the given choice of penalty characters. The latter seems hard to do without knowing T^*. The following lemma implies that we can dismiss some edge subfamilies whose subphylogenies have excessive penalty.

Lemma 1. *A set of species S has a phylogeny of penalty at most q if and only if every subset of S has a subphylogeny of penalty at most q.*

Knowing that some of the states of the connection on non-penalty characters are uniquely determined by P and S is certainly not enough: we still have to compute the states on the penalty characters. Fortunately, we can afford to try all possible penalty assignments. The algorithm tries to construct a minimum-penalty subphylogeny of penalty at most q for each character subfamily. It does so by attempting to solve the problem for each penalty state assignment for the connection; if this is impossible for all guesses, then the character subfamily is dismissed. Since our algorithm finds edge subfamilies by relying on character subfamilies, bad edges pose an obvious problem, because, by definition, their edge subfamilies cannot be found in this way. At first glance it may seem that we cannot afford to have *any* bad edges; we will show that it is possible to handle a bounded number of such edges. Fortunately, as the following lemma indicates, we can also bound the number of bad edges.

Lemma 2. *The number of bad edges in a phylogeny T such that $\text{Penalty}(T) \leq q$ is at most qr.*

Proof. Assume that C_p is the set of penalty characters. Let l_c be the number of edges (u, v) such that $u(c) \neq v(c)$. Then we have that $\sum_{c \in C_p} l_c$ is an upper bound on the number of bad edges and also that $\sum_{c \in C_p} l_c - (r_c - 1) \leq q$. It follows that the number of bad edges is bounded by $q + \sum_{c \in C_p} (r_c - 1) \leq qr$.

4 Bad trees

In what follows, we shall assume that only the first q' characters are penalty characters. Observe that $q' \leq q$; thus, for each fixed q, the number of choices for penalty characters is polynomially-bounded. Our algorithm processes pairs

$(\mathcal{P}, \rho)$, where $\mathcal{P}$ is a character subfamily and ρ is a penalty state assignment, in order of nondecreasing cardinality of the character subfamily. For each such pair, it tries to construct a minimum-penalty subphylogeny for $\mathcal{P}$, where the penalty states of the connection are given by ρ.

Let $\mathcal{P}$ be a character partition which has a minimum-penalty subphylogeny T of penalty at most q. The maximal subtree T' of T that contains the connection of T and only bad edges shall be called a *bad tree* for $\mathcal{P}$. From now on T' shall denote the bad tree. Note that Lemma 2 implies that the number of edges in any bad tree is bounded for any fixed q and r. Note also that, because of the maximality of T', any edge of $T \setminus T'$ which is adjacent to an edge of T' is good.

We will now state some basic properties about the states of the nodes of a bad tree.

Lemma 3. *All nodes in a bad tree have the same states on non-penalty characters.*

Proof. It is, clearly, enough to show that any pair of adjacent nodes in a bad tree have the same states on non-penalty characters. Assume that u and v are two adjacent nodes in the bad tree that differ on the non-penalty character c. Then the edge partition with respect to (u, v) is a character partition with respect to c; i.e., (u, v) is good. But then, by definition, (u, v) cannot be in a bad tree, a contradiction.

Since no two species in a minimum-penalty phylogenetic tree can be identical, we have the following corollary.

Corollary 2 *Every two nodes in a bad tree have different penalty state assignments.*

For each $v \in V(T')$, let $H_T(v)$ denote the set of species that hang from v in T; i.e., $H_T(v)$ consists of the species of $\mathcal{P}$ that belong to the maximal subtree U of T such that $V(U \cap T') = \{v\}$. The idea behind our algorithm is to generate candidate bad trees and use them as "skeletons" from which to "hang" subphylogenies for edge subfamilies $\mathcal{P}' \subseteq \mathcal{P}$ so as to obtain a minimum-penalty subphylogeny for $\mathcal{P}$. As pointed out earlier, since T' is maximal with respect to containment of bad edges, each such $\mathcal{P}'$ is also a character subfamily, which means that it has already been processed by the algorithm. The difficulty is that v can have arbitrarily high degree in T. This implies that even determine $H_T(v)$ is non-trivial, since it cannot be done in polynomial time by simply trying all combinations of character subfamilies. We note that determining $H_T(v)$ is not sufficient: our algorithm will have a second phase where it will try to partition $H_T(v)$ into edge subfamilies.

For each node v, we will determine $H_T(v)$ by enumerating a polynomially-bounded (for fixed q and r) amount of information. The latter will consist of two items for each node v in T': (i) a penalty character subfamily $\tilde{H}(v)$, which will serve as an initial approximation to $H_T(v)$, and (ii) a set $C(v)$ of of character subfamilies. The set $C(v)$ will serve as "correction term" for $\tilde{H}(v)$: in addition to the species that are in $\tilde{H}(v)$, we will add those in the subfamilies contained in $C(v)$; similarly, we will delete the species contained in any subfamily belonging to some $C(x)$, where $x \neq v$. The sets of species that must be added to and deleted from $\tilde{H}(v)$ are,

$$A(v) = \bigcup \{ Q \in C(v) \}$$

$$D(v) = \bigcup \{ Q \mid Q \in C(x), x \in V(T') \setminus \{v\} \}$$

The resulting set is:

$$H(v) = \left(\tilde{H}(v) \cup A(v) \right) \setminus D(v) \qquad (1)$$

To state the next lemma, we need some terminology. A *labeled bad tree* is a triple $(T', \tilde{H}, C)$, where T' is a bad tree $\tilde{H}$ maps each node of T' to a penalty character subfamily, and C maps each node of T to a set of character subfamilies.

Lemma 4. *Let $\mathcal{P}$ a character subfamily with a minimum-penalty subphylogeny T of penalty at most q Then, there exists a labeled bad tree $(T', \tilde{H}, C)$ such that T' is the bad tree of T, each set $C(v)$ contains at most q character subfamilies, and, for every v in T', $H_T(v) = H(v)$.*

Proof. Choose as T' the maximal subtree of T that contains the connection and only bad edges. We will describe how to choose $\tilde{H}$ and C so that $H_T(v) = H(v)$. We first explain how to pick $\tilde{H}$. For each node v of T', let $P(v)$ be the set of penalty state assignments to species in $H_T(v)$. To every penalty state assignment ρ, we shall associate a value $\nu(\rho) \in T'$, chosen as follows. If there is some node v of T' that has ρ as its penalty state assignment, then $\nu(\rho) = v$ (by Corollary 2, node v is unique). Otherwise, if $\rho \in P(v)$ for some node v, let $\nu(\rho)$ be an arbitrary, but fixed, v such that $\rho \in P(v)$. If neither of the above cases applies, we make $\nu(\rho) = \emptyset$. Now, for each node v, let $\tilde{H}(v)$ be the set of all species $u \in \mathcal{P}$ whose penalty state assignment ρ satisfies $\nu(\rho) = v$; clearly, $\tilde{H}(v)$ must be a penalty character subfamily.

We shall now prove that, for each v, there is a set $C(v)$ of at most q character subfamilies such that:

$$\bigcup_{\mathcal{P}' \in C(v)} \mathcal{P}' \subseteq H_T(v) \qquad (2)$$

$$H_T(v) \setminus \tilde{H}(v) \subseteq \bigcup_{\mathcal{P}' \in C(v)} \mathcal{P}' \qquad (3)$$

Since (2) and (3) imply that $H_T(v) = H(v)$, the theorem follows.

Recall that T' is a maximal subtree of T that contains the connection and only bad edges. Thus, for each node v of T' and each $u \in H_T(v)$ there is a character subfamily $\mathcal{P}'$ such that $\mathcal{P}' \subseteq H_T(v)$ and $u \in \mathcal{P}'$. Hence, there is a minimum-size set of character subfamilies $C(v)$ that satisfies (1) and (2). We have yet to bound the size of $C(v)$. For each $\mathcal{P}' \in C(v)$, there must exist a node $v' \neq v$ such that we can choose $u \in \mathcal{P}' \cap (H_T(v) \setminus \tilde{H}(v))$ and $u' \in H_T(v')$, where

u and u' have the same penalty state assignment. Since $u \notin \tilde{H}(v)$ the penalty state assignments of u and v differ on some character, say, c. On the other hand, since u and u' have exactly the same penalty state assignment and $u' \in H_T(v')$, each $\mathcal{P}' \in C(v)$ contributes to the total penalty. We conclude that $|C(v)| \leq q$.

Remark: It is possible to prove a stronger variant of this lemma stating that $\cup \{C(v) : v \in V(T')\}$ has size at most q.

The following three lemmata are useful for the second phase of the algorithm, where we try to partition $H(v)$ into edge subfamilies. The reader may wish to defer reading them until after examining how the previous results are used in the first phase of the algorithm.

If $u \in V(T \setminus T')$ is a child of $v \in V(T')$, then we call the subset of $\mathcal{S}$ contained in T_u an *edge subfamily at v in T*. Note that, since T' is a maximal bad subtree of T, every edge subfamily at v in T is a character subfamily, as well.

Lemma 5. *Let v be a node of T'. If M is a union of edge subfamilies at v in T such that $N = H_T(v) \setminus M$ is not a character subfamily, then the states of the non-penalty characters of v are forced. Moreover, given H_T and M these states can be computed in time $O(m|S|)$.*

Proof. Let c be a non-penalty character. Suppose N is not a character subfamily but $v(c)$ is unforced. Thus, no state on c is shared between N and $\mathcal{S} \setminus N$. This, however, implies that $(N, \mathcal{S} \setminus N)$ is a character partition with respect c, and, hence, N is a character subfamily, a contradiction. We omit the description of the algorithm.

For any species v, let $\sim_v$ be the relation on $\mathcal{S}$ defined by: $s \sim_v s'$ iff $s(c) = s'(c)$ and $s(c) \neq v(c)$ for some character c. Let S_1 and S_2 be edge subfamilies at v. We say that S_1 and S_2 are *state intersecting* if there exist $s_1 \in S_1$ and $s_2 \in S_2$ such that $s_1 \sim_v s_2$.

Lemma 6. *There is a set I of at most $2q$ character subfamilies such that for any node v of T' if $\mathcal{P}_1$ and $\mathcal{P}_2$ are state intersecting edge subfamilies at v in T' then $\{\mathcal{P}_1, \mathcal{P}_2\} \subseteq I$.*

Proof. Let I be the set of all $\mathcal{P}_1$ such that, for some node v and some $\mathcal{P}_2$, $\mathcal{P}_1$ and $\mathcal{P}_2$ are state intersecting edge subfamilies at v in T. Let G be the graph with node set I and such that there is an edge between $\mathcal{P}_1$ and $\mathcal{P}_2$ iff for some node v of T, $\mathcal{P}_1$ and $\mathcal{P}_2$ are intersecting subfamilies at v in T. Let μ be the size of a maximum matching in G. One can verify that

$$q \geq \mu + |I| - 2\mu \geq |I|/2.$$

Lemma 7. *Let S_1 be a non state intersecting edge subfamily at v and let the sets $K_1, \ldots, K_l$ be the equivalence classes in S_1 under the transitive closure of $\sim_v$. If p is the penalty of a minimum-penalty phylogeny for $S_1 \cup \{v\}$ and, for each $1 \leq i \leq l$, p_i is the penalty of a minimum-penalty phylogeny for $K_i \cup \{v\}$, then*

$$p = \sum_{i=1}^{l} p_i.$$

Proof. Omitted.

5 The algorithm

Our algorithm considers each pair $(\mathcal{P}, \rho)$ (where $\mathcal{P}$ is a character subfamily and ρ is a penalty state assignment) and attempts to generate a minimum-penalty subphylogeny T, of penalty at most q, for $\mathcal{P}$ whose connection has penalty state assignment ρ. The information about whether such a subphylogeny exists and, if so, its penalty, is stored in a table, which is indexed by the pair $(\mathcal{P}, \rho)$.

The processing of a pair $(\mathcal{P}, \rho)$ consists of enumerating a polynomially-bounded (for fixed q and r) number of candidate subphylogenies and determining which of these has the least penalty. The generation of a candidate has two phases. In the first, we generate a bad tree T', together with a set $H(v)$ for each node v of this tree. In the second, for each node v, we try to partition $H(v)$ into edge subfamilies and then use this information to obtain the penalty of the best subphylogeny associate with the given bad tree, assuming that such a subphylogeny exists.

Candidates for T' and H are produced by following the steps described below.

1. Generate a labeled bad tree T' with at most qr edges whose root has penalty state assignment ρ. Non-penalty characters of the root that are forced by the character partition have their states set accordingly. All other states of nodes in T' will be assigned later. It is well known that there are $O(4^{qr})$ choices for T'.
2. Choose the states of all penalty characters for every node x of T' other than the root. Note that there is a total of $r^{q^2 r}$ possible choices.
3. For each node v of T', generate a penalty character subfamily $\tilde{H}(v)$. Note that there is a total of $O(2^{q^2 r^2})$ possible choices.
4. For each node u of T', generate a set $C(u) = \{\mathcal{P}_1, \ldots, \mathcal{P}_l\}$ of at most q character subfamilies. Note that there is a total of $O(m^{q^2 r} 2^{q^2 r^2})$ possible choices.
5. For each node v of T', construct the set $H(v)$ defined in equation (1).
6. If there exist distinct nodes u and v in T' such that $H(u) \cap H(v) \neq \emptyset$, the labeled tree generated so far is invalid and is dismissed from further consideration.

7. For every two nodes u and v of T' and every non-penalty character c, if there exist nodes $x \in H(u)$ and $y \in H(v)$ such that $x(c) = y(c) = \sigma$, then for all nodes of T' set character c to σ. We do so because, by Lemma 3, all nodes in T' must have the same states on non-penalty characters. If we discover that two pairs of Hs require conflicting state assignments for a given character, the tree is dismissed as invalid.

In the second phase, we determine which trees to connect to T' to obtain a subphylogeny with least penalty. Note that this will not necessarily be the subphylogeny with the least overall penalty, since other candidates for T' and H may yield better subphylogenies; the fact that we enumerate all the relevant candidates implies that we will eventually find the best subphylogeny.

The main task is to find the edge subfamilies that make up $H(v)$, for any v. We will use an approach similar to the one followed in Agarwala and Fernández-Baca's perfect phylogeny algorithm (and subsequently in Kannan and Warnow's). For each node v in the bad tree T', we choose a set $P(v)$ of at most $2q$ character subfamilies. The total number of choices over all nodes is $O(m^{2q^2 r} 2^{2q^2 r^2})$. By Lemma 6, some choice $P(v)$ contains all state intersecting edge subfamilies at v in T'. Let

$$U(v) = \bigcup \{Q \in P(v)\}.$$

Now consider the remaining set $Q(v) = H(v) \setminus U(v)$. By Lemma 5, only the following two situations are possible:

1. $Q(v)$ is a character subfamily. In this case, for every possible penalty state assignment ρ, we look for $(Q(v), \rho)$ in the table described at the beginning of this section. Similarly, for every $P' \in P(v)$ and every possible penalty state assignment ρ we do a table lookup on (P', ρ). Using this information, we compute the penalty of a minimum-penalty phylogeny for $H(v) \cup \{v\}$.

2. For every non-penalty state c, the state of v is forced. In this case, we first compute the equivalence classes $K_1, \ldots, K_l$ in $Q(v)$ under the transitive closure of $\sim_v$. (This can be done in time $O(|H(v)|m)$; just compute the connected components in the graph where the set of vertices is $H(v) \cup \{\sigma : \sigma$ is a state of some species in $H(v)$ but not of $v\}$ and a species is connected to all its states.) For every $1 \leq i \leq l$, we try all possible penalty state assignments ρ, and do a table lookup on (K_i, ρ). Similarly, for every $P' \in P(v)$ and every possible penalty state assignment ρ we do a table lookup on (P', ρ). By Lemma 7, we can use this information to compute the penalty of a minimum-penalty phylogeny for $H(v) \cup \{v\}$.

Remark: Whenever some state of a node in the bad tree is not determined after this computation it is an easy matter to give it one state that does not increase the penalty.

The running time for one pair (P, ρ) is $|P| m^{O(q^2 r)} 2^{O(q^2 r^2)}$. However, by taking advantage of the full strength of Lemma 6 and using the stronger variant of Lemma 4, mentioned in the remark, we can lower the time used to

$|\mathcal{P}|m^{O(qr)}2^{O(q^2r^2)}$. Our algorithm considers all $O(m2^{q^2r^2+r})$ pairs $(\mathcal{P}, \rho)$ and each of the qm^q possible choices of at most q penalty characters. This gives an overall running time of $|\mathcal{S}|m^{O(qr)}2^{O(q^2r^2)}$, for a given set of species $\mathcal{S}$.

6 Conclusions and open questions

We have shown that a relaxed version of the perfect phylogeny problem can be solved in polynomial time. Since one of the limitations of the perfect phylogeny model is its restrictiveness, our result may have some practical use. The practicality is, however, limited by the fact that the running time of our algorithm is bounded by a polynomial whose degree depends on q; thus we can only expect to be able to use the procedure for small values of the parameters. (The dependency on r is not as large an obstacle, since $r = 4$ gives the most interesting case, that of DNA.) It should be noted that the time bound is based on the perhaps overly pessimistic assumption that all bad edges can occur together in a bad tree. One can ask whether this is likely to happen in practice. Also, is there a parameter that is smaller than q in practice in terms of which to express the time bound? One candidate is the maximum size of a bad tree. In our view, the most important open question raised by our algorithm is whether it is possible to make the degree of the polynomial describing the running time independent of the parameters; i.e., whether there is an algorithm with running time $O(f(q)p(|\mathcal{S}|, m))$, where p is a polynomial whose degree does not depend on q. Alternatively, one could try to show that the case were r is fixed and q is the only parameter is hard for $W[1]$.

References

1. R. Agarwala, D. Fernández-Baca, and G. Slutzki. Fast algorithms for inferring evolutionary trees. To appear in *Journal of Computational Biology*, 1995.
2. R. Agarwala and D. Fernández-Baca. A polynomial-time algorithm for the perfect phylogeny problem when the number of character states is fixed. *SIAM J. Computing* **23** (1994), 1216–1224.
3. S. Arora, C. Lund, R. Motwani, M. Sudan, and M. Szegedy. Proof verification and hardness of approximation problems. In *33rd FOCS*, pages 14–23, 1992.
4. P. Berman and V. Ramayer. Improved approximations for the Steiner tree problem. In *3rd SODA*, pages 1–10, 1992.
5. M. Bern and P. Plassman. The Steiner problem with edge lengths 1 and 2. *Information Processing Letters*, 32:171–176, 1989.
6. H. Bodlaender, M. Fellows, and T. Warnow. Two strikes against perfect phylogeny. In *Proceedings of the 19th International Colloquium on Automata, Languages, and Programming*, pp. 273–283, Springer Verlag, Lecture Notes in Computer Science, 1992.
7. W.H.E. Day, D.S. Johnson, and D. Sankoff. The computational complexity of inferring rooted phylogenies by parsimony. *Mathematical Biosciences*, 81:33–42, 1986.
8. A. Dress and M. Steel. Convex tree realizations of partitions. *Appl. Math. Letters*, Vol. 5, No. 3, 3–6, 1992.

9. G. F. Estabrook, C. S. Johnson Jr., and F. R. McMorris. An idealized concept of the true cladistic character. *Mathematical Biosciences*, 23, 263–272, 1975.
10. G. F. Estabrook. Cladistic methodology: A discussion of the theoretical basis for the induction of evolutionary history. *Annual Review of Ecology and Systematics*, Vol. 3, 427–456, 1972.
11. W. M. Fitch. Aspects of Molecular Evolution. *Annual Reviews of Genetics*, Vol. 7, 343–380, 1973.
12. D. Gusfield. The Steiner tree problem in phylogeny. *Unpublished manuscript.*
13. D. Gusfield. Efficient algorithms for inferring evolutionary trees. *Networks*, Vol. 21, 19–28, 1991.
14. S. Kannan and T. Warnow. Triangulating three-colored graphs. *SIAM J. on Discrete Mathematics*, Vol. 5, 249–258, 1992.
15. S. Kannan and T. Warnow. Inferring evolutionary history from DNA sequences. In *Proceedings of the 31st Annual Symposium on the Foundations of Computer Science*, pp. 362–378, St. Louis, Missouri, 1990.
16. S. Kannan and T. Warnow. A fast algorithm for the computation and enumeration of perfect phylogenies. *Manuscript.* A preliminary version was presented at SODA '95.
17. F.R. McMorris, T.J. Warnow, and T. Wimer. Triangulating vertex colored graphs. To appear in *Proceedings of the 4th Annual Symposium on Discrete Algorithms*, Austin, Texas, 1993.
18. C.H. Papadimitriou and M. Yannakakis. Optimization, approximation, and complexity classes. *Journal of Computer and System Sciences*, 43. Also in STOC '88.
19. M.A. Steel. The complexity of reconstructing trees from qualitative characters and subtrees. *Journal of Classification*, 9:91–116, 1992.
20. Tandy Warnow, Donald Ringe, and Ann Taylor. Reconstructing the evolutionary history of natural languages. To appear in *Proc. SODA 96.*

Author Index

Lecture Notes in Computer Science

For information about Vols. 1–1026

please contact your bookseller or Springer-Verlag